Hydrology and Water Resources in Agriculture and Ecology

Hydrology and Water Resources in Agriculture and Ecology

Editors

Songhao Shang
Qianqian Zhang
Dongqin Yin
Hamza Gabriel
Magdy Mohssen

MDPI

Basel • Beijing • Wuhan • Barcelona • Belgrade • Novi Sad • Cluj • Manchester

Editors

Songhao Shang
Department of Hydraulic Engineering
Tsinghua University
Beijing
China

Qianqian Zhang
Department of Earth System Science
Tsinghua University
Beijing
China

Dongqin Yin
College of Land Science and Technology
China Agricultural University
Beijing
China

Hamza Gabriel
NUST Institute of Civil Engineering
National University of Sciences & Technology
Islamabad
Pakistan

Magdy Mohssen
HydroScience Consultancy
Dunedin
New Zealand

Editorial Office
MDPI
St. Alban-Anlage 66
4052 Basel, Switzerland

This is a reprint of articles from the Topic published online in the open access journals *Remote Sensing* (ISSN 2072-4292), *Water* (ISSN 2073-4441), *Sustainability* (ISSN 2071-1050), and *Hydrology* (ISSN 2306-5338) (available at: https://www.mdpi.com/topics/hydrology_water_agriculture_ecology).

For citation purposes, cite each article independently as indicated on the article page online and as indicated below:

Lastname, A.A.; Lastname, B.B. Article Title. *Journal Name* **Year**, *Volume Number*, Page Range.

ISBN 978-3-7258-0247-0 (Hbk)
ISBN 978-3-7258-0248-7 (PDF)
doi.org/10.3390/books978-3-7258-0248-7

Contents

remote sensing

MDPI

Editorial

Editorial on Hydrology and Water Resources in Agriculture and Ecology

Songhao Shang [1,*], Hamza Farooq Gabriel [2] and Qianqian Zhang [3]

1 State Key Laboratory of HydroScience and Engineering, Department of Hydraulic Engineering, Tsinghua University, Beijing 100084, China
2 NUST Institute of Civil Engineering (NICE), School of Civil & Environmental Engineering (SCEE), National University of Sciences and Technology (NUST), Islamabad 44000, Pakistan; hamza.gabriel@nice.nust.edu.pk
3 Department of Earth System Science, Institute for Global Change Studies, Ministry of Education Ecological Field Station for East Asian Migratory Birds, Tsinghua University, Beijing 100084, China; qqzhang91@tsinghua.edu.cn
* Correspondence: shangsh@tsinghua.edu.cn

Citation: Shang, S.; Gabriel, H.F.; Zhang, Q. Editorial on Hydrology and Water Resources in Agriculture and Ecology. *Remote Sens.* **2024**, *16*, 238. https://doi.org/10.3390/rs16020238

Received: 26 December 2023
Accepted: 2 January 2024
Published: 8 January 2024

The agricultural sector uses the largest share of freshwater, accounting for over 70% of the global freshwater withdrawals, and this proportion can be up to 90% in arid and semi-arid regions [1]. Irrigation uses most of the agricultural water withdrawal and contributes greatly to global food security. Irrigation water diverted or pumped from water sources (rivers, reservoirs, and groundwater aquifers, etc.) is conveyed to croplands through artificial canals or pipe systems, which has a great impact on the hydrological processes in an irrigation district. Consequently, agricultural hydrological processes on the cropland and irrigation district scales are complicated due to the integrated impacts of natural and anthropogenic factors. Salt transport and balance associated with hydrological processes are also key factors influencing crop production on irrigated land. Modeling water flow and salt transport in croplands [2] and irrigation districts [3] has been a major topic in agricultural hydrology, which provides the basis for the spatiotemporal allocation of irrigation water for greater water use efficiency, crop yields, and/or benefits [4].

In recent decades, with the increasing water requirements for domestic and industrial uses, the water available for agriculture and natural ecosystems has been decreasing in most parts of the world, which has been further intensified by climate change. A systemic study on hydrology and water resources in agriculture and ecology will provide a basis for food security and ecosystem security. The main research fields cover water–heat–salt–nutrients transport in the soil–plant–atmosphere continuum (SAPC), agro-hydrological modeling, evapotranspiration modeling in croplands and irrigation district scales, eco-hydrology, water–salt balance and non-point source contamination modeling in an irrigation district, the high-efficient use of water resources for agriculture, interactions among water, agriculture, and natural ecosystems, and remote sensing applications in agricultural and ecological hydrology.

This Editorial refers to the topic "Hydrology and Water Resources in Agriculture and Ecology". This topic highlights new opportunities and challenges for hydrological modeling and the high-efficient use of water resources in agriculture and ecology in a changing environment.

Seventy manuscripts were submitted for the topic, and all of them were subject to the rigorous review process of participating journals. After the review and revision processes, 28 papers were finally accepted for publication and inclusion in this topic, including 10 in *Hydrology*, 6 in *Remote Sensing*, 4 in *Sustainability*, and 8 in *Water*.

As shown in Table 1, the contributions are diverse in the research fields, types of study areas, and geographical regions. The research fields can be classified into crop water requirement (four contributions), drought assessment (three), ecohydrology and environmental hydrology (three), river hydrology (three), forest hydrology (two), groundwater (two), soil water (two), channel leakage (one), cropland hydrology (one), drainage (one), hydrodynamics (one), hydropedology (one), nutrient loss (one), soil physics (one), water balance (one), and water footprint (one).

Table 1. Analysis of the published papers on this topic.

No.	Journal	Research Field	Focus	Type of Study Area	County /Region
1	Water	River hydrology	Impacts of climate change and human activities on streamflow	River Basin	China
2	Water	Groundwater	Recharge channels for sowing water in mountain aquifers	Mountain range	Spain
3	Remote Sensing	Drainage	Effect of controlled tile drainage on growth and yield of spring barley	Experimental fields	Czech
4	Remote Sensing	Environmental hydrology	Relationship between hydrological connectivity and water quality	River Basin	China
5	Sustainability	Ecohydrology	Relationship of zooplankton population growth and environmental factors	Reservoir	China
6	Hydrology	Hydropedology	Impact of hydropedological characteristics on streamflow in mountain catchments	Mountain catchments	South Africa
7	Water	Cropland hydrology	Effects of straw mulching on runoff and soil loss in slope farmland	Experimentalsoil tank	China
8	Remote Sensing	River hydrology	Impact of land use/cover changes on water balance components in plateau watersheds	River Basin	Pakistan
9	Remote Sensing	Drought assessment	Impact of drought on summer maize yield	Region	China
10	Hydrology	Crop water requirement	Water footprint assessment for irrigated paddy cultivation	Irrigation Scheme	Sri Lanka
11	Hydrology	River hydrology	Trends and variabilities in rainfall and streamflow	River Basin	Sri Lanka
12	Remote Sensing	Drought assessment	Remote sensing-based drought monitoring	River Basin	Kingdom of Saudi Arabia
13	Sustainability	Hydrodynamics	Selection of operation mode for irrigation canal headwork	River and canal section	China
14	Remote Sensing	Ecohydrology	Relationship between hydrological processes and ecological evolution	River Basin	China
15	Water	Crop water requirement	Estimation methods for daily crop coefficient of winter wheat	Lysimeter	China
16	Hydrology	Soil physics	Interaction of soil–water–atmosphere on soil crack characteristics	Experimental plot	Brazil
17	Water	Drought assessment	Drought assessment for spring maize	Region	China
18	Hydrology	Forest hydrology	Rainfall partitioning in Amazon Forest	Experimental plot	Brazil

Table 1. *Cont.*

No.	Journal	Research Field	Focus	Type of Study Area	County /Region
19	Sustainability	Nutrient loss	Potential runoff loss risk of nutrients in surface water of saline–alkali paddy	Mesocosm	China
20	Hydrology	Crop water requirement	Sensitivity of reference evapotranspiration to meteorological variables	Island	USA
21	Water	Soil water	Water and salt transport in freeze–thaw soil	Lysimeter	China
22	Hydrology	Forest hydrology	Hydrological properties of litter in different vegetation types	Forest	Brazil
23	Hydrology	Soil water	Photovoltaic-powered soil moisture sensor	Cropland	Brazil
24	Hydrology	Water balance	Large-area water budget analysis and drought monitoring	Countries	USA & Horn of Africa
25	Hydrology	Groundwater	Evolution of tunneling hydro-technology	Globe	Globe
26	Sustainability	Channel leakage	Model for estimating channel leakage	Channel reach	China
27	Water	Water footprint	Water footprint of animal breeding industry	Country	China
28	Water	Reference evapo-transpiration	Reference evapotranspiration estimation method	Lysimeter	USA

The study areas range from experiment sites (lyismeter and experimental plot), river/channel reach, administrative/geographic region, and countries to groups of countries or the globe that cover the globe (one contribution), one country group (one), and nine countries (twenty-six), including Brazil (four), China (thirteen), Czech (one), the Kingdom of Saudi Arabia (one), Pakistan (one), South Africa (one), Spain (one), Sri Lanka (two), and the USA (two).

Moreover, the methods used in these 28 contributions cover laboratory/field experiment analysis, statistical and regression analysis, and conceptual and physical-based hydrological models used in the cropland, regional, watershed, and country scales. There are 27 research papers and 1 review paper (contribution 25) among the 28 published papers on this topic.

Among the 28 contributions, contributions 1 and 11 fall within the scope of river hydrology. Contribution 1 analyzed the spatiotemporal variations of air temperature, precipitation, and potential evapotranspiration in the upper Yongding River Basin in North China based on historical data and assessed the impacts of climate change and human activities on streamflow using the double mass curve method and the Budyko framework for actual evapotranspiration estimation. The results show that human activities contribute more to streamflow changes than climate change in the two studied sub-watersheds. Contribution 11 analyzed the trend of change points in rainfall and streamflow in the Nilwala River Basin of Sri Lanka and explored their linkages. The results are helpful for water resources and hydropower planning.

Contributions 2 and 25 fall within the scope of groundwater. Contribution 2 presented an example of nature-based solutions for water scarcity problems in the Sierra Nevada Range of Spain, a system that uses recharge channels for sowing water in mountain aquifers that is harvested downstream. The authors postulated that this system can be an effective adaptation measure to climate change in similar regions. Contribution 25 reviewed the history of tunneling hydro-technology for groundwater development across the globe over the past several thousand years and discussed emerging trends and challenges of tunneling hydro-technologies in the future.

Contribution 3 evaluated the effect of controlled tile drainage (CTD) on the growth and yield of spring barley at a study site in Central Bohemia, Czech Republic, in 2021 based on vegetation indices calculated from unmanned aerial vehicle (UAV) imagery. The results indicate that CTD can improve spring barley development and grain yield due to higher soil moisture than free tile drainage practice.

Contribution 4 assessed the hydrological connectivity and its influence on water quality in the Bosten Lake Basin of Northwest China. The results indicate that improved hydrological connectivity is beneficial to improving water quality.

Contributions 5 and 14 fall within the scope of ecohydrology. Contribution 5 analyzed the relationship between zooplankton population growth and water environmental factors based on monitoring data in the Shanxi Reservoir in Southeast China. The results show that zooplankton can be taken as an integrated indicator for the assessment of the water environment and ecosystem health. Using an improved SWAT model and comprehensive ecosystem quality (EQ) assessment model, Contribution 14 analyzed the distribution and evolutionary characteristics of hydrological process factors and EQ in the Ulagai River Basin of Northeast China and identified their synergy relationships.

Contribution 6 analyzed the hydropedological characteristics of three mountain catchments in South Africa and their influence on the flow dynamics of the soils. The results indicate that the drying and wetting cycles of a wetland system have a great influence on the baseflow connectivity and the overland flow during wetter periods.

Contribution 7 explored the influences of rainfall pattern, soil structure, and straw mulching on near-surface hydrology and soil erosion in a slope farmland based on a simulated rainfall experiment. The results highlight the roles of rainfall pattern and straw mulching on runoff and soil erosion in slope farmland.

Contribution 8 simulated river flows in the Potohar Plateau of Pakistan using the SWAT model under classified historical and projected future land use/cover maps. Water balance analyses indicate that the land use/cover changes tend to decrease the surface runoff and water yield due to increases in percolation, lateral flow, sub-surface flow, and evapotranspiration.

Contributions 9, 12, and 17 fall within the scope of drought assessment. Contribution 9 obtained the loss risk curve cluster of drought frequency–drought resistance capacity–yield loss rate for summer maize in Benbu of China, which is effective in the quantitative assessment of drought disasters from a physical mechanism perspective. Contribution 12 assessed the drought regime in the Al-Lith Watershed of the Kingdom of Saudi Arabia with Landsat-derived indices and standardized precipitation evapotranspiration index (SPEI). The results indicate that the vegetation health index (VHI) is more appropriate for drought assessment in data-scarce regions. Contribution 17 developed a standardized crop water deficit index based on SPEI and the crop water deficit index (CWDI) and assessed drought for spring maize in the Songnen Plain of Northeast China.

Contributions 10, 15, 20, and 28 fall within the scope of crop water requirements. Contribution 10 assessed the water footprint for irrigated paddy cultivation in the Walawe Irrigation Scheme, Sri Lanka. The results highlight the necessity of improving irrigation practice to reduce excess water usage in the study region. Contribution 15 compared three methods to estimate the stage-wise crop coefficient for winter wheat in East China based on lysimeter measurement, and appropriate method for each growth stage were suggested. Contribution 20 analyzed the sensitivity of meteorological variables in the Penman–Monteith reference evapotranspiration equation for Puerto Rico. Contribution 28 assessed different machine learning (ML) models for reference evapotranspiration estimation in highly advective environments, and the genetic algorithm-optimized extreme learning machine performed better than other models and was recommended for reference evapotranspiration estimation at different time scales.

Contribution 13 simulated water flow and sediment transport in rivers and canal reaches under two operation conditions for irrigation canal headwork through hydrodynamic and sediment modeling. Based on the simulation results, an appropriate operation condition was recommended.

Contribution 16 assessed the soil characteristics and dynamics governing the crack formation and healing processes and quantified the soil moisture limits on soil swelling and shrinking in a vertic soil in a semiarid region of Brazil under natural conditions.

Contributions 18 and 22 fall within the scope of forest hydrology. Contribution 18 analyzed the dynamics and seasonality of litter stocks, water retention capacity, effective water retention, and water content of litter in Amazonian forests. The results are helpful for understanding the impact of sustainable forest management on the hydrological dynamics of litter. Contribution 22 studied the hydrological properties of litter in vegetation covers of *Eucalyptus* sp. plantations, agroforestry, and restoration forests, and highlighted the role of litter composition and species-specific characteristics in the hydrological functions of litter.

Contribution 19 explored the effect of nitrogen-fertilizer types on the potential risks of nitrogen and phosphorus losses through runoff. Carbon-based slow-release fertilizer is recommended for the study region to control nitrogen and phosphorus losses.

Contributions 21 and 23 fall within the scope of soil water. Contribution 20 analyzed the effect of soil texture on soil water flow and salt transport during the freezing-thawing period with a shallow groundwater table based on a lysimeter experiment. Contribution 23 developed an automated soil water tension sensor for soil moisture measurement, which can be used in real-time monitoring of soil moisture and is essential for precision irrigation.

Contribution 24 updated the agro-hydrologic VegET model by considering snow accumulation and melt processes and analyzed water budget in the conterminous United States and the Greater Horn of Africa. The model simulations can be used in drought monitoring and evaluating the impact of changing environments on agriculture and water resources.

Contribution 26 developed a multiple regression model for estimating the channel leakage process by considering the dynamic change in the main driving factors, which provides the basis for irrigation water management and control of the channel flow.

Contribution 27 assessed the water footprint of the animal breeding industry and driving forces at the provincial level in China and identified effective strategies for water footprint reduction.

Several research gaps can be detected from the contributions to this topic.

First, most studies focused on only one or several agro-hydrological processes, while few studies integrated all major hydrological processes into cropland or irrigation district scales that are influenced by both natural and anthropogenic factors. Integrated agro-hydrological models for the cropland and irrigation districts should be further studied.

Second, conflict between water uses for natural ecosystems and humans is unavoidable in areas short of water resources, especially in arid and semiarid regions. How to balance water uses in different sectors and allocate limited water resources optimally are key challenges in water resource management.

Third, nutrient losses associated with water flow from cropland not only waste a large amount of fertilizer and lower the nutrient use efficiency, but also result in pollution in water bodies and groundwater aquifers, together with contaminants from other sources. Controlling non-point source pollution from agriculture requires models for simultaneous water flow and nutrients/contaminants transport.

Fourth, soil salinization is a major threat to crop growth and food security in salinization-prone regions. How to regulate soil water and salt regimes in cropland/irrigation districts through appropriate irrigation and drainage practices will provide suitable soil conditions for crop growth and alleviate the negative influence of salt accumulation in croplands.

Fifth, the combination of data-driven algorithms and physical-based models is a new trend in hydrology and other disciplines. However, no paper used this method on this topic.

Finally, remote sensing technology has provided numerous data for agricultural hydrology research, especially in irrigation districts or on regional scales. Several papers on this topic used remote sensing data in drought assessment, but the application of remote sensing in other fields is less frequently used in this topic. More studies are expected to use remote sensing data in agro-hydrological modeling.

Author Contributions: Conceptualization, S.S., H.F.G. and Q.Z; formal analysis, S.S.; writing—original draft preparation, S.S.; writing—review and editing, H.F.G. and Q.Z. All authors have read and agreed to the published version of the manuscript.

Funding: This research received no external funding.

Data Availability Statement: No new data were created in this editorial.

Conflicts of Interest: The authors declare no conflicts of interest.

List of Contributions:

1. Deng, L.; Guo, Y.; Qi, Y.; Shen, Y.-J.; Shen, Y. Impacts of Climate Change and Human Activities on Streamflow of Upper Yongding River Basin, North China. *Water* **2022**, *14*, 2798. https://doi.org/10.3390/w14182798
2. Jódar, J.; Martos-Rosillo, S.; Custodio, E.; Mateos, L.; Cabello, J.; Casas, J.; Salinas-Bonillo, M.J.; Martín-Civantos, J.M.; González-Ramón, A.; Zakaluk, T.; et al. The Recharge Channels of the Sierra Nevada Range (Spain) and the Peruvian Andes as Ancient Nature-Based Solutions for the Ecological Transition. *Water* **2022**, *14*, 3130. https://doi.org/10.3390/w14193130
3. Duffková, R.; Poláková, L.; Lukas, V.; Fučík, P. The Effect of Controlled Tile Drainage on Growth and Grain Yield of Spring Barley as Detected by UAV Images, Yield Map and Soil Moisture Content. *Remote Sens.* **2022**, *14*, 4959. https://doi.org/10.3390/rs14194959
4. Liu, C.; Chen, Y.; Fang, G.; Zhou, H.; Huang, W.; Liu, Y.; Wang, X.; Li, Z. Hydrological Connectivity Improves the Water-Related Environment in a Typical Arid Inland River Basin in Xinjiang, China. *Remote Sens.* **2022**, *14*, 4977. https://doi.org/10.3390/rs14194977
5. Yin, J.; Xia, J.; Xia, Z.; Cai, W.; Liu, Z.; Xu, K.; Wang, Y.; Zhang, R.; Dong, X. Temporal Variation and Spatial Distribution in the Water Environment Helps Explain Seasonal Dynamics of Zooplankton in River-Type Reservoir. *Sustainability* **2022**, *14*, 13719. https://doi.org/10.3390/su142113719
6. Harrison, R.; van Tol, J.; Amiotte Suchet, P. Hydropedological Characteristics of the Cathedral Peak Research Catchments. *Hydrology* **2022**, *9*, 189. https://doi.org/10.3390/hydrology9110189
7. Yang, L.; Duan, J.; Peng, L.; Zhang, X.; Guo, X.; Yang, J. Effects of Straw Mulching on Near-Surface Hydrological Process and Soil Loss in Slope Farmland of Red Soil. *Water* **2022**, *14*, 3388. https://doi.org/10.3390/w14213388
8. Idrees, M.; Ahmad, S.; Khan, M.W.; Dahri, Z.H.; Ahmad, K.; Azmat, M.; Rana, I.A. Estimation of Water Balance for Anticipated Land Use in the Potohar Plateau of the Indus Basin Using SWAT. *Remote Sens.* **2022**, *14*, 5421. https://doi.org/10.3390/rs14215421
9. Cui, Y.; Tang, H.; Jin, J.; Zhou, Y.; Jiang, S.; Chen, M. System Structure–Based Drought Disaster Risk Assessment Using Remote Sensing and Field Experiment Data. *Remote Sens.* **2022**, *14*, 5700. https://doi.org/10.3390/rs14225700
10. Janani, H.K.; Abeysiriwardana, H.D.; Rathnayake, U.; Sarukkalige, R. Water Footprint Assessment for Irrigated Paddy Cultivation in Walawe Irrigation Scheme, Sri Lanka. *Hydrology* **2022**, *9*, 210. https://doi.org/10.3390/hydrology9120210
11. Panditharathne, R.; Gunathilake, M.B.; Chathuranika, I.M.; Rathnayake, U.; Babel, M.S.; Jha, M.K. Trends and Variabilities in Rainfall and Streamflow: A Case Study of the Nilwala River Basin in Sri Lanka. *Hydrology* **2023**, *10*, 8. https://doi.org/10.3390/hydrology10010008
12. Ejaz, N.; Bahrawi, J.; Alghamdi, K.M.; Rahman, K.U.; Shang, S. Drought Monitoring Using Landsat Derived Indices and Google Earth Engine Platform: A Case Study from Al-Lith Watershed, Kingdom of Saudi Arabia. *Remote Sens.* **2023**, *15*, 984. https://doi.org/10.3390/rs15040984

13. Liu, M.; Lv, S.; Qiao, Q.; Song, L. Design and Numerical Simulation of the Headworks in the Shizuishan Section of the Yellow River. *Sustainability* **2023**, *15*, 4564. https://doi.org/10.3390/su15054564
14. Chen, H.; Meng, F.; Sa, C.; Luo, M.; Zhang, H.; Bao, S.; Liu, G.; Bao, Y. Synergistic Change and Driving Mechanisms of Hydrological Processes and Ecosystem Quality in a Typical Arid and Semi-Arid Inland River Basin, China. *Remote Sens.* **2023**, *15*, 1785. https://doi.org/10.3390/rs15071785
15. Fang, J.; Wang, Y.; Jiang, P.; Ju, Q.; Zhou, C.; Lu, Y.; Gao, P.; Sun, B. Evaluation of Different Methods on the Estimation of the Daily Crop Coefficient of Winter Wheat. *Water* **2023**, *15*, 1395. https://doi.org/10.3390/w15071395
16. Ribeiro Filho, J.C.; Andrade, E.M.d.; Guerreiro, M.S.; Palácio, H.A.d.Q.; Brasil, J.B. Soil–Water–Atmosphere Effects on Soil Crack Characteristics under Field Conditions in a Semiarid Climate. *Hydrology* **2023**, *10*, 83. https://doi.org/10.3390/hydrology10040083
17. Pei, Z.; Wu, B. Spatial-Temporal Characteristics of Spring Maize Drought in Songnen Plain, Northeast China. *Water* **2023**, *15*, 1618. https://doi.org/10.3390/w15081618
18. de Lima, J.A.; Tonello, K.C. Rainfall Partitioning in Amazon Forest: Implications of Reduced Impact Logging on Litter Water Conservation. *Hydrology* **2023**, *10*, 97. https://doi.org/10.3390/hydrology10040097
19. Lyu, J.; Wang, X.; Hou, S.; Zeb, A.; Zhu, H.; Xu, Y. Content Variation and Potential Runoff Loss Risk of Nutrients in Surface Water of Saline-Alkali Paddy in Response to the Application of Different Nitrogen Fertilizer Types. *Sustainability* **2023**, *15*, 7040. https://doi.org/10.3390/su15097040
20. Irizarry-Ortiz, M.; Harmsen, E.W. Sensitivity of the Penman–Monteith Reference Evapotranspiration Equation to Meteorological Variables for Puerto Rico. *Hydrology* **2023**, *10*, 101. https://doi.org/10.3390/hydrology10050101
21. Wang, E.; Chen, J.; Liu, L.; Cui, L.; Xue, J.; Ren, J.; Du, Q. Effect of Soil Texture on Water and Salt Transport in Freeze–thaw Soil in the Shallow Groundwater Area. *Water* **2023**, *15*, 2587. https://doi.org/10.3390/w15142587
22. Castagnolli, L.; Boggiani, F.S.; Lima, J.A.d.; Lima, M.T.; Tonello, K.C. Hydrological Properties of Litter in Different Vegetation Types: Implications for Ecosystem Functioning. *Hydrology* **2023**, *10*, 165. https://doi.org/10.3390/hydrology10080165
23. de Melo, D.A.; Silva, P.C.; da Costa, A.R.; Delmond, J.G.; Ferreira, A.F.A.; de Souza, J.A.; de Oliveira-Júnior, J.F.; da Silva, J.L.B.; da Rosa Ferraz Jardim, A.M.; Giongo, P.R.; et al. Development and Automation of a Photovoltaic-Powered Soil Moisture Sensor for Water Management. *Hydrology* **2023**, *10*, 166. https://doi.org/10.3390/hydrology10080166
24. Senay, G.B.; Kagone, S.; Parrish, G.E.L.; Khand, K.; Boiko, O.; Velpuri, N.M. Improvements and Evaluation of the Agro-Hydrologic VegET Model for Large-Area Water Budget Analysis and Drought Monitoring. *Hydrology* **2023**, *10*, 168. https://doi.org/10.3390/hydrology10080168
25. Angelakis, A.N.; Passchier, C.W.; Valipour, M.; Krasilnikoff, J.A.; Tzanakakis, V.A.; Ahmed, A.T.; Baba, A.; Kumar, R.; Bilgic, E.; Capodaglio, A.G.; et al. Evolution of Tunneling Hydro-Technology: From Ancient Times to Present and Future. *Hydrology* **2023**, *10*, 190. https://doi.org/10.3390/hydrology10090190
26. Ma, J.; Yang, J.; Hao, X.; Cui, B.; Yang, S. Dynamic Simulation Model of Channel Leakage Based on Multiple Regression. *Sustainability* **2023**, *15*, 14904. https://doi.org/10.3390/su152014904
27. Xing, H.; Xie, Y.; Li, B.; Cong, H.; Zheng, W.; Liu, H. Water Footprint of Animal Breeding Industry and Driving Forces at Provincial Level in China. *Water* **2023**, *15*, 4264. https://doi.org/10.3390/w15244264
28. Kiraga, S.; Peters, R.T.; Molaei, B.; Evett, S.R.; Marek, G. Reference Evapotranspiration Estimation Using Genetic Algorithm-Optimized Machine Learning Models and Standardized Penman–Monteith Equation in a Highly Advective Environment. *Water* **2024**, *16*, 12. https://doi.org/10.3390/w16010012

References

1. United Nations. *The United Nations World Water Development Report 2023: Partnerships and Cooperation for Water*; UNESCO: Paris, France, 2023; pp. 12–13.
2. Chen, S.; Mao, X.; Shang, S. Response and contribution of shallow groundwater to soil water/salt budget and crop growth in layered soils. *Agric. Water Manag.* **2022**, *266*, 107574. [CrossRef]

3. Wen, Y.; Wan, H.; Shang, S. A monthly distributed water and salt balance model in irrigated and non-irrigated lands of arid irrigation district with shallow groundwater table. *J. Hydrol.* **2023**, *616*, 128811. [CrossRef]
4. Li, J.; Shang, S.; Jiang, H.; Song, J.; Rahman, K.U.; Adeloye, A.J. Simulation-based optimization for spatiotemporal allocation of irrigation water in arid region. *Agric. Water Manag.* **2021**, *254*, 106952. [CrossRef]

water

MDPI

Article

Impacts of Climate Change and Human Activities on Streamflow of Upper Yongding River Basin, North China

Liqiang Deng, Ying Guo, Yongqing Qi, Yan-Jun Shen and Yanjun Shen *

Key Laboratory of Agricultural Water Resources, Hebei Key Laboratory of Agricultural Water-Saving, Center for Agricultural Resources Research, Institute of Genetics and Developmental Biology, Chinese Academy of Sciences, Shijiazhuang 050021, China

* Correspondence: yjshen@sjziam.ac.cn

Abstract: Streamflow in semiarid areas, especially in North China, was rapidly decreasing, which made it important to analyze the characteristics and influencing factors of streamflow. Using the hydro-meteorological data series of 1961–2017 in the upper Yongding River Basin (UYRB) (including the Yang River Basin (YRB) and Sanggan River Basin (SRB)), spatio-temporal variation characteristics of air temperature, precipitation, and potential evapotranspiration (E_0) were analyzed. The results showed that precipitation has no significant trend; the temperature showed a significant increase of 0.1–0.5 °C per decade; E_0 showed a significant decrease of approximately −2 mm/10yr (in 18 stations); the estimated rates of streamflow change were −7 and −8 mm/10yr for SRB and YRB. As for spatial distribution, the YRB presented a higher E_0 value than the SRB; the mountain areas had more precipitation than the plain areas. The change points of streamflow occurred in 1982 and 2003. Both the Budyko and the DMC methods were used to evaluate the impacts of climate change and human activities on the mean annual streamflow. In variation stage I (1983~2003), impacts of human activities account for 90.6% and 62.7% of the mean annual streamflow changes in YRB and SRB, respectively. In variation stage II (2004~2017), the percentages were 99.5% and 93.5%, respectively. It is also noted that the first change point in streamflow was indeed at the beginning of China's land reform, when the farmers could manage their reallocated lands and, therefore, there was an increase in agricultural water consumption. The second change point coincided with "Capital Water Resources Planning", including water conservation projects and irrigation district construction programs. In general, human activities were mainly responsible for the significant decline in the annual streamflow of UYRB. This paper will provide valuable results for water resources planning and give guidance on the construction of water conservation function areas and ecological environment support areas in the capital.

Keywords: streamflow; climate change; human activities; Yongding River Basin; Budyko

Citation: Deng, L.; Guo, Y.; Qi, Y.; Shen, Y.-J.; Shen, Y. Impacts of Climate Change and Human Activities on Streamflow of Upper Yongding River Basin, North China. *Water* **2022**, *14*, 2798. https://doi.org/10.3390/w14182798

Academic Editors: Songhao Shang, Qianqian Zhang, Dongqin Yin, Hamza Gabriel and Magdy Mohssen

Received: 8 August 2022
Accepted: 5 September 2022
Published: 8 September 2022

1. Introduction

Streamflow, a key link in the water cycle, is of great value to anthropogenic activities and ecosystems [1]. Additionally, it can directly meet social-economic demands as well as the needs of river ecosystems. Nevertheless, such water resources are facing various challenges due to overexploitation and climate change [2–4], leading to diverse variation patterns globally [5]. These phenomena are especially serious in semi-arid areas, and are becoming a pressing topic to be addressed [6]. Streamflow in semi-arid areas has higher spatial and temporal variability due to the complexity of rainfall [7]. Lots of researchers have found that the streamflow in semi-arid areas is more vulnerable to climate change and human activities [8,9]. Therefore, more attention needs to be focused on the semi-arid areas, so as to develop the regional economy and eco-environment in a sustainable way. This paper mainly focuses on the impacts of climate change and human activities on streamflow in the upper Yongding River Basin (UYRB), a semi-arid basin in North China.

Various methods have been applied to quantify the impacts of climate change and human activities on streamflow, which produce lots of valuable results [10,11]. Commonly used methods include hydrological modelling [8,12], conceptual approaches [13,14], analytical approaches [15], and methods based on hydrological observed data [16,17]. All methods mentioned above have advantages and disadvantages summarized by Dey and Mishra [10]. The method used in this paper is based on the Budyko hypothesis (called the Budyko method hereafter), a conceptual approach that describes the relationship between water evaporated and energy available at a basin scale. Additionally, several functions have been derived based on this hypothesis, from generic formulae to basin-specific ones. The Budyko method has a relatively simple structure, and fewer data is needed compared with the hydrological model, which has been applied to separate the influences of climate change from human activities at home and abroad in recent years [18,19]. One of the most popular formulae is Fu's equation with one parameter, which is obtained by dimensional analysis and differential derivation [20]. This research will adopt Fu's equation to analyze the drivers of streamflow change in the UYRB.

The Yongding River, one tributary of the Haihe River, is the mother river of Beijing, and plays a key role in the development of Beijing City [21]. Most of the catchment, however, is located in the UYRB controlled by Guanting Reservoir. The water scarcity in this area has become increasingly severe [22], and several relevant studies have been conducted in this area, which provide certain scientific guidance for improving water resource management. The study conducted by Ren et al. [23] reveals that the increasing amount of water taken from the river course is the direct cause leading to the observed streamflow decrease in the northern part of China. Yang et al. [24] inferred that the key factor leading to the streamflow decreasing in this area in the 1980s was agricultural water consumption. Xia et al. [25] applied a modelling method to the UYRB and found that climate change and human activities were estimated to account for 10.5–12.6% and 87.4–89.5% of the reduction in annual runoff, respectively, indicating that human activities are the main driving factors for the reduction in runoff. The results from Zeng et al. [26] indicate that the water resources in the UYRB will decrease in the early 21st century and increase in the middle of the 21st century, which will further intensify the water resources crisis. Mo et al. [27] conducted a similar study (data 1957–2010) through an elastic coefficient method, and the results showed that the streamflow change rate caused by climate change is 28% and 72% for human activities. However, Zhang's results [28] (data 1957–2010) showed that climate change is the dominant influence factor with a contribution of 65.4%.

To sum up, the UYRB is experiencing a severe water shortage, which makes it a pressing topic to clarify the mechanism of the water cycle. In addition, the research mentioned above mainly focused on the period before 2012, and some views are contradictory. To provide guidance in the construction of water conservation function areas and in ecological environment support areas in the capital (two areas in the capital), it is essential to conduct further studies on the UYRB to clarify the hydrological mechanisms. The objectives of this paper, therefore, are (1) to assess the spatial and temporal variation of hydro-meteorological variables and (2) to quantify the impacts of climate change and human activities on streamflow in the UYRB.

2. Materials and Methods

2.1. Study Area

The UYRB, belonging to Haihe River Basin, covers part of Hebei Province, Shanxi province, and the inner Mongolia Autonomous Region, with a total area of $\approx$ 43,000 km^2 and a population of $\approx$ 9.13 million (2017). The study area lies at an elevation of 479–2852 m above the mean sea level, stretching between longitudes $111°58'$–$116°22'$ E and latitudes $38°50'$–$41°16'$ N (Figure 1). Continental monsoon climate prevails in this area with cold dry winters and hot rainy summers. The annual average precipitation in the basin is 389 mm (1961–2017), approximately 75% falls in rainy months of June–September. Average annual temperature in the study area is 6.8 °C (1961–2017). Main land use/land cover types (2015) are farmland (≈42%), grassland (≈29%), and forest (≈10%), and the forests consist mainly of deciduous broad-leaved forests. Maize is the dominant cultivated crop in the single-crop rotation system in the study area. Rainfall is often unable to meet crop water demand so that irrigated lands are increasingly reliant on groundwater pumping and river water. Hence, as the largest water user, agricultural water consumption has a significant impact on the streamflow. There are two sub-basins in the UYRB, named Yang River Basin (YRB) controlled by Xiangshuibu station and Sanggan River Basin (SRB) controlled by Shixiali station (See Table 1 in detail). This paper will focus on this two sub-basins.

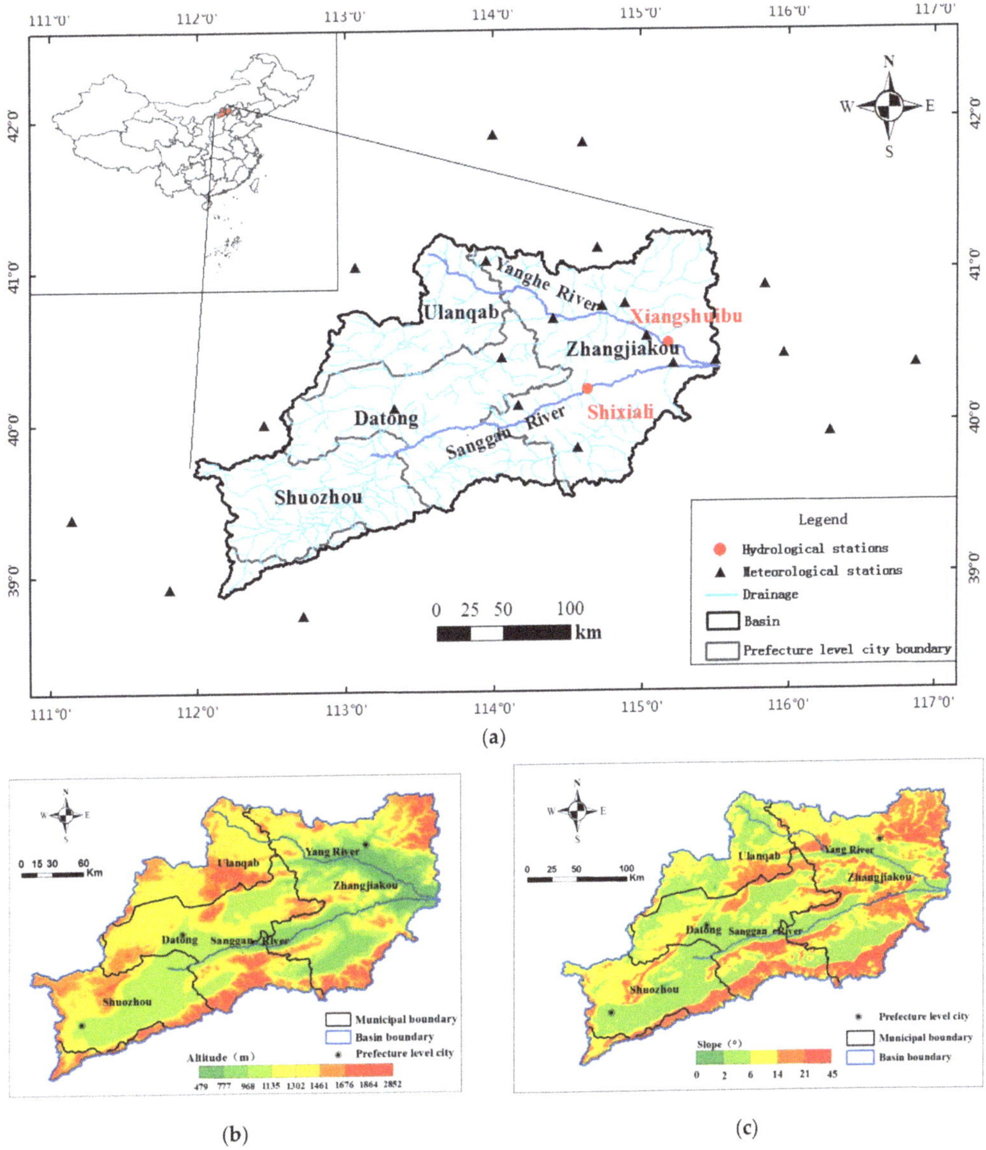

Figure 1. Location of the UYRB showing meteorological and hydrological stations (**a**) and background topography (**b**,**c**).

Table 1. Summary of main features of the two sub-basins in the UYRB.

Basin	Hydrological Station	River Basin Area (km^2)	Average Annual Precipitation (mm)	Average Annual Runoff (mm)	CV
Yang River	Xiangshuibu	14600	389	27	0.59
Sanggan River	Shixiali	23300	433	24	0.59

CV = coefficient of variation for annual runoff.

2.2. Data

Data of 23 meteorological stations from China Meteorological Administration (Figure 1) were selected (among them, 11 within the UYRB and 12 surround it), during 1961–2017 for this study, including mean daily temperature, precipitation, mean relative humidity, mean wind speed, and sunshine duration. Then Penman–Monteith equation was applied to obtain potential evapotranspiration using meteorological data [29]. Kriging interpolation was applied to obtain the spatial distribution of hydro-meteorological variables and the average annual values of the study area were calculated by processing the average grid values. Daily streamflow data were collected for the same period from two hydrological gauge stations situated at the outlets of the YRB and SRB, then, were used to calculate the annual streamflow. The DEM data with a resolution of 30m from ASTER GDEM was used.

2.3. Methods

2.3.1. Trend Detection

The nonparametric Mann–Kendall (M–K) test was applied to detect trends in the hydro-climatic time series [30,31]. For the given time series $X(x_1, x_2, \ldots, x_n)$, the statistic S is defined as:

$$S = \sum_{i=2}^{n}\sum_{j=1}^{i-1} sgn(x_i - x_j) where\ sgn(x_j - x_i) = \begin{cases} 1 & x_j > x_i \\ 0 & x_j = x_i \\ -1 & x_j < x_i \end{cases} \quad (1)$$

Its variance is as follows:

$$\text{var(S)} = \frac{n(n-1)(2n+5)}{18} \quad (2)$$

The standardized statistic is:

$$Z = \begin{cases} \frac{S-1}{\sqrt{var(S)}} & S > 0 \\ 0 & S = 0 \\ \frac{S-1}{\sqrt{var(S)}} & S < 0 \end{cases} \quad (3)$$

The null hypothesis of no trend is rejected if $|Z| > 1.96$ at a 5% significance level. If the Z value is positive then an upward trend exists, otherwise this indicates a downward trend.

2.3.2. Change Point Detection

The double mass curve (DMC) is a simple statistical method to detect the change point of a hydro-meteorological data series [32]. The procedure of plotting the graph with two variable series is as follows. Given the observed variable series X_i and Y_j, the cumulative amount is calculated:

$$X_i' = \sum_{j=1}^{i} X_j \quad (4)$$

$$Y'_i = \sum_{j=1}^{i} Y_j \quad (5)$$

where X_i represents precipitation, and Y_j represents streamflow. Then, plot the curve with X_i and Y_j series on the graph. Finally, the change point can be determined by the slope change in the curve. In order to accurately determine the abrupt change time, this paper also takes into account the field survey data and existing research, making the abrupt change points closer to reality. Therefore, the study period can be divided into baseline and changed period.

2.3.3. Budyko Method

The Budyko assumption describes the water-energy coupling balance in a comparatively long-term scale [33], to put it simply, the water evaporated in a watershed can be expressed as the function of the climate dryness index. The calculation procedures of the Budyko method are as follows:

The basin-scale water balance can be shown as follows:

$$Q = P - E_a - \Delta S \quad (6)$$

where Q is the streamflow depth (mm), P is precipitation (mm), E_a is the actual evapotranspiration (mm), and ΔS is the water storage changes in the basin (mm). According to the Budyko hypothesis, ΔS is assumed to be zero over a comparatively long period. The actual evapotranspiration is computed by Fu's equation [20]:

$$E_a = PF(\phi) \quad (7)$$

$$F(\varphi) = 1 + \varphi - 1 + \phi^{\alpha})^{1/\alpha} \quad (8)$$

where $\varphi = E_0/P$ is the climate dryness index, E_0 is potential evapotranspiration calculated by the Penman–Monteith equation following the procedure outlined in FAO-56 [20]. α is a model parameter fitted by long-term observed data. As one form of the Budyko theoretical framework. Fu's equation has been widely used to model long-term basin-scale water balance [18,34]. A detailed description of Fu's equation is available in [35].

The change of average annual streamflow between baseline and changed period is calculated as:

$$\Delta\overline{Q} = \overline{Q_2} - \overline{Q_1} \quad (9)$$

where $\Delta\overline{Q}$ represents the change in average annual streamflow; $\overline{Q_1}$ and $\overline{Q_2}$ are average annual streamflow during the baseline period and changed period, respectively. The baseline period has no significant human activities, while the changed period is associated with significant human activities.

The total change in the observed mean annual streamflow ΔR_{total} can be decomposed into climate variability ΔR_{clima} and human activities ΔR_{human},

$$\Delta R_{total} = \Delta R_{clima} + \Delta R_{human} \quad (10)$$

Precipitation and E_0 directly determine streamflow in the descriptions of Fu's equation. Therefore, the impact of climate change on streamflow can be calculated by the following formula:

$$\Delta R_{climate} = \frac{\partial R}{\partial P}\Delta P + \frac{\partial R}{\partial ET_0}\Delta E_0 \quad (11)$$

where ΔP and ΔE_0 are the changes in precipitation and E_0, respectively. Addditionally, the partial differential expression in Equation (11) can be derived by the combination of Equations (6)–(8), the results are expressed as:

$$\frac{\partial R}{\partial P} = P^{\alpha-1}(E_0^{\alpha} + P^{\alpha})^{\frac{1}{\alpha}-1} \quad (12)$$

$$\frac{\partial R}{\partial ET_0} = E_0^{\alpha-1}(E_0^{\alpha} + P^{\alpha})^{\frac{1}{\alpha}-1} - 1 \tag{13}$$

2.3.4. Determination of the Parameter α in Fu's Equation

In Budyko's original hypothesis, the Budyko curve was regarded as "universal" for comparatively large basins on a long-term scale, which cannot be applied to small watersheds or watersheds with intensive impacts from human activities. Then, the Fu's equation, one of the most popular Budyko equations, with one parameter, was proposed: each basin has a distinct relationship between precipitation (P), potential evaporation (E_0), and actual evapotranspiration (E_a) on a comparatively long-term scale. The parameter α of Fu's equation represents the integrated effects of catchment attributes (such as climate, vegetation cover, soil properties, and catchment topography) and human activities [35,36]. That is to say, the curve shape parameter α controls how much of the available water will be evaporated given the available energy. The relationship indicates that two factors limit the evapotranspiration: the water supply, when $\varphi > 1$, and the energy supply, when $\varphi < 1$. α can be estimated by minimizing the difference between water-balance-based E_a (P−Q) and simulated E_a with Fu's equation. The parameter α obtained through this method is also called the "basin-specific α" [34]. Consequently, the "basin-specific" value α will provide a good basis for applying Fu's equation.

In this study, the parameter α was fitted through the annual observation data series (Figure 2), and the parameter α is 2.79 and 3.35 for YRB and SRB, respectively. A larger α indicates less water yield capacity for a basin given sufficient energy. Obviously, the water yield capacity of YRB is larger than SRB, in other words, there will be more streamflow in YRB given the same amount of precipitation. For both sub-basins, all annual dryness indices ($\varphi = ET_0/P$) are greater than one, that is to say, the limiting factor to evapotranspiration is water supply.

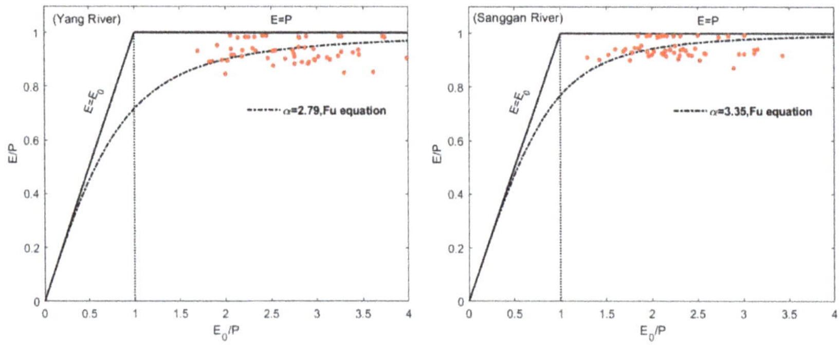

Figure 2. Budyko curve showing the relation between ε (ratio of E_a to P) as a function of φ (E_0/P) in the YRB and SRB. The bold black lines serve as an envelope (energy limit and water limit) to the Budyko curve family. The curve corresponds to α = 2.79 and α = 3.35 for YRB and SRB, respectively.

2.3.5. Assessing the Impacts of Climate Change and Human Activities on Streamflow by the DMC Method

Previous studies have demonstrated that the predominant climate factor to control the streamflow is precipitation in a closed watershed. Therefore, the DMC method, involving two variable series, can determine the impacts of climate change and human activities on streamflow [37]. The calculation procedure of the DMC method is as follows.

$$\Sigma R = k\Sigma P + b \tag{14}$$

R and *P* are annual streamflow and precipitation (mm), respectively, in the baseline period; *k* is slope and *b* is intercept.

$$\delta_{hi} = R_{2m} - R_{2c} \quad (15)$$

$$\delta_{ci} = R_{2c} - R_{1m} \quad (16)$$

$$Q_g = (\delta_{hi} \text{ or } \delta_{ci})/R_d \quad (17)$$

where δ_{hi} and δ_{ci} are streamflow depth variation caused by human activities and climate change, respectively (mm), Q_g represents the contribution rate (%) of human activities and climate change to streamflow reduction. R_{2m} represents measured values in the variation period. R_{2c} represents calculated values in the variation period. R_{1m} represents measured values in the baseline period.

3. Results

3.1. Spatio-Temporal Variation Characteristics of Hydro-Meteorological Factors

The average annual precipitation (for each station) for the UYRB is 319–554 mm and has no significant trend during 1961–2017, as shown in Figure 3a. The Z value of the M–K test shows positive values at 14 stations and negative values at 9 stations, which are not significant at the 5% level. Combining Figures 1b and 3a, we found that precipitation and altitude have roughly similar distribution patterns, which means the mountain area has more precipitation and the plain area has less. Meanwhile, the temperature at 22 stations increased by 0.1–0.5 °C per decade (obtained by linear regression, not shown in this study), which is related to the tendency of global warming.

The annual average E_0 values (for each station during 1961–2017) ranged from 702 to 1112 mm/yr and have a negative trend at 18 stations, which decreased approximately −2 mm/10yr over the study period (Figure 3b). This phenomenon is consistent with the global evaporation paradox [38,39], which may be caused by several climatic variables. The measured data witnessed a significant reduction in wind speed, which may have caused the decline in the E_0 value. The data on wind are not shown in this work. From the perspective of spatial distribution, the northern part of the UYRB presents a higher E_0 value, while the southern part is relatively low.

In this study, characteristics of annual streamflow at two hydrological stations (Shixiali and Xiangshuibu) are described during the period 1961–2017 (Figure 4). The linear regression analysis shows a significant decline in annual streamflow for YRB and SRB (not shown in this work). The estimated rates of change are −7 and −8 mm/10yr for Shixiali and Xiangshuibu, respectively, from 1961 to 2017. From the 1960s to the 1980s, streamflow declined by 20.01% and 22.28% for Xiangshuibu and Shixiali, respectively, and from the 1980s to the 2000s, the percentage was 68.23% and 67.77%, respectively. As Figure 4 shows, the streamflow of both sub-basins decreased rapidly during 1961–2017, which can be obviously divided into three stages. It is worth noting that the first decline appeared approximately in 1985, and the second decline appeared approximately in 2003.

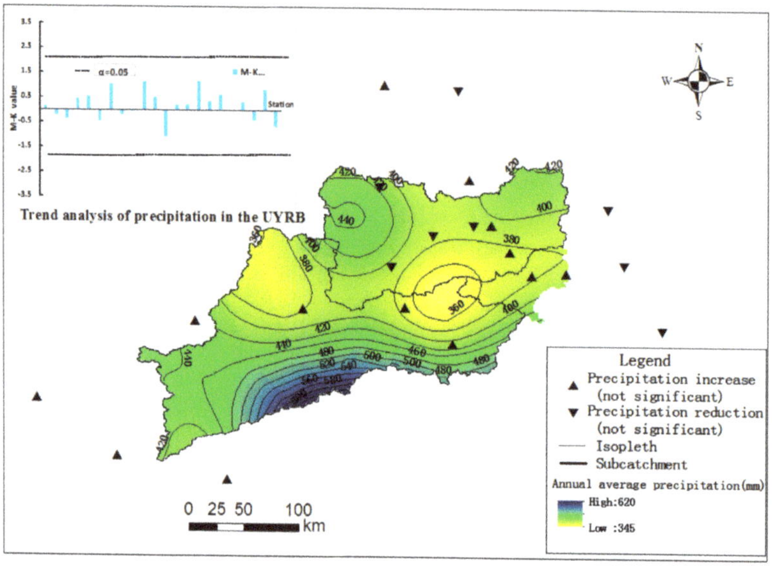

(a)

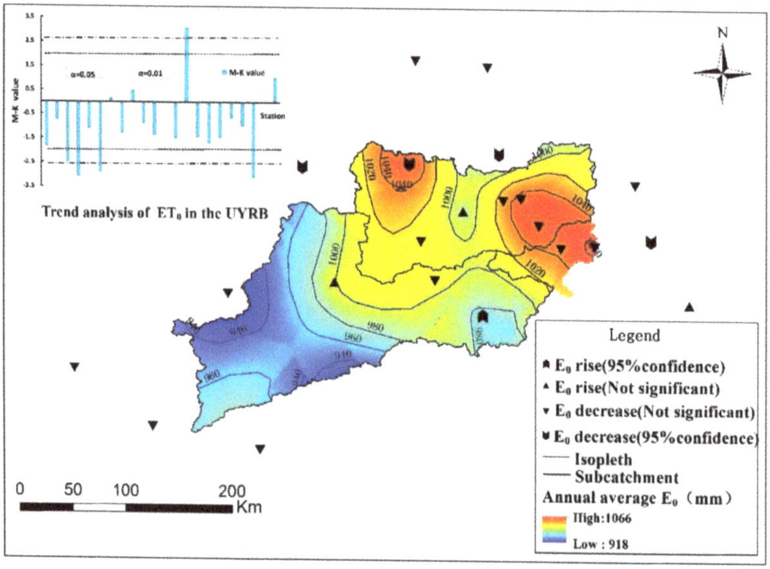

(b)

Figure 3. Trend analysis of precipitation (**a**) and potential evapotranspiration (**b**) in the UYRB. The inset shows the Z values of Mann–Kendall test.

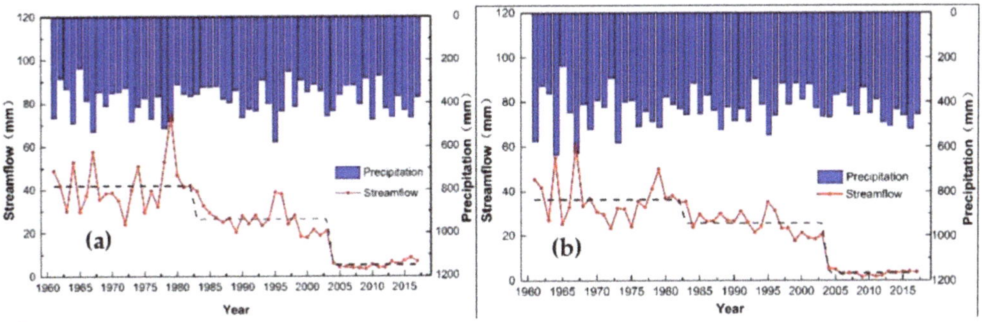

Figure 4. Temporal variations of annual streamflow (red line) and precipitation (blue column) for: (**a**) Xiangshuibu and (**b**) Shixiali.

3.2. Abrupt Changes in Streamflow

As just described, the study period can be divided into three stages. In order to determine the change points more precisely, the method of DMC was applied to examine the change points in annual streamflow at two sub-basins (Figure 5). According to the slope change in the curve, the first abrupt changes of these two sub-basins most likely occurred in the early 1980s; this pattern can be largely attributed to the increased agricultural water consumption [40]. The second abrupt change in the annual streamflow occurred approximately in 2003, where the curve slope changes significantly. This change predominantly resulted from the "Capital water resources planning", which includes large-scale soil and water conservation practices and irrigation district construction. In general, the first change point occurred approximately in 1980. To further prove this claim, we organized a two-week field survey in the UYRB, obtaining the details of water consumption during 1961–2017. We found that it was in 1982 that the household contract responsibility system was implemented in the study area, and the 'Capital Water Resources Planning' was mainly constructed in 2003. Such findings were consistent with the analysis above, therefore, the study period for both sub-basins can be divided into three periods: the baseline period (1961–1982), the variation stage I (1983–2003), and the variation stage II (2004–2017). Previous studies often divided the period into two stages [25,40,41] with one change point. However, we obtained two change points as mentioned above, and the first change point in this paper is roughly consistent with previous studies. Division of study period may be a little subjectivity in our study, nevertheless, the results conform to the region's reality.

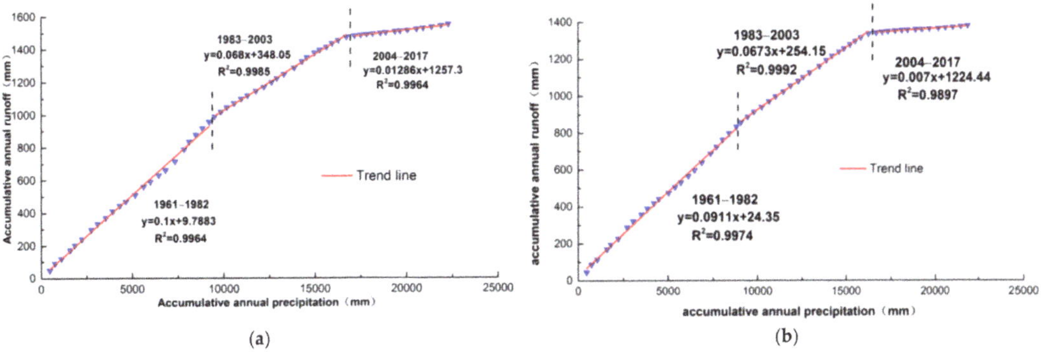

Figure 5. Double mass curve of the annual precipitation and runoff for: (**a**) Yanghe River Basin and (**b**) Sanggan River Basin during the period 1961–2017.

3.3. Effects of Climate Change and Human Activities on Streamflow

The streamflow at the two sub-basins decreased rapidly from 1961–2017, when the streamflow of YRB decreased by 36% (stage I) and 87% (stage II), respectively, compared with the baseline period. Additionally, the percentages for SRB are 30% (stage I) and 91% (stage II). Human activities account for most of this reduction in the mean annual streamflow for both sub-basins (Table 2). In the variation stage I, human activities contribute 90.6% and 62.7% of the mean annual streamflow change in the YRB and SRB, respectively. In the variation stage II, the percentages were 99.5% and 93.5%. This means that the impact of human activities on streamflow increased during the whole period of the UYRB. In general, the contribution rate of human activities to the streamflow of the Yang River is higher than that of the Sanggan River. The impact of climate change on streamflow in the study area is relatively small, which indicates intensive human activities, e.g., agricultural water consumption.

Table 2. Effects of climate change and human activities on streamflow in the UYRB with the Budyko method.

Watershed	Period	P (mm/Year)	E_0 (mm/Year)	R (mm/Year)	ΔR^c (mm)	ΔR^h (mm)	ΔR^c (%)	ΔR^h (%)
Yang River Basin	1961–1982	398	1044	42				
	1983–2003	380	1006	27	−1	−14	9.4	90.6
	2004–2017	389	1024	5	−0	−36	0.5	99.5
Sanggan River Basin	1961–1982	447	940	36				
	1983–2003	416	928	26	−4	−7	37.3	62.7
	2004–2017	438	970	3	−2	−31	6.5	93.5

4. Discussion

4.1. Impacts of Climate Change and Human Activities on Streamflow by the DMC Method

To further clarify the causes of streamflow variation, the DMC method was applied to show the correlation between cumulative annual streamflow and precipitation (Figure 5). In general, the predominant climate variable impacting the streamflow is precipitation, thus, the DMC was roughly a straight line given that streamflow was mainly influenced by climate [42]. As mentioned above, there are slope changes in the DMCs, suggesting that the relationship between streamflow and precipitation was altered by human activities. The results estimated by the DMC method are shown in Table 3, which are roughly consistent with the results of the Budyko method, except for a slight difference.

Table 3. Climate change and human activities on streamflow in the UYRB by the the DMC method.

Watershed	Period	P (mm/Year)	R_{2m} (mm)	R_{2c} (mm)	ΔR^c (mm)	ΔR^c (%)	ΔR^h (mm)	ΔR^h (%)
Yang River Basin	1961–1982	398	42	40				
	1983–2003	380	27	38	−3	21.4	−12	78.6
	2004–2017	389	5	40	−2	5.9	−34	94.1
Sanggan River Basin	1961–1982	447	36	36				
	1983–2003	416	26	34	−3	24.9	−8	75.1
	2004–2017	438	3	36	−1	1.8	−32	98.2

R_{2m} and R_{2c} represent measured values and calculated values in changed period, respectively. ΔR_h and ΔR_c represent runoff depth variation caused by human activities and climate change, respectively. ΔR_h and ΔR_c represent the contribution rate (%) of human activities and climate change to runoff reduction, respectively.

By comparing the results of the two methods, it is obvious that the major climate variable impacting the streamflow is precipitation. The minor differences between the two methods cannot be clarified because the Budyko method describes a nonlinear relationship between the streamflow and climate variables. It is hard to know how many variables are affecting the streamflow. It is also hard to determine the main variable. In any case, the results obtained with the Budyko method are convincing, with a combination of quantitative and qualitative descriptions.

The results in this paper are roughly consistent with Mo et al. [27] (data from 1957–2010), who found that the abrupt change point of the streamflow series occurred in 1983 and human activities were responsible for 72% of the total reduction in mean annual streamflow in the UYRB. The difference may be due to the method adopted and the length of the data series. Hou [43] also obtained a similar result, that human activities account for 80% of the streamflow change with the Budyko method. However, Zhang et al. [28] found that climate change is the major factor influencing the streamflow in the UYRB, which is mainly due to the data series applied (during 1957–2000).

4.2. Impacts of Human Activities on Streamflow in the UYRB

In 1978, the land reform policy was enacted in China, since then, farmers have been progressively managing their own lands [44]. According to the field survey, it was in 1982 that the implementation of this policy in the UYRB led to a rapid increase in agricultural water consumption. The rapid development of the economy also led to the rapid increase in industrial and domestic water use, which further aggravated the reduction in streamflow [41,43]. The "Capital water resources planning" was started in 2002, and it was completed in 2003. The accumulated area of soil and water loss control in the study area is 1841 km^2, and the engineering measures, such as fish scale pit, horizontal ditch, and check dam, play an important role [45]. In addition, forest and grass measures have been extensively taken, and a large number of irrigation areas have been built [46]. According to the field survey, flood irrigation is widely used on farmland in the YRB and SRB, up to 600~800 m^3/mu/year, which greatly reduces the streamflow. On the other hand, the climate dryness index ($\varphi = E_0/P$) increased from 2.627 (baseline period) to 2.648 (variation stage I) and 2.635 (variation stage II) in the YRB, indicating a trend towards a drier climate. The SRB is in the same situation, the value ranges from 2.105 to 2.218 and 2.188. This means that the impact of climate change is increasing, however, covered up by the impacts of strong human activities. Beyond question, human activities are the main driving factor of declining streamflow in the study area, more specifically, agricultural water consumption is the major role in the influence factors of streamflow. Beijing, the capital of China, began to carry out systematic management of the UYRB in 2019, to restore water conservation and ecological environmental support functions. This study can provide a valuable scientific basis for the systematic planning of the UYRB.

5. Conclusions

This study examines the spatial distribution and temporal variation of precipitation, potential evapotranspiration (E_0), temperature, and streamflow using data series from 1961 to 2017 in the UYRB. The impacts of climate change and human activities were investigated, and possible causes of the streamflow changes were analyzed. The conclusion of this study can be summarized as follows:

The results from the M–K test show that a general decrease in the annual E_0 (approximately −2 mm/10yr at 18 stations) and a rising temperature trend (0.1–0.5 °C per decade at 22 stations) have been detected. Nevertheless, the precipitation has no significant trend. The average annual streamflow shows a significant decrease at both YRB and SRB. From the perspective of spatial distribution, the northern part of the UYRB presents a higher E_0 value, while the southern part is relatively low. Additionally, mountain areas have more precipitation than plain areas. Abrupt changes in streamflow occurred in 1982 and 2003, which may be mainly caused by the implementation of land reform policy and "Capital water resources planning".

The impacts of climate change and human activities on streamflow were analyzed. Additionally, consistent results were obtained from both the Budyko method and the DMC method, in spite of a slight discrepancy between the two methods. Generally, the major climate variables impacting the streamflow are precipitation and human activities, where human activities accounted for most of the streamflow changes in the YRB (>78%) and SRB (>62%). The adoption of the household contract responsibility system in 1982 altered the natural streamflow regimes and led to a significant decrease in streamflow. The "Capital water resources planning" further led to a rapid reduction in streamflow circa 2003. All in all, human activities, such as agricultural irrigation (mainly flood irrigation), soil and water conservation measures, and the construction of water control projects seem to be the major causes in the significant decline in the annual streamflow in the UYRB.

Author Contributions: Conceptualization, Y.S. and L.D.; methodology, L.D.; software, Y.G.; validation, Y.-J.S., Y.Q. and Y.G.; formal analysis, Y.S.; investigation, L.D.; resources, Y.Q.; data curation, Y.G.; writing—original draft preparation, L.D.; writing—review and editing, L.D.; visualization, L.D.; supervision, Y.S.; project administration, Y.Q.; funding acquisition, Y.-J.S. All authors have read and agreed to the published version of the manuscript.

Funding: This study was supported by the Hebei Provincial Key R&D Programme (No. 22377001D), the Natural Science Foundation of Hebei Province — China (No. D2021503001) and the pioneer "Hundred Talents Program" of Chinese Academy of Science.

Data Availability Statement: The data may be available to the corresponding author upon request, subject to approval.

Acknowledgments: This paper acknowledges the Zhangjiakou Water Resources Bureau of Hebei Province for partial data support.

Conflicts of Interest: The authors declare no conflict of interest.

References

1. Oki, T.; Kanae, S. Global Hydrological Cycles and World Water Resources. *Science* **2006**, *313*, 1068–1072. [CrossRef] [PubMed]
2. Schiermeier, Q. Increased flood risk linked to global warming. *Nature* **2011**, *470*, 316. [CrossRef] [PubMed]
3. Mekonnen, M.; Hoekstra, A. Four billion people facing severe water scarcity. *Sci. Adv.* **2016**, *2*, e1500323. [CrossRef] [PubMed]
4. Bouimouass, H.; Fakir, Y.; Tweed, S.; Leblanc, M. Groundwater Sustainability in a Semiarid Traditional Irrigation Piedmont Supplied by High Mountain Streamflow. In Proceedings of the EGU General Assembly Conference Abstracts, Online, 19–30 April 2021. EGU21-9925. [CrossRef]
5. Milly, P.C.D.; Dunne, K.A.; Vecchia, A.V. Global pattern of trends in streamflow and water availability in a changing climate. *Nature* **2005**, *438*, 347–350. [CrossRef]
6. Hoekstra, A.Y.; Mekonnen, M.M.; Chapagain, A.K.; Mathews, R.E.; Richter, B.D. Global monthly water scarcity: Blue water footprints versus blue water availability. *PLoS ONE* **2012**, *7*, e32688. [CrossRef]
7. Yakir, H.; Morin, E. Hydrologic response of a semi-arid watershed to spatial and temporal characteristics of convective rain cells. *Hydrol. Earth Syst. Sci.* **2011**, *15*, 393–404. [CrossRef]
8. Kazemi, H.; Hashemi, H.; Maghsood, F.F.; Hosseini, S.H.; Sarukkalige, R.; Jamali, S.; Berndtsson, R. Assessment of streamflow decrease due to climate vs. human influence in a semiarid area. *Hydrol. Earth Syst. Sci. Discuss.* 2020. Available online: https://hess.copernicus.org/preprints/hess-2019-618/ (accessed on 7 August 2022). [CrossRef]
9. Tariku, T.B.; Gan, K.E.; Tan, X.; Gan, T.Y.; Tilmant, A. Global warming impact to river basin of blue nile and the optimum operation of its multi-reservoir system for hydropower production and irrigation. *Sci. Total Environ.* **2021**, *767*, 144863. [CrossRef]
10. Dey, P.; Mishra, A. Separating the impacts of climate change and human activities on streamflow: A review of methodologies and critical assumptions. *J. Hydrol.* **2017**, *548*, 278–290. [CrossRef]
11. Milly, P.; Dunne, K.A. Colorado river flow dwindles as warming-driven loss of reflective snow energizes evaporation. *Science* **2020**, *367*, 1252–1255. [CrossRef]
12. Zhang, H.; Wang, B.; Liu, D.L.; Zhang, M.; Feng, P.; Cheng, L.; Yu, Q.; Eamus, D. Impacts of future climate change on water resource availability of eastern Australia. *J. Hydrol.* **2019**, *573*, 49–59. [CrossRef]
13. Wang, D.; Hejazi, M. Quantifying the relative contribution of the climate and direct human impacts on mean annual streamflow in the contiguous United States. *Water Resour. Res.* **2011**, *47*, 411. [CrossRef]
14. Bai, P.; Liu, W.; Guo, M. Impacts of climate variability and human activities on decrease in streamflow in the Qinhe River, China. *Theor. Appl. Climatol.* **2014**, *117*, 293–301. [CrossRef]
15. Wang, S.; Zhang, Z.; McVicar, T.; Guo, J.; Tang, Y.; Yao, A. Isolating the impacts of climate change and land use change on decadal streamflow variation: Assessing three complementary approaches. *J. Hydrol.* **2013**, *507*, 63–74. [CrossRef]
16. Wang, W.; Shao, Q.; Yang, T.; Peng, S.; Xing, W.; Sun, F.; Luo, Y. Quantitative assessment of the impact of climate variability and human activities on runoff changes: A case study in four catchments of the Haihe River basin, China. *Hydrol. Process.* **2013**, *27*, 1158–1174. [CrossRef]

17. Zhang, L.; Zhao, F.F.; Brown, A.E. Predicting effects of plantation expansion on streamflow regime for catchments in Australia. *Hydrol. Earth Syst. Sci.* **2012**, *16*, 2109–2121. [CrossRef]
18. Zhao, G.; Tian, P.; Mu, X.; Jiao, J.; Wang, F.; Gao, P. Quantifying the impact of climate variability and human activities on streamflow in the middle reaches of the Yellow River basin, China. *J. Hydrol.* **2014**, *519*, 387–398. [CrossRef]
19. Andaryani, S.; Nourani, V.; Ball, J.; Asl, S.J.; Keshtkar, H.; Trolle, D. A comparison of frameworks for separating the impacts of human activities and climate change on river flow in existing records and different near-future scenarios. *Hydrol. Process.* **2021**, *35*, e14301. [CrossRef]
20. Fu, B. On the calculation of land surface evaporation. *Atmos. Sci.* **1981**, *5*, 25–33. (In Chinese)
21. Jinbo, L. Geological background along the banks of the yongding river. *Urban Geol.* **2012**, *7*, 1.
22. Wang, L.; Wang, Z.; Koike, T.; Yin, H.; Yang, D.; He, S. The assessment of surface water resources for the semi-arid Yongding River Basin from 1956 to 2000 and the impact of land use change. *Hydrol. Process.* **2010**, *24*, 1123–1132. [CrossRef]
23. Ren, L.; Wang, M.; Li, C.; Zhang, W. Impacts of human activity on river runoff in the northern area of China. *J. Hydrol.* **2002**, *261*, 204–217. [CrossRef]
24. Yang, Y.; Ren, D.; Yang, Y.; Tian, F.; Hu, Y.; Han, S. Advances in clarification of the driving forces of water shortage in Haihe River Catchment. *Chin. J. Eco-Agric.* **2018**, *26*, 1443–1453.
25. Xia, J.; Zeng, S.; Du, H.; Zhan, C. Quantifying the effects of climate change and human activities on runoff in the water source area of Beijing, China. *Hydrol. Sci. J.* **2014**, *59*, 1794–1807. [CrossRef]
26. Zeng, S.; Zhang, L.; Xia, J.; Yang, Z. Water cycle characteristic and its responses to climate change of the Yongding river basin. *J. Basic Sci. Eng.* **2013**, *21*, 501–511.
27. Mo, C.; Ruan, Y.; Mo, G.; Zhu, X.; Sun, G. Studying Response of Runoff to Climate Change and Human Activities Based on Elastic Coefficient. *J. China Hydrol.* **2018**, *38*, 41–45. (In Chinese)
28. Zhang, L.; Yu, S.; Duan, Y.; Shan, L.; Chen, X.; Xu, Z. Quantitative Assessment of the Effects of Climate Change and Human Activities on Runoff in the Yongding River Basin. *Clim. Change Res.* **2013**, *9*, 391–397. (In Chinese)
29. Allen, R.G.; Pereira, L.S.; Raes, D.; Smith, M. Crop Evapotranspiration Guidelines for Computing Crop Water Requirements. In *FAO Irrigation and Drainage Paper No. 56*; FAO: Rome, Italy, 1998.
30. Mann, H.B. Non-parametric test against trend. *Econometrika* **1945**, *13*, 245–259. [CrossRef]
31. Kendall, M.G. *Rank Correlation Methods*; Charles Griffin: London, UK, 1975.
32. Searcy, J.K.; Hardison, C.H. *Double-Mass Curves. U. S. Geological-Survey-Water-Supply*; U. S. Government Printing Office: Washington, DC, USA, 1960; Volume 1541-B, pp. 31–66.
33. Marlatt, W.E.; Budyko, M.I.; Miller, D.H. Climate and Life. *J. Range Manag.* **1975**, *28*, 160. [CrossRef]
34. Li, D.; Pan, M.; Cong, Z.; Zhang, L.; Wood, E. Vegetation control on water and energy balance within the Budyko framework. *Water Resour. Res.* **2013**, *49*, 969–976. [CrossRef]
35. Zhang, L.; Dawes, W.R.; Walker, G.R. Response of mean annual evapotranspiration to vegetation changes at catchment scale. *Water Resour. Res.* **2001**, *37*, 701–708. [CrossRef]
36. Zhou, G.; Wei, X.; Chen, X.; Zhou, P.; Liu, X.; Xiao, Y.; Sun, G.; Scott, D.F.; Zhou, S.; Han, L.; et al. Global pattern for the effect of climate and land cover on water yield. *Nat. Commun.* **2015**, *6*, 5918. [CrossRef]
37. Pirnia, A.; Golshan, M.; Darabi, H.; Adamowski, J.; Rozbeh, S. Using the Mann–Kendall test and double mass curve method to explore stream flow changes in response to climate and human activities. *J. Water Clim. Chang.* **2018**, *10*, 725–742. [CrossRef]
38. Pockley, P. The evaporation paradox. *Australas. Sci.* **2009**, *396*, 30.
39. Liu, C.; Zhang, D. Temporal and Spatial Change Analysis of the Sensitivity of Potential Evapotranspiration to Meteorological Influencing Factors in China. *Acta Geogr. Sin.* **2011**, *66*, 579–588. (In Chinese)
40. Yang, Y.; Tian, F. Abrupt change of runoff and its major driving factors in haihe river catchment, China. *J. Hydrol.* **2009**, *374*, 373–383. [CrossRef]
41. Ding, A.; Zhao, Y.; Hao, D.; Zhang, S.; Qiao, G. Analysis of variation characteristics of runoff and their influencing factors in the yongding river basin. *South North Water Transf. Water Sci. Technol.* **2013**, *11*, 17–22. (In Chinese)
42. Zhang, S.; Lu, X. Hydrological responses to precipitation variation and diverse human activities in a mountainous tributary of the lower Xijiang, China. *Catena* **2009**, *77*, 130–142. [CrossRef]
43. Hou, L. Study on Mechanism of Ecohydrological Response at the Water resources Shortage Watershed in Northern China—A Case Study of Yongding River. Ph.D. Thesis, China Institute of Water Resources and Hydropower Research, Beijing, China, 2019.
44. Gibson, J. Aggregate and distributional impacts of China's household responsibility system. *Aust. J. Agric. Resour. Econ.* **2020**, *64*, 14–29. [CrossRef]
45. Tan, G. Analysis on the implementation of sustainable utilization planning of water resources in the capital in the early 21st century. *Haihe Water Conserv.* **2004**, *4*, 25–26. (In Chinese)
46. Wang, D. Analysis on the progress and measures of implementing the capital water resources planning project in Zhangjiakou. *Haihe Water Conserv.* **2003**, *5*, 12–13. (In Chinese)

water

MDPI

Opinion

The Recharge Channels of the Sierra Nevada Range (Spain) and the Peruvian Andes as Ancient Nature-Based Solutions for the Ecological Transition

Jorge Jódar [1,*], Sergio Martos-Rosillo [2], Emilio Custodio [3], Luciano Mateos [4], Javier Cabello [5], Jesús Casas [5], María Jacoba Salinas-Bonillo [5], José María Martín-Civantos [6], Antonio González-Ramón [2], Thomas Zakaluk [2], Christian Herrera-Lameli [7], Javier Urrutia [7] and Luis Javier Lambán [1]

1 Instituto Geológico y Minero de España (IGME), CSIC, 50006 Zaragoza, Spain
2 Instituto Geológico y Minero de España (IGME), CSIC, 18006 Granada, Spain
3 Departamento de Ingeniería Civil y Ambiental, Universitat Politècnica de Calalunya, 08034 Barcelona, Spain
4 Instituto de Agricultura Sostenible (IAS), CSIC, 14004 Córdoba, Spain
5 Centro Andaluz Para la Evaluación y Seguimiento del Cambio Global (CAESCG), Universidad de Almería, 04120 Almería, Spain
6 MEMOLab, Laboratorio de Arqueología Biocultural, Universidad de Granada, 18071 Granada, Spain
7 Centro de Investigación y Desarrollo de Ecosistemas Hídricos, Universidad Bernardo O'Higgins, Santiago de Chile 8370993, Chile
* Correspondence: j.jodar@igme.es; Tel.: +34-976-65-0416

Citation: Jódar, J.; Martos-Rosillo, S.; Custodio, E.; Mateos, L.; Cabello, J.; Casas, J.; Salinas-Bonillo, M.J.; Martín-Civantos, J.M.; González-Ramón, A.; Zakaluk, T.; et al. The Recharge Channels of the Sierra Nevada Range (Spain) and the Peruvian Andes as Ancient Nature-Based Solutions for the Ecological Transition. *Water* **2022**, *14*, 3130. https://doi.org/10.3390/w14193130

Academic Editors: Fernando António Leal Pacheco, Songhao Shang, Qianqian Zhang, Dongqin Yin, Hamza Gabriel and Magdy Mohssen

Received: 31 August 2022
Accepted: 30 September 2022
Published: 4 October 2022

Abstract: Nature-Based Solutions for Integrated Water Resources Management (NbS-IWRM) involve natural, or nature-mimicking, processes used to improve water availability in quantity and quality sustainably, reduce the risks of water-related disasters, enhance adaptation to climate change and increase both biodiversity and the social-ecological system's resilience. United Nations and the European Commission promote their research as a cornerstone in the changeover to the Ecological Transition. In the Sierra Nevada range (Spain) and the Andean Cordillera, there is a paradigmatic and ancestral example of NbS-IWRM known as "careo channels" and "amunas", respectively. They recharge slope aquifers in mountain areas and consist of an extensive network of channels that infiltrate the runoff water generated during the snow-thawing and rainy season into the upper parts of the slopes. The passage of water through the aquifers in the slope is used to regulate the water resources of the mountain areas and thus ensure the duration of water availability for the downstream local population and generate multiple ecosystem services. This form of water management is known as Water Sowing and Harvesting (WS&H). As shown in this work, it is a living example of a resilience and climate change adaptation tool that can be qualified as a nature-based solution.

Keywords: careo; amuna; aquifer recharge; nature based solution; water resources management; ecological transition

1. Introduction

Water is an irreplaceable resource for life and human development on the planet. The 2030 Agenda for Sustainable Development, adopted by the United Nations General Assembly, underlines the obligation to ensure the availability and sustainable management of water for all [1]. Furthermore, sustainable water resource management is a guiding principle within the European Union's efforts to achieve an ecological transition towards a circular economy. Sustainable water management depends on water availability and thus on the rate of water renewal, which is marked by exogenous factors (e.g., the hydrological cycle) and to whose changes it is necessary to adapt [2]. Otherwise, the unsustainable use of water resources in economic activities may jeopardize our security and, paradoxically, the economic development it is intended to achieve. Therefore, it is necessary to manage water resources sustainably with solutions that involve their protection while safeguarding the biodiversity of dependent ecosystems and making them more resilient, thus improving human well-being.

Nature-based solutions are "solutions to challenges facing society that are inspired and supported by nature, are cost-effective, provide environmental, social and economic benefits, and help build resilience" [3]. Some solutions can be found among traditional water management practices developed by local communities in drylands (e.g., the inhabitants of the Alpujarras region on the slopes of the Sierra Nevada at the end of the Early Medieval Period [4–6], or the Chavín and Wari pre-Inca cultures in Peru [7]). Such practices aim to ensure sustainable access to water resources in times of low availability and high demand caused by climate and social changes [8,9]. Besides, these practices boost biodiversity conservation [10–14] and the recognition of rural communities' cultural identity and role as custodians of the land [11,15,16]. However, the functioning of many of these water management systems is threatened by globalization and or concentration in urban areas into which people are forced to migrate from rural communities. The reallocation of such people in urban areas may generate a significant impact on both the quantity and quality of water resources in such zones, which are typically reflected in (1) water shortages and (2) water quality issues because of pollution, thus aggravating both intensity and frequency of such water shortages. Nevertheless, the impact of people's migration on water resources is not limited to the urban zones. In rural areas, the abandonment of practices may result in the loss of traditional knowledge systems transmitted from generation to generation. To limit such impact at least in origin, it is essential to protect this knowledge from oblivion, as it provides age-old solutions for sustainable management of water resources to the recurrent problem of water scarcity, even in the most adverse social and climatic circumstances [4,5,17,18].

We present a traditional water management system that uses recharge channels for sowing water in mountain aquifers to be harvested later on downstream, for domestic supply and irrigation. The maintenance of such a system whose maintenance may help in the ecological transition. This highly efficient system [19,20] was developed independently by local communities south of the southeast of the Iberian Peninsula and in the Andes [8,21] to solve problems regarding scarcity. This paper describes the system functioning as a Nature-based Solution for Integrated Water Resources Management (NbS-IWRM) and postulates it as an adaptation measure to climate change.

2. Recharge Channels in Mountain Zones as NbS-IWRM for the Ecological Transition

The ecological transition refers to the process by which humans incorporate nature into society [22]. More recently, in the light of global change, it has developed into a broad set of objectives that seek the transformation of the energy, industrial, and agri-food sector [23,24] to adjust the demand of natural resources derived from human activity to the availability and production capacity of such natural resources. Its implementation seeks to curb the environmental crises threatening humanity's journey on Planet Earth.

Moreover, the European Green Deal, which is assumed as the "new growth strategy" for the European Union (EU), is developed using the ecological transition as one of the main drivers [25,26]. Water resource management is one of the cornerstones, both for the conservation of the environment and as a driver of the circular economy [2]. The sustainable management of water resources plays a crucial role in this ecological transition towards a "green" economy within the EU, and elsewhere. The quest for a society that is coexisting with nature without compromising our future abilities, while balancing the needs of a steadily increasing world economy, strongly depends on whether we will be able to adapt to the changes in the water cycle following climate change [27]. The ecological transition in the EU has obliged to strengthen coordination across the board and integrate all social sectors. All this is to change the management and sustainable use of land and water resources and fight desertification, drought, and non-recoverable resource depletion. This action is critical in the pan-Mediterranean area, where water resource availability is decreasing alarmingly [28].

Integrated water resource management (IWRM) is a concept that emerged in the UN's Mar del Plata conference of 1977 and was defined as a method to provide potable water and sanitation facilities to all and to accelerate political will and investment in the water sector. The transformations needed to implement such a concept were broadly envisaged in Mar del Plata and further elaborated, along with the IWRM concept itself, in Dublin, Rio, The Hague, Bonn Johannesburg, and Kyoto [29]. Currently, IWRM is defined as the method to promote the coordinated development and management of water, land, and related resources to maximize the resulting economic and social welfare equitably without compromising the sustainability of vital ecosystems [30]. This concept has become a paradigm for the UN 2030 Agenda for Sustainable Development, as Target 6.5 of the Sustainable Development Goals calls for the implementation of IWRM at all levels by 2030. IMRW is being embraced by many developed developing and transitional countries [31,32] This paradigm and the climate change context have raised interest among water managers, planners, and stakeholders in the so-called Nature-based Solutions for Water Integrated Resources Management (NbS-IWRM) [33]. This concept consists of the application of actions that mimic natural processes to improve water availability in quantity and quality, reduce water-related disaster risks, enhance adaptation to climate change, and increase socio-ecosystem resilience [34]. Contrary to this, the modern irrigation development where rainfall cannot cover crop growth needs has evolved from the introduction of new physical structures and equipment to a new scheme that looks for a transformation of the management of irrigation water resources, to improve the efficiency and productivity of the resources and services provided to the farmers [35]. This includes the Mediterranean region and other arid and semiarid zones. Unfortunately, this concept of modernization does not respond to the latest challenges of society, which include the depletion of resources, deterioration of the environment, population growth, and climate change [14]. Being aware of this problem, the United Nations Food and Agricultural Organization, the European Commission, and the Spanish Ministry for Ecological Transition and Demographic Challenge, among others, are promoting research on NbS-IWRM [3,36–38]. The Spanish Ministry for Ecological Transition joined World Water Day 2021 with an event entitled "Nature-based solutions for water management in Spain: challenges and opportunities". At this event, the Secretary of State for the Environment underlined the need to look for nature-based solutions to improve the use of water resources by conserving and protecting the headwaters of river basins, and/or by regulating natural flows [39]. Such nature-based solutions can complement conventional infrastructures and reduce the overall costs of water quantity and quality services.

Reported examples of NbS-IWRM applications around the world are scarce and relatively recent. However, in some mountain ranges, such as the Sierra Nevada (Spain) and the Andes (South America), there are good examples of conceivable NbS-IWRM based on the traditional knowledge of local populations [8,40]. The concept of Water Sowing and Harvesting (WS&H) was coined in these areas. WS&H describes the process by which surface runoff water from both snowmelt and rainfall is collected and infiltrated (sown) through a system of channels dug in the upper parts of the mountain basins (Figures 1 and 2) [19,41,42], to be recovered (harvested) elsewhere, sometime later, as a groundwater discharge, for irrigation or domestic use. The delay is due to the slow velocity of groundwater through permeable materials. Such aquifer recharge channels are locally known as "careo channels" in the Sierra Nevada range (Spain) and "amunas" in the Andean Cordillera (South America). The amunas are almost identical to the careo recharge channels in Spain, although developed independently by the pre-Inca cultures in Peru, Chavín initially, and Wari later [7]. The water that is not sown in the area leaves it and goes to the sea or evaporates in flat land downstream where the water cannot be recovered.

Figure 1. Careo recharge channel in the Bérchules watershed, located at Spain's southern slopes of the Sierra Nevada range (Photo: Sergio Martos-Rosillo).

Figure 2. Example of an area where water flowing through a careo channel is released for infiltration. The infiltration zone is 6 km from the beginning of the channel. When the photograph was taken (April 2021), a flow rate of 250 L/s infiltrated. The seepage water generates pastures in the neighboring infiltration zones. (Photo: Blas Ramos).

The careo recharge system has at least three functions:

(i) delaying the transit time of water through the ground to maintain the flow of rivers and springs at lower altitudes during the summers, when radiation and temperature favor crop growth, but rainfall is scarce, and the demand for drinking water increases,
(ii) watering the vegetation on the mountain slopes (Figure 2), favoring the growth of pastures, and enhancing biodiversity, and
(iii) improving water quality by diluting the salinity of evapo-concentrated groundwater and filtering runoff water. Therefore, the spatio-temporal regulation of water resources

for different uses and the associated ecosystem services qualifies this WS&H "green infrastructure" as an NbS-IWRM [43].

The importance of the careo recharge channels and their hydrological, environmental, and socio-economic services, belong to the ancestral knowledge and cultural heritage of the local population, which has kept them operational in the Southwestern Iberian Peninsula since at least the 11th century [4,5]. However, it is only very recently that water and environmental authorities have recognized their importance. The water authorities of the Guadalquivir River basin and of the southern basin of Andalucia (the two basins where recharge channels play an important regulating role), the provincial administration of Granada, the Sierra Nevada National and Natural Parks, and the Association of Historical and Traditional Irrigation Communities of Andalusia have just started to address the careo system by incorporating it into their planning, maintenance, and surveillance activities. Between 2008 and 2011, the Sierra Nevada National Park and the Department of the Environment of the Regional Government of Andalusia invested 5.3 million euros through the project "Conservation of the traditional careo recharge channels of Sierra Nevada". References to the economic interest of the careo channels are also found in the 2nd Plan for the Sustainable Development of the Natural Area for the Sierra Nevada, approved in 2018 by the regional government as a part of the Natural Resources Management Plan. Furthermore, at a national level, the collaboration between water management agencies and researchers has been sought to investigate similar NbS-IWRM in Guadarrama, Gredos and Sierra Morena ranges, and the Canary Islands. Interest in the careo recharge channels has also increased internationally. Examples are (1) the Research Network "Water Sowing and Harvesting in Protected Natural Areas" (https://www.cyted.org/es/syca (accessed on 30 August 2022)), funded by the Ibero-American Programme of Science and Technologies for Development, with additional support for training activities from the INTERCOONENTA program of the Spanish Agency for International Development Cooperation, which unites 76 researchers, water, and environmental planners from eight Ibero-American countries, (2) their relevance to the Focus Group "Nature-Based Solutions for water management under climate change" of the European Innovation Partnership for Agricultural Productivity and Sustainability [44], and (3) by UNESCO's Intergovernmental Hydrological Programme as Demonstration Site in its Global Network of Ecohydrology (http://ecohydrology-ihp.org/demosites (accessed on 31 August 2022)). This action reflects the applicability of such an NbS-IWRM system in many mountain areas with similar climate conditions to those prevailing in the Sierra Nevada range [45]. Such are as the southern slopes of the Alps in France and Italy, the Dinaric Alps in Croatia, Mount Etna in Italy, the Atlas Mountains in Morocco, the Taurus Mountains in Turkey, the Lebanese Cordillera, the Sierra Nevada range in the United States, or the Andes Cordillera in South America.

3. Science to Understand Better Recharge Channels

The first scientific publications on the hydrology and hydrogeology of these ancestral water recharge systems are recent. The earliest paper by Pulido-Bosch and Sbih (1995) [41] described careo recharge channels in the southern slopes of Sierra Nevada (SE-Spain) and measured groundwater residence times from 5 to 10 days by applying dissolved lithium chloride (LiCl) to the water flowing in the Cástaras careo channel, which is located in the Trevélez River basin. In the same basin, Oyonarte et al. (2022) [14] measured for the Busquístar channel a mean infiltration rate per unit length of channel ($\hat{q}$) of 9.32 L/s/km. In the neighboring Bérchules Basin, Martos Rosillo et al. (2019) [19] obtained a $\hat{q}$ value of 20.2 L/s/km in the Espino channel. Here, they measured infiltration rates up to 400 L/s in some channel zones. In the Peruvian Andes, Cárdenas-Panduro (2020) [46] obtained a $\hat{q}$ value of 88.7 L/s/km for the Saywapata channel. Such high $\hat{q}$ values evidence the high infiltration capacity of the recharge channel system. Analyzing the importance of the channel recharge with respect to that of the natural water cycle, Jódar et al. (2022) [20] showed that the total channel recharge in the Bérchules watershed during the hydrological year 2014–2015 was 3.66 hm^3, which is equivalent to 70% of the river water flow at the outlet of the basin (5.3 hm^3) and amounts 48% of the total aquifer recharge for this period (7.62 hm^3). They have demonstrated that this ancestral aquifer recharge system can double natural recharge rates as it increases the average and base groundwater discharge of downstream springs and the mainstream during the summer [6,20,47,48]. Yapa (2016) [21] and Martos-Rosillo et al. (2020a) [8] described similar ancestral methods of water recharge in the Americas since pre-Columbian times. In addition, Ochoa-Tocachi et al. (2019) [7] studied a 1400-year-old rainfall-runoff infiltration enhancement system in the Andes, which is based on amunas. These authors used eosine to trace the recharged water, obtaining that groundwater is held for an average of 45 days before resurfacing, and assessed the effects of this water management technique on the water supply of Lima using a rainfall-runoff model. As research results are very promising, Peruvian local water planners are encouraging the use of this green infrastructure through Mechanisms of Rewards for Ecosystem Services, which allows financing such NbS-IWRM practiced by local peasants but benefiting other downstream water users. However, the greater or lesser impact of these solutions is location-specific and therefore requires deep scientific or traditional knowledge. For instance, Somers et al. (2018) [49] measured scant recharge from recharge channels also in the Peruvian Andes, likely because the water available for recharge in their case study was only from rainfall runoff, which is available only a few days per year compared with snowmelt runoff, that may last months. Nevertheless, the hydrological parameters that control the hydrodynamic, hydrogeochemical, and isotopic responses of the slope aquifers in the Sierra Nevada and the Andes, where traditional recharge channels remain operational, are rather unknown.

The effects of these ancient groundwater recharge systems on terrestrial ecosystems are various and not fully understood. Remote sensing observations show an extension of the growing season and an increase in chlorophyll activity of vegetation in areas where the careo recharge is conducted [50–53]. However, there are no data regarding how the system contributes to increase vegetation productivity and carbon sequestration, nor to characterize its role to sustain threatened drought-intolerant species associated with the channels. Moreover, the positive or negative effects on fluvial and riparian ecosystems appear to vary, depending on the altitude of the river reach considered. Water withdrawal in the upper zone of the basin from the river to feed the recharge channels likely impacts the functioning and biodiversity in the reaches immediately downstream of the diversion site. The regulation of river flow has strong effects on the functional diversity of riparian vegetation [54,55] and aquatic communities such as amphibians and fish [56]. Ecosystem functioning (i.e., primary production, organic matter decomposition, nutrient cycling) can also be impacted by the river flow regulation, particularly in streams from semiarid regions, which are characterized by a high rainfall seasonality [57]. These alterations will worsen in Mediterranean streams as climate change proceeds as suspected (e.g., Salinas et al., 2018 [58]). In either case, there is a need to investigate the magnitude and positive or negative direction of such impacts, in terms of flow regulation as a function of the relative volume and seasonality of water withdrawal.

On the other hand, downstream reaches receiving groundwater from slope aquifers recharged by channel recharge systems could likely provide better conditions for biodiversity and ecosystem functioning and services. In these lower river reaches, groundwater inputs produce higher discharge and relatively low water temperature, improving habitat quality, particularly during summer, for many cold-stenothermal species typical of these rivers (e.g., brown trout). Nevertheless, as this groundwater has passed through slopes with agricultural use (Figure 3), it could transport nutrients (i.e., nitrogen and phosphorous) to the river, favoring eventually eutrophication. Such effects, which may have a positive and/or negative environmental impact, remain unknown. More research is needed to fully understand the behavior of recharge channel systems and their impacts on the associated downstream ecosystems.

Figure 3. Conceptual scheme of hydrogeological behavior of the recharge with careo channels during the snow-melting period in an idealized watershed of Sierra Nevada (Spain). The geological substratum is made up entirely of schists, with a surface alteration zone (Author: Rocío Espín and Sergio Martos-Rosillo).

4. Current Challenges

Demographic and climate changes are threatening the provision of ecosystem services with foreseeable trade-offs that must be considered in the management of the territory's resources. One of the challenges is the exodus of the rural population and the incorporation of new stakeholders. While farmers and shepherds were the lands, water, and careo channel managers for centuries, current actors also seek conservation objectives, such as biodiversity protection, ecotourism, or freshwater provision for growing populations in the lower parts of catchments. In addition, agriculture is intensifying in some areas of the southern slopes of Sierra Nevada (i.e., the Bérchules and Mecina watersheds) in response to favorable market conditions. This may lead to increased demand for irrigation water and the claim for customary water rights, with effects on the availability of water resources similar to those reported in other regions [59,60].

Recharge channels in mountain regions have successfully overcome drastic social [4,5] and climatic changes that have occurred in the Sierra Nevada range since the Middle Ages [17,18]. Further back from as early as the 5th century in the Peruvian Andes [7], recharge channels have played an important role. According to palaeoclimatic reconstructions of the last two millennia, the period from 700 to 1200 was dry and prone to severe drought. After that, the climate became somewhat wetter (Åkesson et al., 2020 [61], and references therein), but not very different from the current climate conditions in the Andean Cordillera at the same latitude, where arid to hyperarid conditions still prevail [62]. The question is how, by delving into the hydrology and hydrogeology of these systems, valuing their ecosystem services, and understanding the effects that socio-economic changes have on traditional organizational structures, we can adapt these WS&H systems to the current context, harnessing their climate change adaptation values and ensuring the resilience that they have shown historically.

The careo channels may become an adaptation measure to climate change. Delineating their role in this regard, when implementing them in dissimilar conditions, including different (1) climate change projections, (2) forcing levels (greenhouse gas emission pathways), (3) socio-economic scenarios, and (4) management alternatives, vulnerability models may be helpful (Joyce y Janowiak, 2011 [63]). They make it possible to analyze the degree to which an ecosystem is affected by climate change and to evaluate the consequences of different adaptation strategies. This information, together with an adequate evaluation and communication of the uncertainties associated with the different scenarios, is the cornerstone for an adequate decision-making process and the implementation of the careo channels as an effective adaptation measure.

5. Conclusions

The careo recharge as NbS-IWRM may enhance biodiversity and ecosystem functioning, both terrestrial and aquatic, at the basin scale. Therefore, to understand the inner workings of the "Recharge channel-Soil-Aquifer-River" system, its social and environmental repercussions, and to maintain and replicate this NbS-IWRM system in other areas with similar characteristics, more in-depth and multidisciplinary research is needed. This research should provide information on (1) how to adapt the careo channels to the new social and climatic scenarios, (2) the hydraulic and hydrogeological variables to take into account when designing new recharge systems in other mountain areas with similar characteristics, and (3) how to maximize the ecosystem services provided. This historical water management system, based on local ecological knowledge and communal practices, in which a balance between land and water use has been attained, should become an adaptation measure to climate change, but also to build a better, more secure, and equitable future through the ecological transition path towards the objectives of the European Green Deal. This is especially important to stabilize the rural population and to preserve the environmental, hydrological, ecological, cultural, and economic conditions in mountainous areas and make compatible the roles of the local economy and "nature gardeners".

Author Contributions: J.J. wrote the paper with further contributions from all authors (i.e., S.M.-R., E.C., L.M., J.C. (Javier Cabello), J.C. (Jesús Casas), M.J.S.-B., J.M.M.-C., A.G.-R., T.Z., C.H.-L., J.U., L.J.L.). All authors were involved and participated in the discussion of ideas, read, and approved the final version of the manuscript. L.J.L., J.C. (Javier Cabello), S.M.-R. and J.J. were principal investigators of research projects that funded this work. All authors have read and agreed to the published version of the manuscript.

Funding: This work was undertaken as part of the projects "Impact, monitoring and assessment of global and climate change on water resources in high-mountain National Parks (CCPM)" (SPIP2021-02741) and "Soluciones basadas en la naturaleza para la gestión resiliente del ciclo hidrológico en zonas de montaña: los sistemas tradicionales de gestión del agua de Sierra Nevada" (NBS4WATER, Ref 2768/2021) funded by Organismo Autónomo Parques Nacionales from the Ministerio para la Transición Ecológica y el Reto Demográfico. The authors thank the Ibero-American Science and Technology for Development Programme (CYTED) for its financial support to the network "Water Sowing and Harvesting in Protected Natural Areas" (419RT0577). Besides, this work was supported by the "Severo Ochoa" extraordinary grants for excellence IGME-CSIC (AECEX2021).

Data Availability Statement: The data supporting reported results can be found in the cited bibliography.

Acknowledgments: We would like to thank the anonymous reviewers for their constructive comments and suggestions which led to a substantial improvement of the paper.

Conflicts of Interest: The authors declare no conflict of interest.

References

1. United Nations. *Transforming Our World: The 2030 Agenda for Sustainable Development(A/RES/70/1)*; UN General Assembly: New York, NY, USA, 2015; Available online: https://sdgs.un.org/2030agenda (accessed on 30 August 2022).
2. Belda, E.T. El Agua en el Contexto de la Transición Ecológica. Entrevista a Pedro Arrojo Agudo [Water in the Context of the Ecological Transition. Interview with Pedro Arrojo Agudo]. *Relac. Int.* **2020**, *45*, 377–383. Available online: https://revistas.uam.es/relacionesinternacionales/article/view/12900 (accessed on 30 August 2022).
3. United Nations. *Nature-Based Solutions for Water 2018: The United Nations World Water Development Report 2018*; UN-Water United Nations World Water Assessment Programme: Geneva, Switzerland, 2018. Available online: https://wedocs.unep.org/20.500.11822/32857 (accessed on 30 August 2022).
4. Martín-Civantos, J.M. *Poblamiento y Territorio Medieval en el Zenete, Granada [Medieval Settlement and Territory in Zenete, Granada]*; Editorial Universidad de Granada: Granada, Spain, 2007; p. 773.
5. Martín Civantos, J.M. Las Aguas del Río Alhama de Guadix y el Sistema de Careos de Sierra Nevada (Granada) en Época Medieval [The Waters of the River Alhama de Guadix and the System of Careos de Sierra Nevada (Granada) in medieval times].

In *El Paisaje y su Dimensión Arqueológica. Estudios Sobre el Sur de la Península Ibérica en la Edad Media*; Alhulia: Granada, Spain, 2010; pp. 79–111. Available online: http://opac.regesta-imperii.de/id/2433415 (accessed on 31 August 2022).
6. Martos-Rosillo, S.; González-Ramón, A.; Marín-Lechado, C.; Cabrera, J.A.; Guardiola-Albert, C.; Jodar, J.; Navarrete, E.; Ruiz-Constán, A.; Moral, F.; Pedrera, A.; et al. Las acequias de careo de Sierra Nevada (Sur de España), un sistema de recarga ancestral en acuíferos de alta montaña [The irrigation channels of the Sierra Nevada (Southern Spain), an ancestral recharge system in high mountain aquifers]. In *Manejo de la Recarga de Acuíferos*; Escolero, O., Gutiérrez, C., Mendoza, E., Eds.; Instituto Mexicano de Tecnología del Agua (IMTA): Jiutepec, Mexico, 2017; pp. 527–563. Available online: https://www.imta.gob.mx/biblioteca/libros_html/manejo-recarga-acuiferos-ehl.pdf (accessed on 30 August 2022)Available online: .
7. Ochoa-Tocachi, B.F.; Bardales, J.D.; Antiporta, J.; Pérez, K.; Acosta, L.; Mao, F.; Zulkafli, Z.; Gil-Ríos, J.; Angulo, O.; Grainger, S.; et al. Potential contributions of pre-Inca infiltration infrastructure to Andean water security. *Nat. Sustain.* **2019**, *2*, 584–593. [CrossRef]
8. Martos-Rosillo, S.; Durán, A.; Castro, M.; Vélez, J.J.; Herrera, G.; Martín-Civantos, J.M.; Mateos, L.; Durán, J.J.; González-Ramón, A.; Ruiz Constán, A.; et al. La Siembra y Cosecha del Agua en Iberoamérica; un sistema ancestral de gestión del agua que utiliza Soluciones Basadas en la Naturaleza [Water Sowing and Harvesting in Ibero-America; an ancestral system of water management using Nature-Based Solutions]. In *Tierra y Tecnología*; Ilustre Colegio Oficial de Geólogos: Madrid, Spain, 2020; Volume 55, p. 15. Available online: https://www.icog.es/TyT/index.php/2020/02/la-siembra-y-cosecha-del-agua-en-iberoamerica-un-sistema-ancestral-de-gestion-del-agua-que-utiliza-soluciones-basadas-en-la-naturaleza (accessed on 30 August 2022).
9. Martos-Rosillo, S.; Durán, A.; Castro, M.; Vélez, J.J.; Herrera, G.; Martín-Civantos, J.M.; Mateos, L.; Durán, J.J.; Jódar, J.; Gutiérrez, C.; et al. Ancestral techniques of Water Sowing and Harvesting in Ibero-America: Examples of hydro-geo-ethical systems. In *Advances in Geoethics and Groundwater Management: Theory and practice for a Sustainable Development*; Abrunhosa, M., Chambel, A., Peppoloni, S., Chaminé, H.I., Eds.; Springer: Cham, Switzerland, 2020; pp. 489–492. [CrossRef]
10. Albarracín, M.; Gaona, J.; Chicharo, L.; Zalewski, M. Ecohydrology and Its Implementation in Ecuador. Original in Spanish 2018. 2019. Available online: https://www.ingeraleza.com/ecohidrologia (accessed on 30 August 2022).
11. Albarracín, M.; Ramón, G.; González, J.; Iñiguez-Armijos, C.; Zakaluk, T.; Martos-Rosillo, S. The ecohydrological approach in water sowing and harvesting systems: The case of the Paltas Catacocha ecohydrology demonstration site, Ecuador. *Ecohydrol. Hydrobiol.* **2021**, *21*, 454–466. [CrossRef]
12. Zalewski, M. Ecohydrology: Process-oriented thinking towards sustainable river basins. *Ecohydrol. Hydrobiol.* **2013**, *13*, 97–103. [CrossRef]
13. Zalewski, M. Ecohidrología como un marco para la mejora del potencial de sostenibilidad de cuencas hidrográficas. In *Ecohidrología y su Implementación en el Ecuador*; Albarracín, E.M., Gaona, J., Chicharo, L., Zalewski, M., Eds.; EDILOJA: Loja, Ecuador, 2018; pp. 51–59. Available online: https://www.scopus.com/record/display.uri?eid=2-s2.0-85113306986&origin=inward (accessed on 30 August 2022).
14. Oyonarte, N.A.; Gómez-Macpherson, H.; Martos-Rosillo, S.; González-Ramón, A.; Mateos, L. Revisiting irrigation efficiency before restoring ancient irrigation canals in multi-functional, nature-based water systems. *Agric. Syst.* **2022**, *203*, 103513. [CrossRef]
15. Ramón, G. Formas Ancestrales de Almacenamiento de Agua en los Andes de Páramo: Una Mirada Histórica. 2008. Available online: http://suia.ambiente.gob.ec/documents/783967/889145/Formas+Ancestrales+De+Almacenamiento+De+Agua+En+Los+Andes+De+Paramo+Una+Mirada+Histórica.pdf/aa9aed68-8cf3-416c-ad48-26aefddd5db0;jsessionid=CopLS8YZbIIAhmqASMJVa7Q+ (accessed on 30 August 2022).
16. Ramón, G. Recuperación de saberes ancestrales de los Paltas para el manejo del agua en Catacocha. In *Ecohidrología y su Implementación en Ecuador*; Albarracín, M., Gaona, J., Chicharo, L., Zalewski, M., Eds.; EDIJOJA: Loja, Ecuador, 2018; pp. 134–135.
17. Ramos-Román, M.J.; Jiménez-Moreno, G.; Anderson, R.S.; García-Alix, A.; Toney, J.L.; Jiménez-Espejo, F.J.; Carrión, J.S. Centennial-scale vegetation and North Atlantic Oscillation changes during the Late Holocene in the southern Iberia. *Quat. Sci. Rev.* **2016**, *143*, 84–95. [CrossRef]
18. García-Alix, A.; Toney, J.L.; Jiménez-Moreno, G.; Pérez-Martínez, C.; Jiménez, L.; Rodrigo-Gámiz, M.; Scott Anderson, R.; Camuera, J.; Jiménez-Espejo, F.J.; Peña-Angulo, D.; et al. Algal lipids reveal unprecedented warming rates in alpine areas of SW Europe during the industrial period. *Clim. Past* **2020**, *16*, 245–263. [CrossRef]
19. Martos-Rosillo, S.; Ruiz-Constán, A.; González-Ramón, A.; Mediavilla, R.; Martín-Civantos, J.M.; Martínez-Moreno, F.J.; Jódar, J.; Marín-Lechado, C.; Medialdea, A.; Galindo-Zaldívar, J.; et al. The oldest managed aquifer recharge system in Europe: New insights from the Espino recharge channel (Sierra Nevada, southern Spain). *J. Hydrol.* **2019**, *578*, 124047. [CrossRef]
20. Jódar, J.; Zakaluk, T.; González-Ramón, A.; Ruiz-Costán, A.; Marín-Lechado, C.; Martín-Civantos, J.M.; Custodio, E.; Urrutia, J.; Herrera, C.; Lambán, L.J.; et al. Artificial recharge by means of careo channels versus natural aquifer recharge in a semi-arid, high-mountain watershed (Sierra Nevada, Spain). *Sci. Total Environ.* **2022**, *825*, 153937. [CrossRef]
21. Yapa, K.A. Nurturing water: Ancestral ground water recharging in the Américas. In Proceedings of the 7th Rural Water Supply Network Forum 2016 Cote d'Ivoire, Abidjan, Cote d'Ivoire, 29 November–2 December 2016; "Water for Everyone". 2016. Available online: https://rwsnforum7.files.wordpress.com/2016/11/full_paper_0067_submitter_0174_a-s-yapa_kashyapa.pdf (accessed on 30 August 2022).
22. Bennet, J.W. The Ecological Transition—Cultural Anthropology and Human Adaptation. 1976. Available online: https://www.elsevier.com/books/the-ecological-transition/bennett/978-0-08-017868-4 (accessed on 30 August 2022).

23. Hopkins, R. The Transition Handbook: From Oil Dependency to Local Resilience. 2008. Available online: https://www.amazon.de/Transition-Handbook-Dependency-Resilience-Guides/dp/1900322188 (accessed on 30 August 2022).
24. Transition-Europe. What is Transition? 2022. Available online: https://www.transition-europe.eu/en/page/definitions-2 (accessed on 30 August 2022).
25. European Commission. The European Green Deal, Communication, COM(2019) 640 final, 11 December 2019. Available online: https://ec.europa.eu/info/sites/info/files/european-green-dealcommunication_en.pdf (accessed on 30 August 2022).
26. COR. European Committee of the Regions: Ecological Transition—What Balance between Social Acceptability and Environmental Imperatives from the Point of View of Cities and Regions with a View to Building Resilient Communities? 150th Plenary Session, 29–30 June 2022. 2022. Available online: https://cor.europa.eu/en/our-work/Pages/OpinionTimeline.aspx?opId=CDR-104-2022 (accessed on 30 August 2022).
27. Ruti, P.M.; Somot, S.; Dubois, C.; Flaounas, E.; Obermann, A.; Dell'Aquila, A.; Pisacane, G.; Harzallah, A.; Lombardi, E.; Ahrens, B.; et al. MED-CORDEX initiative for Mediterranean climate studies. *Bull. Am. Meteorol. Soc.* **2016**, *97*, 1187–1208. [CrossRef]
28. Masseroni, D.; Camici, S.; Cislaghi, A.; Vacchiano, G.; Massari, C.; Brocca, L. The 63-year changes in annual streamflow volumes across Europe with a focus on the Mediterranean basin. *Hydrol. Earth Syst. Sci.* **2021**, *25*, 5589–5601. [CrossRef]
29. Rahaman, M.M.; Varis, O. Integrated water resources management: Evolution, prospects and future challenges. *Sustain. Sci. Pract. Policy* **2005**, *1*, 15–21. [CrossRef]
30. Global Water Partnership (GWP). *Integrated Water Resources Management*; Technical Advisory Committee (TAC): Stockholm, Sweden, 2000.
31. Saravanan, V.S.; McDonald, G.T.; Mollinga, P.P. Critical review of integrated water resources management: Moving beyond polarised discourse. In *Natural Resources Forum*; Blackwell Publishing Ltd.: Oxford, UK, 2009; Volume 33, pp. 76–86.
32. Butterworth, J.; Warner, J.; Moriarty, P.; Smits, S.; Batchelor, C. Finding practical approaches to Integrated Water Resources Management. *Water Altern.* **2010**, *3*, 68–81.
33. WWAP; (United Nations World Water Assessment Programme)/UN-Water. *The United Nations World Water Development Report 2018: Nature-Based Solutions for Water*; UNESCO: Paris, France, 2018. Available online: https://unesdoc.unesco.org/ark:/48223/pf0000261424 (accessed on 30 August 2022).
34. Gunderson, L.H. Ecological Resilience–In Theory and Application. *Annu. Rev. Ecol. Syst.* **2000**, *31*, 425–439. [CrossRef]
35. Playán, E.; Mateos, L. Modernization and optimization of irrigation systems to increase water productivity. *Agric. Water Manag.* **2006**, *80*, 100–105. [CrossRef]
36. Sonneveld, B.G.; Merbis, M.D.; Alfarra, A.; Ünver, O.; Arnal, M.F. Nature-Based Solutions for Agricultural Water Management and Food Security. FAO Land and Water Discussion Paper, p. 12. 2018. Available online: https://www.fao.org/documents/card/es/c/CA2525EN (accessed on 30 August 2022).
37. Bulkeley, H.; Naumann, S.; Vojinovic, Z.; Calfapietra, C.; Whiteoak, K.; Freitas, T.; Vandewoestijne, S.; Wild, T. European Commission, Directorate-General for Research and Innovation. In *Nature-Based Solutions: State of the Art in EU-Funded Projects*; Freitas, T., Vandewoestijne, S., Wild, T., Eds.; Publications Office of the European Union: Luxembourg, 2020. Available online: https://data.europa.eu/doi/10.2777/236007 (accessed on 30 August 2022).
38. MITECO-TNC. Soluciones Basadas en la Naturaleza para la Gestión del Agua en España. Ministerio para la Transición Ecológica [Nature-based solutions for Water Management in Spain. The Nature Conservancy]. 2019. Available online: https://www.miteco.gob.es/es/agua/formacion/soluciones-basadas-en-la-naturaleza_tcm30-496389.pdf (accessed on 30 August 2022).
39. La Moncloa. El Ministerio para la Transición Ecológica se Suma a la Celebración del Día Mundial del Agua. 2021. Available online: https://www.lamoncloa.gob.es/serviciosdeprensa/notasprensa/ecologica/Paginas/2019/210319-diadelagua.aspx (accessed on 30 August 2022).
40. Ribeiro, L. Revisiting ancestral groundwater techniques as nature based solutions for managing water. In *Advances in Geoethics and Groundwater Management: Theory and Practice for a Sustainable Development*; Springer: Cham, Switzerland, 2021; pp. 483–487.
41. Pulido-Bosch, A.; Sbih, Y. Centuries of artificial recharge on the southern edge of the Sierra Nevada (Granada, Spain). *Environ. Geol.* **1995**, *26*, 57–63. [CrossRef]
42. Barberá, J.A.; Jódar, J.; Custodio, E.; González-Ramón, A.; Jiménez-Gavilán, P.; Vadillo, I.; Pedrera, A.; Martos-Rosillo, S. Groundwater dynamics in a hydrologically-modified alpine watershed from an ancient managed recharge system (Sierra Nevada National Park, Southern Spain): Insights from hydrogeochemical and isotopic information. *Sci. Total Environ.* **2018**, *640*, 874–893. [CrossRef]
43. Benedict, M.A.; McMahon, E.T. *Green Infrastructure: Linking Landscapes and Communities*; Island Press: Washington, DC, USA, 2012.
44. EIP-AGRI. European Innovation Partnership for Agricultural productivity and Sustainability. 2022. Available online: https://ec.europa.eu/eip/agriculture/en/focus-groups/nature-based-solutions-water-management-under (accessed on 30 August 2022).
45. Pérez Palazón, M. Análisis de Tendencias en los Flujos de Agua y Energía de la capa de Nieve a Diversas Escalas en Sierra Nevada [Trend Analysis of Snowpack Water and Energy Fluxes at Different Scales in Sierra Nevada]. Ph.D. Thesis, Universidad de Córdoba, Córdoba, Spain, 2019; p. 202. Available online: http://hdl.handle.net/10396/19159 (accessed on 30 August 2022).
46. Cárdenas-Panduro, A. Impacto de las Amunas en la Seguridad Hídrica de Lima. 2020. Available online: https://hdl.handle.net/20.500.12543/4606 (accessed on 30 August 2022).

47. Jódar, J.; Cabrera, J.A.; Martos-Rosillo, S.; Ruiz-Constan, A.; Gonzalez-Ramón, A.; Lambán, L.J.; Herrera, C.; Custodio, E. Groundwater discharge in high-mountain watersheds: A valuable resource for downstream semi-arid zones. The case of the Bérchules River in Sierra Nevada (Southern Spain). *Sci. Total Environ.* **2017**, *593*, 760–772. [CrossRef]
48. Jódar, J.; Carpintero, E.; Martos-Rosillo, S.; Ruiz-Constán, A.; Marín-Lechado, C.; Cabrera-Arrabal, J.A.; Navarrete-Mazariegos, E.; González-Ramón, A.; Lambán, L.J.; Herrera, C.; et al. Combination of lumped hydrological and remote-sensing models to evaluate water resources in a semi-arid high altitude ungauged watershed of Sierra Nevada (Southern Spain). *Sci. Total Environ.* **2018**, *625*, 285–300. [CrossRef]
49. Somers, L.D.; McKenzie, J.M.; Zipper, S.C.; Mark, B.G.; Lagos, P.; Baraer, M. Does hillslope trenching enhance groundwater recharge and baseflow in the Peruvian Andes? *Hydrol. Processes* **2018**, *32*, 318–331. [CrossRef]
50. Cazorla, B.P.; Cabello, J.; de Giles, J.P.; Sánchez, E.G.; Harker, A.R.; Alcaraz-Segura, D. Funcionamiento de la vegetación y diversidad funcional de los ecosistemas de Sierra Nevada [Vegetation functioning and functional diversity of Sierra Nevada ecosystems]. In *Biología de la Conservación de Plantas en Sierra Nevada. Principios y Retos para su Preservación*; Peñas, J.G., Lorite, J.M., Eds.; Editorial Universidad de Granada: Granada, Spain, 2019; pp. 303–321.
51. Cazorla, B.P.; Cabello, J.; Peñas, J.; Garcillán, P.P.; Reyes, A.; Alcaraz-Segura, D. Incorporating ecosystem functional diversity into geographic conservation priorities using remotely-sensed Ecosystem Functional Types. *Ecosystems* **2020**, *24*, 548–564. [CrossRef]
52. Cabello, J.; Alcaraz-Segura, D.; Reyes-Díez, A.; Lourenço, P.; Requena, J.M.; Bonache, J.; Castillo, P.; Valencia, S.; Naya, J.; Ramírez, L.; et al. Sistema para el Seguimiento del funcionamiento de ecosistemas en la Red de Parques Nacionales de España mediante Teledetección. *Rev. De Teledetección* **2016**, *46*, 119–131. [CrossRef]
53. Cabello, J.; Ruiz, J. *Aplicación de la Plataforma REMOTE a la Red de Parques Nacionales de España: Identificación de Procesos y Acciones de Gestión Susceptibles de ser Monitoreadas [Application of the REMOTE Platform to the Spanish National Parks Network: Identification of Processes and Management Actions that can be Monitored]*; TRAGSATEC, OAPN: Madrid, Spain, 2020.
54. Bejarano, M.D.; Jansson, N.; Nilsson, C. The effects of hydropeaking on riverine plants: A review. *Biol. Rev.* **2018**, *93*, 658–673. [CrossRef] [PubMed]
55. Bejarano, M.D.; Nilsson, C.; Aguiar, F.C. Riparian plant guilds become simpler and most likely fewer following flow regulation. *J. Appl. Ecol.* **2018**, *55*, 365–376. [CrossRef]
56. Oliveira, A.G.; Baumgartner, M.T.; Gomes, L.C.; Dias, R.M.; Agostinho, A.A. Long-term effects of flow regulation by dams simplify fish functional diversity. *Freshw. Biol.* **2018**, *63*, 293–305. [CrossRef]
57. Gallart, F.; Llorens, P.; Latron, J.; Regüés, D. Hydrological processes and their seasonal controls in a small Mediterranean mountain catchment in the Pyrenees. *Hydrol. Earth Syst. Sci.* **2002**, *6*, 527–537. [CrossRef]
58. Salinas, M.J.; Casas, J.J.; Rubio-Ríos, J.; López-Carrique, E.; Ramos-Miras, J.J.; Gil, C. Climate-driven changes of riparian plant functional types in permanent headwater streams. Implications for stream food webs. *PLoS ONE* **2018**, *13*, e0199898. [CrossRef]
59. Stahl, K.; Hisdal, H.; Hannaford, J.; Tallaksen, L.M.; Lanen, H.A.J.; van Sauquet, E.; Demuth, S.; Fendekova, M.; Jódar, J. Streamflow trends in Europe: Evidence from a dataset of near-natural catchments. *Hydrol. Earth Syst. Sci.* **2010**, *14*, 2367–2382. [CrossRef]
60. Lorenzo-Lacruz, J.; Vicente-Serrano, S.M.; López-Moreno, J.I.; Morán-Tejeda, E.; Zabalza, J. Recent trends in Iberian streamflows (1945–2005). *J. Hydrol.* **2012**, *414*, 463–475. [CrossRef]
61. Åkesson, C.M.; Matthews-Bird, F.; Bitting, M.; Fennell, C.J.; Church, W.B.; Peterson, L.C.; Valencia, B.G.; Bush, M.B. 2100 years of human adaptation to climate change in the High Andes. *Nat. Ecol. Evol.* **2020**, *4*, 66–74. [CrossRef]
62. Betancourt, J.L.; Latorre, C.; Rech, J.A.; Quade, J.; Rylander, K.A. A 22,000-year record of monsoonal precipitation from northern Chile's Atacama Desert. *Science* **2000**, *289*, 1542–1546. [CrossRef] [PubMed]
63. Joyce, L.A.; Janowiak, M.K. Climate Change Assessments. U.S. Department of Agriculture, Forest Service, Climate Change Resource Center. 2011. Available online: www.fs.usda.gov/ccrc/topics/vulnerability-assessments (accessed on 31 August 2022).

remote sensing

Article

The Effect of Controlled Tile Drainage on Growth and Grain Yield of Spring Barley as Detected by UAV Images, Yield Map and Soil Moisture Content

Renata Duffková [1,*], Lucie Poláková [1,2], Vojtěch Lukas [3] and Petr Fučík [1]

1 Department of Hydrology and Water Conservation, Research Institute for Soil and Water Conservation, Žabovřeská 250, Praha 5, 156 27 Zbraslav, Czech Republic
2 Department of Water Resources and Environmental Modeling, Czech University of Life Sciences Prague, Kamýcká 129, 165 00 Suchdol, Czech Republic
3 Department of Agrosystems and Bioclimatology, Mendel University in Brno, Zemědělská 1, 613 00 Brno, Czech Republic
* Correspondence: duffkova.renata@vumop.cz

Abstract: Controlled tile drainage (CTD) practices are a promising tool for improving water balance, water quality and increasing crop yield by raising shallow groundwater level and capillary rise due to drainage flow retardation. We tested the effect of CTD on growth and grain yield of spring barley, at a study site in central Bohemia using vegetation indices from unmanned aerial vehicle (UAV) imagery and Sentinel-2 satellite imagery. Tile drainage flow was slowed by fixed water level control structures that increased soil moisture in the surrounding area according to the terrain slope. Vegetation indices based on red-edge spectral bands in combination with near-infrared and red bands were selected, of which the Normalized Red Edge-Red Index (NRERI) showed the closest relationships with shoot biomass parameters (dry biomass, nitrogen concentration and uptake, nitrogen nutrition index) from point sampling at the tillering stage. The CTD sites showed significantly more biomass using NRERI compared to free tile drainage (FTD) sites. In contrast, in the period prior to the implementation of CTD practices, Sentinel-2 satellite imagery did not demonstrate higher biomass based on NRERI at CTD sites compared to FTD sites. The grain yields of spring barley as determined from the yield map also increased due to CTD (by 0.3 t/ha, i.e., by 4%). The positive impact of CTD on biomass development and grain yield of spring barley was confirmed by the increase in soil moisture at depths of 20, 40 and 60 cm compared to FTD. The largest increase in soil water content of 3.5 vol% due to CTD occurred at the depth of 40 cm, which also had a higher degree of saturation of available water capacity and the occurrence of crop water stress was delayed by 14 days compared to FTD.

Keywords: controlled tile drainage; UAV images; red-edge vegetation indices; spring barley biomass; grain yield; soil moisture

Citation: Duffková, R.; Poláková, L.; Lukas, V.; Fučík, P. The Effect of Controlled Tile Drainage on Growth and Grain Yield of Spring Barley as Detected by UAV Images, Yield Map and Soil Moisture Content. *Remote Sens.* **2022**, *14*, 4959. https://doi.org/10.3390/rs14194959

Academic Editors: Qianqian Zhang, Songhao Shang, Dongqin Yin, Hamza Gabriel and Magdy Mohssen

Received: 9 September 2022
Accepted: 1 October 2022
Published: 5 October 2022

1. Introduction

In the context of more frequent periods of drought under climate change, it is desirable to reduce water runoff from agricultural land [1–4]. Fields systematically tile-drained to provide suitable conditions for crop production removing excess water from soil profile usually drain water all year round, depending on topography, the crop grown and the hydro-pedological and meteorological conditions [5–8]. Also, nutrients such as nitrogen (N) in the form of nitrates are transported by subsurface drainage at a time when they could be used by growing crops [9–11]. To avoid this inefficient water and nutrients runoff from drained fields, controlled tile drainage (CTD) practices, where drainage flow is retarded by water level control structures (WLCS), are applied especially in some parts of the North America and Europe to retain water and nutrients in drains and related surrounding soil for crop use [12–19]. The purpose of CTD is to increase soil moisture by capillary rise of the regulated drainage water in the surrounding area, being determined by terrain slope (the so-called range of drainage flow control, RDFC). Drainage water under CTD can only drain from the WLCS when its level reaches the height of the gate (board), inserted in WLCS, falls over it and flows downwards in drainage manholes or outlets. Thus, CTD is a promising approach for improving water balance, water quality and increasing crop yields [5,12,20,21]. The agronomic and environmental benefits of CTD are associated with an increase in soil moisture during the growing season, i.e., better availability of water and nutrients (especially N) for the crops grown [22,23]. Likewise, the formation of an anoxic environment in water-saturated soil for plants by CTD is reduced by diverting the water as soon as the water level reaches a critical level set by the WLCS [5].

Drainage discharge in CTD approach can be controlled by several WLCS types; basically, with adjustable or fixed WLCS placed either at drainage outlets, in manholes or directly on collective or conductive drains [6,15,19,21]. The adjustable WLCS pose an advantage to readily respond to precipitation, runoff conditions and the actual crop water requirements by setting the height of WLCS at the requested level [14]. The WLCS with fixed height could be placed directly on drains and raise the water level according to terrain slope. In more sloping conditions (up to 5%), more densely placed WLCS are needed as their effect on water level rise is less than in flat areas. To assess the effect of increased nutrient and water availability imposed by CTD practices on crop productivity, vegetation indices as spectral reflectance indicators derived from multispectral remote sensing can be useful to diagnose the plant nutritional status and environmental stress symptoms [5,24,25]. Vegetation indices are mostly based on spectral reflectance in the red (R, 630–690 nm) and near-infrared (NIR, 770–1300 nm) bands of the electromagnetic spectrum [26]. Reflectance in the NIR spectrum is related to the plant cell walls and rises as the amount of biomass increases, while reflectance in the R region related to the amount of chlorophyll only in the upper leaves is very low and after reaching a certain amount of the biomass it remains at a minimum constant level, which leads to saturation phenomenon of vegetation indices [27,28]. The commonly used vegetation index associated with R and NIR wavebands is Normalized Difference Vegetation Index (NDVI, [29]) mostly recommended in monitoring green vegetation cover. However, when the vegetation cover becomes dense, i.e., when the leaf area index (LAI) is higher than three, NDVI tends to be saturated, leading to underestimated biomass yield predictions [30–33]. Saturation effect can be reduced by using vegetation indices based on reflectance in a narrow band of the red-edge region (RE, 700–750 nm, e.g., REIP, NDRE, NRERI and RENDVI), showing sensitive increase corresponding strongly to vegetation chlorophyll content and the plant N uptake [34–36]. Mittermayer et al. [37] reported REIP (Red Edge Inflection Point) as a suitable vegetation index to identify site-specific N uptake, high N surplus as well as N loss potential. Vegetation indices based on green reflectance, e.g., GNDVI, GRDVI or MTVI2, also show higher sensitivity to changes in chlorophyll content [38–40].

To demonstrate the prediction of crop biomass production using vegetation indices, yield maps produced in precision agriculture as a result of combine harvester yield sensing system appear to be a useful tool for this purpose. In some cases, vegetation indices are involved within the filters used for spatial interpolation of yield maps that improve the management of soil variability on the farm [41]. Vegetation indices and yield maps were also used to determine the spatial variability of N uptake and N balances [37].

The aim of this study was to evaluate the effect of CTD on the growth and grain yield of spring barley at a study site in Central Bohemia in the 2021 growing season using vegetation indices from unmanned aerial vehicle (UAV) imagery and a yield map at harvest. We hypothesized that CTD would positively affect spring barley development and grain yield through increased soil moisture compared to free tile drainage (FTD).

2. Material and Methods

2.1. Study Site

We selected two experimental fields (Za Frajmankou, Pod Hvězdou) at the study site in the Central Bohemia near the village of Hvězda-Malíkovice (50.2247167N, 13.9741050E, Figure 1). The local climate is influenced by the rainfall shadow of the Ore Mountains and according to Quitt [42] is classified as warm with a normal of annual rainfall of 501 mm and air temperature of 8.6 °C. In 2021, annual rainfall was 471 mm (growing season 352 mm), and the average annual temperature was 8.6 °C (growing season 14.5 °C). In the partially tile-drained field Za Frajmankou, we applied CTD practices from autumn 2020 to spring 2021 and examined their impact by measuring soil moisture in the 2021 growing season, when spring barley was grown there. In both fields, two UAV images were taken, and spring barley was sampled from selected points twice during the growing season.

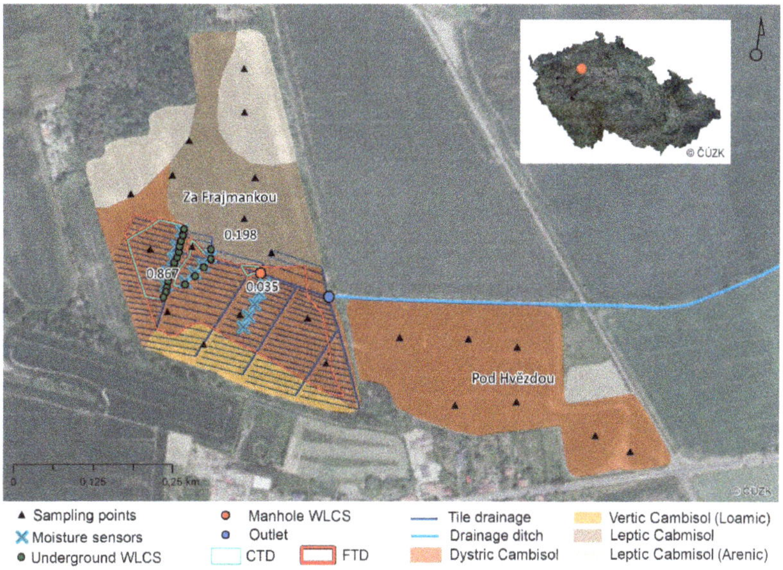

Figure 1. Experimental fields Za Frajmankou and Pod Hvězdou with soil types, sampling points, moisture sensors, underground and manhole water level control structures (WLCS), outlet, drainage ditch and map of the Czech Republic with marked study site. CTD—control tile drainage sites, FTD—free tile drainage sites.

The field Za Frajmankou with an area of 14.6 ha, an average altitude of 364.6 m.a.s.l. and a slope of 3° was partially tile-drained in 1962 (7.18 ha), with drainage spacing 10–14 m and lodgement of drains 0.7–1.1 m below surface, drainage water discharging into a drainage ditch (Figure 1). The average slope of the tile-drained area is 3.15° (0.04–9.10°). The field is soil heterogeneous with different types of Cambisols. The lower part of the experimental field with Vertic Cambisol (Loamic) and Dystric Cambisol [43] is texturally diverse (sandy loam, loam, clay loam, clay), predominantly under tile drainage, which developed on sediments from the Permo-Carboniferous period. The higher parts of the field are dominated by sandstone deposits with texturally lighter Leptic Cambisol and Leptic Cambisol (Arenic) classified as sandy loam and loam.

The field Pod Hvězdou (7.4 ha, Dystric Cambisol), located southeast of the field Za Frajmankou, was not drained, and was used only to expand the number of spring barley sampling points to increase the confidence of the statistical analyses.

In both fields, N-P fertiliser (26% N, 14% P_2O_5) was applied on 1 April 2021 at a rate of 350 kg/ha (i.e., 91 kg N/ha) and spring barley (variety Solist) was sown on 2 April 2021.

2.2. Installation of a Manhole and Water Level Control Structures

The relatively low slope of the field Za Frajmankou provided suitable conditions for slowing drainage flow with CTD practices. In autumn 2020, we installed a total of 20 underground fixed WLCS at 0.7–1.1 m depth below soil surface on selected conductive (main) and collective drains of two drainage groups (Figure 1). The underground fixed WLCS consists of a horizontal PVC pipe with a vertically connected branch (diameter of 110 mm), containing a 6 mm thick polypropylene gate (Figure 2). At all WLCS installation sites, we first removed three original ceramic drainage tiles and replaced them by a PVC drainage pipe with a branch. A gate was inserted into the top hole of the branch and then down leakproof to the PVC pipe. The height of the gate was adjusted so that after inserting the cap at the top of the branch (leaving ca 5 cm for water overflow), approx. 40 cm remained to the soil surface (a safe distance to ensure that the WLCS would not be damaged by ploughing).

Figure 2. Underground fixed WLCS consisting of a drainage T branch (yellow), a polypropylene gate (black) and a PVC pipe vertically connected to the T branch (grey) with a cap (brick-red); before and after installation on tile drainage.

In autumn 2020, we built a control manhole (80 cm in diameter and 1.5 m deep), into which drainage water was connected through 75 mm diameter PVC pipes from CTD and FTD sites and discharged further into the drainage ditch. We installed the same fixed WLCS as on the tile drainage along with a propeller flow meter in the manhole on the PVC pipe bringing water from the CTD sites (Figure 3). This WLCS included a slide valve that, when manually pulled out, would allow the groundwater level (GWL) to be lowered if necessary (as opposed to the underground fixed CWLS). However, it was not needed to control the water level in this way during the monitoring period. The RDFC area of 1.1 ha (0.867 + 0.198 + 0.035, Figure 1) achieved by the installation of both the underground and manhole WLCS and delineated by the known water level (i.e., the height of WLCS) and contours according to DMR 5G (with the declared mean height error 0.18 m in exposed terrain) is considered as an area affected by CTD.

Figure 3. Fixed WLCS installed in the manhole with propeller flow meter (left) and water overflow after a rainfall-runoff event on 14 May 2021.

2.3. Soil Moisture, Soil Texture and Soil Hydrolimits

To determine the differences in soil moisture between CTD and FTD sites, we measured volumetric water content in the field Za Frajmankou between 19 April and 28 July 2021. We installed a total of 25 sensors (7 sensors at a depth of 20 cm and 9 sensors each at depths of 40 and 60 cm) at 9 locations where RDFC occurred because of increased capillary rise via WLCS (Figure 1). At sites where WLCS were not installed, a total of 15 sensors (5 sensors each at 20, 40, and 60 cm depths) were installed at 5 sites (Figure 1). Soil water volumetric content was measured using the SMT-100 soil moisture probes based on a Time Domain Transmission technology and soil water content reflectometer CS650 (Figure 4). Data were stored in a datalogger at hourly intervals. All sensors were located at Dystric Cambisol.

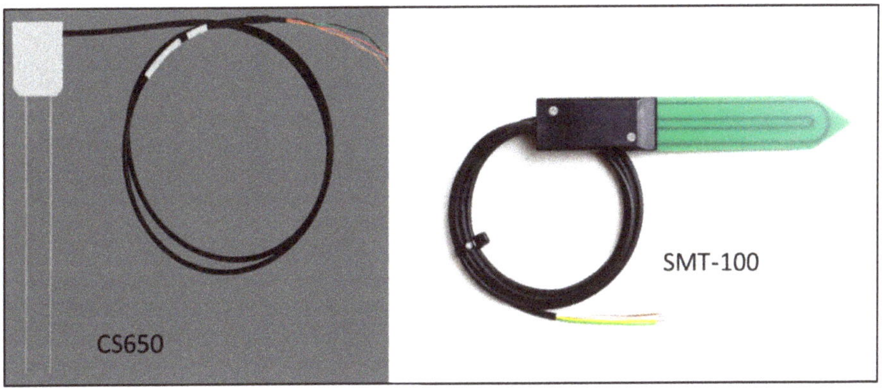

Figure 4. Soil volumetric water sensors CS650 and SMT-100.

For each soil moisture sensor, we collected soil samples to determine the soil texture class using the pipette method [44]. A fine particle size fraction (FPSF, %) < 0.01 mm was used to calculate soil hydrolimits: field capacity (FC), point of decreased availability

(PDA), permanent wilting point (WP) and available water capacity (AWC) using simple pedotransfer functions [45]:

$$FC\ (vol.\%) = 6.66 + 1.03 \times (FPSF) - 0.008 \times (FPSF)^2 \quad (1)$$

$$WP\ (vol.\%) = 2.97 + 0.33 \times (FPSF) - 0.0012 \times (FPSF)^2 \quad (2)$$

$$AWC\ (vol.\%) = FC - WP \quad (3)$$

$$PDA\ (vol.\%) = WP + \%\ AWC \quad (4)$$

FC is the water that remains in the soil after it is thoroughly saturated and free to drain, usually for one to two days. WP is the soil moisture at which plants wilt and do not recover if supplied with sufficient moisture. AWC is the amount of water the soil can retain, and the crops can use. PDA is the minimum soil moisture, expressed as a percentage of AWC, at which plants are still growing and developing successfully (50–60% of AWC for spring barley).

The shallow GWL could have a significant influence on soil water dynamics and is often a profound source of upward water movement by capillary rise. In tile-drained fields across Czechia, based on the experience of the authors from many other sites, the shallow GWL is usually 1.2–1.8 m deep, based on soil morphological and hydrogeological conditions. However, in the trial field, the GWL was not found even at a depth of 2 m in two pits (around the northernmost and southernmost WLCS) which were opened around 3 months prior to WLCS installation. Therefore, we did not measure the GWL depth as it had no effect on soil moisture in the experimental field.

2.4. Point Sampling and Analysis of Spring Barley Biomass

We collected spring barley samples from the selected twenty-two points in the Za Frajmankou and Pod Hvězdou fields (Figure 1) at tillering stage (BBCH 25-29) on 2 June 2021 and prior to harvest on 11 August 2021.

Shoot biomass taken on 2 June of 2021 from an area of 0.25 m^2 was weighed before and after drying at 105 °C and shoot dry weight was converted to t/ha (dry biomass). Subsequently, the N concentration (%) in shoot dry biomass was determined according to the Kjeldahl method [46]. Shoot N uptake was calculated as dry biomass multiplied by N concentration. To assess plant N status, we calculated nitrogen nutrition index (NNI, [47]) as: N_{act}/N_{crit} where N_{act} is the actual and N_{crit} the critical concentration in dry biomass, respectively. Critical N concentration is the minimum concentrations required to achieve maximum shoot growth and was calculated using the power function capturing a typical dilution curve, i.e., decreasing along with increasing shoot dry biomass. N_{crit} (%) was calculated as 5.35 $B^{-0.442}$ [47], where B is shoot dry biomass (t/ha).

Prior to harvest, on 11 August of 2021, 0.2 m^2 of ears were sampled from each sampling point and then, after grain weight was obtained, grain yield (t/ha), grain N concentration according to the Kjeldahl method (%) and grain N uptake (kg/ha) were determined.

2.5. Vegetation Indices Based on UAV and Satellite Imagery

Multispectral images for calculation of vegetation indices and assessment of crop status were acquired by UAV imagery near to the date of plant sampling. We conducted UAV surveys on 2 June (BBCH 25-29, tillering stage) and 30 June 2021 (BBCH 51-57, heading stage) by DJI Phantom 4 Multispectral. This UAV is equipped by multispectral camera which capture five narrow spectral bands—blue (B, center wavelength 450 nm), green (G, 560 nm), R (650 nm), RE (730 nm), NIR (840 nm). Simultaneously, the intensity of incoming radiation is recorded by light sensor installed on the upper part of the UAV for the normalization of incoming light conditions. The survey was carried out at a flight altitude of 140 m; based on the sensor resolution of 1600 × 1300 (2.12 MPx) the final spatial resolution provided by images was 7.56 cm. We ensured radiometric calibration of the multispectral camera by scanning the spectral panel Micasense CRP and using procedures recommended by the manufacturer. Geometric accuracy of acquired images was guaranteed by RTK used in the UAV guidance system and by the placement of 4 ground control points (GCPs) in the observed area.

The orthomosaic of spectral bands was created using the Agisoft Metashape software together with the calculation of the digital surface model (DSM). As a next step, the combined multispectral orthomosaic with all spectral bands was created, from which the set of vegetation indices was subsequently calculated (see Table 1).

Table 1. Vegetation indices calculated from the UAV multispectral images.

Vegetation Index		Equation	Reference
EVI	Enhanced Vegetation Index	2.5 × (NIR − R)/((NIR + 6.0 × R − 7.5 × B) + 1.0)	[48]
EVI2	Enhanced Vegetation Index 2	2.5 × (NIR − R)/(NIR + 2.5 × R + 1)	[49]
GNDVI	Green Normalized Difference Vegetation Index	(NIR − G)/(NIR + G)	[50]
SRI	Simple Ratio Index	NIR/R	[50]
NDRE	Normalized Difference Red Edge Index	(NIR − RE)/(NIR + RE)	[51]
NDVI	Normalized Difference Vegetation Index	(NIR − R)/(NIR + R)	[29]
NRERI	Normalized Red Edge Index	(NIR − RE)/(NIR − R)	[33]
Chl	Chlorophyll Index	(NIR − R)/(RE − R)	[52]
RENDVI	Red-edge NDVI	(RE − R)/(RE + R)	[53]
SAVI	Soil Adjusted Vegetation Index	1.5 × ((NIR − R)/(NIR + R + 0.5))	[54]

To compare crop development prior to the introduction of CTD practices, we used freely available Sentinel-2 satellite imagery, which we obtained from ESA's free data repositories Openhub, CollGS and Google Earth Engine for the period May to June from 2017–2020. Subsequently, we selected cloud-free images and calculated NRERI values using the formula given in Table 1.

2.6. Yield Maps

Crop yield maps were recorded during the harvest of spring barley on 15 August 2021 to analyze the spatial patterns within the field. Raw data were acquired by combine harvester Claas Lexion equipped with sensor system for estimation of grain flow, grain moisture and Differential Global Position System (DGPS) receiver. From the recorded point data, outliers and erroneous values were filtered, followed by spatial interpolation in ESRI ArcGIS using the kriging technique to smooth out the differences at small scale level. The final raster dataset contains information about grain yield in 1 m spatial resolution.

2.7. Statistical Analysis

We used linear regression models to determine the relationships between vegetation indices and shoot or grain parameters of spring barley (shoot parameters: dry biomass, N concentration, N uptake, NNI; grain parameters: yield, N concentration, N uptake) taken from twenty-two sampling points in the Za Frajmankou and Pod Hvězdou fields (Figure 1). We included sampling points from the Pod Hvězdou field in models only to provide more data to increase the power of the test. The closest relationship between vegetation indices and shoot parameters, which was derived from a linear logarithmic regression, was exhibited by the Normalized Red Edge-Red Index (NRERI, [33]) based on the RE band. This vegetation index was further used to test the effect of CTD with defined RDFC on the biomass growth and grain yield of spring barley in the Za Frajmankou field. For this, we selected only the predominantly Dystric Cambisol, i.e., 81% of the tile-drained area of the Za Frajmankou field outside the 29 m strip of headlands to exclude the effect of soil type and agricultural machinery movement on crop development.

The NRERI index from the 2 June 2021 UAV image showed a very wide range of unrealistic values (−476.7 to +266.2) in the 1 × 1 m pixels of the selected tile-drained Dystric Cambisol, mainly caused by the movement of agricultural machinery, low vegetation cover or even bare soil. Hence, 6.26% of the pixels were excluded as outliers, mainly in the track lines. No outliers were identified in the UAV image of 30 June 2021. From the Sentinel-2 imagery, from which two images of the Za Frajmankou field (21 May 2017 and 6 May 2018) with grown winter wheat were selected for comparison before the implementation of CTD practices, outliers of the selected tile-drained Dystric Cambisol were identified only in 2018 (9 pixels of 10 × 10 m at FTD locations).

Outliers were identified using the 25th (Q1) and 75th (Q3) percentiles and the interquartile range (IQR = Q3 − Q1):

$$\text{Outliers} > Q3 + 1.5 \times IQR \text{ or } < Q1 - 1.5 \times IQR \quad (5)$$

To create a balanced data design, we randomly selected two thousand 1 × 1 m pixels with NRERI values from UAV imagery or grain yield values for the CTD and FTD sites using the R script. For Sentinel-2 images, we selected NRERI values from 88 pixels of 10 × 10 m from CTD and FTD sites. For normally distributed data (Sentinel-2 from 6 May 2018), an unpaired two-sample *t*-test was used to identify differences in NRERI values between CTD and FTD sites, and in the case of non-normal data distribution detected by the Shapiro-Wilk test (both UAV imagery and Sentinel-2 from 21 May 2017), the Mann-Whitney U test was used as a non-parametric alternative to the independent two-sample *t*-test.

We used Welch's two-sample *t*-test (unequal variances *t*-test) to test for differences in soil moisture measured at CTD and FTD sites. Each data set contained 101 daily averages from all measuring sensors for each depth (20, 40 and 60 cm).

We conducted all statistical analyses in the R environment [55].

2.8. Creation of Maps

We analyzed the UAV and Sentinel-2 imagery in ArcGIS software (version 10.7.1). The images were first converted to the S-JTSK coordinate system and then cropped with the required layers to the final image using the Extract by Mask function. The Zonal Statistics and Zonal Statistics as Table functions were used to obtain the mean values. For subsequent statistical analyses in the R environment (2.7.), the average pixel values of 1 × 1 m (UAV) or 10 × 10 m (Sentinel-2) were determined by converting the rasters with the required values to a point layer using the Raster to Point function. We added the required attributes to the resulting layers using the join function. The average NRERI values from the UAVs in the vicinity of the sampling points within a 2-m radius area were obtained using the Buffer and Extract by Mask functions. Outlying points were removed from the point layer of the Sentinel-2 image from 6 May 2018 and the UAV image from 2 June 2021, and the layers were then converted back to a raster layer using the Point to Raster function for the resulting visualization.

3. Results

3.1. Vegetation Indices

The three vegetation indices from UAV imagery based on reflectance in either the RE, NIR and R regions (NRERI, Canopy chlorophyll content index Chl) or the RE and NIR regions (Normalized Difference Red Edge NDRE) demonstrated the closest relationships with all shoot parameters from point sampling. Of these, NRERI, based on linear logarithmic regressions with the highest adjusted coefficients of determination (R2adj. in Figure 5a–d), was selected as the best indicator of the effect of CTD on growth and grain yield of spring barley. The outlier value of NRERI (−0.75) was not excluded as it realistically corresponded to a site with poor stand development.

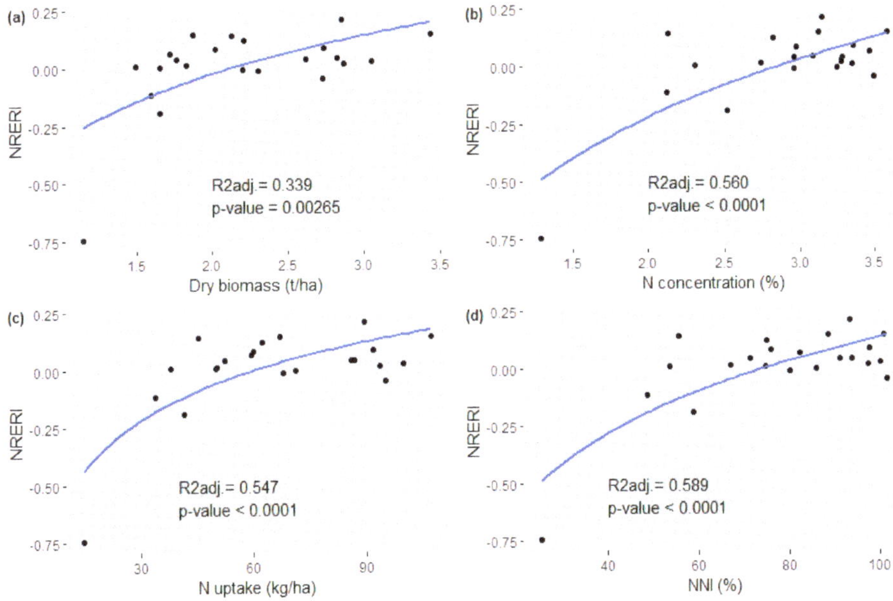

Figure 5. Relationships between NRERI from UAV and spring barley shoot biomass parameters ((**a**) biomass dry weight, (**b**) N concentration, (**c**) N uptake, (**d**) NNI) from point sampling of Za Frajmankou and Pod Hvězdou fields on 2 June 2021 expressed by logarithmic functions with ad-justed coefficient of determination (R2adj.) and *p*-value.

NRERI values differed significantly between the two UAV images depending on the different growth stages of spring barley (Table 2). The low, not fully established crop at the tillering stage (2 June 2021) showed a relatively wide range of values even after excluding outliers (−0.20–0.31). The Mann-Whitney U test revealed statistically significant differences between NRERI values in CTD and FTD sites (*p*-value = 0.0006), although not very clear visually (Figure 6a). On the contrary, considerably higher values with a relatively narrow range even without excluding outliers (0.18–0.60) were shown by the fully established stand at the heading stage (30 June 2021) with a height of 50–70 cm. The differences in NRERI values between CTD and FTD sites were more pronounced compared to the previous UAV image (*p*-value < 0.0001, Figure 6b), as also documented in Figure 7 (i.e., higher NRERI in RDFC).

Table 2. Means, medians and standard deviations of NRERI from UAV images (without outliers from 2 June 2021) and grain yield from the yield map at controlled (CTD) and free (FTD) sites with Dystric Cambisol in the Za Frajmankou field.

Date of UAV Image/Harvest	Means		Medians		Standard Deviations	
	CTD	FTD	CTD	FTD	CTD	FTD
2 June 2021 (NRERI)	0.066	0.055	0.077	0.063	0.084	0.088
30 June 2021 (NRERI)	0.434	0.395	0.435	0.395	0.032	0.033
15 June 2021 (grain yield, t/ha)	7.103	6.807	7.095	6.831	0.193	0.204

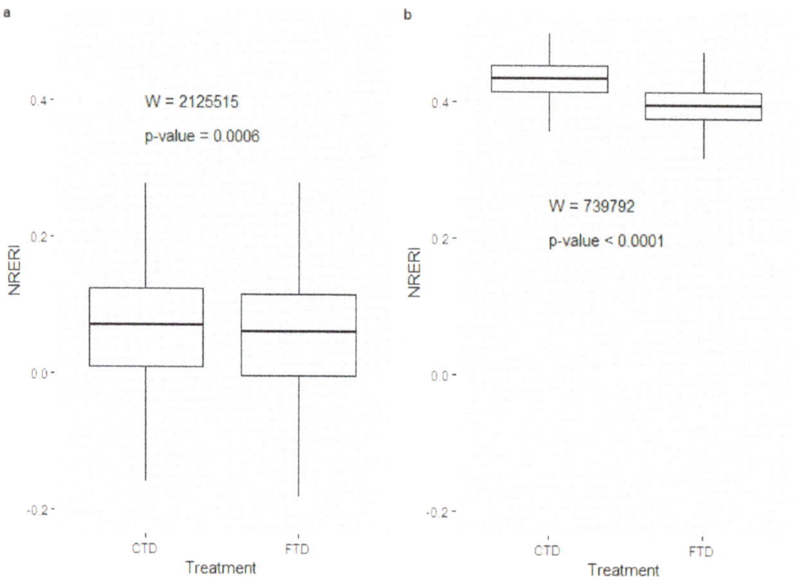

Figure 6. NRERI box plots with Mann-Whitney U test (W, *p*-value) at CTD and FTD sites with Dystric Cambisol in the Za Frajmankou field from UAV imagery on (**a**) 2 June 2021 and (**b**) 30 June 2021.

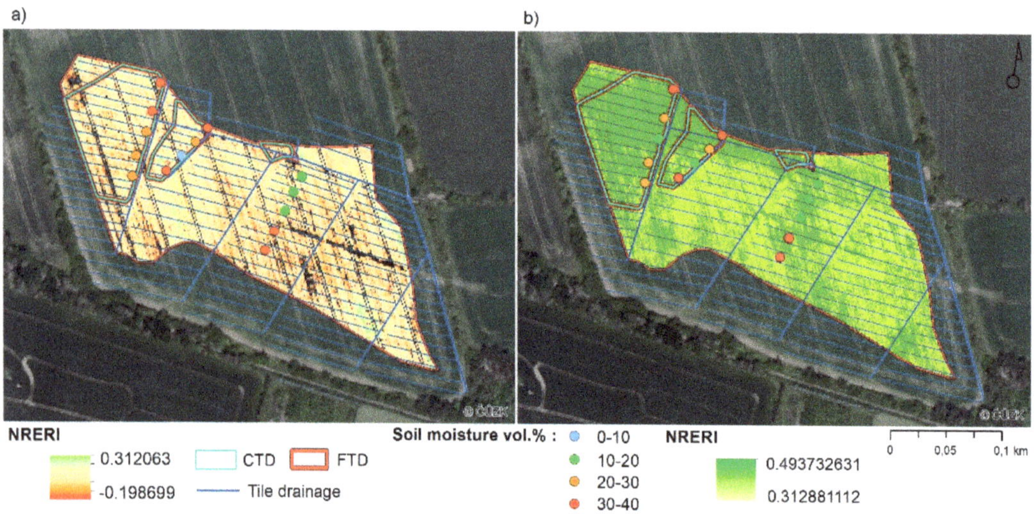

Figure 7. Distribution of NRERI and point soil moisture from 40 cm at CTD and FTD sites with Dystric Cambisol (without headlands) of the Za Frajmankou field taken by UAV on (**a**) 2 June 2021 (ex-cluded outliers in black)and (**b**) 30 June 2021.

We found no differences in NRERI values in the two selected Sentinel-2 images between CTD and FTD sites (Figure 8), as demonstrated by the results of the Mann-Whitney U test for 21 May 2017 (W 14 808, *p*-value = 0.5567) and the unpaired two-sample *t*-test for 6 May 2018 (t = −11,834, *p*-value = 0.2383).

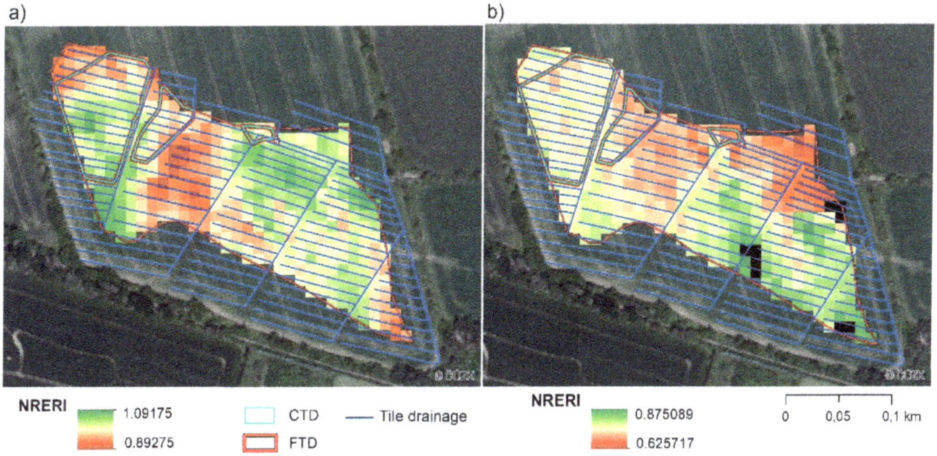

Figure 8. Distribution of NRERI at CTD and FTD sites with Dystric Cambisol (without headlands) of the Za Frajmankou field taken from Sentinel-2 on (**a**) 21 May 2017 and (**b**) 6 May 2018 (excluded outliers in black).

3.2. Grain Yield

Spring barley grain yield as determined from the yield map taken at harvest on 15 August 2021 was, like the NRERI, significantly affected by CTD (Figure 9). Table 2 shows that grain yield at CTD sites was on average 0.3 t/ha higher than in FTD sites, which was due to the largest and most fertile RDFC area in the western part of the field (Figure 10).

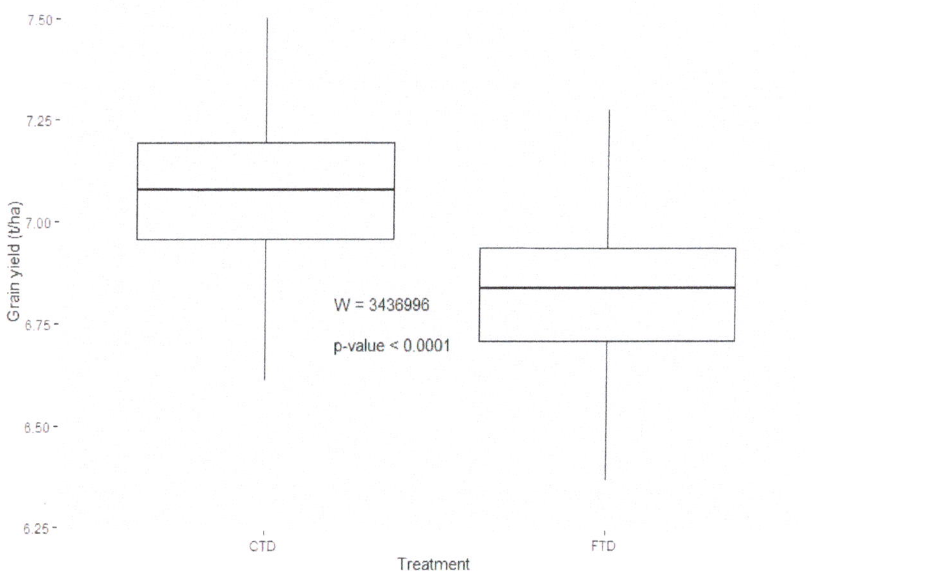

Figure 9. Box plots of grain yield with Mann-Whitney U test (W, *p*-value) at CTD and FTD sites with Dystric Cambisol in the Za Frajmankou field from the yield map taken on 15 August 2021.

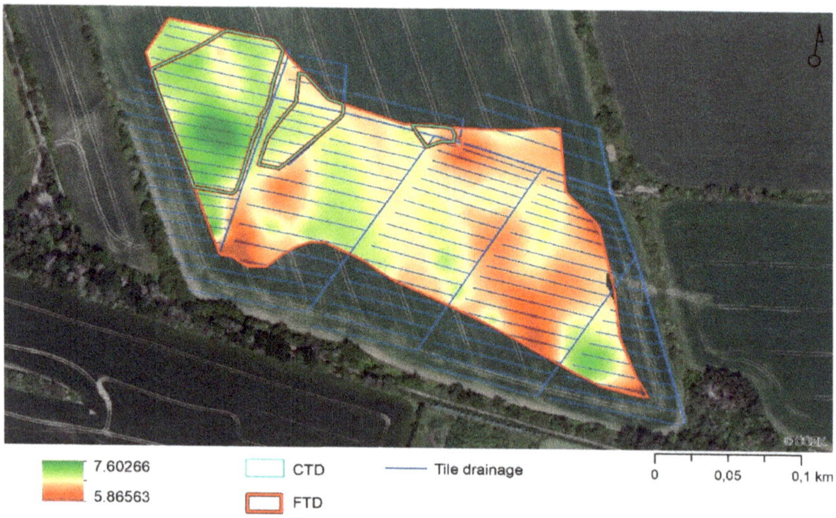

Figure 10. Grain yield map of spring barley (t/ha) at CTD and FTD sites with Dystric Cambisol taken at harvest on 15 August 2021.

Correlation analysis was chosen to investigate the relationship between grain yield/grain N uptake and three selected vegetation indices from the sampling points. At the tillering stage (2 June 2021), grain yield and grain N uptake were not correlated with the three selected vegetation indices (NRERI, Chl, NDRE, Table 3). As stand development progressed (30 June 2021), vegetation indices showed statistically significant results in relation to grain yield. The accuracy of estimating grain N uptake using vegetation indices was also improved, but the significance of the relationship was not demonstrated.

Table 3. Pearson's correlation coefficients between grain yield and vegetation indices (NRERI, Chl and NDRE) from sampling points on 2 and 30 June 2021.

Biomass Parameter	Grain Yield			Grain Nitrogen Uptake		
Date	**NRERI**	**Chl**	**NDRE**	**NRERI**	**Chl**	**NDRE**
2.6.2021	0.22	0.25	0.26	0.21	0.36	0.32
30.6.2021	0.51 *	0.50 *	0.46 *	0.40	0.41	0.36

* Significance p-value < 0.05.

Correlations of the spatial distribution of NRERI and the yield map on tile-drained Dystric Cambisol without headlands provided similar results (no correlation on 2 June 2021 and weak but statistically significant correlation on 30 June 2021, $r = 0.37$, $p < 0.0001$).

3.3. Soil Moisture Content

Soil moisture values at all depths were significantly higher at CTD sites compared to FTD sites (Table 4, Figure 11). At 40 cm depth, differences were evident throughout the whole study period, thus also at the time of UAV imagery, as shown in Figure 7 in the detail of the fourteen measurement locations. At this depth at the CTD sites, the average amount of water in AWC (AWC saturation level in Table 5) was 10.4% higher, and soil moisture exceeding the soil hydrolimit PDA persisted 14 days longer indicating a delayed onset of crop water stress compared to the FTD sites (Table 5). At other depths, soil moisture exceeding PDA was 4% (20 cm) and 11% (60 cm) higher at CTD sites, but of shorter durations of occurrence.

Table 4. Mean soil moisture from each and all depths and Welch´s test parameters (*t*-test, degrees of freedom df, *p*-value) at CTD and FTD sites.

Depths of Soil Moisture Sensors (cm)	Soil Moisture (vol.%) at Sites:		*t*-Test	df	*p*-Value
	CTD	FTD			
20	22.07	20.39	−2.7307	183.13	0.0069
40	25.84	22.41	−6.8521	179.64	<0.0001
60	26.07	24.25	−3.0875	146.92	0.0024
Means of all	24.66	22.35	−4.3867	166.71	<0.0001

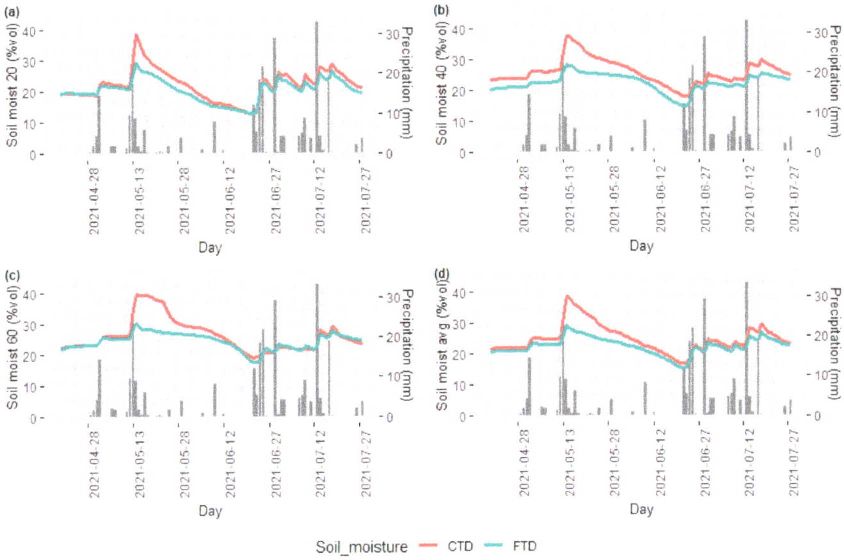

Figure 11. Daily precipitation and soil moisture at depths of (**a**) 20, (**b**) 40 and (**c**) 60 cm and (**d**) averaged over all depths at CTD and FTD sites.

Table 5. Mean soil hydrolimits (field capacity FC, point of decreased availability PDA, wilting point WP, and available water capacity AWC), degree of saturation of AWC and the number of days when soil moisture was equal to or greater than PDA and FC from each and all depths at CTD and FTD sites.

Sites	Depth (cm)	FC (vol.%)	PDA (vol.%)	WP (vol.%)	AWC (vol.%)	AWC Saturation Level (%)	Number of Days with Soil Moisture ≥	
							PDA (% over PDA)	FC
CTD	20	32.6	23.4	13.0	19.6	46.4	39 (14.3)	3
	40	34.3	25.0	14.5	19.7	57.3	52 (16.0)	6
	60	34.1	25.6	15.9	18.2	55.8	46 (20.9)	12
	Means of all	33.7	24.7	14.5	19.2	53.2	46 (17.1)	7
FTD	20	30.0	21.2	11.7	18.3	47.4	46 (10.4)	0
	40	31.9	23.5	14.0	18.0	46.9	38 (7.4)	0
	60	32.6	24.1	14.6	18.0	53.7	53 (9.7)	0
	Means of all	31.5	23.0	13.4	18.1	49.3	46 (9.3)	0

The largest increase in soil moisture content, along with its differences at all depths between CTD and FTD sites, occurred after the heavy rains in early May 2021 (Figure 11). Soil moisture was maintained above PDA until the end of May (at 20 cm) or the end of the first decade of June (at 40 and 60 cm). After a rainfall of 47 mm during 11–15 May, soil moisture exceeded even FC values at CTD sites, which remained at 60 cm until May

24, 2021. During this period, water flowed into the manhole through the drainage pipes and flow rates of up to 1.5 l/s were recorded through the WLCS installed in the manhole (Figure 3). This was the only case in the study period where water flowed through the WLCS in the manhole. At the end of May, crop water stress (soil moisture dropped below PDA) occurred at 20 cm depth, which extended to 40 and 60 cm depths during the first decade of June. Soil moisture differences between the CTD and FTD sites at 20 and 60 cm depth gradually decreased until they disappeared completely by the end of the second decade of June. The wet period from 22 June to 28 July 2021 (total rainfall of 181 mm) induced a renewed increase in soil moisture differences at 20 and 40 cm depth and a gradual removal of crop water stress.

4. Discussion

Vegetation indices are useful tools of remote sensing for identifying trends in crop biomass growth and predicting crop yields [5]. To test the effect of CTD functionality on biomass production of spring barley in our study, we selected the vegetation index NRERI based on spectral reflectance in R, RE and NIR bands as the best indicator of shoot and grain production. The superiority of the NRERI index (similarly, the NRERI and Chl indices), unlike the other indices based on other combinations of spectral bands (NDVI, SAVI, EVI2, SRI, RENDVI, GNDVI), was due to the better relationships in the case of 2 June 2021 with all shoot biomass parameters and in the case of 30 June 2021 with grain yield from point sampling. The advantage of reflectance in the RE band, in contrast to the R band, is that the sensitivity of absorption to chlorophyll content is much higher (i.e., no saturation effect, [56]) and, similarly to the NIR band, a positive correlation with leaf N and biomass exists [57–60]. The combination of RE and NIR bands is recommended for estimating higher biomass with LAI > 2–3, but also to provide insight into N nutritional status (N content, N uptake, NNI) [61–64]. Thus, the NRERI vegetation index provided an opportunity in our study to show the link between biomass development and the N nutritional status of the crop as well as the prediction of grain yield. Similarly, Klem et al. [65] found that NRERI, as affected by water deficit, is the best estimator of N status in both leaves and grain of winter wheat. Klem et al. [66] also confirmed the suitability of RE reflectance for estimating the N nutritional status of malting barley, the accuracy of which can be further improved by using an artificial neural network based on multiple spectral reflectance wavelengths. Holub et al. [67] reported that at the completed heading stage of winter wheat, the NRERI index, as the only one based on reflectance in the R, RE and NIR spectra, had the greatest potential for estimating grain N uptake.

CTD practices increase crop yields by improving soil moisture availability along with the retention of mineral N available to plants [5,14,68,69]. The beneficial effect of CTD practices on soil water availability for spring barley was clearly reflected in a higher degree of saturation of AWC compared to FTD sites. Also, Wesström et al. [70] found increased soil water storage due to CTD in southern Sweden, which they attributed to reduced drainage outflow compared to FTD. Spring barley has a weaker root system than other cereal crops, and most of its roots (ca. 90%) are distributed at a depth of 30–50 cm depending on soil type, with the highest density at depths up to 10 cm [71–73]. Increased soil water supply at 40 cm in relation to CTD practices, which in our case was maintained throughout the study period, demonstrated an improved water supply to barley roots even when water did not flow through the WLCS in the manhole but was only retained in the drainage pipes. This was highly desirable and clearly contributed to increased grain yield.

The probability of retaining plant-available soil water due to CTD is lower in our drier study site than in humid sites, but even a small increase in soil water availability associated with the elimination of drainage runoff is important for stabilizing or slightly increasing crop yields. For instance, Dou et al. [74] identified a delay in groundwater table decline through CTD (drainage depth of 40 cm) in a dryland area, which allowed sunflower to use groundwater at later growth stages, resulting in yield and water use efficiency improvements of 4.52–11.14% and 1.16–10.8%, respectively. Accordingly, the grain yield of spring barley in our study was increased by 4% in relation to CTD.

The prediction of grain yield from the early shoot biomass parameters (2 June 2021) from sampling points was not demonstrated, which contrasts with Křen et al. [75] who estimated grain yield of spring barley based on a strong correlation with dry weight of above-ground biomass per unit area at the early growth stage BBCH 25 ($r = 0.81$). Similarly, the use of selected vegetation indices from the early growth stage for estimating grain yield was not beneficial in our case. As stand development proceeded to the heading stage, the selected vegetation indices based on RE region predicted grain yield at a significant level. Likewise, Erdle et al. [76] demonstrated a close correlation of RE-based vegetation index REIP at later stages of winter wheat development with grain yield. Consistently, Klem et al. [65] considered NRERI, which was measured at the early milk ripening stage, as the best indicator of grain yield in winter wheat. Also, qualitative parameters of harvested crops can be assessed by UAV survey, as shown on the prediction of nutritional values of silage maize using NDVI and NDRE indices by [77].

This study also confirmed the role of UAV multispectral imaging in the monitoring of crop stand and identification of spatial differences in vegetation parameters. The main advantages of UAV in the comparison to the satellite remote sensing, such as free available Sentinel-2 data, are the ultra-high spatial resolution of acquired multispectral data at the few centimeters scale and high operability of drones, which results in the better timing of the survey independent on the cloud condition. Further research is needed for development of collaborative smart drones for fully automated observations [78].

5. Conclusions

The use of vegetation indices from UAV imagery based on a combination of R, RE, and NIR wavebands appears to be a suitable method for determining the effect of CTD on biomass growth and N nutritional status of spring barley, as well as for predicting grain yield. CTD practices have shown a distinctly positive impact on biomass development and increased grain yield, as evidenced by increased soil water storage and delayed crop water stress, especially at 40 cm depth. Although this paper describes the results of a field experiment from only one growing season, the effect of CTD on increased biomass growth was clearly demonstrated by Sentinel-2 imagery from before WLCS installation, when there were no differences between CTD and FTD sites. As showed in many other regions, and now also for the Central Europe, the CTD could be thus considered as a measure with a substantial potential to mitigate or delay crop water stress, enhance crop yields, and reduce the undue water loss from the landscape.

Author Contributions: Conceptualization, R.D.; methodology, R.D., L.P., V.L. and P.F.; validation, R.D. and V.L.; formal analysis, R.D. and P.F.; investigation, R.D. and V.L.; resources, R.D. and V.L.; data curation, R.D., L.P. and V.L.; writing—original draft preparation, R.D., L.P., V.L. and P.F.; writing—review and editing, R.D., V.L. and P.F.; visualization, R.D., L.P. and V.L.; supervision and project administration, R.D., V.L. and P.F.; funding acquisition, R.D. and P.F. All authors have read and agreed to the published version of the manuscript.

Funding: This research was funded by the Technology Agency of the Czech Republic (project No. SS01020309), by The European Union's Horizon 2020 research and innovation program (project No. 818187) and by the Ministry of Agriculture of the Czech Republic (project No. RO0218).

Data Availability Statement: The data used for this work are not publicly available.

Acknowledgments: The authors would like to thank Zbyněk Kulhavý and Vlastimil Osoba for excellent professional and technical assistance, David Šádek for installation of water level control structures, Darina Heřmanovská advice on the R environment and Ondřej Holubík for soil classification.

Conflicts of Interest: The authors declare no conflict of interest.

References

1. Duffková, R.; Holub, J.; Fučík, P.; Rožnovský, J.; Novotný, I. Long-Term Water Balance of Selected Field Crops in Different Agricultural Regions of the Czech Republic Using Fao-56 and Soil Hydrological Approaches. *Sustainability* **2019**, *11*, 5243. [CrossRef]
2. Vicente-Serrano, S.M.; McVicar, T.R.; Miralles, D.G.; Yang, Y.; Tomas-Burguera, M. Unraveling the influence of atmospheric evaporative demand on drought and its response to climate change. *WIREs Clim. Chang.* **2020**, *11*, e632. [CrossRef]
3. Pokhrel, Y.; Felfelani, F.; Satoh, Y.; Boulange, J.; Burek, P.; Gädeke, A.; Gerten, D.; Gosling, S.N.; Grillakis, M.; Gudmundsson, L.; et al. Global terrestrial water storage and drought severity under climate change. *Nat. Clim. Chang.* **2021**, *11*, 226–233. [CrossRef]
4. Bouabdelli, S.; Zeroual, A.; Meddi, M.; Assani, A. Impact of temperature on agricultural drought occurrence under the effects of climate change. *Theor. Appl. Climatol.* **2022**, *148*, 191–209. [CrossRef]
5. Cicek, H.; Sunohara, M.; Wilkes, G.; McNairn, H.; Pick, F.; Topp, E.; Lapen, D.R. Using vegetation indices from satellite remote sensing to assess corn and soybean response to controlled tile drainage. *Agric. Water Manag.* **2010**, *98*, 261–270. [CrossRef]
6. Kulhavý, Z.; Fučík, P. Adaptation Option for Land Drainage Systems Toward Sustainable Agriculture and the Environment: A Czech Perspective. *Pol. J. Environ. Stud.* **2015**, *24*, 1085–1102. [CrossRef]
7. Carstensen, M.V.; Børgesen, C.D.; Ovesen, N.B.; Poulsen, J.R.; Hvid, S.K.; Kronvang, B. Controlled Drainage as a Targeted Mitigation Measure for Nitrogen and Phosphorus. *J. Environ. Qual.* **2019**, *48*, 677–685. [CrossRef]
8. Mourtzinis, S.; Andrade, J.F.; Grassini, P.; Edreira, J.I.R.; Kandel, H.; Naeve, S.; Nelson, K.A.; Helmers, M.; Conley, S.P. Assessing benefits of artificial drainage on soybean yield in the North Central US region. *Agric. Water Manag.* **2021**, *243*, 1–8. [CrossRef]
9. Sunohara, M.D.; Gottschall, N.; Wilkes, G.; Craiovan, E.; Topp, E.; Que, Z.; Seidou, O.; Frey, S.; Lapen, D.R. Long term observations of nitrogen and phosphorus export in paired-agricultural watersheds under controlled and conventional tile drainage management. *J. Environ. Qual.* **2015**, *44*, 1589–1604. [CrossRef]
10. Fučík, P.; Zajíček, A.; Kaplická, M.; Duffková, R.; Peterková, J.; Maxová, J.; Takáčová, Š. Incorporating rainfall-runoff events into nitrate-nitrogen and phosphorus load assessments for small tile-drained catchments. *Water* **2017**, *9*, 712. [CrossRef]
11. Ghane, E.; Askar, M.H.; Skaggs, R.W. Design drainage rates to optimize crop production for subsurface-drained fields. *Agric. Water Manag.* **2021**, *257*, 107045. [CrossRef]
12. Wesström, I.; Ekbohm, G.; Linnér, H.; Messing, I. The effects of controlled drainage on subsurface outflow from level agricultural fields. *Hydrol. Process.* **2003**, *17*, 1525–1538. [CrossRef]
13. Drury, C.; Tan, C.; Reynolds, W.; Welacky, T.; Oloya, T.; Gaynor, J. Managing tile drainage, subirrigation, and nitrogen fertilization to enhance crop yields and reduce nitrate loss. *J. Environ. Qual.* **2009**, *38*, 1193–1204. [CrossRef]
14. Kross, A.; Lapen, D.R.; McNairn, H.; Sunohara, M.; Champagne, C.; Wilkes, G. Satellite and in situ derived corn and soybean biomass and leaf area index: Response to controlled tile drainage under varying weather conditions. *Agric. Water Manag.* **2015**, *160*, 118–131. [CrossRef]
15. Povilaitis, A.; Rudzianskaite, A.; Miseviciene, S.; Gasiunas, V.; Miseckaite, O.; Živatkauskiene, I. Efficiency of drainage practices for improving water quality in Lithuania. *Trans. ASABE* **2018**, *61*, 179–196. [CrossRef]
16. Sojka, M.; Kozłowski, M.; Stasik, R.; Napierała, M.; Kęsicka, B.; Wróżyński, R.; Jaskuła, J.; Liberacki, D.; Bykowski, J. Sustainable Water Management in Agriculture—The Impact of Drainage Water Management on Groundwater Table Dynamics and Subsurface Outflow. *Sustainability* **2019**, *11*, 4201. [CrossRef]
17. Tolomio, M.; Borin, M. Controlled drainage and crop production in a long-term experiment in North-Eastern Italy. *Agric. Water Manag.* **2019**, *222*, 21–29. [CrossRef]
18. Salo, H.; Salla, A.; Koivusalo, H. Seasonal effects of controlled drainage on field water balance and groundwater levels. *Hydrol. Res.* **2021**, *52*, 1633–1647. [CrossRef]
19. King, K.W.; Hanrahan, B.R.; Stinner, J.; Shedekar, V.S. Field scale discharge and water quality response, to drainage water management. *Agric. Water Manag.* **2022**, *264*, 107421. [CrossRef]
20. Almen, K.; Jia, X.; DeSutter, T.; Scherer, T.; Lin, M. Impact of Controlled Drainage and Subirrigation on Water Quality in the Red River Valley. *Water* **2021**, *13*, 308. [CrossRef]
21. Helmers, M.J.; Abendroth, L.; Reinhart, B.; Chighladze, G.; Pease, L.; Bowling, L.; Youssef, M.; Ghane, E.; Ahiablame, L.; Brown, L.; et al. Impact of controlled drainage on subsurface drain flow and nitrate load: A synthesis of studies across the U.S. Midwest and Southeast. *Agric. Water Manag.* **2022**, *259*, 107265. [CrossRef]
22. Yuan, N.; Xiong, Y.; Li, Y.; Xu, B.; Liu, F. Experimental Study of the Effect of Controlled Drainage on Soil Water and Nitrogen Balance. *Water* **2021**, *13*, 2241. [CrossRef]
23. Chi, B.; Shi, H.; Xu, D.; Jiao, P. Distribution and variation of water and salt in soil profile under controlling subsurface drainage. *Trans. Chin. Soc. Agric. Eng.* **2021**, *37*, 148–158. [CrossRef]
24. Viña, A.; Gitelson, A.A.; Rundquist, D.C.; Keydan, G.; Leavitt, B.; Schepers, J. Monitoring maize (*Zea mays* L.) phenology with remote sensing. *Agron. J.* **2004**, *96*, 1139–1147. [CrossRef]
25. Jackson, T.J.; Chen, D.; Cosh, M.; Li, F.; Anderson, M.; Walthall, C.; Doriaswamy, P.; Hunt, E.R. Vegetation water content mapping using Landsat data derived normalized difference water index for corn and soybeans. *Remote Sens. Environ.* **2004**, *92*, 475–482. [CrossRef]
26. Wiegand, C.L.; Richardson, A.J.; Escobar, A.J.; Gerbermann, A.H. Vegetation indices in crop assessments. *Remote Sens. Environ.* **1991**, *35*, 105–119. [CrossRef]

27. Wang, C.; Feng, M.C.; Yang, W.D.; Ding, G.W.; Sun, H.; Liang, Z.Y.; Xie, Y.K.; Qiao, X.X. Impact of spectral saturation on leaf area index and aboveground biomass estimation of winter wheat. *Spectrosc. Lett.* **2016**, *49*, 241–248. [CrossRef]
28. Tesfaye, A.A.; Awoke, B.G. Evaluation of the saturation property of vegetation indices derived from sentinel-2 in mixed crop-forest ecosystem. *Spat. Inf. Res.* **2021**, *29*, 109–121. [CrossRef]
29. Rouse, J.W.; Haas, R.H.; Schell, J.A.; Deering, D.W. Monitoring vegetation systems in the Great Plains with ERTS. *NASA Spec. Publ.* **1974**, *351*, 309.
30. Asrar, G.; Fuchs, M.; Kanemasu, E.T.; Hatfield, J.L. Estimating absorbed photosynthetic radiation and leaf area index from spectral reflectance in wheat. *Agron. J.* **1984**, *76*, 300–306. [CrossRef]
31. Sellers, P.J. Canopy reflectance, photosynthesis, and transpiration, II. The role of biophysics in the linearity of their interdependence. *Remote Sens. Environ.* **1987**, *21*, 143–183. [CrossRef]
32. Chen, P.Y.; Fedosejevs, G.; Tiscareño-López, M.; Arnold, J.G. Assessment of MODIS-EVI, MODIS-NDVI and VEGETATION-NDVI composite data using agricultural measurements: An example at corn fields in western Mexico. *Environ. Monit. Assess.* **2006**, *119*, 69–82. [CrossRef] [PubMed]
33. Klem, K.; Rajsnerová, P.; Novotná, K.; Míša, P.; Křen, J. Changes in Vertical Distribution of Spectral Reflectance within Spring Barley Canopy as an Indicator of Nitrogen Nutrition, Canopy Structure and Yield Parameters. *Agriculture* **2014**, *60*, 50–59. [CrossRef]
34. Li, F.; Mistele, B.; Hu, Y.; Yue, X.; Yue, S.; Miao, Y.; Chen, X.; Cui, Z.; Meng, X.; Schmidhalter, U. Remotely estimating aerial N status of phenologically differing winter wheat cultivars grown in contrasting climatic and geographic zones in China and Germany. *Field Crop Res.* **2012**, *138*, 21–32. [CrossRef]
35. Li, F.; Mistele, B.; Hu, Y.; Chen, X.; Schmidhalter, U. Comparing hyperspectral index optimization algorithms to estimate aerial N uptake using multi-temporal winter wheat datasets from contrasting climatic and geographic zones in China and Germany. *Agr. Forest Meteorol.* **2013**, *180*, 44–57. [CrossRef]
36. Lin, S.; Li, J.; Liu, Q.; Li, L.; Zhao, J.; Yu, W. Evaluating the Effectiveness of Using Vegetation Indices Based on Red-Edge Reflectance from Sentinel-2 to Estimate Gross Primary Productivity. *Remote Sens.* **2019**, *11*, 1303. [CrossRef]
37. Mittermayer, M.; Maidl, F.-X.; Nätscher, L.; Hülsbergen, K.-J. Analysis of site-specific N balances in heterogeneous croplands using digital methods. *Eur. J. Agron.* **2022**, *133*, 126442. [CrossRef]
38. Li, F.; Mistele, B.; Hu, Y.; Chen, X.; Schmidhalter, U. Optimising three–band spectral indices to assess aerial N concentration, N uptake and aboveground biomass of winter wheat remotely in China and Germany. *ISPRS J. Photogramm.* **2014**, *92*, 112–123. [CrossRef]
39. Cao, Q.; Miao, Y.; Feng, G.; Gao, X.; Li, F.; Liu, B.; Yue, S.; Cheng, S.; Ustin, S.L.; Khosla, R. Active canopy sensing of winter wheat nitrogen status: An evaluation of two sensor systems. *Comput. Electron. Agr.* **2015**, *112*, 54–67. [CrossRef]
40. Huang, S.; Miao, Y.; Zhao, G.; Yuan, F.; Ma, X.; Tan, C.; Yu, W.; Gnyp, M.L.; Lenz–Wiedemann, V.I.S.; Rascher, U.; et al. Satellite Remote Sensing–Based In–Season Diagnosis of Rice Nitrogen Status in Northeast China. *Remote Sens.* **2015**, *7*, 10646–10667. [CrossRef]
41. Vega, A.; Córdoba, M.; Castro-Franco, M.; Balzarini, M. Protocol for automating error removal from yield maps. *Precis. Agric.* **2019**, *20*, 1030–1044. [CrossRef]
42. Tolasz, R.; Kveton, V. *Climate Atlas of Czechia*, 1st ed.; Czech Hydrometeorological Institute: Prague, Czech, 2007.
43. WRB. World reference base for soil resources 2014. In *International Soil Classification System for Naming Soils and Creating Legends for Soil Maps*; World Soil Resources Reports No. 106; FAO: Rome, Italy, 2015.
44. *ISO 11277*; Soil Quality–Determination of Particle Size Distribution in Mineral Soil Material—Method by Sieving and Sedimentation. International Organization for Standardization: Geneva, Switzerland, 2009.
45. Novotný, M.; Kervališvili, D.M.; Šanta, M. *Irrigation of Field and Special Crops*, 1st ed.; Príroda: Bratislava, Slovakia, 2000.
46. *ISO 11261*; Soil Quality—Determination of Total Nitrogen—Modified Kjeldahl Method. International Organization for Standardization: Geneva, Switzerland, 1995.
47. Justes, E.; Mary, B.; Meynard, J.M.; Machet, J.M.; Thelierhuche, L. Determination of a critical nitrogen dilution curve for winter–wheat crops. *Ann. Bot.* **1994**, *74*, 397–407. [CrossRef]
48. Huete, A.; Didan, K.; Miura, T.; Rodriguez, E.P.; Gao, X.; Ferreira, L.G. Overview of the radiometric and biophysical performance of the MODIS vegetation indices. *Remote Sens. Environ.* **2002**, *83*, 195–213. [CrossRef]
49. Jiang, Z.; Huete, A.R.; Didan, K.; Miura, T. Development of a two-band Enhanced Vegetation Index without a blue band. *Remote Sens. Environ.* **2008**, *112*, 3833–3845. [CrossRef]
50. Gitelson, A.A.; Kaufman, Y.J.; Merzlyak, M.N. Use of a green channel in remote sensing of global vegetation from EOS-MODIS. *Remote Sens. Environ.* **1996**, *58*, 289–298. [CrossRef]
51. Barnes, E.; Clarke, T.; Richards, S.; Colaizzi, P.; Haberland, J.; Kostrzewski, M.; Waller, P.; Choi, C.; Riley, E.; Thompson, T. Coincident detection of crop water stress, nitrogen status and canopy density using ground based multispectral data. In Proceedings of the 5th International Conference on Precision Agriculture, Bloomington, MN, USA, 16–19 July 2000; pp. 16–19.
52. Dash, J.; Curran, P.J. The MERIS terrestrial chlorophyll index. *Int. J. Remote Sens.* **2004**, *25*, 5403–5413. [CrossRef]
53. Gitelson, A.; Merzlyak, M. Spectral Reflectance Changes Associated with Autumn Senescence of *Aesculus hippocastanum* L. and *Acer platanoides* L. leaves. *J Plant Physiol.* **1994**, *143*, 286–292. [CrossRef]
54. Huete, A.R. A Soil-Adjusted Vegetation Index (SAVI). *Remote Sens. Environ.* **1988**, *25*, 295–309. [CrossRef]

55. R Core Team. *R: A Language and Environment for Statistical Computing*; R Foundation for Statistical Computing: Vienna, Austria, 2021.
56. Lichtenthaler, H.K. Chlorophylls and carotenoids: Pigments of photosynthetic biomembranes. *Method. Enzymol.* **1987**, *148*, 350–382. [CrossRef]
57. Guyot, G.; Baret, F.; Major, D.J. High spectral resolution: Determination of spectral shifts between the red and the near infrared. *Int. Arch. Photogram. Rem. Sens.* **1988**, *11*, 750–760.
58. Campbell, J.B. *Introduction to Remote Sensing*; The Guilford Press: New York, NY, USA, 2002.
59. Shaver, T.M.; Khosla, R.; Westfall, D.G. Evaluation of two ground-based active crop canopy sensors in maize: Growth stage, row spacing, and sensor movement speed. *Soil Sci. Soc. Am. J.* **2010**, *74*, 2101–2108. [CrossRef]
60. Clevers, J.G.P.W.; Gitelson, A.A. Remote estimation of crop and grass chlorophyll and nitrogen content using red-edge bands on Sentinel-2 and -3. *Int. J. Appl. Earth Obs. Geoinf.* **2013**, *23*, 334–343. [CrossRef]
61. Curran, P.J.; Dungan, J.L.; Gholz, H.L. Exploring the relationship between reflectance red edge and chlorophyll content in slash pine. *Tree Physiol.* **1991**, *7*, 33–48. [CrossRef] [PubMed]
62. Mistele, B.; Gutser, R.; Schmidhalter, U. Validation of field-scaled spectral measurements of nitrogen status in winter wheat. In Proceedings of the 7th International Conference on Precision Agriculture and Other Precision Resources Management, Hyatt Regency, Minneapolis, MN, USA, 25–28 July 2004; pp. 1187–1195.
63. Erdle, K.; Mistele, B.; Schmidhalter, U. Comparison of active and passive spectral sensors in discriminating biomass parameters and nitrogen status in wheat cultivars. *Field Crop Res.* **2011**, *124*, 74–84. [CrossRef]
64. Cisneros, A.; Fiorio, P.; Menezes, P.; Pasqualotto, N.; van Wittenberghe, S.; Bayma, G.; Nogueira, S.F. Mapping Productivity and Essential Biophysical Parameters of Cultivated Tropical Grasslands from Sentinel-2 Imagery. *Agronomy* **2020**, *10*, 711. [CrossRef]
65. Klem, K.; Záhora, J.; Zemek, F.; Trunda, P.; Tůma, I.; Novotná, K.; Hodaňová, P.; Rapantová, B.; Hanuš, J.; Vavříková, J.; et al. Interactive effects of water deficit and nitrogen nutrition on winter wheat. Remote sensing methods for their detection. *Agr. Water Manage.* **2018**, *210*, 171–184. [CrossRef]
66. Klem, K.; Křen, J.; Šimor, J.; Kováč, D.; Holub, P.; Míša, P.; Svobodová, I.; Lukas, V.; Lukeš, P.; Findurová, H.; et al. Improving Nitrogen Status Estimation in Malting Barley Based on Hyperspectral Reflectance and Artificial Neural Networks. *Agronomy* **2021**, *11*, 2592. [CrossRef]
67. Holub, P.; Klem, K.; Tůma, I.; Vavříková, J.; Surá, K.; Veselá, B.; Urban, O.; Záhora, J. Application of organic carbon affects mineral nitrogen uptake by winter wheat and leaching in subsoil: Proximal sensing as a tool for agronomic practice. *Sci. Total Environ.* **2020**, *717*, 137058. [CrossRef]
68. Ayars, J.E.; Christen, E.W.; Hornbuckle, J.W. Controlled drainage for improved water management in arid regions irrigated agriculture. *Agr. Water Manage.* **2006**, *86*, 128–139. [CrossRef]
69. Wesström, I.; Messing, I. Effects of controlled drainage on N and P losses and N dynamics in a loamy sand with spring crops. *Agr. Water Manage.* **2007**, *87*, 229–240. [CrossRef]
70. Wesström, I.; Messing, I.; Linnér, H.; Lindström, J. Controlled drainage—effects on drain outflow and water quality. *Agr. Water Manage.* **2001**, *47*, 85–100. [CrossRef]
71. Madsen, H.B. Distribution of spring barley roots in Danish soils, of different texture and under different climatic conditions. *Plant Soil* **1985**, *88*, 31–43. [CrossRef]
72. Gregory, P.J. *Plant Roots: Growth, Activity, and Interaction with Soils*, 1st ed.; Blackwell Publishing: Oxford, UK, 2006; p. 318.
73. Wahlström, E.M.; Kristensen, H.L.; Thomsen, I.K.; Labouriau, R.; Pulido-Moncada, M.; Nielsen, J.A.; Munkholm, L.J. Subsoil compaction effect on spatio-temporal root growth, reuse of biopores and crop yield of spring barley. *Eur. J. Agron.* **2021**, *123*, 126225. [CrossRef]
74. Dou, X.; Shi, H.; Li, R.; Miao, Q.; Tian, F.; Yu, D.; Zhou, L.; Wang, B. Effects of Controlled Drainage on the Content Change and Migration of Moisture, Nutrients, and Salts in Soil and the Yield of Oilseed Sunflower in the Hetao Irrigation District. *Sustainability* **2021**, *13*, 9835. [CrossRef]
75. Křen, J.; Klem, K.; Svobodová, I.; Míša, P.; Neudert, L. Yield and grain quality of spring barley as affected by biomass formation at early growth stages. *Plant Soil Environ.* **2014**, *60*, 221–227. [CrossRef]
76. Erdle, K.; Mistele, B.; Schmidhalter, U. Spectral high-throughput assessments of phenotypic differences in biomass and nitrogen partitioning during grain filling of wheat under high yielding Western European conditions. *Field Crop Res.* **2013**, *141*, 16–26. [CrossRef]
77. Janoušek, J.; Jambor, V.; Marcoň, P.; Dohnal, P.; Synková, H.; Fiala, P. Using UAV-Based Photogrammetry to Obtain Correlation between the Vegetation Indices and Chemical Analysis of Agricultural Crops. *Remote Sens.* **2021**, *13*, 1878. [CrossRef]
78. Alsamhi, S.H.; Ma, O.; Ansari, M.S.; Almalki, F.A. *Survey on Collaborative Smart Drones and Internet of Things for Improving Smartness of Smart Cities*; IEEE Access: Piscataway, NJ, USA, 2019; Volume 7, pp. 128125–128152. [CrossRef]

remote sensing

MDPI

Article

Hydrological Connectivity Improves the Water-Related Environment in a Typical Arid Inland River Basin in Xinjiang, China

Chuanxiu Liu [1,2], Yaning Chen [1,*], Gonghuan Fang [1], Honghua Zhou [1], Wenjing Huang [1,2], Yongchang Liu [1,2], Xuanxuan Wang [1,2] and Zhi Li [1]

1 State Key Laboratory of Desert and Oasis Ecology, Xinjiang Institute of Ecology and Geography, Chinese Academy of Sciences, Urumqi 830011, China
2 Department of Resources and Environment, University of China Academy of Sciences, Beijing 100049, China
* Correspondence: chenyn@ms.xjb.ac.cn

Abstract: Hydrological connectivity directly affects aquatic ecological processes, water environment and wetland ecological security, which is essential to the stability of arid ecosystems. However, the mechanism between hydrological connectivity and water-related environment has not been revealed completely. To address these issues, we use a landscape connectivity approach to assess the connectivity of water patches for analyzing the hydrological connectivity of the Bosten Lake Basin (BLB), as well as its response to human activities and climate change, based on the Joint Research Centre (JRC) global surface water dataset. It shows that the integral index of connectivity (IIC) of the BLB is low (ranging from 0 to 0.2) from 1990 to 2019, with an increasing interannual trend. The connectivity is higher in wet periods and in oases compared with dry periods and high-altitude mountain regions. Correlation and regression analyses indicate that hydrological connectivity has a strong correlation ($r > 0.5$, $p \leq 0.05$) with water area and water level. The interannual and seasonal trends of eight hydrochemical indices in the Bosten Lake have been investigated to systematically elaborate the complex relationships between hydrological connectivity and water quality in the BLB. Results indicated that better hydrological connectivity can improve water quality, and the minimum of pollutants were observed in high hydrological connectivity period, covering approximately 75% of the high-water quality period. These findings could provide scientific support for the water management in the BLB.

Keywords: hydrological connectivity; water quality; arid region; Bosten Lake; inland river-lake systems

Citation: Liu, C.; Chen, Y.; Fang, G.; Zhou, H.; Huang, W.; Liu, Y.; Wang, X.; Li, Z. Hydrological Connectivity Improves the Water-Related Environment in a Typical Arid Inland River Basin in Xinjiang, China. *Remote Sens.* **2022**, *14*, 4977. https://doi.org/10.3390/rs14194977

Academic Editors: Songhao Shang, Qianqian Zhang, Dongqin Yin and Hamza Gabriel

Received: 6 September 2022
Accepted: 30 September 2022
Published: 6 October 2022

1. Introduction

Rivers fluctuate back and forth between dry and wet periods in arid regions [1]. The emergence and disappearance of water patches provide necessary connectivity for river and lake ecosystems, which also provide connecting pathways for biological habitats in both space and time. However, climate change and human activities bring great uncertainty to the hydrological processes and ecological changes in arid zones [2–5], which may increase the extreme precipitation and the degree and frequency of droughts [6]. Thus, they will change the flow state and river morphology of inland rivers [7,8], and weaken or even isolate the hydrological connectivity processes between rivers and lakes [9,10]. That will lead to deterioration of hydrological environment and water quality as well as a rapid decline in essential ecosystem services [11], such as lake shrinkage, water pollution, and loss of biodiversity [12–14].

Hydrological connectivity is widely defined as the water-mediated transfer of material, energy, and/or organisms within or between elements of the water cycle [15–18]. Recently, with the concept of hydrological connectivity having attracted much attention among geoscientists [19,20], theories and methods related to landscape connectivity have been introduced into hydrology and widely used in current hydrological connectivity studies [21–23]. Hydrological connectivity describes the connectivity of lake ecosystems to terrestrial ecosystems through sedimentation, soil leaching, diffusion in wetlands, lake and river inputs (receiving water) and outputs (discharging water) [24–26]. The ability of water to exchange pollutants between different water patches depends on their connectivity with each other, and the rate of water exchange depends on the degree of connectivity between patches [25,27–30]. Hydrological connectivity determines the ability of species spread and gene flow [31,32], which is an important driver for maintaining healthy ecosystem function and social development in arid areas [33–35], and it is also important to protect biodiversity and maintain the stability as well as the integrity of natural ecosystems [36].

The fragmentation of water systems in arid inland river basins has caused the decline in hydrological connectivity, which has directly resulted in deterioration of water environment and ecological conditions. Previous studies have investigated the effects of topography, geomorphology, and human activities on hydrologic connectivity by means of graph theory, hydrologic models, and connectivity indices [21,25,37–39]. However, the relationships between fragile and irreversible ecosystems and hydrological connectivity in arid zones are still unclear due to discontinuous monitoring data and incomplete monitoring networks [6]. Meanwhile, only a few studies have focused on the correlation between water quality and hydrological connectivity [25,29,40], and they all concentrate on the relationship between hydrological connectivity and a single environmental factor [37,41–43]. Few studies attempt to explore the complex relationship between hydrological connectivity and multiple environmental factors [27,44,45]. Therefore, it is particularly urgent to investigate the response between water quality and hydrological connectivity in arid areas.

The Bosten Lake, together with the Kaidu River and the Konqi River, constitutes the Bosten Lake Basin (BLB), which is a typical inland river basin in the arid zone and one of the headwaters of the Tarim River Basin, the largest inland river basin in China [46,47]. With climate change and increased human activity, the Bosten Lake has gradually become a mesotrophic lake due to extensive salinization caused by large-scale agricultural irrigation and water diversion projects for industrial development [28,48,49]. To enhance the capacity of the water environment, a series of water system connectivity projects was implemented in recent years to improve the water quality of the Bosten Lake. As most of the surrounding tributaries of the lake are seasonal rivers, and the change in hydrological connectivity may have a significant impact on the Bosten Lake, which connects with the upstream and downstream of the basin, but these effects have never been assessed. Therefore, this study aims to (1) assess the interannual and seasonal variability characteristics of the connectivity among water patches in the BLB from 1990 to 2019; (2) analyze the characteristics of hydrochemistry changes; (3) clarify the relationship between hydrological connectivity and its water-related environmental effects. It is intended to provide a comprehensive framework to explain the hydrological connectivity and ecological response. The results of the study can provide a reference for the implementation of hydraulic engineering in inland river basins of arid regions, which is important for maintaining the stability of oasis-desert ecosystems and the integrity of river functions, and promoting the sustainable development of the region.

2. Material and Methods

2.1. Study Area

The BLB mainly includes the Kaidu River, the Konqi River, and the Bosten Lake, which is the largest inland freshwater lake in China [50]. The basin is located in the arid and semi-arid region in China. It spans an area of approximately 90,944.24 km^2 [51]. The total annual precipitation is only 76.1 mm; however, evaporation amounts to 2000 mm year^{-1} [50]. A total of 65% of rainfall and 70% of evaporation happen between May and August [52]. As shown in Figure 1, the watershed has a complex topography, with mountains, oases, and desert interspersed, which is representative of a typical mountain-oasis-desert complex ecosystem. Based on the extent and characteristics of the study area, we divided it into three sub-basins: the upper and middle reaches of the Kaidu River (KDH), the lower reaches of the Kaidu River and upper reaches of the Konqi River (KKH), and the middle and lower reaches of the Konqi River (KQH). In terms of primary topographical features, KDH is a high-altitude mountainous area, KKH is a densely populated area, and KQH includes part of the desert area.

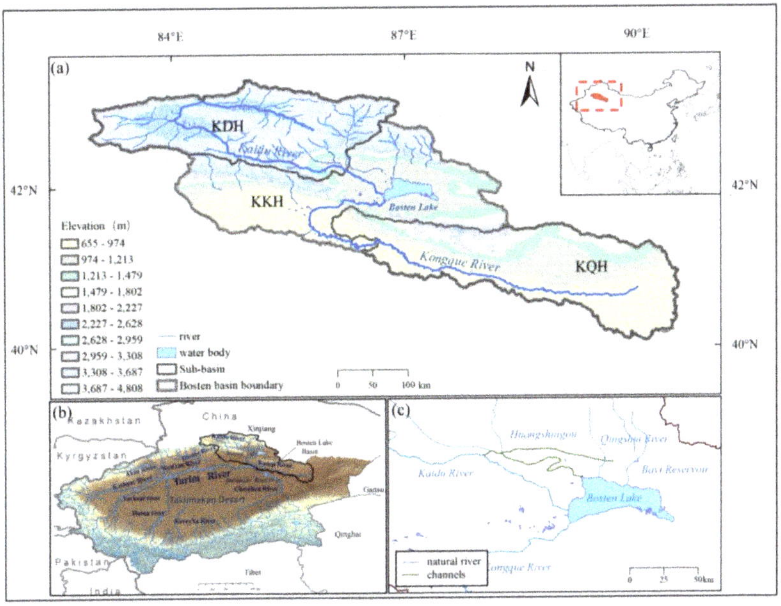

Figure 1. Location and overview of the Bosten Lake Basin. Scale and overview of the whole Bosten Lake Basin (**a**), location of the catchment within Tarim River Basin (**b**), enlarged view of the Bosten Lake (**c**).

2.2. Data

2.2.1. Remote Sensing Data

JRC Yearly Water Classification History, v1.3 (JRC-Yearly), and JRC Monthly Water Classification History, v1.3 (JRC-Monthly) are used to extract the yearly and monthly surface water body information. Both datasets contain information on the location and temporal distribution of surface water from 1984 to 2020, which provides statistical data about the extent and variability of these water surfaces [53]. These data were generated based on Landsat 5–8 imagery data, using expert systems, visual analytics, and evidential reasoning to classify each pixel individually as water body or non-water body. The results were then collated into monthly-scale for two time periods to detect water body changes, which are widely used in studies related to terrestrial hydrology [54,55].

The JRC Yearly Water Classification History, v1.3 (JRC-Yearly) collection preserves yearly water body distributions from 1984 to 2019 and contains 36 images in total, which has a spatial resolution of 30 m. Among them, there are many missing patches in the study area before 1990, so we selected all the 30 images based on the JRC-Yearly from 1990 to 2019 to identify the water area in the BLB. The JRC-Yearly dataset classifies land as permanent water, seasonal water, and others throughout the year.

The JRC Monthly Water Classification History, v1.3 (JRC-Monthly) collection preserves monthly water body distributions from 1984 to 2020 and contains 442 images in total, which have a spatial resolution of 30 m. Among them, monthly-scale water bodies have a large number of missing cases prior to 2000, and open water bodies in November-March show as ice in BLB [51], resulting in a large bias in water area statistics. To avoid large biases skewing our results, we extracted the distribution of the water area for the months of April to October from 2000 to 2019 based on the JRC-Monthly. The JRC-Monthly The JRC-Monthly dataset classifies land as water, not water and others for the month.

2.2.2. Ecological and Hydrochemical Data

The water level and water quality data for the Bosten Lake were provided by the Bosten Lake Administration of Bayingol Mongolian Autonomous Prefecture (Xinjiang). The hydrological data includes the Bosten Lake level (BLL) data from 1990 to 2019. To investigate the relationship between hydrological connectivity and the water environmental quality of the Bosten Lake comprehensively, eight pollutions indices from 2001 to 2019 are collected from the local ecological and environmental bureau. The eight indices are: dissolved oxygen (DO), permanganate index (COD_{MN}), chemical oxygen demand (COD), five-day biochemical oxygen demand (BOD_5), total phosphorus (TP), total nitrogen (TN), ammonia nitrogen (NH3-N), and total dissolved salts (TDS) or mineralization.

2.2.3. Hydro-Meteorological Data

The hydro-meteorological data used in this study are from the Terra Climate (TC) dataset (http://www.climatologylab.org/, accessed on 1 September 2021). Three variables are selected: temperature (TMP), actual evapotranspiration (ETa), and precipitation (PREC). The TC dataset, which compiles global land surface monthly-scale climate data covering 1958–2019, is a high-precision climate dataset with 4 km (1/24 degree) spatial resolution. It combines the high spatial resolution WorldClim dataset with the low spatial resolution CRU Ts4.0 and the Japanese 55-year Reanalysis (JRA55) meteorological data [56].

2.2.4. Socioeconomic Data

The population data come from the WorldPop Global Project Population Data dataset [57]. By using a machine learning approach that decomposes the population size into 100 × 100 m grid cells, it generates data on the spatial distribution of the global population from 2000–2021 utilizing the relationship between population density and a series of geospatial covariate layers.

Croplands data have been derived from MCD12Q1.006 MODIS Land Cover Type Yearly Global 500 m (MCD12Q1 V6) [58], which has a spatial resolution of 0.5 km. This dataset contains global land cover types for 2001–2020 and is based on MODIS Terra and Aqua reflectance data derived from a supervised classification.

Details of the data source for this article can be found in Supplementary Materials Table S1.

2.3. Data Processing

2.3.1. Hydrological Connectivity Index

According to the landscape connectivity theory (Figure 2), the integral index of connectivity (IIC) is used to evaluate the hydrological connectivity of BLB using Conefor Sensinode 2.6 and ArcGIS 10.2 software [59,60]. The interannual hydrological connectivity (Yearly-IIC) is calculated based on the permanent water patches in JRC-Yearly datasets, and the seasonal hydrological connectivity (Monthly-IIC) is calculated using the water patches in JRC-Monthly datasets. The interannual variability of hydrological connectivity (dIIC) serves as an index to identify the significant patches of water [22]. The threshold value for Conefor Sensinode 2.6 is determined in Supplementary Materials Text S1 and Supplementary Materials Figure S5.

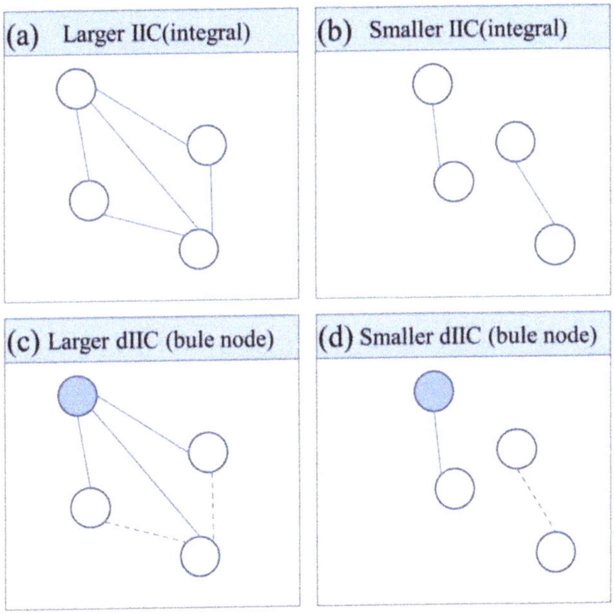

Figure 2. Different cases illustrating these two connectivity indexes. (**a**) Larger IIC: any two nodes are connecting. (**b**) Smaller IIC: pairwise connecting. (**c**) Larger dIIC for bule node: key node. (**d**) Smaller dIIC for bule node: non-key node.

Calculation of hydrological connectivity index. The integral index of connectivity (IIC) value is used to assess complex traffic topological networks, which is widely applied in the evaluation of landscape connectivity [22,61–63]. Since the index is highly sensitive to connectivity, it has also been widely used in previous studies on hydrological connectivity [22,64–66]. Therefore, IIC and dIIC are used to assess the dynamics of interannual hydrological connectivity dynamics of the BLB from 1990 to 2019 and the seasonal hydrological connectivity dynamics from 2000 to 2019 for April to October, respectively.

The IIC indicates the integral index of connectivity of the basin which is based on a binary connectivity model, indicating direct connectivity or disconnection and intuitive structural connectivity between two patches. The higher the connectivity of the study area, the higher the IIC value. The connectivity index is calculated as:

$$IIC = \frac{\sum_{i=1}^{n}\sum_{j=1}^{n}\frac{a_i \times a_j}{1 + nl_{ij}}}{A_L^2} = \frac{\sum_{i=1}^{n}\sum_{j=1}^{n} a_i \times a_j \times P_{ij}^*}{A_L^2}, \ (0 \leq IIC \leq 1) \quad (1)$$

where, n is the total number of water patches; a_i and a_j represent the area of patch i and j, respectively; nl_{ij} denotes the number of links in the shortest path between patch i and j; and A_L the area of the BLB. P_{ij}^* is the maximum multiplication probability of all possible paths between patches i and j. P_{ij}^* = 0 means that the two patches are completely isolated from each other. The values of IIC range from 0 to 1: when IIC = 0, there is no connection between patches; when IIC = 1, the whole landscape is actually one habitat patch. Meanwhile, in order to analyze the relationship between hydrological connectivity and water-related environment, we divided the years into high hydrological connectivity years (i.e., IIC is above the mean value) and low hydrological connectivity years (i.e., IIC is above the mean value) from 1990 to 2019.

Although, the IIC can assess the overall degree of hydrological connectivity, it lacks the ability to assess the importance of individual water patches [67]. Therefore, to identify the importance of specific patches in the BLB and better support water conservation, the contribution of each node to the overall index was measured by removing each specific node and recalculating the IIC [22] to take into account the percentage loss. The calculation is as follows:

$$dIIC = \frac{IIC - IIC'}{IIC} \times 10 \quad (2)$$

where, IIC and $dIIC'$ correspond to the IIC value before and after the removal of a certain patch, respectively.

The Jenks natural breakpoint method [68] has been used to classify the important index of water patches for each year, and the *dIIC* is classified in six levels as unimportant (I. 0~0.99), low importance (II. 1~4.99), medium importance (III. 5~9.99), high importance (IV. 10~14.99), higher importance (V. 15~19.00) and highest importance (VI. 20~100). Based on the grading results of *dIIC*, the image maps of each year are superimposed by ArcGIS10.2 software raster calculator and classified into five categories, which are less important (importance index fluctuates between gradient I and II), important (importance index fluctuates between gradient II and III), more important (importance index fluctuates between gradient III and IV), very important (importance index fluctuates between gradient IV and V), and most important (importance index is between gradient V and VI).

2.3.2. Statistical Analysis

For discussing the effect of water abundance and depletion patterns on interannual and seasonal hydrological connectivity, based on the runoff data of the Kaidu River outlet, the years were classified into wet year, dry year, and normal year according to the typical year method (Table S2).

Trend analysis was performed using Sen's slope estimator and Pearson correlation in this study. Sen's slope estimator is often used in trend analysis of long time series data as a robust non-parametric statistical method of trend calculation that is computationally efficient and insensitive to measurement errors and outliers [69]. Pearson correlation was used to analyze the relationship between two different factors. Standard deviation (SD) and coefficient of variation (CV) were used to characterize the data's degree of dispersion.

The multiple stepwise regression model (MSRM) is used to quantify the effects of climate factors and human activities on *IIC* [70,71]. Slope > 0 is defined as positive correlation and slope < 0 is defined as negative correlation (95% confidence level). The MSRM equation is shown below:

$$IIC = aX1 + bX2 + cX3 + dX4 + eX5 + k \quad (3)$$

where, $X1$ is the total annual precipitation (mm), $X2$ is the average annual temperature (°C), $X3$ is the total annual actual evapotranspiration (mm), $X4$ is the croplands area (km^2), $X5$ is the average annual total population (person), and k is a constant.

3. Results

3.1. Characteristics of Multi-Scale Changes in Hydrological Connectivity

3.1.1. Inter-Annual Variation Characteristics of Hydrological Connectivity

The hydrological connectivity of the BLB basin in the past 30 years is low overall, with an annual average yearly-IIC index of only 0.169. In addition, three obvious fluctuating trends have emerged (Figure 3a) in three distinct corresponding periods. First, the yearly-IIC index increased at 0.004/yr from 1990 to 2000, giving an average yearly-IIC of 0.170. Secondly, from 2000 to 2007, it decreased at a rate of 0.008/yr, with an average yearly-IIC of 0.164, and then plunged to a 30-year low value of 0.146 in 2007, which was 13.59% lower than the average. In the third period (2007 to 2019), the yearly-IIC index increased again at 0.005/yr, with an average yearly-IIC of 0.173, which is the highest mean value among the three periods. It reached a maximum in 2018 of up to 0.203, which is 19.68% higher than the yearly mean value.

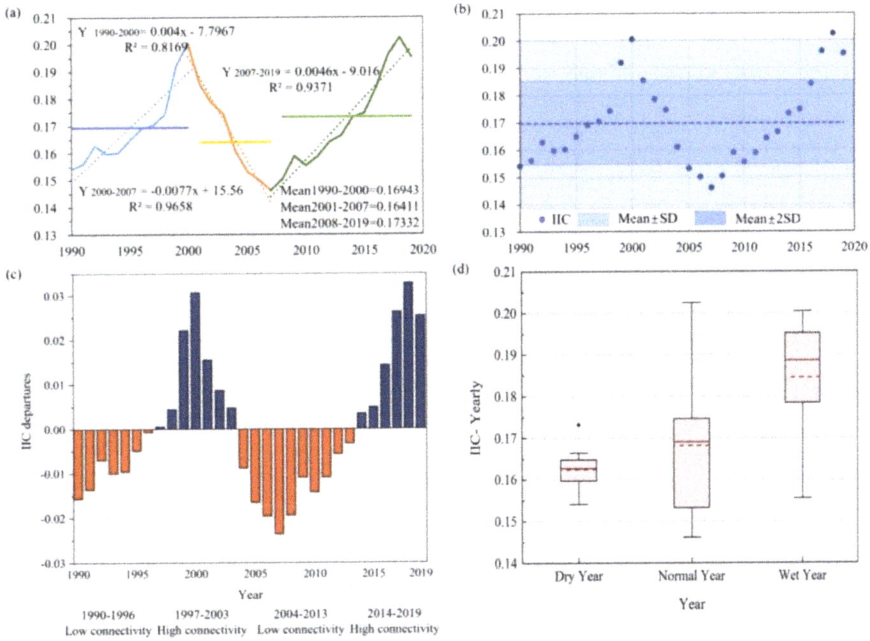

Figure 3. (**a**) Time series, trends, and staged changes of yearly-IIC of the Bosten Lake Basin from 1990–2019; (**b**) The discrete situation of IIC in the Bosten Lake Basin from 1990–2019; (**c**) IIC departures in the Bosten Lake Basin, 1990–2019, based on the average of 30-year period; (**d**) the distribution of yearly-IIC index in different years. The top and bottom of boxes represent the 75th and 25th percentiles, respectively, while the top and bottom whiskers represent the 90th and 10th percentiles, respectively. Solid red lines in the boxes are median values and dotted red lines represent mean value.

Yearly-IIC in BLB has four phases of fluctuation with surpluses and deficits in the range of ± 0.03 over the last 30 years (Figure 3c). The yearly-IIC shows two high value phases (1997–2003, 2014–2019) and two low value phases (1990–1996, 2004–2013). In 1990–1996 and 2004–2013, the yearly-IIC index was lower than the 30-year average value, and the average deficit was -0.114. These two periods had an average value of yearly-IIC of 0.161 and 0.156, respectively. On the other hand, 1997–2003 and 2014–2019 were above-average periods of higher hydrological connectivity. The average surplus of these two phases was 0.149, and the mean values of yearly-IIC were 0.181 and 0.191, respectively.

The hydrological connectivity of the BLB has obvious spatial heterogeneity. The annual average yearly-IIC of KDH, KKH and KQH are 0.021, 0.032 and 0, respectively (Supplementary Figure S2). The yearly-IIC of KQH is almost zero, which is due to the prominent disconnection of the Konqi River channel [72] and serious desertification downstream, which resulted in few and scattered patches of water. The average annual IIC of KKH is higher than that of KDH and less discrete, which is mainly because KDH is a high-altitude mountainous area with unstable connectivity of water patches and is vulnerable to climate change factors, while KKH is an oasis area with more stable and higher connectivity of water patches under human management.

Figure 3d shows that the yearly-IIC is the largest in wet years (0.185) and the smallest in dry years (0.163). These data indicate that the hydrological connectivity of the basin was influenced by runoff, the hydrological connectivity was higher in wet years than in dry ones, but the variability of the yearly-IIC index is higher in wet years.

3.1.2. Seasonal Variations in Hydrological Connectivity

As shown in Figure 4a,c, there was a clear seasonal pattern in the hydrological connectivity of BLB, with an average monthly-IIC value of 0.157 from April to October. The average monthly-IIC was highest in April at 0.159 and lowest in October at 0.151. Monthly-IIC was increased at a rate of 0.0011/yr in October, while the rest of the months experienced a decreasing trend, with hydrological connectivity decreased the most (−0.0026/month) in May.

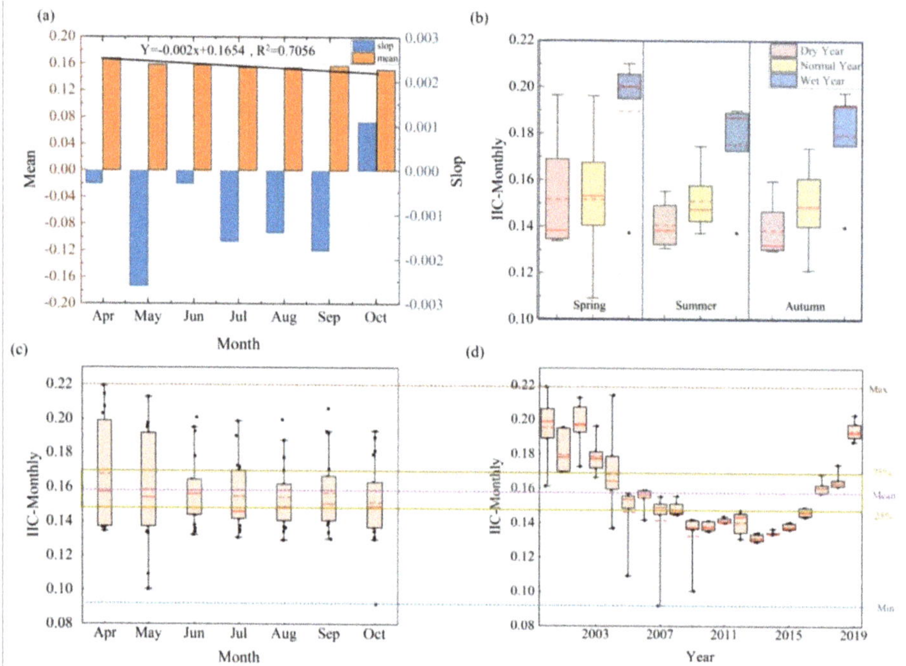

Figure 4. (**a**) Distribution of monthly-IIC; (**b**) Monthly-IIC under different seasons for different year types; (**c**) Box plot of monthly-IIC distribution from April to October in 2000 to 2019; (**d**) Monthly-IIC Distribution boxplots from April to October in 2000 to 2019. The meaning of the boxes and lines are the same with Figure 3d.

The interannual trend of monthly-IIC is similar to yearly-IIC, a 'U' shaped trend has been observed between 2000 and 2019 (Figure 4d), the data show that the monthly-IIC is lower than the yearly-IIC. To be specific, the lowest monthly-IIC value of 0.114 occurring in 2010, and the maximum value occurs in 2019 with an average monthly-IIC of 0.195. In seasonal (Figure 4b), it can be found that the monthly-IIC reached 0.160 in spring, which is higher than summer (monthly-IIC = 0.155) and autumn (monthly-IIC = 0.154). Furthermore, monthly-IIC is higher in wet years (yearly-IIC = 0.182) than in dry years (yearly-IIC = 0.144). The highest monthly-IIC occurred in the spring during the wet period (monthly-IIC = 0.190) and the lowest in the fall during the dry period (monthly-IIC = 0.139).

The monthly-IIC in the BLB is higher in the wet period than that in the dry, and higher in spring than in summer and autumn. The reason for this phenomenon may be that with the increase in temperature during the year, the water flux in the basin gets larger and vegetation grows vigorously, leading to an increase in evapotranspiration and soil evaporation, which may in turn reduce the hydrological connectivity.

3.1.3. Identification of Key Nodes in Hydrological Connectivity

As shown in Figure 5, the key nodes of hydrological connectivity BLB were mainly distributed around the Bosten Lake and the upper reaches of the Kaidu River. The dIIC value of the Bosten Lake is always more than 99, which is the largest hydrological connectivity node in the BLB. The Bosten Lake is the central node of the water network which is of great significance to the regional ecological environment. The other key nodes are primarily distributed along the Kaidu River, mainly in the small Yuldus wetland and the large Yuldus wetland, with an average dIIC of 4.142. The small Yuldus wetland and the large Yuldus wetland, important biodiversity reserves in the BLB, have a particularly prominent water conservation function.

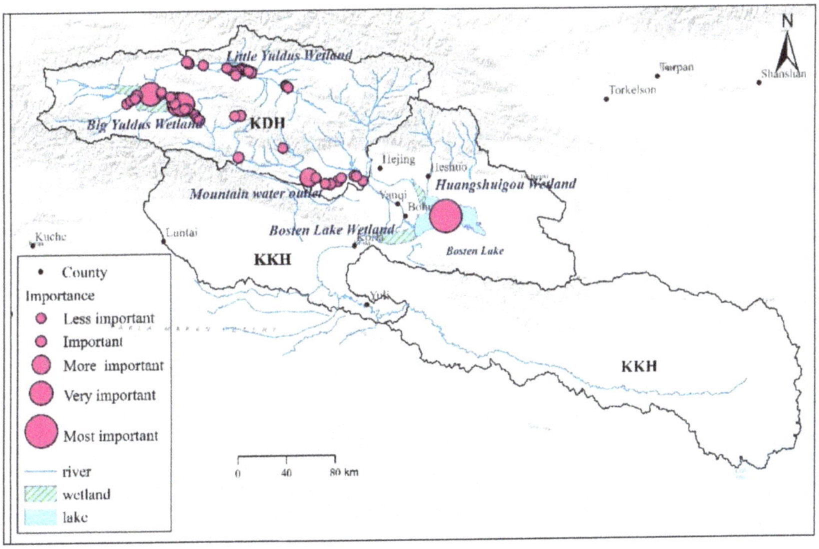

Figure 5. Distribution of key hydrological connectivity nodes in BLB.

3.2. *Characteristics and Dynamics of Water Quality in the Bosten Lake*

Increased hydrological connectivity will promote water exchange capacity, which will enhance the water cycle and improve the water quality. The water environmental quality of the Bosten Lake shows an overall improvement from 2001 to 2019 (Figure 6). Specifically, all indicators reveal improved quality of the Bosten Lake (except for TN), with the compliance rate of all indicators up to the Chinese Environmental Quality for Surface Water III Standard (Standard-III) (Tables S3 and S4), except for TDS, COD and TN.

Among these indices, the annual average concentration of COD is 23.41 mg/L (Figure 6b), which already exceeds the Standard III ($\leq$20 mg/L), the attainment rate only 36.84%. The maximum value was as high as 28.68 mg/L in 2013, and only in 2006 and 2018 were the concentrations lower than 20 mg/L. The annual average concentration of TN was 0.83 mg/L, which achieved the Standard-III, and the attainment rate is 94.74%; the concentration was only exceeded in 2011, with a concentration of 1.01 mg/L (Figure 6c).

The water environmental quality of the Bosten Lake varies under different hydrological connectivity periods (Figure 7a). The annual average concentrations of DO, COD_{MN}, BOD_5, TP, TN, NH3-N and TDS in high hydrological connectivity periods are smaller than those in low periods, and the dispersion of BOD_5, TP, NH3-N and TDS has a higher degree (CV > 20%) (Table S5). However, the maximum value and the CV (CV = 9–58%) of COD_{MN}, COD, BOD_5, TN and TDS in high hydrological connectivity periods are larger than those in low hydrological connectivity periods (Figure 7a). The reason is that water movement is enhanced during high hydrological connectivity, leading to an increase in the water quality exchange capacity, which results in fluctuating changes in water quality indices and a subsequent increase in the dispersion of water quality indicators. Among them are COD and TDS, which have the lowest rate for fulfilling the Standard III. Most of their compliance occur in high hydrological connectivity (75%). Moreover, only in one year (2011) during the low hydrological connectivity periods did TN not reach the standard.

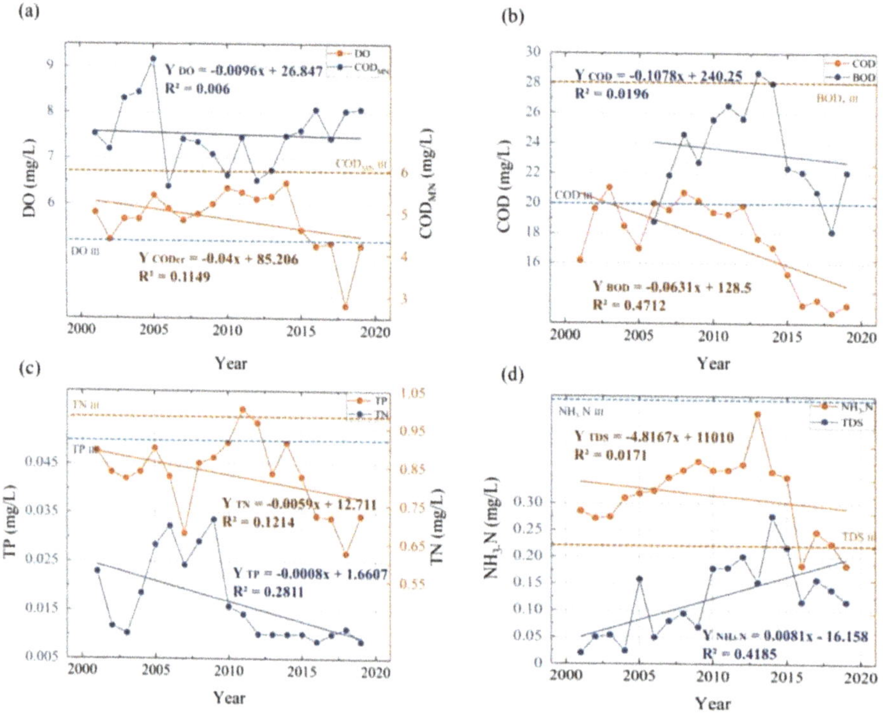

Figure 6. Interannual variation curves of eight water quality indicators, (**a**) DO and COD_{MN} (**b**) COD and BOD_5 (**c**) TP and TN (**d**) NH3-N and TDS, where the dashed lines represent the Class III standard values and freshwater standard values (TDS) of each indicator.

There are pronounced seasonal differences in the water environmental quality of the Bosten Lake. As shown in Figure 7b, concentrations of COD_{MN}, COD, BOD_5, TN, NH3-N and TDS were higher in autumn and summer than in spring, while DO concentrations were slightly higher in spring than in summer and autumn. This indicates that seasonal water quality was significantly better in the spring when hydrological connectivity was higher than in the summer and autumn when it was lower. The seasonal variation of TP was more discrete than the rest of the indicators, with CVs as high as 99%, 61%, and 80% in spring, summer, and autumn, respectively, which indicates that the higher the inter-seasonal hydrologic connectivity, the higher the variability of hydro-chemical concentration. The seasonal variation of NH3-N varies greatly (CV > 50%), while that of DO, COD, TN and TDS is less so (CV < 20%). From these data, we can see that the water quality of the Bosten Lake during high hydrological connectivity periods is better than during lower hydrological connectivity periods. However, the degree of dispersion is higher, and the seasonal variations in hydrological connectivity majorly impact the quality of the water environment.

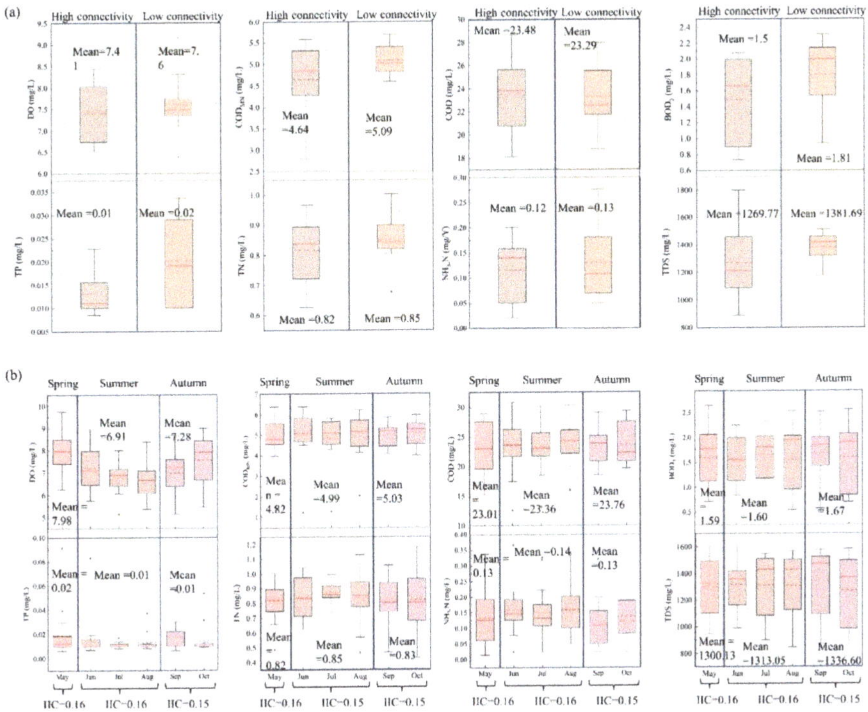

Figure 7. (**a**) Box line diagrams of 8 water quality indicators during high connectivity (green) and low connectivity (yellow) periods, respectively, dots represent the average concentration of each indicator per year; (**b**) Boxplots of 8 water quality indicators during May to October, respectively, where May is spring (red), June–August is summer (green), and September–October is autumn (blue). The meaning of the boxes and lines are the same with Figure 3d, dots represent the average concentration of each indicator per mounth.

3.3. *Hydrological Connectivity and Its Water-Related Environmental Relationship*

3.3.1. Anthropogenic and Climatic Drivers for IIC Dynamics

The results of the multi-step regression model used to quantify the effects of climate change and human activities on hydrological connectivity in the BLB are shown in Table S6. IIC is significantly and positively correlated with precipitation (PREC) ($p < 0.05$) and Population (POP) ($p < 0.05$). With the increase of PREC, water patch area of the basin increases, which obviously resulted in an increase in the inter-patch connectivity. As the regional population increases, the government has implemented a series of water projects and management measures (Table S7) in order to achieve sustainable water resources in the region, which may contribute to a positive correlation between population and hydrological connectivity. Simultaneously, IIC is significantly negatively correlated with actual evapotranspiration (ETa) ($p < 0.05$) and cropland ($p < 0.05$), which is also attributed to the increase in water consumption, decrease in water patch area and increase in fragmentation. The absolute value of the correlation coefficient between ETa and IIC is greater than the other indexes (Table S6), which indicates that the connectivity of water patches is mainly negatively influenced by ETa in BLB, which is different from the humid area [73–75]. The water resources in inland river basins in arid zones are formed in the mountainous areas, and precipitation in plains, which is almost not hydrologically significant, is too subtle to recharge the streamflow. Temperature (TEM) is negatively correlated with IIC, but not statistically significant.

3.3.2. Relationship between Water Quality and Hydrological Characteristics

To examine the interactions between hydrological related characteristics and hydrochemistry, a correlation study has been conducted in this paper (Figure 8). The results show that COD_{MN}, BOD_5, TP, TN, and TDS are significantly negatively correlated with IIC ($p \leq 0.05$), while DO is significantly positively correlated with IIC. The results indicated that the increase in hydrological connectivity improves water quality. At the same time, the area of total water body (TWA) and the area of the Bosten Lake area (BLA) have significant negative correlations ($p \leq 0.05$) with COD_{MN}, BOD_5 and TP, whereas the Bosten Lake level (BLL) has significant positive correlations ($p \leq 0.05$) with COD_{MN}, COD, BOD_5, TN and TDS, indicating that hydrological connectivity increases with increases in watershed area and lake level, thus improving the water environment quality.

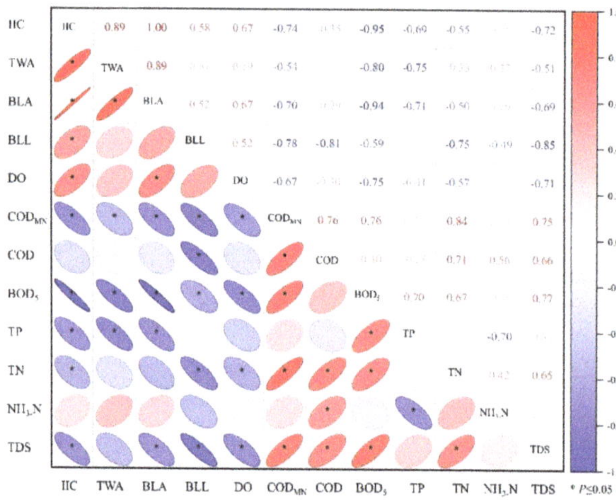

Figure 8. Correlation test of each index. IIC: integral index of connectivity; TWA: total water area; BLA: Bosten Lake area; BLL: Bosten Lake level.

Overall, there is a strong correlation between the water quality of the water environment and the hydrological characteristics. The results show that DO has a very strong positive correlation with IIC and the Bosten Lake area (BLA), ($|r| > 0.6$, $p \leq 0.05$); COD_{MN} has a very strong negative correlation with IIC, BLA and the Bosten Lake level (BLL) ($|r| > 0.6$, $p \leq 0.05$); BOD5 and TP have a very strong negative correlation with IIC, total water area (TWA) and BLA, ($|r| > 0.6$, $p \leq 0.05$); TN has a very strong negative correlation with BLL ($|r| > 0.6$, $p \leq 0.05$); TDS has a very strong negative correlation with IIC, BLA and BLL ($|r| > 0.6$, $p \leq 0.05$); and TDS has a very strong negative correlation with IIC, BLA and BLL ($|r| > 0.6$, $p \leq 0.05$). Furthermore, the results indicate that there is a strong negative correlation between TDS and IIC, and between BLA and BLL ($|r| > 0.6$, $p \leq 0.05$). This shows that the quality of the water environment will be effectively improved with increases in hydrological connectivity and water body area.

The changes and interactions between water environmental quality indicators are more complex, as shown by the significant negative correlations between DO and COD_{MN}, BOD_5, TN, TDS ($p \leq 0.05$) and negative correlations with COD and TP. The correlations with NH3-N are weak and statistically insignificant. COD_{MN} has a strong positive correlation with COD, BOD_5, TN, and TDS ($|r| > 0.6$, $p \leq 0.05$); COD has a strong positive correlation with TN, NH3-N, and TDS ($p \leq 0.05$); BOD_5 has a strong positive correlation with TP, TN, and TDS ($|r| > 0.6$, $p \leq 0.05$); and TN has a strong positive correlation with TDS ($|r| > 0.6$, $p \leq 0.05$). This phenomenon indicates that the increase in DO has a positive effect on the self-cleaning ability of the water environment. All the other water quality indicators show a synergistic effect, such that when one water quality indicator starts to deteriorate, it causes the deterioration of other indicators as well.

4. Discussion

4.1. Anthropogenic and Climatic Drivers for Hydrological Connectivity Dynamics

Previous studies [73,75] have demonstrated that hydraulic measures such as river dredging and land use changes are the main drivers of hydrological connectivity changes. According to the statistics (Figure S3), the population and cropland in BLB increased continuously from 2001 to 2019, which likewise increased water consumption and affected terrestrial water storage [76]. This, in turn, had an impact on hydrological connectivity. Table S7 shows that most of the BLB channels have been dredged after 1990, and that several artificial channels have been built since 2000 to satisfy the increasing irrigation demand. The construction of these channels has increased the flow paths and significantly improved the possibility of water system connectivity.

Ecological water conveyance, as an effective measure to restore and protect the natural vegetation and water environment in inland river basins, is commonly used in northwest China [76–78]. Around 2010, several water system connectivity projects and hydraulic facilities were carried out in the BLB, including artificial dredging and construction channels and water transfer projects. All these hydraulic measures have effectively increased the water system connectivity in the basin, and improved the fractal structure of the water system and artificial water system connectivity. It has enabled the basin to achieve supplementary water resources and boost the complementary water resources between river-lake and reservoir. These water management measures have enhanced of the exchange of water and matter within and outside the water system, leading to an improvement of the carrying capacity of water resources in the basin, and an increase in the stability of the water system network. In cases where protective behaviors are stronger than destructive behaviors, the hydrological connectivity of the basin will improve. Otherwise, the connectivity will be reduced. Therefore, regulating all types of water use activities and construction projects in inland river basins in arid zones is essential, which can improve the structure and pattern of hydrological connectivity, as well as the function and connectivity of river and lake systems.

Our results show that the increase in hydrological connectivity caused by precipitation and temperature is not significant in the arid zone. The increase in TMP accelerate glacier melt, which led to an increase in runoff and hydrological connectivity to some extent. In contrast, ETa has a negative impact on the watershed, which will reduce hydrological connectivity.

Beel's [78] argued that the increase in river function in summer and autumn in the Arctic highlands could increase terrestrial hydrological connectivity. However, we find that the seasonal variation in hydrological connectivity is higher in spring than in summer and autumn (Figure 4), as surface water in the arid inland river basins is mainly influenced by ETa. Hydrological connectivity is higher in spring with snowmelt, but as temperature rises, vegetation grows and regional evapotranspiration increases, the water flux becomes larger, resulting in an amount of water surface evapotranspiration and the connectivity of water patches is reduced.

4.2. Ecological and Hydrochemical Characteristics of Lakes in Response to Hydrological Characteristics

As illustrated in Figure 8, there is a strong positive relationship among hydrological connectivity, water body area, lake area and water level. When the water level of the lake increases, the water area likewise increases, and as total water body area increases, there is an obvious increase in hydrological connectivity. As the largest throughput freshwater lake in China, the Bosten Lake is a broad and shallow basin, with rises in water level, the surface area of the lake will increase, isolated patches of water will be connected with each other, thus increasing hydrological connectivity [79,80]. However, as the lake area increases, the evaporation of water from the lake surface will increase, as well as the ineffective water loss of the lake. While when the water level is too low, the exposed surface area increases, and the large area of exposed fine sand of the lake basin, that is in the northwest and south shore of the lake, will directly aggravate the wind and dust storms around the Bosten Lake, and affect the environment suitable for people to live. Therefore, there may be an inverted U-shaped relationship between hydrological connectivity, lake level, water area, and ecology that is similar to an environmental Kuznets curve [40].

As the biggest inland freshwater lake in China, the largest hydrologically connected node in the basin (Figure 5), the soil salinization in the surrounding areas is seriously overloaded with nutrients. Despite the local government implementing a series of water conservation policies and projects, it still has not been able to change the fact that the Bosten Lake is gradually transforming into a micro-salt lake (TDS > 1000 mg/L) [2,28,81]. The continued increase of salinity not only adversely affects the lake ecosystem, regional ecology and water resources utilization, but also has become a serious environmental problem in the lake itself [48]. Our study found that between 1990 and 2019, the Bosten Lake had met freshwater standards for TDS concentrations twice only—in 2016 and 2019 (TDS $\leq$ 1000 mg/L) (Figure 6d).

From 2016 to 2020, in response to the problems of broken streams and deteriorating water environment around the Bosten Lake, along with poor water circulation and water environment quality in the northern part of the lake, three water system connection projects were implemented. A total of about 4.79×10^8 m^3 (Table S7) of water was ecologically transferred to the Bosten Lake, which provided an important basis for the improvement of the lake's water environment quality. According to the correlation test (Figure 8), TDS shows a strong negative correlation with IIC, BLA, and BLL ($|r| > 0.6$, $p \leq 0.05$). This phenomenon suggests that the hydrological connectivity, BLL and BWA play an important role in the dynamics of lake water salinity. Our results also show that from 2007 to 2014, hydrological connectivity increased (Figure 4a), while BLL decreased (Figure S1). At the same time, however, TDS also increased (Figure 6d). The correlation between TDS and BLL was greater than that between IIC and BLA based on the correlation test, indicating a more significant regulation of TDS by BLL [28,47]. This may be because as the lake level increases, the decrease in hydrochemical concentration caused by the increase in lake volume is more efficient than that of water exchange. However, the existing hydrological connectivity evaluation models lack the representation of hydrological connectivity processes, so it is difficult to reveal the kinetics of water exchange quantitatively, which will be the focus of our next research. Also, we will improve the method of assessing hydrological connectivity, which is influenced by the area of water bodies and the distance of patches.

5. Conclusions

In this study, we applied the concept of landscape connectivity and used landscape graph theory to evaluate the connectivity characteristics of water patches in BLB from 1990 to 2019, and analyzed the water quality changes in Lake Bosten over the last 20 years, emphasizing the regulation effects of hydrological connectivity on water quality. The results indicate that the hydrological connectivity of BLB is low (IIC = 0 ~ 0.2025, IIC_{MAX} = 1) and cyclical, and the hydrological connectivity is higher in oasis than mountain and desert areas. At the same time, the water resources in inland river basins in arid zones are formed in mountainous areas, and precipitation in plains is almost not hydrologically significant. Temperature (TEM) is negatively correlated with IIC, but not statistically significant. Furthermore, seasonal hydrological connectivity is highest in spring and lowest in autumn.

From 2000 to 2019, the water environmental quality of the lake gradually improved, and the pollution indicators mainly related to TDS, COD and TN. Most importantly, there is a significant negative correlation between COD_{MN}, BOD_5, TP, TN, and TDS and IIC ($p \leq 0.05$), and DO was significantly positively correlated with IIC, and the annual average concentrations of COD_{MN}, BOD_5, TP, TN, NH3-N and TDS in high hydrological connectivity periods are smaller than those in low periods, and most of their compliance periods occur during high hydrological connectivity (75%), it can be found that hydrological connectivity plays a key role in improving the water quality of the Bosten Lake, the minimum value have occurred during periods of high hydrological connectivity.

Supplementary Materials: The following supporting information can be downloaded at: https://www.mdpi.com/article/10.3390/rs14194977/s1, Text S1. Selection of hydrological connectivity thresholds, Text S2. Characteristics of water body changes in the BLB, Text S3. Water quality change characteristics from 2001–2019, Figure S1. Change trend of Bosten lake water area and level change from 1990 to 2019, Figure S2. Time series and trends of yearly IIC from 1990–2019 of each sub-basin, Figure S3. Arable land area and population change of BLB, Figure S4. Temporal and spatial variations of PREC, ETa and TMP and their significance tests. Black dots represent passing the $p < 0.05$ test, Figure S5. Variation of IIC at different distance thresholds, Figure S6. 1990–2019 BLB (a) variations of TWA, PWA, SWA, (b) percentage of PWA and SWA, and (c) conversion between different types of water bodies, Figure S7. (a) Water area distribution in 1990, (b) water area distribution in 2019, (c) water area change from 1990 to 1996, (d) water area change from 1996 to 2000, (e) water area change from 2000 to 2005, (f) water area change from 2005 to 2010, (g) water area change from 2010 to 2015, (h) water area change from 2015 to 2019, Table S1. Details of the data source for this article, Table S2. Classification results of wet, dry and normal years from 1990 to 2019, Table S3. Chinese Environmental Quality for Surface Water III Standard(GB3838-2002), Table S4. Standard for classification of fresh and salt water, Table S5. Characteristic values of interannual variation of water quality indexes, Table S6. Multiple stepwise regression results, Table S7. Water System connectivity Project of BLB. References [82–85] are cited in Supplementary Materials.

Author Contributions: C.L. and Y.C. conceived of the original design of this paper. G.F., H.Z. and W.H. improve the structure of the paper. Z.L., Y.L. and X.W. provided comments on this paper. All authors have read and agreed to the published version of the manuscript.

Funding: This work is supported by the Natural Science Foundation of Xinjiang (2021D01D01) and the Key Research Program of the Chinese Academy of Sciences (ZDRWZS-2019-3).

Institutional Review Board Statement: Not applicable.

Informed Consent Statement: Not applicable.

Data Availability Statement: JRC Yearly Water Classification History, v1.3 and JRC Monthly Water Classification History, v1.3 are openly available via https://earthengine.google.com/ (accessed on 20 September 2021).

Conflicts of Interest: The authors declare that they have no known competing financial interests or personal relationships that could have appeared to influence the work reported in this paper.

References

1. Boulton, A.J.; Rolls, R.J.; Jaeger, K.L.; Datry, T. Chapter 2.3—Hydrological Connectivity in Intermittent Rivers and Ephemeral Streams. In *Intermittent Rivers and Ephemeral Streams*; Datry, T., Bonada, N., Boulton, A., Eds.; Academic Press: Cambridge, MA, USA, 2017; pp. 79–108; ISBN 978-0-12-803835-2.
2. Hassani, A.; Azapagic, A.; Shokri, N. Global Predictions of Primary Soil Salinization under Changing Climate in the 21st Century. *Nat. Commun.* **2021**, *12*, 6663. [CrossRef] [PubMed]
3. Keeley, A.T.H.; Beier, P.; Jenness, J.S. Connectivity Metrics for Conservation Planning and Monitoring. *Biol. Conserv.* **2021**, *255*, 109008. [CrossRef]
4. Kraaijenbrink, P.D.A.; Stigter, E.E.; Yao, T.; Immerzeel, W.W. Climate Change Decisive for Asia's Snow Meltwater Supply. *Nat. Clim. Chang.* **2021**, *11*, 591–597. [CrossRef]
5. Osman, M.B.; Tierney, J.E.; Zhu, J.; Tardif, R.; Hakim, G.J.; King, J.; Poulsen, C.J. Globally Resolved Surface Temperatures since the Last Glacial Maximum. *Nature* **2021**, *599*, 239–244. [CrossRef] [PubMed]
6. Yao, J.; Chen, Y.; Guan, X.; Zhao, Y.; Chen, J.; Mao, W. Recent Climate and Hydrological Changes in a Mountain–Basin System in Xinjiang, China. *Earth-Sci. Rev.* **2022**, *226*, 103957. [CrossRef]
7. Messager, M.L.; Lehner, B.; Cockburn, C.; Lamouroux, N.; Pella, H.; Snelder, T.; Tockner, K.; Trautmann, T.; Watt, C.; Datry, T. Global Prevalence of Non-Perennial Rivers and Streams. *Nature* **2021**, *594*, 391–397. [CrossRef]
8. Grill, G.; Lehner, B.; Thieme, M.; Geenen, B.; Tickner, D.; Antonelli, F.; Babu, S.; Borrelli, P.; Cheng, L.; Crochetiere, H.; et al. Mapping the World's Free-Flowing Rivers. *Nature* **2019**, *569*, 215–221. [CrossRef]

9. Cooley, S.W.; Ryan, J.C.; Smith, L.C. Human Alteration of Global Surface Water Storage Variability. *Nature* **2021**, *591*, 78–81. [CrossRef]
10. Schmitt, R.J.P.; Bizzi, S.; Castelletti, A.; Kondolf, G.M. Improved Trade-Offs of Hydropower and Sand Connectivity by Strategic Dam Planning in the Mekong. *Nat. Sustain.* **2018**, *1*, 96–104. [CrossRef]
11. Tan, Z.; Li, Y.; Zhang, Q.; Liu, X.; Song, Y.; Xue, C.; Lu, J. Assessing Effective Hydrological Connectivity for Floodplains with a Framework Integrating Habitat Suitability and Sediment Suspension Behavior. *Water Res.* **2021**, *201*, 117253. [CrossRef]
12. Belletti, B.; Garcia de Leaniz, C.; Jones, J.; Bizzi, S.; Börger, L.; Segura, G.; Castelletti, A.; van de Bund, W.; Aarestrup, K.; Barry, J.; et al. More than One Million Barriers Fragment Europe's Rivers. *Nature* **2020**, *588*, 436–441. [CrossRef] [PubMed]
13. Lynch, L.M.; Sutfin, N.A.; Fegel, T.S.; Boot, C.M.; Covino, T.P.; Wallenstein, M.D. River Channel Connectivity Shifts Metabolite Composition and Dissolved Organic Matter Chemistry. *Nat. Commun.* **2019**, *10*, 459. [CrossRef] [PubMed]
14. Ali, G.; English, C. Phytoplankton Blooms in Lake Winnipeg Linked to Selective Water-Gatekeeper Connectivity. *Sci. Rep.* **2019**, *9*, 8395. [CrossRef] [PubMed]
15. Anderson, E.P.; Jenkins, C.N.; Heilpern, S.; Maldonado-Ocampo, J.A.; Carvajal-Vallejos, F.M.; Encalada, A.C.; Rivadeneira, J.F.; Hidalgo, M.; Cañas, C.M.; Ortega, H.; et al. Fragmentation of Andes-to-Amazon Connectivity by Hydropower Dams. *Sci. Adv.* **2018**, *4*, eaao1642. [CrossRef] [PubMed]
16. Lam, N.S.-N.; Cheng, W.; Zou, L.; Cai, H. Effects of Landscape Fragmentation on Land Loss. *Remote Sens. Environ.* **2018**, *209*, 253–262. [CrossRef]
17. Bracken, L.J.; Croke, J. The Concept of Hydrological Connectivity and Its Contribution to Understanding Runoff-Dominated Geomorphic Systems. *Hydrol. Processes* **2007**, *21*, 1749–1763. [CrossRef]
18. Pringle, C. What Is Hydrologic Connectivity and Why Is It Ecologically Important? *Hydrol. Processes* **2003**, *17*, 2685–2689. [CrossRef]
19. Wohl, E.; Brierley, G.; Cadol, D.; Coulthard, T.J.; Covino, T.; Fryirs, K.A.; Grant, G.; Hilton, R.G.; Lane, S.N.; Magilligan, F.J.; et al. Connectivity as an Emergent Property of Geomorphic Systems. *Earth Surf. Processes Landf.* **2019**, *44*, 4–26. [CrossRef]
20. Coulthard, T.J.; Van De Wiel, M.J. Modelling Long Term Basin Scale Sediment Connectivity, Driven by Spatial Land Use Changes. *Geomorphology* **2017**, *277*, 265–281. [CrossRef]
21. Jahanishakib, F.; Salmanmahiny, A.; Mirkarimi, S.H.; Poodat, F. Hydrological Connectivity Assessment of Landscape Ecological Network to Mitigate Development Impacts. *J. Environ. Manag.* **2021**, *296*, 113169. [CrossRef]
22. Sun, C.; Chen, L.; Zhu, H.; Xie, H.; Qi, S.; Shen, Z. New Framework for Natural-Artificial Transport Paths and Hydrological Connectivity Analysis in an Agriculture-Intensive Catchment. *Water Res.* **2021**, *196*, 117015. [CrossRef] [PubMed]
23. Liu, J.; Engel, B.A.; Dai, L.; Wang, Y.; Wu, Y.; Yan, G.; Cong, L.; Zhai, J.; Zhang, Z.; Zhang, M. Capturing Hydrological Connectivity Structure of Wetlands with Indices Based on Graph Theory: A Case Study in Yellow River Delta. *J. Clean. Prod.* **2019**, *239*, 118059. [CrossRef]
24. Yu, Z.; Wang, Q.; Xu, Y.; Lu, M.; Lin, Z.; Gao, B. Dynamic Impacts of Changes in River Structure and Connectivity on Water Quality under Urbanization in the Yangtze River Delta Plain. *Ecol. Indic.* **2022**, *135*, 108582. [CrossRef]
25. Li, Y.; Zhang, Q.; Cai, Y.; Tan, Z.; Wu, H.; Liu, X.; Yao, J. Hydrodynamic Investigation of Surface Hydrological Connectivity and Its Effects on the Water Quality of Seasonal Lakes: Insights from a Complex Floodplain Setting (Poyang Lake, China). *Sci. Total Environ.* **2019**, *660*, 245–259. [CrossRef] [PubMed]
26. Goodwell, A.E.; Kumar, P.; Fellows, A.W.; Flerchinger, G.N. Dynamic Process Connectivity Explains Ecohydrologic Responses to Rainfall Pulses and Drought. *Proc. Natl. Acad. Sci. USA* **2018**, *115*, E8604–E8613. [CrossRef]
27. Deng, X. Correlations between Water Quality and the Structure and Connectivity of the River Network in the Southern Jiangsu Plain, Eastern China. *Sci. Total Environ.* **2019**, *664*, 583–594. [CrossRef]
28. Zhou, L.; Zhou, Y.; Hu, Y.; Cai, J.; Bai, C.; Shao, K.; Gao, G.; Zhang, Y.; Jeppesen, E.; Tang, X. Hydraulic Connectivity and Evaporation Control the Water Quality and Sources of Chromophoric Dissolved Organic Matter in Lake Bosten in Arid Northwest China. *Chemosphere* **2017**, *188*, 608–617. [CrossRef]
29. Yu, X.; Hawley-Howard, J.; Pitt, A.L.; Wang, J.-J.; Baldwin, R.F.; Chow, A.T. Water Quality of Small Seasonal Wetlands in the Piedmont Ecoregion, South Carolina, USA: Effects of Land Use and Hydrological Connectivity. *Water Res.* **2015**, *73*, 98–108. [CrossRef]
30. Peršić, V.; Čerba, D.; Bogut, I.; Horvatić, J. Trophic State and Water Quality in the Danube Floodplain Lake (Kopački Rit Nature Park, Croatia) in Relation to Hydrological Connectivity. In *Eutrophication: Causes, Consequences and Control*; Ansari, A.A., Singh Gill, S., Lanza, G.R., Rast, W., Eds.; Springer Netherlands: Dordrecht, The Netherlands, 2011; pp. 109–129. ISBN 978-90-481-9625-8.
31. Maes, D.; Van Dyck, H. Climate-Driven Range Expansion through Anthropogenic Landscapes: Landscape Connectivity Matters. *Glob. Change Biol.* **2022**, *28*, 4920–4922. [CrossRef]
32. Damschen, E.I.; Brudvig, L.A.; Burt, M.A.; Fletcher, R.J.; Haddad, N.M.; Levey, D.J.; Orrock, J.L.; Resasco, J.; Tewksbury, J.J. Ongoing Accumulation of Plant Diversity through Habitat Connectivity in an 18-Year Experiment. *Science* **2019**, *365*, 1478–1480. [CrossRef]
33. Saco, P.M.; Rodríguez, J.F.; Moreno-de las Heras, M.; Keesstra, S.; Azadi, S.; Sandi, S.; Baartman, J.; Rodrigo-Comino, J.; Rossi, M.J. Using Hydrological Connectivity to Detect Transitions and Degradation Thresholds: Applications to Dryland Systems. *Catena* **2020**, *186*, 104354. [CrossRef]

34. McLaughlin, D.L.; Diamond, J.S.; Quintero, C.; Heffernan, J.; Cohen, M.J. Wetland Connectivity Thresholds and Flow Dynamics from Stage Measurements. *Water Resour. Res.* **2019**, *55*, 6018–6032. [CrossRef]
35. Okin, G.S.; Sala, O.E.; Vivoni, E.R.; Zhang, J.; Bhattachan, A. The Interactive Role of Wind and Water in Functioning of Drylands: What Does the Future Hold? *BioScience* **2018**, *68*, 670–677. [CrossRef]
36. Ryser, R.; Hirt, M.R.; Häussler, J.; Gravel, D.; Brose, U. Landscape Heterogeneity Buffers Biodiversity of Simulated Meta-Food-Webs under Global Change through Rescue and Drainage Effects. *Nat. Commun.* **2021**, *12*, 4716. [CrossRef] [PubMed]
37. Liu, J.; Engel, B.A.; Zhang, G.; Wang, Y.; Wu, Y.; Zhang, M.; Zhang, Z. Hydrological Connectivity: One of the Driving Factors of Plant Communities in the Yellow River Delta. *Ecol. Indic.* **2020**, *112*, 106150. [CrossRef]
38. Liu, Y.; Cui, B.; Du, J.; Wang, Q.; Yu, S.; Yang, W. A Method for Evaluating the Longitudinal Functional Connectivity of a River–Lake–Marsh System and Its Application in China. *Hydrol. Processes* **2020**, *34*, 5278–5297. [CrossRef]
39. Zuecco, G.; Rinderer, M.; Penna, D.; Borga, M.; van Meerveld, H.J. Quantification of Subsurface Hydrologic Connectivity in Four Headwater Catchments Using Graph Theory. *Sci. Total Environ.* **2019**, *646*, 1265–1280. [CrossRef]
40. Deng, X. Influence of Water Body Area on Water Quality in the Southern Jiangsu Plain, Eastern China. *J. Clean. Prod.* **2020**, *254*, 120136. [CrossRef]
41. Norton, A.J.; Rayner, P.J.; Wang, Y.-P.; Parazoo, N.C.; Baskaran, L.; Briggs, P.R.; Haverd, V.; Doughty, R. Hydrologic Connectivity Drives Extremes and High Variability in Vegetation Productivity across Australian Arid and Semi-Arid Ecosystems. *Remote Sens. Environ.* **2022**, *272*, 112937. [CrossRef]
42. Souza, J.; Hooke, J. Influence of Seasonal Vegetation Dynamics on Hydrological Connectivity in Tropical Drylands. *Hydrol. Processes* **2021**, *35*, e14427. [CrossRef]
43. Wu, Y.; Zhang, Y.; Dai, L.; Xie, L.; Zhao, S.; Liu, Y.; Zhang, Z. Hydrological Connectivity Improves Soil Nutrients and Root Architecture at the Soil Profile Scale in a Wetland Ecosystem. *Sci. Total Environ.* **2021**, *762*, 143162. [CrossRef] [PubMed]
44. Mainali, J.; Chang, H. Landscape and Anthropogenic Factors Affecting Spatial Patterns of Water Quality Trends in a Large River Basin, South Korea. *J. Hydrol.* **2018**, *564*, 26–40. [CrossRef]
45. Dai, X.; Zhou, Y.; Ma, W.; Zhou, L. Influence of Spatial Variation in Land-Use Patterns and Topography on Water Quality of the Rivers Inflowing to Fuxian Lake, a Large Deep Lake in the Plateau of Southwestern China. *Ecol. Eng.* **2017**, *99*, 417–428. [CrossRef]
46. Wang, Y.; Zhou, X.; Engel, B. Water Environment Carrying Capacity in Bosten Lake Basin. *J. Clean. Prod.* **2018**, *199*, 574–583. [CrossRef]
47. Yao, J.; Chen, Y.; Zhao, Y.; Yu, X. Hydroclimatic Changes of Lake Bosten in Northwest China during the Last Decades. *Sci. Rep.* **2018**, *8*, 9118. [CrossRef] [PubMed]
48. Guo, M.; Wu, W.; Zhou, X.; Chen, Y.; Li, J. Investigation of the Dramatic Changes in Lake Level of the Bosten Lake in Northwestern China. *Theor. Appl. Climatol.* **2015**, *119*, 341–351. [CrossRef]
49. Dai, J.; Tang, X.; Gao, G.; Chen, D.; Shao, K.; Cai, X.; Zhang, L. Effects of Salinity and Nutrients on Sedimentary Bacterial Communities in Oligosaline Lake Bosten, Northwestern China. *Aquat. Microb. Ecol.* **2013**, *69*, 123–134. [CrossRef]
50. Ma, L.; Abuduwaili, J.; Liu, W. Environmentally Sensitive Grain-Size Component Records and Its Response to Climatic and Anthropogenic Influences in Bosten Lake Region, China. *Sci. Rep.* **2020**, *10*, 942. [CrossRef] [PubMed]
51. Zhang, L.; Shen, T.; Cheng, Y.; Zhao, T.; Li, L.; Qi, P. Temporal and Spatial Variations in the Bacterial Community Composition in Lake Bosten, a Large, Brackish Lake in China. *Sci. Rep.* **2020**, *10*, 304. [CrossRef] [PubMed]
52. Li, N.; Kinzelbach, W.; Li, W.; Dong, X. Box Model and 1D Longitudinal Model of Flow and Transport in Bosten Lake, China. *J. Hydrol.* **2015**, *524*, 62–71. [CrossRef]
53. Pekel, J.-F.; Cottam, A.; Gorelick, N.; Belward, A.S. High-Resolution Mapping of Global Surface Water and Its Long-Term Changes. *Nature* **2016**, *540*, 418–422. [CrossRef] [PubMed]
54. Worden, J.; de Beurs, K.M.; Koch, J.; Owsley, B.C. Application of Spectral Index-Based Logistic Regression to Detect Inland Water in the South Caucasus. *Remote Sens.* **2021**, *13*, 5099. [CrossRef]
55. Xi, Y.; Peng, S.; Ciais, P.; Chen, Y. Future Impacts of Climate Change on Inland Ramsar Wetlands. *Nat. Clim. Chang.* **2021**, *11*, 45–51. [CrossRef]
56. Abatzoglou, J.T.; Dobrowski, S.Z.; Parks, S.A.; Hegewisch, K.C. TerraClimate, a High-Resolution Global Dataset of Monthly Climate and Climatic Water Balance from 1958–2015. *Sci. Data* **2018**, *5*, 170191. [CrossRef] [PubMed]
57. Tatem, A.J. WorldPop, Open Data for Spatial Demography. *Sci. Data* **2017**, *4*, 170004. [CrossRef] [PubMed]
58. Chirachawala, C.; Shrestha, S.; Babel, M.S.; Virdis, S.G.P.; Wichakul, S. Evaluation of Global Land Use/Land Cover Products for Hydrologic Simulation in the Upper Yom River Basin, Thailand. *Sci. Total Environ.* **2020**, *708*, 135148. [CrossRef] [PubMed]
59. Pascual-Hortal, L.; Saura, S. Comparison and Development of New Graph-Based Landscape Connectivity Indices: Towards the Priorization of Habitat Patches and Corridors for Conservation. *Landsc. Ecol.* **2006**, *21*, 959–967. [CrossRef]
60. Saura, S.; Torné, J. Conefor Sensinode 2.2: A Software Package for Quantifying the Importance of Habitat Patches for Landscape Connectivity. *Environ. Model. Softw.* **2009**, *24*, 135–139. [CrossRef]
61. Justeau-Allaire, D.; Vieilledent, G.; Rinck, N.; Vismara, P.; Lorca, X.; Birnbaum, P. Constrained Optimization of Landscape Indices in Conservation Planning to Support Ecological Restoration in New Caledonia. *J. Appl. Ecol.* **2021**, *58*, 744–754. [CrossRef]
62. Guo, X.; Zhang, X.; Du, S.; Li, C.; Siu, Y.L.; Rong, Y.; Yang, H. The Impact of Onshore Wind Power Projects on Ecological Corridors and Landscape Connectivity in Shanxi, China. *J. Clean. Prod.* **2020**, *254*, 120075. [CrossRef]

63. Babí Almenar, J.; Bolowich, A.; Elliot, T.; Geneletti, D.; Sonnemann, G.; Rugani, B. Assessing Habitat Loss, Fragmentation and Ecological Connectivity in Luxembourg to Support Spatial Planning. *Landsc. Urban Plan.* **2019**, *189*, 335–351. [CrossRef]
64. Heckmann, T.; Cavalli, M.; Cerdan, O.; Foerster, S.; Javaux, M.; Lode, E.; Smetanová, A.; Vericat, D.; Brardinoni, F. Indices of Sediment Connectivity: Opportunities, Challenges and Limitations. *Earth-Sci. Rev.* **2018**, *187*, 77–108. [CrossRef]
65. Lehotský, M.; Rusnák, M.; Kidová, A.; Dudžák, J. Multitemporal Assessment of Coarse Sediment Connectivity along a Braided-Wandering River. *Land Degrad. Dev.* **2018**, *29*, 1249–1261. [CrossRef]
66. Erős, T.; Schmera, D.; Schick, R.S. Network Thinking in Riverscape Conservation—A Graph-Based Approach. *Biol. Conserv.* **2011**, *144*, 184–192. [CrossRef]
67. Ribeiro, R.; Carretero, M.A.; Sillero, N.; Alarcos, G.; Ortiz-Santaliestra, M.; Lizana, M.; Llorente, G.A. The Pond Network: Can Structural Connectivity Reflect on (Amphibian) Biodiversity Patterns? *Landsc. Ecol.* **2011**, *26*, 673–682. [CrossRef]
68. Xie, Y.; Li, Z.; Zhu, L.; Zhou, Y.; Liu, H.; Yu, T. Dynamic Monitoring of Desertification in Response to Climatic Factors: A Case Study from the Gelintan Oasis on the Southeastern Edge of the Tengger Desert, China. *Geocarto Int.* **2021**, 1–21. [CrossRef]
69. Notarnicola, C. Hotspots of Snow Cover Changes in Global Mountain Regions over 2000–2018. *Remote Sens. Environ.* **2020**, *243*, 111781. [CrossRef]
70. Finkler, N.R.; Gücker, B.; Boëchat, I.G.; Ferreira, M.S.; Tanaka, M.O.; Cunha, D.G.F. Riparian Land Use and Hydrological Connectivity Influence Nutrient Retention in Tropical Rivers Receiving Wastewater Treatment Plant Discharge. *Front. Environ. Sci.* **2021**, *9*, 709922. [CrossRef]
71. Niu, L.; Cai, H.; Jia, L.; Luo, X.; Tao, W.; Dong, Y.; Yang, Q. Metal Pollution in the Pearl River Estuary and Implications for Estuary Management: The Influence of Hydrological Connectivity Associated with Estuarine Mixing. *Ecotoxicol. Environ. Saf.* **2021**, *225*, 112747. [CrossRef] [PubMed]
72. Fu, A.; Li, W.; Chen, Y.; Liu, Y. Suitable Oasis Scales under a Government Plan in the Kaidu-Konqi River Basin of Northwest Arid Region, China. *PeerJ* **2018**, *6*, e4943. [CrossRef] [PubMed]
73. Xia, Y.; Fang, C.; Lin, H.; Li, H.; Wu, B. Spatiotemporal Evolution of Wetland Eco-Hydrological Connectivity in the Poyang Lake Area Based on Long Time-Series Remote Sensing Images. *Remote Sens.* **2021**, *13*, 4812. [CrossRef]
74. Liu, X.; Zhang, Q.; Li, Y.; Tan, Z.; Werner, A.D. Satellite Image-Based Investigation of the Seasonal Variations in the Hydrological Connectivity of a Large Floodplain (Poyang Lake, China). *J. Hydrol.* **2020**, *585*, 124810. [CrossRef]
75. Deng, X.; Xu, Y.; Han, L.; Song, S.; Xu, G.; Xiang, J. Spatial-Temporal Changes in the Longitudinal Functional Connectivity of River Systems in the Taihu Plain, China. *J. Hydrol.* **2018**, *566*, 846–859. [CrossRef]
76. Wang, X.; Xiao, X.; Zou, Z.; Dong, J.; Qin, Y.; Doughty, R.B.; Menarguez, M.A.; Chen, B.; Wang, J.; Ye, H.; et al. Gainers and Losers of Surface and Terrestrial Water Resources in China during 1989–2016. *Nat. Commun.* **2020**, *11*, 3471. [CrossRef] [PubMed]
77. Hu, S.; Ma, R.; Sun, Z.; Ge, M.; Zeng, L.; Huang, F.; Bu, J.; Wang, Z. Determination of the Optimal Ecological Water Conveyance Volume for Vegetation Restoration in an Arid Inland River Basin, Northwestern China. *Sci. Total Environ.* **2021**, *788*, 147775. [CrossRef] [PubMed]
78. Beel, C.R.; Heslop, J.K.; Orwin, J.F.; Pope, M.A.; Schevers, A.J.; Hung, J.K.Y.; Lafrenière, M.J.; Lamoureux, S.F. Emerging Dominance of Summer Rainfall Driving High Arctic Terrestrial-Aquatic Connectivity. *Nat. Commun.* **2021**, *12*, 1448. [CrossRef]
79. Li, Z.; Sun, W.; Chen, H.; Xue, B.; Yu, J.; Tian, Z. Interannual and Seasonal Variations of Hydrological Connectivity in a Large Shallow Wetland of North China Estimated from Landsat 8 Images. *Remote Sens.* **2021**, *13*, 1214. [CrossRef]
80. Deng, X.; Xu, Y.; Han, L. Impacts of Human Activities on the Structural and Functional Connectivity of a River Network in the Taihu Plain. *Land Degrad. Dev.* **2018**, *29*, 2575–2588. [CrossRef]
81. Tang, X.; Guijuan, X.; Shao, K.; Bayartu, S.; Chen, Y.; Gao, G. Influence of Salinity on the Bacterial Community Composition in Lake Bosten, a Large Oligosaline Lake in Arid Northwestern China. *Appl. Environ. Microbiol.* **2012**, *78*, 4748–4751. [CrossRef]
82. Ariken, M.; Zhang, F.; Liu, K.; Fang, C.; Kung, H.-T. Coupling coordination analysis of urbanization and eco-environment in Yanqi Basin based on multi-source remote sensing data. *Ecol. Indic.* **2020**, *114*, 106331. [CrossRef]
83. Deng, H.; Chen, Y.; Wang, H.; Zhang, S. Climate change with elevation and its potential impact on water resources in the Tianshan Mountains, Central Asia. *Glob. Planet. Change* **2015**, *135*, 28–37. [CrossRef]
84. Fan, M.; Xu, J.; Chen, Y.; Li, W. Modeling streamflow driven by climate change in data-scarce mountainous basins. *Sci. Total Environ.* **2021**, *790*, 148256. [CrossRef]
85. Li, N.; McLaughlin, D.; Kinzelbach, W.; Li, W.; Dong, X. Using an ensemble smoother to evaluate parameter uncertainty of an integrated hydrological model of Yanqi basin. *J. Hydrol.* **2015**, *529*, 146–158. [CrossRef]

sustainability

MDPI

Article

Temporal Variation and Spatial Distribution in the Water Environment Helps Explain Seasonal Dynamics of Zooplankton in River-Type Reservoir

Jingyun Yin [1,*], Jihong Xia [1,*], Zhichang Xia [2], Wangwei Cai [1], Zewen Liu [1], Kejun Xu [1], Yue Wang [1], Rongzhen Zhang [2] and Xu Dong [2]

1 College of Agricultural Science and Engineering, Hohai University, Nanjing 210098, China
2 Wenzhou Water Resources and Conservancy Bureau, Wenzhou 325000, China
* Correspondence: 211310020002@hhu.edu.cn (J.Y.); syjhxia@hhu.edu.cn (J.X.)

Abstract: Integrated assessment of the water environment has become widespread in many rivers, lakes, and reservoirs; however, aquatic organisms in freshwater are often overlooked in this process. Zooplankton, as primary consumers, are sensitive and responsive to changes in the water environment. Water and zooplankton samples were collected on-site at Shanxi Reservoir quarterly to determine 12 water environmental indicators and to quantify the abundance of zooplankton of Cladocera, Copepoda and Rotifera by using the ZooScan zooplankton image-scanning analysis system, combined with OLYMPUS BX51 using machine learning recognition classification. The aim was to explore the relationship between water environmental factors and zooplankton through their spatial and temporal heterogeneity. Through principal component analysis, redundancy analysis and cluster analysis, variations in the factors driving zooplankton population growth in different seasons could be identified. At the same time, different taxa of zooplankton can form clusters with related water environmental factors during the abundant water period in summer and the dry water period in winter. Based on long-term monitoring, zooplankton can be used as a comprehensive indicator for water environment and water ecological health evaluation, as well as providing scientific support for regional water resources deployment and management.

Keywords: Cladocera; Copepoda; freshwater reservoir; water environment; Rotifera; ZooScan

Citation: Yin, J.; Xia, J.; Xia, Z.; Cai, W.; Liu, Z.; Xu, K.; Wang, Y.; Zhang, R.; Dong, X. Temporal Variation and Spatial Distribution in the Water Environment Helps Explain Seasonal Dynamics of Zooplankton in River-Type Reservoir. *Sustainability* **2022**, *14*, 13719. https://doi.org/10.3390/su142113719

Academic Editors: Songhao Shang, Qianqian Zhang, Dongqin Yin, Hamza Gabriel and Magdy Mohssen

Received: 6 September 2022
Accepted: 19 October 2022
Published: 22 October 2022

1. Introduction

Water is one of the most important natural resources on which all life depends. The earth's freshwater resources play an important role in the survival of mankind, the development of society, and contemporary progress that cannot be ignored [1,2]. In recent years, however, increasing population and climate change have caused serious pollution and damage to limited freshwater resources [3,4]. Therefore, many water-poor countries around the world, like China, have adopted the construction of reservoirs to collect, store, and utilize their limited freshwater resources [5]. Freshwater reservoirs not only provide water for drinking, agricultural irrigation, and industrial production for nearby residents, but also prevent flooding by regulating the water level downstream through storage and discharge processes [6], and generating electricity through the drainage process can effectively reduce the production of greenhouse gases in the thermal power generation process [7]. Nevertheless, due to human activities, the water environment of freshwater reservoirs in different regions of the world has deteriorated to different degrees [8], which makes the monitoring of the water environment of freshwater reservoirs and the water environment and ecological restoration of freshwater reservoirs a hot topic of current research [9,10]. As freshwater reservoirs are closely related to human production and life, they have been attracting the attention of scholars worldwide in recent years. For a long time, people have been monitoring the water environment and water ecology of freshwater reservoirs through various water environment indicators to ensure water safety and avoiding the crisis caused by water pollution and water environment degradation [11,12]. With the development of science and technology, more and more intelligent equipment and high-tech research methods are applied to water environment monitoring. Sagan, et al. [13] used the potential of remote sensing and the limitations of spectral indices, bio-optical simulations, machine learning, and cloud computing to monitor inland water quality. Shi, et al. [14] used spectroscopic analysis and ultra-high resolution mass spectrometry to monitor dissolved organic matter in wastewater and drinking water treatment.

In addition to conventional water quality indicators, the inclusion of zooplankton-like ecological indicators is an important reference for the comprehensive consideration of the health of the water environment. Zooplankton is an important part of aquatic ecosystems [15,16]. Zooplankton is mostly in the second trophic level of the food web and is the key intermediate link in the transfer of material and energy from primary producers to higher trophic levels [17]. The ecological niche of zooplankton and its specificity has led many researchers to work on integrating it with water environmental indicators. Sousa, et al. [18] found that water quality changes had a significant effect on zooplankton community structure. Due to the widespread use of zooplankton to monitor water quality in productive life as well as in scientific research, in 2011 scholars called for the inclusion of zooplankton in the ecological quality assessment of lakes according to the European Water Framework Directive (WFD) [19]. With further research, it was found that the changes in zooplankton abundance are not only directly related to water environment indicators, but also respond to some extent to changes in the abundance of phytoplankton such as algae [20,21]. Zooplankton species composition and quantity changes are not only the basic content of water ecological health research but can also accurately reflect the quality of the water ecological environment [22,23]. Its biodiversity index is one of the important evaluation indicators of water ecological environment health [4,24].

With the development of computer and electronic technology, more and more researchers rely on the ZooScan zooplankton image scanning and analysis system for scientific and rapid identification, measurement, and identification of zooplankton in the water body [25,26]. Naito, et al. [27] used ZooScan, Optical Plankton Recorder (OPC), and microscopic study methods to study the sea surface zooplankton respectively, and ZooScan has significant advantages among these three quantitative methods. Wang, et al. [28] studied the spatial variation of the size structure of medium-sized plankton and its relationship with environmental factors with the help of the ZooScan system. Maas, et al. [29] used image-based observation systems in marine ecosystems, and with the help of ZooScan, the ecological zonation of zooplankton can be quantitatively analyzed. Noyon, et al. [30] scanned mesozooplankton samples by ZooScan to get the distribution of medium-sized zooplankton communities on the Agulhas Bank in autumn to predict scale structure and production. Garcia-Herrera, et al. [31] used ZooScan in combination with the web-based platform EcoTaxa 2.0 system to study differences in integrated zooplankton abundance, biovolume, and biomass. In addition, the microscope enables accurate species identification and classification of zooplankton [32].

The present study area was conducted in a mountainous riverine freshwater reservoir located in a subtropical monsoonal zone with a mild climate in southeastern China. The study area is in a typhoon area with many inhabitants, and the water environment and water ecology are affected by the combination of extreme typhoon climate and surface pollution from residential areas [33]. During the rainy season, the initial rainwater is mixed with surface pollutants from residential areas into the reservoir due to the catchment effect of the valley [34]. The typhoon season causes drastic changes in the water environment in the reservoir due to the wind disturbance effect [35]. The study area is the main freshwater water source in southern Zhejiang province, which assumes multiple roles as drinking water, domestic water, and industrial water [36], so it is of great importance to maintain water security, guarantee the sustainability of water resources use, and protect the water environment and water ecological stability of the study area.

In this study, zooplankton was linked to its surroundings based on the most important theory in ecology (ecological niche theory) [37], and the spatial and temporal heterogeneity of typical freshwater zooplankton and common water environmental indicators was used to analyze mountain river-type reservoirs. Zooplankton, as an integral part of freshwater aquatic ecosystems, plays an important role in the monitoring and assessment of the water environment [38]. However, studies of the spatial heterogeneity of freshwater zooplankton communities and their correlation with the water environment often take researchers a great deal of time due to the lack of rapid and accurate plankton statistics [39]. Therefore, a systematic study on the spatial distribution of ecological niches of freshwater zooplankton, with the help of OLYMPUS BX51 and ZooScan, a zooplankton image acquisition and analysis system, and an artificial intelligence recognition and classification system, will help to enhance the dimensions of freshwater water environment monitoring. It can also help to explore the spatial and temporal heterogeneity of zooplankton and water environment, identifying the response mechanism of water environment changes to specific zooplankton, and providing new ideas for the sustainable and healthy operation and maintenance of the same type of freshwater reservoirs.

2. Materials and Methods

2.1. Description of the Study Area

Shanxi Reservoir (latitude 27°36′–27°50′; longitude 119°47′–120°15′), a large freshwater reservoir in the south of Zhejiang Province, is in the upper reaches of the Feiyun River, a typical river-type reservoir. It has a total reservoir capacity of 1.82×10^9 m^3 in a rainfall catchment area of 1529 km^2, controlling nearly 80% of Wenzhou's water resources, providing 1.34×10^9 m^3 of freshwater to Wenzhou's urban area every year, supplying five million people in the water supply area and providing 2.20×10^5 kW of peaking power to Wenzhou's power grid, with obvious environmental and economic benefits. The Shanxi Reservoir is located in the subtropical monsoon climate zone, with an average multi-year precipitation of 1843.3 mm and an average multi-year temperature of 18.7 °C.

2.2. Sampling

For this study, based on the habitat and ecological characteristics of zooplankton [30], four seasons were selected for sampling; in April (spring), July (summer), October (autumn) in 2021, and January (winter) in 2022. As shown in Figure 1, 25 monitoring points were set up in the study area, according to the distribution of the main Shanxi Reservoir area and the tributary catchment area. According to the width of the section of different monitoring points set up with the bank perpendicular to the monitoring section, each monitoring section selected the mid-point and near-shore points for sampling.

According to the experience of relevant researchers, the distribution of the thermocline in lakes or oceans is generally used to determine the depth of samples to be collected at the corresponding sampling sites [40–42]. Water samples were collected from the upper layer (1 m), middle layer (4–7 m), and lower layer (10–20 m) using a 5 L water collector, considering the distribution of the thermocline in the study area.

Water environment indicators were obtained as follows. For the collected water samples on-site using DS5X (Hach Corporation) to measure water temperature (Tem), pH, dissolved oxygen (DO), electrical conductivity (EC), oxidation-reduction potential (ORP), and chlorophyll a (Chl-a), which are conventional indicators. For each layer of the sample retention volume of 10 L of water samples for the determination of routine water pollution indicators in the laboratory; according to the national standard (GB11892-89) [43] for the determination of permanganate index (COD_{Mn}) of environmental water quality. For the determination of total phosphorus (TP), the ammonium molybdate spectrophotometric method was used according to the national standard (GB11893-89) [44]. According to this, the determination of total nitrogen (TN) was carried out by UV spectrophotometric method using alkaline potassium persulfate elimination. The determination of ammonia nitrogen (NH_3-N) in water samples was made according to the national standard (HJ535-2009) [45].

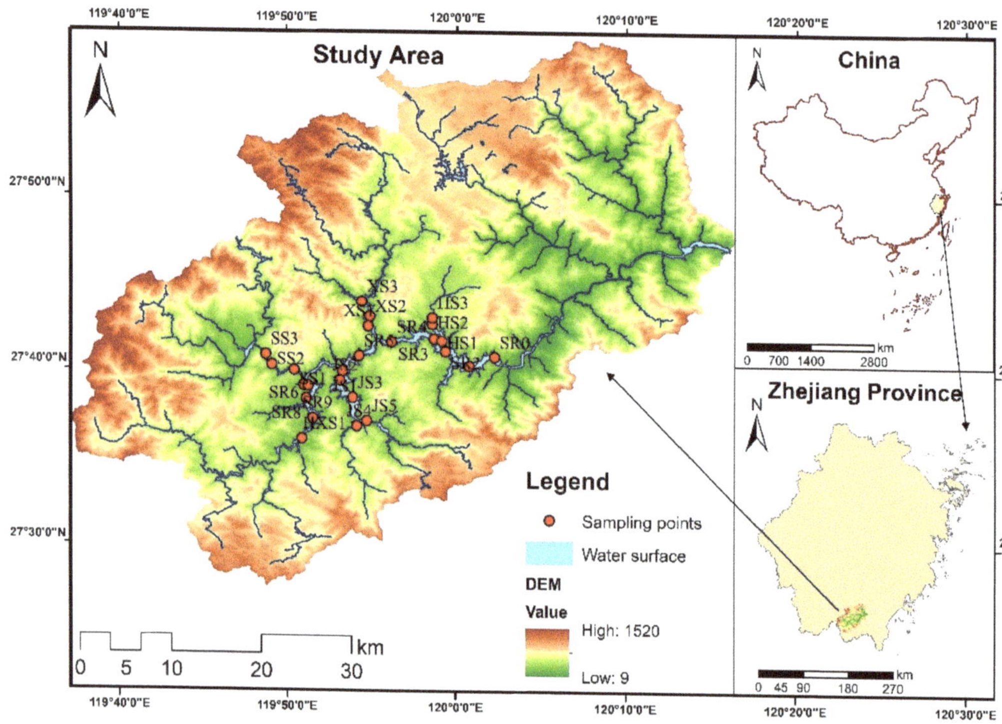

Figure 1. Sampling points in the Shanxi reservoir.

The zooplankton samples were obtained as follows: 10 L of water samples were collected from each layer, filtered using a 64μm pore size filter according to the national standard (SCT9402-2010) [46], the filter rinsed to obtain about 250 mL of zooplankton concentrate, and formaldehyde solution with a volume fraction of 5% was added for sample fixation. Zooplankton samples were classified and data collected in the laboratory using ZooScan, a zooplankton image acquisition system, in combination with imageJ image processing software and EcoTaxa 2.0 image analysis and processing website. Considering the accuracy of the ZooScan system for zooplankton monitoring and the related habits of zooplankton [47], the identification of zooplankton in the study area was completed with the assistance of OLYMPUS BX51 in this study.

2.3. Data Processing

In order to investigate the response between zooplankton and the water environment in Shanxi Freshwater Reservoir, the study area was divided into six sub-regions: the Shanxi Reservoir (SR), the Huangtankeng Stream (HS), the Xuezuokou Stream (XS), the Jujiangxi Stream (JS), the Sanchaxi Stream (SS), and the Hongkouxi Stream (HXS) according to the characteristics of the catchments in the study area. The screening of three representative zooplankton species based on the collected zooplankton samples gave results on the spatial and temporal heterogeneity of zooplankton in freshwater reservoirs. The results of the spatial and temporal heterogeneity of the water environmental factors of freshwater reservoirs were obtained by dividing them according to their physical, chemical, and ecological properties.

As the data in this study came from field sampling, there were a certain number of outliers, and to ensure better representativeness, the abnormal data were not presented. Correlations between three representative zooplankton species and 14 water environmental factors were analyzed using SPSS25 and Origin2021 using ANOVA and W-MERT methods respectively [48], and principal component analysis (PCA) was performed on each area of freshwater reservoirs in the study area to ensure that covariance matrices were characteristically decomposed with no or little loss of information, and multiple water environmental factors were generalized into representative major influences [49,50]. In addition, the relationship between zooplankton and water environmental factors was studied using detrended correspondence analysis (DCA) and redundancy analysis (RDA) using Canoco 5 software [51]. A cluster analysis, based on Euclidian distance, was used to check how Cladocera, Copepoda, and Rotifera were distributed along the study area using the sampling points and seasons of the reservoir as attributes [52]. A chi-square test was performed on the data for the water environment factors. One-way analysis of variance (ANOVA) was used for the normally distributed factors Tem, COD_{Mn}, pH, and TN to obtain significant differences in the spatial and temporal distribution of TN (ANOVA, $p < 0.05$). A Welch's mean equality robustness test was applied to the non-normally distributed factors with uneven variances [53]. Finally, the results of spatial and temporal heterogeneity analysis of zooplankton and water environment factors were combined with data analysis methods to obtain the status of zooplankton and water environment in Shanxi Reservoir and to analyze their causes, and to explore the response relationship between typical zooplankton and water environment indicators.

3. Results

3.1. Spatial and Temporal Heterogeneity of Reservoir's Water Environment

The variability of 12 water environmental factors in different areas of Shanxi Reservoir in terms of water temperature (Tem), electrical conductivity (EC), dissolved oxygen (DO), redox potential (ORP), pH, picocyanobacteria (PCY), turbidity (Tur), chlorophyll α (Chl-a), total nitrogen (TN), permanganate index (COD_{Mn}), total phosphorus (TP) and ammonia nitrogen (NH_3-N) are presented in Table 1. After chi-square test and Welch's mean equality robustness test, significant differences in the spatial and temporal distributions of EC, ORP, PCY, Chl-a, TP, and NH_3-N (W-MERT, $p < 0.05$) were obtained. The 12 water environmental factors are further classified and discussed concerning their different properties.

3.1.1. Physicochemical Indicators of Water Bodies

The physical and chemical properties of water bodies are the basic components of the water environment and are a fundamental indication of the health of the water environment. Water temperature (Tem), turbidity (Tur), electrical conductivity (EC), pH, dissolved oxygen (DO), and redox potential (ORP)—these physical and chemical properties of water bodies are the basic components of the water environment and are a fundamental indication of the health of the water environment.

The study area is generally characterized by high water temperatures in summer and low water temperatures in winter. Combined with Figure 2 and the results presented by the data, the temporal and spatial differences in turbidity (Tur) and electrical conductivity (EC) in the study area were significant. The pH of the water bodies in the study area fluctuated significantly more in spring and summer than in autumn and winter, and the values increased significantly in summer. The spatial and temporal variability of dissolved oxygen (DO) in the water bodies of the study area was greatest in summer. The redox potential (ORP) also produced significant fluctuations and differences during the summer months.

Table 1. Water Environment in the Study Area.

Water Environment \ Region	Shanxi Reservoir (SR)	Huangtankeng Stream (HS)	Xuezuokou Stream (XS)	Jujiang Stream (JS)	Sanchaxi Stream (SS)	Hongkouxi Stream (HKS)
Tem (°C)	15.30~31.22 ^ (22.67 ± 5.06) *	15.47~29.08 (22.39 ± 5.11)	15.59~30.91 (22.80 ± 5.34)	15.84~31.00 (23.43 ± 4.99)	15.84~31.00 (22.95 ± 4.96)	16.21~31.02 (23.90 ± 6.06)
EC (mS/cm)	31.16~47.60 (42.60 ± 4.05)	40.70~66.60 (50.60 ± 4.11)	31.42~46.20 (41.01 ± 5.08)	32.90~70.00 (46.71 ± 10.37)	35.13~52.20 (43.19 ± 5.32)	37.40~47.40 (44.70 ± 4.88)
DO (mg/L)	5.07~8.16 (6.78 ± 0.86)	5.14~8.32 (6.88 ± 1.06)	5.84~7.69 (6.99 ± 0.70)	5.25~7.75 (6.84 ± 0.81)	5.26~8.48 (7.02 ± 0.82)	6.08~8.48 (7.30 ± 1.00)
ORP (V)	0.19~0.46 (0.37 ± 0.07)	0.34~0.52 (0.40 ± 0.05)	0.18~0.44 (0.37 ± 0.08)	0.14~0.43 (0.33 ± 0.10)	0.16~0.46 (0.35 ± 0.11)	0.20~0.43 (0.35 ± 0.10)
pH	6.43~8.17 (7.21 ± 0.43)	6.81~7.88 (7.29 ± 0.30)	6.49~8.23 (7.20 ± 0.50)	6.54~8.02 (7.34 ± 0.43)	6.68~8.13 (7.36 ± 0.49)	7.15~8.10 (7.43 ± 0.45)
PCY ($\times 10^3$ cell/L)	0.38~19.38 (2.05 ± 1.56)	0.38~19.37 (2.05 ± 1.56)	0.53~3.99 (2.02 ± 1.47)	0.67~19.38 (5.27 ± 4.74)	0.74~10.03 (3.87 ± 3.04)	0.95~4.90 (3.33 ± 1.71)
Tur (NTU)	1.10~8.00 (4.34 ± 1.75)	3.40~6.20 (4.71 ± 0.89)	2.60~5.30 (4.36 ± 0.87)	1.50~7.95 (4.14 ± 1.79)	1.70~11.10 (5.13 ± 2.55)	1.90~9.90 (5.44 ± 3.62)
Chl-a (mg/L)	0.60~4.16 (1.71 ± 1.11)	1.24~13.66 (4.09 ± 3.50)	0.63~7.54 (2.13 ± 2.43)	0.50~4.79 (2.31 ± 1.52)	1.20~6.28 (2.85 ± 1.61)	1.05~3.58 (2.43 ± 1.25)
TN (mg/L)	0.10~0.63 (0.38 ± 0.12)	0.30~0.67 (0.44 ± 0.12)	0.2282~0.4754 (0.35 ± 0.065)	0.14~0.47 (0.33 ± 0.096)	0.07~0.46 (0.31 ± 0.14)	0.17~0.49 (0.36 ± 0.14)
COD_{Mn} (mg/L)	0.37~2.00 1.36 ± 0.43	0.72~2.30 (1.58 ± 0.53)	0.73~1.88 (1.42 ± 0.37)	0.75~1.80 (1.42 ± 0.35)	0.63~1.70 (1.35 ± 0.43)	0.78~1.89 (1.53 ± 0.51)
TP (μg/L)	6.10~128.40 (23.90 ± 23.80)	4.40~40.00 (20.90 ± 12.30)	6.10~128.40 (23.91 ± 23.82)	6.50~36.00 (22.90 ± 9.50)	8.30~38.80 (23.20 ± 11.10)	10.60~30.30 (21.30 ± 9.70)
NH_3N (μg/L)	13.10~174.30 (50.20 ± 42.40)	7.90~235.10 (78.30 ± 74.90)	11.80~118.60 (37.70 ± 36.30)	10.20~155.80 (52.40 ± 54.30)	3.20~143.00 (60.30 ± 46.10)	3.00~170.00 (76.20 ± 66.50)

^ indicates the range of an indicator. * Mean values and standard deviations (SD) are shown in parentheses.

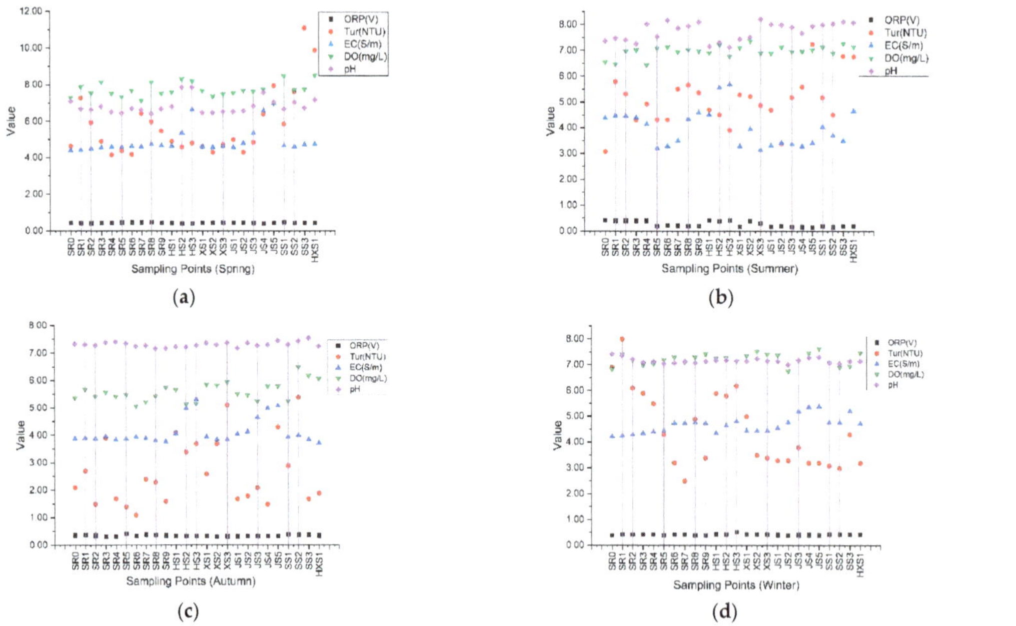

Figure 2. Plumbline diagrams of the spatial and temporal distribution of physical and chemical properties of water bodies in the study area in different seasons: (**a**) Spring; (**b**) Summer; (**c**) Autumn; (**d**) Winter.

3.1.2. Biological Indicators of Water Bodies

Similar to chlorophyll a (Chl-a), which is mainly derived from phytoplankton in the water, PCY values reflect the abundance of picocyanobacteria, which are typical of primary producers in the water environment of the study area.

According to the results of the Welch's mean equality robustness test (W-MERT), Chl-a (W-MERT, $p = 0.0001367 < 0.05$) and PCY (W-MERT, $p = 2.6499 \times 10^{-13} < 0.05$) showed highly significant differences in spatial and temporal distributions. In terms of temporal distribution, Chl-a and PCY were significantly less in winter than in other seasons; in terms of spatial distribution, Chl-a and PCY in spring showed higher levels in Huangtankeng Stream and Jujiangxi Stream than in other study areas. However, the Chl-a in summer showed a higher phenomenon in the Shanxi Reservoir than in the tributaries. In contrast, PCY was further elevated in number but spatially distributed more evenly than Chl-a.

3.1.3. Pollution Indicators of Water Bodies

Ammonia nitrogen (NH_3-N) reflects the elemental nitrogen in water in the form of NH_3 and NH_4^+. Total nitrogen (TN) contains various forms of nitrogen in the water body, and various forms of phosphorus elements in the water body are indicated by total phosphorus (TP). NH_3-N, TN, and TP are mainly from domestic sewage and agricultural surface source pollution, and are important indicators of the health of water bodies. The permanganate index is a common indicator of oxidizable pollutants in water bodies. In terms of spatial and temporal distribution, TN (ANOVA, $p = 0.0089 < 0.05$), NH_3-N (W-MERT, $p = 0.006 < 0.05$), and TP (W-MERT, $p = 0.016 < 0.05$) showed highly significant differences in spatial and temporal distribution.

In Figure 3, the amount of NH_3-N in the water was significantly more in summer than in other seasons. TN and TP were higher than the yearly average in spring and summer, and COD_{Mn} was significantly less in winter than in other seasons. In addition, TN and NH_3-N showed an increase in summer and a higher level in autumn. For the spatial distribution of pollution indicators, both TN and NH_3-N showed extreme values in the Huangtankeng Stream, and fluctuated more between sub-regions within the same season; both COD_{Mn} and TP showed extreme values in the Shanxi Reservoir, but TP fluctuated more than COD_{Mn} within each sub-region.

3.2. Spatial and Temporal Heterogeneity of Reservoir's Zooplankton

Through microscopic examination of zooplankton samples using OLYMPUS BX51, combined rapid statistical and technical work with the ZooScan zooplankton image scanning analysis system, 46 genus of zooplankton were identified at the genus level in accordance with the Atlas of Major Freshwater Zooplankton of Zhejiang Province (Drinking Water Sources) [54]. The detailed composition, frequency of recurrence and degree of dominance are shown in Table 2.

According to the data in Table 2, three of the genus *Sinocalanus*, *Sinodiaptomus* and *Nitocra* are prominent in terms of dominance of Copepoda, and Cladocera's *Macrothrix*, *Daphnia*, *Bosmina* and *Diaphanosoma* all have a numerical dominance of more than 4%, significantly more than the other species, which further suggests that it is Cladocera and Copepoda that dominate the study area. Combining the information in Figure 4, in terms of temporal distribution, the number of zooplankton in the water column showed a gradual increase from spring to autumn and reached a maximum in summer and autumn, with a significant decrease after winter. In spring, the abundance of Cladocera fluctuated more in different subregions than Rotifera and Copepoda. All three orders in the same sub-region showed a greater increase in zooplankton abundance in summer compared to spring, with the mean abundance reaching its highest throughout the year. Copepoda was the dominant species within each subregion in summer under its abundance. The abundance of Cladocera in each subregion decayed with the onset of autumn but was generally much higher than in spring. The abundance of Rotifera and Copepoda in the Huangtanke Stream and Sanchaxi Stream sub-regions showed an increase compared to summer. The zooplankton abundance of the three orders showed a dramatic decrease in winter. The abundance of Rotifera and Cladocera in the Shanxi Reservoir sub-region decreased dramatically and was lower than the spring average. The abundance in the tributary sub-region decreased but overall was higher than the spring average. The Copepoda, the typical dominant zooplankton species in the study area during summer and autumn, had higher mean values in the study area than in spring, even though its abundance showed a dramatic decrease in winter.

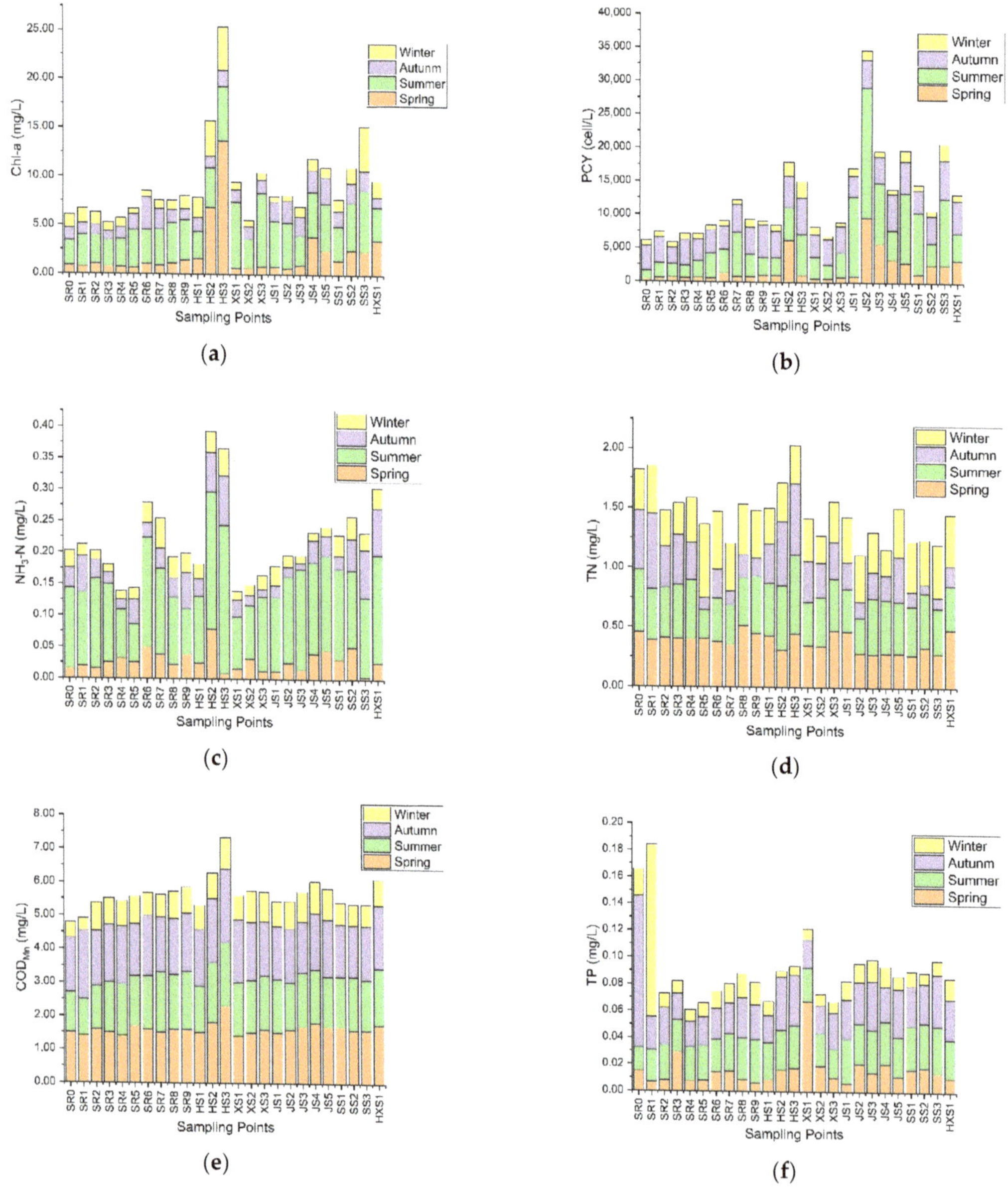

Figure 3. Stacked bar charts of the spatial and temporal distribution of water environmental factors: (**a**) Chl-a; (**b**) PCY; (**c**) NH_3-N; (**d**) TN; (**e**) COD_{Mn}; (**f**) TP.

Table 2. The species composition and individual dominance of the study area.

Categories (Orders)	Name of the Species (Genus)	Frequency of Recurrence *	Degree of Dominance	Name of the Species (Genus)	Frequency of Recurrence	Degree of Dominance
Rotifera	*Asplanchna*	100%	2.60%	*Keratella*	40%	0.20%
	Polyarthra	80%	4.31%	*Testudinalla*	40%	0.60%
	Trichocerca	60%	0.25%	*Filinia*	40%	0.53%
	Gastropus	60%	0.44%	*Ascomorpha*	40%	0.33%
	Mytilina	60%	0.49%	*Eosphora*	40%	0.63%
	Brachiomus	60%	3.07%	*A.fissa*	20%	0.02%
	Cephalodella	60%	3.72%	*Rotaria*	20%	0.26%
	Pompholyx	60%	1.16%	*Epiphanes*	20%	0.42%
	Synchaeta	60%	0.34%	*Euchlanis*	20%	0.02%
	Notholeca	40%	0.74%	*Ploesoma*	20%	0.06%
Copepoda	*Nitocra*	100%	6.90%	*Paracyclops*	80%	2.58%
	Sinocalanus	100%	11.80%	*Mesocyclops*	80%	1.99%
	Sinodiaptomus	100%	9.00%	*Heliodiaptomus*	80%	5.39%
	Cyclops	100%	1.51%	*Onchocamptus*	60%	0.15%
	Tropocyclops	80%	2.12%	*Limnoithona*	60%	1.06%
	Themocyclops	80%	2.23%	*Neodiaptomus*	60%	1.64%
	Canthocamptus	80%	1.47%	*Macrocyclops*	40%	0.05%
	Mongolodiaptpmus	80%	1.00%			
Cladocera	*Macrothrix*	100%	4.20%	*Diaphanosoma*	60%	5.60%
	Daphnia	80%	6.12%	*Camptocercus*	40%	0.49%
	Bosminopsis	60%	1.37%	*Leydigia*	20%	0.24%
	Chydorus	60%	3.54%	*Alona*	20%	0.32%
	Simocephalus	60%	2.11%	*Moinodaphnia*	20%	0.08%
	Bosmina	60%	6.92%			

* Probability of being observable in a single survey cycle.

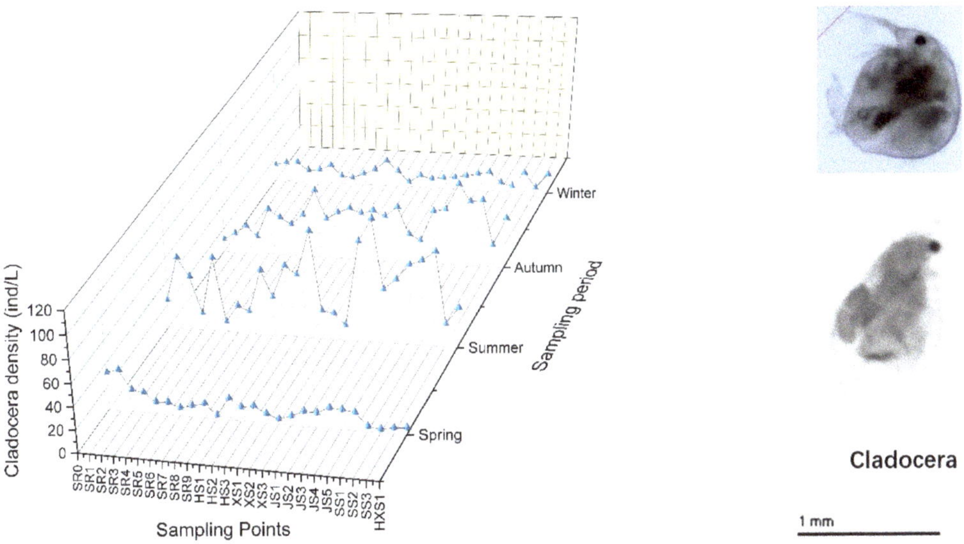

(**a**)

Figure 4. *Cont.*

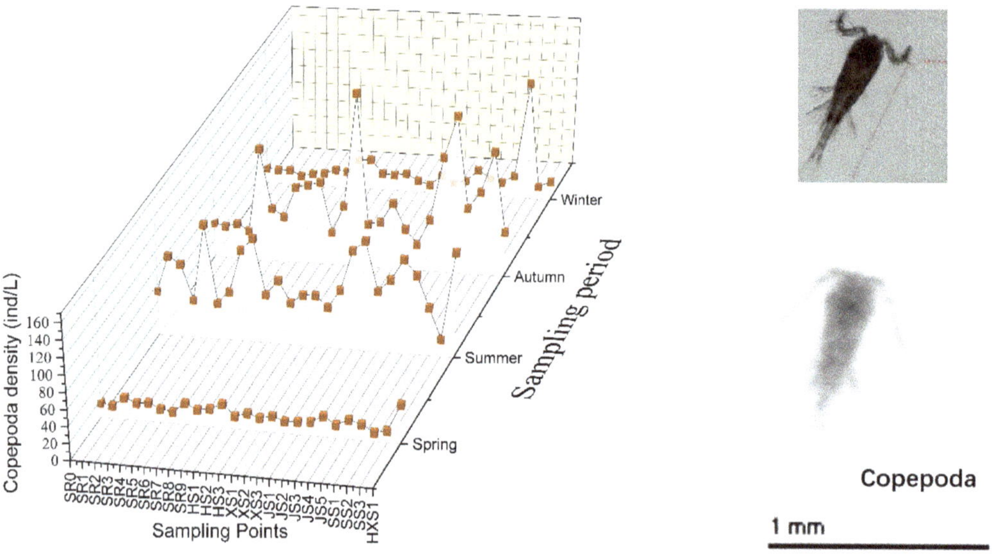

(**b**)

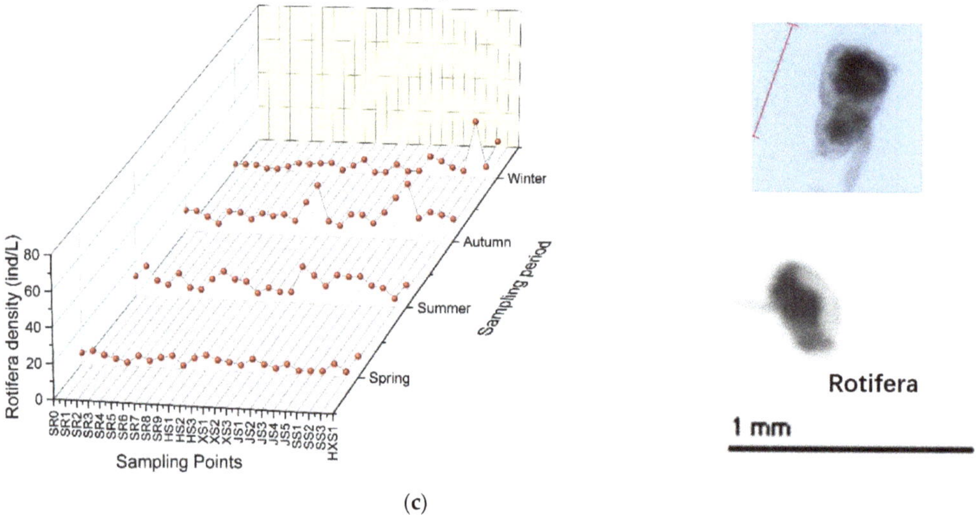

(**c**)

Figure 4. Three-dimensional dotted waterfall maps of the spatial and temporal distribution of plankton and images from OLYMPUS BX51 and ZooScan scans of (**a**) Cladocera; (**b**) Copepoda; (**c**) Rotifera.

In terms of spatial distribution, the following results can be obtained by comparing typical zooplankton abundances at different times during the same season. In spring, the average number of Rotifera in Shanxi Reservoir was lower than in the other five sub-regions. By contrast, the average number of Cladocera and Copepoda was higher than in the other five sub-regions. In summer, the average number of Rotifera and Cladocera in Shanxi Reservoir was lower than in the other five sub-regions, and the average number of Copepoda remained higher than the other five sub-regions. The average number of zooplankton decreased slightly in autumn compared to summer and generally showed a lower average number in Shanxi Reservoir than in the sub-region. The average number of zooplankton decreased further in winter, and the average number of Rotifera and Copepoda in Shanxi Reservoir was lower than the other five sub-regions, whereas the average number of Cladocera remained higher than the other five sub-regions.

3.3. *Relationships between Water Environmental Factors and Zooplankton*

3.3.1. Principal Component Analysis (PCA)

Due to the large number of water environmental factors involved in this study, in order to facilitate the study of the response relationship between water environmental factors and zooplankton, after using the KMO (Kaiser-Meyer-Olkin) test for judgement ($0 < 0.735 < 1$), the dimensionality of the water environmental factors could be reduced using principal component analysis [55,56], and the loadings plot for the principal component analysis and the ellipses corresponding to the four 95% confidence levels for spring, summer, autumn and winter are shown in Figure 5.

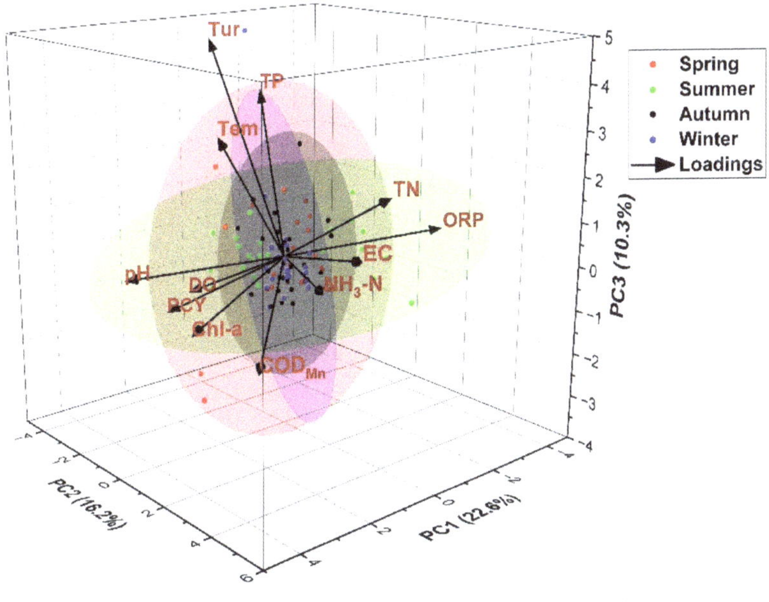

Figure 5. Loadings of principal component analysis of water environment factors.

For the water environment, the PCA of the 12 correlated variables resulted in three principal components (PC) that had eigenvalues >1 and accounted for 49.10% of the variance in the data. The first principal component (PC1) eigenvalue was 2.71 with a contribution rate of 22.60%, which accounted for the largest proportion of the variance, indicating that it had the strongest ability to combine the original variables. The water environment factors pH, Chl-a, COD_{Mn}, and Tem had larger weight coefficients, and pH had the largest weight coefficient and, very significantly, was positively correlated with the other three variables ($p < 0.01$). Therefore, PC1 could be synthesized as a chemical biological factor in water bodies. The second principal component (PC2) with an eigenvalue of 1.95 and a contribution of 16.22%, had the highest EC weight and the second highest TN weight. Very significantly, they were positively correlated ($p < 0.01$), so PC2 could be resolved as a physiochemical factor. The third principal component (PC3) had the largest Tur and TP weight, so PC3 could be resolved as a contamination risk factor.

3.3.2. Redundancy and Correlation Analysis

Before the constrained ranking analysis, the detrended correspondence analysis (DCA) of the principal water environmental factors was conducted. The results of the DCA analysis of the principal water environmental factors in all four seasons showed that the gradient length was less than 3, so the choice of redundancy analysis (RDA) was more reasonable [53]. Three major zooplankton taxa were used as forecast objects in the redundancy analysis, as well as eight principal water environmental factors as forecast factors. Clusters were determined based on a similarity matrix using Euclidian distance, with distances calculated by group-average sorting and ranked similarities. The water environmental factors involved in the analysis were screened using a Monte Carlo Permutation test ($p < 0.05$) and, combined with the results of the principal component analysis, the water environmental factors with explanatory value for each season were identified as Tem, Chl-a, pH, NH_3-N, Tur, DO, TP and EC in spring, Tur, NH_3-N, pH, Chl-a, DO, TP, EC, in summer ORP, EC, TN, pH, Chl-a, COD_{Mn}, DO, Tem in autumn, EC, Tur, pH, Chl-a, COD_{Mn}, Tem, DO in winter.

In spring, the eigenvalues for the first two axes of the RDA analysis were 0.247 and 0.121. From Figure 6a, it can be seen that EC, TN, COD_{Mn}, and NH_3-N were positively correlated with Cladocera and Copepoda. The pH, Chl-a, and Tem were positively correlated with Rotifera, and ORP was negatively correlated with Rotifera. Cladocera formed good clusters with pH, Chl-a and EC, and Copepoda formed good clusters with Tur, TP and DO together with Rotifera. As the time enters summer, the total variation by RDA was 18.47, where the explanatory variables account for 62.80%. The pH, Chl-a, Tur, TP, Tem were positively correlated with Copepoda and Rotifera, and EC, TN were negatively correlated with them. pH, Chl-a, EC, TN, Tem were positively correlated with Cladocera. As shown in Figure 6b, Cladocera, Copepoda and Rotifera together formed good clusters with pH, Chl-a, Tur, TP, Tem. The three populations formed clusters with pH, Chl-a, Tur, TP, and Tem in summer to the results presented in the clustering heatmap in Figure 6b. In autumn, the eigenvalues for the first two axes of the RDA were 0.454 and 0.112, explaining 95.2% of the zooplankton variation. The pH, EC, Chl-a, TN, COD_{Mn}, Tem, NH_3-N were positively correlated with Copepoda and Rotifera. pH, Chl-a, CODMn were positively correlated with Cladocera and EC, TN were negatively correlated with it. Cladocera, Copepoda, and Rotifera together formed good clusters with pH, Chl-a, TP, Tem, DO in Figure 6c. In winter, the species-environment correlation coefficients for RDA analysis were 0.757 and 0.419, respectively, indicating a significant correlation between water environmental factors and zooplankton. tur and TP were positively correlated with Cladocera, whereas EC, TN and COD_{Mn}, which were negatively correlated with Cladocera, were positively correlated with Copepoda and Rotifera. In addition, NH_3-N was positively correlated with Copepoda, and pH and Chl-a were positively correlated with Rotifera. In Figure 6d, Cladocera, Copepoda and Rotifera together formed good clusters with Tur, TP and DO.

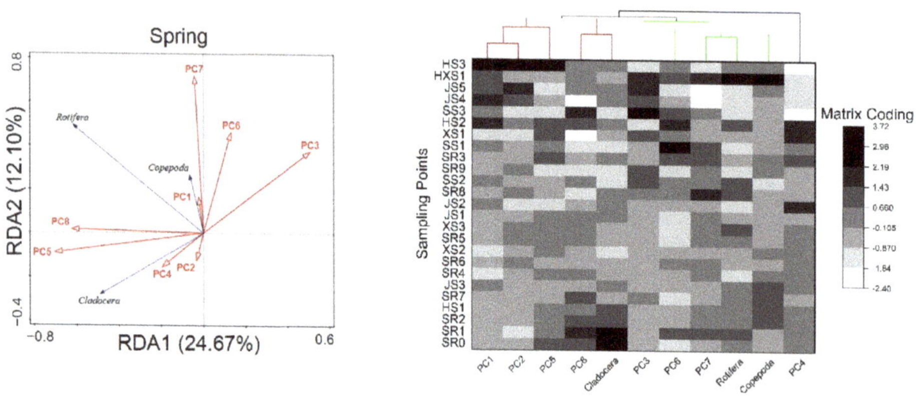

(a)

Figure 6. *Cont.*

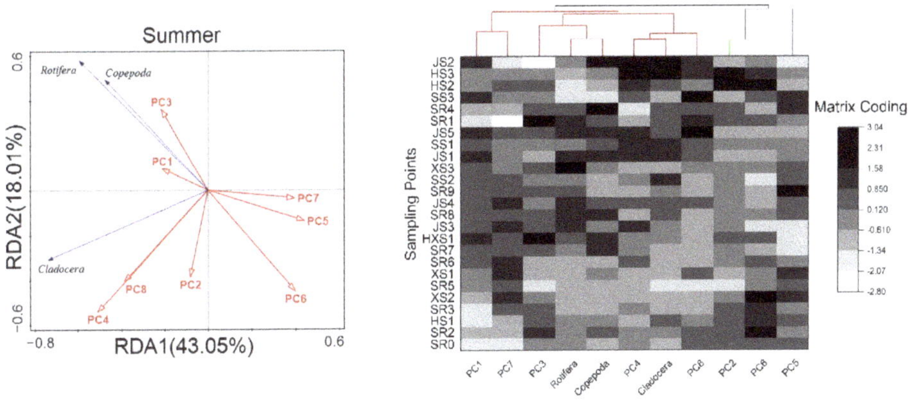

(**b**)

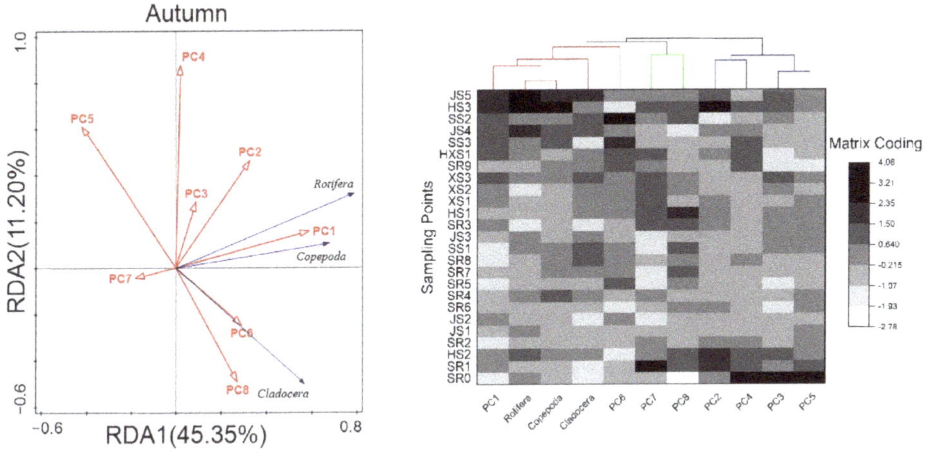

(**c**)

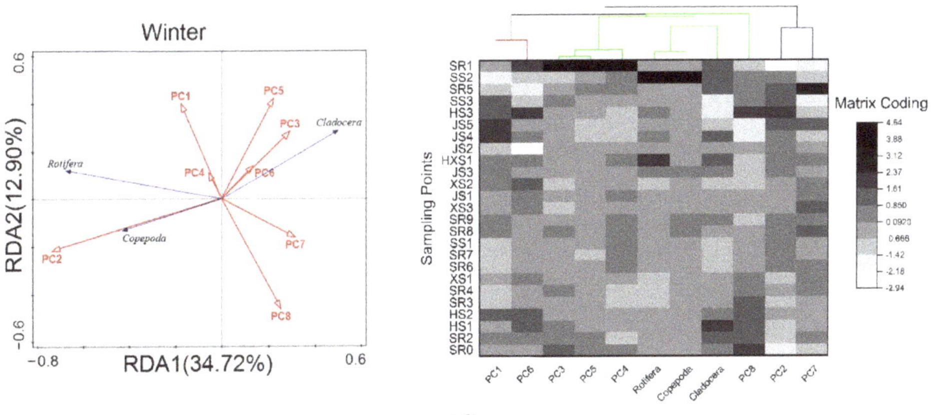

(**d**)

Figure 6. Plot of principal components with zooplankton redundancy analysis and clustering heatmap in different seasons: (**a**) Spring; (**b**) Summer; (**c**) Autumn; (**d**) Winter.

4. Discussion

The significant differences between the water level variation, water flow, and water connectivity of river-type reservoirs and natural water bodies lead to the aquatic ecosystems of river-type reservoirs being different from those of natural water bodies, and these differences also make the aquatic environment of reservoirs very fragile [50]. Because of this, a comprehensive evaluation of the water environment of river-type reservoirs and the health of the water ecology is essential to ensure water security and protect the ecosystem.

4.1. Factors Influencing Zooplankton Distribution

Our study found that the Shanxi Reservoir, a source of drinking water, is at some risk of contamination at certain times and locations. In fact, damming not only changes the connectivity of the river, but also alters the hydrodynamic conditions of natural water bodies, affecting their renewal cycles and making them more vulnerable to eutrophication risks [57,58]. Unlike previous studies [36], our study found that the PCY values in Huangtankeng Stream and Jujiang Stream in the study area were too high in summer, and the higher the PCY value, the greater the risk of algal blooms. According to research [59], local algal blooms will form in subtropical reservoirs when the density of blue-green algae is steadily higher than 50,000 cells/L. In Figure 3a,b, peaks in Chl-a and PCY are observed in Huangtankeng Stream and Jujiangxi Stream in spring and summer. And pH in these regions tended to be lower than in other regions during the same season. This is because, in addition to the obvious regional differences in pH distribution due to zonal factors, the horizontal distribution of pH is not entirely uniform in the same lake or water body due to environmental conditions such as incoming runoff, intensity of water exchange, and the number of biological populations in the lake [60]. Generally, in freshwater bodies, the pH level is related to the free carbon dioxide and dissolved carbonate ions in the lake water. Where the free carbon dioxide is high, the pH is low; the pH is correspondingly high in lakes with high content of bicarbonate ions and carbonate ions [61]. At the same time, lake algae generally consume free carbon dioxide in the water during their photosynthesis, increasing in pH response [62]. Our study also found that Tur and EC showed a large number of variations between seasons and spaces, as presented in Table 1 and Figure 2. Since the conductive particles in water bodies are mainly K^+, Na^+, Ca^{2+}, Mg^{2+}, Cl^-, SO_4^{2-}, HCO_4^{2-} and CO_3^{2-}, the total amount of these ions is close to the degree of mineralization of the water body, so the electrical conductivity is a side reflection of the degree of mineralization of the water body [63]. Turbidity can visually reflect the degree of turbidity of water bodies, which is mainly caused by insoluble sediment, humus, planktonic algae, colloidal particles, and other substances in the water bodies [64,65]. Dissolved oxygen did not show a large variability in spatial distribution in our study area, but showing fluctuations between seasons. Oxygen in water comes mainly from atmospheric dissolution and photosynthesis by aquatic organisms, and its content in water is influenced by temperature [66]. Notably, as presented in Figure 3c,d, our study found a high degree of variability in the spatial and temporal distribution of ammonia and total nitrogen, and the detection of ammonia and total nitrogen indicators compensated for the lack of relevant studies conducted by researchers on phytoplankton groups in the same area [36]. Nitrogenous compounds in natural water bodies often exist in three forms: ammonium nitrogen, nitrite nitrogen, and nitrate nitrogen, all of which can be used by aquatic organisms. When the water body is polluted or aquatic organisms die, organic nitrogen undergoes a series of decomposition to ammonia nitrogen form, then ammonia nitrogen is further oxidized to nitrite and finally to nitrate form [67]. Temperature is likewise an indicator of the large spatial and temporal variability within our study area. For the northern hemisphere, the lowest water temperature often occurs in January to February and the highest water temperature often occurs in July to August each year. The physiological and biochemical reactions in organisms change with temperature [68]. Within the appropriate temperature range, the physiological and biochemical reactions in organisms are accelerated with the increase in temperature, which promotes growth and development.

The results of our study indicate that the factors limiting the increase in zooplankton abundance in spring are mainly Tem and nutrient salts, which can be represented by PC1, PC2 in Figure 5. Comparing the observed data with the results of the RDA analysis, it is easy to see that the abundance of Rotifera is well explained by pH, Chl-a, and Tem, and that the abundance of Cladocera and Copepoda increases with EC, TN, NH_3-N. In contrast to the previous study [69], the main factors limiting the increase in zooplankton abundance during summer are nutrient salinity, PCY and DO. The abundance of Rotifera and Copepoda increases with increasing pH, Chl-a, Tur, TP, Tem, and decreases with increasing EC, TN. The situation in autumn and winter is similar to previous studies [70], the main factors limiting the increase in zooplankton abundance in autumn are nutrient salinity and DO, which can be expressed by PC1, PC2. The abundance of Cladocera increases with pH, Chl-a. Similarly, within the appropriate range, increases in pH, EC, Chl-a, TN, COD_{Mn}, Tem, NH_3-N result in a growing number of Copepoda and Rotifera. The main factors limiting the increase in abundance conferred on the animals in winter are nutrient salinity and PCY. The pH, Chl-a, EC, TN, and Tem predicted the abundance of Copepoda and Rotifera, and EC, TN, Tur, and TP were good predictors of Cladocera abundance. Another important finding was that zooplankton populations are influenced by different water environmental factors in different seasons, and they can also form good clusters with the corresponding water environmental factors under different conditions. Comparing the observed data with the results of the cluster analysis, it is easy to see that in spring the three zooplankton species form clusters with different water environmental factors, with Rotifera forming good clusters with NH_3-N and DO, Cladocera with Tur, and Copepoda with EC, pH, and TP. In summer, when environmental conditions are favorable, the three zooplankton species form good clusters with pH, Chl-a, Tur, NH_3-N, and DO. In autumn, the three zooplankton species form good clusters with pH, Chl-a, COD_{Mn}, ORP, and DO. In winter, when environmental conditions were harsh, the three zooplankton species formed good clusters with Tur, TP, Tem, Chl-a, and NH_3-N, respectively.

4.2. *Variation of Zooplankton Distribution*

As shown in Table 2, we identified a variety of zooplankton in the study area, including 20 genus of Rotifera, 15 genus of Copepoda and 11 genus of Cladocera, all of which are common freshwater zooplankton in southeast China [71]. Although Rotifera species were more abundant, the crustaceans (Copepoda and Cladocera) were overwhelmingly dominant in terms of the number of individuals (79.8% of the microscopic samples), with a total of 30 genus with a reproduction rate of 60% or more, including 9 genus of Rotifera, 14 genus of Copepoda and 7 genus of Cladocera. The number of species with a numerical dominance of more than 1% was 5 genus of Rotifera, 13 genus of Copepoda, and 7 genus of Cladocera, respectively.

Previous research has focused more on zooplankton as indicators of water quality and for ecological assessment [19]. The formation of clusters between zooplankton and water environmental factors could be of interest in the future for ecological restoration and environmental monitoring. Similar to the results of other related studies [59,72], the nutrient levels in the study area were lower in spring than in summer and autumn, and levels of the water environment factors PCY and Chl-a, which measure biological indicators, were low in spring, suggesting that the primary producers, picocyanobacteria, as well as other algae, were less abundant in the water column. Tur, also a measure of suspended matter in the water column [65], was lower in spring than in other seasons, suggesting that the low levels of suspended particulate matter and organic detritus, in the water column limited the food sources of Copepoda and Cladocera, thus making differences in zooplankton abundance on a spatial scale less pronounced in spring.

The water temperature and solar radiation in the study area were significantly higher in summer and autumn than in spring and winter, and the suitable water temperature and solar radiation provided a good environment for the growth and development of algae, the producers located in the first trophic level in the study area [73]. Along with the algal blooms, Chl-a and PCY that could be detected in the water environment factors are increasing and zooplankton that feeds on algae and their metabolites are experiencing a peak in growth and reproduction [71], making the abundance of the three typical zooplankton species reach their maximums in spring and summer. The higher NH_3-N content in the sub-regions of Huangtankeng Stream and Jujiangxi Stream provided sufficient nutrients for the growth and development of algae, making the Chl-a content in the sub-region of Huantankeng Stream much higher than that in other sub-regions, and likewise making the PCY content in the sub-region of Jujiangxi Stream much higher than that in other sub-regions.

In our study, variations of physicochemical indicators changed considerably with the seasons. The water temperature in autumn is between spring and summer, with more nutrient salts than in spring, making it more suitable for the growth and reproduction of producers such as algae. The values of PCY and Chl-a were correspondingly lower in Jujiangxi Stream, which had less TN and NH_3-N, and higher in Huangtangkeng Stream, which had more TN and NH_3-N. In the Shanxi Reservoir, which has a high TN content, the PCY values were higher than in the other sub-regions. Cladocera also occurs in abundance in the Jujiangxi Stream and Huangtangkeng Stream, areas of fertile water. Algae-feeding Copepoda were more abundant in Shanxi Reservoir, Jujiangxi Stream, and Huangtangkeng Stream than in the other sub-regions. Copepoda, which has a wide range of food sources [74], was significantly more abundant in Huangtankeng Stream than in the other regions.

Biological indicators in the study area were lower in winter than in summer and autumn, due to the amount of solar radiation, PCY and Chl-a were the lowest throughout the year. The life activities of aquatic organisms in the water column are somewhat restricted at this temperature [73], and the PCY and Chl-a in the water column are reduced to their lowest values throughout the year. Similarly, the decomposition of microorganisms is also restricted due to the temperature, causing the NH_3-N content in the water column to decrease to some extent. More Copepoda spend the winter in the Sanchaxi Stream and Hongkouxi Stream, where Chl-a levels are high, and Cladocera spend the winter in sub-regions with slow flow conditions and rich nutrient salinity.

5. Conclusions

In the study area of Shanxi Reservoir, zooplankton showed the same spatial and temporal heterogeneity as the water environmental factors. Freshwater zooplankton, represented by Cladocera, Copepoda, and Rotifera showed not only clear seasonal differences in time, but also spatially variable response relationships with changes in key water environmental factors such as Tem, DO, nutrient salinity and EC. The main response relationships between water environmental factors and zooplankton exist in different seasons as follows. In spring, Tem and TP affect the number of Cladocera. EC, pH, and Tem affect the number of Copepoda, and pH, EC, and Chl-a affect the number of Rotifera. In summer, Tur, TP, and DO affect the abundance of Copepoda and Rotifera, and NH_3-N affects the amount of Cladocera in the water. In autumn, Tem has an effect on the numbers of all three. Cladocera will be affected mainly by DO, and pH, EC will affect Copepoda numbers. Rotifera will also be affected mainly by EC in autumn. In winter, EC, pH, DO, TP affect the numbers of Copepoda and Rotifera. EC and Tur affect the number of Cladocera. Meanwhile, the three zooplankton species formed good clusters with pH, DO, Chl-a, and NH_3-N during the summer months in abundant water, and with Tem, Tur, TP, and NH_3-N during the winter months in dry water. This correspondence between zooplankton and relevant water environment factors in the respective seasons can be used not only to predict trends in water environment development but also as a key indicator of regional water environment deployment in the process of water diversion and storage during periods of abundance and drought.

Overall, the observations and analyses in this study suggest that temperature is a key factor influencing the differences in the spatial and temporal distribution of water environment and water ecology in riverine reservoirs. On the one hand, water temperature is a direct reflection of the intensity of solar radiation on the water body, and changes in water temperature can lead to changes in DO, EC, ORP, pH, Chl-a, PCY and other physicochemical and biological indicators. On the other hand, in the catchment area of river-type reservoirs, the change of temperature also indicates the change of seasons, which fits with the growth and development cycle of economic crops near the catchment area and economic fish in the reservoir. Nutrients such as nitrogen and phosphorus, as well as organic pollutants, will enter the reservoir with surface runoff, causing changes in chemical indicators. Changes in physical and chemical indicators will cause changes in biological indicators.

Due to the time and scale constraints, the analysis method in this study affected the generalizability of the results to a certain extent. In future studies, the assessment of the water environment and water ecological health of river-type reservoirs requires not only data support for a longer time series, but also data mining and the establishment and validation of models with relevant data in order to provide constructive evaluation and analysis of the water environment and water ecological health of other reservoirs with similar conditions using accurate and generalized models.

Author Contributions: Conceptualization, J.Y. and J.X.; methodology, W.C. and J.X.; software, J.Y.; validation, Y.W., R.Z. and X.D.; formal analysis, W.C.; investigation, J.Y., X.D. and Z.L.; resources, K.X.; data curation, Z.L.; writing—original draft preparation, J.Y. and J.X.; writing—review and editing, K.X.; visualization, Y.W.; supervision, Z.X.; project administration, Z.X., R.Z. and X.D.; funding acquisition, J.X. All authors have read and agreed to the published version of the manuscript.

Funding: This research was funded by the National Science and Technology Basic Resources Survey Special Project [Grant No. 2022FY100404], National Key Research and Development Program of China [Grant No. 2018YFD0900805], Key Program of Water Conservancy Science and Technology of Zhejiang Province [Grant No. RB1915], National Natural Science Foundation of China [Grant No. 41471069], Postgraduate Research and Practice Innovation Program of Jiangsu Province [Grant No. B200203137, KYCX20_0493], Fundamental Research Funds for the Central Universities [Grant No. B210203028].

Institutional Review Board Statement: Not applicable.

Informed Consent Statement: Not applicable.

Data Availability Statement: Not applicable.

Acknowledgments: The authors would like to thank BA students Shuyi Ji, Yuanshuo Lu, Zhengxin Wang, Chuwen Wang, MSc student Peng Zhang, Hao Wang, and Xiaohan Zhu of Hohai University for their field and laboratory contributions to this study. The authors are particularly grateful to Valerie Lopes and Tihana Mirkovic from York University, and Han Wang from Hong Kong University of Science and Technology for constructive advice during data analysis and visualization.

Conflicts of Interest: The authors declare no conflict of interest.

References

1. Foley, J.A.; Ramankutty, N.; Brauman, K.A.; Cassidy, E.S.; Gerber, J.S.; Johnston, M.; Mueller, N.D.; O'Connell, C.; Ray, D.K.; West, P.C.; et al. Solutions for a Cultivated Planet. *Nature* **2011**, *478*, 337–342. [CrossRef] [PubMed]
2. Steffen, W.; Richardson, K.; Rockstrom, J.; Cornell, S.E.; Fetzer, I.; Bennett, E.M.; Biggs, R.; Carpenter, S.R.; de Vries, W.; de Wit, C.A.; et al. Planetary Boundaries: Guiding Human Development on a Changing Planet. *Science* **2015**, *347*, 11. [CrossRef] [PubMed]
3. Li, C.R.; Busquets, R.; Campos, L.C. Assessment of Microplastics in Freshwater Systems: A Review. *Sci. Total Environ.* **2020**, *707*, 12. [CrossRef] [PubMed]
4. Acuna-Alonso, C.; Alvarez, X.; Lorenzo, O.; Cancela, A.; Valero, E.; Sanchez, A. Water Toxicity in Reservoirs after Freshwater Algae Harvest. *J. Clean. Prod.* **2021**, *283*, 104560. [CrossRef]
5. Soares, L.M.V.; Calijuri, M.D. Deterministic Modelling of Freshwater Lakes and Reservoirs: Current Trends and Recent Progress. *Environ. Model. Softw.* **2021**, *144*, 16. [CrossRef]
6. Zhang, Y.G.; Zhang, Z.; Xue, S.; Wang, R.J.; Xiao, M. Stability Analysis of a Typical Landslide Mass in the Three Gorges Reservoir under Varying Reservoir Water Levels. *Environ. Earth Sci.* **2020**, *79*, 14. [CrossRef]

7. Hossain, M.; Huda, A.S.N.; Mekhilef, S.; Seyedmahmoudian, M.; Horan, B.; Stojcevski, A.; Ahmed, M. A State-of-the-Art Review of Hydropower in Malaysia as Renewable Energy: Current Status and Future Prospects. *Energy Strategy Rev.* **2018**, *22*, 426–437. [CrossRef]
8. Kosek, K.; Ruman, M. Arctic Freshwater Environment Altered by the Accumulation of Commonly Determined and Potentially New Pops. *Water* **2021**, *13*, 1739. [CrossRef]
9. Ji, B.; Liang, J.C.; Chen, R. Bacterial Eutrophic Index for Potential Water Quality Evaluation of a Freshwater Ecosystem. *Environ. Sci. Pollut. Res.* **2020**, *27*, 32449–32455. [CrossRef]
10. Park, Y.; Lee, H.K.; Shin, J.K.; Chon, K.; Kim, S.; Cho, K.H.; Kim, J.H.; Baek, S.S. A Machine Learning Approach for Early Warning of Cyanobacterial Bloom Outbreaks in a Freshwater Reservoir. *J. Environ. Manag.* **2021**, *288*, 9. [CrossRef] [PubMed]
11. Brack, W.; Ait-Aissa, S.; Burgess, R.M.; Busch, W.; Creusot, N.; Di Paolo, C.; Escher, B.I.; Hewitt, L.M.; Hilscherova, K.; Hollender, J.; et al. Effect-Directed Analysis Supporting Monitoring of Aquatic Environments—An in-Depth Overview. *Sci. Total Environ.* **2016**, *544*, 1073–1118. [CrossRef] [PubMed]
12. Altenburger, R.; Brack, W.; Burgess, R.M.; Busch, W.; Escher, B.I.; Focks, A.; Hewitt, L.M.; Jacobsen, B.N.; de Alda, M.L.; Ait-Aissa, S.; et al. Future Water Quality Monitoring: Improving the Balance between Exposure and Toxicity Assessments of Real-World Pollutant Mixtures. *Environ. Sci. Eur.* **2019**, *31*, 17. [CrossRef]
13. Sagan, V.; Peterson, K.T.; Maimaitijiang, M.; Sidike, P.; Sloan, J.; Greeling, B.A.; Maalouf, S.; Adams, C. Monitoring Inland Water Quality Using Remote Sensing: Potential and Limitations of Spectral Indices, Bio-Optical Simulations, Machine Learning, and Cloud Computing. *Earth-Sci. Rev.* **2020**, *205*, 31. [CrossRef]
14. Shi, W.X.; Zhuang, W.E.; Hur, J.; Yang, L.Y. Monitoring Dissolved Organic Matter in Wastewater and Drinking Water Treatments Using Spectroscopic Analysis and Ultra-High Resolution Mass Spectrometry. *Water Res.* **2021**, *188*, 16. [CrossRef] [PubMed]
15. Brooks, J.L.; Dodson, S.I. Predation, Body Size, and Composition of Plankton. *Science* **1965**, *150*, 28–35. [CrossRef] [PubMed]
16. Elser, J.J.; Fagan, W.F.; Denno, R.F.; Dobberfuhl, D.R.; Folarin, A.; Huberty, A.; Interlandi, S.; Kilham, S.S.; McCauley, E.; Schulz, K.L.; et al. Nutritional Constraints in Terrestrial and Freshwater Food Webs. *Nature* **2000**, *408*, 578–580. [CrossRef]
17. Frank, K.T.; Petrie, B.; Choi, J.S.; Leggett, W.C. Trophic Cascades in a Formerly Cod-Dominated Ecosystem. *Science* **2005**, *308*, 1621–1623. [CrossRef]
18. Sousa, W.; Attayde, J.L.; Rocha, E.D.; Eskinazi-Sant'Anna, E.M. The Response of Zooplankton Assemblages to Variations in the Water Quality of Four Man-Made Lakes in Semi-Arid Northeastern Brazil. *J. Plankton Res.* **2008**, *30*, 699–708. [CrossRef]
19. Jeppesen, E.; Noges, P.; Davidson, T.A.; Haberman, J.; Noges, T.; Blank, K.; Lauridsen, T.L.; Sondergaard, M.; Sayer, C.; Laugaste, R.; et al. Zooplankton as Indicators in Lakes: A Scientific-Based Plea for Including Zooplankton in the Ecological Quality Assessment of Lakes According to the European Water Framework Directive (Wfd). *Hydrobiologia* **2011**, *676*, 279–297. [CrossRef]
20. Winder, M.; Schindler, D.E. Climate Change Uncouples Trophic Interactions in an Aquatic Ecosystem. *Ecology* **2004**, *85*, 2100–2106. [CrossRef]
21. Ger, K.A.; Urrutia-Cordero, P.; Frost, P.C.; Hansson, L.A.; Sarnelle, O.; Wilson, A.E.; Lurling, M. The Interaction between Cyanobacteria and Zooplankton in a More Eutrophic World. *Harmful Algae* **2016**, *54*, 128–144. [CrossRef] [PubMed]
22. Gallardo, B.; Clavero, M.; Sanchez, M.I.; Vila, M. Global Ecological Impacts of Invasive Species in Aquatic Ecosystems. *Glob. Chang. Biol.* **2016**, *22*, 151–163. [CrossRef] [PubMed]
23. Jeppesen, E.; Brucet, S.; Naselli-Flores, L.; Papastergiadou, E.; Stefanidis, K.; Noges, T.; Noges, P.; Attayde, J.L.; Zohary, T.; Coppens, J.; et al. Ecological Impacts of Global Warming and Water Abstraction on Lakes and Reservoirs Due to Changes in Water Level and Related Changes in Salinity. *Hydrobiologia* **2015**, *750*, 201–227. [CrossRef]
24. Miloslavich, P.; Bax, N.J.; Simmons, S.E.; Klein, E.; Appeltans, W.; Aburto-Oropeza, O.; Garcia, M.A.; Batten, S.D.; Benedetti-Cecchi, L.; Checkley, D.M.; et al. Essential Ocean Variables for Global Sustained Observations of Biodiversity and Ecosystem Changes. *Glob. Chang. Biol.* **2018**, *24*, 2416–2433. [CrossRef]
25. Marcolin, C.R.; Lopes, R.M.; Jackson, G.A. Estimating Zooplankton Vertical Distribution from Combined Lopc and Zooscan Observations on the Brazilian Coast. *Mar. Biol.* **2015**, *162*, 2171–2186. [CrossRef]
26. Wang, W.C.; Sun, S.; Zhang, F.; Sun, X.X.; Zhang, G.T. Zooplankton Community Structure, Abundance and Biovolume in Jiaozhou Bay and the Adjacent Coastal Yellow Sea During Summers of 2005-2012: Relationships with Increasing Water Temperature. *J. Oceanol. Limnol.* **2018**, *36*, 1655–1670. [CrossRef]
27. Naito, A.; Abe, Y.; Matsuno, K.; Nishizawa, B.; Kanna, N.; Sugiyama, S.; Yamaguchi, A. Surface Zooplankton Size and Taxonomic Composition in Bow Doi N Fjord, North-Western Greenland: A Comparison of Zooscan, Opc and Microscopic Analyses. *Polar Sci.* **2019**, *19*, 120–129. [CrossRef]
28. Wang, W.C.; Sun, S.; Sun, X.X.; Zhang, G.T.; Zhang, F. Spatial Patterns of Zooplankton Size Structure in Relation to Environmental Factors in Jiaozhou Bay, South Yellow Sea. *Mar. Pollut. Bull.* **2020**, *150*, 10. [CrossRef] [PubMed]
29. Maas, A.E.; Gossner, H.; Smith, M.J.; Blanco-Bercial, L.; Irigoien, X. Use of Optical Imaging Datasets to Assess Biogeochemical Contributions of the Mesozooplankton. *J. Plankton Res.* **2021**, *43*, 475–491. [CrossRef]
30. Noyon, M.; Poulton, A.J.; Asdar, S.; Weitz, R.; Giering, S.L.C. Mesozooplankton Community Distribution on the Agulhas Bank in Autumn: Size Structure and Production. *Deep-Sea Res. Part Ii-Top. Stud. Oceanogr.* **2022**, *195*, 15. [CrossRef]
31. Garcia-Herrera, N.; Cornils, A.; Laudien, J.; Niehoff, B.; Hofer, J.; Forsterra, G.; Gonzalez, H.E.; Richter, C. Seasonal and Diel Variations in the Vertical Distribution, Composition, Abundance and Biomass of Zooplankton in a Deep Chilean Patagonian Fjord. *Peerj* **2022**, *10*, 31. [CrossRef] [PubMed]

32. Schultes, S.; Lopes, R.M. Laser Optical Plankton Counter and Zooscan Intercomparison in Tropical and Subtropical Marine Ecosystems. *Limnol. Oceanogr.-Methods* **2009**, *7*, 771–784. [CrossRef]
33. Dong, X.; Zhou, H. Calculation and Study of Total Nitrogen Influx Fluxes in Tributaries of Shanxi Reservoir. *Zhejiang Hydrotech.* **2017**, *45*, 22–24. [CrossRef]
34. Li, A.L.; Haitao, C.; Yuanyuan, L.; Qiu, L.; Wenchuan, W. Simulation of Nitrogen Pollution in the Shanxi Reservoir Watershed Based on Swat Model. *Nat. Environ. Pollut. Technol.* **2020**, *19*, 1265–1272. [CrossRef]
35. Chen, H.T.; Chen, J.; Liu, Y.Y.; He, J. Study of Nitrogen Pollution Simulation and Management Measures on Swat Model in Typhoon Period of Shanxi Reservoir Watershed, Zhejiang Province, China. *Pol. J. Environ. Stud.* **2021**, *30*, 2499–2507. [CrossRef]
36. Yang, M.Z.; Xia, J.H.; Cai, W.W.; Zhou, Z.Y.; Yang, L.B.; Zhu, X.X.; Li, C.D. Seasonal and Spatial Distributions of Morpho-Functional Phytoplankton Groups and the Role of Environmental Factors in a Subtropical River-Type Reservoir. *Water Sci. Technol.* **2020**, *82*, 2316–2330. [CrossRef] [PubMed]
37. Pianka, E.R. Niche Overlap and Diffuse Competition. *Proc. Natl. Acad. Sci. USA* **1974**, *71*, 2141–2145. [CrossRef] [PubMed]
38. Xiong, W.; Huang, X.N.; Chen, Y.Y.; Fu, R.Y.; Du, X.; Chen, X.Y.; Zhan, A.B. Zooplankton Biodiversity Monitoring in Polluted Freshwater Ecosystems: A Technical Review. *Environ. Sci. Ecotechnol.* **2020**, *1*, 11. [CrossRef]
39. Warren, J.D.; Leach, T.H.; Williamson, C.E. Measuring the Distribution, Abundance, and Biovolume of Zooplankton in an Oligotrophic Freshwater Lake with a 710 Khz Scientific Echosounder. *Limnol. Oceanogr.-Methods* **2016**, *14*, 231–244. [CrossRef]
40. Sastri, A.R.; Gauthier, J.; Juneau, P.; Beisner, B.E. Biomass and Productivity Responses of Zooplankton Communities to Experimental Thermocline Deepening. *Limnol. Oceanogr.* **2014**, *59*, 1–16. [CrossRef]
41. Gauthier, J.; Prairie, Y.T.; Beisner, B.E. Thermocline Deepening and Mixing Alter Zooplankton Phenology, Biomass and Body Size in a Whole-Lake Experiment. *Freshw. Biol.* **2014**, *59*, 998–1011. [CrossRef]
42. Yaseen, T.; Bhat, S.U.; Bhat, F.A. Study of Vertical Distribution Dynamics of Zooplankton in a Thermally Stratified Warm Monomictic Lake of Kashmir Himalaya. *Ecohydrology* **2022**, *15*, 18. [CrossRef]
43. *GB11892-89*; Water Quality-Determination of Permanganate index. Ministry of Environmental Protection of the People's Republic of China: Beijing, China, 1989. (In Chinese)
44. *GB11893-89*; Water Quality-Determination of Total Phosphorus-Ammonium Molybdate Spectrophotometric Method. Ministry of Environmental Protection of the People's Republic of China: Beijing, China, 1989. (In Chinese)
45. *HJ535-2009*; Water Quality-Determination of Ammonia Nitrogen-Nessler's Reagent Spectrophotometry. Ministry of Environmental Protection of the People's Republic of China: Beijing, China, 2009. (In Chinese)
46. *SCT9402-2010*; Specifications for Freshwater Plankton Surveys. Ministry of Agriculture of the People's Republic of China: Beijing, China, 2010. (In Chinese)
47. Vogelmann, C.; Teichert, M.; Schubert, M.; Martens, A.; Schultes, S.; Stibor, H. The Usage of a Zooplankton Digitization Software to Study Plankton Dynamics in Freshwater Fisheries. *Fish. Res.* **2022**, *251*, 9. [CrossRef]
48. Kumar, S.; Singh, R.; Singh, T.P.; Batish, A. On Mechanical Characterization of 3-D Printed Pla-Pvc-Wood Dust-Fe3o4 Composite. *J. Thermoplast. Compos. Mater.* **2022**, *35*, 36–53. [CrossRef]
49. Sharma, A.S.; Gupta, S.; Singh, N.R. Zooplankton Community of Keibul Lamjao National Park (Klnp) Manipur, India in Relation to the Physico-Chemical Variables of the Water. *Chin. J. Oceanol. Limnol.* **2017**, *35*, 469–480. [CrossRef]
50. Xiang, R.; Wang, L.J.; Li, H.; Tian, Z.B.; Zheng, B.H. Water Quality Variation in Tributaries of the Three Gorges Reservoir from 2000 to 2015. *Water Res.* **2021**, *195*, 12. [CrossRef]
51. Tian, Z.Y.; Zheng, S.; Guo, S.J.; Zhu, M.L.; Liang, J.H.; Du, J.; Sun, X.X. Relationship between Zooplankton Community Characteristics and Environmental Conditions in the Surface Waters of the Western Pacific Ocean During the Winter of 2014. *J. Ocean Univ. China* **2021**, *20*, 706–720. [CrossRef]
52. Lima, A.R.A.; Costa, M.F.; Barletta, M. Distribution Patterns of Microplastics within the Plankton of a Tropical Estuary. *Environ. Res.* **2014**, *132*, 146–155. [CrossRef]
53. Wu, S.; Hua, P.; Gui, D.; Zhang, J.; Ying, G.; Krebs, P. Occurrences, Transport Drivers, and Risk Assessments of Antibiotics in Typical Oasis Surface and Groundwater. *Water Res.* **2022**, *225*, 119138. [CrossRef]
54. Committee, E. *Atlas of the Main Freshwater Zooplankton of Zhejiang Province (Drinking Water Sources)*; China Environment Publishing House: Beijing, China, 2013; Volume 445. (In Chinese)
55. Olsen, R.L.; Chappell, R.W.; Loftis, J.C. Water Quality Sample Collection, Data Treatment and Results Presentation for Principal Components Analysis—Literature Review and Illinois River Watershed Case Study. *Water Res.* **2012**, *46*, 3110–3122. [CrossRef]
56. ZerfaSs, C.; Lehmann, R.; Ueberschaar, N.; Sanchez-Arcos, C.; Totsche, K.U.; Pohnert, G. Groundwater Metabolome Responds to Recharge in Fractured Sedimentary Strata. *Water Res.* **2022**, *223*, 118998. [CrossRef] [PubMed]
57. Yang, N.; Li, Y.; Zhang, W.L.; Lin, L.; Qian, B.; Wang, L.F.; Niu, L.H.; Zhang, H.J. Cascade Dam Impoundments Restrain the Trophic Transfer Efficiencies in Benthic Microbial Food Web. *Water Res.* **2020**, *170*, 115351. [CrossRef] [PubMed]
58. Shi, X.; Sun, J.; Xiao, Z.J. Investigation on River Thermal Regime under Dam Influence by Integrating Remote Sensing and Water Temperature Model. *Water* **2021**, *13*, 133. [CrossRef]
59. Yao, L.G.; Zhao, X.M.; Zhou, G.J.; Liang, R.C.; Gou, T.; Xia, B.C.; Li, S.Y.; Liu, C. Seasonal Succession of Phytoplankton Functional Groups and Driving Factors of Cyanobacterial Blooms in a Subtropical Reservoir in South China. *Water* **2020**, *12*, 1167. [CrossRef]
60. Akis, S.; Ozcimen, D. Optimization of Ph Induced Flocculation of Marine and Freshwater Microalgae Via Central Composite Design. *Biotechnol. Prog.* **2019**, *35*, 6. [CrossRef]

61. Gollnisch, R.; Alling, T.; Stockenreiter, M.; Ahren, D.; Grabowska, M.; Rengefors, K. Calcium and Ph Interaction Limits Bloom Formation and Expansion of a Nuisance Microalga. *Limnol. Oceanogr.* **2021**, *66*, 3523–3534. [CrossRef]
62. Price, G.A.V.; Stauber, J.L.; Holland, A.; Koppel, D.J.; Van Genderen, E.J.; Ryan, A.C.; Jolley, D.F. The Influence of Ph on Zinc Lability and Toxicity to a Tropical Freshwater Microalga. *Environ. Toxicol. Chem.* **2021**, *40*, 2836–2845. [CrossRef]
63. Assuncao, A.W.D.; Souza, B.P.; da Cunha-Santino, M.B.; Bianchini, I. Formation and Mineralization Kinetics of Dissolved Humic Substances from Aquatic Macrophytes Decomposition. *J. Soils Sediments* **2018**, *18*, 1252–1264. [CrossRef]
64. Sosa-Aranda, I.; Zambrano, L. Relationship between Turbidity and the Benthic Community in the Preserved Montebello Lakes in Chiapas, Mexico. *Mar. Freshw. Res.* **2020**, *71*, 824–831. [CrossRef]
65. Gozdziejewska, A.M.; Kruk, M. Zooplankton Network Conditioned by Turbidity Gradient in Small Anthropogenic Reservoirs. *Sci. Rep.* **2022**, *12*, 12. [CrossRef]
66. Vad, C.F.; Horvath, Z.; Kiss, K.T.; Toth, B.; Pentek, A.L.; Acs, E. Vertical Distribution of Zooplankton in a Shallow Peatland Pond: The Limiting Role of Dissolved Oxygen. *Ann. Limnol.-Int. J. Limnol.* **2013**, *49*, 275–285. [CrossRef]
67. Ding, S.; Dan, S.F.; Liu, Y.; He, J.; Zhu, D.D.; Jiao, L.X. Importance of Ammonia Nitrogen Potentially Released from Sediments to the Development of Eutrophication in a Plateau Lake. *Environ. Pollut.* **2022**, *305*, 11. [CrossRef] [PubMed]
68. Pulsifer, J.; Laws, E. Temperature Dependence of Freshwater Phytoplankton Growth Rates and Zooplankton Grazing Rates. *Water* **2021**, *13*, 1591. [CrossRef]
69. Amorim, C.A.; Moura, A.D. Ecological Impacts of Freshwater Algal Blooms on Water Quality, Plankton Biodiversity, Structure, and Ecosystem Functioning. *Sci. Total Environ.* **2021**, *758*, 13. [CrossRef] [PubMed]
70. Burns, C.W.; Galbraith, L.M. Relating Planktonic Microbial Food Web Structure in Lentic Freshwater Ecosystems to Water Quality and Land Use. *J. Plankton Res.* **2007**, *29*, 127–139. [CrossRef]
71. Li, C.C.; Feng, W.Y.; Chen, H.Y.; Li, X.F.; Song, F.H.; Guo, W.J.; Giesy, J.P.; Sun, F.H. Temporal Variation in Zooplankton and Phytoplankton Community Species Composition and the Affecting Factors in Lake Taihu-a Large Freshwater Lake in China. *Environ. Pollut.* **2019**, *245*, 1050–1057. [CrossRef]
72. Huang, J.Y.; Wang, X.; Wang, X.Y.; Chen, Y.J.; Yang, Z.W.; Xie, S.G.; Li, T.T.; Song, S. Distribution Characteristics of Ammonia-Oxidizing Microorganisms and Their Responses to External Nitrogen and Carbon in Sediments of a Freshwater Reservoir, China. *Aquat. Ecol.* **2022**, *56*, 841–857. [CrossRef]
73. Rhode, S.C.; Pawlowski, M.; Tollrian, R. The Impact of Ultraviolet Radiation on the Vertical Distribution of Zooplankton of the Genus Daphnia. *Nature* **2001**, *412*, 69–72. [CrossRef]
74. Diniz, L.P.; Franca, E.J.; Bonecker, C.C.; Marcolin, C.R.; De Melo, M. Non-Predatory Mortality of Planktonic Microcrustaceans (Cladocera and Copepoda) in Neotropical Semiarid Reservoirs. *An. Acad. Bras. Cienc.* **2021**, *93*, 16. [CrossRef]

hydrology

MDPI

Article

Hydropedological Characteristics of the Cathedral Peak Research Catchments

Rowena Harrison [1,2,*], Johan van Tol [2] and Philippe Amiotte Suchet [3]

[1] South African Environmental Observation Network (SAEON), Grasslands, Forests, Wetlands Node, Montrose, Pietermaritzburg 3201, South Africa
[2] Department of Soil, Crop and Climate Sciences, University of the Free State, Bloemfontein 9300, South Africa
[3] Biogéosciences, UMR 6282 CNRS, Université Bourgogne, Franche-Comté, 6 Boulevard Gabriel, 21000 Dijon, France
* Correspondence: 2018258745@ufs4life.ac.za

Abstract: It has long been recognised that the role of soils is critical to the understanding of the way catchments store and release water. This study aimed to gain an understanding of the hydropedological characteristics and flow dynamics of the soils of three mountain catchment areas. Digital soil maps of the hydropedological characteristics of the catchments were interpreted and a conceptual response of these watersheds to precipitation was formed. This conceptual response was then tested with the use of site-specific precipitation and streamflow data. Furthermore, piezometers were installed in soils classified as the interflow hydropedological soil group as well as the saturated responsive hydropedological soil group and water table depth data for the three catchments were analysed. Climatic data indicated that there is a lag time effect in the quantity of precipitation that falls in the catchment and the corresponding rise in streamflow value. This lag time effect coupled with data obtained from the piezometers show that the various hydropedological soil groups play a pivotal role in the flow dynamics. Of importance is the unique influence of different wetland systems on the streamflow dynamics of the catchments. The drying and wetting cycles of individual wetland systems influenced both the baseflow connectivity and the overland flow during wetter periods. They are the key focus in understanding the connectivity between the hydropedological flow paths and the contribution of soil water to the stream networks of the three catchments.

Keywords: hydropedology; soil science; catchment hydrology; hydropedology soil maps; soil flow paths

Citation: Harrison, R.; van Tol, J.; Amiotte Suchet, P. Hydropedological Characteristics of the Cathedral Peak Research Catchments. *Hydrology* **2022**, 9, 189. https://doi.org/10.3390/hydrology9110189

Academic Editors: Songhao Shang, Qianqian Zhang, Dongqin Yin, Hamza Gabriel and Magdy Mohssen

Received: 27 September 2022
Accepted: 20 October 2022
Published: 25 October 2022

1. Introduction

Understanding how catchments store and release water and the resulting ecosystem services they provide is a crucial element in improving the management of these resources [1]. It has long been recognised that the role of soils is critical to these processes. The study of hydropedology as an intertwined branch of soil science and hydrology is used at multiple scales to gain a better understanding of the variability of saturated and unsaturated surface and subsurface environments and how these influence rainfall-runoff processes [2]. Hydropedology has therefore gained popularity in establishing the role of soils in the storage, flow dynamics and connectivity between hillslopes and streams of watersheds [3,4].

Soils are three-dimensional bodies in the landscape with different arrangements of vertical horizons and lateral variability of soil properties [5]. Ref. [6] showed that the quantity and type of soil macropores are variable across short distances, but spatial patterns of preferential flow at the landscape scale are far from being completely random. They instead show a clear pattern comprised of recognizable diagnostic soil horizons, soil materials, and pedons which all display characteristic flow and transport arrangements. Soil water processes can therefore be described in terms of content (volumetric or gravitational), potential (matric, osmotic, and gravitational potentials) and movement (subsurface flows in quantity or in speed). All of these descriptions are variable in time and space, creating multiple differences in the temporal structure of how precipitation moves through a landscape and is then discharged [7].

Despite these variations in soil patterns, the range of specific soil types within a catchment is generally restricted based on location [8]. This distribution of different soil characteristics over a landscape is the key to connecting the pedon scale to the landscape scale [2,9]. These soil patterns are expressed in the different soil forms identified within a catchment area. The periodicity of water movement through a soil causes distinct processes of oxidation and reduction. For example, the vertical and lateral percolation of water through a soil profile can cause the leaching out of iron and manganese, creating a unique set of characteristics that pertain to a particular soil form. In other areas, where there is excess water, soil forms are expressed by an accumulation of organic matter and/or a reduction process within the soil horizons. These specific morphological features in the soil profile are indicators of landscape processes including percolation, lateral flow, and water storage [10]. These different types of flow paths within a catchment area may be isolated or connect the flow paths to a stream network [11]. Thus, the characteristics of a soil profile can be utilised to gain an understanding of hydrological dynamics at landscape scale. A further contributing factor is soil thickness, as this is a key factor in the storage and redistribution of rainfall within the soil profile. It therefore plays an important role in controlling the types of various runoff processes and is often a decisive factor in the processes that generate baseflow as well as overland flow [12].

In mountainous regions, changes within the landscape occur over short distances, and this creates a marked internal (i.e., subsurface) heterogeneity within soils, as well as heterogeneity in the catchment conditions. This makes it difficult to determine the direct measure of how much water is stored within particular areas of the catchment as well as the internal flow dynamics [1]. This is particularly so given the added interrelated influence of climate, geology, topography, and vegetation characteristics on the flow dynamics of these watersheds [3]. The understanding of these processes is important as mountainous headwater catchments provide key water-related services for downstream ecosystems, and the regulation of streamflow by these catchments is highly influenced by their capacity to store and release water [1]. Recent studies have shown that the way in which water is stored and transferred within catchment areas is furthermore a crucial link in generating both base flows and storm flows during precipitation events as well as influences the sediment yield [3,13].

In South Africa, the uKhahlamba-Drakensberg Mountain range is one such area in which the spatial heterogeneity of catchments allows for the study of these various processes over a relatively short distance. Utilising this area, one can gain a deeper understanding of the way in which soil landscape functions control the movement of water in these areas and influence streamflow discharge. This is an initial and important component in understanding how streamflow discharge from these areas impacts the downstream ecosystems.

The aims of this chapter are therefore to gain an understanding of the hydropedological characteristics and flow dynamics of the soils of three mountain catchments within the uKhahlamba-Drakensberg Mountain range. This is achieved through (1) interpreting hydropedological soil maps to conceptualize the hydrological functioning of the catchments in terms of dominant flow paths and storage mechanisms and how these influence the streamflow dynamics and (2) to test the conceptual understanding of the hydropedological character of the catchment areas through a series of site-specific measurements taken within the catchment areas.

2. Materials and Methods

2.1. Study Area

The Cathedral Peak experimental research catchment site forms part of the uKhahlamba-Drakensberg escarpment within the Ezemvelo KZN Wildlife Maloti-Drakensberg Park.

The Park is a World Heritage Site and is situated in the northern part of the uKhahlamba-Drakensberg escarpment, KwaZulu-Natal, South Africa. The South African National Environment Observatory Network (SAEON) undertakes the monitoring of the catchment site. There are fifteen research catchments within this site, and these are situated at the head of three isolated Little Berg spurs and are underlain by basaltic lavas, which overlie Clarens Sandstone [14,15]. The catchment areas are mainly covered by mesic grasslands of the uKhahlamba Basalt Grassland vegetation type interspersed with Northern Afrotemperate Forest patches and wetlands [16,17]. The fifteen catchments range in altitude from 1820 m.a.s.l. to 2463 m.a.s.l. Topography varies from relatively flat to very steep (1–39°) with the aspect ranging from north to south facing [18].

Three catchments were selected from the fifteen for this study and are named CP-III, CP-VI, and CP-IX (Figure 1). These catchment areas have similar soil properties, but different historic and current land management practices. General details of the three catchments as well as the climatic hydrologic properties during the study period are provided in Table 1. The study period was from September 2019 to June 2021.

Table 1. General details of the three catchment areas during the study period (adapted from [19]).

Catchment Name	Size (ha) and Altitude Range (m.a.s.l.)	Rainfall Dynamics during Study Period	Streamflow Discharge Dynamics during Study Period	Description of Catchment
CP-III	138.9 1847–2323	Mean (mm) 4.34 Max (mm) 65.23 Min (mm) 0.00 Annual PCP 2019—1095 mm 2020—1572 mm 2021—1664 mm	Mean (mm): 2.33 Max (mm): 13.39 Min (mm): 0.22	The catchment is degraded as a result of a forestry experiment in which *Pinus patula* was planted throughout the catchment in the 1950s and 1960s as well as accidental fires which led to the removal of these trees in 1981. The catchment was rehabilitated with *Eragrostis curvula* following the removal of the trees [15]. There is, however, erosion throughout the catchment area, with large portions of the catchment covered by Pteridium sp. (Bracken).
CP-VI	67.7 1844–2073	Mean PCP (mm) 3.62 Max PCP (mm) 65.28 Min (PCP) (mm) 0.00 Annual PCP 2019—829 mm 2020—1261 mm 2021—1472 mm	Mean (mm): 1.84 Max (mm): 18.90 Min (mm): 0.00	This catchment is covered by mesic grassland of the uKhahlamba Basalt Grassland type which is burned biennially during spring. CP-VI is considered the core catchment with focused, detailed monitoring ongoing in this catchment. A full array of evaporation, soil moisture and groundwater monitoring is undertaken.
CP-IX	64.5 1823–1966	Mean PCP (mm) 3.70 Max PCP (mm) 68.34 Min (PCP) (mm) 0.00 Annual PCP 2019—884.94 mm 2020—1274 mm 2021—1378 mm	Mean (mm): 1.28 Max (mm): 11.81 Min (mm): 0.09	This catchment has been completely protected from fire since 1952 but has experienced accidental burns and wildfires in some years. As a result of fire exclusion, this catchment is dominated by woody scrub (*Leucasidea serica* and *Buddleia salvifolia*).

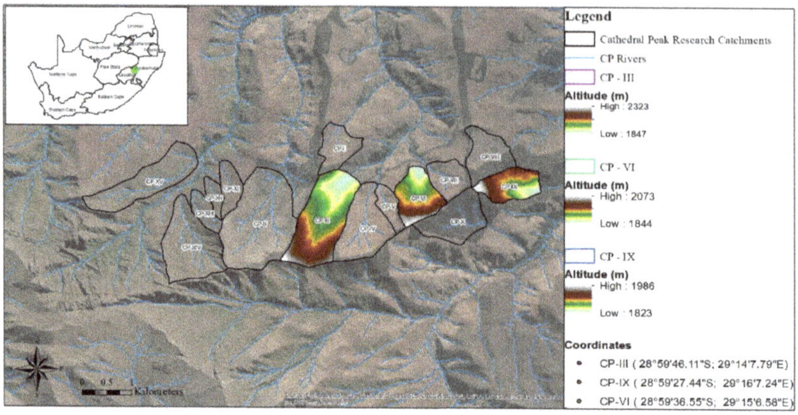

Figure 1. Locality of the catchments selected for the study.

2.2. *Climate and Hydrological Monitoring*

The Cathedral Peak research catchments fall within the summer rainfall region of South Africa. The mean annual precipitation (MAP) for the area is approximately 1400 mm with a gradient of increasing rain between the south-eastern areas (which receive approximately 1300 mm) to the western areas (receive approximately 1700 mm). CP-III has a MAP of 1564 mm, CP-VI has a MAP of 1340 mm, and CP-IX has a MAP of 1257 mm [15]. Rainfall is measured with tipping bucket rain gauges installed in the mid position of each of the catchments. Half of the rainfall events in the catchments are brought about by localised thunderstorms which fall during the spring and summer months (September to March), with occasional snowfall received during winter (May to August). The clouds forming these thunderstorms come from the west of the catchment areas. Orographic rainfall produced from clouds forming in the east of the catchments also creates longer periods of softer rainfall which can fall for several days [14–17]. Mean monthly temperatures range from 17.1 °C to 10 °C with frost common in autumn and winter (April to August) [15,17,18,20].

Streamflow monitoring was initiated in the three catchment areas during the late 1940's and 1950's [15]. At the outlet of each catchment, a concrete weir and stilling hut with 90-degree V Notches were installed. These V Notches are 45.72 cm deep and are surmounted by 1.82 meter-wide rectangular notches of varying depth. Details of how early measurements were taken, error checked and processed are given in [15]. The water stage-height at each weir is currently monitored using an Orpheus Mini (Ott Hydromet GmbH, Kempten, Germany) at CP-VI weir and a CS451 Stainless steel SDI-12 Pressure Transducers with CR200 loggers at weirs CP-III, CP-VI, and CP-IX [15].

Catchment-specific rainfall and streamflow data were therefore utilised for this study period (September 2019 to June 2021). However, in CP-III and CP-IX accidental fires, weir silting, and equipment problems have led to periods of missing streamflow discharge data. In CP-III there is no streamflow discharge data between February and November 2020, while in CP-IX there is no streamflow discharge data in October 2019 as well as between August and November 2020. These periods of missing data were removed from the database.

2.3. *Hydropedology and Soil Mapping*

The development of hydropedology studies in South Africa has led to the classification of hydropedological soil types and how these are distributed down a hillslope catena [21]. A digital soil mapping exercise was undertaken for the three catchment areas utilising these hydropedological soil classifications [22]. The procedure used for the digital soil maps (DSMs) is detailed in [19] and is briefly described here. The soils of the three catchment areas were mapped and classified as per the South African classification system [23] and then regrouped into hydropedological soil types, namely shallow recharge soils, deep recharge soils, interflow soils, and saturated responsive soils. The dominant properties of these soils are provided in Table 2.

Table 2. Dominant properties of the dominant hydropedological soil groups [19].

Hydropedological Soil Group	Characteristics of the Soils
Recharge Shallow	These are soils that are freely drained and do not show any indication of saturation. They are typically shallow in nature (<500 mm). The freely drained B horizon merges with fractured rock or a lithic horizon. These soils typically occur on steeper convex slopes in the higher lying or steeper parts of the catchments.
Recharge Deep	These are soils that are freely drained and do not show any indication of saturation. They are typically deeper than the Recharge Shallow Soils (>500 mm). The freely drained B horizon merges into fractured rock or a lithic horizon. These soils where identified throughout the catchments on gentler convex and concave slopes and away from wetlands and watercourses.
Interflow	These soils have a freely drained upper solum which overlies relatively impermeable bedrock. Hydromorphic properties are identified at this interface and signify periodic saturation associated with a water table. They typically occur on gentler concave slopes in areas delineated as wetlands as well as adjacent to watercourses.
Responsive Saturated	These soils display morphological indications of long-term saturation. They characteristically respond quickly to rainfall events and generate overland flow as they are typically close to saturation during the wet season and therefore any additional precipitation will flow overland due to saturation excess. These soils were identified in the valley bottom positions of the catchments, in permanently saturated wetlands. They typically occur on gentle concave slopes.

The ArcSIE (Soil Inference Engine) version 10.2.105 was used to create the DSMs. A rules-based approach was first utilised based on knowledge of the catchments as well as the outcomes of the creation of Digital Terrain Models (DTMs) with the following environmental control variables applied to the rules: wetness index, slope, elevation, and planform curvature. The rules applied were aimed at producing the optimal relationships between soil type and a particular DTM [24,25]. The initial maps created following the rules-based approach were then validated based on the information gained during soil surveys undertaken within each of the catchment areas. The maps were refined according to the validation points taken during these surveys. The final hydropedological soil group maps are displayed in Figure 2. The performance of the ArcSIE interface to create the combined hydropedological maps for each of the catchments was analysed using the Kappa coefficient of agreement. The Kappa coefficient for CP-III is 0.57, for CP-VI is 0.59, and for CP-IX is 0.74, showing that there are some discrepancies between the hydropedological soil maps created and the site-specific soils identified within the catchment areas.

2.4. Dominant Hydropedological Soil Groups of the Catchments

Comparison of the hydropedological soil group maps revealed that each catchment had a different percentage of the various hydropedological soil groups. This is based on the different topographies of the catchments as well as the various soil characteristics of each hydropedological soil group [19]. Table 3 gives an indication of the dominant hydropedological soil groups in CP-III, CP-VI, and CP-IX.

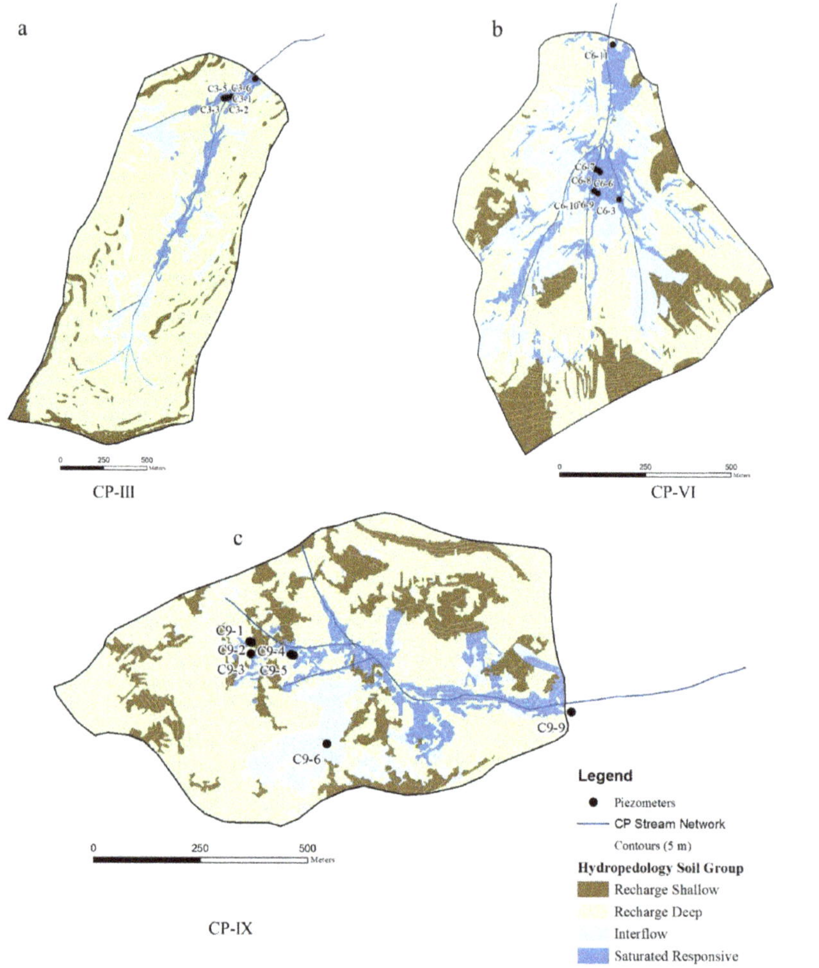

Figure 2. Location of the piezometers in relation to the hydropedological soil group in (**a**) CP-III, (**b**) CP-VI, and (**c**) CP-IX.

Table 3. Percentage of the catchment area covered by each hydropedological soil group.

	CP-III	CP-VI	CP-IX
Hydropedological soil group	Percentage of catchment covered by each hydropedological soil group		
Recharge Shallow	18.3	17.1	27.6
Recharge Deep	43.3	33.8	38.4
Interflow	24.1	28.7	15.9
Responsive Saturated	14.3	20.4	18.1

As shown in Table 3, CP-III, CP-VI, and CP-IX are dominated by the recharge deep hydropedological soil group (43.3, 38.8 and 38.4% of the catchment area, respectively), followed by the interflow soil group in CP-III (24.1%) and CP-VI (28.7%), and the recharge shallow group in CP-IX (27.6%). CP-VI has a greater area classified as responsive saturated soils (20.4%) as compared to CP-III (14.3%) and CP-IX (18.1%).

By utilising the hydropedological soil group maps as well as the dominant groups identified in each catchment, a theoretical interpretation of the various flow paths for each catchment was identified and described.

2.5. Piezometer Installations

Piezometers were installed within the three catchments: six piezometers in CP-III, twelve in CP-VI, and nine in CP-IX. The piezometers were installed in clusters of two or three within a location, with this location chosen to represent the upper, mid, and lower portions of the catchments. Furthermore, the position of the piezometers was chosen within wetland and seepage areas of the catchments. In CP-III, all piezometers were, however, installed in the lower sections of the catchment area as a result of a lack of seepage areas within the upper portions of the catchment. This is due to the shallow nature of soils within the upper reaches of this catchment.

Soil profiles were dug using an extension Dutch auger to refusal with signs of a gleyic or gley horizon noted within the profiles. These horizons display gleying and are considered indicators of the redox state of the soil. Gley horizons are recognised by low chroma grey matrix colours which may contain blue or green tints. The gleyic horizon displays low chroma, grey and light-yellow colours, with the morphology of this horizon indicating less reduction and shorter duration of water saturation compared to the gley horizon [21]. A PVC pipe with slits cut around the end of the pipe to a height of 30 cm were then installed into the auger holes. The diameter of the PVC pipe utilised ensured a close fit with the hole. The piezometers were then capped, and measurements taken once a month between January 2019 to June 2021; however, due to a drought within the region, the majority of piezometers only received water in September 2019 and thus this was chosen as the start point for comparison of water levels.

As a result of the drought conditions, some of the piezometers had to be discontinued, and thus water was sampled and water heights were recorded each month in five piezometers in CP-III, seven piezometers in CP-VI, and seven piezometers in CP-IX (Figure 2). The height of the water within the piezometer was calculated from the surface of the soil to the depth of the water table.

3. Results and Discussion

3.1. Conceptual Response Based on Hydropedological Interpretations

From the hydropedological soil maps created for each catchment coupled with the descriptions of these dominant soil groups, the principal hillslopes and flow paths could be conceptually described. These conceptual descriptions were used as a working hypothesis of the catchments' function. The conceptual descriptions are then evaluated against site-specific measurements.

When precipitation falls in the upper reaches of the three catchment areas, it will enter the hydropedological recharge soil group. The dominant flow direction in recharge soils is the vertical flow of water through and out of the profile into the underlying bedrock. In the three catchment areas, this hydropedological soil group is separated into the recharge shallow soils and the recharge deep soils. Recharge shallow soils occur in the steeper areas of the catchments, and this forms their shallow nature (<500 mm). In this soil group, the freely drained B horizon merges with fractured rock or a lithic horizon. The recharge deep soils are similar to the recharge shallow soils, but the thickness of the profile is far greater (>500 mm). This is largely due to their position within gentler topographical areas of the catchments. Water that moves through these soils would recharge the deeper aquifers associated with the catchment areas, or if it encounters less permeable rock such as non-weathered and compacted sandstone or basaltic outcrops; it will flow laterally, and recharge shallow aquifers associated with seasonal hillslope seepage areas.

Interflow soils located downgradient of the recharge soils are associated with two dominant flow paths. Precipitation would first flow vertically through the free-draining upper profile of these soils before it encounters relatively impermeable bedrock. Hydromorphic properties have developed at this point in the soil profile, signifying periodic saturation associated with a water table. At this soil, bedrock interface water will move laterally into the stream network or downgradient.

The responsive saturated soils are located in the permanently saturated wetlands of the catchment areas. These soils show morphological evidence of long periods of saturation such as a gleyed matrix as well as mottling. They are close to saturation, particularly during the wet season, and once saturated and incapable of attenuating any more water they will generate overland flow to the stream network.

3.2. Precipitation and Streamflow Dynamics

Precipitation data for CP-III, CP-VI, and CP-IX during the study period September 2019 to June 2021 showed that precipitation largely falls within the spring–summer months (September to March) with little to no rain within the autumn and winter months (April to August) (Figure 3). There is a decline in annual rainfall from CP-III to CP-VI to CP-IX. Furthermore, a greater quantity of precipitation was recorded in all three catchments for the spring–summer season of 2020–2021 (CP-III = 1492 mm, CP-VI = 1307 mm and CP-IX = 1045 mm) as compared to the same season within the preceding year (2019–2020) as a result of the drought conditions experienced in 2019 (CP-III = 1150 mm, CP-VI = 842 mm, and CP-IX = 771 mm).

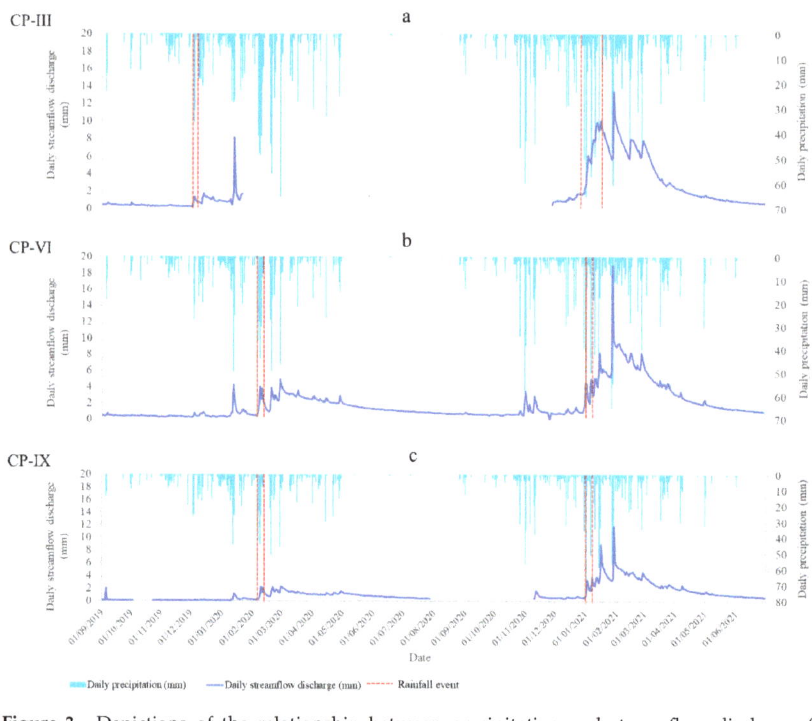

Figure 3. Depictions of the relationship between precipitation and streamflow discharge for: (**a**) CP-III; (**b**) CP-VI; (**c**) CP-IX.

Streamflow discharge values, like the precipitation values, were highest during the spring–summer months, and lowest during the autumn–winter months. Streamflow discharge for the study period also varied between catchment areas, with the greatest values obtained in CP-VI (ranged from 0.0 mm to 18.89 mm), followed by CP-III (ranged from 0.2 mm to 13.39 mm) and CP-IX (ranged from 0.0 mm to 11.81 mm).

The correlation between rainfall and streamflow values is non-linear, particularly during the drier period associated with the drought conditions in 2019 as well as the seasonal variations in the quantity of rainfall received. This is due to a lag time effect in the quantity of precipitation that falls in the catchment and the corresponding rise in streamflow value that is noticeable when comparing daily precipitation and daily streamflow discharge values within all three catchment areas over the study period. As shown in Figure 3, a lag time effect occurs in all three catchments from when a rainfall event occurs to when there is a corresponding increase in streamflow discharge. This lag time differs depending on the pre-rainfall event hydrological conditions of the catchment. For example, following the end of the drought conditions experienced in the catchment areas in 2019, the first large rainfall event took place between 06/02/2020 and 11/02/2020 in which 153.67 mm fell into the CP-VI. Given the largely desiccated conditions of the soils within CP-VI at the time, there is little effect of this rainfall event on the streamflow discharge values during the same time period (streamflow discharge has a combined value of 15.34 mm over the 5 days). No corresponding increase in streamflow discharge takes place during the time of the rainfall event as well as within the following month after the rainfall event. When a similar rainfall event took place in CP-VI but during the wetter season from 01/01/2021 to 06/01/2021, in which 138.94 mm of rain fell, there was a corresponding increase in streamflow discharge approximately 1 month after the event from 29/01/2021 to 04/02/2021 (streamflow discharge has a combined value of 65.34 mm for the time period).

In CP-IX, a similar trend was noticed. Just after the drought of 2019, the same larger rainfall event between 06/02/2020 and 11/02/2020 in which 139.45 mm fell had little effect on the streamflow discharge both at the time of the event and within the following month after the event. Again, the desiccated soils were becoming saturated before they could contribute to the streamflow. During the same rainfall event as in CP-VI, which occurred between 01/01/2021 to 06/01/2021 in which 136.91 mm of rain fell, there was a corresponding increase in the streamflow discharge approximately one month after the event where the streamflow discharge had a combined value of 45.43 mm for the time period from 29/01/2021 to 04/02/2021. Given the wetter time in which the storm event occurred, the soils in the catchment were already partially saturated, thus storm events which occurred during this time could lead to oversaturation of the wetlands and the subsequent creation of overland and shallow subsurface flow which contributed to the increase in streamflow discharge values.

Given the limited streamflow data available for CP-III, obtaining correlation examples between rainfall and streamflow discharge were not possible. However, a similar trend was noted in comparison to CP-VI and CP-IX, particularly in the time after the drought period. Following the drought period, a rainfall event occurred between 03/12/2019 and 06/12/2019 in which 69.08 mm of rain fell. Little effect on the streamflow discharge was observed during the event as well as within the following month after the event. The wetlands in this catchment were, as in CP-VI and CP-IX, becoming saturated again. However, unlike CP-VI and CP-IX, once the wetlands were saturated, the corresponding rise in streamflow discharge values following a rainfall event responded at a much quicker rate. For example, a rainfall event takes place from 28/12/2020 to 15/01/2021. There is an immediate increase in streamflow discharge values both during the event and in the following days after the event, with the streamflow values peaking (13.39 mm) on the 29/01/2021 following a 60.45 mm rainfall event the preceding day.

So, while the hydrological preconditions of the soil groups in all three catchments play a pivotal role in the storage and runoff dynamics of the catchment areas, in CP-III there is a far more immediate response in streamflow discharge following a rainfall event. This could be attributed to the topography of the catchment, the streamflow network or the shallower soils within this catchment, which have largely been created as a result of erosion brought about by the use of the catchment as a *Pinus patula* plantation.

3.3. Piezometer Data and Flow Paths

The water table height within the areas where piezometers were installed varied throughout the study period and was dependent on the depth of the soil profile, the location of the piezometer within the catchment (i.e., the topographical position) as well as seasonal climatic variations. Average depths to the water table showed that, following the end of the drought conditions, the saturated responsive soils became saturated and remained so throughout the study period, while the depth to the water table within the interflow soil group showed greater variation in all three catchment areas. The three catchment areas are explained in more detail in the following sections.

3.3.1. CP-III

In CP-III, the average depth to the water table for the saturated responsive soil group decreased from 530 mm following the end of the drought in September 2019 to 70 mm in November 2019 and remained between 30 mm and 150 mm for the rest of the study period depending on the seasonal variations in the rainfall received. In comparison, the average depth to the water table for the interflow soil group remained at a depth of 1200 mm until January 2020 where it decreased to 768 mm and then increased again to over 1000 mm during the drier period of 2020 (March to September). Following the onset of the spring rains in October 2020, the average depth to the water table decreased to 900 mm where it fluctuated throughout the wetter summer period (between 800 mm and 1100 mm) depending on the rainfall received. With the onset of the drier autumn to winter period from April 2021, the depth to the water table increased again (1100 mm to 1200 mm) (Figure 4).

Figure 4. Comparisons of piezometer data installed in the (**a**) interflow soils, (**b**) saturated responsive soils, (**c**) monthly streamflow discharge and (**d**) monthly rainfall CP-III.

Individual piezometers followed a similar pattern to the average depth to the water table with the piezometers located in the saturated responsive soil group becoming saturated in December 2019 and remaining at or near saturation for the entire study period, depending on the seasonal rainfall received. This saturation level showed that when rainfall was received in the catchment, the wetland systems became oversaturated and contributed more to overland and shallow subsurface flow toward the stream network (Figure 4). With regards to the piezometers installed in the interflow soil group, C3-2 and C3-3 (average water depth is 1051 mm and 985 mm, respectively), which were situated higher in the catchment, received more water compared to C3-1 (average water depth of 1241 mm). C3-1 is situated in close proximity to the stream network. All three piezometer locations are associated with deep water table depths, and this could be attributed to this location contributing more to the baseflows of the stream discharge values and not to overland flow. The more water in C3-2 (average depth of 1051 mm) and C3-3 (average depth of 985 mm) compared to C3-1 (average depth of 1241 mm) shows a down gradient flow path from the upslope recharge soil group through the interflow soils where the piezometers are located, and then laterally into the stream network.

Given the quicker rate at which the streamflow discharge values responded to rainfall events, particularly once the wetland systems were saturated, and taking into account the deep-water table depths of the interflow soils as well as the small size of the wetland systems in which the saturated responsive soil piezometers were located, it is apparent that infiltration of precipitation does not occur on the recharge soils during larger rainfall events but that rather overland or shallow subsurface flow occurs and water reaches the streamflow network at a much quicker rate. This is most likely a result of erosion, particularly from the upper reaches of the catchment, and the resultant shallow nature of these recharge soils and therefore the reduced recharge properties that these soil profiles display. Figure 5 shows a diagram of these flow paths during both the drier and wetter seasons.

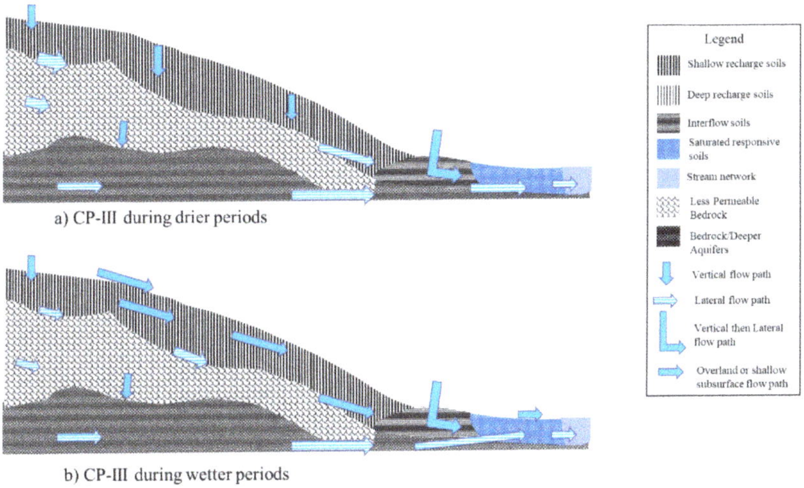

Figure 5. Flow path diagrams for (**a**) the drier periods and (**b**) wetter periods for CP-III.

3.3.2. CP-VI

In this catchment, the average depth to the water table within the piezometers installed in the saturated responsive soils decreased from 720 mm in December 2019 following the onset of rains after the drought period and fluctuated between 200 mm and 18 mm for the remainder of the study period. The average depth to the water table for the interflow soil group fluctuated between 700 mm (December 2019) and 287 mm (December 2020). The average depth to the water table for the interflow soil group furthermore fluctuated depending on the rainfall conditions, with an increase in the depth during the drier autumn to winter months and a decrease in depth in the wetter spring to summer months (Figure 6).

As can be seen in Figure 6, each piezometer had a varied fluctuation in the depth to the water table, with some piezometers remaining more saturated compared to others. In the saturated responsive soil group, the C6-7 piezometer remained more saturated compared to the remaining piezometers in this group, particularly during the drought conditions. The C6-3, C6-9, and C6-10 piezometers which are situated higher in the catchment compared to the C6-7 piezometer became far drier during the drought conditions. With the onset of rains and the end of the drought period, these piezometers became saturated and then fluctuated slightly depending on the seasons and associated rainfall conditions. Once saturated, during rainfall events, the wetlands in which the piezometers were located would become oversaturated and contribute to overland and shallow subsurface flow (Figure 7).

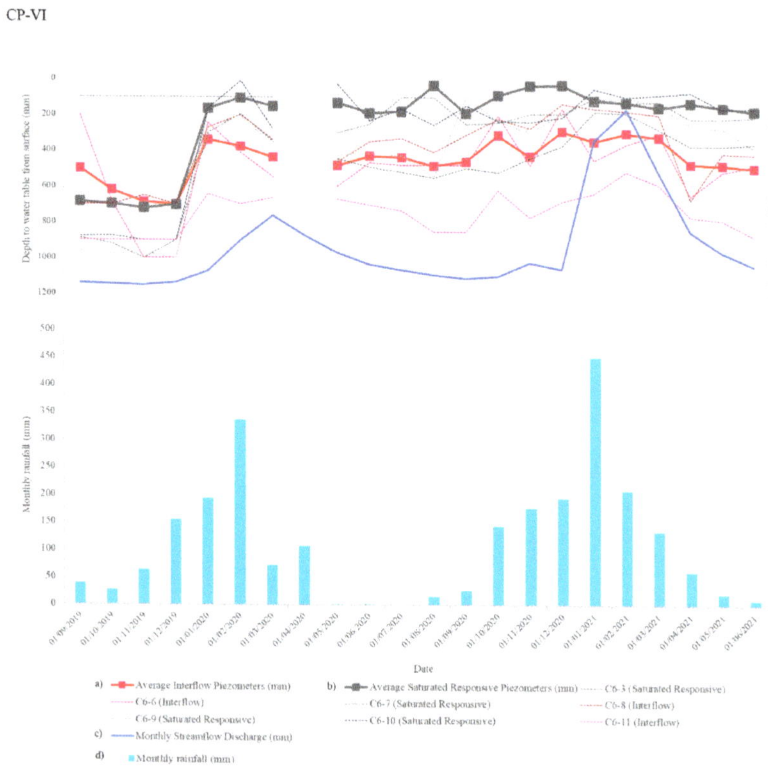

Figure 6. Comparisons of piezometer data installed in the (**a**) interflow soils, (**b**) saturated responsive soils, (**c**) monthly streamflow discharge and (**d**) monthly rainfall in CP-VI.

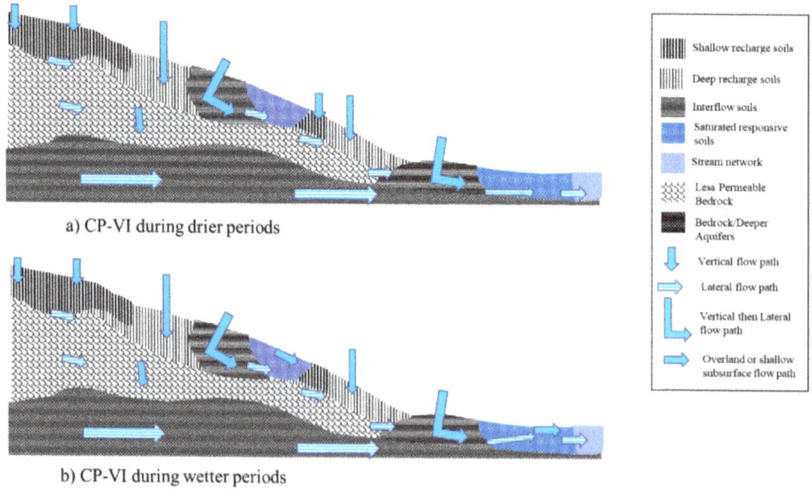

Figure 7. Flow path diagrams for (**a**) the drier periods and (**b**) wetter periods for CP-VI.

Individual piezometers in the interflow soil group also responded differently. Saturation content of the piezometers decreased from C6-8 to C6-6 to C6-11, with the C6-8 piezometer consistently more saturated than the C6-6 and C6-11 piezometers. This piezometer was situated in close proximity but outside of the permanently saturated areas of the C6-7 piezometer. The C6-6 piezometer was furthermore located on the edge of the same wetland system. The C6-11 piezometer was located at the lower end of the catchment, adjacent to the weir and remained drier throughout the study period in comparison to the other interflow piezometers.

These saturation levels of the piezometers show that the interflow soils largely contribute to the baseflow of the streams following a downgradient movement of water from the higher reaches of the catchment before moving laterally into the stream network (Figure 7). During the drought, the wetland system located where the piezometers C6-6, C6-7, and C6-8 (average depth to water table from September 2019 to February 2020 was 436 mm) were installed attenuated more water compared to other wetland systems (average depth to water table from September 2019 to February 2020 was 682 mm) within the catchment. Water moved downgradient from the upper reaches of the catchment and was attenuated within this wetland before moving further downgradient toward the outlet of the catchment area. This movement of water within the drier phase of the study period contributed to the baseflow of the stream network. Once the rains began, the wetland in which the C6-6, C6-7, and C6-8 piezometers were installed became wetter (average depth to water table increased to between 50 mm and 270 mm in January 2020) at a quicker rate than other wetlands within the catchment and started contributing to overland and shallow subsurface flow. The wetlands in which C6-3, C6-9, and C6-10 piezometers were installed became saturated in January/February 2020 (average depth to the water table increased to between 10 mm and 210 mm in February 2020) and then contributed to overland and shallow subsurface flow (Figure 7).

3.3.3. CP-IX

In CP-IX, the average depth to the water table for the saturated responsive soil group decreased from 487 mm to 102 mm following the onset of the rains by January 2020. The average depth to the water table then remained between 200 mm and 75 mm depending on the seasonal variation of rainfall received. The average depth to the water table for the piezometers installed in the interflow soil group also decreased following the onset of rains from 1000 mm in September 2019 to 326 mm in January 2020. The fluctuation of the average depth to the water table then also followed the seasonal variation in the rainfall received, but this variation was more pronounced in comparison to the saturated responsive soil group (depths ranged from 636 mm at the start of spring in October 2020 to 211 mm in February 2021) (Figure 8).

The depth to the water table was different in the individual piezometers. In the saturated responsive soil group, C9-3 remained more saturated even during the drought conditions compared to the other piezometers (water table depth remained at 10 mm until January 2020), followed by C9-4 (water table depth fluctuated between 400 mm and 75 mm until January 2020). C9-5 and C9-9 dried out in comparison and became saturated again in January 2020 with a decrease in water table depth from 750 mm to 140 mm in C9-5 and a decrease from 630 mm to 185 mm in C9-9 (Figure 8).

Piezometers located in the interflow soil group had a greater depth to the water table during the drought conditions, with this depth decreasing following the onset of rains until they reached a peak depth in January and February 2020. The C9-1 and C9-2 piezometers (average water table depth of 330 mm and 460 mm, respectively) which are situated higher up in the catchment remained more saturated compared to C9-6 (average water table depth of 797 mm) which is situated mid catchment.

As was the case in CP-III and CP-VI, the interflow soils contribute more to the lateral flow of water in the sub-horizons of the soil profile (average water table depth ranges from 1000 mm) and the base flow of the streams within the catchment. The saturated responsive soils, which become saturated and remain so, contribute both to the baseflow of the streams and storm flow in the form of overland and shallow subsurface flow once they become saturated. Figure 9 shows a diagram of these flow paths during both the drier and wetter seasons.

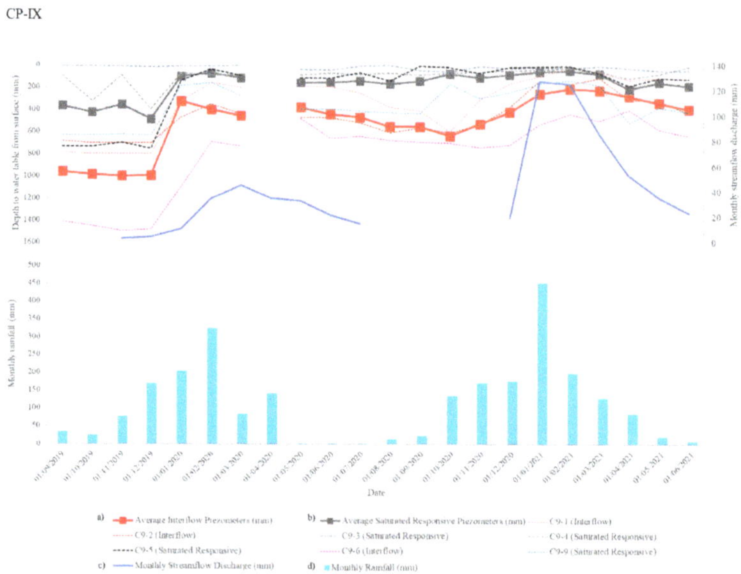

Figure 8. Comparisons of piezometer data installed in the (**a**) interflow soils, (**b**) saturated responsive soils, (**c**) monthly streamflow discharge and (**d**) monthly rainfall in CP-IX.

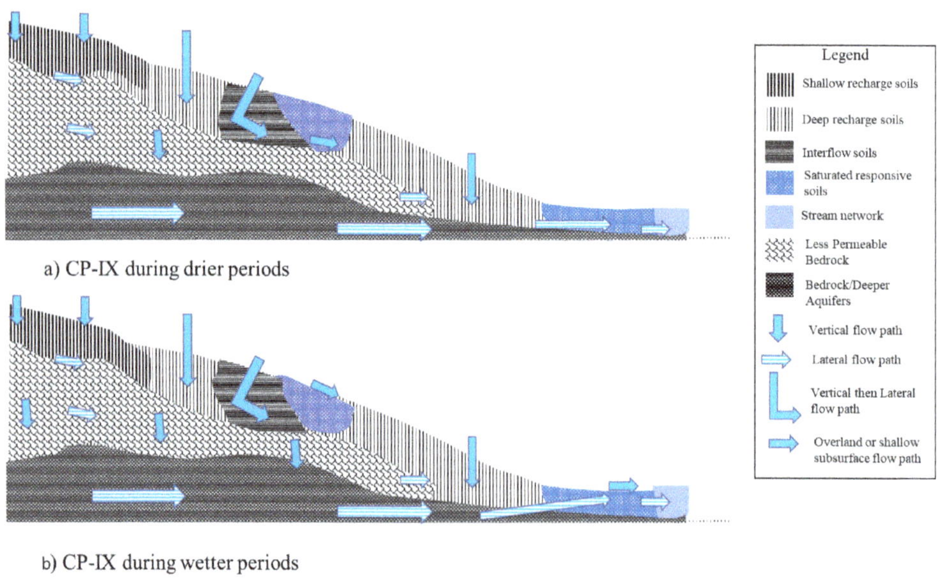

Figure 9. Flow path diagrams for (**a**) the drier periods and (**b**) wetter periods for CP-IX.

3.4. Catchment-Specific Attributes Affecting the Hydropedological Flow Paths

Data obtained from the climatic and hydrologic variables (rainfall and streamflow discharge) as well as the piezometers shows that water moves through the soils of the catchment areas before contributing to the streamflow. The various hydropedological soil groups which affect the flow rates of water before it contributes to streamflow therefore play a pivotal role in the flow dynamics of the catchment areas.

These hydropedological soil groups were mapped following a digital soil mapping process, and while this allowed us to gain a general understanding of the dominant flow paths of these catchment areas, the more detailed investigation of the climatic, hydrologic and water table depth fluctuations have shown that the hydropedological characteristics of the catchment areas are both specific to the catchment and created as a result of various interrelated factors.

The various interactions between the flow dynamics of the hydropedological soil groups of an area and how they become disconnected and reconnected to each other during drying and wetting cycles is unique to the various landscapes in which the flow paths are situated. A number of studies have been conducted in a variety of landscapes [10,26,27]. However, studies conducted in mountainous landscapes highlight the effect of the lower reaches of catchment areas continuously receiving water from the steeper surrounding hillslopes and these flow dynamics contributing to the baseflow of streams. During the wetter periods, the connectivity between the various hydropedological soil groups becomes more established and this allows for the generation of greater overland and shallow subsurface fluxes of water, particularly during larger rainfall events. These flows contribute to peaks within the streamflow hydrographs during storm events [4,28,29].

The connectivity between the hydropedological soil groups is furthermore influenced by the topography of the mountain catchments. In CP-VI for example, the wetland in which the C6-6, C6-7, and C6-8 piezometers were installed and that remained more saturated when compared to other wetland systems in the catchment is situated in an area with a gentler slope as well as a concave landform. Various studies have shown similar findings with catchment areas that have gentler terrain resulting in poorer drainage conditions and therefore the storage of higher volumes of water. Areas with steeper terrain increase the hydraulic gradients of the soils thus increasing the flow between the different hydropedological soil groups and reducing the storage capacity of these soils [1,29]. This influence of topography on the flow dynamics should be studied further within these research catchments.

A further effect on the hydropedological characteristics of the catchment areas is both the historic and current land management practices. The hydropedological dynamics of a site in a certain time is not the result of present processes and events but is related to and strongly influenced by the land use management history as well as the natural plant succession patterns [30].

In CP-III, the historic use of the area as a *Pinus patula* plantation and the subsequent lack of rehabilitation has led to a decrease in basal cover and erosion, particularly in the upper reaches of this catchment. This erosion has created shallower soils and therefore reduced the storage capacity and infiltration rates of the recharge hydropedological soil group. This reduced infiltration capacity has likely changed the flow dynamics of the catchment compared to what would have been historically present, and this is evident in the quicker response of the streamflow discharge values following a rainfall event during the wetter periods of this study. The impact of erosion on the hydropedological characteristics of a catchment area has been highlighted in other studies, with [31] identifying that soils with a degraded structure tend to have increased bulk density and consequently a decrease in soil porosity. This impacts the water movement in the soil profile, having knock-on effects at the landscape scale. Ref. [12] showed that rainfall infiltration into shallower soils will reach the bedrock interface quickly and flow along this interface as preferential flow. The slope runoff from these areas will therefore appear to occur as subsurface stormflow (i.e., similar to overland flow) but occurring at the shallow soil/bedrock interface. In areas of thicker soil profiles, as is the case in CP-VI and CP-IX, rainfall infiltration into the deeper recharge soils will supply the stream network largely as shallow groundwater and contribute more to the baseflow.

The erosion of the upper reaches has also led to deeper deposition areas within the lower reaches of CP-III. It is within these deposition areas that the interflow soil group piezometers were installed. When comparing the depth to the water table in all interflow soil group piezometers from the three catchment areas, CP-III has the deepest water table depths. This is due to the burying of the original soil profile by sediment which has been eroded from the top of the catchment; this has implications for the flow dynamics of this area of the catchment. Refs. [32,33] showed that in areas of deposition, soil particles have been mixed, causing changes in the pore structure of the soil matrix resulting in pore clogging and the reduction in the soil hydraulic capabilities. Thus, in these areas of CP-III, there is likely to be a slow reconnection of the subsurface flow paths following dry periods and these flow paths reconnecting to the stream network. This is demonstrated by the fact that the interflow soil group piezometers located in the depositional areas did not have a substantial increase in water table depth throughout the study period in comparison to the interflow soil group piezometers in CP-VI and CP-IX. They are thus likely to contribute slowly to the baseflow of the stream network and not to the stormflow peaks of the hydrograph of this catchment even following large rainfall events.

CP-VI is managed as a mesic grassland interspersed with wetland systems, while the fire exclusion since 1952 in CP-IX has led to this catchment becoming a woody dominated area. When comparing CP-VI and CP-IX, fluctuations in the average water table depths of the saturated responsive soil groups show that in both catchments the wetland areas dried out to an extent during the drought period, and these became re-saturated in January/February 2020 and then remained saturated throughout the study period. The wetlands in both catchments contributed to shallow sub-surface flow and at times overland flow depending on rainfall conditions. The average water table depth of the interflow soil group also followed similar patterns when comparing the two catchment areas. The effect of plant cover on the hydropedological characteristics of catchment areas has been reported in different environmental settings [11,30,34–36], with these studies showing that woody cover areas have greater infiltration rates compared to pasture areas and that tree canopies can reduce the interception of rainfall within catchment areas, influencing infiltration rates within soils [36]. These studies were conducted in commercial plantations, fallow pastures, and old wood forests. The results of this study suggest that the flow dynamics of each catchment area are not a product of land cover but a factor of a combination of interrelated components.

The pivotal role that the wetland systems play in the streamflow dynamics of the catchment areas has been highlighted in this study. The drying and wetting cycles of individual wetland systems as well as specific saturation zones of these wetlands influenced both the baseflow connectivity and the overland flow during wetter periods. Ref. [37] identified similar findings utilising remote sensing techniques to show how a wetland system has different internal saturation compartments and how these both differ in saturation content depending on the climatic conditions and in providing lateral flow and overland flow to downgradient environments. Ref. [29] furthermore utilised isotopes to show that baseflow within the stream network consists predominantly of pre-event water (or dryer cycles) with larger rainfall events (particularly during the wetter cycles) displacing this water within the wetland systems and moving it as overland flow to the stream networks. The contribution of wetland systems to the stream network is therefore a heterogeneous and complex interaction of the soil physical properties, the climatic conditions, and the land management of an area. This has an impact on areas within the catchments classified as saturated responsive soils as these areas do not always contribute to overland flow, but rather the timing of their contribution to both baseflow and overland flow is specific to the wetland system, its location within the catchment, and climatic variables. Future isotopic studies within the Cathedral Peak research catchments are recommended to help gain a deeper understanding of flow dynamics from the wetland systems to the stream networks.

4. Conclusions

This study has highlighted the effects of climate, hydrologic conditions, land management and soil properties on the hydropedological characteristics of three montane catchment areas. The results suggest that a number of factors which are interrelated play a key role in determining the flow paths and the connection between flow paths in these areas. These factors are dominated by antecedent soil moisture, rainfall intensity, the duration of dry and wet periods as well as the depth of soil profiles.

The conceptual interpretation of the hydropedological flow paths of each catchment area following the creation of the digital soil maps provided a general understanding of the flow paths and storage areas of these watersheds. However, utilising catchment-specific climate and streamflow data coupled with water table depth measurements as well as an understanding of how historic and current land management practices have influenced the soil properties, we were able to gain a more accurate interpretation of the response of each hydropedological soil group following a rainfall event. The dominant role of wetland systems and how these have drying and wetting cycles (the average water table depth ranged from 520 mm to 20 mm in CP-III, from 720 mm to 28 mm in CP-VI and from 487 mm to 51 mm in CP-IX) are the key focus in understanding the connectivity between the hydropedological flow paths and the contribution of soil water to the stream networks of the three catchments.

Given the importance of small mountain watersheds in maintaining water supplies to downgradient systems, the understanding of how streamflow is generated and maintained in these headwater catchments is of importance. This is particularly so in understanding the importance of the water storage capacity and water flux rate of the wetlands of the catchments in creating a buffering capacity against hydroclimatic variability which is becoming an ever-increasing reality [33]. The health of the wetland systems in storing water during droughts and their capacity to become saturated quickly and then contribute to the stream network is an important consideration in the ecological services these mountain headwater catchments provide.

Author Contributions: Conceptualization, R.H. and J.v.T.; methodology, R.H. and J.v.T.; formal analysis, R.H. and J.v.T.; funding acquisition, J.v.T.; resources, J.v.T. and P.A.S.; supervision, J.v.T. and P.A.S.; writing—original draft preparation, R.H.; writing—review and editing, R.H., J.v.T. and P.A.S.; visualization, R.H., J.v.T. and P.A.S.; validation, P.A.S. All authors have read and agreed to the published version of the manuscript.

Funding: This research was funded by the IPHAKADE EARTH STEWARDSHIP SCIENCE BURSARY PROGRAMME.

Data Availability Statement: Not applicable.

Acknowledgments: The authors wish to thank the South African Environmental Observation Network (SAEON) for all the logistical support in gathering the data for this study.

Conflicts of Interest: The authors declare no conflict of interest.

References

1. Lazo, P.; Mosquera, G.; McDonnell, J.; Crespo, P. The role of vegetation, soils, and precipitation on water storage and hydrological services in Andean Páramo catchments. *J. Hydrol.* **2019**, *572*, 805–819. [CrossRef]
2. Lin, H.; Brooks, E.; Mcdaniel, P.; Boll, J. Hydropedology and Surface/Subsurface Runoff Processes. In *Encyclopedia of Hydrological Sciences*; John Wiley & Sons, Ltd.: New York, NY, USA, 2008. [CrossRef]
3. Geris, J.; Tetzlaff, D. Resistance and resilience to droughts: Hydropedological controls on catchment storage and run-off response. *Hydrol. Process.* **2015**, *29*, 4579–4593. [CrossRef]
4. Pinto, L.; de Mello, C.; Darrell, N.; Poggere, G.; Owens, P.; Curi, N. A hydropedological approach to a mountainous Clayey Humic Dystrudept. *Sci. Agric.* **2018**, *75*, 60–69. [CrossRef]
5. Lin, H. *Hydropedology Synergistic Integration of Soil Science and Hydrology*, 1st ed.; Academic Press: Cambridge, MA, USA, 2012.

6. Jarvis, N.J. A review of non-equilibrium water flow and solute transport in soil macropores: Principles, controlling factors and consequences for water quality. *Eur. J. Soil Sci.* **2007**, *58*, 523–546. [CrossRef]
7. Juez, C.; Nadal-Romero, E. Long-term time-scale bonds between discharge regime and catchment specific landscape traits in the Spanish Pyrenees. *Environ. Res.* **2020**, *191*, 110158. [CrossRef]
8. Zhu, Q.; Castellano, M.; Yang, G. Coupling soil water processes and nitrogen cycle across spatial scales: Potentials, bottlenecks and solutions. *Earth. Sci. Rev.* **2018**, *187*, 248–258. [CrossRef]
9. Lin, H. Temporal stability of soil moisture spatial pattern and subsurface preferential flow pathways in the Shale Hills Catchment. *Vadose Zone J.* **2006**, *5*, 317–340. [CrossRef]
10. Novak, P. Soil Genetics in Hydropedological Survey in Mountain Areas. *Soil Sci. Annu.* **1986**, *37*, 215–223.
11. Zuecco, G.; Rinderer, M.; Penna, D.; Borga, M.; van Meerveld, H.J. Quantification of subsurface hydrologic connectivity in four headwater catchments using graph theory. *Sci. Total Environ.* **2019**, *646*, 1265–1280. [CrossRef]
12. Fu, Z.; Li, Z.; Cai, C.; Shi, Z.-H.; Xu, Q.; Wang, X. Soil thickness effect on hydrological and erosion characteristics under sloping lands: A hydropedological perspective. *Geoderma* **2011**, *167*, 41–53. [CrossRef]
13. Juez, C.; Nadal-Romero, E.; Cammeraat, E.; Regüés, D. Spatial and temporal variability of water table dynamics in an afforested catchment of the Central Spanish Pyrenees. *Hydrol. Process.* **2021**, *35*, e14311. [CrossRef]
14. Nanni, U.W. Forest Hydrological Research at the Cathedral Peak Research Station. *J. South Afr. For. Assoc.* **1956**, *27*, 2–35. [CrossRef]
15. Toucher, M.L.; Clulow, A.; van Rensburg, S.; Morris, F.; Gray, B.; Majozi, S.; Everson, C.E.; Jewitt, G.P.W.; Taylor, M.A.; Mfeka, S.; et al. *Establishment of a more Robust Observation Network to Improve Understanding of Global Change in the Sensitive and Critical Water Supply Area of the Drakensberg*; WRC Report No. 2236/1/16; Water Research Commission: Pretoria, South Africa, 2016.
16. Mucina, L.; Rutherford, M.C.; Powrie, L.W. *Vegetation Map of South Africa, Lesotho and Swaziland*, 2nd ed.; South African National Biodiversity Institute: Pretoria, South Africa, 2017.
17. Bosch, J.M. Treatment effects on annual and dry period streamflow at Cathedral Peak. *S. Afr. For. J.* **1979**, *108*, 29–38. [CrossRef]
18. Gordijn, P.J.; Everson, T.M.; O'Connor, T.G. Resistance of Drakensberg grasslands to compositional change depends on the influence of fire-return interval and grassland structure on richness and spatial turnover. *Perspect. Plant Ecol. Evol.* **2018**, *34*, 26–36. [CrossRef]
19. Harrison, R.; van Tol, J. Digital Soil Mapping for Hydropedological Purposes of the Cathedral Peak Research Catchments, South Africa. In *Remote Sensing of African Mountains*, 1st ed.; Adelabu, S., Ramoelo, A., Olusola, A., Adagbasa, E., Eds.; Springer: Cham, Switzerland, 2022. [CrossRef]
20. Everson, C.E.; Molefe, G.L.; Everson, T.M. *Monitoring and Modelling Components of the Water Balance in a Grassland Catchment in the Summer Rainfall Area of South Africa*; WRC Report No. 493/1/98.1998; Water Research Commission: Pretoria, South Africa, 1998.
21. van Tol, J. Hydropedology in South Africa: Advances, applications and research opportunities. *S. Afr. J. Plant Soil.* **2020**, *37*, 23–33. [CrossRef]
22. van Tol, J.J.; Roux, P.A.L. Hydropedological grouping of South African soil forms. *S. Afr. J. Plant Soil.* **2019**, *36*, 233–235. [CrossRef]
23. Soil Classification Working Group. *Soil Classification: A Natural and Anthropogenic System for South Africa*; ARC-Institute for Soil, Climate and Water: Pretoria, South Africa, 2018.
24. de Menezes, M.D.; Silva, S.H.G.; Owens, P.R.; Curi, N. Solum depth spatial prediction comparing conventional with knowledge-based digital soil mapping approaches. *Sci. Agric.* **2014**, *71*, 4. [CrossRef]
25. Zhu, A.X.; Band, L.; Vertessy, R.; Dutton, B. Derivation of soil properties using a soil land inference model (SoLIM). *Soil. Sci. Soc. Am. J.* **1997**, *61*, 523–533. [CrossRef]
26. D'Amore, D.V.; Fellman, J.B.; Edwards, R.T.; Hood, E.W.; Ping, C.L. Hydropedology of the North American coastal temperate rainforest. In *Hydropedology: Synergistic Integration of Soil Science and Hydrology*; Lin, H., Ed.; Academic Press: Cambridge, MA, USA, 2012.
27. MacEwan, R.; Dahlhaus, P.; Fawcett, J. Hydropedology, Geomorphology, and Groundwater Processes in Land Degradation: Case Studies in South West Victoria, Australia. In *Hydropedology: Synergistic integration of Soil Science and Hydrology*; Lin, H., Ed.; Academic Press: Cambridge, MA, USA, 2012; pp. 449–481. [CrossRef]
28. van Tol, J.J.; Le Roux, P.; Hensley, M.; Lorentz, S. Soil as indicator of hillslope hydrological behaviour in the Weatherley Catchment, Eastern Cape, South Africa. *Water SA* **2010**, *36*, 5. [CrossRef]
29. Tetzlaff, D.; Birkel, C.; Dick, J.; Geris, J.; Soulsby, C. Storage dynamics in hydropedological units control hillslope connectivity, runoff generation, and the evolution of catchment transit time distributions. *Water Resour. Res.* **2014**, *50*, 969–985. [CrossRef]
30. Gómez-Tagle, A. Linking hydropedology and ecosystem services: Differential controls of surface field saturated hydraulic conductivity in a volcanic setting in central Mexico. *Hydrol. Earth. Syst. Sci. Discuss.* **2009**, *6*, 2499–2536. [CrossRef]
31. Pinto, L.C.; de Mello, C.R.; Owens, P.R.; Norton, L.D.; Curi, N. Role of inceptisols in the hydrology of mountainous catchments in Southeastern Brazil. *J. Hydrol. Eng.* **2016**, *75*, 60–69. [CrossRef]
32. Dikinya, O.; Hinz, C.; Aylmore, G. Dispersion and re-deposition of fine particles and their effects on saturated hydraulic conductivity. *Aust. J. Soil Res.* **2006**, *44*, 47–56. [CrossRef]
33. Blumstock, M.; Tetzlaff, D.; Dick, J.J.; Nuetzmann, G.; Soulsby, C. Spatial organization of groundwater dynamics and streamflow response from different hydropedological units in a montane catchment. *Hydrol. Process.* **2016**, *30*, 3735–3753. [CrossRef]

34. Zimmermann, B.; Elsenbeer, H.; De Moraes, J.M. The influence of land-use changes on soil hydraulic properties: Implications for runoff generation. *For. Ecol. Manag.* **2006**, *222*, 29–38. [CrossRef]
35. Zou, C.; Barron-Gafford, G.; Breshears, D. Effects of Topography and Woody Plant Canopy Cover on Near-Ground Solar Radiation: Relevant Energy Inputs for Ecohydrology and Hydropedology. *Geophys. Res. Lett.* **2007**, *34*, 24. [CrossRef]
36. Granados, M.E.; Vilagrosa, A.; Chirino, E.; Vallejo, R. Reforestation with resprouter species to increase diversity and resilience in Mediterranean pine forests. *For. Ecol. Manag.* **2016**, *362*, 231–240. [CrossRef]
37. Furlan, L.; Rosolen, V.; Sales, J.; Moreira, C.; Ferreira, M.; Bueno, G.; Coelho, C.; Mounier, S. Natural superficial water-storage and aquifer recharge assessment in Brazilian savanna wetland using unmanned aerial vehicle and geophysical survey. *J. Unmanned Veh. Syst.* **2020**, *8*, 224–244. [CrossRef]

water

MDPI

Article

Effects of Straw Mulching on Near-Surface Hydrological Process and Soil Loss in Slope Farmland of Red Soil

Luyang Yang [1,2,3,4], Jian Duan [3,4], Lang Peng [3,4], Xinyin Zhang [3,4], Xiaomin Guo [1,2,*] and Jie Yang [3,4,*]

1 Key Laboratory of National Forestry and Grassland Administration on Forest Ecosystem Protection and Restoration of Poyang Lake Watershed, Nanchang 330045, China
2 College of Forestry, Jiangxi Agricultural University, Nanchang 330045, China
3 Jiangxi Academy of Water Science and Engineering, Nanchang 330029, China
4 Jiangxi Provincial Key Laboratory of Soil Erosion and Prevention, Nanchang 330029, China
* Correspondence: gxmjxau@163.com (X.G.); zljyj@126.com (J.Y.)

Abstract: Slope farmland is prone to soil erosion, especially in sub/tropical regions. However, our understanding of near-surface hydrology characteristics and their controlled factors in red soil sloping farmland remains limited. Here, we conducted simulated rainfall experiments to assess the impact of rainfall pattern, straw mulching, and soil structure on near-surface hydrological processes of red soil sloping farmland of southern China. Results showed that: (1) short duration-high intensity rain caused greater surface runoff and sediment production than did long duration-low intensity rain, whereas the variation pattern of subsurface flow exhibited the opposite trend; (2) tillage behavior could weaken the surface runoff intensity and promote the development of subsurface flow; (3) straw mulching increased the water infiltration rate and associated subsurface flow production (increased by 1.33~12.71 times), and thus reduced the surface runoff production (reduced by 99.68~100%). These findings highlight the crucial roles of rainfall pattern and straw mulching in regulating the spatial distribution pattern of rainwater and suggest that straw mulching can effectively reduce soil erosion via accelerating water infiltration and subsurface flow form in slope farmland of soil erosion in southern China.

Keywords: slope farmland; simulated rainfall; subsurface flow; soil structure; rape straw mulching

Citation: Yang, L.; Duan, J.; Peng, L.; Zhang, X.; Guo, X.; Yang, J. Effects of Straw Mulching on Near-Surface Hydrological Process and Soil Loss in Slope Farmland of Red Soil. *Water* **2022**, *14*, 3388. https://doi.org/10.3390/w14213388

Academic Editors: Songhao Shang, Qianqian Zhang, Dongqin Yin, Hamza Gabriel and Magdy Mohssen

Received: 26 September 2022
Accepted: 23 October 2022
Published: 26 October 2022

1. Introduction

Cultivated land is a scarce resource for the survival of people in various developing countries, including China [1,2]. As an important part of cultivated land, slope farmland has attracted more and more attention. However, due to unreasonable land use, the soil and water loss of slope farmland has been aggravated, making slope farmland become an important source of soil and water loss in many regions [3–5], which has become an ecological environment and agricultural resource problem that has attracted much attention worldwide. To ensure the normal life needs of people in hilly areas of developing countries, it is urgent to control soil erosion in sloping farmland.

Soil structure is an important factor affecting the near-surface hydrology of sloping farmland. Some scholars had preliminarily discussed the influence of soil configuration in black soil [6], purple soil [7,8] and karst topography [9] on near-surface hydrology. For example, in the report of Zhang [9], the surface runoff in the karst area after rainfall was less, and it was usually characterized by full runoff. Subsurface flow and deep leakage are important hydrological processes in the region. Zhang [6] found that in the black soil region, the soil layer was thick, the vertical stratification was obvious, the upper layer was loose, the lower layer was dense, and the infiltration performance was poor. The special soil configuration was easy to form the upper stagnant water. Long [7] found that the purple soil area had steep slope, thin soil layer, high gravel content, large porosity and strong infiltration capacity. Subsurface flow occurs in the middle and late stages of rainfall after soil saturation, and the runoff yield was large. However, the response of near-surface hydrology to rainfall caused by special tillage soil structure was still unclear. Long-term tillage of red soil slope farmland makes the surface soil loose and permeable. The bottom soil of the tillage layer was compacted by tillage equipment, which increased soil density and thus reduced soil permeability. Porosity, especially the content of macropores, is low, and the permeability is significantly lower than that of the loose plow layer, forming a typical soil structure of "upper loose-lower tight" [10]. Under sufficient precipitation and special soil configuration, red soil slope farmland makes the development of subsurface flow more active. In some extreme rainfall events, the outflow of subsurface flow even exceeds the surface runoff, which dominates the rainfall-flow process on the slope [11]. Surface runoff and subsurface flow of sloping farmland are important components of watershed runoff, and directly affect the hydrological process of the whole watershed [12,13]. Both of them play an equally important role in soil erosion.

As one of the measures to control soil erosion, straw mulching cannot only improve soil properties but also play an active role in reducing the surface erosion of sloping farmland [14–16]. Some scholars had preliminarily discussed the hydrodynamic process of straw mulch. Compared with bare land, straw mulching can increase surface roughness [17], significantly reduce flow velocity [18,19], reduce runoff [20,21], and reduce sediment. For example, Prosdocimi [21] reported a significant decrease in soil erosion rates due to reduced surface runoff due to straw mulching. Straw can affect the process of surface runoff and sediment production by increasing the roughness of the surface. Mulumba [22] and Guo [23] showed that straw mulching could effectively reduce surface temperature and water evaporation. Straw mulching can reduce soil bulk density and significantly promote soil water use efficiency [24]. In addition, straw mulching can also protect the soil surface from the splashing effect caused by raindrops, prevent the crusting effect of the soil surface during rainfall, and reduce the surface runoff and the amount of sediment covered by runoff [25,26]. However, few studies have considered the interaction between straw mulching on surface runoff and subsurface flow.

The southern red soil region is an important food production area in China, and it is also one of the most densely populated areas. As an important cultivated land resource in the region, slope farmland accounts for 26.3% of the cultivated land area there [27]. However, due to the unique natural conditions (such as subtropical monsoon rainfall patterns and landforms, soil parent materials, etc.) and social factors, this region has become one of the regions with serious soil and water loss in China, and agricultural development is greatly restricted. In recent years the region has used a series of means to solve the problem of soil and water loss, and has achieved remarkable phased results. However, there is still insufficient research on the response of near-surface hydrology to rainfall under straw mulching on sloping farmland in this region. Because of this, the purpose of this study is (1) to explore the influence of different rainfall patterns on near-surface hydrology and soil erosion; (2) to explore the influence of slope farmland soil structure on near-surface hydrology and soil erosion; and (3) to explore the impacts of straw mulching on near-surface hydrology and soil erosion for different rainfall patterns.

2. Materials and Methods

2.1. Study Site

The experiment was performed in the simulated rainfall hall of Jiangxi Soil and Water Conservation Ecological Science and Technology Park (115°43′42″, 29°16′55″) from July to October 2020. The science and technology park is located in the south of Dean County, Jiujiang City, Jiangxi Province. The study site is located in the small watershed of Yangou and the west bank of Boyang River in the Poyang Lake system (Figure 1). This site belongs to the subtropical monsoon climate zone with abundant rainfall. The average annual rainfall is 865.6~1807.7 mm, accounting for nearly 50% of the annual rainfall from June to October. The annual average temperature is 16.7 °C, annual sunshine hours range from 650~2100 h, the frost-free period 245~260 d, the soil parent material is Quaternary red clay, and the zonal vegetation is subtropical evergreen broad-leaved forest. The landform in the region is shallow hilly land with an elevation of 30~100 m and a slope of less than 25°. The science and technology park is located in the red soil center of China, and the terrain and soil conditions are typically representative.

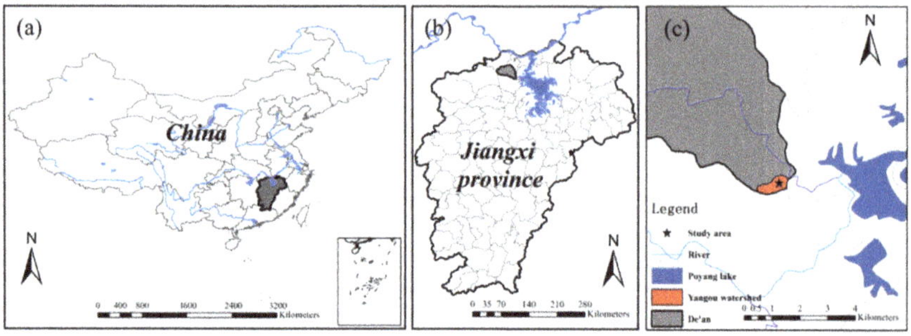

Figure 1. Location map of test area (**a**) Location of Jiangxi Province in China; (**b**) Location of De'an County in Jiangxi Province; (**c**) The location relationship between the test site and Yangou watershed, De'an County and Poyang Lake.

2.2. Test Materials

In this experiment, we used the steel soil tank with 1.5 m × 0.5 m × 0.5 m (length × width × height) to collect surface runoff, subsurface flow, and deep leakage in soils (Figure 2). According to the field investigation, the test slope was treated as 10° [28,29], representing the slope of most red soil slope farmland. The filling soil was a typically quaternary red soil collected from the study area (Table 1). The soil was naturally dried and screened to remove plant roots and gravel for standby. A layer of 5 cm thick graded gravel was filled at the bottom of the soil tank, and a layer of geotextile was covered on the gravel to prevent the loss of the upper soil and facilitate water permeability.

Table 1. Basic physical and chemical properties of soils.

Sand/(%)	Silt/(%)	Clay/(%)	pH	Organic Matter/(g kg^{-1})
13.700	53.055	33.245	5.533	13.24

Note: Sand (2.0 mm ≥ D > 0.02 mm). Silt (0.02 mm ≥ D > 0.002 mm). Clay (D < 0.002 mm).

In this study, the soils with the same bulk density were regarded as homogeneous soils (HoS) and filled uniformly to 5~45 cm of the soil tank. However, heterogeneous soils (HeS) were composed of two soil layers with different bulk densities. The details are exhibited in Table 2. For each treatment, soil was filled into the soil tank every 5 cm. And breaking the soil surface before filling the next layer. The soil moisture content of the soil tank test was controlled at 10 ± 1%, and the test was carried out after the completion of the soil tank filling.

Table 2. Backfilling bulk density of soil tank.

Soil Depth/(cm)	HoS/(g cm^{-3})	HeS/(g cm^{-3})	HeS + SM/(g cm^{-3})
0~15	1.10	1.10	1.10
15~40	1.10	1.45	1.45

Note: HoS, Homogeneous soil. HeS, Heterogeneous soil. HeS + SM, Heterogeneous soil + Straw mulch.

Oilseed Rape (*Brassica napus* L.), as a common oil crop in the south of China, is widely planted in red soil sloping fields. Rapeseed leaves a large amount of straw after harvesting, which is an excellent material for straw mulching. The rape straw was taken from the rape planted in the slope farmland of the study area, and naturally dried after mature harvest. Before the experiment, the rape straw was crushed to 5~10 cm for covering. The covering amount was 10 t hm^{-2}, and the soil tank area in this experiment was 0.75 m^2. Straw laying thickness was about 3–5 cm.

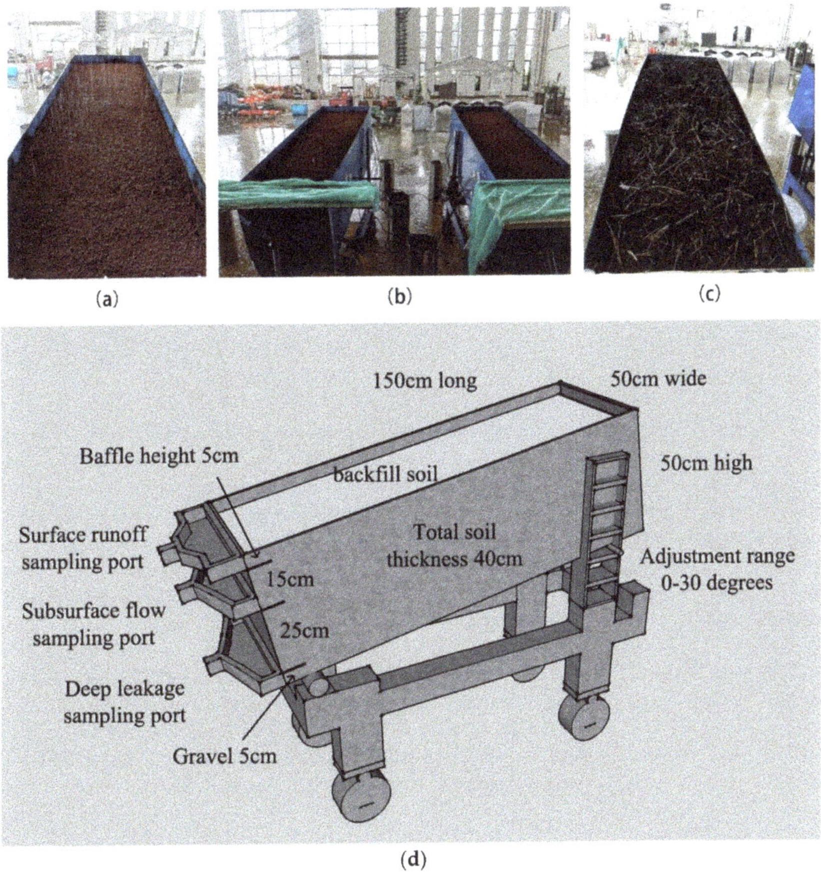

Figure 2. Experimental demonstration and soil tank schematic. (**a**) Uncovered soil tank. (**b**) Soil tanks during rainfall experiment. (**c**) Straw–covered soil tank. (**d**) Schematic diagram of steel soil tank.

2.3. Test Method

To simulate natural precipitation, this work was conducted in a rainfall hall, in which effective rainfall area, rainfall height, and rainfall evenness were 786 m^2, 18 m, and 0.85, respectively. The whole rainfall hall was divided into four independent rainfall test areas, of which areas 1, 2 and 3 were the lower spraying area, and area 4 was the side spraying area. The whole process of this test was completed in zone 2, which adopts FULLJET rotary down jet nozzle. The variable range of rainfall intensity in the spraying area is 10~200 mm h^{-1}.

Under the condition of consistent total rainfall control, two typical rainfall patterns in the southern red soil region were selected for the simulation test [30]. Rainfall pattern A was designed to simulate the rainfall with short duration-high intensity rain in summer, which lasts 1.5 h, and the rainfall intensity was 90 mm h^{-1}. Rainfall pattern B was designed to simulate the rainfall with long duration-low intensity rain in spring, which lasted for 4.5 h and the rainfall intensity was 30 mm h^{-1}. The designed rainfall field of this test was 3 (treatment) $\times$ 3 (repeat) $\times$ 2 (rain pattern) = 18. In the actual process of the test, there were 18 rainfalls, including 0 invalid rainfalls and 18 effective rainfalls. The test results were taken as the average of three repetitions.

In this experiment, water samples were collected every three minutes from the beginning of the rainfall and measured until the end of surface runoff. Among them, due to the mixed sediment in the surface runoff sample, the samples were weighed and filtered with a fast filter paper. The paper-wrapped sediment was put into the oven, the temperature of the oven was set to 105 °C, and the sediment was baked for more than 8 h and read after drying.

2.4. Analysis Method

All statistical analyses were performed with R version 4.1.1 [31], and the accepted significance level was set at $\alpha = 0.05$. A one-way ANOVA of Tukey's HSD comparison was used to determine the differences in the runoff, sediment yield, and subsurface flow among different treatments with the 'agricolae' package. Excel 2019 and Origin pro 2022 were used to process data and draw charts.

3. Results

3.1. Surface Runoff Generation Characteristics

Analysis of surface runoff start time and flow under straw mulching on red soil slope farmland is shown in Table 3. It can be seen that the surface runoff generation time of HoS (homogeneous soil) and HeS (heterogeneous soil) was similar under long duration-low intensity rain. The runoff generation time was ranked as HoS < HeS < HeS + SM from early to late. HoS and HeS began to produce runoff after 157.00 min and 160.42 min, respectively, and the HeS + SM with straw mulch did not produce surface runoff. The total surface runoff yields of HoS and HeS were 15.72 L and 9.83 L, respectively. The surface runoff of HoS was 1.6 times higher than HeS. The peak flow and average flow showed the same regularity. When the short duration-high intensity rain, the surface runoff of red soil slope land was faster, and the order was HoS < HeS < HeS + SM from early to late. HoS and HeS started runoff at 6.18 and 7.83 min after the beginning of rainfall, respectively. HeS + SM began runoff at 19.03 min. By comparing the two rainfall patterns, we can see that the short duration-high intensity rain causes early surface runoff generation, large total runoff, and high peak flow. Straw mulching significantly reduced surface runoff in red soil slopes.

Table 3. Effect of straw mulching on surface runoff of Red Soil Sloping Farmland.

Rain Pattern	Test Treatment	Runoff Generation Time/(min)	Total Flow/(L)	Peak Flow/(mL)	Average Flow/(mL)
Long duration -low intensity	HoS	157.00 ± 4.25 a	15.72 ± 1.60 a	740.00 ± 3.00 a	413.71 ± 21.05 a
	HeS	160.42 ± 3.35 a	9.83 ± 1.63 b	500.00 ± 45.00 b	265.54 ± 7.65 b
	HeS + SM	/	0	0	0
Short duration -high intensity	HoS	6.18 ± 0.03 c	81.41 ± 3.82 a	3300.00 ± 155.00 a	3015.19 ± 127.41 a
	HeS	7.83 ± 0.31 b	66.70 ± 7.52 b	2800.00 ± 55.00 b	2382.14 ± 90.60 b
	HeS + SM	19.03 ± 1.20 a	0.21 ± 0.02 c	18.00 ± 5.00 c	9.22 ± 0.84 c

Note: For the same column, different lowercase letters indicate significant differences among different treatments under the same rain pattern ($p < 0.05$). ± denotes standard deviation. HoS, Homogeneous soil. HeS, Heterogeneous soil. HeS + SM, Heterogeneous soil + Straw mulch.

It can be seen that the surface runoff intensity of HoS was greater than that of HeS under the long duration-low intensity rain (Figure 3a). Generally, the runoff intensity of Hos and HeS increased during rainfall. The runoff intensity reached its peak at the end of rainfall, and the runoff immediately ended after the end of rainfall. Differently, under short duration-high intensity rain (Figure 3b), the runoff intensity of soil tanks without straw mulching increased rapidly after runoff generation and formed superosmotic runoff. Between 20~25 min after the beginning of rainfall, the runoff intensity stabilized until the end of rainfall. Straw mulching leads to low surface runoff intensity and stable runoff yield on red soil slope.

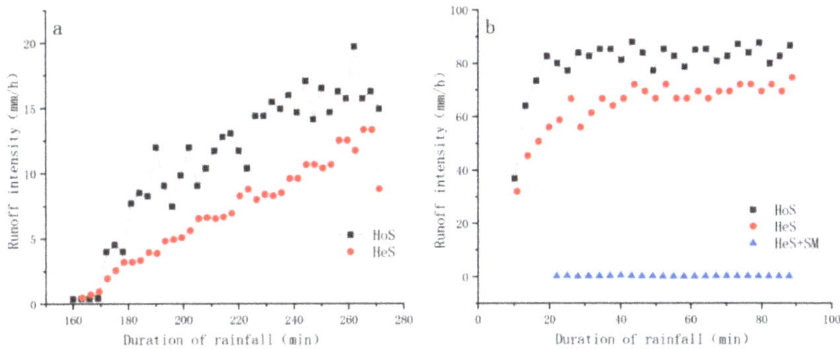

Figure 3. Surface runoff process under different rainfall patterns (**a**) Long duration-low intensity rain; (**b**) Short duration-high intensity rain. HoS, Homogeneous soil. HeS, Heterogeneous soil. HeS + SM, Heterogeneous soil + Straw mulch.

3.2. Sediment Yield Characteristics

The soil loss analysis of red soil slope farmland under straw mulching conditions was shown in Table 4. We can know that the total erosion of HoS was 1.4 times of HeS at long duration-low intensity rain. However, due to the large surface runoff of HoS, the sediment concentration was diluted, and the sediment concentration of HoS was 0.8 times that of HeS. Under short duration-high intensity rain, the total erosion amount was 1.2 times that of HeS. The sediment concentrations of the two were close, and HoS was 0.98 times that of HeS. It is worth noting that the total amount of soil erosion caused by short duration-high intensity rain was 12.12~12.39 times that of long duration-low intensity rain, and the peak erosion reached 35.29~44.29 times. In the case of constant rainfall, short duration-high intensity rain can form greater soil erosion.

Table 4. Effect of straw mulching on soil loss in red soil sloping farmland.

Rain Pattern	Test Treatment	Sediment Concentration (g L^{-1})	Total Sediment Yield (g)	Peak Sediment Yield (g)	Average Sediment Yield (g)
Long duration -low intensity	HoS	2.58 ± 0.06 [a]	40.43 ± 5.94 [a]	1.54 ± 0.05 [a]	1.19 ± 0.041 [a]
	HeS	3.23 ± 0.19 [a]	28.25 ± 3.17 [b]	1.19 ± 0.03 [a]	0.94 ± 0.039 [a]
	HeS + SM	0	0	0	0
Short duration -high intensity	HoS	5.96 ± 0.23 [b]	490.06 ± 20.08 [a]	54.35 ± 1.40 [a]	17.50 ± 1.07 [a]
	HeS	6.09 ± 0.02 [a]	406.52 ± 10.70 [a]	52.70 ± 0.95 [b]	14.52 ± 0.50 [b]
	HeS + SM	0	0	0	0

Note: For the same column, different lowercase letters indicate significant differences among different treatments under the same rain pattern ($p < 0.05$). ± denotes standard deviation. HoS, Homogeneous soil. HeS, Heterogeneous soil. HeS + SM, Heterogeneous soil + Straw mulch.

The analysis of soil loss processes caused by different rainfall patterns on sloping arable land was shown in Figure 4. We know that the erosion amount of red soil slope fluctuates with the continuous rainfall under the long duration-low intensity rain (Figure 4a). At the same rainfall time, the erosion amount generated by HoS was lower than that generated by HeS. Under short duration-high intensity rain, the trend of the HoS and HeS erosion process line was close (Figure 4b). Erosion of HoS and HeS increased rapidly from the beginning of runoff to peak in about 20 min and then decreased rapidly. The runoff intensity tends to level off until the end of the rainfall, about 50 min after the start of the rainfall.

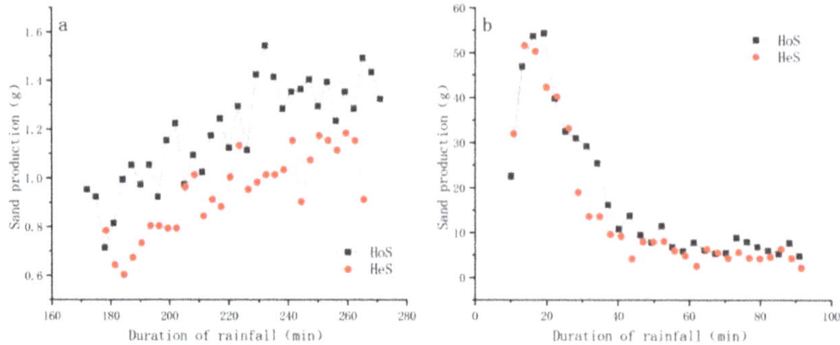

Figure 4. Surface sediment production process under different rain patterns (**a**) Long duration-low intensity rain; (**b**) Short duration-high intensity rain. HoS, Homogeneous soil. HeS, Heterogeneous soil. HeS + SM, Heterogeneous soil + Straw mulch.

3.3. Subsurface Flow and Deep Infiltration Characteristics

Analysis of the beginning time and flow rate of subsurface flow in red soil sloping farmland under straw mulching was shown in Table 5. It can be seen that under long duration-low intensity rain, the runoff generation time of subsurface flow of HoS and HeS was earlier than that of surface runoff. The subsurface flow of HoS was 14.5 min earlier than that of surface runoff, and the subsurface flow of HeS was 19.19 min earlier than that of surface runoff. The total flow in soil from high to low was HeS + SM > HeS > HoS, but the peak flow production was HeS + SM > HoS > HeS. Under long duration-low intensity rain, the beginning time of HoS leakage (40 cm) runoff was close to that subsurface flow. However, the leakage of HeS and HeS + SM was significantly different from the runoff generation time of subsurface flow. The time of subsurface flow and leakage of HeS + SM was earlier than that of HeS. We found that the total runoff showed HeS + SM > HeS > HoS in the long duration-low intensity rain. This indicates that the flow rate was positively correlated with the flow time. Under short duration-high intensity rain, the runoff generation time of subsurface flow was later than that of surface flow, and the runoff generation time from early to late was HoS < HeS < HeS + SM, which was opposite to that under long duration-low intensity rain. The total flow of subsurface flow was inversely proportional to the runoff generation time, which was HoS < HeS < HeS + SM. Straw mulch had a great influence on leakage. The leakage runoff of the soil tank covered by straw mulch started early and the runoff was large.

The response process of subsurface flow to different rainfall patterns in red soil sloping farmland under straw mulching was shown in Figure 5. We can see that the HeS + SM flow generation time was the earliest (Figure 5a). With continuous rainfall, the runoff intensity was increased. Runoff intensity was proportional to time. The runoff intensity was proportional to time until the peak intensity is reached at the end of rainfall, and then rapidly decreased to the end of runoff. The runoff generation time of HoS was close to that of HeS. After HoS began to produce runoff, the runoff intensity increased rapidly, reached the peak at 169.50 min, and then decreased rapidly. The runoff stopped at 208.5 min. The runoff intensity increased rapidly after the beginning of HeS runoff, remained stable at about 180 min, and gradually decreased to the end of rainfall after 210 min. Under the condition of short duration-high intensity rain, the runoff intensity of HoS and HeS was higher than that of HeS + SM at the beginning of runoff generation (Figure 5b). The runoff generation time of HeS + SM was relatively late. With continuous rainfall, the runoff intensity of subsurface flow increased and reached the peak intensity at the end of rainfall.

Table 5. Effects of straw mulching on subsurface flow and deep leakage in red soil sloping farmland.

Rain Pattern	Test Treatment	Subsurface Flow				Deep Leakage			
		Runoff Generation Time (min)	Lag Time Compared with the Surface (min)	Total Flow (L)	Peak Flow (mL)	Runoff Generation Time (min)	Lag Time Compared with the Surface (min)	Total Flow (mL)	Peak Flow (mL)
Long duration -low intensity	HoS	142.50 ± 1.75 [a]	−14.50 ± 2.50 [a]	1.55 ± 0.09 [c]	175.00 ± 7.00 [a]	141.50 ± 4.95 [b]	−15.5 ± 0.7 [b]	34.43 ± 2.30 [a]	1467.00 ± 75.00 [a]
	HeS	141.23 ± 1.52 [a]	−19.19 ± 1.83 [a]	3.92 ± 0.19 [b]	152.00 ± 11.00 [b]	168.75 ± 3.67 [a]	8.33 ± 0.32 [a]	23.24 ± 3.20 [b]	480.00 ± 56.00 [b]
	HeS + SM	101.50 ± 2.74 [b]	/	5.23 ± 0.30 [a]	180.00 ± 10.00 [a]	161.30 ± 2.12 [a]	/	29.03 ± 3.55 [ab]	600.00 ± 60.00 [b]
Short duration -high intensity	HoS	10.85 ± 0.37 [c]	4.67 ± 0.34 [b]	0.35 ± 0.04 [c]	130.00 ± 5.00 [b]	92.12 ± 11.28 [a]	85.94 ± 11.25 [a]	1.35 ± 0.11 [b]	30.00 ± 5.00 [b]
	HeS	13.00 ± 0.80 [b]	5.17 ± 0.49 [a]	2.18 ± 0.14 [b]	140.00 ± 4.00 [b]	/	/	/	/
	HeS + SM	23.02 ± 1.28 [a]	3.99 ± 0.08 [c]	27.71 ± 0.93 [a]	1790.00 ± 120.00 [a]	39.65 ± 1.75 [b]	20.62 ± 0.55 [b]	41.42 ± 5.03 [a]	1280.00 ± 80.00 [a]

Note: For the same column, different lowercase letters indicate significant differences among different treatments under the same rain pattern ($p < 0.05$). ± denotes standard deviation. HoS, Homogeneous soil. HeS, Heterogeneous soil. HeS + SM, Heterogeneous soil + Straw mulch.

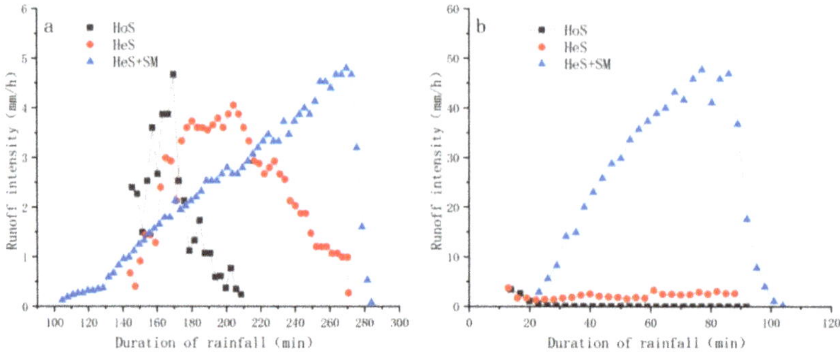

Figure 5. Subsurface flow process under different rainfall patterns (**a**) Long duration-low intensity rain; (**b**) Short duration-high intensity rain. HoS, Homogeneous soil. HeS, Heterogeneous soil. HeS + SM, Heterogeneous soil + Straw mulch.

4. Discussion

4.1. Effects of Rainfall Pattern on Near-Surface Hydrological Process and Soil Loss

In this study, the surface runoff generated by short duration-high intensity rain in uncovered sloping farmland was 5.18~6.79 times that of long duration-low intensity rain, and soil erosion was 12.12~14.39 times as much. For short duration-high intensity rain, the surface runoff accounted for 66~80% of the total rainfall water, and the surface runoff caused by long duration-low intensity rain accounted for only 10~16% of the rainfall water (Figure 6). Previous [32,33] studies showed that short duration-high intensity rain is an important rainfall pattern that causes erosion in various regions compared to long duration-low intensity rain. Ma [34] found that the mean runoff depth of short duration-high intensity rain was 3.4 times that of long duration-low intensity rain. Short duration-high intensity rain is the main pattern of rainfall that causes runoff development on purple soil slopes. Fang [32] analyzed natural rainfall and found that the total runoff caused by short duration-high intensity rain was the highest among all rainfall patterns in karst areas. Ma [33] reported that the rill erosion of red soil slope farmland occurred in the 33rd minute under the condition of heavy rain and heavy rain, and in the 160th minute under the condition of light rain and heavy rain. Generally, short duration-high intensity rain in a certain period rainfall, more than surface soil moisture infiltration rate to form surface runoff. The water not absorbed by the soil forms surface runoff, which takes away the surface soil and leads to soil erosion. Therefore, short duration-high intensity rain is more likely to cause soil erosion than long duration-low intensity rain under the same value of rainfall.

Different rainfall patterns have a great influence on subsurface flow beginning time and runoff yield. Specifically, the subsurface flow of long duration-low intensity rain started earlier than that of short duration-high intensity rain. Moreover, the subsurface flow produced by long duration-low intensity rain was 1.80~4.43 times that of short duration-high intensity rain in uncovered sloping farmland. Long duration-low intensity rain produced a subsurface flow of 2~4% of total rainfall, while short duration-high intensity rain produced 0~2% (Figure 6). We believe that in long duration-low intensity rain, the rainfall intensity per unit time is small, and the water has sufficient time to infiltrate, thus forming subsurface flow. In short duration-high intensity rain, the infiltration rate of surface soil was lower than that of rainfall intensity, forming infiltration excess runoff [35]. Moreover, short duration-high intensity rain forms surface crusts after raindrops hit the surface soil [36,37], further reducing the soil infiltration rate and subsurface flow. This is consistent with most previous studies [38]. These findings indicate that long duration-low intensity rain is more likely to promote the development of subsurface flow than short duration-high intensity rain.

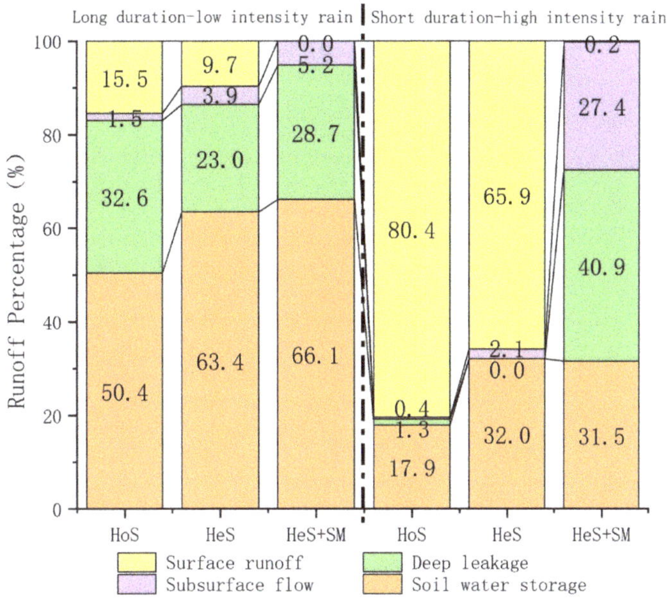

Figure 6. Surface runoff and subsurface flow under different rainfall patterns. HoS, Homogeneous soil. HeS, Heterogeneous soil. HeS + SM, Heterogeneous soil + Straw mulch. (The proportion is less than 1%, and the figure is not marked).

4.2. Effect of Soil Structure on Near-Surface Hydrological Process and Soil Loss

In red soil slope farmlands, soil structure is an important factor affecting surface runoff and erosion [39]. Our results showed that the surface runoff generated by HeS (heterogeneous) was lower than that done by HoS (homogeneous), regardless of rainfall pattern. Because the surface infiltration rate of soil with heterogeneous is higher than that of homogeneous soil. Part of the water was absorbed by the surface soil infiltration, thereby reducing the generation of surface runoff. This is consistent with previous research results [40]. Nanda [40] report that the spatial heterogeneity of soil properties affects the spatial distribution of runoff. Relatedly, less surface runoff can cause less erosion. This is consistent with our results that HeS has lower surface soil erosion than HoS. These phenomena indicate that the soil structure of red soil slope farmland can effectively reduce the yield of surface runoff and weaken the surface erosion caused by rainfall.

The subsurface flow is also affected by the soil structure. Different from surface runoff, the subsurface flow of HeS was 2.53~6.23 times of HoS under two rainfall patterns. Under HeS conditions, the proportion of subsurface flow in rainfall water increased by about 2%. Soil structure increases the flow of subsurface flow, which was consistent with previous studies [41]. Soil heterogeneity caused by long-term tillage of sloping farmland can effectively reduce the intensity of surface runoff [42]. When the water infiltrates into the heterogeneous soil, it encounters the plow pan with poor permeability, forming a lateral subsurface flow. This will reduce the water pressure in the surface soil, increase the water infiltration rate and reduce the surface runoff. Aldo [41] report that heterogeneity is more sensitive to rainfall inputs. The increase of subsurface flow and saturation is faster than that of a homogeneous case. In summary, the soil structure of red soil slope farmland can increase the water infiltration rate and increase the proportion of subsurface flow in the overall water. Therefore, soil structure can affect near-surface hydrology, reduce soil erosion, and achieve the effect of water and fertilizer conservation.

4.3. Effect of Straw Mulching on Near-Surface Hydrological Process and Soil Loss

Straw mulching is an effective means of dealing with soil erosion caused by rainfall. Straw mulching reduced 99.7% (Long duration-low intensity rain)~100% (Short duration-high intensity rain) surface runoff and 100% surface erosion under different rainfall patterns, respectively. Compared with bare land, surface runoff and erosion in straw mulching experiments decreased significantly, which was similar to the results of several authors [43–46]. Misagh [43] found that a higher straw application rate would produce less runoff and reduce soil erosion. Wang [44] also found that straw mulching had a significant effect on the average infiltration rate, cumulative runoff and cumulative sediment yield on the slope. We believe that the physical and biochemical properties of surface soil change when the straw is covered on sloping farmland. Straw mulching can reduce the impact of rainfall on surface soil and reduce surface crust. Straw mulching can also reduce surface soil temperature [47], decrease soil bulk density [48,49], increase soil porosity [50] and increase soil water storage. For example, Liu [47] proposed that soil porosity gradually increased with the continuous coverage of straw. Increasing soil porosity can improve soil infiltration, thus affecting surface runoff and subsurface flow. These results suggest that straw mulching can regulate the formation of surface runoff and reduce surface soil erosion.

In this study, the subsurface flow was significantly affected by straw mulching (Table 4). Specifically, in long duration-low intensity rain, the initial runoff time of subsurface flow caused by straw mulching was 39.73 min earlier than that without mulching, and the total subsurface flow runoff increased by 33.42%. After straw mulching, the proportion of total precipitation in subsurface flow runoff increased by 1~25%, and the amount of deep leakage also increased by 6~41% (Figure 6). These findings indicate that straw mulching can effectively promote the development of subsurface flow [38]. The report of Gao [46] suggested that straw mulching could significantly increase subsurface flow under various rainfall intensities. Duan [38] reported that straw mulching significantly reduced surface runoff and significantly increased subsurface flow at a depth of 30 cm. Different from the previous research results, in short duration-high intensity rain, straw mulching increased the total amount of subsurface flow runoff to 12.71 times, but the initial runoff time-lagged 10.02 min. We believe that heavy rainfall has compacted and splashed the surface of sloping farmland without straw mulching in this experiment. This leads to the change of surface micro-topography, and some surface soil forms a crust effect. Therefore, a large amount of water carried by rainfall cannot be infiltrated, forming runoff on the surface. Due to the large surface runoff water potential, the water rapidly infiltrates the large gap to form a preferential flow, and the subsurface flow is formed at the change of soil structure. Straw mulching reduced the splash erosion of raindrops on surface soil, resulting in loose surface soil and a high infiltration rate. Moreover, straw mulching reduced the flow velocity of surface runoff and made it infiltrate nearby. The straw mulching treatment made the surface soil moisture reach saturation before the formation of subsurface flow, resulting in the formation of subsurface flow later. Straw mulching enables the water carried by short duration-high intensity rain to be fully fixed by the surface soil. After the surface soil is saturated, subsurface flow is formed to recharge groundwater resources. This plays a positive role in water and fertilizer conservation and reduces soil erosion in slope farmland.

This study provides theoretical support for the application of straw mulching on red soil slope farmland. The protective effect of rape straw mulching on the surface soil of red soil slope farmland under short duration-high intensity rain was proposed, which was conducive to the popularization and application of this measure. However, the conclusion of this study was only obtained under the condition of single rape straw coverage and slope, which may vary with different coverage and slope. It will be the focus of our future research to select different coverage and find the optimal straw coverage of red soil slope farmland to expand the application scope and practicability of the research results.

5. Conclusions

In this study, the effects of rainfall pattern, soil structure, and straw mulching on near-surface hydrology and erosion were investigated in the sloping cropland of red soil. The main conclusions are as follows: (1) The intensity of surface runoff formed by short duration-high intensity rain is greater than that of long duration-low intensity rain, and the surface soil is more eroded. At the same time, short duration-high intensity rain is not conducive to the development of subsurface flow and cannot make full use of the water brought by rainfall; (2) Under the two rainfall patterns, the surface runoff intensity of HeS (heterogeneous) is smaller than that of HoS (homogeneous), while the subsurface flow runoff intensity is the opposite. This shows that the soil structure of red soil slope farmland can affect the development process of near surface hydrology, increase the infiltration of rainfall water, and reduce the surface runoff and erosion of the slope; (3) After mulching with rape straw, the intensity of surface runoff decreased significantly, and the intensity of subsurface flow increased significantly. The improvement of coverage on short duration-high intensity rain is particularly evident. Straw mulching can make the precipitation water fully fixed by the surface soil, which is conducive to the absorption and utilization of water by crops, and reduce the loss of soil. These findings will help to understand the impact of rainfall on the near-surface hydrology of sloping farmland in red soil and provide a theoretical basis for production practice.

Author Contributions: Study design: X.G. and J.Y.; Investigation: L.Y., L.P. and X.Z.; Data analysis and mapping: L.Y. and L.P.; Funding acquisition: J.Y.; Writing—original: L.Y. and J.D.; Writing—editing: X.G. and J.Y. All authors have read and agreed to the published version of the manuscript.

Funding: This work was supported by the project of National Natural Science Foundation of China (41761065), Supported by the Cultivation Plan for Reserved Project of National Science and Technology Award (Grants No. 20212AEI91011) and Key Science and Technology Project of Jiangxi Province (20213AAG01012; 2022KSG01001).

Data Availability Statement: The data reported in this study are available on request from the corresponding authors. The data are not publicly available yet due to the authors are writing some other papers by mining these data.

Conflicts of Interest: The authors declare no conflict of interests.

References

1. Tong, S.; Zhiming, F.; Yanzhao, Y.; Yumei, L.; Yanjuan, W. Research on Land Resource Carrying Capacity: Progress and Prospects. *J. Resour. Ecol.* **2018**, *9*, 331–340. [CrossRef]
2. Mishra, V. Review of Arable Land-Use Problems in Present-Day China. *AMBIO J. Hum. Environ.* **2009**, *38*, 110–112. [CrossRef] [PubMed]
3. Grum, B.; Assefa, D.; Hessel, R.; Woldearegay, K.; Kessler, A.; Ritsema, C.; Geissen, V. Effect of In Situ Water Harvesting Techniques on Soil and Nutrient Losses in Semi-Arid Northern Ethiopia. *Land Degrad. Dev.* **2017**, *28*, 1016–1027. [CrossRef]
4. Hacısalihoğlu, S. Variations of Measured and Simulated Soil-Loss Amounts in a Semiarid Area in Turkey. *Environ. Monit. Assess.* **2010**, *165*, 255–262. [CrossRef] [PubMed]
5. Wairiu, M. Soil Organic Carbon in Relation to Cultivation and Topsoil Removal on Sloping Lands of Kolombangara, Solomon Islands. *Soil Tillage Res.* **2003**, *70*, 19–27, (In Chinese with English Abstract). [CrossRef]
6. Zhang, G.H.; Yang, Y.; Liu, Y.N.; Wang, Z.Q. Advances and Prospects of Soil Erosion Research in the Black Soil Region of Northeast China. *J. Soil Water Conserv.* **2022**, *36*, 1–12, (In Chinese with English Abstract). [CrossRef]
7. Long, T.Y.; Liu, X.Z.; Liu, J. Modeling Spatial and Temporal Variation of Nitrate Loss via Interflow in a Sloping Field of Purple Soil. *J. Agro-Environ. Sci.* **2015**, *34*, 1973–1978, (In Chinese with English Abstract). [CrossRef]
8. Fu, Z.Y.; Li, Z.X.; Cai, C.F.; Guo, Z.L. Impact of boundary conditions on soil water and subsurface flow in shallow purple soil slope. *J. Hydraul. Eng.* **2011**, *42*, 899–907, (In Chinese with English Abstract). [CrossRef]
9. Zhang, X.; Wang, K.L.; Fu, Z.Y.; Chen, H.S.; Zhang, W.; Shi, Z.H. Hydrological characteristics of calcareous soil with contrasting architecture on dolomite slope of Northwest Guangxi, China. *J. Appl. Ecol.* **2017**, *28*, 2186–2196, (In Chinese with English Abstract). [CrossRef]
10. Lal, R.; Mahboubi, A.A.; Fausey, N.R. Long-Term Tillage and Rotation Effects on Properties of a Central Ohio Soil. *Soil Sci. Soc. Am. J.* **1994**, *58*, 517–522. [CrossRef]

11. Liu, Y.-J.; Yang, J.; Hu, J.-M.; Tang, C.-J.; Zheng, H.-J. Characteristics of the Surface-Subsurface Flow Generation and Sediment Yield to the Rainfall Regime and Land-Cover by Long-Term in-Situ Observation in the Red Soil Region, Southern China. *J. Hydrol.* **2016**, *539*, 457–467. [CrossRef]
12. Wilson, G.V.; Cullum, R.F.; Roemkens, M.J.M. Ephemeral Gully Erosion by Preferential Flow through a Discontinuous Soil-Pipe. *Catena* **2008**, *73*, 98–106. [CrossRef]
13. Wang, T.; Zhu, B. Nitrate Loss via Overland Flow and Interflow from a Sloped Farmland in the Hilly Area of Purple Soil, China. *Nutr. Cycl. Agroecosyst.* **2011**, *90*, 309–319. [CrossRef]
14. Getahun, G.T.; Kätterer, T.; Munkholm, L.J.; Rychel, K.; Kirchmann, H. Effects of Loosening Combined with Straw Incorporation into the Upper Subsoil on Soil Properties and Crop Yield in a Three-Year Field Experiment. *Soil Tillage Res.* **2022**, *223*, 105466. [CrossRef]
15. Wang, J.L.; Shi, X.X.; Li, Z.Z.; Zhang, Y.; Liu, Y.Q.; Peng, Y.X. Responses of Runoff and Soil Erosion to Planting Pattern, Row Direction, and Straw Mulching on Sloped Farmland in the Corn Belt of Northeast China. *Agric. Water Manag.* **2021**, *253*, 106935. [CrossRef]
16. Keesstra, S.D.; Rodrigo-Comino, J.; Novara, A.; Gimenez-Morera, A.; Pulido, M.; Di Prima, S.; Cerda, A. Straw Mulch as a Sustainable Solution to Decrease Runoff and Erosion in Glyphosate-Treated Clementine Plantations in Eastern Spain. An Assessment Using Rainfall Simulation Experiments. *Catena* **2019**, *174*, 95–103. [CrossRef]
17. Jordan, A.; Zavala, L.M.; Gil, J. Effects of Mulching on Soil Physical Properties and Runoff under Semi-Arid Conditions in Southern Spain. *Catena* **2010**, *81*, 77–85. [CrossRef]
18. Rahma, A.E.; Warrington, D.N.; Lei, T. Efficiency of Wheat Straw Mulching in Reducing Soil and Water Losses from Three Typical Soils of the Loess Plateau, China. *Int. Soil Water Conserv. Res.* **2019**, *7*, 335–345. [CrossRef]
19. Rahma, A.E.; Lei, T.; Shi, X.; Dong, Y.; Zhou, S.; Zhao, J. Measuring Flow Velocity under Straw Mulch Using the Improved Electrolyte Tracer Method. *J. Hydrol.* **2013**, *495*, 121–125. [CrossRef]
20. Jourgholami, M.; Abari, M.E. Effectiveness of Sawdust and Straw Mulching on Postharvest Runoff and Soil Erosion of a Skid Trail in a Mixed Forest. *Ecol. Eng.* **2017**, *109*, 15–24. [CrossRef]
21. Prosdocimi, M.; Jordán, A.; Tarolli, P.; Keesstra, S.; Novara, A.; Cerdà, A. The Immediate Effectiveness of Barley Straw Mulch in Reducing Soil Erodibility and Surface Runoff Generation in Mediterranean Vineyards. *Sci. Total Environ.* **2016**, *547*, 323–330. [CrossRef] [PubMed]
22. Mulumba, L.N.; Lal, R. Mulching Effects on Selected Soil Physical Properties. *Soil Tillage Res.* **2008**, *98*, 106–111. [CrossRef]
23. Guo, T.; Wang, Q.; Li, D.; Zhuang, J.; Wu, L. Flow Hydraulic Characteristic Effect on Sediment and Solute Transport on Slope Erosion. *Catena* **2013**, *107*, 145–153. [CrossRef]
24. Fan, W.; Wu, J.G.; Li, J.M.; He, R.C.; Yao, Y.Y.; Wang, D.C.; Sun, L.; Wang, C.Y. Effects of Straw Return on Soil Physico-chemical Properties of Chernozem in Northeast China and Maize Yield Therein. *Acta Pedol. Sin.* **2018**, *55*, 835–846, (In Chinese with English Abstract). [CrossRef]
25. Materechera, S.A. Aggregation in a Surface Layer of a Hardsetting and Crusting Soil as Influenced by the Application of Amendments and Grass Mulch in a South African Semi-Arid Environment. *Soil Tillage Res.* **2009**, *105*, 251–259. [CrossRef]
26. Gicheru, P.; Gachene, C.; Mbuvi, J.; Mare, E. Effects of Soil Management Practices and Tillage Systems on Surface Soil Water Conservation and Crust Formation on a Sandy Loam in Semi-Arid Kenya. *Soil Tillage Res.* **2004**, *75*, 173–184. [CrossRef]
27. Jiang, G.Y.; Zen, X.Y.; Shi, D.M.; Zhang, J.L. Minimal Date Set of Effect Evalutaion and Measure Classiflcation of Soil and Water Conservation on Slope Farmland in South of China. *Soil Water Conserv. China* (In Chinese with English Abstract). **2021**, 5, 56–59. [CrossRef]
28. Yang, J.; Zheng, H.J.; Chen, X.A. Effects of Tillage Practices on Nutrient Loss and Soybean Growth in Red-Soil Slope Farmland. *Int. Soil Water Conserv. Res.* **2013**, *1*, 49–55. [CrossRef]
29. Chen, X.; Liang, Z.; Zhang, Z.; Zhang, L. Effects of Soil and Water Conservation Measures on Runoff and Sediment Yield in Red Soil Slope Farmland under Natural Rainfall. *Sustainability* **2020**, *12*, 3417. [CrossRef]
30. Wang, B. LinHo Rainy Season of the Asian-Pacific Summer Monsoon. *J. Clim.* **2002**, *15*, 386–398. [CrossRef]
31. R Team. R: A Language and Environment for Statistical Computing. R Foundation for Statistical Computing: Vienna, Austria. *Computing* **2009**, *14*, 12–21. [CrossRef]
32. Nu, F.F.; Shi, Z.H.; Li, L.; Guo, Z.L.; Liu, Q.J.; Ai, L. The Effects of Rainfall Regimes and Land Use Changes on Runoff and Soil Loss in a Small Mountainous Watershed. *Catena* **2012**, *99*, 1–8. [CrossRef]
33. Ma, J.Y.; Ma, B.; Wang, Y.X.; Wang, C.G.; Li, C.D.; Xiao, J.B. Combined Effects of the Wheat Straw Length and Incorporation Rate on Reducing Runoff and Sediment Yields. *Catena* **2022**, *215*, 106310. [CrossRef]
34. Ma, X.; Zheng, J.K.; Wang, W.W.; Chen, X.; Qin, W.; Shan, Z.J.; Lang, D.X.; Liao, F. Characteristics of the Runoff and Sediment Yield of Sloping Farmland in the Purple Soil Area Under Different Rainfall Patterns. *J. Soil Water Conserv.* **2017**, *31*, 17–21, (In Chinese with English Abstract). [CrossRef]
35. Zhang, X.Y.; Shi, X.Z.; Yu, D.S.; Wang, H.J.; Liu, Z.B.; Zhang, W.T. Effects of antecedent soil moistrue on hill slope runoff-generation and soil erosion over red soil-mantled landscapes. *Adv. Water Sci.* **2010**, *21*, 23–29, (In Chinese with English Abstract). [CrossRef]
36. Giménez, R.; Govers, G. Effects of Freshly Incorporated Straw Residue on Rill Erosion and Hydraulics. *Catena* **2008**, *72*, 214–223. [CrossRef]

37. Knapen, A.; Poesen, J.; Govers, G.; De Baets, S. The Effect of Conservation Tillage on Runoff Erosivity and Soil Erodibility during Concentrated Flow. *Hydrol. Process.* **2008**, *22*, 1497–1508. [CrossRef]
38. Duan, J.; Yang, J.; Tang, C.; Chen, L.; Liu, Y.; Wang, L. Effects of Rainfall Patterns and Land Cover on the Subsurface Flow Generation of Sloping Ferralsols in Southern China. *PLoS ONE* **2017**, *12*, e0182706. [CrossRef]
39. Chen, X.A.; Yang, J.; Tang, C.J.; Zheng, T.H.; Li, L.F. Effects of rainfall intensity and slope on surface and subsurface runoff in red soil slope farmland. Trans. *Chin. Soc. Agric. Eng.* **2017**, *33*, 141–146, (In Chinese with English Abstract). [CrossRef]
40. Nanda, A.; Sen, S.; Mcnamara, J.P. How Spatiotemporal Variation of Soil Moisture Can Explain Hydrological Connectivity of Infiltration-Excess Dominated Hillslope: Observations from Lesser Himalayan Landscape. *J. Hydrol.* **2019**, *579*, 124146. [CrossRef]
41. Fiori, A.; Russo, D. Numerical Analyses of Subsurface Flow in a Steep Hillslope under Rainfall: The Role of the Spatial Heterogeneity of the Formation Hydraulic Properties. *Water Resour. Res.* **2007**, *43*, W07445. [CrossRef]
42. Shi, D.M.; Jiang, G.Y.; Jiang, P.; Lou, Y.B.; Ding, W.B.; Jin, H.F. Effects of soil erosion factors on cultivated-layer quality of slope farmland in purple hilly area. *Trans. Chin. Soc. Agric. Eng.* **2017**, *33*, 270–279, (In Chinese with English Abstract). [CrossRef]
43. Parhizkar, M.; Shabanpour, M.; Lucas-Borja, M.E.; Zema, D.A.; Li, S.; Tanaka, N.; Cerdà, A. Effects of Length and Application Rate of Rice Straw Mulch on Surface Runoff and Soil Loss under Laboratory Simulated Rainfall. *Int. J. Sediment Res.* **2021**, *36*, 468–478. [CrossRef]
44. Wang, C.G.; Ma, J.Y.; Wang, Y.X.; Li, Z.B.; Ma, B. The Influence of Wheat Straw Mulching and Straw Length on Infiltration, Runoff and Soil Loss. *Hydrol. Process.* **2022**, *36*, e14561. [CrossRef]
45. Won, C.H.; Choi, Y.H.; Shin, M.H.; Lim, K.J.; Choi, J.D. Effects of Rice Straw Mats on Runoff and Sediment Discharge in a Laboratory Rainfall Simulation. *Geoderma* **2012**, *189*, 164–169. [CrossRef]
46. Gao, Z.C.; Xu, Q.X.; Si, Q.; Zhang, S.P.; Fu, Z.Y.; Chen, H.S. Effects of Different Straw Mulch Rates on the Runoff and Sediment Yield of Young Citrus Orchards with Lime Soil and Red Soil under Simulated Rainfall Conditions in Southwest China. *Water* **2022**, *14*, 1119. [CrossRef]
47. Liu, L.; Zhang, L.; Liu, J.; Fu, Q.; Xu, Q. Idimesheva Soil Water and Temperature Characteristics under Different Straw Mulching and Tillage Measures in the Black Soil Region of China. *J. Soil Water Conserv.* **2021**, *76*, 256–262. [CrossRef]
48. Ram, H.; Dadhwal, V.; Vashist, K.K.; Kaur, H. Grain Yield and Water Use Efficiency of Wheat (*Triticum Aestivum* L.) in Relation to Irrigation Levels and Rice Straw Mulching in North West India. *Agric. Water Manag.* **2013**, *128*, 92–101. [CrossRef]
49. Sun, J.N.; Yang, R.Y.; Zhu, J.J.; Zhou, C.X.; Yang, M.; Pan, Y.H.; Zhang, Z.H. Contrasting Effects of Corn Straw Biochar on Soil Water Infiltration and Retention at Tilled and Compacted Bulk Densities in the Yellow River Delta. *Can. J. Soil Sci.* **2019**, *99*, 357–366. [CrossRef]
50. Glab, T.; Kulig, B. Effect of Mulch and Tillage System on Soil Porosity under Wheat (*Triticum Aestivum*). *Soil Tillage Res.* **2008**, *99*, 169–178. [CrossRef]

remote sensing

MDPI

Article

Estimation of Water Balance for Anticipated Land Use in the Potohar Plateau of the Indus Basin Using SWAT

Muhammad Idrees [1], Shakil Ahmad [1,*], Muhammad Wasif Khan [1], Zakir Hussain Dahri [2], Khalil Ahmad [3], Muhammad Azmat [1] and Irfan Ahmad Rana [1]

[1] School of Civil and Environmental Engineering (SCEE), National University of Sciences and Technology (NUST), Sector H-12, Islamabad 44000, Pakistan
[2] Pakistan Agricultural Research Council, Sector G-5, Islamabad 44000, Pakistan
[3] Department of Civil and Environmental Engineering, King Abdulaziz University, Jeddah 22254, Saudi Arabia
* Correspondence: shakilahmad@nice.nust.edu.pk; Tel.: +92-51-9085-4614

Abstract: Land Use/Land Cover (LU/LC) change is among the dominant driving factors that directly influence water balance by transforming hydrological responses. Consequently, a thorough comprehension of its impacts is imperative for sustainable water resource planning and development, notably in developing worlds such as Pakistan, where agriculture is a major livelihood. This research intends to assess the continuing changes in LU/LC and evaluate their probable repercussions on the hydrological regime of the Potohar Plateau. The maximum likelihood classification (MLC) algorithm for land use classification of the high-resolution satellite imageries, the Cellular-Automata Markov Chain Model (CA-MCM) for the projection of LU/LC maps, and the Soil and Water Assessment Tool (SWAT) in tandem with SWAT-CUP for hydrological modeling were employed in this research. The high-resolution climatic dataset (10 × 10 km) was used in SWAT. The LU/LC analysis revealed a continual propagation of agricultural and built-up lands at the detriment of forest and barren land during the last three decades, which is anticipated to continue in the future, too. Hydrological model calibrations and validations were performed on multi-basins, and the performance was evaluated using different statistical coefficients, e.g., the coefficient of determination (R^2), Nash–Sutcliffe Efficiency (NSE), Kling–Gupta Efficiency (KGE), and Percent Bias (PBIAS). The results yielded that the model performed very well and demonstrated the model's robustness in reproducing the flow regime. The water balance study revealed that the anticipated LU/LC changes are projected to decrease the mean annual surface runoff, water yield, and streamflow due to an increase in percolation, lateral flow, sub-surface flow, and evapotranspiration. More significant variations of the water balance components were observed at the sub-basin level, owing to the heterogeneous spatial distribution of LU/LC, than at the basin level. The outcome of this study will provide pragmatic details to legislative bodies to develop land and water management ameliorative strategies to harness hydrological changes sustainably.

Keywords: LU/LC; MLC; CA-Markov; SWAT; Potohar Plateau; water balance; Indus basin

Citation: Idrees, M.; Ahmad, S.; Khan, M.W.; Dahri, Z.H.; Ahmad, K.; Azmat, M.; Rana, I.A. Estimation of Water Balance for Anticipated Land Use in the Potohar Plateau of the Indus Basin Using SWAT. *Remote Sens.* **2022**, *14*, 5421. https://doi.org/10.3390/rs14215421

Academic Editor: Magaly Koch

Received: 10 September 2022
Accepted: 24 October 2022
Published: 28 October 2022

1. Introduction

Sustainable management of watersheds and environmental systems is receiving thriving attention from local, continental, and intercontinental institutes because they are essential for ecosystem preservation, poverty alleviation, and food security in developing countries, where agriculture is a major livelihood [1]. Additionally, Sustainable Development Goals (SDGs) specify essential socioeconomic, environmental, and hydrological processes characterized by performance indices [2].

Numerous unidentified anthropogenic biomes affect ecosystems at various spatial-temporal scales. The cumulative effects of environmental change, which integrates multi-scale climate change (CC), and disruptions to natural and socio-ecosystems, alter and deteriorate ecosystem characteristics [3]. Climate change and LU/LC seem to be the most overriding drivers of the hydrological processes, impacting flow regimes and water balance components in watersheds worldwide [4]. LU/LC change is a significant attribute of global environmental change due to its diverse implications. LU/LC change is intrinsically accompanied by an expansion of economic activities, machine-intensive agriculture, unregulated infrastructural sprawl, and a substantial demographic transition from remote rural areas to urban centers, a characteristic feature of LU/LC change in the Anthropocene era [5]. Furthermore, interactions between these parameters at the basin scale might thus have a confounding impact, resulting in spatial-temporal variation of hydrological components [6,7].

Changes in LU/LC are the most prominent driving agents of hydrological variation at spatio-temporal scales. Various researchers have identified the effects of LU/LC change on the flow regime. For example, an increase in flow during the wet season and a decrease in flow during the dry season in Hoeya River Basin, Korea, from 1975–2050 were attributed to urban sprawl [8]. The growth of agricultural and built-up lands at the cost of vegetation in the Andassa watershed, Ethiopia, from 1985 to 2045 caused an increase in annual surface runoff and water yield [9]. An increase in infrastructural development and agricultural activities in the upper Athi basin of the Nairobi metropolitan area from 1985 to 2055 corresponds to the increasing surface runoff and evapotranspiration [7]. Other studies in the East African watershed [10], Big Sioux River watershed [11], Krishna river basin [12], Xinanjiang basin [13], and Bhavani basin [14] have also revealed significant implications of LU/LC changes to exacerbate hydrological components. [15] The quantification of water balance helps to better comprehend the dynamics of physical processes in the upper Chao Phraya River Basin, Thailand. The author of [16] evaluated the trends in the various hydrological variables to rationalize the intensification of the global hydrological cycle. Other studies in Peninsular India [17], 32 global basins [18], and 24 global basins [19] presented uncertainties in the quantification of water balance.

Spatio-temporal LU/LC projections are useful tools for identifying the relationships between different simulated changes in LU/LC dynamics and their underlying causes [20]. The CA-Markov Chain Model (CA-MCM) embedded into MOLUSCE was applied to project changes in LU/LC to 2050 and performed very well with a kappa coefficient of 0.72 in the Astore watershed [21]. Other studies such as those in the Nashe basin, Ethiopia [22], the Wuhan metropolitan area, Central China [23], Pu county, Shanxi province, Chinese Loess Plateau [24], the city of Faisalabad, Pakistan [25], Heihe river basin China [26], and in Shiyang river basin, China [20] have recommended an integrated CA-Markov model for LU/LC projection.

Hydrological models are essential in quantifying the implications of LU/LC changes in water balance [12]. The SWAT model was recommended since it had been commonly employed in large-scale modeling and simulations to evaluate the plausible ramifications of land management practices and LU/LC changes related to hydrologic components. Previous investigations have suggested the efficacy of the SWAT model in quantifying the implications of LU/LC changes on hydrological components. For instance, studies in the Potohar Plateau [27], in Europe [28], Ib river watershed, India [29], the Upper Sind river basin, India [30], Tons river basin, India [31], Ghataprabha basin, India [32], and in the eastern Baltic Sea region [33] have recommended the SWAT model to simulate the water balance components of the basins.

Multiple research works have been carried out to investigate the implications of climate change on flow regimes since the publication of the 5th Assessment Report of IPCC [34]. Investigations on the Haro and Soan Basin have noted that climate change will, in turn, intensify uncertainties regarding water availability [35,36]. However, the degree to which the flow regime responds to LU/LC change has not been thoroughly studied, and this response differs between watersheds and LU/LC scenarios. Due to demographic growth, agricultural and built-up lands have significantly increased, causing a dramatic change in flow regime at spatial-temporal scales. To manage water resources sustainably, it is crucial to evaluate how climate change will alter the hydrological processes and how different LU/LC scenarios will impact the flow regime [13]. Though a number of researchers have identified the heightened problem of LU/LC change [11] and have investigated LU/LC changes in the region [37,38], there is a lack of quantifying the impact of these changes on the hydrological regime. Thus, this research is one of the few to evaluate the LU/LC change implications on hydrologic components in a vast rainfed agricultural-dominated region, which might be essential for sustainable LU/LC planning and water resource management.

The study targeted formulating a pivotal framework to comprehend historical and projected LU/LC changes and their implications on water balance in the Plateau. Specifically, the historical LU/LC maps projected by CA-MCM were employed to drive the SWAT model to simulate flow regimes under changing scenarios. With this in mind, the overarching objectives of this research are: (1) To assess the spatiotemporal LU/LC changes in the Plateau, (2) To setup and parameterize a high-resolution hydrological model, and (3) To quantify spatiotemporal water balance under changing LU/LC scenarios. Using the results of this study, water managers and policymakers would be able to develop ameliorative strategies to mitigate water availability on spatial and temporal scales.

2. Materials and Methods

2.1. Study Area

The Potohar Plateau is the largest rainfed tract in Pakistan and is situated in the northern Punjab province, which is in the northeastern part of Pakistan. The Plateau consists of five administrative units, namely Attock, Chakwal, Jhelum, Rawalpindi, and Islamabad Capital Territory. Its area is approximately 22,254 km^2 with an elevation of 133 to 2221 m a.m.s.l. The region geographically lies in Sindh Sagar Doab, which is bound by the Indus River on the west and the Jhelum River on the east and stretches southward from the foothills of the Himalayas to the salt range [27]. The region's topography is highly undulating, which is constituted by rolling plains in the east, lofty mountains to the west, and dissected ravine belts (Figure 1). The Potohar Plateau mainly includes six river basins, namely Bhuna, Haro (the hatched portion incorporated in hydrological modeling, which lies outside the Potohar Plateau), Kahan, Kanshi, Reshi, and Soan. Out of six river basins, only three gauged river basins, i.e., Soan, Haro, and Kanshi (Figure 2), have been analyzed to simulate the hydrological regime under current and projected LU/LC [39].

The climate of the Plateau varies from semi-arid to sub-humid, with scorching summers and relatively cold winters. Generally, moist and sub-humid climates predominate in the northern portion of the region, whereas dry and semi-arid environments predominate in the central and southern regions, respectively. The yearly precipitation varies from 1750 mm in the northwest to 450 mm in the southwest. Precipitation distribution in the region is bi-modular in nature because of two precipitation patterns, i.e., westerlies and Monsoon. The westerly aggravation triggers the first peak, while Monsoon triggers the second peak of higher magnitude. The temperature ranges from 4 to 40 °C, but it can increase beyond its normal limits due to natural processes and non-environment-friendly anthropogenic activities [40–42].

The Potohar Plateau is the fastest-growing region of Pakistan, with a population of around 12 million in 2017 [43]. Due to the rapid economic and demographic growth, water is becoming increasingly scarce due to the increasing water demands. Therefore, it is imperative to determine the LU/LC changes and their hydrological implications.

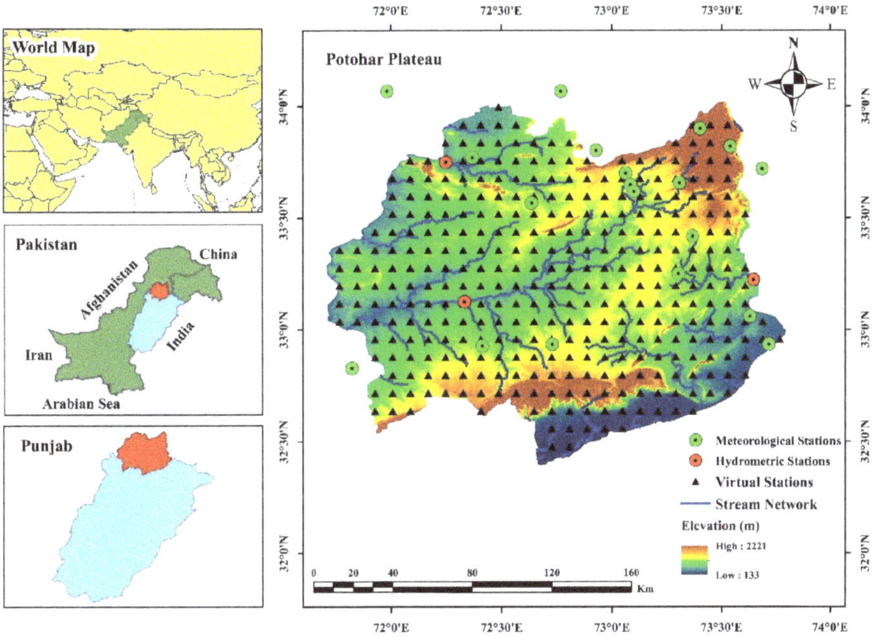

Figure 1. Study area map of the Potohar Plateau with the topography, hydrometric stations, meteorological stations, and gridded virtual stations.

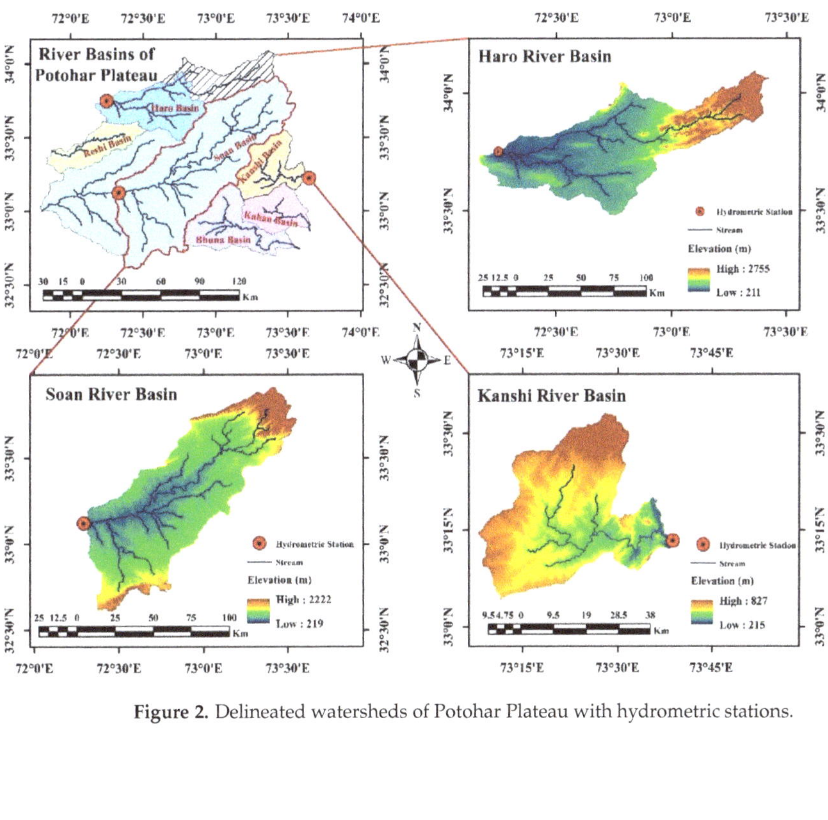

Figure 2. Delineated watersheds of Potohar Plateau with hydrometric stations.

2.2. Datasets

Datasets required for the setup of a hydrological model are geo-hydro-climatic datasets, for instance, soil map, topography (Figure S2), i.e., Digital Elevation Model (DEM), land use map, climatic data, namely precipitation, maximum and minimum temperature, and hydrometric data such as streamflow data (Table S2). The concise description and sources of these datasets are displayed in Table 1.

Table 1. Details of geo-hydro-climatic input datasets.

Data Type	Data Name	Description	Time	Resolution	Source
Spatial Data	DEM	ALOS PALSAR	NA	12.5 × 12.5 m	NASA Earth-Data [44]
	LU/LC	Landsat 5, 7	1990, 2000, 2010	30 × 30 m	USGS [45]
		Sentinel-2A	2020	10 × 10 m	USGS [45]
	Soil Map	DSMW	NA	30 Arc Second	FAO [46]
Hydro-Climatic Data	Climatic Data	Precipitation	1991–2019	10 × 10 Km	Submitted and Unpublished [47]
		Temperature	1991–2019	10 × 10 Km	Submitted and Unpublished [47]
	Hydrometric Data	Flow Data	1991–2007	NA	Surface Water Hydrology Project (WAPDA) [48]

2.3. Methodological Framework

The primary goal of the research is to estimate the implications of LU/LC changes on water balance under historical and projected scenarios, and the detailed methodology employed for this research is given in Figure 3. This study is structured into four major components: (a) To perform supervised classification of satellite imagery for historical LU/LC, (b) To project LU/LC, (c) To set up a hydrological model, (d) To quantify water balance under LU/LC scenarios.

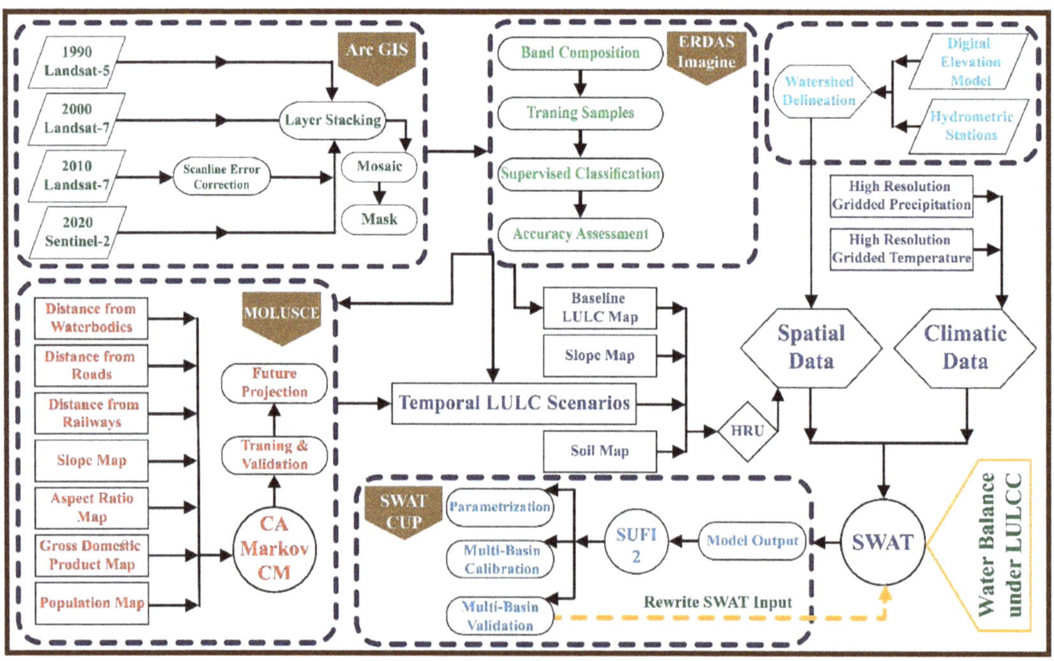

Figure 3. Methodological framework of the study.

2.4. Cellular Automata Markov Chain Model (CA-MCM)

The Cellular Automata Markov Chain Model (CA-MCM) is a widely accepted LU/LC change modeling algorithm for projecting spatiotemporal changes. The CA-MCM combines Cellular Automata (CA, a spatial model) and the Markov Chain (MCM, a quantitative temporal model) to project LU/LC change trends and characteristics over time. The rationale for integrating the models is that the MCM describes the likelihood of cells changing from one form to another but does not provide spatial distribution. Therefore, to distribute the projected changes spatially, the MCM is hybridized with CA [20,23,24].

MCM is a stochastic model that utilizes two historical LU/LC maps, which are the main inputs for determining the transition area matrix and transition probabilities matrix [49]. For instance, it will provide information on how many cells are likely to transform from one LU/LC class to another based on a transition probability matrix [22,50]. MCM is based on Bayes' Equation (1) to project LU/LC change, and P is determined using Equation (2):

$$S_{(t+1)} = P_{mn} \times S_{(t)} \tag{1}$$

$$P = \begin{Vmatrix} P_{11} & P_{12} & P_{13} & \dots & P_{1n} \\ P_{21} & P_{22} & P_{23} & \dots & P_{2n} \\ P_{31} & P_{32} & P_{33} & \dots & P_{3n} \\ \vdots & \vdots & \vdots & \vdots & \vdots \\ P_{m1} & P_{m2} & P_{m3} & \dots & P_{mn} \end{Vmatrix} \tag{2}$$

where $\{0 \leq P_{mn} \leq 1$ and $\sum_{m,n=1}^{j}(P_{mn}) = 1\ (m,n = 1,2,3,\ldots,j)\}$.

MCM is unable to recognize spatial variability in LU/LC due to a spatial constraint. So, by coupling MCM with CA, spatial and temporal LU/LC changes can be simulated. CA can be expressed as Equation (3):

$$S_{(t,t+1)} = f\left\{P_{(t)}, N\right\} \tag{3}$$

where S_t and S_{t+1} are the LU/LC condition at the time of t and t + 1, respectively; P is the transition probability matrix, j is the number of LU/LC classes, N is the cellular field, and f is the transformation rule of cellular states [13,22,51].

2.5. SWAT Model

SWAT is a physically based, continuous time series, a semi-distributed hydrological model that functions on sub-daily/daily time steps. It was devised to determine the influence of land management strategies on water, sediment, and agricultural chemical yields over extended time periods in a large-scale, complex watershed with heterogeneous soil, slope, and LU/LC scenarios [2,34,52]. The SWAT model simulates a hydrological cycle using the water budget Equation (4):

$$SW_f = SW_i + \sum_{n=1}^{t}\left(R_{day} - Q_{surface} - W_{seep} - E_{ET} - Q_{gw}\right) \tag{4}$$

The SWAT model employs the Soil Conservation Service curve number (SCS-CN) to characterize the surface runoff [52], which can be defined as Equation (5):

$$Q_{surface} = \frac{\left(R_{day} - I_a\right)^2}{R_{day} - I_a + S'}, \quad I_a = \lambda S \tag{5}$$

where SW_f is the final water content of the soil, SW_i is the initial soil water content, t is time (days), R_{day}, $Q_{surface}$, E_{ET}, W_{seep}, Q_{gw}, I_a, and S are the daily amount of precipitation, surface flow, evapotranspiration, percolation, return flow, initial abstraction, and the retention parameter after runoff begins, respectively; all the parameters are measured in mm of H_2O. When I_a is taken to be 0.2S, Equation (6) becomes:

$$Q_{surface} = \frac{\left(R_{day} - 0.2S\right)^2}{R_{day} + 0.8S}, \quad \text{for } P > 0 \text{ else } Q = 0 \tag{6}$$

The retention parameter fluctuates geo-spatially owing to the changes in LU/LC, soil, slope, and soil water content, and it may be computed using Equation (7) [31,53,54].

$$S = \frac{25,400}{CN} - 254 \tag{7}$$

Evapotranspiration of the basin is evaluated using the Penman–Monteith method, which can be defined as Equation (8).

$$\lambda * ET = \frac{\Delta(R_n - G) + \rho_a c_p * \left(\frac{e_s - e_a}{r_a}\right)}{\Delta + \gamma * \left(1 + \frac{e_s}{r_a}\right)} \tag{8}$$

where R_n, G, $e_s - e_a$, ρ_a, r_s, r_a, c_p, and g are the lateral flow, transmission losses, net radiation, soil heat flux, air vapor pressure deficit, mean air density, surface resistance, aerodynamic resistance, the specific heat of air, and psychrometric constant, respectively.

2.6. Evaluation of Historical Land Use/Land Cover

The assessment of LU/LC was accomplished by employing four satellite imagery, three (3) from Landsat with a spatial resolution of 30 m and one (1) from Sentinel with a spatial resolution of 10 m, namely, Landsat-5 TM 1990, Landsat-7 ETM+ 2000, Landsat-7 ETM+ 2010, and Sentinel-2A 2020, respectively (Table S1). Landsat 7 ETM+ had scan line errors which were collected after 31 May 2003 due to a malfunction of the scan line corrector (SLC). Firstly, these scan line errors were corrected using the Landsat tool. After this operation, the next step was to perform the supervised classification, which was executed using Imagine ERDAS software. For this purpose, more than 700 training samples were acquired to create signature files that were further used to execute supervised classification based on the Maximum Likelihood Classification (MLC) algorithm [34,35]. After successful classification, the accuracy assessment of the classified images was performed using the ground truth data. For this purpose, 100 points for each LU/LC type were taken using Google Earth maps and satellite imageries. After ground truthing, Imagine ERDAS automatically calculated the kappa coefficient, overall, producer's and user's accuracy.

2.7. Land Use/Land Cover Projection

CA-MCM is available in MOLUSCE (Modules for Land Use Change Simulations), an extension of QGIS, and is a comprehensive model that projects the trends and geospatial configuration of LU/LC classes based upon historical LU/LC thematic maps, transition probability matrices, and suitability matrices [34,55]. Before proceeding towards the projection of LU/LC, inputs were required. The first input was the creation of three discrete LU/LC classified maps, two maps for the preparation of the transition probability matrix, and one map to validate the model. The other input was the driving parameters, which include the spatial and socio-economic factors (Table S4). Several parameters can control the growth in each LU/LC class, and these parameters can interact and establish an intricate relationship. Thus, major driving parameters such as topography, aspect, slope, Euclidean distance from roads, Euclidean distance from railways, Euclidean distance from water bodies, gross domestic product (GDP), and population density (Figure S1) were chosen depending on their availability and impact on LU/LC changes [14,24,56]. These driving factors were used to prepare the suitability matrix. The classified LU/LC maps of 2000 and 2010 were employed to generate a transition probability matrix and driving parameters to create a suitability matrix. MLP-ANN algorithm used these matrices to project the LU/LC map of 2020. This projected map was validated with a supervised classified LU/LC 2020 map [14,57]. After successful validation, the trained model was used to project LU/LC to 2030, 2040, and 2050.

2.8. Setup, Sensitivity Analysis, Calibration and Validation of SWAT

A SWAT model was employed to simulate the implications of the historical and projected LU/LC change scenarios on water balance components in the Potohar Plateau. The Arc-SWAT 2012 model delineated the whole region into three major river basins, namely Soan, Haro, and Kanshi. These basins were divided into multiple sub-basins by choosing a threshold value of 5005 hectares. Since these sub-basins had heterogeneity, they were further segmented into hydrologic response units (HRUs) associated with distinct sets of topography, soil, and LU/LC class with the threshold value of 10%, 20%, and 10%, respectively, as put forth by Kiprotich [7]. These HRUs were homogeneous geospatial units with identical hydrological and geomorphological characteristics. In our case, this region had 3 river basins, 57 sub-basins, and 486 HRUs for the baseline model. The model performed a simulation using high-resolution gridded climatic data such as precipitation, and minimum and maximum temperature for a span of 29 years (1991–2019), encapsulating a spin-up period of an initial three years [28,58–60].

The SWAT model was refined by coupling it with SWAT Calibration and Uncertainty Programs (SWAT-CUP). The SUFI-2 algorithm enables parameterization using sensitivity analysis, calibration, and validation [61,62]. In order to carry out this operation, available observed data was divided into three-time windows, with the middle time window used for model calibration and rest two windows used for backward and forward validations. These windows were defined such that the monthly average and standard deviation were nearly equal for all the windows.

Based on past research on similar catchments, thirty-eight of the most influential parameters were chosen for sensitivity analysis. Global Sensitivity Analysis (GSA) was employed to perform sensitivity analysis. The sensitivity of the parameters was determined using a multiple regression approach that regresses Latin hypercube-generated parameters against an objective function. Due to the diverse number of parameters, the calibration process becomes complex and computationally extensive. So, to reduce the number of parameters, local sensitivity analysis was also performed, which not only reduced the number of parameters but also provided upper and lower bounds for an expeditious calibration process [63,64]. In order to analyze the SWAT performance during the simulation, we applied four statistical performance indicators given in Table 2 [52,65].

Table 2. List of employed statistical parameters.

Coefficient	Formula	Performance Rating
R^2: Coefficient of determination	$R^2 = \frac{\left[\sum_{i=1}^{n}(Q_{obs,i}-\overline{Q_{obs}})(Q_{sim,i}-\overline{Q_{sim}})\right]^2}{\sum_{i=1}^{n}(Q_{obs,i}-\overline{Q_{obs}})^2\sum_{i=1}^{n}(Q_{sim,i}-\overline{Q_{sim}})^2}$	$0 \leq R^2 \leq 1$ >0.5 Satisfactory >0.65 Good
NSE: Nash–Sutcliffe Efficiency	$NSE = 1 - \frac{\sum_{i=1}^{n}(Q_{obs}-Q_{sim})^2}{\sum_{i=1}^{n}(Q_{obs,i}-\overline{Q_{obs}})^2}$	$0 \leq NSE \leq 1$ >0.5 Satisfactory >0.65 Good
KGE: Kling–Gupta efficiency	$KGE = 1 - \sqrt{(r-1)^2 + (\alpha-1)^2 + (\beta-1)^2}$	$0 \leq KGE \leq 1$ >0.5 Satisfactory >0.65 Good
PBIAS: Percent bias	$PBIAS = 100 \times \frac{\sum_{i=1}^{n}(Q_{obs}-Q_{sim})_i}{\sum_{i=1}^{n}Q_{obs,i}}$	$-\infty \leq PBIAS \leq +\infty$ $<\pm 25$ Satisfactory $<\pm 15$ Good

3. Results

3.1. Spatio-temporal Changes in Historical LU/LC

The LU/LC maps were classified into five classes, i.e., water bodies, agricultural lands, forest lands, barren lands, and built-up lands. Principally, it was conceived that there was an escalating anthropogenic biome in the region owing to the infrastructural development and high land-intensive agriculture, and abatement of barren and forest lands, resulting in a major loss of natural cover. The spatiotemporal distribution of LU/LC change in the Plateau is depicted in Figure 4. It demonstrates that built-up lands were drastically increased in the northern areas by transforming agricultural, forest, and barren lands into non-porous lands between 1990 and 2020 (Figure 5). Concurrently, barren land was converted to machine-intensive agricultural land in the central and southern areas. A similar scenario prevailed in the subsequent decades, with the sprawl of unregulated settlements and agricultural activities at the detriment of barren and forest lands.

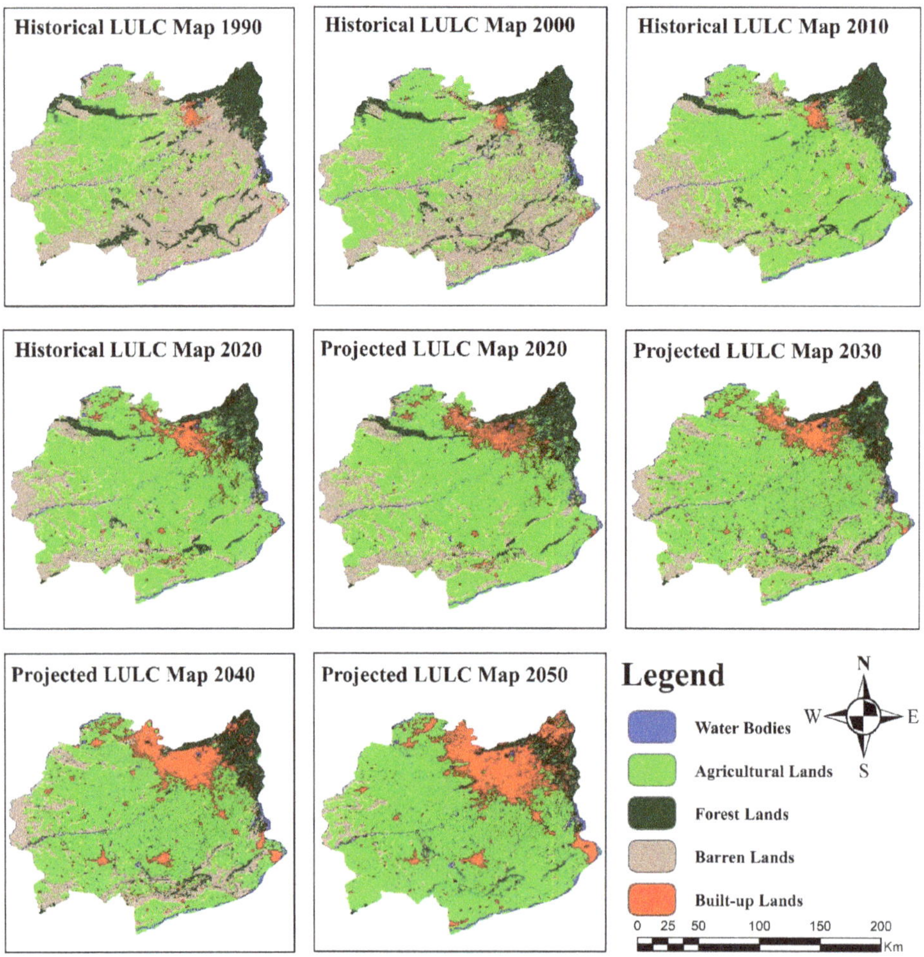

Figure 4. Spatio-temporal variation of LU/LCs, historical and projected.

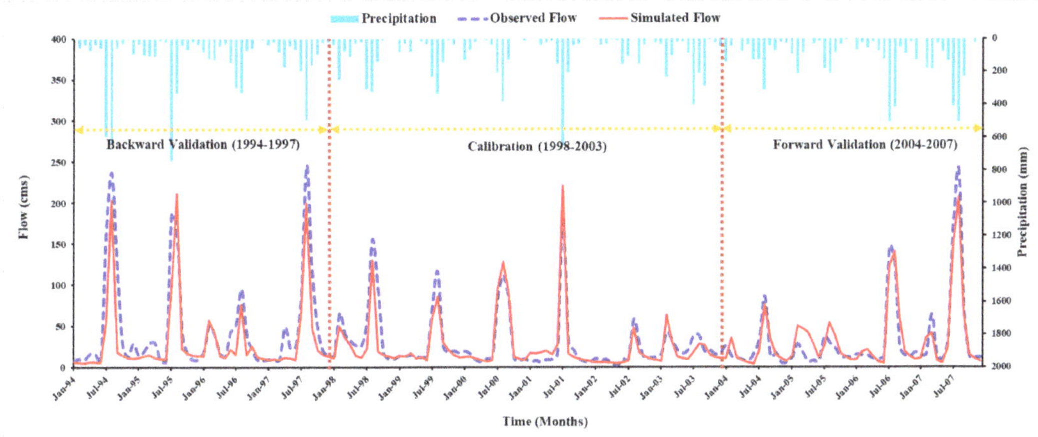

Figure 5. Calibration and validations of stream flows, Soan River Basin.

The historical LU/LC change (Table 3) was dominated by the change in a barren land with a total loss of 7895.47 km^2, followed by urban and agriculture. Agricultural land and built-up land exhibited an increasing trend gaining 6864.94 km^2 and 1585.67 km^2, respectively, while water changed the least. The highest decline in barren lands was primarily owing to the proliferation of agricultural and built-up lands. The agricultural lands were expanded to fulfill the agrarian needs and generate livelihood for the increasing population, while the built-up lands were expanded to accommodate migration from rural communities to major urban centers.

Table 3. Historical LU/LC composition in the Plateau.

LU/LC	Classified 1990		Classified 2000		Classified 2010		Classified 2020	
	Km^2	%	Km^2	%	Km^2	%	Km^2	%
Water Bodies	252.04	1.08	316.56	1.36	324.59	1.40	338.76	1.46
Agriculture Lands	5790.60	24.91	6631.58	28.58	11,102.87	47.85	12,655.54	54.43
Forest Lands	2859.59	12.30	2786.42	12.01	2585.14	11.14	2478.51	10.66
Barren Lands	14,086.34	60.59	13,159.87	56.71	8637.84	37.23	6190.87	26.63
Built-up Lands	260.79	1.12	309.55	1.33	553.63	2.39	1585.67	6.82

However, the LU/LC change is trivial compared to the Plateau's total geographical area. This signifies that the LU/LC change has been more substantial at the sub-basin level than at the Plateau level. At the Plateau level, positive LU/LC changes in one class have been counterbalanced by negative LU/LC changes in another class. For example, an increase in agricultural activity has been observed in central and downstream areas, simultaneously, the transition of agricultural land to built-up land in upstream areas led to a decline in agricultural land within the Plateau. The findings of this research are congruent with those of others conducted on and around the Plateau. For instance, in the Potohar Plateau, [37,38] concluded that there was a dramatic increase in agricultural and infrastructural development. The authors of [57] also summarized that the built-up land are increasing, which is causing the loss of vegetation cover in the northern part of the region.

3.2. Accuracy Assessment of Supervised Classified LU/LC Maps

Accuracy assessment was used to determine the reliability of classified images. In this regard, a confusion matrix was employed to determine the correctness of the classified image. Randomly selected sampling points were analyzed on the classified maps from

Google Earth maps and mosaicked satellite imagery as a reference. Around 500 ground truthing points for each classified image, with 100 sampling points for each class, were selected to validate images from 1990, 2000, 2010, and 2020. The results (Table S3) exhibit that the average kappa coefficient and overall accuracy for 1990, 2000, 2010, and 2020 maps were 0.79, 0.81, 0.82, and 0.84; 83.8%, 85.2%, 85.88%, and 87.8%, respectively. The producer's accuracy ranged between 75% and 94%, while the user's accuracy ranged between 79% and 97%. According to Monserud and Landis [66,67], a Kappa coefficient above 0.75 is an acceptable indicator of the classified image. Therefore, the outcome of the assessment demonstrated that the classified image and ground truths were in good agreement.

3.3. LU/LC Projections

LU/LC maps of 2000 and 2010 functioned as observed data for training the CA-MCM, whereas the LU/LC map of 2020 was utilized to evaluate the reliability of the projected map for 2020. A comparative analysis of classified and projected LU/LC maps for 2020 was conducted. The pictorial comparison demonstrates that the CA-MCM was an effective LU/LC projection tool for the Potohar Plateau (Figure 4). An assessment of conformity between classified and projected maps for 2020 was performed to ascertain the adaptability of the deployed model to reliably simulate the distribution of LU/LC changes (Table S7). The Kappa coefficient (K) was determined to comprehend the resemblance between the classified and projected LU/LC maps for 2020. Since all the K statistics ($K_{overall}$ = 0.77, $K_{historical}$ = 0.79, and K_{local} = 0.82) have been above the value of 0.75 [63,64], the CA-MCM trained for LU/LC projection was deemed to be acceptable.

Furthermore, since the hydrological model was taken into account to estimate the implications of LU/LC change on the water balance of the study area, a comparison between the simulations performed using the classified LU/LC 2020 map, and the CA-MCM projected LU/LC 2020 map (Figure S4) was necessary. The graphical analysis reveals that the flow regime simulated by the SWAT model at the river basins changed minimally for the two employed LU/LC maps (Figure S3), restating the good match between the classified and simulated LU/LC maps. The trained CA-MCM was then utilized to project LU/LC maps for 2030, 2040, and 2050 (Table 4).

Table 4. Future projected LU/LC composition in the Plateau.

LU/LC	Projected 2020		Projected 2030		Projected 2040		Projected 2050	
	Km²	%	Km²	%	Km²	%	Km²	%
Water Bodies	334.69	1.46	353.53	1.52	350.57	1.51	348.44	1.50
Agriculture Lands	12,439.41	54.43	12,831.91	55.19	13,040.56	56.09	14,586.03	62.74
Forest Lands	2335.59	10.66	2314.06	9.95	2189.18	9.42	2054.69	8.84
Barren Lands	6289.47	26.63	5119.34	22.02	4043.06	17.39	1522.11	6.55
Built-up Lands	1800.17	6.82	2630.50	11.31	3627.24	15.60	4738.08	20.38

The LU/LC change detection of the Potohar Plateau indicates that agricultural lands hold the dominant portion of the area (Figure 4). From the analysis, it is clear that the study area has assimilated significant LU/LC change, and the dominant LU/LC change has occurred in the expansion of built-up and agricultural lands. Additionally, LU/LC change has been occurring at the expense of a reduction in barren and forest lands (Figure S2).

3.4. Hydrological Model Calibration and Validation

The initial SWAT model setup utilizing default parameter ranges was unable to successfully simulate the flow regime at all hydrometric gauges because the base flow and peak flow were overestimated. As a result, parameter calibration was required to precisely model the hydrologic regime. A global sensitivity analysis (GSA) for the simulated flow was undertaken using a monthly hydrometric flow to recognize the most sensitive parameters which affect model response. Incipiently soil, groundwater, evapotranspiration, surface runoff, and geomorphology parameters were considered, and 29 parameters were recognized as sensitive parameters. The greater the absolute value of the t-stat and the lower the p-value, the more sensitive a parameter, and the sensitivity ranks of parameters after GSA are shown in Table S5. Subsequently, the calibration was performed incorporating 29 parameters [28,63], which was laborious and computationally intensive. Therefore, local sensitivity analysis (LSA) was also performed, reducing the number of parameters to 22 and providing upper and lower bounds for swift calibration (Table S6). The model was calibrated to flow series from 1998 to 2003. The calibrated and validated parameter values were integrated into the SWAT database to estimate the implications of LU/LC on the water balance.

The SWAT model performed well during calibration (1998–2003) and validation (1994–1997 and 2004–2007) windows. According to the visual analysis of the stream flows (Figures 5, S5 and S6) and statistical evaluation of the model using R^2, NSE, KGE, and PBIAS with reference to criteria defined by Moriasi [65], all hydrometric gauges demonstrated good or very good performance during calibration and validation windows, and the statistical performance indices are shown in Table 5. Furthermore, R-factor and p-value statistics criteria are governed by Karim Abbaspour's [61] model exhibits a good agreement with the calibration and validations. Overall, the performance rating indices computed during the calibration and validation windows confirmed that the model performed quite well [27,68], which suggests that it has the potential to model the implications of LU/LC changes on the hydrological regime.

Table 5. Statistical performance of model during calibration and validation windows.

River Basin	Calibration				Backward Validation				Forward Validation			
	R^2	NSE	KGE	PBIAS	R^2	NSE	KGE	PBIAS	R^2	NSE	KGE	PBIAS
Soan	0.81	0.79	0.77	9.8	0.78	0.76	0.75	−12.9	0.78	0.76	0.76	−17.8
Haro	0.80	0.77	0.78	8.7	0.76	0.74	0.76	8.7	0.76	0.76	0.77	12.7
Kanshi	0.77	0.79	0.73	19.7	0.77	0.74	0.74	14.7	0.75	0.74	0.73	10.6

3.5. Plausible Impacts of LU/LC Changes on Hydrological Regime

Four supervised classified LU/LC maps of the Potohar Plateau (1990, 2000, 2010, and 2020) and three future projected LU/LC maps (2030, 2040, and 2050) were deployed in the calibrated SWAT model under the assumption that the climate was immutable. The impact of LU/LC alterations on the hydrological responses was assessed using a baseline model over the span of 29 years (1991–2019), including a three-year spin-up period (1991–1993) and with varied LU/LC maps, for instance, LU/LC 2000, LU/LC 2010, LU/LC 2020, LU/LC 2030, LU/LC 2040, and LU/LC 2050.

The hydrological regime was investigated in terms of how hydrological processes respond to evapotranspiration, water yield, surface, and sub-surface flows, as exhibited in Table 6. As a consequence of the anticipated drastic transformation of LU/LC from 1990 to 2020, mean annual evapotranspiration (+11%) and groundwater flow (+15%) tend to increase, while mean annual surface flow (−19%) and total water yield (−12%) decreased. Nevertheless, the rate of increase and decrease is proportional to the rate of change in LU/LC. In general, the exacerbating ramifications of the LU/LC changes are accompanied by increased agricultural and built-up lands in conjunction with decreased forest and barren lands.

According to the LU/LC scenarios, agricultural lands are anticipated to increase from 24.91% to 62.74%, while built-up lands are expected to increase from 1.12% to 20.38%, leading to an increase in evapotranspiration (+15.54%) and a decrease in surface runoff (−24%), as shown in Table 7 [12,51]. In the research area, from 1990 to 2050, agricultural expansion (+37.84%) dominated infrastructural development (+19.26%). Therefore, it may neutralize the implications of built-up lands on hydrological components, decreasing surface flow. The expansion of machine-intensive agriculture with the loss of barren land may be the cause of the flow reduction.

Table 6. Mean annual water balance components of the river basins.

Components (mm)	1990	2000	2010	2020	2030	2040	2050
Precipitation	845.80	845.80	845.80	845.80	845.80	845.80	845.80
Surface Runoff	394.09	372.96	348.94	319.82	309.34	307.98	303.33
Evapotranspiration	406.65	419.67	433.96	452.66	464.60	468.66	471.80
Percolation	52.83	54.32	63.55	72.60	73.22	74.94	76.08
Groundwater Flow	38.43	39.45	42.71	44.35	44.73	46.61	48.02
Return Flow	5.14	6.59	6.94	7.04	7.15	7.32	8.61
Lateral Flow	8.63	9.43	9.46	9.70	10.03	10.09	10.16
Water Yield	432.87	419.37	402.11	378.84	366.61	366.27	357.95

Table 7. Comparison between water balance components under LU/LC scenarios.

Components (mm)	Baseline Scenario (1990)	Recent Scenario (2020)	Mid-Century Scenario (2050)	Baseline to Recent (%)	Recent to Mid-Century (%)
Surface Runoff	394.09	319.82	303.33	−74.28 (−18.85%)	−16.49 (−5.15%)
Evapotranspiration	406.65	452.66	471.80	46.01 (11.31%)	19.14 (4.23%)
Water Yield	432.87	378.84	357.95	−54.03 (−12.48%)	−20.90 (−5.52%) *

* The number in parentheses indicates a percent change.

4. Discussion

The integration of LU/LC with the SWAT model can enhance the performance of the model in simulating the processes happening in the basins. In this study, the impacts of LU/LC changes on water balance were estimated for the Potohar Plateau. In this context, supervised classification was performed using ERDAS Imagine for the satellite imageries for the years 1990, 2000, 2010, and 2020. Subsequently, the future of LU/LC was projected using CA-MCM for the years 2030, 2040, and 2050. After that, the SWAT model was set up for the base year and then calibrated and validated to enhance its efficiency of the model. A SWAT model with optimized parameters was utilized to quantify the plausible impact of LU/LC changes in the water balance on the sub-basin level.

The analysis of LU/LC maps of 1990, 2000, 2010, 2020, 2030, 2040, and 2050 revealed that an increase in agricultural and built-up areas from 24.91% to 54.43% and 1.12% to 6.82% and likely to be increased to 62.74% and 20.38%, respectively, in the mid-century due to an increase in demographic and socio-economic growth. The LU/LC analysis also showed that a decrease in forest and barren lands from 12.30% to 10.66% and 60.59% to 26.63% are likely to be increased to 8.84% and 6.55%, respectively, as presented in Tables 3 and 4. The authors of [37] showed that there is an increase in agricultural and built-up areas from 11% to 29% and 6% to 11%, respectively, while forest and barren areas reduced from 69% to 43% and 16% to 10%, respectively, in the Simly watershed. The researchers of [69] also showed that there is an increase in agricultural and built-up areas from 33.44% to 63.1% and 1.77% to 5.78%, respectively, while forest and waste lands reduced from 29.81% to 11.32% and 9.27% to 2%, respectively, in the Narmada watershed. This built-up area and forest shift are more prominent in the northern regions and urban centers, especially in the twin cities. Similarly, an augmentation in agriculture and the associated loss of barren land in the central and southern regions of the Plateau as shown in Figure 4. An analysis from the authors of [38] showed similar patterns in the Potohar Plateau, confirming that there is an increase in agriculture in the central and south-eastern regions and increased urbanization in north-western and developed areas. The findings of other studies in the Astore basin, Pakistan [21], Bogota basin, Colombia [3], Ganga basin, India [4], and Geba catchment, Ethiopia [70] are also in agreement with this study.

Hydrological models are crucial for evaluating the implications of LU/LC changes on water balance. The model is selected in conformity with the problem complexity, data availability, computing cost, and model robustness [12,27,68]. A comparison of different hydrological models was conducted on those models that are suitable for evaluating the influence of LU/LCC on the water balance. The SWAT model was the best for the punctilious modeling of LU/LC changes on the flow regime. The SWAT model was also recommended since it had been commonly employed in large-scale modeling and simulations to evaluate the plausible ramifications of land management practices and LU/LC changes on the hydrologic components [52,62]. In this study, the SWAT model setup for the baseline scenario was calibrated for the period 1998 to 2003, backward validated from 1994–1997, and forward validated from 2004–2007, as this approach was followed by Dahri et al., 2021. The result of statistical indices, i.e., R^2, NSE, KGE, and PBIAS for calibration and validation windows ranged between 0.75–0.81, 0.74–0.79, 0.73–0.78, and −17.8–+14.7%, respectively. Overall, the performance rating indices computed during calibration, and validation windows revealed that the model performed quite well [52,62,65], which suggests that it has the potential to model the implications of LU/LC changes on the hydrological regime.

The quantitative analysis of water balance components revealed that the decrease in surface runoff and water yield from 394.09 mm to 319.82 mm and 432.87 mm to 432.87 mm are likely to be increased to 303.33 mm and 357.95 mm, respectively, in mid-century due to LU/LC changes [4,71]. It also showed an increase in evapotranspiration, percolation, groundwater flow, and lateral flow from 406.65 mm to 452.66 mm, 52.83 mm to 72.60 mm, 38.43 mm to 44.35 mm, and 8.63 mm to 9.70 mm, and are likely to be increased to 471.80 mm, 76.08 mm, 48.02 mm, and 10.16 mm, respectively, as presented in Table 6 [51,72]. The potential of the model to simulate the geospatial distribution of hydrological components provides a substantial contribution to sustainable watershed management. At HRU and sub-basin level, the spatial distribution of hydrological components, namely evapotranspiration, water yield, surface runoff, percolation, lateral flow, and groundwater flow, have been computed as shown in Figures 6, 7 and S7–S10. Evapotranspiration of the Plateau was lesser in the northern parts and higher in the central and southern parts (Figure 7). Conversely, besides climatic and orographic variations, surface runoff, and water yield had a high concentration in the northern region due to the transformation of agricultural and forest lands into built-up land, while a lower concentration was found in the central and southern regions due to the conversion of barren land into agricultural land (Figures 6 and S7).

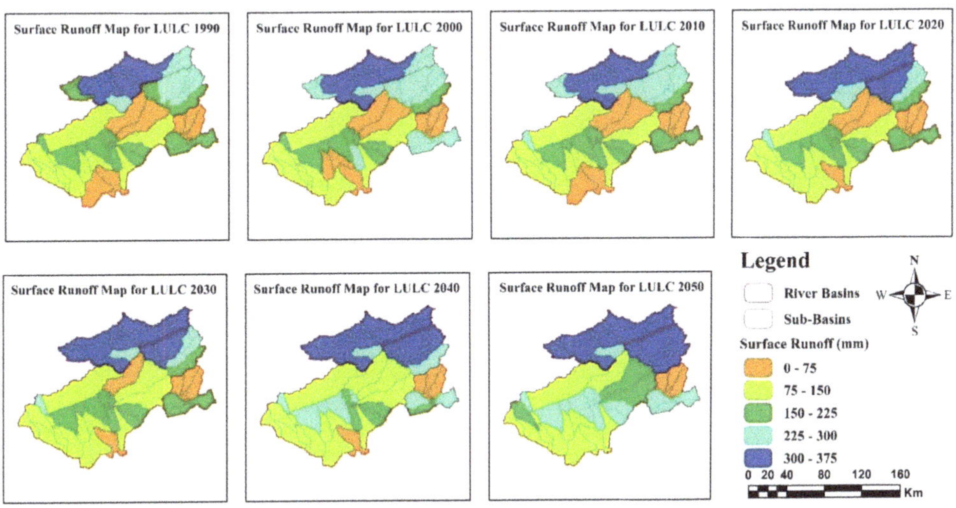

Figure 6. Spatial and temporal distribution of variations in simulated surface runoff.

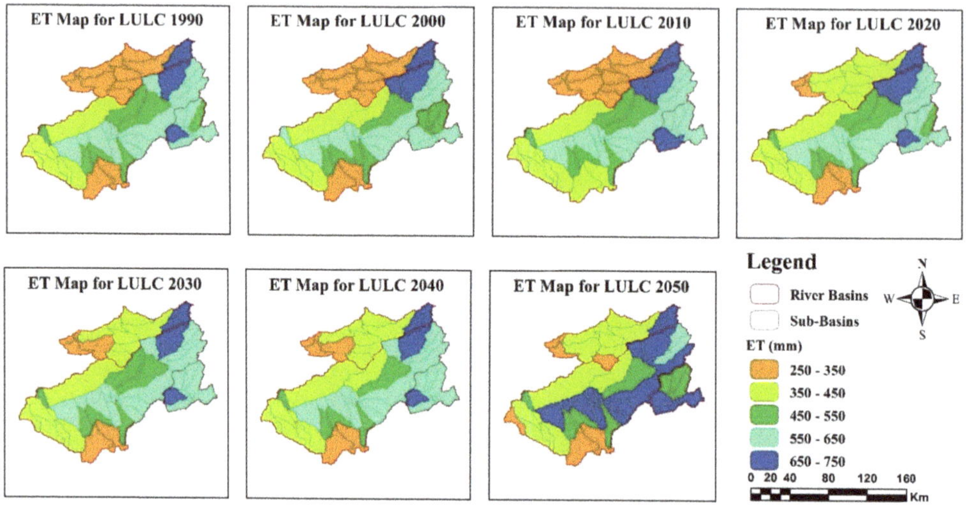

Figure 7. Spatial and temporal distribution of variations in simulated evapotranspiration.

The decreased surface runoff may be accompanied by an increase in evapotranspiration, percolation, lateral flow, and decreased runoff, signifying the augmentation of soil water storage. Groundwater flow varied comparatively less in the proximity of the northern districts and higher in the southern districts. An increase in groundwater flow was simulated on plain terrain, and there was relatively no change in the hilly terrain (Figure S8). Unlike surface runoff, the lateral flow was higher in plain terrain and low infiltration areas. The southern region was found to contribute significant lateral flow to streamflow. Seasonal streamflow variability was also evaluated for the wet and dry seasons (Figure S11). The streamflow was slightly increased in the wet season and decreased in the dry season under the LU/LC scenarios. This increase and decrease in flow might be due to increased urbanization and agricultural activities, respectively. However, it was concluded from the quantitative outputs of the SWAT model that the implications of LU/LC changes were more pronounced at the sub-basin level than at the basin level [4,64,73,74]. The findings are in agreement with other studies conducted in watersheds around the globe. For instance, [72] in the Nam Rom catchment, Vietnam concluded that LU/LC changes have reduced surface runoff and increased evapotranspiration. The authors of [51] in the Beiluo river basin on the Loess Plateau of China found that on an annual scale, surface runoff and water yield may gradually decrease, but evapotranspiration may increase. The authors of [69] also showed a decline of surface runoff and water yield but higher ET due to the presence of more vegetation and forest areas between 1990–2050 in the Narmada river basin, India, which was attributed to the LU/LCC of the catchment.

Despite the fact that the Soan, Haro, and Kanshi river basins encompass 78% of the Potohar Plateau, this still does not meticulously characterize the Plateau, as it constitutes three other river basins. Further studies on the other river basins are required to completely comprehend the dynamics of water balance components in the Plateau. Overall, the SWAT model simulated the flow regimes and water balance components efficiently. However, the accuracy of the simulations is constrained by the coarse resolution of soil data, uncertainties in hydrometric data [75], unavailability of reservoirs data (small dams), evapotranspiration data, and groundwater flow data for the calibration and validation of the model, these should be incorporated in subsequent studies. Moreover, climate change impacts should also be studied in conjunction with LU/LC changes to develop rational strategies for sustainable water management.

5. Conclusions

The enclosed research executed a methodical framework that comprised LU/LC scenarios, hydrological modeling, and quantification of hydrological components alteration in the Potohar Plateau, which were attributed to the effects of human-induced LU/LC changes. Seven distinct LU/LC scenarios were accomplished and evaluated by means of the hydrological SWAT model. Furthermore, the outcomes of each scenario were compared to those of the baseline scenario. According to historical LU/LC maps from 1990 to 2020, anthropogenic pressure increased agriculture and built-up lands by 29.52 and 7.7%, respectively, while barren and forest lands decreased by 33.96 and 1.64%, respectively. LU/LC projections for 2030, 2040, and 2050 were simulated using CA-MCM, which showed an increase in agricultural and built-up lands with decreased barren and forest lands. Among all LU/LC scenarios from 1990 to 2050, it is anticipated that the agricultural and built-up lands will have the greatest proliferation.

The hydrological regime was modeled through the SWAT model, which was then compared to observed hydrometric flows. The SWAT model was calibrated and validated by simulating flow time series from 1994 to 2007. After calibration and validations, the obtained results of the four most recommended coefficients R^2, NSE, KGE, and PBIAS for calibration ranged between 0.77–0.81, 0.77–0.79, 0.73–0.78, and +9.8–+19.7%, respectively, while for forward and backward validation, values ranged between 0.75–0.78, 0.74–0.76, 0.73–0.77, and −17.8–+14.7%, respectively. The result of statistical indices and visualization of streamflow yields a good degree of agreement between the simulated and observed flow regimes, which indicates that the calibrated model, in conjunction with optimized parameters, has the ability to simulate the water balance under LU/LC scenarios.

A distinct outcome of this study is that the transformation in intrinsic water balance and flow regime is substantially sensitive to alteration in LU/LC. The analysis signifies that the influence of LU/LC change has been diverse for peculiar river basins and more prominent at the sub-basin level. The anticipated results indicate that LU/LC attributes to dramatic decrease of surface runoff and water yield by 19% and 12%, respectively, which is expected to exacerbate. Conversely, there is an increase in groundwater flow, lateral flow, and evapotranspiration of 15%, 12%, and 11%, respectively, largely due to the transformation of barren land into agricultural land.

The findings of this study lead to a greater comprehension of the plausible effects of LU/LC changes on water balance in the Potohar Plateau, which will substantially aid decision-makers in planning and executing potential adaptation strategies for watersheds under changing LU/LC scenarios. Moreover, the methodological framework of this research can be useful for any other watershed to evaluate the effects of anthropogenic biomes on the hydrological regime.

Supplementary Materials: The following supporting information can be downloaded at: https://www.mdpi.com/article/10.3390/rs14215421/s1, Figure S1. Spatial maps of driving parameters of LU/LC change, Figure S2. Temporal variance of LU/LC, historical and projected, Figure S3. Soil map of Potohar Plateau, Figure S4. Streamflow Evaluation of Classified and Simulated LU/LC 2020, Figure S5. Calibration and validations of stream flows, Haro River Basin, Figure S6. Calibration and validations of stream flows, Kanshi River Basin, Figure S7. Spatial and temporal distribution of variations in simulated water yield, Figure S8. Spatiotemporal patterns of percolation under LU/LC scenarios, Figure S9. Spatiotemporal patterns of groundwater flow under LU/LC scenarios, Figure S10. Spatiotemporal patterns of lateral flow under LU/LC scenarios, Figure S11. Seasonal variation of flow regime under LU/LC scenarios at different hydrometric stations, Table S1. Description of satellite imagery acquisition, Table S2. Details of hydrometric data acquisition, Table S3. Driving factors used to prepare the suitability matrix for LU/LC projection, Table S4. Accuracy assessment of historical LU/LCs, Table S5. Global sensitivity values and ranks, Table S6. List of most sensitive parameters with best-fitted values, Table S7. Quantitative analysis of classified and simulated LU/LC 2020.

Author Contributions: Conceptualization, S.A.; data curation, M.I. and M.W.K.; formal analysis, Z.H.D.; investigation, M.I.; methodology, S.A. and Z.H.D.; resources, I.A.R. and M.W.K.; software, M.A. and M.I.; writing—original draft, M.I. and S.A.; writing—review and editing, K.A. All authors have read and agreed to the published version of the manuscript.

Funding: This research received no external funding.

Data Availability Statement: Data sharing is not applicable.

Conflicts of Interest: The authors declare no conflict of interest.

References

1. Idrissou, M.; Diekkrüger, B.; Tischbein, B.; de Hipt, F.O.; Näschen, K.; Poméon, T.; Yira, Y.; Ibrahim, B. Modeling the Impact of Climate and Land Use/Land Cover Change on Water Availability in an Inland Valley Catchment in Burkina Faso. *Hydrology* **2022**, *9*, 12. [CrossRef]
2. Ha, L.T.; Bastiaanssen, W.G.M.; Van Griensven, A.; Van Dijk, A.I.J.M.; Senay, G.B. Calibration of Spatially Distributed Hydrological Processes and Model Parameters in SWAT Using Remote Sensing Data and an Auto-Calibration Procedure: A Case Study in a Vietnamese River Basin. *Water* **2018**, *10*, 212. [CrossRef]
3. Clerici, N.; Cote-Navarro, F.; Escobedo, F.J.; Rubiano, K.; Villegas, J.C. Spatio-temporal and cumulative effects of land use-land cover and climate change on two ecosystem services in the Colombian Andes. *Sci. Total Environ.* **2019**, *685*, 1181–1192. [CrossRef]
4. Anand, J.; Gosain, A.; Khosa, R. Prediction of land use changes based on Land Change Modeler and attribution of changes in the water balance of Ganga basin to land use change using the SWAT model. *Sci. Total Environ.* **2018**, *644*, 503–519. [CrossRef]
5. Ellis, E.C.; Beusen, A.H.; Goldewijk, K.K. Anthropogenic Biomes: 10,000 BCE to 2015 CE. *Land* **2020**, *9*, 129. [CrossRef]
6. Verburg, P.H.; Neumann, K.; Nol, L. Challenges in using land use and land cover data for global change studies. *Glob. Chang. Biol.* **2011**, *17*, 974–989. [CrossRef]
7. Kiprotich, P.; Wei, X.; Zhang, Z.; Ngigi, T.; Qiu, F.; Wang, L. Assessing the Impact of Land Use and Climate Change on Surface Runoff Response Using Gridded Observations and SWAT+. *Hydrology* **2021**, *8*, 48. [CrossRef]
8. Kim, J.; Choi, J.; Choi, C.; Park, S. Impacts of changes in climate and land use/land cover under IPCC RCP scenarios on streamflow in the Hoeya River Basin, Korea. *Sci. Total Environ.* **2013**, *452–453*, 181–195. [CrossRef]
9. Gashaw, T.; Tulu, T.; Argaw, M.; Worqlul, A.W. Modeling the hydrological impacts of land use/land cover changes in the Andassa watershed, Blue Nile Basin, Ethiopia. *Sci. Total Environ.* **2018**, *619–620*, 1394–1408. [CrossRef] [PubMed]
10. Baker, T.J.; Miller, S.N. Using the Soil and Water Assessment Tool (SWAT) to assess land use impact on water resources in an East African watershed. *J. Hydrol.* **2013**, *486*, 100–111. [CrossRef]
11. Neupane, R.P.; Kumar, S. Estimating the effects of potential climate and land use changes on hydrologic processes of a large agriculture dominated watershed. *J. Hydrol.* **2015**, *529*, 418–429. [CrossRef]
12. Chanapathi, T.; Thatikonda, S. Investigating the impact of climate and land-use land cover changes on hydrological predictions over the Krishna river basin under present and future scenarios. *Sci. Total Environ.* **2020**, *721*, 137736. [CrossRef]
13. Guo, Y.; Fang, G.; Xu, Y.-P.; Tian, X.; Xie, J. Identifying how future climate and land use/cover changes impact streamflow in Xinanjiang Basin, East China. *Sci. Total Environ.* **2019**, *710*, 136275. [CrossRef]
14. Kamaraj, M.; Rangarajan, S. Predicting the future land use and land cover changes for Bhavani basin, Tamil Nadu, India, using QGIS MOLUSCE plugin. *Environ. Sci. Pollut. Res.* **2022**, *2022*, 1–12. [CrossRef]
15. Abhishek; Kinouchi, T.; Abolafia-Rosenzweig, R.; Ito, M. Water Budget Closure in the Upper Chao Phraya River Basin, Thailand Using Multisource Data. *Remote Sens.* **2021**, *14*, 173. [CrossRef]
16. Huntington, T.G. Evidence for intensification of the global water cycle: Review and synthesis. *J. Hydrol.* **2006**, *319*, 83–95. [CrossRef]
17. Kinouchi, T. Synergetic application of GRACE gravity data, global hydrological model, and in-situ observations to quantify water storage dynamics over Peninsular India during 2002–2017. *J. Hydrol.* **2021**, *596*, 126069. [CrossRef]
18. Pan, M.; Sahoo, A.K.; Troy, T.J.; Vinukollu, R.K.; Sheffield, J.; Wood, A.E.F. Multisource Estimation of Long-Term Terrestrial Water Budget for Major Global River Basins. *J. Clim.* **2012**, *25*, 3191–3206. [CrossRef]
19. Abolafia-Rosenzweig, R.; Pan, M.; Zeng, J.; Livneh, B. Remotely sensed ensembles of the terrestrial water budget over major global river basins: An assessment of three closure techniques. *Remote Sens. Environ.* **2020**, *252*, 112191. [CrossRef]
20. Wang, Q.; Guan, Q.; Lin, J.; Luo, H.; Tan, Z.; Ma, Y. Simulating land use/land cover change in an arid region with the coupling models. *Ecol. Indic.* **2020**, *122*, 107231. [CrossRef]
21. Haleem, K.; Khan, A.U.; Ahmad, S.; Khan, M.; Khan, F.A.; Khan, W.; Khan, J. Hydrological impacts of climate and land-use change on flow regime variations in upper Indus basin. *J. Water Clim. Chang.* **2021**, *13*, 758–770. [CrossRef]
22. Leta, M.; Demissie, T.; Tränckner, J. Modeling and Prediction of Land Use Land Cover Change Dynamics Based on Land Change Modeler (LCM) in Nashe Watershed, Upper Blue Nile Basin, Ethiopia. *Sustainability* **2021**, *13*, 3740. [CrossRef]
23. Wang, Q.; Wang, H. An integrated approach of logistic-MCE-CA-Markov to predict the land use structure and their micro-spatial characteristics analysis in Wuhan metropolitan area, Central China. *Environ. Sci. Pollut. Res.* **2022**, *29*, 30030–30053. [CrossRef] [PubMed]

24. Huang, H.; Zhou, Y.; Qian, M.; Zeng, Z. Land Use Transition and Driving Forces in Chinese Loess Plateau: A Case Study from Pu County, Shanxi Province. *Land* **2021**, *10*, 67. [CrossRef]
25. Tariq, A.; Shu, H. CA-Markov Chain Analysis of Seasonal Land Surface Temperature and Land Use Landcover Change Using Optical Multi-Temporal Satellite Data of Faisalabad, Pakistan. *Remote Sens.* **2020**, *12*, 3402. [CrossRef]
26. Zhao, M.; He, Z.; Du, J.; Chen, L.; Lin, P.; Fang, S. Assessing the effects of ecological engineering on carbon storage by linking the CA-Markov and InVEST models. *Ecol. Indic.* **2018**, *98*, 29–38. [CrossRef]
27. Rahman, K.U.; Shang, S.; Shahid, M.; Wen, Y. Hydrological evaluation of merged satellite precipitation datasets for streamflow simulation using SWAT: A case study of Potohar Plateau, Pakistan. *J. Hydrol.* **2020**, *587*, 125040. [CrossRef]
28. Abbaspour, K.C.; Rouholahnejad, E.; Vaghefi, S.; Srinivasan, R.; Yang, H.; Kløve, B. A continental-scale hydrology and water quality model for Europe: Calibration and uncertainty of a high-resolution large-scale SWAT model. *J. Hydrol.* **2015**, *524*, 733–752. [CrossRef]
29. Singh, L.; Saravanan, S. Simulation of monthly streamflow using the SWAT model of the Ib River watershed, India. *J. Hydro-Environ. Res.* **2020**, *3*, 95–105. [CrossRef]
30. Narsimlu, B.; Gosain, A.K.; Chahar, B.R. Assessment of Future Climate Change Impacts on Water Resources of Upper Sind River Basin, India Using SWAT Model. *Water Resour. Manag.* **2013**, *27*, 3647–3662. [CrossRef]
31. Kumar, N.; Singh, S.K.; Singh, V.G.; Dzwairo, B. Investigation of impacts of land use/land cover change on water availability of Tons River Basin, Madhya Pradesh, India. *Model. Earth Syst. Environ.* **2018**, *4*, 295–310. [CrossRef]
32. Tanksali, A.; Soraganvi, V.S. Assessment of impacts of land use/land cover changes upstream of a dam in a semi-arid watershed using QSWAT. *Model. Earth Syst. Environ.* **2020**, *7*, 2391–2406. [CrossRef]
33. Tamm, O.; Maasikamäe, S.; Padari, A.; Tamm, T. Modelling the effects of land use and climate change on the water resources in the eastern Baltic Sea region using the SWAT model. *CATENA* **2018**, *167*, 78–89. [CrossRef]
34. Getachew, B.; Manjunatha, B.; Bhat, H.G. Modeling projected impacts of climate and land use/land cover changes on hydrological responses in the Lake Tana Basin, upper Blue Nile River Basin, Ethiopia. *J. Hydrol.* **2021**, *595*, 125974. [CrossRef]
35. Nauman, S.; Zulkafli, Z.; Bin Ghazali, A.H.; Yusuf, B. Impact Assessment of Future Climate Change on Streamflows Upstream of Khanpur Dam, Pakistan using Soil and Water Assessment Tool. *Water* **2019**, *11*, 1090. [CrossRef]
36. Usman, M.; Ndehedehe, C.; Manzanas, R.; Ahmad, B.; Adeyeri, O. Impacts of Climate Change on the Hydrometeorological Characteristics of the Soan River Basin, Pakistan. *Atmosphere* **2021**, *12*, 792. [CrossRef]
37. Butt, A.; Shabbir, R.; Ahmad, S.S.; Aziz, N. Land use change mapping and analysis using Remote Sensing and GIS: A case study of Simly watershed, Islamabad, Pakistan. *Egypt. J. Remote Sens. Space Sci.* **2015**, *18*, 251–259. [CrossRef]
38. Tariq, A.; Riaz, I.; Ahmad, Z.; Yang, B.; Amin, M.; Kausar, R.; Andleeb, S.; Farooqi, M.A.; Rafiq, M. Land surface temperature relation with normalized satellite indices for the estimation of spatio-temporal trends in temperature among various land use land cover classes of an arid Potohar region using Landsat data. *Environ. Earth Sci.* **2019**, *79*, 40. [CrossRef]
39. Waseem Ghani, M.; Arshad, M.; Shabbir, A.; Shakoor, A.; Mehmood, N.; Ahmad, I. Investigation of Potential Water Harvesting Sites at Potohar Using Modeling Approach. *Pakistan J. Agric. Sci.* **2013**, *50*, 723–729.
40. Khan, M.T.; Shoaib, M.; Hammad, M.; Salahudin, H.; Ahmad, F.; Ahmad, S. Application of Machine Learning Techniques in Rainfall–Runoff Modelling of the Soan River Basin, Pakistan. *Water* **2021**, *13*, 3528. [CrossRef]
41. Hussain, F.; Nabi, G.; Wu, R.-S. Spatiotemporal Rainfall Distribution of Soan River Basin, Pothwar Region, Pakistan. *Adv. Meteorol.* **2021**, *2021*, 6656732. [CrossRef]
42. Nusrat, A.; Gabriel, H.F.; e Habiba, U.; Rehman, H.U.; Haider, S.; Ahmad, S.; Shahid, M.; Jamal, S.A.; Ali, J. Plausible Precipitation Trends over the Large River Basins of Pakistan in Twenty First Century. *Atmosphere* **2022**, *13*, 190. [CrossRef]
43. Final Results (Census-2017) | Pakistan Bureau of Statistics. Available online: https://www.pbs.gov.pk/content/final-results-census-2017 (accessed on 2 August 2022).
44. ALOS PALSAR—ASF. Available online: https://asf.alaska.edu/data-sets/sar-data-sets/alos-palsar/ (accessed on 13 August 2022).
45. USGS.Gov | Science for a Changing World. Available online: https://www.usgs.gov/ (accessed on 13 August 2022).
46. FAO/UNESCO Soil Map of the World | FAO SOILS PORTAL | Food and Agriculture Organization of the United Nations. Available online: https://www.fao.org/soils-portal/data-hub/soil-maps-and-databases/faounesco-soil-map-of-the-world/en/ (accessed on 13 August 2022).
47. Muhammad, W.K.; Shakil, A.; Zakir, H.D.; Zain, S.; Khalil Ahmad, F.K.M.A. Development of High Resolution Daily Gridded Precipitation and Temperature Dataset for Potohar Plateau of Indus Basin. *Remote Sens.* **2022**, *in press*.
48. Water & Power Development Authority. Available online: http://www.wapda.gov.pk/ (accessed on 31 August 2022).
49. Firozjaei, M.K.; Sedighi, A.; Argany, M.; Jelokhani-Niaraki, M.; Arsanjani, J.J. A geographical direction-based approach for capturing the local variation of urban expansion in the application of CA-Markov model. *Cities* **2019**, *93*, 120–135. [CrossRef]
50. Tadese, S.; Soromessa, T.; Bekele, T. Analysis of the Current and Future Prediction of Land Use/Land Cover Change Using Remote Sensing and the CA-Markov Model in Majang Forest Biosphere Reserves of Gambella, Southwestern Ethiopia. *Sci. World J.* **2021**, *2021*, 6685045. [CrossRef] [PubMed]
51. Yan, R.; Cai, Y.; Li, C.; Wang, X.; Liu, Q. Hydrological Responses to Climate and Land Use Changes in a Watershed of the Loess Plateau, China. *Sustainability* **2019**, *11*, 1443. [CrossRef]
52. Arnold, J.G.; Moriasi, D.N.; Gassman, P.W.; Abbaspour, K.C.; White, M.J.; Srinivasan, R.; Santhi, C.; Harmel, R.D.; van Griensven, A.; Van Liew, M.W.; et al. SWAT: Model Use, Calibration, and Validation. *Trans. ASABE* **2012**, *55*, 1491–1508. [CrossRef]

53. Shahid, M.; Rahman, K.U.; Haider, S.; Gabriel, H.F.; Khan, A.J.; Pham, Q.B.; Pande, C.B.; Linh, N.T.T.; Anh, D.T. Quantitative assessment of regional land use and climate change impact on runoff across Gilgit watershed. *Environ. Earth Sci.* **2021**, *80*, 743. [CrossRef]
54. Abbas, T.; Nabi, G.; Boota, M.W.; Hussain, F.; Faisal, M.; Ahsan, H.; Lahore, T.; Lahore, T. Impacts of Landuse Changes on Runoff Generation in Simly. *Sci. Int.* **2015**, *27*, 4083–4089.
55. Dibaba, W.T.; Demissie, T.A.; Miegel, K. Watershed Hydrological Response to Combined Land Use/Land Cover and Climate Change in Highland Ethiopia: Finchaa Catchment. *Water* **2020**, *12*, 1801. [CrossRef]
56. Zhang, S.; Yang, P.; Xia, J.; Wang, W.; Cai, W.; Chen, N.; Hu, S.; Luo, X.; Li, J.; Zhan, C. Land use/land cover prediction and analysis of the middle reaches of the Yangtze River under different scenarios. *Sci. Total Environ.* **2022**, *833*, 155238. [CrossRef] [PubMed]
57. Hakim, A.M.Y.; Baja, S.; Rampisela, A.D.; Arif, S. Spatial dynamic prediction of landuse/landcover change (case study: Tamalanrea sub-district, makassar city). *IOP Conf. Ser. Earth Environ. Sci.* **2019**, *280*, 012023. [CrossRef]
58. Anand, J.; Gosain, A.; Khosa, R.; Srinivasan, R. Regional scale hydrologic modeling for prediction of water balance, analysis of trends in streamflow and variations in streamflow: The case study of the Ganga River basin. *J. Hydrol. Reg. Stud.* **2018**, *16*, 32–53. [CrossRef]
59. Desai, S.; Singh, D.; Islam, A.; Sarangi, A. Multi-site calibration of hydrological model and assessment of water balance in a semi-arid river basin of India. *Quat. Int.* **2020**, *571*, 136–149. [CrossRef]
60. Nusrat, A.; Gabriel, H.; Haider, S.; Ahmad, S.; Shahid, M.; Jamal, S.A. Application of Machine Learning Techniques to Delineate Homogeneous Climate Zones in River Basins of Pakistan for Hydro-Climatic Change Impact Studies. *Appl. Sci.* **2020**, *10*, 6878. [CrossRef]
61. Abbaspour, K.C. Swat-Cup 2012. In *SWAT Calibration Uncertain. Program—A User Man*; Swiss Federal Institute of Aquatic Science and Technology: Dübendorf, Switzerland, 2012; p. 106.
62. Abbaspour, K.C.; Yang, J.; Maximov, I.; Siber, R.; Bogner, K.; Mieleitner, J.; Zobrist, J.; Srinivasan, R. Modelling hydrology and water quality in the pre-alpine/alpine Thur watershed using SWAT. *J. Hydrol.* **2007**, *333*, 413–430. [CrossRef]
63. Shrestha, M.K.; Recknagel, F.; Frizenschaf, J.; Meyer, W. Assessing SWAT models based on single and multi-site calibration for the simulation of flow and nutrient loads in the semi-arid Onkaparinga catchment in South Australia. *Agric. Water Manag.* **2016**, *175*, 61–71. [CrossRef]
64. Zhang, H.; Wang, B.; Liu, D.L.; Zhang, M.; Leslie, L.M.; Yu, Q. Using an improved SWAT model to simulate hydrological responses to land use change: A case study of a catchment in tropical Australia. *J. Hydrol.* **2020**, *585*, 124822. [CrossRef]
65. Moriasi, D.N.; Gitau, M.W.; Pai, N.; Daggupati, P. Hydrologic and Water Quality Models: Performance Measures and Evaluation Criteria. *Trans. ASABE* **2015**, *58*, 1763–1785. [CrossRef]
66. Monserud, R.A.; Leemans, R. Comparing global vegetation maps with the Kappa statistic. *Ecol. Model.* **1992**, *62*, 275–293. [CrossRef]
67. Landis, J.R.; Koch, G.G. The Measurement of Observer Agreement for Categorical Data. *Biometrics* **1977**, *33*, 159–174. [CrossRef] [PubMed]
68. Syed, Z.; Ahmad, S.; Dahri, Z.H.; Azmat, M.; Shoaib, M.; Inam, A.; Qamar, M.U.; Hussain, S.Z.; Ahmad, S. Hydroclimatology of the Chitral River in the Indus Basin under Changing Climate. *Atmosphere* **2022**, *13*, 295. [CrossRef]
69. Kundu, S.; Khare, D.; Mondal, A. Individual and combined impacts of future climate and land use changes on the water balance. *Ecol. Eng.* **2017**, *105*, 42–57. [CrossRef]
70. Gebremicael, T.; Mohamed, Y.; Van der Zaag, P. Attributing the hydrological impact of different land use types and their long-term dynamics through combining parsimonious hydrological modelling, alteration analysis and PLSR analysis. *Sci. Total Environ.* **2019**, *660*, 1155–1167. [CrossRef]
71. Spruill, C.A.; Workman, S.R.; Taraba, J.L. Simulation of daily stream discharge from small watersheds using the SWAT model. *Am. Soc. Agric. Biol. Eng.* **2000**, *1*, 1431–1439. [CrossRef]
72. Son, N.T.; Le Huong, H.; Loc, N.D.; Phuong, T.T. Application of SWAT model to assess land use change and climate variability impacts on hydrology of Nam Rom Catchment in Northwestern Vietnam. *Environ. Dev. Sustain.* **2022**, *24*, 3091–3109. [CrossRef]
73. Garg, V.; Aggarwal, S.P.; Gupta, P.K.; Nikam, B.R.; Thakur, P.K.; Srivastav, S.K.; Kumar, A.S. Assessment of land use land cover change impact on hydrological regime of a basin. *Environ. Earth Sci.* **2017**, *76*, 635. [CrossRef]
74. Samal, D.R.; Gedam, S. Assessing the impacts of land use and land cover change on water resources in the Upper Bhima river basin, India. *Environ. Chall.* **2021**, *5*, 100251. [CrossRef]
75. Dahri, Z.H.; Ludwig, F.; Moors, E.; Ahmad, S.; Ahmad, B.; Ahmad, S.; Riaz, M.; Kabat, P. Climate change and hydrological regime of the high-altitude Indus basin under extreme climate scenarios. *Sci. Total Environ.* **2021**, *768*, 144467. [CrossRef]

remote sensing

Article

System Structure–Based Drought Disaster Risk Assessment Using Remote Sensing and Field Experiment Data

Yi Cui [1,2], Huiyan Tang [1,2], Juliang Jin [1,2,*], Yuliang Zhou [1,2], Shangming Jiang [3] and Menglu Chen [1,2]

1 School of Civil Engineering, Hefei University of Technology, Hefei 230009, China
2 Institute of Water Resources and Environmental Systems Engineering, Hefei University of Technology, Hefei 230009, China
3 Key Laboratory of Water Conservancy and Water Resources of Anhui Province, Water Resources Research Institute of Anhui Province and Huaihe River Commission, Ministry of Water Resources, Hefei 230088, China
* Correspondence: jinjl66@hfut.edu.cn

Abstract: With the impacts of climate change and human activities, agricultural drought disaster losses have increased remarkably. Drought disaster risk assessment is a prerequisite for formulating disaster reduction strategies and ensuring food security. However, drought disaster risk is a complex system, and quantitative assessment methods reflecting the risk formation mechanism are still rarely reported. This study presented a chain transmission system structure of drought disaster risk, which meant that drought disaster loss risk R was derived from drought hazard H by the transformation of drought disaster vulnerability V. Based on this point, firstly, a drought hazard curve between drought intensity and drought frequency was established using remote sensing data and the copula function. Then, a crop loss calculation approach under various drought events and drought resistance capacity scenarios was achieved by two-season field experiments and the AquaCrop model. Finally, a loss risk curve cluster of "drought frequency–drought resistance capacity–yield loss rate" was proposed by the composition of the above two quantitative relationships. The results of the case study for summer maize in Bengbu City indicated that the average yield loss rate under 19 droughts occurring during the growth period of maize from 1982 to 2017 was 24.51%. High risk happened in 1988, 1992, 1994, 2001, and 2004, with the largest loss rate in 2001, up to 65.58%. Overall, droughts with a return period less than two years occurred frequently during the growth period of summer maize in Bengbu, though the loss risk was still controllable. In conclusion, the results suggest that the loss risk curve provides a new effective method of drought disaster risk quantitative assessment from a physical mechanism perspective, which lays a scientific foundation for decision-making in risk management.

Keywords: drought disaster risk assessment; disaster loss risk curve; system structure; GSMaP precipitation product; AquaCrop model; field experiment; scenario simulation

Citation: Cui, Y.; Tang, H.; Jin, J.; Zhou, Y.; Jiang, S.; Chen, M. System Structure–Based Drought Disaster Risk Assessment Using Remote Sensing and Field Experiment Data. *Remote Sens.* **2022**, *14*, 5700. https://doi.org/10.3390/rs14225700

Academic Editors: Songhao Shang, Magdy Mohssen, Qianqian Zhang, Dongqin Yin and Hamza Gabriel

Received: 29 September 2022
Accepted: 9 November 2022
Published: 11 November 2022

1. Introduction

Food security is an important issue internationally [1]. Drought disaster is one of the natural disasters that has the widest range of influence and causes the largest agricultural losses [2]. Nearly half of the countries in the world suffer from drought disasters [3]. In recent years, with global climate change and the effect of human activities, the frequency, intensity, and influence of drought disaster has witnessed a significant increase [4,5]. Correspondingly, the drought disaster losses and risk have remarkably increased, which poses a serious threat to world food security and agricultural sustainable development. Risk management is an effective means of dealing with extreme climate events, and risk assessment is the scientific foundation for risk management [6]. Therefore, the quantitative assessment of drought disaster risk is of great significance to improving human's ability to cope with drought, reducing agricultural losses and guaranteeing food security.

Drought disaster risk assessment is a difficult issue and a hot spot in the current research field of natural disasters [7]. At present, there are three main assessment methods [8,9]: (1) probability statistics–based method [10,11], (2) comprehensive evaluation–based method [12,13], and (3) system structure–based method [14,15]. Furthermore, the drought disaster risk assessment method from a system structure perspective can reflect the element structure and relational structure of drought disaster risk systems. It can simulate various uncertainties in the process of drought disaster risk formation. Therefore, this method has become the main direction of drought disaster risk study.

Drought disaster risk in this study is defined as the quantitative relationship between the possibility of future drought events with different intensities and the corresponding possible losses of drought disaster-bearing bodies. Further to the perspective of the drought disaster risk system structure, drought disaster loss risk (R) is a system output resulting from system input, i.e., the hazard of drought disaster–causing factor (H) by system transformation, i.e., the vulnerability of the drought disaster–bearing body (V), expressed as $R = f\ (H, V)$ [16,17]. Thus, it can be considered that drought disaster risk has a chain transmission system structure, which consists of the element structure of H, V, and R and the relation structure among these three. Drought hazard can be described quantitatively by the relation curve between drought intensity and the occurrence frequency of the corresponding drought event [18,19]. Drought disaster vulnerability can be described quantitatively by the relation curve between drought intensity and the corresponding losses of the disaster-bearing body [20,21]. Therefore, the drought disaster loss risk curve between drought possibility and the corresponding losses is obtained from the transformation of the drought hazard curve through to the drought disaster vulnerability curve. This loss risk curve method can fully reflect the drought disaster risk system structure and the risk formation mechanism [22]. However, there are a few studies focusing on this. In addition, the scenario simulation of crop modelling is an effective approach to establishing the drought disaster loss risk curve [23,24]. Crop models simulate the growth process of crops under drought stress by using statistical and dynamic methods, which can reveal the quantitative response relationship of crop growth and yield formation to drought.

AquaCrop is a crop model developed by the FAO in 2009 based on the water driving principle [25]. Due to advantages such as few input parameters, simple interface, strong intuition, and high precision, the AquaCrop model has been used in crop yield simulation [26], crop response to drought [27], and irrigation system optimization [28]. At present, crop models such as DSSAT [15], EPIC [23], and APSIM [24] are used for drought disaster risk assessment. Nevertheless, there are few studies on the application of the AquaCrop model, with its stronger adaptability and higher efficiency in drought disaster risk.

Maize is one of the worlds' three major food crops [29]. With the upgrading of the industrial structure and the improvement of people's living standards, the proportion of maize as fodder and industrial product has gradually increased [30]. Thus maize production plays a key role in global food security and industrial development [31]. According to the statistics released by the FAO, China's maize yield in 2020 was 2.61×10^{11} kg, accounting for 22.90% of the world's total maize yield and ranking No. 2 in the world [32,33]. However, maize requires a large amount of water during the growth period, and the main water source is precipitation. Therefore, precipitation has a great influence on the growth and yield of maize. China is located in the southeast of Eurasia; it has a significant monsoon climate and extremely uneven interannual and annual distributions of precipitation [22,34]. Moreover, due to the high temperature, strong evaporation, and frequent droughts during the growth period of summer maize, drought disaster has become an important factor affecting China's maize production [35,36]. To sum up, the drought disaster risk assessment method based on the system structure is less reported; there is a lack of studies on risk assessment that combine the scenario simulation of crop models with the disaster loss risk curve. In addition, the quantitative assessment of maize drought disaster risk in China's major production areas is urgently needed. Therefore, this study chose Bengbu City in the Huaibei Plain of Anhui Province in China as the study region. Using the field experiment data of summer maize implemented at Xinmaqiao Agriculture and Water Conservancy Comprehensive Experimental Station in Bengbu during 2018 and 2019 seasons, this study calibrated and verified the parameters of the AquaCrop model. Furthermore, drought index SPI and run theory were used to identify the drought events and characteristic variables during the growth period of summer maize in Bengbu from 1982 to 2017. Run theory is a method of time series analysis. For SPI series, the drought processes were identified by truncation level. The length of negative run (when SPI value was lower than the truncation level) was drought duration, and the degree of negative run (the cumulative deviation of SPI value below the truncation level) was drought intensity. Moreover, the copula function was applied to calculate the drought frequency of double variables. Then, the calibrated AquaCrop model was used to simulate the yield of maize under normal and drought conditions in various irrigation scenarios. Finally, the drought disaster loss risk curve between drought frequency and the corresponding yield loss rate of maize under different drought resistance capacities were established, and the drought disaster risk of summer maize in Bengbu was assessed quantitatively to provide decision support for guaranteeing regional maize production and food security.

2. Materials and Methods

2.1. Study Area

Bengbu City is the central city in Northern Anhui Province, China (32°43′N–33°30′N, 116°45′E–118°04′E) (Figure 1), with a land area of 5.95 × 10^3 km^2 and a population of 3.30 million. Meanwhile, Bengbu is a major production area of maize, belonging to the Huang-Huai-Hai planting region, as well as an important commodity grain base in China [37], with a cultivated area of 3.79 × 10^5 hm^2. According to statistics, in 2020, the sown area of summer maize in Bengbu was 1.31 × 10^5 hm^2 and the yield was 6.57 × 10^8 kg [38]. Furthermore, the summer maize in this region is mainly rain-fed. However, because Bengbu is located in the transition zone between temperate and subtropical monsoon climates, the annual distribution of precipitation is extremely uneven. Meanwhile, the monsoon climate causes a large inter-annual variation and frequent droughts [22,39,40]. For instance, during July and August 2013, the precipitation was 91% less than that of a normal year. There were 4.26 × 10^4 hm^2 of crops suffering from drought, which accounted for 13.62% of the planting area [41]. In addition, from September to November 2019, there was high temperature and little precipitation. Most regions suffered from mild drought, while some suffered from moderate drought. The affected area was 9.87 × 10^4 hm^2, and the area not sown was 1.05 × 10^4 hm^2 [42]. Therefore, assessing the summer maize drought disaster risk in Bengbu City is of great significance for stable grain production.

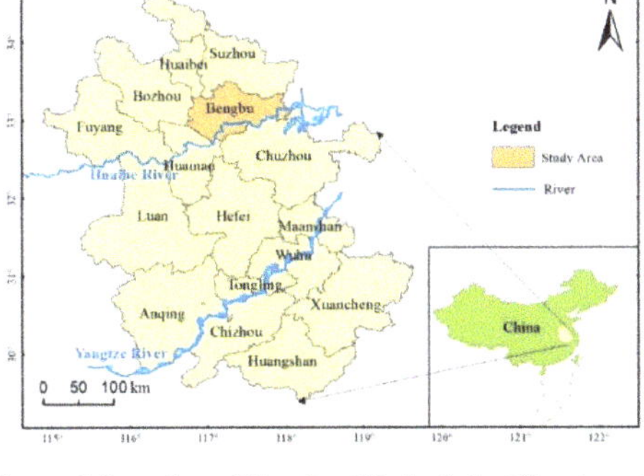

Figure 1. Location of Bengbu City in Anhui Province, China.

2.2. Establishment of Summer Maize Drought Disaster Risk Assessment Model

The establishment of the summer maize drought disaster risk assessment model in Bengbu City is as follows (Figure 2):

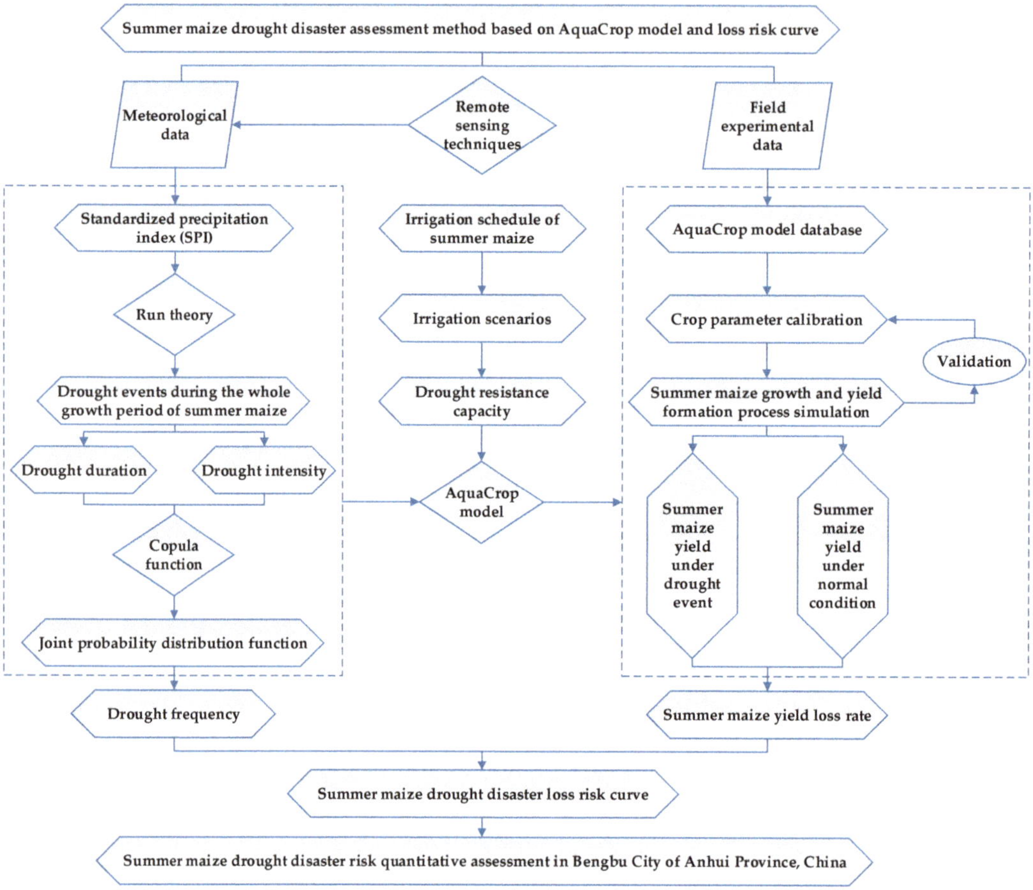

Figure 2. Establishment of summer maize drought disaster risk assessment model in Bengbu City.

(a) Drought frequency calculation

A standardized precipitation index with one-month SPI_1 was chosen as the drought index. Based on the precipitation data in Bengbu City from 1982 to 2017, the SPI_1 values were obtained. Furthermore, run theory was used to identify the drought events during the growth period of summer maize for each year. The two variables of drought duration and drought intensity were extracted. Then, the copula function was used to calculate the drought frequency corresponding to the joint probability distribution of duration and intensity.

(b) Drought resistance capacity simulation

The irrigation water amount at each growth stage of maize was selected as a quantitative index describing drought resistance capacity. Various irrigation scenarios were set to simulate different resistance capacities combined with the actual irrigation schedule.

(c) AquaCrop model parameter calibration

The measured field experiment data of summer maize conducted in 2018 and 2019 were used to calibrate and verify the key parameters of the AquaCrop model, so as to ensure the simulation precision of maize growth, development, and yield formation in Bengbu.

(d) Summer maize drought loss assessment

The drought loss of summer maize was represented by the yield loss rate. According to the identified drought events during the growth period of maize, the meteorological data in the year corresponding to each drought event and various irrigation scenarios were input into the calibrated AquaCrop model. Then, the yields of maize with different drought resistance capacities in the drought year were obtained, and the yield loss rate relative to the yield in normal year was assessed.

(e) Summer maize drought disaster loss risk curve establishment

The drought disaster loss risk in this study referred to the crop yield loss rate under a certain level of drought disaster–causing factors and a certain level of drought resistance capacity (assuming that the crop was fully exposed to a drought environment). The loss risk curve of drought frequency–drought resistance capacity–crop yield loss rate was established to quantitatively assess the drought disaster risk of summer maize in Bengbu.

2.3. Remote Sensing Data and Techniques

The daily precipitation data from 1982 to 2017 in Bengbu City were mainly from the ground-based Bengbu National Meteorological Station, which can be accessed from https://data.cma.cn (accessed on 12 September 2020). In addition, the missing and abnormal observations from the meteorological station were replaced with the corrected remote sensing data, guaranteeing the precision of precipitation data and drought disaster risk assessment results.

NASA (National Aeronautics and Space Administration) and JAXA (Japan Aerospace Exploration Agency) jointly implement the Global Precipitation Measurement (GPM). The satellite precipitation products under the GPM have a wider coverage and a higher spatial and temporal resolution; the satellite data and inversion algorithm used are enhanced compared with previous products, and the accuracy is improved. Furthermore, JAXA develops the Global Satellite Mapping of Precipitation (GSMaP), which fully integrates the observation data of GPM satellites and continuously optimizes the inversion algorithm, with the precision and resolution being further improved. In addition, GSMaP_Gauge is a high precision product with 1 h and 0.1°, which is adjusted by the Climate Prediction Center (CPC) precipitation gauge dataset (daily precipitation data derived from more than 30,000 gauges worldwide). This study replaced the missing and abnormal precipitation observations from the Bengbu National Meteorological Station with the matching GSMaP_Gauge product, which can be accessed from https://sharaku.eorc.jaxa.jp/GSMaP/ (accessed on 26 May 2022). Hence the remote sensing techniques provide an effective precipitation data supplement of the ground-based meteorological station, which greatly supports and promotes the drought identification and drought disaster risk quantitative assessment in this study.

2.4. Drought Frenquency Calculation

Bengbu City mainly belongs to the rain-fed agricultural region; thus, the meteorological drought index was more appropriate to reflect drought events. In this study, the standardized precipitation index with one-month SPI_1 [39] was selected as the drought index. Then, the run theory [15] was used to identify the drought events during the growth period of summer maize, and the drought duration and drought intensity of each event were extracted.

The process of drought event identification using SPI_1 and run theory is as follows (Figure 3); there were three truncation levels, R_0, R_1, and R_2, set in the process.

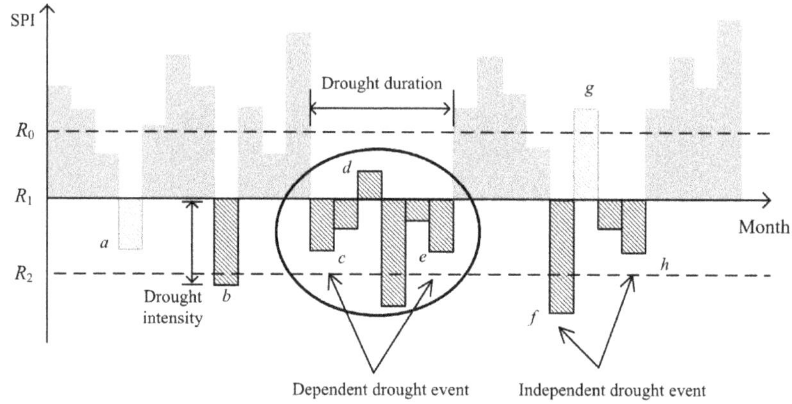

Figure 3. Drought event identification process based on standardized precipitation index (SPI) and run theory.

(1) When the SPI value was lower than R_1, it was preliminarily determined that a drought event occurred in this month (such as *a*, *b*, *c*, *e*, *f*, and *h* in Figure 3). Otherwise, there were no droughts (such as *g*).
(2) Then, for the drought event that lasted for only one month, when the SPI value was less than R_2, it was finally considered that there was a drought event in that month (such as *b* and *f*). Otherwise, there were no droughts (such as *a*).
(3) Furthermore, for two adjacent drought events with an interval of only one month, when the SPI value in the month of the interval was lower than R_0, these two adjacent droughts were merged into one event. The drought duration was the sum of these two drought durations plus 1; the drought intensity was the sum of two drought intensities (such as *c*, *d*, and *e*). Otherwise, there were two independent droughts (such as *f* and *h*).

Drought duration and drought intensity can generally be described by exponential distribution and gamma distribution, respectively. However, some studies have indicated that the fitting effects of exponential and gamma distributions were not adequate [43,44]. Therefore, in this study, the empirical frequency of drought duration and drought intensity were calculated by the formula of mathematical expectation, and the P-III distribution function was used to fit the empirical frequency points.

The probability density function of P-III distribution is shown as follows [39]:

$$f(x) = \frac{\beta^{\alpha}}{\Gamma(\alpha)}(x - a_0)^{\alpha-1}e^{-\beta(x-a_0)} \tag{1}$$

where α, β, and a_0 are the parameters of the probability density function of P-III distribution, which were obtained by the population parameters [39]:

$$\begin{cases} \alpha = \frac{4}{C_s^2} \\ \beta = \frac{2}{C_v C_s \overline{x}} \\ a_0 = \overline{x}\left(1 - \frac{2C_v}{C_s}\right) \end{cases} \tag{2}$$

where x, C_v, and C_s are the mean value, coefficient of variation, and coefficient of skew for the drought characteristic variable series (drought duration or intensity), respectively.

The G-H (Gumbel-Hougaard) function [45] in Archimedean copula was used to describe the correlation between drought duration and drought intensity. The copula joint distribution function is shown by the following formula [45,46]:

$$C_\theta(u,v) = e^{\{-\{[-\ln(u)]^\theta + [-\ln(v)]^\theta\}^{\frac{1}{\theta}}\}} \tag{3}$$

where $u = F_T(t)$, $v = F_D(d)$, and $F_T(t)$ and $F_D(d)$ are the probability distribution functions of drought duration and drought intensity, respectively. θ ($\theta \geq 1$) is the parameter of the G-H function and was calculated as follows [45,46]:

$$\theta = \frac{1}{1-\tau} \tag{4}$$

where τ is the Kendall rank correlation coefficient, and its calculation formula is as follows [45,46]:

$$\tau = \frac{2}{n(n-2)} \sum_{k>j} \mathrm{sgn}\left[(t_k - t_j)(d_k - d_j)\right] \tag{5}$$

where t_k, d_k and t_j, d_j represent the duration and intensity of the k-th and the j-th drought events, respectively. sgn(z) represents a sign function, that is, when $z > 0$, sgn(z) = 1, when $z = 0$, sgn(z) = 0, and when $z < 0$, sgn(z) = −1 [45,46].

When the drought duration was longer than t and the drought intensity was larger than d, the corresponding probability of exceedance was as follows [46]:

$$P(T > t \cap D > d) = 1 - F_T(t) - F_D(d) + C_\theta(t,d) \tag{6}$$

In addition, when the drought duration was longer than t or the drought intensity was larger than d, the corresponding probability of exceedance was as follows [46]:

$$P(T > t \cup D > d) = 1 - C_\theta(t,d) \tag{7}$$

2.5. Summer Maize Field Experiments

The field experiments were conducted at Xinmaqiao Agriculture and Water Conservancy Comprehensive Experimental Station (33°09′N, 117°22′E) in 2018 and 2019. This station was located in Xinmaqiao Town, Bengbu City, with an average altitude of 19.7 m, average temperature of 14.3 °C, average precipitation of 911 mm, and average evaporation of 917 mm. In each year, the experiment was implemented in a field with an area of about 600 m^2, and the field was divided into three plots (10 m × 15 m) for repeated tests. To avoid the influence caused by lateral migration of water, a 3 m long isolation zone was set between plots. The summer maize was sown on 15 June 2018 and 12 June 2019, and the variety was "Longping 206". Furthermore, according to the field planting density of summer maize in the Huaibei Plain, the density in each plot was 65,000 plants/hm^2. The application rate of compound fertilizer was 750 kg/hm^2, and the urea was 300 kg/hm^2.

The field experiments aimed to verify the simulation effect of the growth and yield formation process of summer maize under natural drought conditions by the AquaCrop model. Therefore, in the experiments during the 2018 and 2019 seasons, a completely rain-fed mode without irrigation for summer maize was designed. The soil in the field tillage layer was the typical Shajiang black soil of the Huaibei Plain [22]. In addition, during the experiment period, the field management measures in all experimental plots were consistent, so as to ensure the normal growth and development of summer maize plants.

According to the growth records of summer maize at the station over many years, and the studies on growth stage division of maize [30,35,36], the whole growth period of summer maize in the experiments was divided into four stages, i.e., the seedling stage, jointing stage, tasseling stage, and filling and ripening stage. Moreover, the whole growth process of maize was monitored, and some plants with uniform growth were randomly selected at each stage from each plot for destructive tests. The test items are described as follows.

2.5.1. Canopy Cover Degree

The WinFOLIA leaf image analysis system (Version: 2007b Basic, Regent Instruments Inc., Canada, Quebec, QC) and a scanner (CanoScan LiDE 90, Canon Inc., Que Vo, Vietnam) were used to scan the leaf area per maize plant sample. The leaf area index of the whole plant was obtained by accumulation. The average value of leaf areas for all samples in a plot was regarded as the leaf area index per plant. Furthermore, the canopy cover degree of each maize plant is shown as follows:

$$CC = 1.005 \times \left[1 - e^{(-0.6LAI)}\right]^{1.2} \tag{8}$$

where *LAI* and *CC* represent the leaf area index and canopy cover degree per plant.

2.5.2. Aboveground Biomass

The bottom of the maize stem was cut off to obtain the complete aboveground part. Each part (leaf, stem, spike, and seed) was separated and cleaned and put into a drying oven at 105 °C for one hour. The sample was dried at 80 °C until the weight was constant. Then, an electronic balance (TD30K-0.1, Tianjin Balance Instrument Co., Ltd., Tianjin, China) was used to weigh the total amount of aboveground dry matter for each plant. The average value of all samples in each plot was the aboveground biomass.

2.5.3. Seed Yield

All maize fruits in each plot were taken down and put into a numbered yarn bag and then were exposed uniformly outdoors until the weighs were constant. After the extraction of maize seeds, an electronic balance (TD30K-0.1, Tianjin Balance Instrument Co., Ltd., Tianjin, China) was used to weigh the total amount of seeds in each plot. The average value of all samples in each plot was the seed yield of maize.

2.6. Summer Maize Drought Loss Assessment

2.6.1. Input of AquaCrop Model

The input parameters of the AquaCrop model mainly included meteorological parameters, crop parameters, soil parameters, and field management parameters.

(a) Meteorological parameters

Meteorological parameters mainly included daily maximum air temperature, minimum air temperature, precipitation, CO_2 concentration, and reference crop evapotranspiration ET_0. The daily meteorological data in this study were obtained from the automatic meteorological station at Xinmaqiao station. The daily maximum and minimum air temperature during the whole growth period of summer maize at the station in 2018 and 2019 are shown in Figure 4. Furthermore, ET_0 was obtained by the "ET_0 calculator" recommended by the FAO [25]. The parameters required for the calculator, including daily maximum and minimum air temperature, relative humidity, wind speed, and hours of sunshine, were also provided by the automatic meteorological station. The daily precipitation and ET_0 during the whole growth period of maize in 2018 and 2019 are shown in Figure 5. The CO_2 concentration was obtained from the data of Mauna Loa CO_2 in the AquaCrop model [25].

(b) Crop parameters

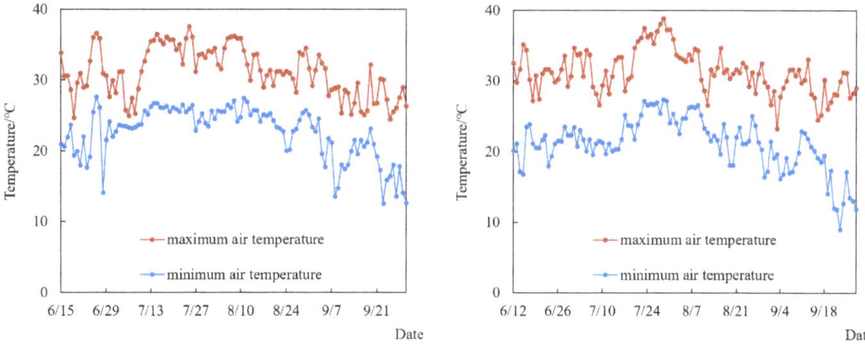

(**a**) (**b**)

Figure 4. Daily maximum and minimum air temperature during the whole growth period of summer maize at the experimental station. (**a**) 2018 season; (**b**) 2019 season.

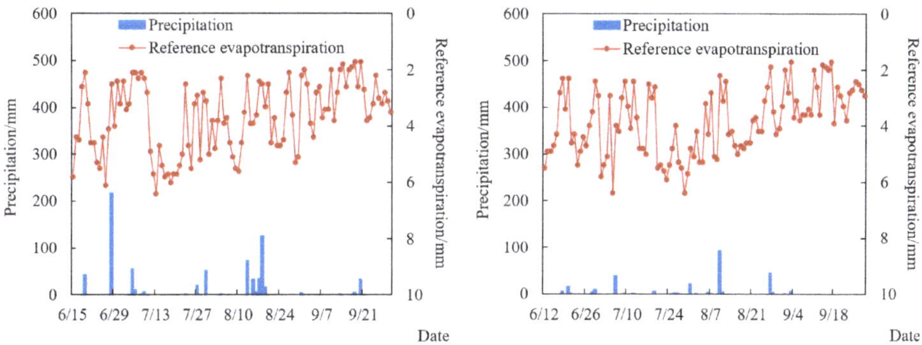

(**a**) (**b**)

Figure 5. Daily precipitation and reference evapotranspiration during the whole growth period of summer maize at the experimental station. (**a**) 2018 season; (**b**) 2019 season.

Crop parameters were determined according to the crop types (C3 or C4 crops, grain, tuber, or vegetable crops) and the actual situation of crop planting. These mainly included planting density, growth stage, canopy expansion, and root growth parameters, as well as those parameters that reflected the influences of various stress conditions (water, fertility, and temperature) on crop growth. The AquaCrop model manual provides the reference values or reference range of some physiological parameters for summer maize [25,47]. Specifically, for some physiological parameters that did not change with the planting time and location, this study adopted the reference values in the manual. However, for those that changed with the planting conditions, this study further calibrated and verified them based on the data of summer maize field experiments, i.e., the localization of the AquaCrop model.

(c) Soil parameters

Soil parameters mainly included the number of soil layers as well as the thickness, texture, and physical and chemical properties of each layer. The experimental soil in this study was Shajiang black soil, a kind of medium and low yield field soil with poor physical and chemical properties, adhesive texture, and permeability [22]. The summer maize selected in this study had a developed root system. In addition, a previous study reported that the proportion of maize root weight in the soil layer of 0–10 cm is 75.6%, and 94.4%

of root weight is distributed in the tillage layer of 0–40 cm [48]. Some have proposed that the maize root is mainly distributed in the soil layer of 0–40 cm, with little in the layer of 60–80 cm [49]. Some have found that the root system of maize during the seedling stage is distributed in a shallow soil layer, mainly between 0–20 cm. After the jointing stage, the root quickly reached about 100 cm underground during the tasseling stage [50]. Therefore, this study set three soil layers in combination with previous studies of the properties of Shajiang black soil [51,52] and input the soil parameters into the AquaCrop model (Table 1).

(d) Field management parameters

Table 1. Main soil parameters input into AquaCrop model.

Soil Layer	Thickness	Bulk Density ($g\ cm^{-3}$)	Wilting Water Content ($cm^3\ cm^{-3}$)	Field Capacity ($cm^3\ cm^{-3}$)	Saturated Water Content ($cm^3\ cm^{-3}$)
1	0–40 cm	1.18	0.13	0.33	0.42
2	40–80 cm	1.35	0.14	0.35	0.39
3	80–120 cm	1.48	0.17	0.37	0.41

Field management parameters included field management and irrigation management measures. Field management measures included fertilization level, farmland coverage, and covering materials, which were input according to the actual experimental situation. Irrigation management measures included irrigation method and irrigation schedule. In this study, the field experiments of maize were conducted under rain-fed conditions without irrigation; thus, the irrigation management parameters did not need to be set.

2.6.2. AquaCrop Model Parameter Calibration

The AquaCrop model manual listed the reference values or reference range of some physiological parameters of summer maize. However, some parameters that changed with planting conditions needed to be optimized according to the actual experimental situation. In this study, the data of maize field experiments in the 2019 season were used for calibration of AquaCrop model parameters, and those in 2018 were used for parameter verification. The process of AquaCrop model parameter calibration is shown in Figure 6.

In the process of AquaCrop model parameter calibration (Figure 6), canopy cover degree, aboveground biomass, and seed yield were selected as the objective functions. Firstly, the required meteorological parameters, soil parameters, field management parameters, and appropriate initial values of partial crop parameters referring to the AquaCrop model manual [47] were input into the model. Then, the values of crop parameters were adjusted continuously until the simulated canopy cover degrees approached the field measured results. Secondly, the remaining crop parameters were input and adjusted until the simulated aboveground biomass and seed yield were close to the measured values.

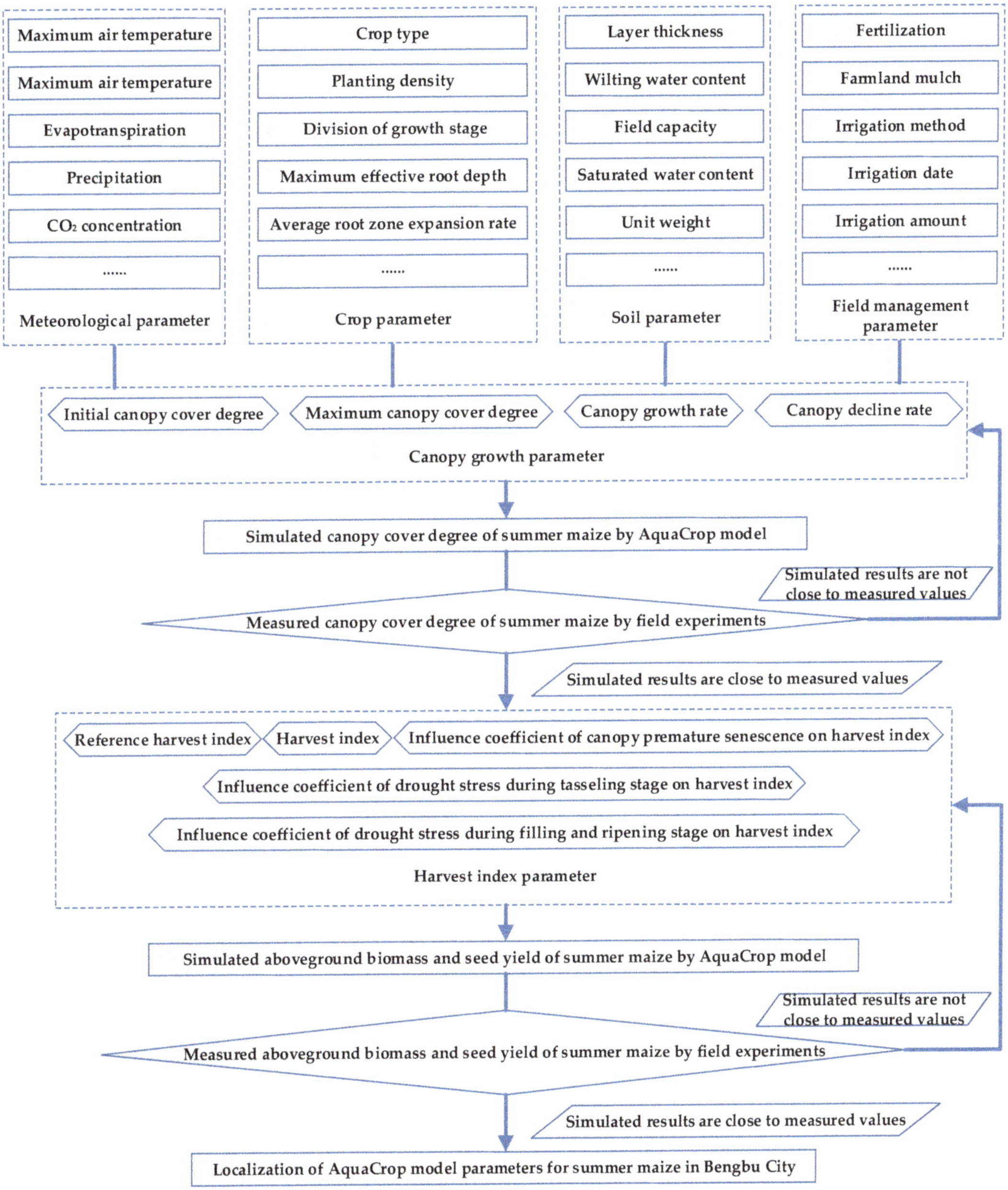

Figure 6. Process of AquaCrop model parameter calibration and verification.

Partially calibrated crop parameters in the AquaCrop model are shown in Table 2. Other crop parameters adopt the reference values given in the AquaCrop model manual.

Table 2. Partially calibrated crop parameters input into AquaCrop model.

Parameter Type	Parameter Meaning	Unit	Value
Canopy	initial canopy cover degree	%	0.35
	soil cover degree of single plant when germination rate reaches 90%	cm^2 per plant	5.0
	planting density	plant per hm^2	0.65×10^5
	canopy growth rate	% per day, relative increase in canopy cover	14.2
	maximum canopy cover degree	%	78
	canopy decline rate	% per day, relative decrease in canopy cover	12.5
Crop transpiration	crop transpiration coefficient		1.00
Doot zone	maximum effective root depth	m	1.20
	root zone expansion rate	cm per day	1.9
Yield	reference harvest index	%	48
	duration of harvest index	day	35
	duration of flowering period	day	13
Drought stress	lower threshold for influence coefficient of drought stress on canopy growth		0.12
	upper threshold for influence coefficient of drought stress on canopy growth		0.70
	influence coefficient of drought stress on stomatal control		5.5
	influence coefficient of drought stress on canopy senescence		3.5
	upper threshold for influence coefficient of drought stress on pollination		0.68

2.6.3. Validity Test of AquaCrop Model

To reasonably evaluate the fitting degree between the simulated results of the calibrated AquaCrop model and the field measured values, relative error (*RE*) and root mean squared error (*RMSE*) were used to quantify the simulation accuracy of the AquaCrop model:

$$RE = \frac{OV_n - SV_n}{OV_n} \times 100\% \tag{9}$$

$$RMSE = \sqrt{\frac{\sum\limits_{n=1}^{N} (OV_n - SV_n)^2}{N}} \tag{10}$$

where OV_n and SV_n are the measured and simulated values of canopy cover degree (aboveground biomass or seed yield), respectively. N is the number of maize samples.

2.6.4. Summer Maize Yield Loss Rate Calculation

The average value of many years for each meteorological parameter from the Bengbu National Meteorological Station was used as the input under normal meteorological conditions, while the meteorological data in each year corresponding to the actual drought events (during the growth period of summer maize) was used as the input under drought conditions. Meanwhile, combined with the irrigation scenarios that had been set, various irrigation management measures were input into the calibrated AquaCrop model to simulate the maize yields under normal and drought conditions. Thus, the yield loss rate *LS* of maize under different drought resistance capacities is calculated as follows:

$$LS = \frac{Y_N - Y_D}{Y_N} \times 100\% \tag{11}$$

where Y_N is the summer maize yield under the normal meteorological conditions, and Y_D is the maize yield under the actual drought events and different drought resistance capacities.

2.7. Summer Maize Drought Diasater Loss Risk Curve Establishment

In this study, the summer maize drought disaster loss risk curve was a set of curves between drought frequency and the corresponding maize yield loss rate under various drought resistance capacities (Figure 7). According to the water requirement characteristics and irrigation schedule of summer maize in Bengbu City, the amount of irrigation water, which was 0%, 50%, and 100% of the actual irrigation quota, was set and input into the calibrated AquaCrop model to simulate the maize yields under different resistance capacities during the same drought event period. Furthermore, the "drought frequency–drought resistance capacity–yield loss rate" summer maize drought disaster loss risk curve was established using the following formula:

$$LS = a\ln(P) + b \tag{12}$$

where P is the drought frequency, and a and b are the parameters of summer maize drought disaster loss risk curve.

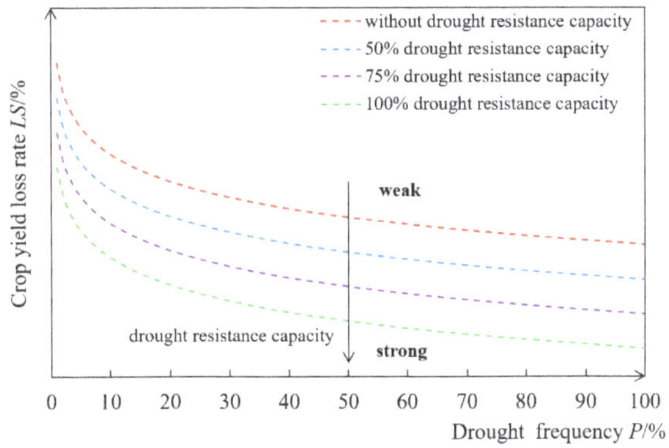

Figure 7. Drought disaster loss risk curve between drought frequency and the corresponding crop yield loss rate under various drought resistance capacities.

3. Results

3.1. Drought Frequency Analysis

The drought index SPI_1 and run theory were used to identify the drought events in Bengbu City from 1982 to 2017, and 19 drought events during the growth period of summer maize were selected. Then, the drought duration and drought intensity of each event were extracted. The empirical frequency points of drought duration and drought intensity of these 19 drought events were fitted, the results are shown in Figure 8 and Table 3. The degree of fitting between the P-III distribution curve and empirical frequency points were both higher than 0.90, indicating that the P-III distribution curve method had a good fit with the univariate probability distributions of drought duration and drought intensity for the drought events during the growth period of summer maize in Bengbu City.

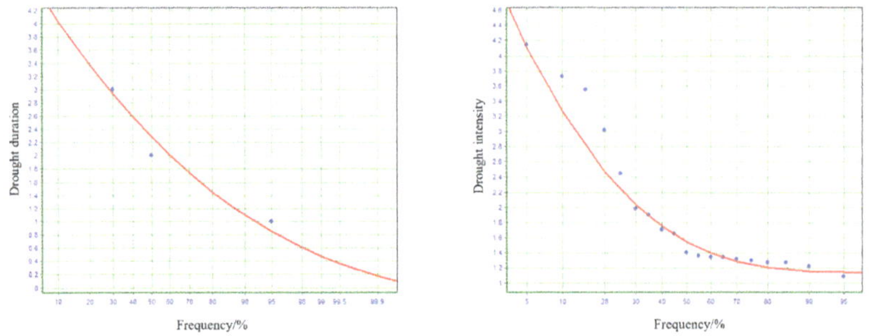

(a) (b)

Figure 8. P-III distribution curves of drought duration and drought intensity for drought events during the growth period of summer maize in Bengbu. (**a**) Drought duration; (**b**) drought intensity.

Table 3. P-III function parameters of drought duration and drought intensity for drought events during the growth period of summer maize in Bengbu.

Drought Characteristic Variable	Drought Duration				Drought Intensity			
	Mean Value	C_v	C_s	R^2	Mean Value	C_v	C_s	R^2
P-III function parameter value	2.46	0.48	0.87	0.96	1.95	0.55	2.65	0.93

The joint probability distribution function for drought duration and drought intensity was calculated by the G-H copula as shown in Figure 9; $\theta = 1.02$.

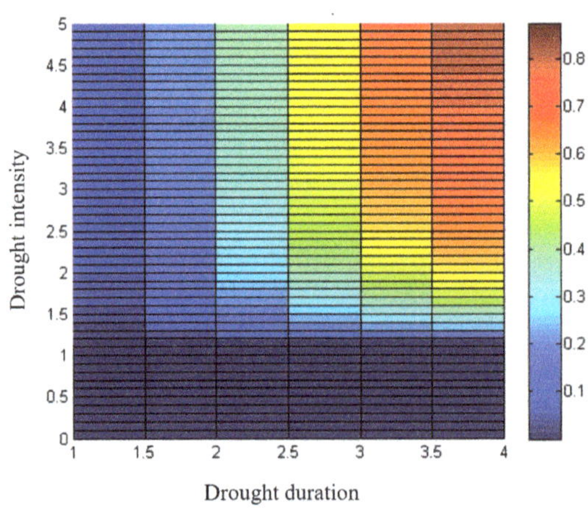

Figure 9. The joint probability distribution of drought duration and drought intensity for drought events during the growth period of summer maize in Bengbu.

The drought duration–drought intensity joint probability was the x-coordinate, and the drought intensity of the corresponding drought event was the y-coordinate. The power function, exponential function, and semilogarithmic function were used to fit the relationship, and the results are shown in Table 4. The degree of fitting of the semilogarithmic function was relatively high ($R^2 = 0.94$). Therefore, semilogarithmic function was selected to establish the quantitative relationship between drought duration–drought intensity joint drought frequency and drought intensity for 19 drought events during the growth period of summer maize in Bengbu City, i.e., drought hazard curve, as shown in Figure 10.

Table 4. Parameters of drought hazard curve in Bengbu.

Function Type	Power Fucntion $D = aP^b$			Exponential Function $D = ae^{bP}$			Semilogarithmic Function $D = a\ln(P) + b$		
	a	b	R^2	a	b	R^2	a	b	R^2
Function parameter value	8.17	−0.44	0.91	3.51	−0.02	0.83	−1.02	5.49	0.94

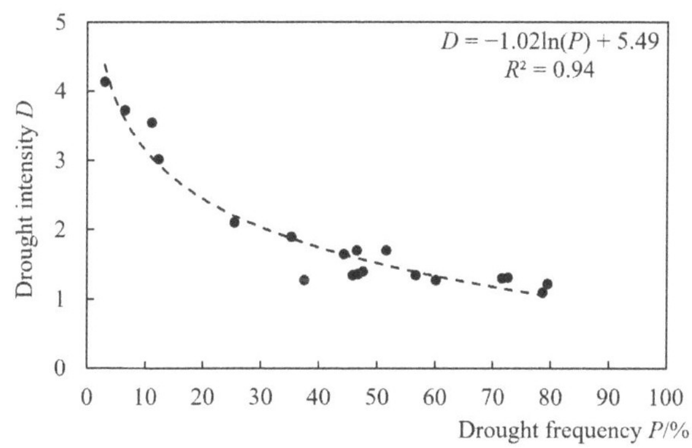

Figure 10. The drought hazard curve between drought frequency and drought intensity in Bengbu.

From Figure 10, as the drought duration–drought intensity joint frequency increased, drought intensity declined. There was a strong correlation between these two. Among 19 drought events during the growth period of maize in Bengbu from 1982 and 2017, there were 15 events whose drought frequency was between 0.2–0.8. Furthermore, the drought intensity was basically below 2.0; the events whose intensity were larger than 3.0 were fewer. This reflected that droughts occurred frequently during the growth period of summer maize in Bengbu, with most of them mild droughts with low intensity. These were consistent with the drought identification results in Cheng et al. [37] and Duan et al. [39].

3.2. *AquaCrop Model Simulation Analysis*

3.2.1. Canopy Growth Process Simulation Analysis

Canopy cover degree was used to describe the growth of summer maize leaves. The canopy cover degree of maize measured in the field experiments and the simulated results of the AquaCrop model are shown in Figure 11. According to Figure 11a, the canopy growth processes of maize plants under natural conditions without irrigation in 2018 and 2019 were basically consistent, and the maximum canopy cover degree in 2019 was slightly higher than that in 2018. In addition, the calibrated AquaCrop model accurately simulated the dynamic change of canopy cover degree with the growth of maize plants in 2018 and 2019. From Figure 11, the degree of fitting for low canopy cover was slightly worse than that in the second stage, i.e., the stable stage, which was the rapid growth period of the canopy. This was mainly influenced by the canopy growth rate in the AquaCrop model parameters. Furthermore, there were only a few observed canopy cover degrees during the seedling and tasseling stages in 2018, which also caused a certain impact on simulation accuracy. The *RMSE* of simulated canopy cover degree relative to the measured values in the field experiments in 2018 and 2019 was 0.067 and 0.079, respectively. In addition, there was a satisfactory correlation between the simulated and measured values (R^2 = 0.99) (Figure 11b). The linear regression fitting function passed through the point of origin, while the simulated results were overall less than the observed values, especially during the period of low canopy cover. Therefore, it is considered that a good degree of fit was achieved between the simulated canopy growth by the AquaCrop model and the field measured results.

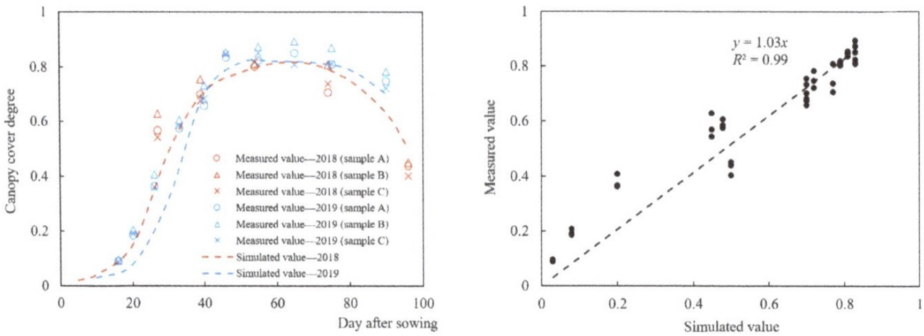

(a) (b)

Figure 11. Simulated canopy growth processes of summer maize during 2018 and 2019 seasons in Bengbu by the calibrated AquaCrop model. (**a**) Results comparison; (**b**) results correlation.

3.2.2. Aboveground Biomass Accumulative Process Simulation Analysis

The aboveground biomass accumulative process of plants is an important index to measure the productivity and development of maize. From Figure 12a, after parameter localization, the AquaCrop model well simulated the aboveground biomass accumulative process of summer maize plants in 2018 and 2019. As the number of planting days increased, maize plants gradually grew up, and the total amount of aboveground biomass continuously increased. Moreover, the accumulative rate firstly increased and then declined. Similar to canopy growth, the accumulative process of aboveground biomass under natural conditions without irrigation in 2018 and 2019 was basically consistent in the early growth period. However, when the maize plant grew to about 80 days, the aboveground biomass in 2019 was larger than that in 2018 until harvest. The *RMSE* of simulated aboveground biomass by the AquaCrop model relative to the field observed values in 2018 and 2019 was 0.755 and 0.966, respectively. There was a satisfactory correlation between simulated and measured results (R^2 = 0.99), and the linear regression fitting function passed through the point of origin (Figure 12b). This indicated that there was an accurate fitting between the simulated accumulative process of aboveground biomass for summer maize by the AquaCrop model and the measured values in field experiments.

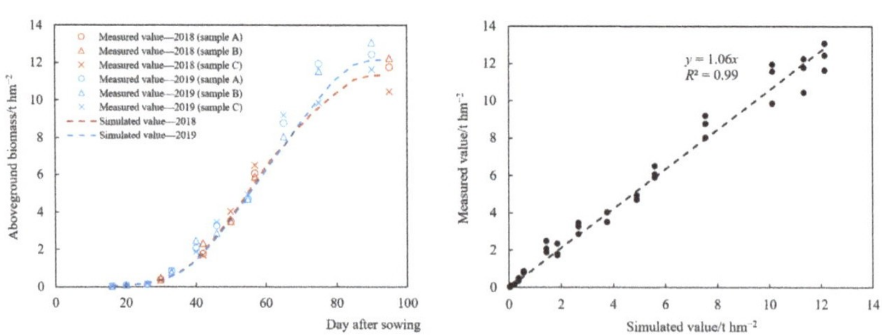

(a) (b)

Figure 12. Simulated accumulative processes of aboveground biomass for summer maize during 2018 and 2019 seasons in Bengbu by the calibrated AquaCrop model. (**a**) Results comparison; (**b**) results correlation.

3.2.3. Yield Simulation Analysis

The comparison between the simulated aboveground biomass at harvest and yield per unit area obtained by the AquaCrop model and the field observed values is shown in Table 5. The simulated aboveground biomass and yield in 2018 were both lower than those in 2019, which accorded with the measured results in field experiments. Furthermore, the relative error (*RE*) between the simulated and measured values of aboveground biomass at harvest in 2018 and 2019 were 1.38% and 1.79%, respectively, and the *RE* of yield per unit area was 0.67% and 0.31%. Overall, the simulated results were slightly larger than the measured values. The relative errors were quite small, indicating that the calibrated and verified AquaCrop model well simulated the growth and yield formation process of summer maize in Bengbu, and the crop parameters after localization were reasonable and accurate. Therefore, the drought disaster loss simulation of summer maize in Bengbu can be effectively implemented based on the calibrated AquaCrop model.

Table 5. Comparison between simulated aboveground biomass at harvest and yield per unit area by the calibrated AquaCrop model and field measured values of summer maize during 2018 and 2019 seasons in Bengbu.

Simulation Variable	2018 Season			2019 Season		
	Simulated Value	Measured Value	Relative Effor (*RE*)	Simulated Value	Measured Value	Relative Effor (*RE*)
Aboveground biomass at harvest (t hm^{-2})	11.33	11.49	1.38%	12.17	12.39	1.79%
Yield per unit area (t hm^{-2})	5.50	5.54	0.67%	5.86	5.88	0.31%

3.3. Summer Maize Drought Loss Simulation Analysis

3.3.1. Determination of Data under Normal Meteorological Conditions

The daily maximum and minimum temperature under normal meteorological conditions took the average values of daily data from the Bengbu National Meteorological Station from 1982 to 2017. The daily reference crop evapotranspiration was calculated by the "ET_0 calculator"; the required parameters took the average values of daily data from the Bengbu Station from 1982 to 2017. CO_2 concentration was obtained from the data of Mauna Loa CO_2 in the AquaCrop model. In addition, the daily precipitation was constructed from each month during the growth period of summer maize from the Bengbu Station from 1982 to 2017 combined with the corrected GSMap_Gauge precipitation product. Taking June as an example, the average value of monthly precipitation in June from 1982 to 2017 was obtained, and the precipitation in June of each year was compared with the average value; the daily precipitation that was closest to the average value was selected as the data under normal meteorological conditions. The daily maximum and minimum air temperature, precipitation, and reference crop evapotranspiration under normal meteorological conditions during the growth period of summer maize are shown in Figure 13.

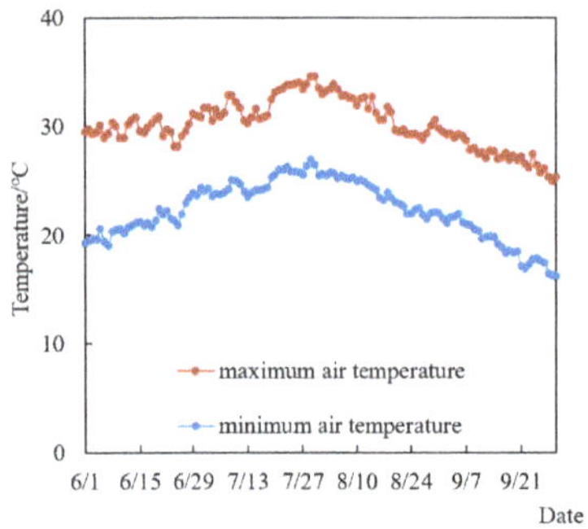

(a)

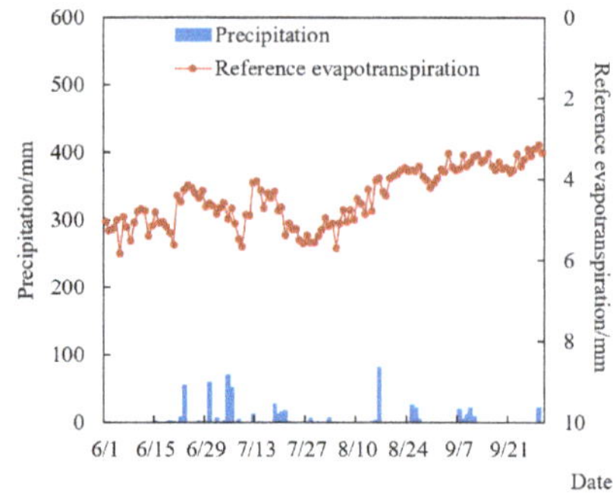

(b)

Figure 13. Daily maximum and minimum air temperature, precipitation, and reference evapotranspiration under normal meteorological conditions during the growth period of maize in Bengbu. (**a**) Maximum and minimum air temperature; (**b**) precipitation and reference evapotranspiration.

3.3.2. Summer Maize Yield Loss Simulation Analysis

The data under the normal meteorological conditions were input to the calibrated AquaCrop model, and the yield per unit area of summer maize under normal meteorological conditions in Bengbu was obtained. Then, the meteorological data under the actual drought events during the growth period of maize from 1982 to 2017 were input into the AquaCrop model, and the yield per unit area of maize during each drought period was obtained. Furthermore, compared with the yield under the normal meteorological conditions, the corresponding yield loss rate caused by each drought event was determined, as shown in Table 6.

Table 6. Simulated yield losses of summer maize under each drought event in Bengbu from 1982 to 2017 by the calibrated AquaCrop model.

Drought Event		Yied Per Unit Area during Drought Period/t hm^{-2}	Yield Loss Per Unit Area/t hm^{-2}	Yield Loss Rate/%
Year	**Month**			
1983	June, July, August	5.199	0.739	12.44
1985	August	4.967	0.971	16.35
1986	August	5.183	0.755	12.71
1987	September	5.098	0.840	14.14
1988	June, July, August	3.233	2.705	45.55
1992	June, July, August	3.557	2.381	40.10
1994	June, July, August	3.375	2.563	43.16
1996	August	4.897	1.041	17.53
1998	September	4.266	1.672	28.16
1999	September	5.246	0.692	11.65
2000	July, August	4.741	1.197	20.15
2001	June, July, September	2.044	3.894	65.58
2004	June, July, August	4.373	1.565	26.35
2010	July	4.715	1.223	20.59
2011	June, July	4.836	1.102	18.56
2012	June, July	4.579	1.359	22.88
2014	July	5.167	0.771	12.98
2015	July	5.094	0.844	14.21
2016	August, September	4.598	1.340	22.56
Mean value		4.483	1.455	24.51

The average yield per unit area under drought events for summer maize in Bengbu was 4.483 t/hm^2, and the yield loss was 1.455 t/hm^2. The average yield loss rate per unit area was 24.51%, which accounted for approximately 1/4 of the total yield. This reflected that droughts caused a severe impact on the summer maize growth in Bengbu, resulting in significant yield losses. In addition, serious yield reduction happened in 1988, 1992, 1994, 2001, and 2004, which is in agreement with the findings of Zhang et al. [53], Sun et al. [54], and Gao et al. [40]. The largest yield loss rate was in 2001, at up to 65.58%. According to the historical drought data in Bengbu, the period of 1990–1992 was three continuous drought years, 1994–1995 were the most severe drought years, and 2000–2001 was another period of serious drought, following 1978 and 1994 [15]. For instance, in 2001, the precipitation in Anhui Province was low; the flood season encountered an empty plum rain period [55]. On 27 July, the upstream water level of Bengbu Sluice declined to the lowest value for the same period in history [40]. The government of Bengbu City took emergency measures and gave priority to the urban water supply, which caused a great reduction of maize production. This indicates that the maize yield losses simulated by the AquaCrop model were basically consistent with the actual drought situations in agricultural production for Bengbu.

3.4. Summer Maize Drought Disaster Risk Loss Curve Analysis

According to the analysis of crop irrigation experiments in the Huaibei Plain of Anhui Province, water consumption during the growth period of the main crops in this region, such as maize [56], wheat [57], and soybean [58], were basically equal to the mean annual precipitation in the same period. However, due to the uneven distribution of precipitation, additional irrigation was required in most cases. Based on the experiments of crop water production function conducted at Xinmaqiao station from 1996 to 2001, some studies established models of optimal irrigation schedules for the four main crops in various types of hydrological years in the Huaibei Plain of Anhui Province [56]. The optimal irrigation schedule of summer maize in Bengbu City is shown in Table 7.

Table 7. Optimal irrigation schedule and economical irrigation quota of summer maize in Bengbu.

Hydrological Year Type	Irrigation Times	Irrigation Amount at Each Growth Stage/mm				Total Irrigation Amount/mm	Total Water Consumption/mm	Economical Irrigation Quota/mm
		Seedling Stage	Jointing Stage	Tasseling Stage	Filling and Ripening Stage			
50% (wet year)	0	0	0	0	0	0	375	45
	1	0	45	0	0	45	405	
75% (normal year)	0	0	0	0	0	0	320	125
	1	0	0	45	0	45	365	
	2	0	90	0	0	90	405	
	3	35	45	45	0	125	425	
95% (dry year)	0	0	0	0	0	0	250	225
	1	0	0	45	0	45	300	
	2	0	45	45	0	90	345	
	3	0	45	90	0	135	390	
	4	0	135	45	0	180	435	
	5	0	135	90	0	225	470	
	6	45	90	135	0	270	500	

From Table 7, the irrigation times and irrigation amounts during the growth period of summer maize increased with the reduction of precipitation, which can effectively increase the water supply and reduce the adverse impact on maize plants in drought years. For the hydrological years of 50%, 75%, and 95%, the corresponding economical irrigation quota of summer maize in Bengbu was 45 mm, 125 mm, and 225 mm, respectively. The optimal irrigation schedule of maize indicated that the irrigation time was basically concentrated in the jointing and tasseling stages. This reflected that the precipitation during this period in Bengbu was relatively low and could not meet the large water demand of maize plants for growth and development. In addition, the jointing stage and the tasseling stage were two key stages in which the growth of maize was quite sensitive to drought stress. The maize drought sensitivity results are consistent with the study of Wei et al. [59].

Based on the above research results and Industry Water-use Quota for Anhui Province (DB34/T 679—2019) and considering the practical field planting situation of summer maize in the Huaibei Plain of Anhui Province, the irrigation management measures in the AquaCrop model were set. Specifically, the irrigation method was flood irrigation, and irrigation dates were July 5 (jointing stage) and August 5 (tasseling stage). Moreover, three irrigation levels, which were 100% irrigation (45 mm on each irrigation date), 50% irrigation (22.5 mm on each irrigation date), and without irrigation, were set. Correspondingly, three irrigation scenarios (100%, 50%, and without drought resistance capacity scenarios) were arranged, as shown in Table 8. Consequently, the meteorological data of each actual drought event during the growth period of summer maize were input to the calibrated AquaCrop model to obtain the yield loss rates caused by each event under various irrigation levels. Then, the semilogarithmic function was adopted to fit the summer maize drought disaster loss risk curve between drought frequency and the corresponding yield loss rate under different irrigation levels (i.e., different drought resistance capacities) in Bengbu, as shown in Figure 14 and Table 9.

Table 8. Irrigation scenarios set for simulating different drought resistance capacities.

Irrigation Scenarios	Irrigation Times	Irrigation Amount at Each Growth Stage/mm				Total Irrigation Amount/mm
		Seedling Stage	Jointing Stage	Tasseling Stage	Filling and Ripening Stage	
Without drought resistance capacity	0	0	0	0	0	0
50% drought resistance capacity	2	0	22.5	22.5	0	45.0
100% drought resistance capacity	2	0	45.0	45.0	0	90.0

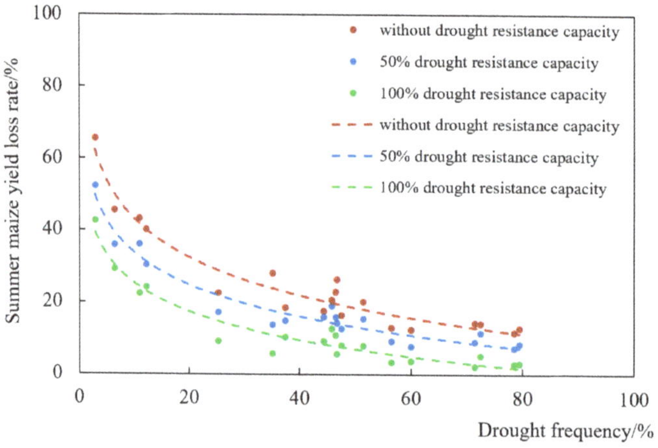

Figure 14. Summer maize drought disaster loss risk curve in Bengbu City.

Table 9. Function parameters of summer maize drought disaster loss risk curve in Bengbu City.

Drought Resistance Capacity Level	$LS = a\ln(P) + b$								
	Without Drought Resistance Capacity			50% Drought Resistance Capacity			100% Drought Resistance Capacity		
	a	b	R^2	a	b	R^2	a	b	R^2
Function parameter value	−15.37	78.55	0.94	−12.81	63.27	0.95	−11.43	51.64	0.94

According to Table 9, the semilogarithmic function was used to fit the quantitative relationships between drought frequency and the corresponding yield loss rate of maize under different irrigation levels, and the determination coefficients R^2 were all higher than 0.90. This indicates that there was a significant correlation between the drought frequency of the identified drought events during the growth period of summer maize in Bengbu and the corresponding yield loss rate simulated by the calibrated AquaCrop model. From Figure 14, as the drought frequency declined, the yield loss rate of maize continuously increased. That is, the drought disaster loss risk of maize continuously increased with the increase of drought hazard; the quantitative relationship between these two as fitted by the semilogarithmic function is reliable and precise, which reflects the drought-causing disaster mechanism and the system structure of drought disaster risk.

From Figure 14, the loss risk curve of maize without irrigation was significantly higher than that with irrigation, and the curve with 100% irrigation was lower than that with 50% irrigation. That is, under the same drought hazard, as the drought resistance capacity increases, the loss risk reduces significantly, which is in accordance with the physical mechanism of drought disaster risk [9,22]. This suggests that irrigation during the drought period can effectively relieve the yield loss of maize, and the stronger the drought resistance capacity, the smaller the loss [59]. For the droughts with frequency between 0.2–0.8, 100% irrigation reduced the yield loss rate to below 10%. Especially when the drought frequency was higher than 0.8, 100% irrigation basically prevented the drought loss. This indicates that for the droughts with high frequency and low intensity, adding timely irrigation is of great significance for avoiding yield loss of summer maize [30]. Nevertheless, for the droughts with frequency lower than 0.2, the loss reduction effect of irrigation was far less than that for droughts with high frequency. This phenomenon may be related to the fact that the irrigation water amount for summer maize set in this study is lower than the economical irrigation quota in Table 7, which cannot compensate for the water demand of plants after suffering from drought stress. Alternatively, this may be due to the fact that the drought intensity is too high and that drought stress causes irreversible damage to summer maize plants; thus, the loss mitigation effect of irrigation decreases. These are consistent with the compensation effect of irrigation for different drought degrees in the studies of Cui et al. [58,60].

For droughts with frequency between 0.3–0.5, as seen in Figure 14, the yield loss rate also increased with the adding of drought frequency, which was related to the occurrence time and intensity of drought events. For the identified drought events in Bengbu City, some differ in the occurrence time but have similar drought frequency between 0.3–0.5. However, the sensitivity of maize growth and development to drought stress at different growth stages is markedly different, so that the recovery effects of irrigation at different stages are different. These are in accordance with the phenomenon that the maize sensitivity to drought stress at different growth stages is significantly different, as obtained by Wang et al. [20] and Wei et al. [59]. Therefore, although there is similar drought frequency with various occurrence times, the yield losses are different. As a whole, the curves under three irrigation levels still meet the physical mechanism of drought disaster risk. That is, under the same drought resistance capacity, as the drought frequency increases, the yield loss declines.

4. Discussions

Drought disaster risk has a clear chain transmission system structure, which includes an element structure of drought hazard (H), drought disaster vulnerability (V), and drought disaster loss risk (R), and a relation structure that R is derived from H by the transformation of V. For the specific functional relationship, the drought disaster loss risk curve between drought frequency and crop losses is obtained from the transformation of the drought hazard curve between drought frequency and drought intensity, by the drought disaster vulnerability curve between drought intensity and crop losses (Figure 15). Moreover, drought intensity is the key intermediate variable in the process. Hence, the established loss risk curve in this study fully reflects the formation mechanism of drought disaster risk.

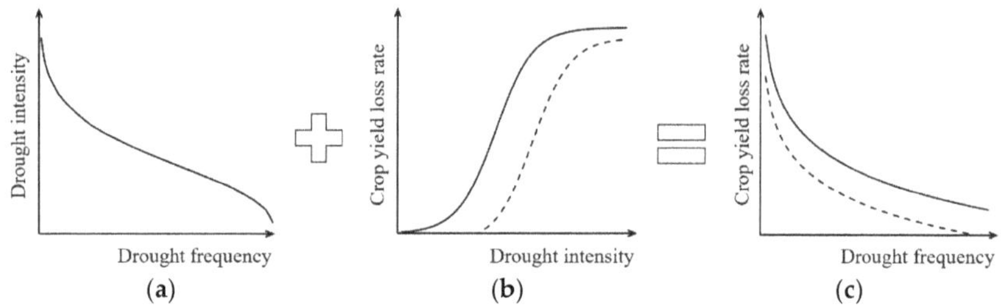

Figure 15. Chain transmission process of drought disaster loss risk curve. (**a**) Drought hazard curve; (**b**) drought disaster vulnerability curve; (**c**) drought disaster loss risk curve.

The loss risk curve established in this study provides effective support for the system structure of drought disaster risk, i.e., the drought hazard is transformed into drought disaster loss risk by the vulnerability of the drought disaster–bearing body. This is the same as the chain transmission theory of drought disaster risk proposed by Jin et al. [61]. Furthermore, this study shows that the semilogarithmic risk curve between drought frequency and the corresponding yield loss rate quantitatively describes the drought disaster risk and basically agrees with the relevant research on drought disaster risk assessment by Zhang et al. [14], Yin et al. [23], and Wang et al. [24]. In addition, from the perspective of practical significance, in Figure 14, the x-coordinate represents drought frequency, and the y-coordinate represents crop yield loss rate under different irrigation levels, which visually reflects the physical meaning of drought disaster risk [15,22]. Moreover, Figure 14 can be used to quickly estimate the possible yield loss of summer maize in various drought and irrigation scenarios, which provides a scientific guarantee to conduct a reasonable assessment of drought loss and an effective response to drought disaster risk for Bengbu City.

The drought disaster risk curve cluster under various drought resistance capacities built in this study can accurately evaluate the potential losses when encountering droughts with different frequencies in the future. Then, appropriate drought resistance measures can be taken in advance or not, according to the acceptable drought disaster risk (crop yield loss rate) threshold, which provides key decision support for risk prevention and control and effectively reduces losses. Furthermore, this curve cluster is a fundamental work for assessing drought disaster risk under the actual drought resistance capacity.

In fact, the drought resistance capacity is generally not constant but decreases with the increase of drought severity. The lower the drought frequency, the lower the available water resource amount, and the weaker the drought resistance capacity. Therefore, to assess the drought disaster risk under an actual drought resistance capacity, firstly, the quantitative relationship between drought resistance capacity and drought frequency should be built. For a given frequency, the actual resistance capacity can be obtained. Then, according to the established loss risk curve cluster under various resistance capacities in this study (Figure 7), the crop yield loss rate under the given frequency and resistance capacity can be obtained by interpolation. Thus, the loss risk curve between drought frequency and the corresponding crop yield loss rate under the actual drought resistance capacity is established, which may represent important future work on the basis of this study.

Precipitation is the most direct drought disaster-inducing factor and is usually adopted to construct the drought index; thus, the precision of precipitation data markedly affects the results of drought event identification and drought disaster risk assessment. Normally, precipitation data use the observations from ground-based meteorological stations. However, due to the influences of geographic, economic, external environment, and other factors, the station network is usually sparsely and unevenly distributed, lacking good temporal continuity and spatial consistency. In addition, precipitation has a large variability of temporal and spatial distributions and a strong uncertainty, especially for extreme precipitation events, such as droughts. Therefore, for the ground-based meteorological station, it is difficult to provide precipitation information with high temporal and spatial resolutions in a large range. The missing and abnormal precipitation observations from meteorological stations are usually difficult to obtain by spatial interpolation. Thus, when the available stations are sparse, the calculated drought index according to the station observations cannot accurately depict the actual drought situations on a regional scale. In recent years, with the rapid development of remote sensing and data inversion techniques, a range of precipitation products based on satellite remote sensing inversion have been released, which have a wide scale coverage and high temporal and spatial resolutions. These remote sensing data effectively make up for the lack of ground station spatial distribution, provide a new data source for the calculation of drought index, which improves the precision of drought identification and drought disaster risk assessment.

This study replaced the missing and abnormal precipitation observations from 1982 to 2017 in Bengbu City from the Bengbu National Meteorological Station with the corrected GSMaP_Gauge satellite products. According to the results of drought event identification (Figure 8 and Table 3) and the drought hazard curve (Figure 9), the adjusted GSMaP_Gauge data were well matched with the ground station observations. The fused precipitation data were accurately used to identify the drought process combined with drought index SPI in Bengbu. This may be due to the fact that the GSMaP_Gauge is a satellite precipitation product adjusted by the CPC global gauge dataset. Therefore, the satellite remote sensing data play an important role in this study. GSMaP_Gauge provides a valid precipitation data source for supplementing the Bengbu station, which lays the data foundation for system structure–based drought disaster risk quantitative assessment in Bengbu. In addition, this study verifies the effectiveness of fusion between station precipitation data and the GSMaP_Gauge product, providing an effective way to further conduct regional or large-scale drought disaster risk study using remote sensing data.

This study uses the field experimental data, meteorological data, and soil and crop parameters provided by Xinmaqiao station to calibrate the AquaCrop model and then calculate the drought disaster loss risk of summer maize in Bengbu City. Firstly, the Xinmaqiao experimental station (33°09′N, 117°22′E) is located in the center of Bengbu City (32°43′N–33°30′N, 116°45′E–118°04′E) (Figure 16), close to the Huaihe River; it has the typical climate characteristics of temperate and subtropical monsoon transition zones, like Bengbu. Specifically, the mean monthly precipitation and reference crop evapotranspiration in Xinmaqiao station and Bengbu City are highly consistent (Figure 17). Meanwhile, the main daily meteorological indexes during the growth period of summer maize (June to September) for the 2018 season in Xinmaqiao station are all very close to those in Bengbu (Figure 18). Furthermore, the soil (Shajiang black soil) and maize variety (Longping 206) used in Xinmaqiao station are the main types in Bengbu. Therefore, it can be considered that the data provided by Xinmaqiao station are representative for the whole Bengbu area. In addition, if the experimental conditions are sufficient, multiple stations' data will be further applied to the drought disaster risk assessment of Bengbu City in future work.

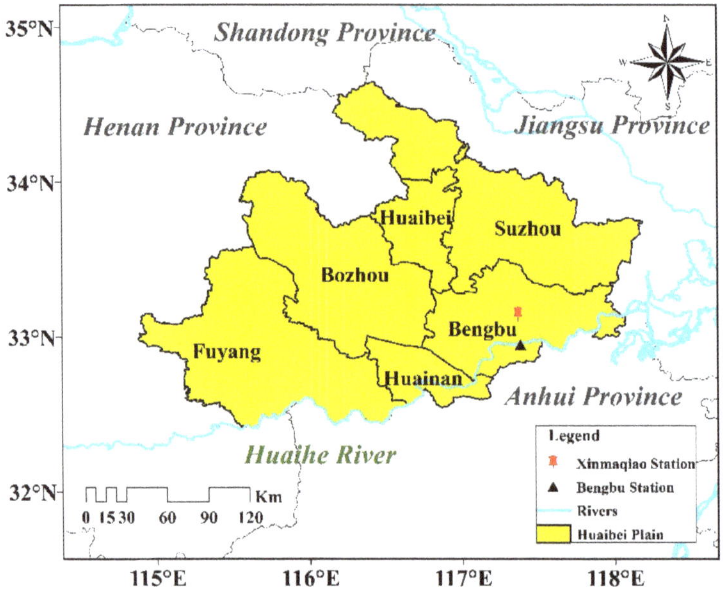

Figure 16. Location of Xinmaqiao experimental station in Bengbu City.

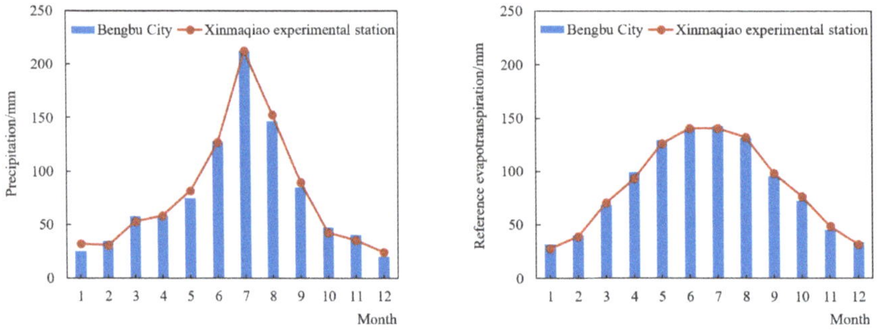

(a) (b)

Figure 17. Comparison between mean monthly precipitation and reference evapotranspiration in Xinmaqiao station and Bengbu City. (**a**) Precipitation; (**b**) reference evapotranspiration.

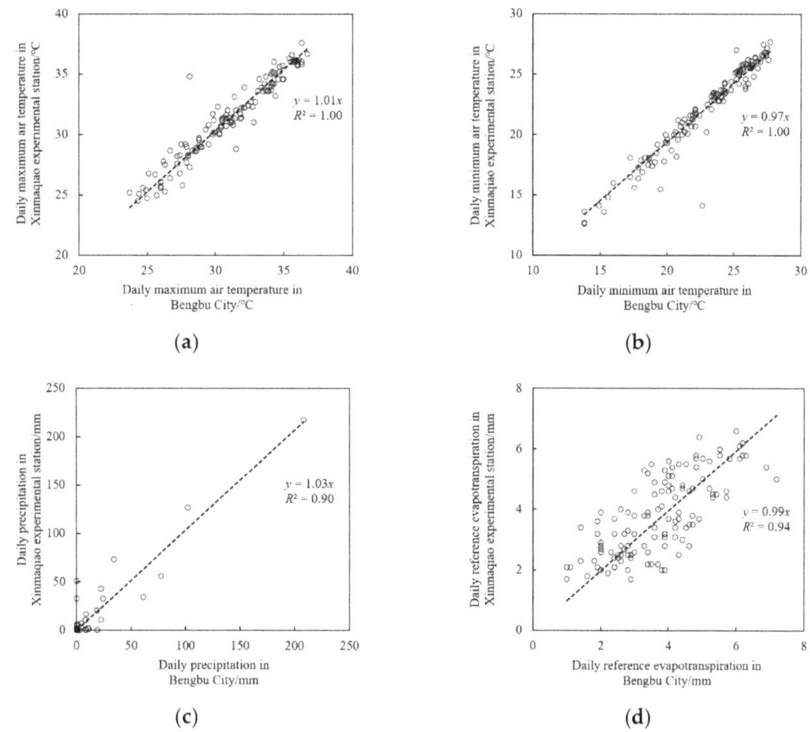

Figure 18. Correlations between main daily meteorological indexes during the growth period of summer maize for the 2018 season in Xinmaqiao station and Bengbu City. (**a**) Maximum air temperature; (**b**) minimum air temperature; (**c**) precipitation; (**d**) reference evapotranspiration.

Based on the reference values of maize crop parameters in the AquaCrop model manual [47], this study further adopts two-season field experiment data to calibrate and verify partial crop parameters, which change with the actual planting conditions (Table 2). Firstly, the simulated canopy cover degree (Figure 11), aboveground biomass (Figure 12), and biomass yield (Table 5) of summer maize in Bengbu using the calibrated parameters indicate that the simulated results are all highly consistent with the field measured values; the simulation accuracy meets the requirements. Furthermore, the simulated yield losses of maize caused by severe droughts in Bengbu from 1982 to 2017 (Table 6) are all in accordance with the historical drought situations [41,55] and relevant studies [40,52,53]. In addition, the calibrated maize crop parameters in this study (Table 2) are basically consistent with the studies of Han et al. [62], Wolka et al. [63], and Wu et al. [64], who obtained the parameters by field experiments in the Heihe River Basin of China, the BokoleKartha watershed of southwest Ethiopia, and Wuwei City of northwest China, respectively. Therefore, it can be considered that the obtained crop parameters of summer maize for the AquaCrop model in Table 2 are reasonable. Moreover, these parameters can be further verified and modified by continuous field experiments in future work.

The crop parameter of maximum canopy cover in the AquaCrop model for summer maize in this study is 78%. First, the simulated canopy cover degrees are highly consistent with the field measured results (Figure 11). Then, the measured samples of maize maximum canopy cover degree in 2018 and 2019 seasons are only 80.15% and 85.00%, respectively. The field experiments in this study are conducted under a rain-fed condition; the water demand of maize plants cannot be fully met. According to the measured gravimetric soil water content during the growth period of maize (Figure 19), the values are basically lower than 75% field capacity (19.65%). Furthermore, the measured yields per unit area in 2018 and 2019 seasons are 5.54 t/hm^2 and 5.88 t/hm^2 (Table 5), respectively, which are both less than that under normal meteorological conditions (5.94 t/hm^2). This reflects that there is significant drought stress for maize plant growth in the experiments, resulting in a maximum canopy cover degree of only about 80%. Similar results were proposed by Abedinpour et al. [65], who found that the maximum canopy cover degree of maize decreased with the declining of soil water content in New Delhi, India; those under full irrigation and rain-fed conditions were about 90% and 80%, respectively. Moreover, the maximum canopy cover parameter of 78% may be related to the planting density, meteorological conditions, and maize variety [66]. A relatively low maximum canopy cover parameter of maize in the AquaCrop model was also obtained in some studies. Nyakudya et al. [66] calibrated the parameter of maximum canopy cover CC_x for rain-fed maize in a semi-arid region of Zimbabwe by field experiments in various sites, and the CC_x in the Magaranhewe site and Chongma site was 70% and 65%, respectively. Ran et al. [67] presented that the calibrated CC_x for summer maize in an arid region of northwest China during the 2012 and 2013 seasons was 85%, and the field measured maximum canopy cover degree was about 80%. Furthermore, in the study of Li et al. [68], the calibrated CC_x for summer maize in the Shijin irrigation district of North China was 80%. Therefore, it can be considered that the CC_x of 78% for maize in this study is reasonable.

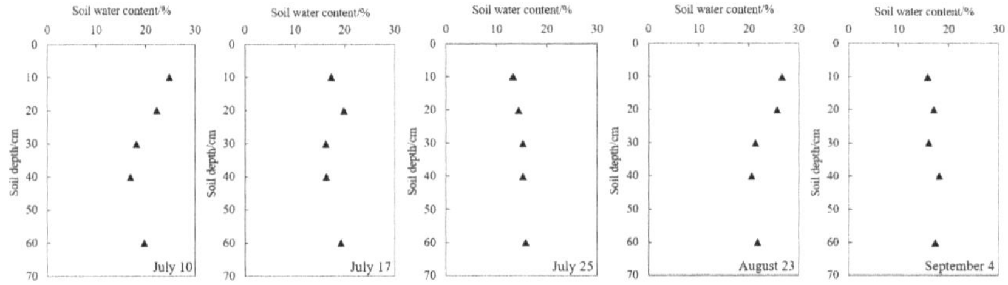

Figure 19. Field measured gravimetric soil water content during the growth period of summer maize in 2018 season.

5. Conclusions

In this study, the drought events and characteristic variables during the growth period of summer maize in Bengbu from 1982 to 2017 were identified, and the drought frequency of double variables was calculated. In addition, the AquaCrop model was used to simulate the yield loss of maize based on field experiments in 2018 and 2019. Finally, the loss risk curves between drought frequency and the corresponding yield loss rate under different drought resistance capacities were established. The conclusions are as follows:

(1) The P-III distribution curve method well fitted the univariate probability distributions of drought duration and drought intensity. Furthermore, the semilogarithmic function quantitatively described the drought hazard curve between drought duration–drought intensity joint drought frequency and drought intensity. Among 19 drought events, there were 15 events whose drought frequency was 0.2–0.8. Moreover, the drought intensity was basically below 2.0; the events whose intensity was larger than 3.0 were less. Therefore, droughts occurred frequently during the growth period of summer maize in Bengbu, though most of them were mild droughts with low intensity.

(2) The *RMSE* of simulated maize canopy cover degree by the AquaCrop model relative to the field measured results in 2018 and 2019 was 0.067 and 0.079, respectively. The *RMSE* of aboveground biomass was 0.755 and 0.966. In addition, the relative error (*RE*) between the simulated and measured aboveground biomass at harvest in 2018 and 2019 was 1.38% and 1.79%, respectively. The *RE* of yield per unit area was 0.67% and 0.31%. Therefore, the optimized crop parameters were effective; the calibrated

AquaCrop model accurately simulated the growth and yield formation process of summer maize in Bengbu.

(3) The simulated average yield loss per unit area under 19 drought events identified during the growth period of summer maize in Bengbu from 1982 to 2017 was 1.455 t/hm^2, and the yield loss rate was 24.51%. Droughts caused a severe impact on the summer maize production in Bengbu, resulting in significant yield losses. In addition, serious yield reduction happened in 1988, 1992, 1994, 2001, and 2004, and the largest yield loss rate was in 2001, at up to 65.58%. Therefore, the simulated yield losses were consistent with the actual drought situations in maize production for Bengbu.

(4) The semilogarithmic function accurately depicted the summer maize drought disaster loss risk curve in Bengbu City. Under the same drought hazard condition, as the drought resistance capacity increased, the maize loss risk reduced significantly. Furthermore, for the droughts with frequency between 0.2–0.8, 100% irrigation reduced the yield loss rate of maize to below 10%. Especially when the frequency was higher than 0.8, 100% irrigation basically prevented the loss. Therefore, for the droughts with high frequency and low intensity in Bengbu, adding timely irrigation was a key measure to reduce the yield loss of maize. Nevertheless, for the droughts with frequency lower than 0.2, the loss reduction effect of irrigation was far less than that for the droughts with high frequency. This study provides an effective approach for quantifying the regional drought disaster loss risk and supporting the decisions of regional drought disaster risk management.

Author Contributions: Conceptualization, Y.C. and J.J.; methodology, Y.C. and J.J.; validation, H.T., Y.Z. and M.C.; resources, S.J.; writing—original draft preparation, Y.C.; writing—review and editing, H.T., J.J. and Y.Z.; visualization, S.J. and M.C.; funding acquisition, Y.C. and J.J. All authors have read and agreed to the published version of the manuscript.

Funding: This research was funded by the National Natural Science Foundation of China (grant Nos. 52109009, U2240223), the Natural Science Foundation of Anhui Province, China (grant Nos. 2108085QE254, 2208085QE179), and the Fundamental Research Funds for the Central Universities (grant No. JZ2021HGTA0165).

Data Availability Statement: The raw data supporting the conclusion of this article will be made available by the authors, without undue reservation.

Conflicts of Interest: The authors declare no conflict of interest.

References

1. Kchouk, S.; Melsen, L.A.; Walker, D.W.; van Oel, P.R. A geography of drought indices: Mismatch between indicators of drought and its impacts on water and food securities. *Nat. Hazards Earth Syst. Sci.* **2022**, *22*, 323–344. [CrossRef]
2. Hameed, M.; Ahmadalipour, A.; Moradkhani, H. Drought and food security in the Middle East: An analytical framework. *Agric. For. Meteorol.* **2020**, *281*, 107816. [CrossRef]
3. Farrukh, M.U.; Bashir, M.K.; Hassan, S.; Adil, S.A.; Kragt, M.E. Mapping the food security studies in India, Pakistan and Bangladesh: Review of research priorities and gaps. *Glob. Food Secur.* **2020**, *26*, 100370. [CrossRef]
4. Sheffield, J.; Wood, E.F.; Roderick, M.L. Little change in global drought over the past 60 years. *Nature* **2012**, *491*, 435–438. [CrossRef] [PubMed]
5. Mccabe, G.J.; Wolock, D.M. Variability and trends in global drought. *Earth Space Sci.* **2015**, *2*, 223–228. [CrossRef]
6. IPCC (Intergovernmental Panel on Climate Change). *Managing the Risks of Extreme Events and Disasters to Advance Climate Change Adaptation: Special Report of the Intergovernmental Panel on Climate Change;* Cambridge University Press: Cambridge, UK, 2012.
7. Kim, J.S.; Park, S.Y.; Hong, H.P.; Chen, J.; Choi, S.J.; Kim, T.W.; Lee, J.H. Drought risk assessment for future climate projections in the Nakdong River Basin, Korea. *Int. J. Climatol.* **2020**, *40*, 4528–4540. [CrossRef]
8. Qu, Y.P.; Hui, G.; Lv, J.; Su, Z.C.; Cheng, X.T.; Sun, H.Q. Agricultural drought disaster risk assessment in China based on the regional disaster system theory. *J. Hydraul. Eng.* **2015**, *46*, 908–917.
9. Cui, Y.; Jin, J.L.; Bai, X.; Ning, S.W.; Zhang, L.B.; Wu, C.G.; Zhang, Y.L. Quantitative evaluation and obstacle factor diagnosis of agricultural drought disaster risk using connection number and information entropy. *Entropy* **2022**, *24*, 872. [CrossRef]
10. Xie, Z.T.; Xu, J.P.; Deng, Y.F. Risk analysis and evaluation of agricultural drought disaster in the major grain-producing areas, China. *Geomat. Nat. Hazards Risk* **2016**, *7*, 1691–1706. [CrossRef]

11. Yu, X.B.; Li, C.L.; Huo, T.Z.; Ji, Z.H. Information diffusion theory-based approach for the risk assessment of meteorological disasters in the Yangtze River Basin. *Nat. Hazards* **2021**, *107*, 2337–2362.
12. Ma, Y.N.; Guga, S.; Xu, J.; Liu, X.P.; Tong, Z.J.; Zhang, J.Q. Assessment of maize drought risk in Midwestern Jilin Province: A comparative analysis of TOPSIS and VIKOR models. *Remote Sens.* **2022**, *14*, 2399. [CrossRef]
13. Villani, L.; Castelli, G.; Piemontese, L.; Penna, D.; Bresci, E. Drought risk assessment in Mediterranean agricultural watersheds: A case study in Central Italy. *Agric. Water Manag.* **2022**, *271*, 107748. [CrossRef]
14. Zhang, F.; Chen, Y.N.; Zhang, J.Q.; Guo, E.L.; Wang, R.; Li, D.J. Dynamic drought risk assessment for maize based on crop simulation model and multi-source drought indices. *J. Clean. Prod.* **2019**, *233*, 100–114. [CrossRef]
15. Wei, Y.Q.; Jin, J.L.; Cui, Y.; Ning, S.W.; Fei, Z.Y.; Wu, C.G.; Zhou, Y.L.; Zhang, L.B.; Liu, L.; Tong, F. Quantitative assessment of soybean drought risk in Bengbu city based on disaster loss risk curve and DSSAT. *Int. J. Disaster Risk Reduct.* **2021**, *56*, 102126. [CrossRef]
16. Schwarz, M.; Landmann, T.; Cornish, N.; Wetzel, K.-F.; Siebert, S.; Franke, J. A spatially transferable drought hazard and drought risk modeling approach based on remote sensing data. *Remote Sens.* **2020**, *12*, 237. [CrossRef]
17. Zhang, Q.; Zhang, J.Q.; Wang, C.Y. Risk assessment of drought disaster in typical area of corn cultivation in China. *Theor. Appl. Climatol.* **2017**, *128*, 533–540. [CrossRef]
18. Yoo, J.; Kim, D.; Kim, H.; Kim, T.-W. Application of copula functions to construct confidence intervals of bivariate drought frequency curve. *J. Hydro-Environ. Res.* **2016**, *11*, 113–122. [CrossRef]
19. Sun, P.; Zhang, Q.; Singh, V.P.; Xiao, M.Z.; Zhang, X.Y. Transitional variations and risk of hydro-meteorological droughts in the Tarim River basin, China. *Stoch. Environ. Res. Risk Assess.* **2017**, *31*, 1515–1526. [CrossRef]
20. Wang, Z.Q.; Jiang, J.Y.; Ma, Q. The drought risk of maize in the farming–pastoral ecotone in Northern China based on physical vulnerability assessment. *Nat. Hazards Earth Syst. Sci.* **2016**, *16*, 2697–2711. [CrossRef]
21. Zhong, S.B.; Wang, X.L.; Yang, Y.S.; Huang, Q.Y. Risk assessment of drought in Yun-Gui-Guang of China jointly using the Standardized Precipitation Index and vulnerability curves. *Geomat. Nat. Hazards Risk* **2018**, *9*, 892–918. [CrossRef]
22. Cui, Y.; Jiang, S.M.; Jin, J.L.; Ning, S.W.; Feng, P. Quantitative assessment of soybean drought loss sensitivity at different growth stages based on S-shaped damage curve. *Agric. Water Manag.* **2019**, *213*, 821–832. [CrossRef]
23. Yin, Y.Y.; Zhang, X.M.; Lin, D.G.; Yu, H.; Wang, J.A.; Shi, P.J. GEPIC-V-R model: A GIS-based tool for regional crop drought risk assessment. *Agric. Water Manag.* **2014**, *144*, 107–119. [CrossRef]
24. Wang, Y.X.; Lv, J.; Wang, Y.C.; Sun, H.Q.; Hannaford, J.; Su, Z.C.; Barker, L.J.; Qu, Y.P. Drought risk assessment of spring maize based on APSIM crop model in Liaoning province, China. *Int. J. Disaster Risk Reduct.* **2020**, *45*, 101483. [CrossRef]
25. Steduto, P.; Hsiao, T.C.; Raes, D.; Fereres, E. AquaCrop-the FAO crop model to simulate yield response to water: I. concepts and underlying principles. *Agron. J.* **2009**, *101*, 426–437. [CrossRef]
26. Paredesa, P.; Wei, Z.; Liu, Y.; Xu, D.; Xin, Y.; Zhang, B.; Pereira, L.S. Performance assessment of the FAO AquaCrop model for soil water, soil evaporation, biomass and yield of soybeans in North China Plain. *Agric. Water Manag.* **2015**, *152*, 57–71. [CrossRef]
27. Katerji, N.; Campi, P.; Mastrorilli, M. Productivity, evapotranspiration, and water use efficiency of corn and tomato crops simulated by AquaCrop under contrasting water stress conditions in the Mediterranean region. *Agric. Water Manag.* **2015**, *152*, 57–71. [CrossRef]
28. Ahmadi, S.H.; Mosallaeepour, E.; Kamgar-Haghighi, A.A.; Sepaskhah, A.R. Modeling maize yield and soil water content with AquaCrop under full and deficit irrigation managements. *Water Resour. Manag.* **2015**, *29*, 2837–2853. [CrossRef]
29. Mashaba-Munghemezulu, Z.; Chirima, G.J.; Munghemezulu, C. Mapping smallholder maize farms using multi-temporal Sentinel-1 data in support of the sustainable development goals. *Remote Sens.* **2021**, *13*, 1666. [CrossRef]
30. Jia, H.C.; Wang, J.A.; Cao, C.X.; Pan, D.H.; Shi, P.J. Maize drought disaster risk assessment of China based on EPIC model. *Int. J. Digit. Earth* **2012**, *5*, 488–515. [CrossRef]
31. Zampieri, M.; Ceglar, A.; Dentener, F.; Dosio, A.; Naumann, G.; van den Berg, M.; Toreti, A. When will current climate extremes affecting maize production become the norm? *Earth's Future* **2019**, *7*, 113–122. [CrossRef]
32. Zhang, S.; Bai, Y.; Zhang, J.H. Remote sensing-based quantification of the summer maize yield gap induced by suboptimum sowing dates over North China Plain. *Remote Sens.* **2021**, *13*, 3582. [CrossRef]
33. Zhu, B.X.; Chen, S.B.; Cao, Y.J.; Xu, Z.Y.; Yu, Y.; Han, C. A regional maize yield hierarchical linear model combining Landsat 8 vegetative indices and meteorological data: Case study in Jilin Province. *Remote Sens.* **2021**, *13*, 356. [CrossRef]
34. He, Q.J.; Zhou, G.S.; Lü, X.M.; Zhou, M.Z. Climatic suitability and spatial distribution for summer maize cultivation in China at 1.5 and 2.0 °C global warming. *Sci. Bull.* **2019**, *64*, 690–697. [CrossRef]
35. Wang, R.; Zhang, J.Q.; Wang, C.Y.; Guo, E.L. Characteristic analysis of droughts and waterlogging events for maize based on a new comprehensive index through coupling of multisource data in Midwestern Jilin Province, China. *Remote Sens.* **2020**, *12*, 60. [CrossRef]
36. Wu, X.; Wang, P.J.; Ma, Y.P.; Gong, Y.D.; Wu, D.R.; Yang, J.Y.; Huo, Z.G. Standardized relative humidity index can be used to identify agricultural drought for summer maize in the Huang-Huai-Hai Plain, China. *Ecol. Indic.* **2021**, *131*, 108222. [CrossRef]
37. Cheng, H.; Wang, W.; van Oel, P.R.; Lu, J.X.; Wang, G.; Wang, H.J. Impacts of different human activities on hydrological drought in the Huaihe River Basin based on scenario comparison. *J. Hydrol. Reg. Stud.* **2021**, *37*, 100909. [CrossRef]
38. Anhui Statistical Bureau; NBS Survey Office in Anhui. *Anhui Statistical Yearbook 2021*; China Statistics Press: Beijing, China, 2021.

39. Duan, K.; Mei, Y.D. Comparison of meteorological, hydrological and agricultural drought responses to climate change and uncertainty assessment. *Water Resour. Manag.* **2014**, *28*, 5039–5054. [CrossRef]
40. Gao, C.; Li, X.W. Precipitation thresholds of drought disaster for maize in areas in front of Bengbu Sluice, Huaihe River Basin, China. *Water* **2018**, *10*, 1395. [CrossRef]
41. China Meteorological Administration. *Yearbook of Meteorological Disasters in China 2014*; China Meteorological Press: Beijing, China, 2015.
42. China Meteorological Administration. *Yearbook of Meteorological Disasters in China 2020*; China Meteorological Press: Beijing, China, 2021.
43. Wei, T.; Song, S.B. Copula-based composite likelihood approach for frequency analysis of short annual precipitation records. *Hydrol. Res.* **2018**, *49*, 1498–1512. [CrossRef]
44. Du, M.C.; Zhang, J.Y.; Yang, Q.L.; Wang, Z.L.; Bao, Z.X.; Liu, Y.L.; Jin, J.L.; Liu, C.S.; Wang, G.Q. Spatial and temporal variation of rainfall extremes for the North Anhui Province Plain of China over 1976–2018. *Nat. Hazards* **2021**, *105*, 2777–2797. [CrossRef]
45. Tosunoglu, F.; Kisi, O. Joint modelling of annual maximum drought severity and corresponding duration. *J. Hydrol.* **2016**, *543*, 406–422. [CrossRef]
46. Lu, G.H.; Yan, G.X.; Wu, Z.Y.; Kang, Y.X. Regional drought analysis approach based on copula function. *Adv. Water Sci.* **2010**, *21*, 188–193.
47. Hsiao, T.C.; Heng, L.; Steduto, P.; Rojas-Lara, B.; Raes, D.; Fereres, E. AquaCrop-the FAO crop model to simulate yield response to water: III. parameterization and testing for maize. *Agron. J.* **2009**, *101*, 448–459. [CrossRef]
48. Dai, J.Y.; Gu, W.L. Studies on the relationship between root growth and yield in maize II: The interaction of root system and leaves of maize and its relation with yield. *Acta Agron. Sin.* **1998**, *14*, 310–314.
49. Song, H.X.; Li, S.X. Effect of root growing space of on maize its absorbing characteristics. *Sci. Agric. Sin.* **2003**, *36*, 899–904.
50. Yang, Q.H.; Gao, E.M.; Ma, X.M. Study on growing dynamic of maize root system in Shajiang black soil. *Acta Agron. Sin.* **2000**, *26*, 587–593.
51. Zhang, Y.F.; Wang, Y.F.; Liu, L.X. Function mechanism between the drought and waterlogging disaster and the soil-structure of the Shajiang soil in Huaibei Plain. *Prog. Geogr.* **2001**, *20*, 169–176.
52. Wang, Z.L. Study on "Four Waters" Transformation Mechanism in the Huaibei Plain. Ph.D. Thesis, Hohai University, Nanjing, China, 2017.
53. Zhang, C.L.; Ye, J.Y.; Qiu, X.M. Characteristics of drought and flood in Bengbu City in recent 45 years based on standardized precipitation index. *Water Resour. Power* **2013**, *31*, 9–12, 135.
54. Sun, P.; Zhang, Q.; Cheng, C.; Singh, V.P.; Shi, P.J. ENSO-induced drought hazards and wet spells and related agricultural losses across Anhui province, China. *Nat. Hazards* **2017**, *89*, 963–983. [CrossRef]
55. Zhou, Y.L.; Zuo, Y.L.; Zhang, Y.L.; Jin, J.L.; Zhou, P.; Wu, C.G.; Cui, Y.; Ning, S.W. Identification and characteristics analysis of Meiyu in Anhui Province based on the National Standard of Meiyu monitoring indices. *Hydrol. Res.* **2021**, *52*, 975–989. [CrossRef]
56. Tang, G.M.; Wang, Y.Z. Optimization irrigation procedure and economic irrigation norm of main arid crops. *J. Irrig. Drain.* **2006**, *25*, 24–29.
57. Cui, Y.; Jiang, S.M.; Feng, P.; Jin, J.L.; Yuan, H.W. Winter wheat evapotranspiration estimation under drought stress during several growth stages in Huaibei Plain, China. *Water* **2018**, *10*, 1208. [CrossRef]
58. Cui, Y.; Ning, S.W.; Jin, J.L.; Jiang, S.M.; Zhou, Y.L.; Wu, C.G. Quantitative lasting effects of drought stress at a growth stage on soybean evapotranspiration and aboveground biomass. *Water* **2021**, *13*, 18. [CrossRef]
59. Wei, Y.Q.; Jin, J.L.; Jiang, S.M.; Ning, S.W.; Cui, Y.; Zhou, Y.L. Simulated assessment of summer maize drought loss sensitivity in Huaibei Plain, China. *Agronomy* **2019**, *9*, 78. [CrossRef]
60. Cui, Y.; Jin, J.L.; Jiang, S.M.; Zhou, Y.L.; Wu, C.G.; Ning, S.W. Study on response relationship between evapotranspiration and aboveground growth matter of soybean under different drought stress. *J. Catastrophology* **2021**, *36*, 79–87.
61. Jin, J.L.; Ma, Q.; Cui, Y.; Bai, X.; Jiang, S.M.; Zhou, Y.L.; Zhou, R.X. Analysis on comprehensive prevention mechanism of actual risk of regional drought based on ternary chain transmission structure. *J. Catastrophology* **2022**, *37*, 6–12.
62. Han, C.Y.; Zhang, B.Z.; Chen, H.; Liu, Y.; Wei, Z. Novel approach of upscaling the FAO AquaCrop model into regional scale by using distributed crop parameters derived from remote sensing data. *Agric. Water Manag.* **2020**, *240*, 106288. [CrossRef]
63. Wolka, K.; Biazin, B.; Martinsen, V.; Mulder, J. Soil and water conservation management on hill slopes in southwest Ethiopia. II. Modeling effects of soil bunds on surface runoff and maize yield using AquaCrop. *J. Environ. Manag.* **2021**, *296*, 113187. [CrossRef]
64. Wu, H.; Yue, Q.; Guo, P.; Xu, X.Y.; Huang, X. Improving the AquaCrop model to achieve direct simulation of evapotranspiration under nitrogen stress and joint simulation-optimization of irrigation and fertilizer schedules. *Agric. Water Manag.* **2022**, *266*, 107599. [CrossRef]
65. Abedinpour, M.; Sarangi, A.; Rajput, T.B.S.; Singh, M.; Pathak, H.; Ahmad, T. Performance evaluation of AquaCrop model for maize crop in a semi-arid environment. *Agric. Water Manag.* **2012**, *110*, 55–66. [CrossRef]
66. Nyakudya, I.W.; Stroosnijder, L. Effect of rooting depth, plant density and planting date on maize (*Zea mays* L.) yield and water use efficiency in semi-arid Zimbabwe: Modelling with AquaCrop. *Agric. Water Manag.* **2014**, *146*, 280–296. [CrossRef]

67. Ran, H.; Kang, S.Z.; Li, F.S.; Tong, L.; Ding, R.S.; Du, T.S.; Li, S.E.; Zhang, X.T. Performance of AquaCrop and SIMDualKc models in evapotranspiration partitioning on full and deficit irrigated maize for seed production under plastic film-mulch in an arid region of China. *Agric. Syst.* **2017**, *151*, 20–32. [CrossRef]
68. Li, F.W.; Liu, Y.; Yan, W.H.; Zhao, Y.; Jiang, R.G. Effect of future climate change on summer maize growth in Shijin Irrigation District. *Theor. Appl. Climatol.* **2020**, *139*, 33–44. [CrossRef]

hydrology

MDPI

Article

Water Footprint Assessment for Irrigated Paddy Cultivation in Walawe Irrigation Scheme, Sri Lanka

Higgoda K. Janani [1], Himasha Dilshani Abeysiriwardana [1], Upaka Rathnayake [1] and Ranjan Sarukkalige [2,*]

[1] Department of Civil Engineering, Faculty of Engineering, Sri Lanka Institute of Information Technology, Malabe 10115, Sri Lanka
[2] School of Civil and Mechanical Engineering, Faculty of Science and Engineering, Curtin University, Bentley, Perth, WA 6102, Australia
* Correspondence: p.sarukkalige@curtin.edu.au

Abstract: Water footprint (WF) is a comprehensive summation of the volume of freshwater consumed directly and indirectly in all the steps of the production chain of a product. The water footprint concept has been widely used in agricultural water resources management. Water for irrigation is supplied in Sri Lanka to farmers at no cost, and thus the question is arising, whether the current management strategies the authorities and the farmers follow are appropriate to achieve productive water utilization. Therefore, this study aims at evaluating the water footprint of rice production in an irrigation scheme in the dry zone of Sri Lanka, the Walawe irrigation scheme. Due to the unreliability of the rainfall in the study area paddy cultivation depends entirely on irrigation, thus, the WF_{blue}, in other terms the volume of water evaporated from the irrigation water supply is considered as the total WF (WF_{tot}) in this study. Actual crop evapotranspiration (equivalent to ET_{blue}) was estimated based on the Penman-Monteith (P-M) model integrating effective rainfall, and crop coefficient published in Sri Lankan Irrigation Design Guidelines. The study spanned for three irrigation years from 2018–2021. Actual irrigation water issued to the field was estimated based on the data recorded by the government body responsible for irrigation water management of the area—Mahaweli Authority of Sri Lanka. The total volume of percolated water was computed employing the water balance method while assuming runoff is negligible. Results show that the average annual WF_{blue} found to be 2.27 m^3/kg, which is higher than global and national WF_{tot}. As the crop yield in the study area (6.5 ton/ha) is also higher than the global (4.49 ton/ha) and national (3.5 ton/ha) yields, a conclusion was drawn that the irrigation water usage ($CWUT_{blue}$) in the area may be significantly higher. It was then noted the higher $CWUT_{blue}$ was due to relatively higher evapotranspiration in the area. Thus, it is vital to reduce excess water usage by shifting irrigation practices from flooded irrigation to the System of Rice Intensification (SRI).

Keywords: blue water footprint; rice production; evapotranspiration; irrigation water management

Citation: Janani, H.K.; Abeysiriwardana, H.D.; Rathnayake, U.; Sarukkalige, R. Water Footprint Assessment for Irrigated Paddy Cultivation in Walawe Irrigation Scheme, Sri Lanka. *Hydrology* **2022**, 9, 210. https://doi.org/10.3390/hydrology9120210

Academic Editors: Songhao Shang, Qianqian Zhang, Dongqin Yin, Hamza Gabriel and Magdy Mohssen

Received: 13 October 2022
Accepted: 23 November 2022
Published: 25 November 2022

1. Introduction

Agriculture, the main consumer of freshwater globally [1,2] plays a major roll in freshwater scarcity [3]. About 70% of total direct extractions and 90% of indirect consumption (e.g., evapotranspiration) of the freshwater resources are accounted for agricultural water demand, which is predicted to be increased by 55% over the next three decades [2] due to factors such as rapid population growth, economic development, urbanization and climatic changes [3]. Therefore, both future food and water security are at risk. This necessitates increased crop water use efficiencies and sustainable irrigation water management. Characterisation and quantification of both direct and indirect water use in agriculture are paramount to managing agricultural water allocations sustainably to ensure future water security [4].

The concept of water footprint (WF) has been considered a valuable tool by many scholars recently [2,5] in estimating water consumption in agriculture. WF of a product is the comprehensive summation of the volume of water consumed directly and indirectly in all the steps of the production chain of a product. WF is an indicator that characterises consumed water volumes by source and polluted volumes by type of pollution and specifies all the components of consumed water spatially and temporally. WF has three components: Blue WF (WF_{blue}), Green WF (WF_{green}), and Grey WF (WF_{grey}), which, respectively refer to total evapotranspiration from water withdrawn from surface and groundwater resources (in other terms evaporation from irrigated water), total evapotranspiration from rainwater utilised in crop production and volume of freshwater required to dilute the load of pollutants to the agreed maximum permissible levels [6].

Rice is one of the major crops that feed the global population, with a total annual production of approximately 600 million metric tons, of which about 30% being produced in the South Asian region [7]. In South Asia, a larger portion of agricultural WF could be attributed to paddy due to two factors: (1) rice, being the staple food of South Asians, needs significantly higher production than any other single crop, and (2) rice, being a water intense crop, requires paddy fields to be saturated (in many cases this is achieved by flooding the fields). There are two major rice production systems namely wetland systems and upland systems. Wetland systems can either be rainfed or irrigated. In South Asia, rice occupies about 30–35% of all the irrigated croplands [8], which has shown a substantial expansion over the last few decades [9]. In order to feed the rapidly growing population in the region more rice needed to be produced. Since the rainwater is already scant and unreliable in the study area, with the impacts of climate change, it could be expected have even higher dependability on irrigated rice production. Hence, the pressure on the water resources in the region will be on the rise, as the total the WF of irrigated rice is almost two times that of rain-fed rice [10], demanding a comprehensive and continuous assessment of WF of paddy for efficient irrigation water resources management in the region.

Sri Lanka (refer to Figure 1) an island nation in the Indian Ocean, located between 5° N to 10° N latitudes and 79° E to 82° E longitudes, with a total geographical area of 65,610 km^2, is home to approximately 21 million people. Although Sri Lanka represents only a tiny portion of South Asia in terms of both the landmass and the population, because of relative similarities between the rest of the South Asian countries in terms of irrigation practices; policies; socio-economic and environmental conditions, this case study has a considerable potential for up- and out scaling of findings in the region. Although Sri Lanka has agriculture-based economy and some areas in the country are facing water scarcity, there seemed to be little or no attempts in Sri Lanka to evaluate WF, which allows sound overall assessment of water utilizing in crop production as the first step to a proper irrigation water management. This study was formulated to address this identified and vital research gap.

Approximately, 10% of the total land area of Sri Lanka (7080 km^2) has been allocated as Paddy fields [11]. Sri Lanka records the highest annual rice yield (about 3885 kg/ha) in South Asia [9]. However, the current rice yields are already reaching the maximum yield that could be achieved with the current sown area, which does not seem to be extending [12]. In Sri Lanka, rice is produced in two cultivation seasons namely, namely Yala (April to August) and Maha (November to March next year), which are synonymous with two monsoons: Southwest monsoon (SWM) from March to September and Northeast monsoon (NEM) from December to February [13]. So that, the irrigation year spans from October to September of next year. There are two major climatic zones delineated based on the spatial heterogeneity of rainfall. The two zones, i.e., the Wet zone, and Dry zone, are separated by the 2000 mm annual average rainfall isohyet. Paddy cultivation in Sri Lanka is broadly categorised as (1) wet zone paddy and dry zone paddy based on the local agroecology and (2) low land paddy and upland paddy based on the topography. About 60% of the annual rice production is coming from the dry zone, which receives most of its rain from NEM [14]. Therefore, for this study, a dry zone irrigation scheme was selected to evaluate the WF_{blue} of paddy cultivation in the dry zone of Sri Lanka.

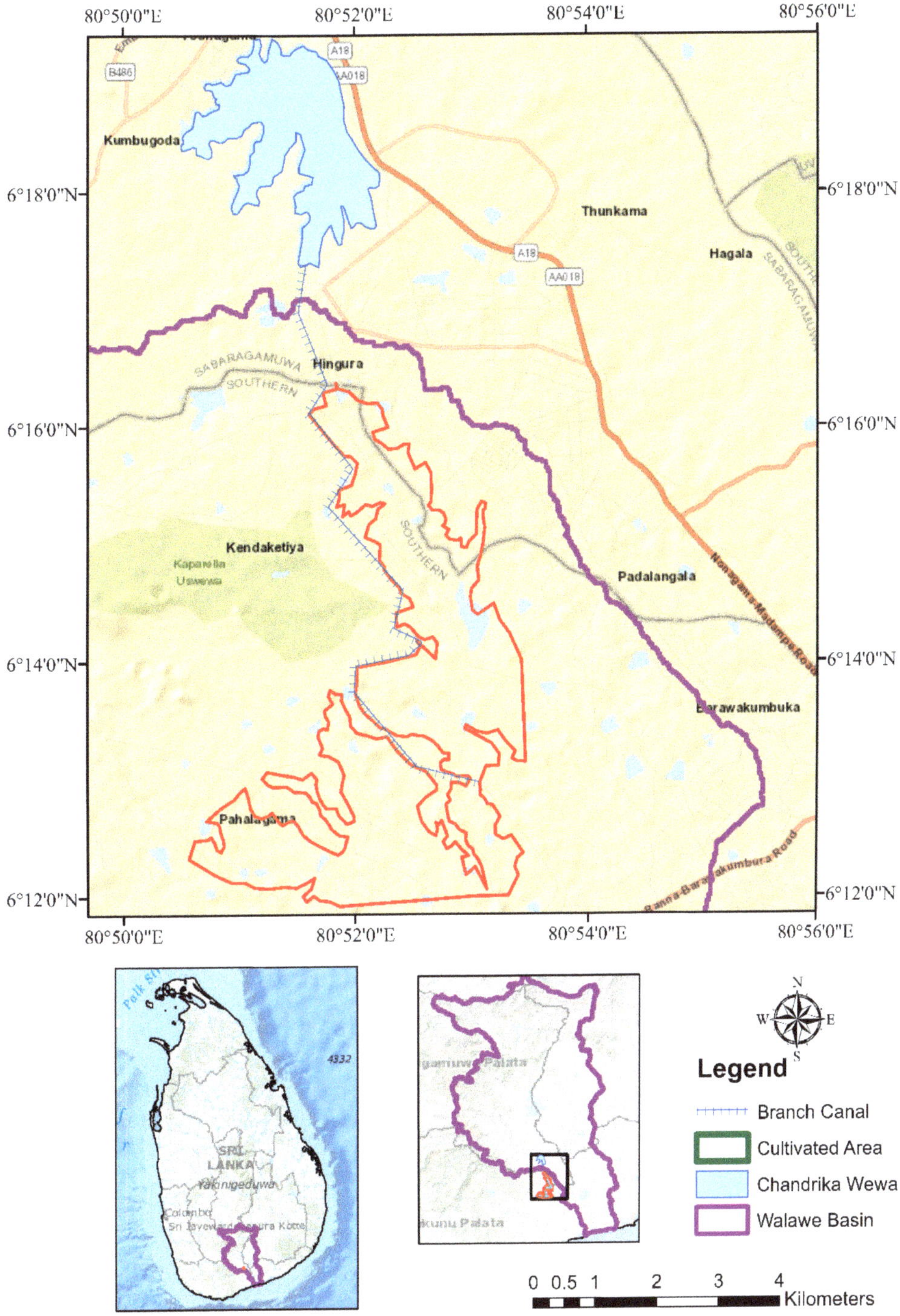

Figure 1. Cultivated area fed by Chandrikawewa Branch Canal. Source: ESRI ArcGIS Basemap.

2. Study Area

The case study was conducted in an irrigation scheme named, Walawe Special Area of which the administrative activities are being carried out by the Mahaweli Authority of Sri Lanka (MASL), which plays a key role in managing large reservoirs and irrigation distribution systems in Sri Lanka. Walawe Special Area is located in the Walawe river basin in the dry zone of Sri Lanka. "Chandrika wewa" is one of the three major reservoirs in the Walawe basin with a capacity of 27 million cubic meters (MCM). Chandrika wewa has three main canals, namely the right bank canal, left bank canal, and branch canal, irrigating downstream paddy fields. Paddy fields of 1549 ha which are located between 6°12′00″ N to 6°17′00″ N and 80°50′00″ E to 80°54′00″ E) and being fed by branch canal of Chandrika wewa were considered for this study (refer to Figure 1). Irrigation water is supplied to paddy fields by 30 sub canals, both lined and unlined and all having 1 m width, connected to the branch canal of Chandrika wewa (Refer to Table 1 for more details on canals). Discharge from each canal is quantified by measuring water level at the gate to each canal and substituting to a discharge equation derived and calibrated for the scheme. The area is characterised by an average annual rainfall of 1120 mm, with a majority coming from NEM, a flat terrain, with elevation ranging between 60–70 m, and reddish-brown soil. Reddish brown soil shows low hydraulic conductivity values of its bottom layers suggesting possible impedance to water movement under saturation. The profile saturated hydraulic conductivity range between 0.4 and 2.1 cm/hr. The bulk density values range from 1.06 gcm^3 to 2.15 gcm^3 at 10–20 cm soil depth and from 1.07 gcm^3 to 2.18 gcm^3 at 40–50 cm depth with averages of 1.68 gcm^3 and 1.69 gcm^3, respectively.

Table 1. Details of canals in the study area—length and command area of each canal; number water supplying days per week during the crop growth stage as per MASL rotation plan. L—lined canals, UL—unlined.

Canal	Length (m)	Command Area (ha)	Number of Days Per Week	Discharge Capacity (10^{-3} m/s)	Canal Lining	Canal	Length (m)	Command Area (ha)	Number of Days Per Week	Discharge Capacity (10^{-3} m/s)	Canal Lining
D1	598	6.5	2	15	L	D16 FC1	12	24.7	2	16	L
D2	482	10.4	2	19	UL	D17	2068	48	2	25	L
D3	543	48.8	3	20	L	CW FC1	581	8.4	2	12	L
D4	695	14.4	3	20	L	CWFC2	244	10	2	12	UL
D5 FC1	20	28.9	3	20	UL	CWFC3	385	11.2	2	15	UL
D6	3181	208.6	4	27	L	D18	361	38.4	3	12	L
D7 FC1	12	27	2	20	UL	CWFC4	639	11.1	2	12	UL
D8	2205	149.5	4	28	L	CWFC5	425	10.5	2	12	UL
D9	530	36.9	3	23	UL	D20	2769	131	4	22	UL
D10	4922	234.5	5	28	L	CWFC6	217	4.7	2	10	UL
D11	587	27.7	3	15	L	CWFC7	388	7.2	2	12	UL
D12	782	27.8	3	15.5	L	CWFC8	693	13	2	15	UL
D13	600	22.6	3	15.5	L	CWFC9	306	6.1	2	15	UL
D14	140	28.4	3	15	L	D21	742	50.7	3	18	UL
D15	3448	233.6	5	26	L	D22	2687	68.4	3	27	UL

2.1. Irrigation Water Allocation

As per ID-SL guidelines, in each cultivation season, the total crop production process can be broadly categorised into two major stages: Land Preparation and Crop growth. The crop growth stage can be further divided into four substages: initial stage; crop development stage; mid-season (mid) stage and late stage. Table 2 shows the duration (in weeks) of each stage for 105 days of paddy cultivation in both seasons. As per MASL, water is issued continuously (24 h everyday) via all sub canals during the Land Preparation, so that the issued water would amount to a field water depth (FWD) of 5 inches (122.5 mm) over 7 days standing in the field. During the crop growth period, irrigation water is supplied according to a rotation plan (given in Table 1 and the same for Yala and Maha seasons) implemented by MASL. The rotation plan is prepared, so that irrigation water issued would sum up to an FWD of 73.5 mm over 7 days during the initial and mid stages, and 49 mm over 7 days for crop development and late stages in both Yala and Maha seasons. These are the depths of water needed to meet the water output from the irrigation system as evapotranspiration during each stage. Thus, the actual water issued to a canal is greater than FWD by 60–70% after accounting for canal losses and field percolation losses.

Table 2. Duration (in weeks) of land preparation stage and each crop growth stage for 105 days paddy cultivation in Yala and Maha seasons. Lp—Land preparation stage, Initial—Initial stage, Crop dev.—Crop development stage, Mid—Mid season stage and Late—Late stage.

Stage	Number of Weeks									
	Yala					Maha				
	Apr	May	Jun	Jul	Aug	Nov	Dec	Jan	Feb	Mar
Lp	4					4				
Initial		3					3			
Crop dev.		1	3				1	3		
Mid			1	4				1	4	
Late					3					3

3. Methodology

3.1. Estimation of Water Footprint

WF of a crop product can be expressed in terms of water volume per unit time of production process (m^3/yr) or water volume per unit mass of product (m^3/kg) [5]. Due to the unreliability of the rainfall in this study area (Walawe irrigation scheme), rainfed irrigation is uncommon, hence all the paddy fields are irrigated—supplemental irrigation. However, as per MASL, all the water requirement of paddy cultivation is released from Chandrika Wewa according to the rotation plan. Thus, it is reasonable to state that paddy cultivation in the study area is based entirely on irrigation, which also means a loss of green water component. Therefore, from the three components of water footprint; WF_{blue}, WF_{green}, and WF_{grey}, this assessment used WF_{blue}, which refers to the volume of water evaporated from the irrigation water supply is considered as the total WF (WF_{tot}) in this study. The WF_{blue} for rice in the study area was calculated as water volume per unit mass of product as shown in Equation (1), where WF_{blue} is the blue water footprint (m^3/kg), $CWUA_{blue}$ is the crop blue water use per area (m^3/ha) and Y is the crop yield for rice production in the study area (kg/ha). The WF_{blue} was estimated for the three most recent irrigation years (2018/19, 2019/20, 2020/21). As per MASL, the average seasonal crop yield of rice in the study area was 6500 kg/ha in all three irrigation years.

$$WF_{blue} = \frac{CWUA_{blue}}{Y} \quad (1)$$

$CWUA_{blue}$ was calculated as in Equation (2), where $CWUT_{blue}$ is the total crop blue water use in the field (m^3) and A is the cultivation area (ha). The $CWUT_{blue}$ was considered as the summation of evapotranspiration from the paddy fields and evaporation from canal water surfaces. $CWUT_{blue}$ for each stage was computed using the relationship given by Equation (3), where i is the index denoting a canal, ET_{blue} is the blue water evapotranspiration (mm/day), A_i is the command area (ha) under the canal i, digit 10 is the unit conversion factor, n is the number of days in each stage, E_c is the evaporation from a water surface (mm/day), L_i is the length of the canal i, L_i is the length of the canal i and m_i is the number of days irrigation water issued via the canal i during a given stage. Pan (Class A) evaporation values recorded in 2018, 2019, 2020 and 2021 at the nearest weather station (Hambantota), which were obtained from the Department of Meteorology of Sri Lanka, were considered as E_c.

$$CWUA_{blue} = \frac{CWUT_{blue}}{A} \tag{2}$$

$$CWUT_{blue} = \sum_{i=1}^{30}\left\{(ET_{blue} \times A_i \times 10 \times n) + \left(\frac{E_c}{1000} \times L_i \times W_i \times m_i\right)\right\} \tag{3}$$

Total ET_{blue} for a given stage was estimated using Equation (4), where ET_c is the actual crop evapotranspiration (mm/day), which was estimated by using Equation (5) and P_{eff} is the effective rainfall (mm/day), estimated based on the relationship explained in Ponrajah [15], which provides the guidelines for irrigation designs in Sri Lanka and hereinafter referred to as ID-SL guidelines (as expressed in Equation (6)).

$$ET_{blue} = ET_c - P_{eff} \tag{4}$$

$$ET_c = k_c \times ET_o \tag{5}$$

where, k_c is the crop coefficient for a given stage and ET_o is the potential crop evapotranspiration (mm/day). k_c values were abstracted from the ID-SL guidelines; k_c = 1.00, 1.00, 1.15, 1.20 and 0.90 for Land Preparation, initial, crop development, mid-season and late stages, respectively. ET_o was estimated using the Penman-Monteith (P-M) model standardized by Food and Agricultural Organization (FAO) [16]. Reason for this selection is that the FAO P-M model considered to be applicable worldwide without any extra adjustment to its parameters [16].

$$P_{eff} = \begin{cases} \frac{0.67(R-25.4)}{30}, & R > 25.4 \\ 0, & R \leq 25.4 \end{cases} \tag{6}$$

where R is the monthly rainfall (mm/month) and digit 30 the unit conversion factor. Equation (7) is the FAO P-M model as expressed in [17].

$$ET_o = \frac{0.408\Delta(R_n - G) + \gamma\frac{900}{T+273}U_2(e_s - e_a)}{\Delta + \gamma(1 + 0.34U_2)} \tag{7}$$

where, ET_o is the total daily PET (mm/day), Δ is the slope of saturation vapour pressure curve (kPa °C^{-1}), R_n is the net incoming radiation (kPa), G is the soil heat flux (MJ m^{-2}), γ is the psychrometric constant (kPa °C^{-1}), T is the average daily temperature (°C), U_2 is the wind speed at 2 m height (m s^{-1}), e_s and e_a are the saturation vapour pressure (kPa) and the ambient vapour pressure (kPa), respectively. Rainfall and hydro-meteorological data required by FAO P-M model (i.e., solar radiation, wind speed, relative humidity and temperature) were obtained from the Department of Meteorology of Sri Lanka for the years 2018, 2019, 2020 and 2021. Figure 2 provides a comparison of E_c and ET_o. E_c is lower than ET_o, and this is in agreement with the findings of Abeysiriwardana [16] who have reported a similar relationship over the whole Sri Lanka. Furthermore, Nandagiri and Kovoor [18] have identified similar relationship between the two parameters in several south Indian states with both paraments ranging within the same range estimated in this study.

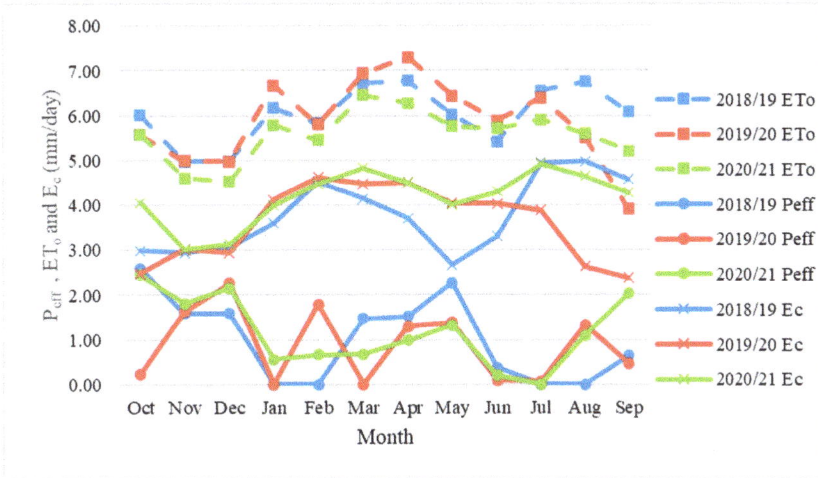

Figure 2. Temporal variations of ET_o estimated by FAO P-M model, E_c and effective rainfall.

3.2. Computation of Total Irrigation Water Issued

In order to find the fraction of $CWUT_{blue}$ in the total crop water use, the total irrigation water issued ($CWUT_{tot}$) via the canals were computed. To account for canal losses (evaporation + percolation) and field percolation losses, FWDs (Section 2.1) were multiplied by respective loss factors relevant to the stage of the cultivation process. Thus, the total IWS from a canal for crop production in a given cultivation season was estimated as in Equation (8).

$$(CWUT_{tot})_i = A_i \times \left\{ \sum_{j=1}^{5} \left(U_f \right)_j \times n_j \times FWD_j \right\} \times 10 \tag{8}$$

where, i is the index denoting a canal, $(CWUT_{tot})_i$ is the total irrigation water issued via the canal i for crop production (m^3), A_i is the area cultivated under the canal i (ha), j is the index denoting a stage in crop production cycle, U_f is the loss factor for stage j, n is the number of weeks in the stage j, FWD is the required field water depth in the stage j (mm/week) and digit 10 is the unit conversion factor. U_f for land preparation stage is 1.15 and for all the crop growth stages 1.42 (factors used by MASL for estimating discharge volumes).

3.3. Comparison of Irrigation Water Issued and Irrigation Water Requirement Estimated Based on Irrigation Guidelines for the Crop Growth Period

Further, to quantify excess irrigation water usage for paddy production, a comparison was made between the actual irrigation water issues and design irrigation requirement (DIR), which was estimated as per the ID-SL guidelines. Equations (9)–(13) describe the estimation of irrigation requirements and these equations are expressed in this paper as described in the ID-SL guidelines [15].

$$ET_c = n \times \frac{7\ ET_o}{30} \times k_c \tag{9}$$

$$FIR = \frac{ET_c}{E_a} \tag{10}$$

$$P_e = \frac{0.67 \times (R-1) \times 7}{30} \tag{11}$$

$$IR = \frac{FIR - P_e}{E_c} \quad (12)$$

where, ET_c is crop water requirement (mm), n is the number of weeks in the respective stage, ET_o is reference crop evapotranspiration (mm/month), k_c is the crop factor which is specific for the crop and growth stage, FIR is the field irrigation requirement (mm), E_a is application efficiency (%), P_e is effective rainfall (mm/week), R is the monthly 75% probable rainfall (mm), IR is the irrigation requirement (mm/week) and E_c is the conveyance loss (%). ET_c is defined as the "depth of water needed to meet the water loss through evapotranspiration" (ID-SL guidelines). Monthly ET_o, R values (Table 3), and k_c values for each stage (Table 4) were extracted from the ID-SL guidelines. E_a for lowland farms and E_c for a supply canal are given as 60% and 70%, respectively (ID-SL guidelines).

$$DIR_i = A_i \times \left\{ \sum_{j=1}^{5} n_j \times IR_j \right\} \times 10 \quad (13)$$

Finally, the total design irrigation water requirement (DIR) of a given canal for a given season was calculated as per Equation (8), where i is the index denoting a canal, A is the cultivated area under the canal i (ha), j is the index denoting a stage in crop production cycle, n is the number of weeks in the stage j and digit 10 is the unit conversion factor.

Table 3. Reference crop evapotranspiration values and monthly 75% probability rainfall. Source: ID-SL guidelines [14]. All the values are given in inches in the source, and the values are converted to SI units by multiplying by 25.4.

Month	ET_o (mm/month)	75% Probable Rainfall (mm/Month)	Month	ET_o (mm/Month)	75% Probable Rainfall (mm/Month)
January	96.520	304.8	July	154.94	12.7
February	114.300	114.3	August	161.29	12.7
March	129.032	88.9	September	161.29	50.8
April	129.032	114.3	October	129.032	152.4
May	135.382	38.1	November	91.44	228.6
June	148.336	12.7	December	91.44	330.2

Table 4. Crop growth stages and crop factors for lowland paddy.

	Initial	Development	Mid	Late
k_c	1.00	1.15	1.20	0.90
No of days	30 (21)	40 (21)	45 (21)	20 (14)

4. Results

4.1. Water Footprint and Crop Water Use

Using the data collected from MASL, the crop water use and water footprint were calculated for the study period. Stagewise, Seasonal, and annual $CWUA_{blue}$ estimates are given in Table 5. When each stage was considered separately, mid-season stage required the highest $CWUA_{blue}$ in both seasons in all three irrigation years. This can be associated with ET_{blue} and k_c. Mid stage of Yala is in July and that of Maha season is in February. Both months have highest ET_{blue} values with respect to the other months in respective season. Additionally, the k_c is highest in the mid stage.

Table 5. Stagewise, Seasonal, and annual crop blue water usage per area.

Year		$CWUA_{blue}$ (m^3/ha)						
		Stage					Seasonal	Annual
		Lp	Initial	Crop Dev	Mid	Late		
2018/19	Yala	1471.9	788.0	1556.0	2630.9	1277.2	7724.0	15,265.5
	Maha	1396.9	713.1	1780.8	2477.7	1172.9	7541.4	
2019/20	Yala	2045.5	1063.0	1823.4	2618.4	766.1	8316.4	15,101.8
	Maha	1399.7	572.9	1855.5	2009.3	948.1	6785.4	
2020/21	Yala	1759.8	930.5	1708.2	2449.7	829.6	7677.8	13,977.1
	Maha	1292.6	500.3	1497.7	2096.2	912.4	6299.3	

The initial stage had accounted for the lowest $CWUA_{blue}$ in the Maha season in all three years, whereas, in the Yala season, association of the lowest $CWUA_{blue}$ with cultivation stages was not consistent. $CWUA_{blue}$ of land preparation stage and Late stage were recorded to be higher in the Yala all the time. Though $CWUA_{blue}$ of other stages did not exhibit a similar trend, when the total seasonal $CWUA_{blue}$ considered, $CWUA_{blue}$ of Yala always exceeded that of Maha. This can be clarified by comparing (1) total ET_o in Yala season (893.68 mm) and Maha Season (810.78 mm) and (2) total P_{eff} in Yala Season (109.97 mm) and Maha (177.43 mm). It is clear when the entire season is considered Maha records lesser ET_o and higher P_{eff}, making it the wetter season of the two. Thus, requirement for irrigation is lower in Maha, hence lower $CWUA_{blue}$. As per Figure 2, estimated ET_o follows a similar pattern in all three irrigation years; however, the P_{eff} displays great differences in each year, and has resulted in above mentioned inconsistencies in $CWUA_{blue}$ values. Average annual $CWUA_{blue}$ of the study area was 14,781.4 m^3/ha, with the lowest (13,977.1 m^3/ha) was recorded in 2020/21, which can be explained by the fact that the said irrigation year was wetter than the other two years.

Table 6 tabulates seasonal and annual WF_{blue} of rice in all three years. Annual WF_{blue} were calculated to compare our results with previous studies. Though the annual figures in this study is for an irrigation year defined earlier, estimating annual figures provide a common base to compare with the available literature as cultivation seasons in other countries may differ to those of Sri Lanka. Except in Maha 2020/21, WF_{blue} ranges between 1.0–1.3 m^3/kg. Considering all three Yala seasons and all three Maha seasons, the average Yala and Maha WF_{blue} were found to be 1.22 m^3/kg and 1.06 m^3/kg, respectively. Average annual WF_{blue} of rice production in the study area was 2.27 m^3/kg. The spatial variability of WF within the scheme was not significant in all three years in all three years. When WF was estimated separately for each canal, maximum and minimum WF values varied only percentages less than 1% in both seasons.

Table 6. Seasonal and annual blue water foot print for rice production.

Year	Season	WF_{blue} (m^3/kg)
2018/19	Yala	1.19
	Maha	1.16
	Annual	2.35
2019/20	Yala	1.28
	Maha	1.04
	Annual	2.32
2020/21	Yala	1.18
	Maha	0.97
	Annual	2.15

4.2. Total Irrigation Water Issued and Percolation

Seasonal and annual total irrigation water supplied ($CWUT_{tot}$) to the entire command area, which includes both blue water and percolated water, are given in Table 7. As

$CWUT_{tot}$, duration of each stage and cultivated land area under each canal remain the same for both seasons, the $CWUT_{tot}$, values are the same for both seasons. As per Table 7, the maximum $CWUT_{tot}$ over a stage was 9.049 million m^3 and has been issued for the land preparation stage, while the minimum $CWUT_{tot}$ was recorded during a late stage at 3.352 million m^3. Total $CWUT_{tot}$ for the entire cultivated land during the initial, crop development stage and mid stages are 5.028 million m^3, 4.469 million m^3, and 8.380 million m^3, respectively. Accordingly, the $CWUT_{tot}$ for the entire command area under consideration, was 30.279 million m^3, hence resulted in $CWUA_{tot}$ of 19,547 m^3/ha. $CWUA_{tot}$ of each sub canal was found to be as same as the total command area (19,547 m^3/ha).

Table 7. Seasonal and annual total irrigation water supplied (blue water plus percolated water) to the total command area fed by all the 30 canals considered. Lp—Land preparation stage, Initial—Initial stage, Crop dev.—Crop development stage, Mid—Mid season stage and Late—Late stage.

	$CWUT_{tot}$ (BW + Percolation)											
	Lp		**Initial**		**Crop Dev**		**Mid**		**Late**		**Total**	
	10^6 m^3	10^3 m^3/ha	10^6 m^3	10^3 m^3/ha	10^6 m^3	10^3 m^3/ha	10^6 m^3	10^3 m^3/ha	10^6 m^3	10^3 m^3/ha	10^6 m^3	10^3 m^3/ha
Yala	9.049	5.842	5.028	3.246	4.470	2.885	8.380	5.410	3.352	2.164	30.280	19.550
Maha	9.049	5.842	5.028	3.246	4.470	2.885	8.380	5.410	3.352	2.164	30.280	19.550
Annual	18.099	11.680	10.056	6.492	8.939	5.771	16.761	10.820	6.704	4.328	60.559	3.910

Assuming no runoff occurred (this is the actual field situation according to MASL), the total amount of water percolated $CWUT_{per}$ was estimated as the difference between $CWUT_{tot}$ and $CWUT_{blue}$. The annual $CWUT_{tot}$, $CWUT_{blue}$ and $CWUT_{per}$ are given in Table 8. Considering the annual estimated percolated amount, the annual average percolation rates were calculated to be 6.53, 6.57 and 6.88 mm/day, respectively in 2018/19, 2019/20 and 2020/21 irrigation years, resulting in average percolation rate of 6.66 mm/day.

Table 8. Summary of annual total irrigation water supply, total annual blue water volume, total volume of irrigated water that percolated annually and average percolation rate. Notations are as same as described in the text.

Year	**$CWUT_{tot}$ (10^6 m^3)**	**$CWUT_{blue}$ (10^6 m^3)**	**$CWUT_{per}$ (10^6 m^3)**	**Average Annual Percolation Rate (mm/Day)**
2018/19	60.559	23.646	36.913	6.53
2019/20	60.559	23.393	37.167	6.57
2020/21	60.559	21.651	38.909	6.88

4.3. Comparison of Irrigation Water Requirement and Issued

A summary of design irrigation requirement (DIR) estimated based on the ID-SL guidelines is shown in Table 9. During the Yala season, DIR varies from the lowest of 3.840 million m^3 during the land preparation stage to a maximum of 8.258 million m^3 during the mid-season stage. When Maha season is considered, the minimum DIR, which is 0.395 million m^3 is recorded in the crop development stage while the maximum DIR of 2.587 million m^3 is recorded in the late stage. For all crop growth stages except the initial stage, DIR values in the Maha season are relatively lower than those of the Yala season. Yala season total DIR 29.03 million m^3 exceeds the Maha season total DIR (7.37 million m^3) by nearly four times.

Table 9. Summary of design irrigation requirement estimated based on ID-SL guidelines and summed up stagewise, seasonally and annually. Lp—Land preparation stage, Initial—Initial stage, Crop dev.—Crop development stage, Mid—Mid season stage and Late—Late stage.

	DIR (BW + Percolation) (10^6 m^3)					
	Lp	**Initial**	**Crop Dev**	**Mid**	**Late**	**Total**
Yala	3.840	3.931	7.277	8.285	5.721	29.028
Maha	2.428	5.077	0.395	1.455	2.587	7.3722
Annual	6.268	4.439	7.672	9.713	8.308	36.400

Figure 3 illustrates a comparison of the stagewise breakdown of $CWUT_{tot}$ and DIR estimated based on, in Yala and Maha, respectively. It was evident that the total $CWUT_{tot}$ values were higher than DIR, in all stages of Maha. However, during crop development and late stages in Yala, the $CWUT_{tot}$ values are lesser than DIR. Furthermore, the seasonal sum of $CWUT_{tot}$ (Table 5) is always greater than DIR (Table 9). Annual $CWUT_{tot}$ is almost twice the DIR.

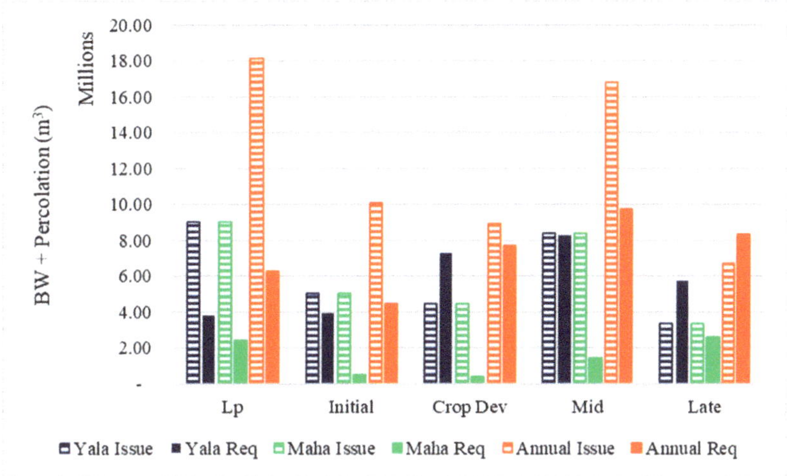

Figure 3. Stage wise comparison of total irrigation water issued to the field and the computed irrigation water requirement based on ID-SL guidelines.

5. Discussion

When having a Look at previous studies which focused on estimating the WF_{tot} of crops, some researchers ([10,19,20]) have considered all three components of WF_{tot} (WF_{green}, WF_{blue} and WF_{grey}) while some others ([21–23]) have omitted WF_{grey}. WF_{green} and WF_{blue} have always been quantified [24]. However, since this study area receives only a very little rainfall compared to the rest of the country and the cultivation in the area largely depends on irrigation, it can be assumed that the green water component associated with the rice production in the study area is minimal compared to blue water fraction. This assumption was later confirmed by estimated ET_c and P_{eff} values, of which the relationship can be expressed as $ET_c \gg P_{eff}$. Based on the same assumption this study evaluated only the WF_{blue} of rice production in the area. Therefore, despite the fact the WF_{tot} should be slightly higher than the WF_{blue}, the study considered computed WF_{blue} as the WF_{tot} when comparing the results of this study with the previous studies conducted worldwide. As mentioned by some researchers intensified precipitation do not change the WF_{blue} substantially, with only about 1% change in WF_{blue} when the rainfall increased by 200 mm, neglecting WF_{green} component under this study can be justified [5]. Though the

P_{eff} estimated under this illustrated significant variations in the three years considered, $CWUT_{blue}$ did not fluctuate as much over the three years, showcasing a lesser correlation between P_{eff} and $CWUT_{blue}$, hence, further supporting the study procedure.

As per the literature, the WF_{tot} of rice production ranges mostly between 1.0 to 2.0 m^3/kg [5], with some outliers going high about 6 m^3/kg [20] and low about 0.8 m^3/kg [25]. Average Yala, Maha, and Annual WF_{blue} of rice production in the present study area were found to be 1.22, 1.06, and 2.27 m^3/kg, respectively. Thus, it can be seen that, in the study area, the WF_{tot} of rice production in Yala and Maha falls within said range, and the annual WF_{tot} is about 11% higher. Mekonnen and Hoekstra [10] have found the global annual WF_{tot} (including WF_{grey}. WF_{green} and WF_{blue}) of rice as 1.673 m^3/kg, which is about 1.4 times lesser than the WF_{tot} of rice production in the study area of the present study. WF_{blue} component in the global annual WF_{tot} was only about 0.341 m^3/kg. Thus the global annual WF_{blue} was about 85% lesser than the annual WF_{blue} in the study area. The average rice yield in the study area is about 6.5 ton/ha, which is nearly 1.5 times higher than the global irrigated rice yield is 4.26 ton/ha [10]. Therefore, the substantial difference in global and study area annual WF_{blue} is due to a significantly large irrigation water usage in the study area. As the study area is characterised by a tropical climate (all year long high temperature, humidity, and wind) and receives approximately 12 h of direct sunlight all year long with the sun rising directly above Sri Lanka twice a year, the evapotranspiration in the study area experiences high ET_o (6.8–5.00 mm/day). The irrigation practice in the study area can be labelled as full irrigation, where full evapotranspiration demand is met, no water stress occurs and the highest ET_c compared to other irrigation methods takes place [26]. This evapotranspiration demand is fully met via flooded paddy fields, which enhances the rate of evapotranspiration. As WF is influenced to a larger extent by the irrigation practice rather than agro climate [10], said irrigation practice in the study area may play a huge role in higher WF_{blue} in the area compared to the global WF_{blue}.

A past study conducted in Indonesia resulted in WF_{tot} of 3.473 m^3/kg [19]. Though this figure is 53% higher than the WF_{tot} of present study area, given that Indonesia is also a tropical island, it can be expected to have similar crop water demand as in Sri Lanka. However, Bulsink et al. [19] claimed only about 21% of WF_{tot} was WF_{blue} (0.733 m^3/kg) and the majority of WF_{tot} (around 73%) was composed of WF_{green}. This suggests that in Indonesia paddy cultivation is mainly rainfed, which is true for the wet zone of Sri Lanka., However, it is in contrast to the dry zone of Sri Lanka, where with the less frequent, less reliable little rain, farmers tend to irrigate the paddy fields instead of utilising the direct rain. Thus, it would be reasonable to compare WF_{green} (which have been considered as WF_{tot}) computed by [19] with current study's WF_{blue} (which was considered as WF_{tot} in the area). As per this study in Indonesia, WF_{green} was 2.528 m^3/kg in their study area, which was only about 11.4% higher than the WF_{blue} of the present study area [9]. Crop yield in Indonesia is below the global yield [5], hence explain the slightly larger WF. WF_{tot} ($WF_{blue} + WF_{green}$) computed by [22] for Thailand (2.005 m^3/kg) was about 11.7% lower than the outcome of the present study. They also found that the WF_{green} was higher than the WF_{blue}. Although Thailand has similar temperature and evapotranspiration as in Sri Lanka, the difference can be the differences in the distribution of rainfall, the length of growing periods, size of the cultivated areas and agricultural practices.

As stated in Chapagain and Hoekstra [7], average WF_{tot} of Sri Lanka was 1.32 m^3/kg and average rice yield was 3.5 ton/ha—both values are considerably lower than this study's estimates and data provided by MASL. As per Chapagain and Hoekstra [7], out of the average WF_{tot}, only about 44% is WF_{blue} and 47% is WF_{green}. These percentages are surprising, as only a small fraction (30%) of Sri Lankan rice production depends entirely on the rain (rainfed) [11]. In other terms it would be expected to have the larger faction from WF_{blue}. Since Chapagain and Hoekstra [7] have conducted their study for 2000–2004, the discrepancies between the national values obtained by Chapagain and Hoekstra [7] and regional values obtained by this study area may be expected due to many reasons including: (1) increase in total cultivation area, (2) enhanced crop yield and (3) increased evapotranspiration due climate changes. On further note, the seasonal WF_{blue} in the study area (Yala–1.22 m^3/kg and Maha—1.06 m^3/kg) however were closer to the national value of 1.32 m^3/kg, with only 7.6% and 20% of deviations in Yala and Maha, respectively. Whereas the annual WF_{blue} in the study area exceeded the national value by 72%. In addition to aforementioned, another reason for this may be because in some areas rice cultivation is taken place only in one season (Yala or Maha), whereas in the study area fields are cultivated in both seasons.

WF_{tot} and composition of three components (Blue, Green, and Grey) show great differences in different studies [5,10,24]. In addition to before mentioned reasons, the identified differences could be due to various other factors including the type of models used for computations, period considered, crop parameters, climate [10], geographical and geomorphological conditions [23]. The variations also can be due to different crop yield which might have been strongly influenced by factors such as seed quality, agricultural practices, nutrition and fertilizer application. Though many studies [10,19,22] have noted that WF_{green} was greater than WF_{blue}, a study [7] showed that in countries like the USA and Pakistan, which depend entirely on irrigation for rice cultivation, WF_{blue} is noticeably larger than WF_{green}. This strengthens the argument of the current study, i.e., the contribution from rainfall can be neglected, hence WF_{blue} can be considered as WF_{tot} with much confidence.

Different studies have used various methods and models to estimate WF_{blue} and WF_{green}, however the widely applied method was Penman-Monteith (P-M) equation [19,20,22], under which it was assumed that crop water requirements can be satisfied [5]. Without following previous studies, [5] have measured field water balance and claimed their estimations of WF are closer to the actual situation on the ground than previous studies. Following the widely applied method, this study employed FAO P m method to estimate $CWUT_{blue}$. However, the actual total water use in any crop production is the WF plus percolation [7]. As WF concepts considers only the loss of freshwater to the catchment and percolation is not really a loss, percolated amount is never considered under WF. In this study, based on the data provided by MASL, the actual amount of irrigation water issued to the field ($CWUT_{tot}$) was estimated. Then, based on $CWUT_{tot}$ and $CWUT_{blue}$, field percolation volume ($CWUT_{per}$) and average percolation rate were also computed. The average percolation rate (6.66 mm/day) in the area is well below the range (10 mm/day–20 mm/day) which percolation rate can be expected to vary in reddish brown soil [27]. This supports the assumption of not occurring runoff in the fields, as the percolation rate still can escalate—also explains the highest percolated volume in the wettest irrigation year (2020/21).

Previous studies [2,28] have discussed the sustainability and unsustainability of WF_{blue} of crop production. The sustainability indicator is the ratio between the total WF_{blue} and blue water availability [2]. Since, in this study area, Chandrika wewa is not only collecting rainwater but receiving transboundary inflow from another large reservoir located in the wet zone, which hardly undergoes water stresses. This water trade between the two basins is irregular making the computation of blue water availability extremely complex. Therefore, this study did not attempt to find the sustainability indicator of rice production in the area.

As annual WF_{tot} (or WF_{blue}) in the study area slightly surpasses global and national annual WF_{tot} and some occasions, annual WF_{tot} of nearby countries. Therefore, the study proposes to implement water saving measures in the study area for more sustainable agriculture in the area. Given the higher evapotranspiration in the area, the application of the System of Rice Intensification (SRI) is highly recommended. Unlike flooded irrigation, SRI suggests only keeping the soil moisture at saturation level all the time instead of standing water in the field [7]. Unlike flooded irrigation, SRI suggests only keeping the soil moisture at saturation level all the time instead of standing water in the field [7]. In addition to SRI, several other methods such as drip irrigation, rainwater harvesting and alternate wetting and drying (AWD) have been applied worldwide as water saving methods [29]. Most of these methods, do not reduce the potential evapotranspiration, but allow utilization of seasonal rainfall [29], which in this study area goes to waste. Although WF is an excellent tool for overall assessment of water utilization and the need for water saving, as it only accounts for evaporation, evaluation of aforementioned water saving methods is not feasible with WF concept alone. Thus, it requires shifting to or combining WF concept with other methods such as water balancing to estimate other component of local water budget. On further note, this requirement is also one of the identified research gap in Sri Lankan context and would be a part of the future extensions of this project. Change of cultivation calendar and introduction of less water demanding variants of rice are also recommended to save water use. Additionally, it was also observed that the computation of actual irrigation water discharges ($CWUT_{tot}$) does not incorporate actual meteorological and percolation data. Instead, the practice is to issue the same amount of water each year each season, indicating poor management of irrigation water. Thus, it is highly recommended to use field observed meteorological and percolation data when preparing an irrigation water release schedule. This action in fact could reduce the $CWUT_{tot}$. Further, the comparison of total water issued ($CWUT_{tot}$) and estimated total irrigation water requirement (DIR) displayed significant variations, suggesting a necessity for re-evaluation of the parameter (such as ET_0, k_c, duration of stages and loss coefficients) given in the ID-SL guidelines, as it is the only guideline in the country to be used in irrigation water management practices. In addition, as mentioned earlier, due to the limitations inherited by WF method, it should be supplemented with other approaches to estimate the efficiencies of each and every changes made to the irrigation scheme.

6. Conclusions

WF_{blue} of rice production in an irrigation scheme in the dry zone of Sri Lanka was calculated based on FAO P-M model. The actual irrigation water issued from the reservoir was computed based on the MASL records. The volume of irrigated water percolated was estimated based on the water balance method. Irrigation water requirement was estimated following the guidelines of Sri Lanka ID-SL guidelines. Three recent irrigation years 2018/19, 2019/20, and 2020/21 were considered in this study and the averaged WF_{blue} of rice for three years was about 2.27 m^3/kg. As the rice production in the study area entirely depends on irrigation, assuming WF_{green} is negligible, WF_{blue} was considered as the WF_{tot}, which was larger than the global and national WF_{tot} of rice as noted by previous other studies. This outcome was attributed to relatively high evapotranspiration, much lower effective rainfall in the area, and the irrigation practice (full and flooded). Therefore, the study highlights the requirement of changing irrigation practice in to System of Rice Intensification to reduce the ET_{blue}. Though, no clear pattern of annual WF_{tot} could be observed, the findings revealed no significant variations to WF_{blue} in wetter years. Furthermore, $CWUT_{blue}$ shows only slight variations in the study years, whereas effective rainfall (P_{eff}) fluctuates significantly during the study period. Thus, the study found no correlation (qualitatively) between the effective rainfall (P_{eff}) and $CWUT_{blue}$, and concluded that if the rice production is 100% irrigated, there would be no considerable effects of rainfall on the WF_{blue} of rice production. The average percolation rate was about 6.66 mm/day, which is below the capacity of the soil type in the area. Thus, it was confirmed that no overflow of irrigated water happened in the field. However, the estimated annual irrigation requirement was substantially lesser than the actual annual irrigation water releases. This indicates that either there is an excessive wastage of freshwater or there is a need to revise ID-SL guideline parameters or both.

It is recommended to change the rotation system to match seasonal needs and considering climate conditions such as rainfall instead of adhering to same pattern both seasons. To reduce water losses through percolation, the study recommends all the canals to be lined in the future, however this may result in other implications on the surrounding ecosystems. Common water savings methods such as rainwater harvesting, drip irrigation, SRI and AWD are suggested by this study. As a further improvement to the current project, it is planned at supplementing WF concept with water balancing to evaluate efficiencies of proposed water savings method.

Author Contributions: Conceptualization, R.S.; methodology, H.K.J.; software, H.K.J. and H.D.A.; validation, H.K.J. and H.D.A.; formal analysis, H.K.J. and H.D.A.; resources, U.R.; data curation, H.K.J. and H.D.A.; writing—original draft preparation, H.K.J.; writing—review and editing, H.D.A., U.R. and R.S.; visualization, H.K.J. and H.D.A.; supervision, U.R. and R.S.; project administration, U.R. and R.S. All authors have read and agreed to the published version of the manuscript.

Funding: This research received no external funding.

Data Availability Statement: Data used in this study can be requested only for research purposes from the corresponding author.

Acknowledgments: The authors would like to acknowledge the Mahaweli Authority, Udawalawe, Sri Lanka for their support in conducting this research (allowing the authors to use their sensitive data in agriculture).

Conflicts of Interest: The authors declare no conflict of interest.

References

1. Chu, Y.; Shen, Y.; Yuan, Z. Water footprint of crop production for different crop structures in the Hebei southern plain, North China. *Hydrol. Earth Syst. Sci.* **2017**, *21*, 3061–3069. [CrossRef]
2. Novoa, V.; Ahumada-Rudolph, R.; Rojas, O.; Munizaga, J.; Sáez, K.; Arumí, J. Sustainability assessment of the agricultural water footprint in the Cachapoal River basin, Chile. *Ecol. Indic.* **2019**, *98*, 19–28. [CrossRef]
3. Sidhu, B.; Sharda, R.; Singh, S. Water Footprint of Crop Production: A Review. *Indian J. Ecol.* **2021**, *48*, 358–366.
4. Wanniarachchi, S.; Sarukkalige, R. A Review on Evapotranspiration Estimation in Agricultural Water Management: Past, Present, and Future. *Hydrology* **2022**, *9*, 123. [CrossRef]
5. Cao, X.; Wu, M.; Shu, R.; Zhuo, L.; Chen, D.; Shao, Z.; Guo, X.; Wang, W.; Tang, S. Water footprint assessment for crop production based on field measurements: A case study of irrigated paddy rice in East China. *Sci. Total Environ.* **2018**, *610–611*, 84–93. [CrossRef]
6. Hoekstra, A.; Chapagain, A.; Aldaya, M.; Mekonnen, M. *Water Footprint Manual*; Water Footprint Network: Enschede, The Netherlands, 2009.
7. Chapagain, A.; Hoekstra, A. The blue, green and grey water footprint of rice from production and consumption perspectives. *Ecol. Econ.* **2011**, *70*, 749–758. [CrossRef]
8. Bouman, B.; Humphreys, E.; Tuong, T.; Barker, R. Rice and Water. *Adv. Agron.* **2007**, *92*, 187–237. [CrossRef]
9. Morita. Chapter 7—Past growth in agricultural productivity in South Asia. In *Water Productivity and Food Security: Global Trends and Regional Patterns*; Kumar, D.M., Ed.; Elsevier: Amsterdam, The Netherlands, 2021; Volume 3, pp. 137–156. [CrossRef]
10. Mekonnen, M.; Hoekstra, A. The green, blue and grey water footprint of crops and derived crop products. *Hydrol. Earth Syst. Sci.* **2011**, *15*, 1577–1600. [CrossRef]
11. Department of Census and Statistics (DoCS). Agriculture. 2022. Available online: http://www.statistics.gov.lk/Agriculture/StaticalInformation/rubpaddy (accessed on 11 September 2021).
12. Davis, K.; Gephart, J.; Gunda, T. Sustaining food self-sufficiency of a nation: The case of Sri Lankan rice production and related water and fertilizer demands. *Ambio* **2015**, *45*, 302–312. [CrossRef] [PubMed]
13. Department of Agriculture Sri Lanka (DOA). 2022. Available online: https://doa.gov.lk/home-page/ (accessed on 11 September 2021).
14. Ratnasiri, S.; Walisinghe, R.; Rohde, N.; Guest, R. The effects of climatic variation on rice production in Sri Lanka. *Appl. Econ.* **2019**, *51*, 4700–4710. [CrossRef]
15. Ponrajah, A.J.P. *Design of Irrigation Headworks for Small Catchments*; Department of Irrigation: Colombo, Sri Lanka, 1984.
16. Abeysiriwardana, H.D.; Muttil, N.; Rathnayake, U.A. Comparative Study of Potential Evapotranspiration Estimation by Three Methods with FAO Penman–Monteith Method across Sri Lanka. *Hydrology* **2022**, *9*, 206. [CrossRef]
17. Allen, R.; Pereira, L.; Raes, D.; Smith, M. *Crop Evapotranspiration (Guidelines for Computing Crop Water Requirements)*; FAO Irrigation and Drainage Paper No. 56; FAO—Food and Agriculture Organization of the United Nations: Rome, Italy, 1998; pp. 65–275.
18. Nandagiri, L.; Kovoor, G.M. Performance Evaluation of Reference Evapotranspiration Equations across a Range of Indian Climates. *J. Irrig. Drain. Eng.* **2006**, *132*, 238–249. [CrossRef]
19. Bulsink, F.; Hoekstra, A.; Booij, M. The water footprint of Indonesian provinces related to the consumption of crop products. *Hydrol. Earth Syst. Sci.* **2010**, *14*, 119–128. [CrossRef]
20. Zhuo, L.; Mekonnen, M.; Hoekstra, A. The effect of inter-annual variability of consumption, production, trade and climate on crop-related green and blue water footprints and inter-regional virtual water trade: A study for China (1978–2008). *Water Res.* **2016**, *94*, 73–85. [CrossRef] [PubMed]
21. Liu, J.; Zehnder, A.; Yang, H. Historical Trends in China's Virtual Water Trade. *Water Int.* **2007**, *32*, 78–90. [CrossRef]

22. Gheewala, S.; Silalertruksa, T.; Nilsalab, P.; Mungkung, R.; Perret, S.; Chaiyawannakarn, N. Water Footprint and Impact of Water Consumption for Food, Feed, Fuel Crops Production in Thailand. *Water* **2014**, *6*, 1698–1718. [CrossRef]
23. Zhuo, L.; Mekonnen, M.; Hoekstra, A. Sensitivity and uncertainty in crop water footprint accounting: A case study for the Yellow River basin. *Hydrol. Earth Syst. Sci.* **2014**, *18*, 2219–2234. [CrossRef]
24. Lovarelli, D.; Bacenetti, J.; Fiala, M. Water Footprint of crop productions: A review. *Sci. Total Environ.* **2016**, *548–549*, 236–251. [CrossRef]
25. Yoo, S.; Choi, J.; Lee, S.; Kim, T. Estimating water footprint of paddy rice in Korea. *Paddy Water Environ.* **2013**, *12*, 43–54. [CrossRef]
26. Chukalla, A.; Krol, M.; Hoekstra, A. Green and blue water footprint reduction in irrigated agriculture: Effect of irrigation techniques, irrigation strategies and mulching. *Hydrol. Earth Syst. Sci.* **2015**, *19*, 4877–4891. [CrossRef]
27. Jayawardane, S.; Weerasena, L. Crop diversification in Sri Lanka. In Proceedings of the Crop Diversification in the Asia-Pacific Region, Bangkok, Thailand, 4–6 July 2000; Food and Agriculture Organization of the United Nations Regional Office for Asia and the Pacific: Bangkok, Thailand, 2000. Available online: https://www.fao.org/3/x6906e/x6906e00.htm#Contents (accessed on 4 December 2021).
28. Mekonnen, M.; Hoekstra, A. Sustainability of blue water footprint of crops. *Adv. Water Resour.* **2020**, *143*, 103679. [CrossRef]
29. Alauddin, M.; Sarker, A.R.; Islam, Z.; Tisdell, C. Adoption of alternate wetting and drying (AWD) irrigation as a water-saving technology in Bangladesh: Economic and environmental considerations. *Land Use Policy* **2019**, *91*, 104430. [CrossRef]

hydrology

MDPI

Article

Trends and Variabilities in Rainfall and Streamflow: A Case Study of the Nilwala River Basin in Sri Lanka

Ravindu Panditharathne [1,2], Miyuru B. Gunathilake [3,4,*], Imiya M. Chathuranika [1], Upaka Rathnayake [1], Mukand S. Babel [5] and Manoj K. Jha [6]

1 Department of Civil Engineering, Faculty of Engineering, Sri Lanka Institute of Information Technology, New Kandy Road, Malabe 10115, Sri Lanka
2 Faculty of Technology, Wayamba University of Sri Lanka, Kuliyapitiya 27411, Sri Lanka
3 Hydrology and Aquatic Environment, Division of Environment and Natural Resources, Norwegian Institute of Bioeconomy and Research, 1430 Ås, Norway
4 Water, Energy and Environmental Engineering, Faculty of Technology, University of Oulu, 90570 Oulu, Finland
5 Water Engineering and Management, Asian Institute of Technology, Pathumthani 12120, Thailand
6 Civil, Architectural and Environmental Engineering, 456 McNair Hall, North Carolina Agricultural and Technical State University, Greensboro, NC 27411, USA
* Correspondence: miyuru.gunathilake@nibio.no; Tel.: +47-922-65-434

Abstract: Rainfall is one of the dominating climatic parameters that affect water availability. Trend analysis is of paramount significance to understand the behavior of hydrological and climatic variables over a long timescale. The main aim of the present study was to identify trends and analyze existing linkages between rainfall and streamflow in the Nilwala River Basin (NRB) of Southern Sri Lanka. An investigation of the trends, detection of change points and streamflow alteration, and linkage between rainfall and streamflow were carried out using the Mann–Kendall test, Sen's slope test, Pettitt's test, indicators of hydrological alteration (IHA), and Pearson's correlation test. Selected rainfall-related extreme climatic indices, namely, CDD, CWD, PRCPTOT, R25, and Rx5, were calculated using the RClimdex software. Trend analysis of rainfall data and extreme rainfall indices demonstrated few statistically significant trends at the monthly, seasonal, and annual scales, while streamflow data showed non-significant trends, except for December. Pettitt's test showed that Dampahala had a higher number of statistically significant change points among the six rainfall stations. The Pearson coefficient correlation showed a strong-to–very-strong positive relationship between rainfall and streamflow. Generally, both rainfall and streamflow showed non-significant trend patterns in the NRB, suggesting that rainfall had a higher impact on streamflow patterns in the basin. The historical trends of extreme climatic indices suggested that the NRB did not experience extreme climates. The results of the present study will provide valuable information for water resource planning, flood and disaster mitigation, agricultural operations planning, and hydropower generation in the NRB.

Keywords: Mann–Kendall test; Nilwala River Basin; indicators of hydrological alteration (IHA); Pettitt's test; rainfall trends

Citation: Panditharathne, R.; Gunathilake, M.B.; Chathuranika, I.M.; Rathnayake, U.; Babel, M.S.; Jha, M.K. Trends and Variabilities in Rainfall and Streamflow: A Case Study of the Nilwala River Basin in Sri Lanka. *Hydrology* **2023**, *10*, 8. https://doi.org/10.3390/hydrology10010008

Academic Editors: Songhao Shang, Qianqian Zhang, Dongqin Yin, Hamza Gabriel and Magdy Mohssen

Received: 6 November 2022
Revised: 21 December 2022
Accepted: 23 December 2022
Published: 29 December 2022

1. Introduction

Climate is a key factor that affects environmental systems, socioeconomic conditions, and water resource availability [1]. The changes in rainfall patterns will directly affect streamflow and thereby domestic, agricultural, and industrial water needs [2]. Moreover, streamflow will also be affected by anthropogenic activities [2,3], such as land-use change, operation of dams and reservoirs, and direct water extraction from surface water and groundwater systems [4]. Hence, identifying and analyzing the long-term trends of meteorological and hydrologic data will be useful for water resource planning and management [5], flood protection and disaster mitigation [3,6], and agricultural operations [2].

Trend analysis will be valuable to eliminate errors in approximations in designing hydraulic structures under assumed fixed hydrometeorological variables [2].

Many studies in different geographic regions of the world were directed toward identifying trends and variabilities in rainfall and streamflow and their associated linkages [2,3,6–11]. Mersin et al. [10] stated that the variations in the frequency and magnitude of rainfall caused biotic and abiotic disturbances in the environment. Kastridis et al. [11] investigated the relationship between climate and tree growth for a tree species called A. Borisii-regis in the Mediterranean. They found that rainfall was the key driving factor for tree growth during the study period. Ademe et al. [2] demonstrated that the change in the water flow of the Birr River in Ethiopia was not only influenced by the change in rainfall but was also due to changes in land cover and land use, as well as human interventions, such as upstream water abstraction. In another study by Chaluka et al. [3], it was found that changes in rainfall influenced the alterations in streamflow patterns. Bellabas et al. [8] used a climate elasticity model and a hydrologic model to examine the effects of anthropogenic activities and changes in climate on streamflow. The results revealed that anthropogenic reasons were the dominant causes for the alterations in streamflow. In contrast to the above results, several others, such as Hannaford [12] and Wang et al. [13], found that the variations in rainfall significantly influenced streamflow patterns. Moreover, studies such as those by Azari et al. [14], Dey and Mishra [15], and Xu et al. [16] found that climate change had impacts on streamflow changes to varying degrees. Most of the trend analysis studies [2,3,17] used the Mann–Kendall test and Sen's slope estimator to study rainfall and streamflow trends. Pettitt's test was used for the detection of changing points in a hydrometeorological time series [18–21]. Other trend analysis methods, such as Spearman's rho and the linear regression test, were used by Fentaw et al. [9] and Coloiero [22]. However, some of these should be performed under certain assumptions, for instance, when the data is normally distributed and there are specific criteria on the length of the data series [22]. The indicators of hydrologic alteration (IHA) are commonly used to identify the hydrological impacts of human activities and to provide recommendations for environmental flow management [23–25].

Sri Lanka is an agrarian country that is highly dependent on rainfed and irrigation water. According to the Annual Report of the Central Bank of Sri Lanka, in 2021, the agriculture sector contributed 6.9% of the gross domestic product. Nearly 27.3% of Sri Lankans' engage in the agricultural sector as their livelihood. Sri Lanka experiences two major monsoon periods, which are the northeast monsoon (NEM; December to February) and the southwest monsoon (SWM; May to September). The two inter-monsoon periods are the first inter-monsoon (FIM; March to April) and the second inter-monsoon (SIM; October to November) [26].

Several studies, including Abeysingha [27], Perera et al. [28], Alahacoon and Edirisinghe [29], and Ruwangika et al. [30], studied rainfall and streamflow trends in Sri Lanka. These studies identified an increasing rainfall trend over the country that was most prominent in the eastern, southeastern, north, and north–central areas. Jayasekara and Abeysingha [17] found that there was a significant association between streamflow and rainfall variations for 70% of gauging stations in the Kelani River Basin. Chathuranika et al. [26] found that the climate and streamflow conditions of the Nilwala River Basin are expected to change in the future relative to the current conditions. Even though rainfall and streamflow trends studies were carried out in Sri Lanka, a handful of them focused on extreme rainfall indices, while none of the document studies used IHA parameters to assess the shifts in hydrologic regimes. Rainfall and streamflow trends and variabilities and their existing linkages have not been assessed for the Nilwala River Basin (NRB), which is one of the major river basins in the southern part of the island. Therefore, this study aimed to analyze long-term rainfall and streamflow trends, detect change points, and identify hydrological variables and their linkage over the NRB. The findings of this study will be helpful for both public and private sectors that are involved in water resource planning and development, disaster management, agricultural development, etc.

2. Materials and Methods

2.1. Study Area

The Nilwala River Basin (NRB) is in the southern part of the country between latitudes 5°55′ N and 6°13′ N and between longitudes 80°25′ E and 80°38′ E [4]. The river originates from Panilkanda in Deniyaya at an altitude of 988 m above the MSL (mean sea level), flows about 72 km through agricultural, urban, and other land uses, and finally drains into the Indian Ocean in Matara [31]. The total basin area is about 1010 km^2. The annual discharge of the river is more than 800 million cubic meters (MCM). The mean annual precipitation of the upper part of the NRB is about 3000 mm, while in the lower part, it is about 1900 mm [26]. Figure 1 demonstrates rainfall and hydrological stations in the NRB.

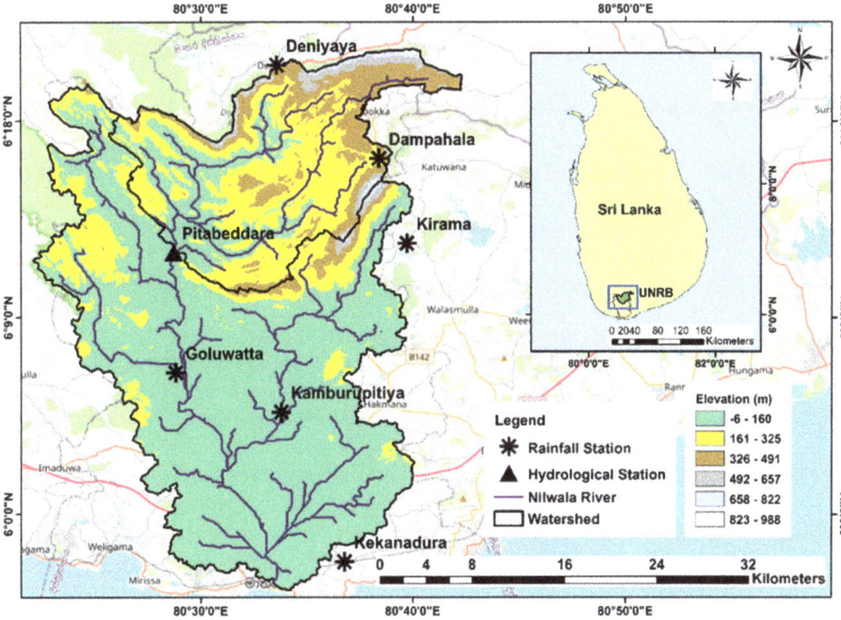

Figure 1. Location map of Nilwala River Basin with rainfall and hydrological stations.

2.2. Rainfall and Streamflow Data

2.2.1. Observed Data

Daily observed rainfall data from 1991–2014 (24 years) for six stations, namely, Dampahala, Kamburupitiya, Kekenadura, Kirama, Goluwawatta, and Deniyaya, were collected from the Department of Meteorology of Sri Lanka. Daily observed discharge data for the Pitabeddara hydrological station was collected from the Department of Irrigation of Sri Lanka for the same period.

2.2.2. Gridded Data

Missing rainfall data were filled using two versions of the Asian Precipitation-Highly Resolved Observational Data Integration towards Evaluation of Water Resources (APHRODITE) products, namely, V1101_MA and V1901_MA, with the same resolutions of 0.25° × 0.25°. APHRODITE products are available in three main geographical domains: Monsoon Asia (MA), Middle East (ME), and Russia (RU) (http://aphrodite.st.hirosaki-u.ac.jp/, accessed on 2 August 2022) [26]. The high temporal and spatial resolution in APHRODITE when compared with other gridded-based products and well-developed quality control methods influenced researchers to use this gridded product. Yatagai et al. [32], Yatagai et al. [33], and many other studies used APHRODITE data products for climatological studies, to validate satellite data, and to downscale low-resolution model data. Table 1 provides the general information of the stations used in the present study.

Table 1. General information of the stations used.

Station	Latitude (N)	Longitude (E)	Elevation (m.a.s.l)	Period	Missing %
Meteorological stations					
Dampahala	6.27	80.64	176		25.03
Kamburupitiya	6.08	80.56	244	1991–2014	5.91
Kekenadura	5.97	80.57	49		3.49
Kirama	6.22	80.67	122		3.79
Goluwawatta	6.10	80.48	16		17.10
Deniyaya	6.33	80.55	399		24.66
Hydrological station					
Pitabaddara	6.20	80.48	27	1991–2014	0.205

2.3. Methodology

Initially, the data quality and consistency were checked, and missing rainfall data were filled in using the APHRODITE precipitation data. Both hydrological and meteorological data were categorized into monthly, seasonal, and annual timescales. Mann–Kendall, modified Mann–Kendall, Sen's slope estimator, and Pettitt's tests were performed to identify the trends in the rainfall data and streamflow data to compute their magnitudes and to detect change points in the time series data using XLSTAT software (available at https://www.xlstat.com/en/, accessed on 5 August 2022) [34]. In addition, 5 extreme rainfall indices were calculated using the RClimdex software (available at https://www.climdex.org/, accessed on 9 August 2022) [35] and computed trends and magnitudes were found using MK and Sen's slope tests. The correlation between rainfall and streamflow was analyzed using Pearson's correlation coefficient. Thereafter, indicators of hydrological alteration (IHA) software was used to analyze the variations of the 16 selected hydrological parameters. Figure 2 below demonstrates the overall methodology of the study.

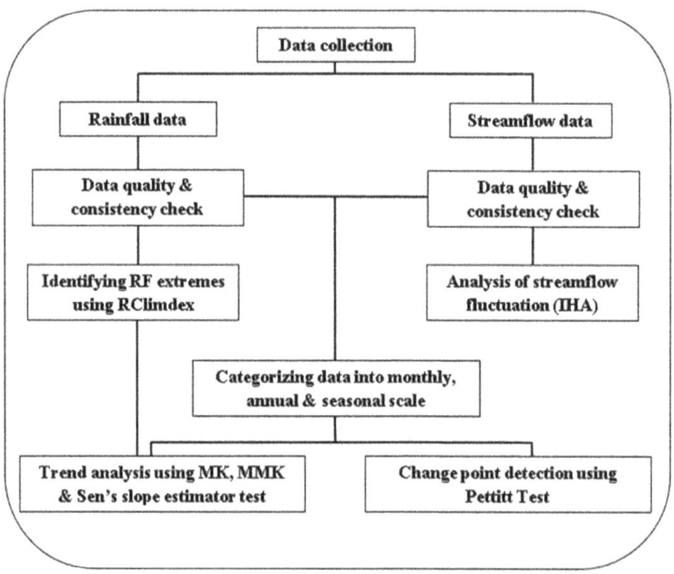

Figure 2. Overall methodology.

2.4. Trend Analysis Methods

2.4.1. Mann–Kendall Test

The non-parametric Mann-Kendall test [36,37] was used to identify the rainfall trends. The Mann–Kendall test static S was calculated using the following Equations (1) and (2):

$$S = \sum_{i=1}^{n-1} \sum_{j=i+1}^{n} sgn(x_j - x_i), \tag{1}$$

$$sgn(x_j - x_i) = \begin{cases} +1 & if\ x_j > x_i \\ 0 & if\ x_j = x_i \\ -1 & if\ x_j < x_i \end{cases} \tag{2}$$

where x_j and x_i are the data values at times j and i ($j > i$) and n is the length of the data series.

The Mann–Kendall Z statics and the variance *Var(S)* were calculated using the following Equations (3) and (4):

$$Z = \begin{cases} \frac{s-1}{\sqrt{Var(S)}} \\ 0 \\ \frac{s+1}{\sqrt{Var(S)}} \end{cases}, \qquad if \begin{bmatrix} S > 0 \\ S = 0 \\ S < 0 \end{bmatrix} \tag{3}$$

The Mann–Kendall test accepts the null hypothesis if $-Z \leq Zcr \leq Z$ is the critical value of the normalized statics Z at a 5% confidence level. The negative and positive values of Z and S statistics indicate decreasing and increasing tendencies, respectively.

$$Var(S) = \frac{1}{18}[n(n-1)(2n+5)] \tag{4}$$

2.4.2. Modified Mann–Kendall Test

Importantly, the influence of the accuracy of values in time series data on each other is highly affected by the serial correlation. The serial correlation is present in hydrologic data sets, such as streamflow and water levels [38–41]. The presence of a positive serial correlation will increase the ability to show trends from a certain level other than the actual status, and sometimes it shows trends that have no significant trends in the actual scenario [42]. Therefore, a serial correlation check should be conducted before applying the MK test. If a serial correlation is present, the modified Mann–Kendall test proposed by Hamed and Rao [42] should be applied to the time series data to eliminate the effects of serial correlation. Trend analysis of the hydrological data was checked using the modified Mann–Kendall (MMK) test and the modified variance can be calculated using the following Equations (5) and (6):

$$Var(S)^* = Var(S)\frac{n}{n^{*\prime}} \tag{5}$$

where *Var(S)** is the modified variance and the correction factor $n/n^{*\prime}$ is given by the following Equation (6):

$$\frac{n}{n^{*\prime}} = 1 + \frac{2}{n(n-1)(n-2)} \sum_{j=1}^{n-1} (n-k)(n-k-1)(n-k-2) r_{k'}^{R} \tag{6}$$

where n is the actual number of observations, n^* is the effective number of observations to count for the auto-correlation data, and $r_{k'}^{R}$ is the lag k autocorrelation coefficient of the rank of data.

2.5. Sen's Slope Estimator

Sen's slope estimator [43] is also a non-parametric test that is widely used to compute the magnitude of trend series. The slope of the data series is computed using the following Equations (7) and (8):

$$T_i = \frac{X_j - X_k}{j - k} \quad For i = 1,\ 2,\ 3 \ldots N \tag{7}$$

$$\beta = \begin{cases} T_{\frac{N+1}{2}} & N \text{ is odd} \\ \frac{1}{2}\left(\frac{T_N}{2} + \frac{T_{N+1}}{2}\right) & N \text{ is even} \end{cases} \tag{8}$$

where x_j and x_i are data values at the time j and k ($j < k$). N is the number of data pairs (x_j, x_k), $1 \leq j < k \leq n$. β defines Sen's slope value. Positive β values signify an increasing trend and negative β values signify a decreasing trend

2.6. Change Point Detection

The nonparametric Pettitt's test developed by Pettitt [44] was used to identify abrupt changes in time rainfall data within the study area. The method was derived using the Mann–Whitney statistic $U_{t,\,n}$ and two test samples from the same population, namely, $x_1, \ldots, x_t$ and $x_{t+1}, \ldots, x_n$. The test statistic $U_{t,\,n}$ can be obtained using the following Equations (9) and (10):

$$U_{t,\,n} = U_{t-1,\,n} + \sum_{J=1}^{n} sgn\left(x_t - x_j\right) \tag{9}$$

where $t = 2, 3, \ldots, n$; n is the length of the time series; and

$$\begin{array}{c} \text{If } \left(x_t - x_j\right) > 0,\ sgn\left(x_t - x_j\right) = 1 \\ \text{If } \left(x_t - x_j\right) = 0,\ sgn\left(x_t - x\right) = 0 \\ \text{If } \left(x_t - x_j\right) < 0,\ sgn\left(x_t - x_j\right) < -1 \end{array} \tag{10}$$

The test statistic quantifies the number of times when the first sample exceeds the second sample. The null hypothesis test indicates there are no changes in the data series, while the alternative hypothesis indicates the existence of changing points in the data series. The commonly used 5% significant level was used for the analysis. The test statistic K_n and the associate probability P used in the test can be obtained using the following Equations (11) and (12), respectively:

$$K_n = max_{1 \leq t \leq n}\left|U_{t,\,n}\right| \tag{11}$$

$$P \cong 2exp\left\{\frac{-6(K_n)^2}{(n^3 + n^2)}\right\} \tag{12}$$

2.7. Indicators of Hydrological Alteration (IHA)

IHA software was developed by the US Nature Conservancy to measure the extent of hydrological changes due to climatic and human influences [21]. This tool has the ability to calculate 33 IHA parameters under five categories: magnitudes of monthly water conditions, magnitude and duration of annual extreme water conditions, the timing of annual extreme water conditions, frequency and duration of high/low pulses, and rate/frequency of water condition changes. A 5% significance level p-value was used to evaluate and compare the parameters' consistency.

2.8. RClimdex

RClimdex software was developed and maintained by Xuebin Zhang and Feng Yang at the Climate Research Branch of the Meteorological Service of Canada and designed to obtain 27 climate extreme indices that are recommended by the Expert Team for Climate Change Detection Monitoring and Indices (ETCCDMI) [4]. This study used five selected extreme precipitation indices: consecutive dry days (CDD), consecutive wet days (CWD), annual total wet day precipitation (PRCPTOT), number of days above 25 mm (R25), and maximum 5-day precipitation amount (Rx5) [45].

3. Results and Discussion

3.1. Correlation between Observed and APHRODITE Data

Since rainfall was missing from some of the rainfall stations, gridded precipitation data were used to fill them. To check the reliability, the correlation coefficient was calculated between the observed data and APHRODITE products V1901 (from January 2004 to March 2015) and V1101 (from February 1991 to September 1991). The longest and continuous data periods that were common for both the observed and APHRODITE data were selected on this basis. The results of the correlation are shown in Table 2 below. Most of the stations showed a moderate-to-very-strong relationship in the correlation analysis. Hence, we could justify the use of APHRODITE data for our study. The types of correlations were classified as per the rule of thumb for interpreting the correlation coefficient. The classification for the Pearson correlation coefficient was as follows: a positive very strong correlation was 0.8–1.0, a positive strong correlation was 0.6–0.8, a positive strong moderate correlation was 0.4–0.6, a positive moderate correlation was 0.2–0.4, and a positive weak or insignificant correlation was 0–0.2 [3].

Table 2. Results of correlation check—observed data vs. APHRODITE data.

Station	V1901 (January 2004–March 2005)		V1101 (February 1991–September 1991)	
	Correlation	Correlation Type	Correlation	Correlation Type
Dampahala	0.47	Positive moderate	0.12	Positive weak
Kamburupitiya	0.79	Positive very strong	0.69	Positive strong
Kekenadura	0.85	Positive very strong	0.61	Positive strong
Kirama	0.28	Positive weak	0.02	Positive weak
Goluwawatta	0.77	Positive very strong	0.38	Positive moderate
Deniyaya	0.45	Positive moderate	0.75	Positive very strong

3.2. Trend Analysis of the Rainfall

Trend analysis was carried out for the monthly, seasonal, and annual scales using Mann–Kendall and Sen's slope tests. The significant trends are denoted in bold font in Table 3. Dampahala station showed significant increasing trends in March, September, November, and December, with magnitudes of 12.2 mm/yr, 8.96 mm/yr, 14.26 mm/yr, and 11.65 mm/yr, respectively. Kamburupitiya showed significant decreasing trends of 3.97 mm/yr in July. Kirama demonstrated significant decreasing trends in April, June, and July, with magnitudes of 7.23 mm/yr, 4.63 mm/yr, and 4.61 mm/yr, respectively. Deniyaya also revealed a significant 13.83 mm/yr decreasing trend in May, while Kekenadura and Goluwatta did not show any significant trends at the monthly scale. In general, most of the stations showed decreasing patterns at the monthly scale. Interestingly, Deniyaya showed only decreasing trends and Dampahala showed only increasing trends. Dampahala showed a significant increasing trend of 73.85 mm/yr, while Deniyaya showed a significant decreasing trend of 70.3 mm/yr at the annual scale. At the seasonal scale, Dampahala station revealed significant increasing trends during the NEM and FIM, with magnitudes of 23.43 mm/yr and 17.42 mm/yr, respectively. During the SWM, Kamburupitiya, Kirama, and Deniyaya showed significant 16.83 mm/yr, 11.68 mm/yr, and 33.19 mm/yr decreasing trends, respectively. During the SIM period, only the Deniyaya station revealed a significant decreasing trend, with a magnitude of 13.89 mm/yr. The highest magnitudes of increasing trends at the monthly (November) and annual scales and for the NEM and FIM were identified in the Dampahala station. Deniyaya exhibited the highest significant decreasing trends at the monthly (May) and annual scales and for the SWM and SIM. Other stations, except for Dampahala, experienced decreasing trends at the monthly and annual scales and for the SWM and FIM.

Table 3. Trend analysis results for the rainfall data.

Timescale	Dampahala			Kamburupitiya			Kekenadura			Kirama			Goluwatta			Deniyaya		
	p-Value	Sen's Slope	Trend Type	*p*-Value	Sen's Slope	Trend Type	*p*-Value	Sen's Slope	Trend Type	*p*-Value	Sen's Slope	Trend Type	*p*-Value	Sen's Slope	Trend Type	*p*-Value	Sen's Slope	Trend Type
January	0.188	2.88	NSIT	0.941	0.11	NSIT	0.747	0.53	NSIT	0.551	1.39	NSIT	0.941	0.37	NSIT	0.637	−1.33	NSDT
February	0.108	7.08	NSIT	0.823	−0.40	NSDT	0.747	0.89	NSIT	0.823	−0.19	NSDT	0.980	0.49	NSIT	0.063	−9.49	NSDT
March	**0.005**	**12.20**	**SIT**	0.248	1.72	NSIT	0.136	1.57	NSIT	0.205	2.55	NSIT	0.941	0.50	NSIT	0.333	−2.78	NSDT
April	0.132	8.46	NSIT	0.248	−3.93	NSDT	0.209	−2.18	NSDT	**0.039**	**−7.23**	**SDT**	0.713	−1.25	NSDT	0.188	−10.12	NSDT
May	0.338	4.86	NSIT	0.268	−3.57	NSDT	0.447	−3.08	NSDT	0.980	−0.20	NSDT	0.160	−4.61	NSDT	**0.039**	**−13.83**	**SDT**
June	0.677	1.59	NSIT	0.063	−4.45	NSDT	0.788	0.91	NSIT	**0.050**	**−4.63**	**SDT**	0.864	−0.44	NSDT	0.118	−8.26	NSDT
July	0.677	1.24	NSIT	**0.044**	**−3.97**	**SDT**	0.175	−2.16	NSDT	**0.002**	**−4.61**	**SDT**	0.192	−3.38	NSDT	0.078	−3.01	NSDT
August	0.390	3.77	NSIT	0.750	0.69	NSIT	0.244	2.73	NSIT	0.785	−0.42	NSDT	0.540	1.51	NSIT	0.535	−2.22	NSDT
September	**0.020**	**8.96**	**SIT**	0.192	−4.33	NSDT	0.573	−1.31	NSDT	0.338	−2.02	NSDT	0.192	−5.17	NSDT	0.823	−0.72	NSDT
October	0.364	2.62	NSIT	0.903	0.69	NSIT	0.602	2.44	NSIT	0.903	0.66	NSIT	0.607	−2.09	NSDT	0.078	−7.56	NSDT
November	**0.017**	**14.26**	**SIT**	0.826	−0.62	NSDT	0.677	−2.28	NSDT	0.338	3.55	NSIT	0.418	−2.58	NSDT	0.188	−4.32	NSDT
December	**0.003**	**11.65**	**SIT**	0.120	5.60	NSIT	0.108	4.66	NSIT	0.338	3.49	NSIT	0.573	1.50	NSIT	0.980	−0.06	NSDT
Annual	**0.034**	**73.85**	**SIT**	0.314	−13.94	NSDT	0.713	5.03	NSIT	0.677	−8.62	NSDT	0.228	−17.06	NSDT	**0.002**	**−70.30**	**SDT**
NEM	**0.013**	**23.43**	**SIT**	0.268	6.49	NSIT	0.145	4.05	NSIT	0.750	4.93	NSIT	0.750	−1.81	NSDT	0.160	−15.75	NSDT
FIM	**0.015**	**17.42**	**SIT**	0.477	−2.96	NSDT	0.826	−0.74	NSDT	0.314	−6.48	NSDT	0.747	−2.10	NSDT	0.056	−17.21	NSDT
SWM	0.268	17.64	NSIT	**0.007**	**−16.83**	**SDT**	0.338	−8.45	NSDT	**0.050**	**−11.68**	**SDT**	0.088	−14.58	NSDT	**0.003**	**−33.19**	**SDT**
SIM	0.063	16.44	NSIT	0.941	−0.40	NSDT	0.637	3.22	NSIT	0.477	3.47	NSIT	0.447	−6.36	NSDT	**0.031**	**−13.98**	**SDT**

* NSIT refers to a non-significant increasing trend, NSDT refers to a non-significant decreasing trend, SIT refers to a significant increasing trend, SDT refers to a significant decreasing trend.

3.3. Trend Analysis of the Streamflow

Trend analysis of the streamflow at the Pitabeddara hydrologic station was computed for the monthly, seasonal, and annual scales. According to the results in Table 4, only one significant trend was observed, which was in December with a magnitude of 0.59 m^3s^{-1}/yr. During the other months, non-significant increasing and decreasing trends were observed. Significant trends were not observed on the annual or seasonal scale. However, non-significant increasing trends were observed for the annual scale and the NEM and FIM. Non-significant decreasing trends during the SWM and SIM were seen.

Table 4. Trend analysis results for the streamflow data.

Timescale	Kendall's Tau	*p*-Value	Sen's Slope	Trend Type
January	0.152	0.298	0.17	NSIT
February	0.123	0.399	0.13	NSIT
March	0.217	0.137	0.25	NSIT
April	0.123	0.399	0.15	NSIT
May	−0.058	0.691	−0.15	NSDT
June	−0.080	0.567	−0.15	NSDT
July	−0.130	0.372	−0.18	NSDT
August	−0.051	0.728	−0.02	NSDT
September	−0.014	0.921	−0.10	NSDT
October	−0.196	0.067	−0.30	NSDT
November	0.101	0.487	0.26	NSIT
December	0.319	**0.029**	**0.59**	**SIT**
Annual	0.094	0.519	0.10	NSIT
NEM	0.275	0.059	0.31	NSIT
FIM	0.203	0.165	0.20	NSIT
SWM	−0.159	0.275	−0.12	NSDT
SIM	−0.036	0.804	−0.03	NSDT

* NSIT refers to a non-significant increasing trend, NSDT refers to a non-significant decreasing trend, SIT refers to a significant increasing trend, SDT refers to a significant decreasing trend.

3.4. Trend Analysis of the Extreme Rainfall Indices

Five selected extreme rainfall indices, namely, consecutive dry days (CDD), consecutive wet days (CWD), annual total wet day precipitation (PRCPTOT), number of days above 25 mm (R25), and maximum 5-day precipitation amount (Rx5), were computed using the RClimdex software. According to the results shown in Table 5, Dampahala station revealed significant increasing trends in PRCPTOT and R25, with magnitudes of 74.55 mm/yr and 1.64 days/yr. In Deniyaya, significant decreasing trends in PRCPTOT and R25 were observed, with magnitudes of 71.75 mm/yr and 1.33 days/yr. Kamburupitiya showed a 0.7 days/yr significant decreasing trend for CWD. No significant trends were observed for the Kekenadura, Kirama, and Goluwatta stations for CDD, CWD, PRCPTOT, and R25.

Table 5. Trend analysis results for the extreme rainfall indices.

Station	Extremes	Kendall's Tau	*p*-Value	Sen's Slope	Trend Type
Dampahala	CDD	−0.115	0.440	−0.14	NSDT
	CWD	−0.268	0.070	−0.38	NSDT
	PRCPTOT	0.312	**0.034**	**74.55**	**SIT**
	R25	0.396	**0.007**	**1.64**	**SIT**
Kamburupitiya	CDD	0.129	0.384	0.20	NSIT
	CWD	−0.462	**0.002**	**−0.70**	**SDT**
	PRCPTOT	−0.145	0.338	−14.53	NSDT
	R25	0.106	0.471	0.23	NSIT

Table 5. *Cont.*

Station	Extremes	Kendall's Tau	*p*-Value	Sen's Slope	Trend Type
Kekenadura	CDD	−0.172	0.243	−0.31	NSDT
	CWD	−0.061	0.688	0.00	No trend
	PRCPTOT	0.065	0.677	5.60	NSIT
	R25	0.030	0.842	0.00	No trend
Kirama	CDD	0.128	0.384	0.57	NSIT
	CWD	−0.282	0.065	−0.12	NSDT
	PRCPTOT	−0.058	0.713	−8.45	NSDT
	R25	−0.022	0.881	0.00	No trend
Goluwatta	CDD	−0.069	0.637	−0.10	NSDT
	CWD	−0.134	0.369	−0.15	NSDT
	PRCPTOT	−0.174	0.248	−18.10	NSDT
	R25	−0.274	0.065	−0.60	NSDT
Deniyaya	CDD	0.101	0.500	0.09	NSIT
	CWD	−0.044	0.765	−0.04	NSDT
	PRCPTOT	−0.454	**0.002**	**−71.75**	**SDT**
	R25	−0.442	**0.003**	**−1.33**	**SDT**

* NSIT refers to a non-significant increasing trend, NSDT refers to a non-significant decreasing trend, SIT refers to a significant increasing trend, SDT refers to a significant decreasing trend.

Tables 6 and 7 present the trend results of Rx5. According to the trend results of Rx5, significant increasing trends in March, September, and December were found for the Dampahala station, with values of 5.76 mm/yr, 4.92 mm/yr, and 5.25 mm/yr. Kamburupitiya revealed a 3.38 mm/yr significant increasing trend in December, while in the annual scale analysis, Kamburupitya revealed a significant decreasing trend, with a magnitude of 3.63 mm/yr. Kirama and Goluwatta stations also revealed decreasing trends in July and May, with magnitudes of 3.34 mm/yr and 4.0 mm/yr, respectively. Deniyaya revealed a comparatively high number of significant decreasing trend events in February, April, May, June, and November, and at the annual scale, with magnitudes of 4.24 mm/yr, 3.13 mm/yr, 5.24 mm/yr, 4,88 mm/yr, 4.81 mm/yr, and 6.45 mm/yr respectively. Generally, Rx5 showed more significant decreasing trends in both monthly and annual scales at most stations.

Table 6. Trend analysis results for the extreme rainfall indices (Rx5).

Timescale	Dampahala			Kamburupitiya			Kekenadura		
	p-Value	Sen's Slope	Trend Type	*p*-Value	Sen's Slope	Trend Type	*p*-Value	Sen's Slope	Trend Type
January	0.082	2.64	NSIT	0.642	0.56	NSIT	0.607	0.54	NSIT
February	0.314	2.48	NSIT	0.903	0.12	NSIT	0.215	1.49	NSIT
March	**0.002**	**5.76**	**SIT**	0.710	0.39	NSIT	0.228	1.10	NSIT
April	0.132	4.51	NSIT	0.096	−2.56	NSDT	0.508	−0.81	NDIT
May	0.096	3.05	NSIT	0.268	−1.99	NSDT	0.145	−2.45	NDIT
June	0.338	1.88	NSIT	0.143	−2.58	NSDT	0.864	0.27	NSIT
July	0.607	0.70	NSIT	0.447	−0.44	NSDT	0.079	−1.46	NDIT
August	0.442	1.38	NSIT	0.607	0.75	NSIT	0.124	1.56	NSIT
September	**0.008**	**4.92**	**SIT**	0.447	−1.43	NSDT	0.359	−1.78	NDIT
October	0.070	2.56	NSIT	0.710	−0.35	NSDT	0.785	−0.22	NDIT
November	0.078	3.87	NSIT	0.573	−0.59	NSDT	0.921	0.07	NSIT
December	**0.007**	**5.25**	**SIT**	**0.030**	**3.38**	**SIT**	0.192	2.16	NSIT
Annual	0.244	4.32	NSIT	**0.030**	**−3.63**	**SDT**	0.228	−1.11	NDIT

* NSIT refers to a non-significant increasing trend, NSDT refers to a non-significant decreasing trend, SIT refers to a significant increasing trend, SDT refers to a significant decreasing trend.

Table 7. Trend analysis results for the extreme rainfall indices (Rx5).

Timescale	Kirama			Goluwatta			Deniyaya		
	p-Value	Sen's Slope	Trend Type	p-Value	Sen's Slope	Trend Type	p-Value	Sen's Slope	Trend Type
January	0.385	1.35	NSIT	0.788	−0.25	NSDT	0.508	−1.10	NDIT
February	0.901	0.12	NSIT	0.750	−0.98	NSDT	**0.027**	**−4.24**	**SDT**
March	0.413	0.80	NSIT	0.980	−0.01	NSDT	0.070	−3.19	NDIT
April	0.160	−2.68	NSDT	0.980	0.09	NSIT	**0.050**	**−3.13**	**SDT**
May	0.901	0.27	NSIT	**0.014**	**−4.00**	**SDT**	**0.024**	**−5.24**	**SDT**
June	0.130	−1.62	NSDT	0.447	−1.65	NSDT	**0.050**	**−4.88**	**SDT**
July	**0.003**	**−3.34**	**SDT**	0.607	−0.95	NSDT	0.172	−1.08	NDIT
August	0.766	0.31	NSIT	0.941	−0.05	NSDT	0.087	−2.05	NDIT
September	0.290	−1.57	NSDT	0.107	−3.35	NSDT	0.286	−1.59	NDIT
October	0.677	0.78	NSIT	0.175	−2.68	NSDT	0.941	0.11	NSIT
November	0.862	0.10	NSIT	0.314	−2.32	NSDT	**0.039**	**−4.81**	**SDT**
December	0.160	1.84	NSIT	0.673	0.66	NSIT	0.862	0.62	NSIT
Annual	0.209	−1.41	NSDT	0.070	−5.52	NSDT	**0.014**	**−6.45**	**SDT**

* NSIT refers to a non-significant increasing trend, NSDT refers to a non-significant decreasing trend, SIT refers to a significant increasing trend, SDT refers to a significant decreasing trend.

3.5. Change Point Detection in the Rainfall Data

Pettitt's test results, which showed statistically significant changes at the annual and seasonal scales, are shown in Figure 3. Most stations did not show significant increasing or decreasing changes at both scales. The annual scale results of the Pettit's test at Dampahala revealed a significant increasing shift in the year 1998 and Deniyaya revealed a significant decreasing shift in the year 2000. However, the Dampahala station showed significant increasing shifts in 1998 during the NEM, FIM, and SIM. Moreover, Kamburupitiya and Deniyaya showed significant decreasing changes in 1998 and 1999, respectively.

Table 8 shows a few significant changes in the monthly Pettitt's test results for the Dampahala, Kamburupitiya, Kirama, and Deniyaya rainfall stations. The Dampahala station revealed significant increasing changes during March, September, and December in 1999, 1999, and 1997, respectively. The Kamburupitiya station revealed a significant decreasing shift in July 1998. Moreover, the Kirama station showed significant decreasing shifts in June 2000 and July 2005, and the Deniyaya station showed a significant decreasing change in October 1997.

Table 8. Pettitt's test results of monthly rainfall data.

Station	Month	p-Value	Change Point	Shift
Dampahala	March	0.014	1999	Upward
	September	0.040	1999	Upward
	December	0.020	1997	Upward
Kamburupitiya	July	0.031	1998	Downward
Kirama	June	0.013	2000	Downward
	July	0.007	2005	Downward
Deniyaya	October	0.034	1997	Downward

3.6. Linkage between Rainfall and Streamflow

Pearson's correlation was used to analyze the relationship between rainfall and streamflow during 1991–2014 at the monthly, annual, and seasonal timescales. Considering the contribution of rainfall to the Pitabeddara streamflow station, only the Dampahala, Kirama, and Deniyaya stations were chosen to find the existing linkages between rainfall and streamflow. The results of the correlation analysis are given in Table 9.

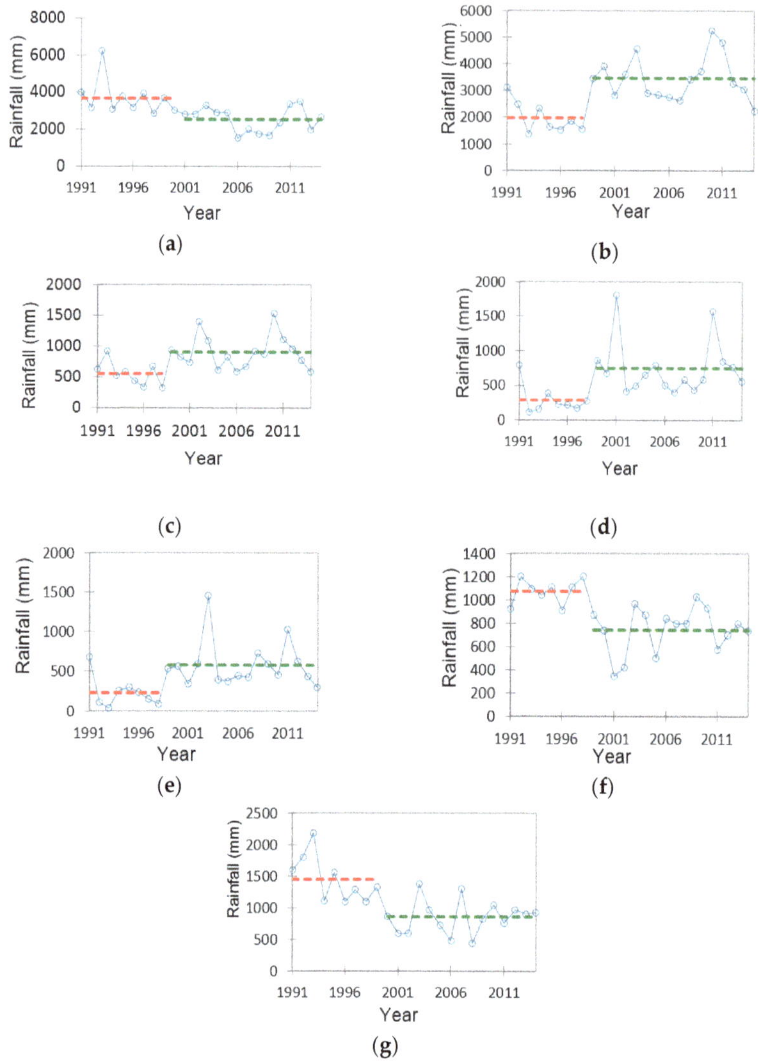

Figure 3. Pettitt's test results of the annual and seasonal rainfall data: (**a**) Deniyaya annual, (**b**) Dampahala annual, (**c**) Dampahala SIM, (**d**) Dampahala NEM, (**e**) Dampahala FIM, (**f**) Kamburupitiya SWM, and (**g**) Deniyaya SWM.

The Pearson correlation results indicated that February, March, May, and September showed positive very strong correlations, while January, June, November, and December showed positive strong correlations between rainfall and streamflow. Moreover, at the monthly scale, April, July, and October showed positive strong moderate correlations between rainfall and streamflow, while August had a positive moderate correlation. The annual scale results showed a positive strong moderate correlation between rainfall and streamflow. At the seasonal scales, a positive very strong correlation was observed for the NEM, and the FIM showed a positive strong moderate correlation. Both the SWM and SIM periods showed positive strong correlations between the rainfall and streamflow data.

Table 9. Pearson correlation results.

Timescale	Correlation	Linkage between Rainfall and Streamflow
January	0.79	*PSC*
February	0.85	*PSVC*
March	0.85	*PSVC*
April	0.40	*PSMC*
May	0.85	*PSVC*
June	0.72	*PSC*
July	0.46	*PSMC*
August	0.33	*PMC*
September	0.80	*PSVC*
October	0.56	*PSMC*
November	0.61	*PSC*
December	0.71	*PSC*
Annual	0.55	*PSMC*
NEM	0.81	*PSVC*
FIM	0.58	*PSMC*
SWM	0.62	*PSC*
SIM	0.67	*PSC*

PSC—positive strong correlation; *PVSC*—positive very strong correlation; *PSMC*—positive strong moderate correlation; *PMC*—positive moderate correlation.

3.7. Indicators of Hydrological Alteration (IHA)

According to the literature based on the NRB, no major obstruction or dam was built. However, we decided to investigate the changes in the flow regime before and after 2003. This year was selected purely arbitrarily. Two-period parametric analysis was used in this IHA method. The pre-impact period was chosen from 1991–2003 and the post-impact period was selected from 2004–2014. This section discusses 16 selected hydrological parameters out of the 33 that fall under the main IHA parameter groups of magnitude and duration of annual extreme water conditions, timing of annual extreme water conditions, and frequency and duration of high and low pulses. According to the IHA analysis, the mean annual flow during the post-impact period increased slightly from 15.87 m^3s^{-1} to 16.23 m^3s^{-1}. Table 10 given below demonstrates the statistics for the pre-impact and post-impact periods for the selected IHA parameters.

Table 10. IHA results.

IHA Parameters	Means			Coefficient of Variation		
	Pre-Impact	Post-Impact	Deviation Factor (%)	Pre-Impact	Post-Impact	Deviation of C.V (%)
Magnitude and Duration (IHA Group 2)						
1-day minimum	2.375	2.322	−2.223	0.4769	0.3378	−29.17
3-day minimum	2.466	2.424	−1.71	0.4594	0.3298	−28.22
7-day minimum	2.766	2.723	−1.564	0.449	0.3073	−31.56
30-day minimum	4.171	4.224	1.266	0.4171	0.3226	−22.67
90-day minimum	6.564	7.762	18.26	0.3865	0.2785	−27.96
1-day maximum	189.2	136.3	−27.94	0.5865	0.3681	−37.24
3-day maximum	129.9	89.85	−30.82	0.7325	0.4233	−42.21
7-day maximum	82.79	63.03	−23.87	0.5919	0.334	−43.56
30-day maximum	43.27	40.66	−6.036	0.3191	0.3589	12.48
90-day maximum	28.46	27	−5.138	0.2447	0.2198	−10.2
No. of zero days	0	0	-	0	0	-
Baseflow index	0.1722	0.1737	0.8578	0.401	0.3362	−16.14
Timing (IHA Group 3)						
Date of each annual 1-day minimum	113	199.5	47.24	0.1857	0.2228	19.96

Table 10. *Cont.*

IHA Parameters	Means			Coefficient of Variation		
	Pre-Impact	Post-Impact	Deviation Factor (%)	Pre-Impact	Post-Impact	Deviation of C.V (%)
Date of each annual 1-day maximum	211.5	261.5	27.28	0.2693	0.2606	−3.227
Frequency (IHA Group 4)						
No. of low pulses in each year	14.54	13.36	−8.081	0.2731	0.4628	69.54
No. of high pulses in each year	12.54	13.73	9.481	0.4253	0.3092	−27.29

Although notable changes were not observed for minimum flows in IHA group 2, maximum flows decreased significantly during the post-impact period. For instance, the 1-day maximum, 3-day maximum, and 7-day maximum flows decreased by 27.94%, 30.82%, and 23.87%, respectively. The only notable change in minimum flow was the 90-day minimum flow with an increase of 18.26%.

Results of the IHA group 3 parameters in Table 9 showed that the annual 1-day minimum flow was recorded on the 113th Julian date in the calendar for the pre-impact period and the 119th and 120th days for the post-impact period; this showed that the date shifted a little bit forward in the post-impact period. Julian's date of each annual 1-day maximum for the pre-impact was recorded during the 211th and 212th days, and for the post-impact period, it was recorded during the 261st and 262nd days.

According to the results in Table 10 under the IHA group 3 category, 14.56 low pulses were found in the pre-impact period, which decreased to 13.36 per year during the post-impact period. The coefficient of variation for the low pulses was increased by 69.54% from the pre-impact to post-impact period. The number of high pulses for each year in the pre-impact period was found to be 12.54 per year, which increased in the post-impact period up to 13.73 per year. The deviation of the coefficient of variation also decreased by 27.29% from the pre-impact to post-impact period.

3.8. Discussion

According to the results found in the present study, most of the stations showed decreasing rainfall trend patterns at the monthly scale. Generally, the annual scale and the FIM and SWM exhibited decreasing rainfall trend patterns, NEM exhibited an increasing tendency and SIM exhibited average results. A previous study that was carried out during 1987–2017 by Nisansala et al. [46] reported similar results for the NEM, SWM, and SIM seasons. Wickramagamage [47] also reported an increasing rainfall trend during the NEM season and decreasing trends in the SWM for Sri Lanka during 1981–2010. However, Nisansala et al. [46] and Wickramagamage [47] also showed different results compared with the present study at the annual and seasonal scales for rainfall trends by showing an increasing tendency. In the present study, the Dampahala station showed increasing trends for rainfall, while all the other stations showed decreasing trend patterns. Similar to the present study, contrasting directions of the magnitude of trends in nearby stations of the same basin were reported by Khaniya et al. [48] and Pawar and Rathnayake [49]. These contrasting results might have been because of local rainfall events. Mehta and Yadav [50] demonstrated that the magnitude of climate variability change across spatial scales. According to the results of the present study, the streamflow trend analysis did not show significant trends, except in December. However, non-significant increasing trends were demonstrated at the annual scale and the NEM and FIM, but not for the SWM SIM. Dinethra and Basnayake [51] showed that the discharge of the Nilwala River increased during 2004–2013. When considering the extreme rainfall indices, they also showed few significant trends. However, extreme rainfall indices trend patterns also showed similar contrasting results, especially for the Dampahala and Deniyaya stations. These contrasting results might have been due to variations in elevations, as explained by Bizuneh [52]. The NRB comprises lowlands and mountains. Due to the windward and leeward sides of the mountains, these types of contrasting results in the direction of rainfall trends can happen. This is because the windward side normally receives higher rainfall, while the leeward side of the mountain gets lower rainfall. Considering rainfall, extreme rainfall, and streamflow trend results, we concluded that rainfall was not only the influencing factor for the changes in streamflow patterns for the NRB. Other factors, such as the density of physical features, watershed characteristics, and vegetation cover, can be influential as well [2].

4. Conclusions

The main objective of this study was to analyze the trends that were present in the rainfall and streamflow records and to identify the correlation between the rainfall and streamflow over the Nilwala River Basin. Six rainfall stations and one hydrological gauge station were chosen based on the data availability. There was a considerably good correlation present between the rainfall and streamflow in the upper Nilwala Basin, indicating that rainfall was the main driver for the changes in the streamflow. This study showed that the NRB did not face extreme climatic events from 1990 to 2014. Variations in the topographical features in the basin might lead to contrasting results in rainfall trends within the NRB. Five out of six stations showed decreasing trends in rainfall, suggesting that, in general, rainfall had been decreasing during the 25 years between 1990 and 2014. The insignificant trends in rainfall and streamflow suggested that the climate and hydrologic regimes were not altered during this period. Future research is advocated for in the NRB considering that other possible factors, such as the impacts of land-use change, could also be influential.

Author Contributions: Conceptualization, M.B.G.; methodology, R.P. and M.B.G.; software, R.P.; formal analysis, R.P.; writing—original draft preparation, R.P.; writing—review and editing, M.B.G., R.P., I.M.C., M.K.J., M.S.B. and U.R.; supervision, M.B.G., M.S.B., U.R. and M.K.J.; funding acquisition, M.B.G. and U.R. All authors have read and agreed to the published version of the manuscript.

Funding: This research was carried out under the financial support of a SLIIT Research Grant: FGSR/RG/FE/2022/02.

Institutional Review Board Statement: Not applicable.

Informed Consent Statement: Not applicable.

Data Availability Statement: The climatic data used in this research study are available upon request for research purposes.

Acknowledgments: The authors of this study would like to thank the Sri Lanka Institute of Information Technology (SLIIT), Wayamba University of Sri Lanka, and the Norwegian Institute of Bioeconomy Research (NIBIO) for providing the environment to conduct this research.

Conflicts of Interest: The authors declare no conflict of interest.

References

1. Fentaw, F.; Hailu, D.; Nigussie, A. Trend and Variability Analysis of Rainfall & Stream Flow Series at Tekeze River Basin, Ethiopia. *Int. J. Sci. Eng. Res.* **2017**, *8*, 665–680.
2. Ademe, D.; Alamirew, T.; Rotich, J.; Thi, N.; Linh, T.; Gebrie, T. Analysis of Rainfall and Streamflow Trend and Variability over Birr River Watershed, Abbay Basin, Ethiopia. *Environ. Chall.* **2022**, *7*, 100528. [CrossRef]
3. Chauluka, F.; Singh, S.; Kumar, R. Rainfall and Streamflow Trends of Thuchila River, Southern Malawi. *Mater. Today Proc.* **2021**, *34*, 846–855. [CrossRef]
4. Zhai, R.; Tao, F. Contributions of Climate Change and Human Activities to Runoff Change in Seven Typical Catchments across China. *Sci. Total Environ.* **2017**, *605–606*, 219–229. [CrossRef] [PubMed]
5. Alemu, M.M.; Bawoke, G.T. Analysis of Spatial Variability and Temporal Trends of Rainfall in Amhara Region, Ethiopia. *J. Water Clim. Chang.* **2020**, *11*, 1505–1520. [CrossRef]
6. Kliment, Z.; Matoušková, M. Long-Term Trends of Rainfall and Runoff Regime in Upper Otava River Basin. *Soil Water Res.* **2008**, *3*, 155–167. [CrossRef]
7. Arrieta-Castro, M.; Donado-Rodríguez, A.; Acuña, G.J.; Canales, F.A.; Teegavarapu, R.S.V.; Kaźmierczak, B. Analysis of Streamflow Variability and Trends in the Meta River, Colombia. *Water* **2020**, *12*, 1451. [CrossRef]
8. Charifi Bellabas, S.; Benmamar, S.; Dehni, A. Study and Analysis of the Streamflow Decline in North Algeria. *J. Appl. Water Eng. Res.* **2021**, *9*, 20–44. [CrossRef]

9. Fentaw, F.; Melesse, A.M.; Hailu, D.; Nigussie, A. Precipitation and Streamflow Variability in Tekeze River Basin, Ethiopia. In *Extreme Hydrology and Climate Variability*; Elsevier: Amsterdam, The Netherlands, 2019; pp. 103–121. [CrossRef]
10. Mersin, D.; Tayfur, G.; Vaheddoost, B.; Safari, M.J.S. Historical Trends Associated with Annual Temperature and Precipitation in Aegean Turkey, Where Are We Heading? *Sustainability* **2022**, *14*, 3380. [CrossRef]
11. Kastridis, A.; Kamperidou, V.; Stathis, D. Dendroclimatological Analysis of Fir (A. Borisii-Regis) in Greece in the Frame of Climate Change Investigation. *Forests* **2022**, *13*, 879. [CrossRef]
12. Hannaford, J. Climate-Driven Changes in UK River Flows: A Review of the Evidence. *Prog. Phys. Geogr.* **2015**, *39*, 29–48. [CrossRef]
13. Wang, G.; Xia, J.; Chen, J. Quantification of Effects of Climate Variations and Human Activities on Runoff by a Monthly Water Balance Model: A Case Study of the Chaobai River Basin in Northern China. *Water Resour. Res.* **2009**, *45*, 1–12. [CrossRef]
14. Azari, M.; Moradi, H.R.; Saghafian, B.; Faramarzi, M. Climate Change Impacts on Streamflow and Sediment Yield in the North of Iran. *Hydrol. Sci. J.* **2016**, *61*, 123–133. [CrossRef]
15. Dey, P.; Mishra, A. Separating the Impacts of Climate Change and Human Activities on Streamflow: A Review of Methodologies and Critical Assumptions. *J. Hydrol.* **2017**, *548*, 278–290. [CrossRef]
16. Xu, F.; Zhao, L.; Niu, C.; Qiu, Y. Effect of Climate Change and Anthropogenic Activities on Streamflow Indicators in a Tropical River Basin in Southern China. *Water* **2022**, *14*, 304. [CrossRef]
17. Jayasekara, S.M.; Abeysingha, N.S.; Meegastenna, T. Streamflow Trends of Kelani River Basin in Sri Lanka (1983–2013). *J. Natl. Sci. Found. Sri Lanka* **2020**, *48*, 449. [CrossRef]
18. Kale, S.; Ejder, T.; Hisar, O.; Mutlu, F. Climate Change Impacts on Streamflow of Karamenderes River (Çanakkale, Turkey). *Mar. Sci. Technol. Bull.* **2016**, *5*, 1–6. [CrossRef]
19. Abungba, J.A.; Khare, D.; Pingale, S.M.; Adjei, K.A.; Gyamfi, C.; Odai, S.N. Assessment of Hydro-Climatic Trends and Variability over the Black Volta Basin in Ghana. *Earth Syst. Environ.* **2020**, *4*, 739–755. [CrossRef]
20. Salarijazi, M. Trend and Change-Point Detection for the Annual Stream-Flow Series of the Karun River at the Ahvaz Hydrometric Station. *Afr. J. Agric. Res.* **2012**, *7*, 4540–4552. [CrossRef]
21. Zhang, W.; Yan, Y.; Zheng, J.; Li, L.; Dong, X.; Cai, H. Temporal and Spatial Variability of Annual Extreme Water Level in the Pearl River Delta Region, China. *Glob. Planet. Chang.* **2009**, *69*, 35–47. [CrossRef]
22. Caloiero, T.; Coscarelli, R.; Ferrari, E. Application of the Innovative Trend Analysis Method for the Trend Analysis of Rainfall Anomalies in Southern Italy. *Water Resour. Manag.* **2018**, *32*, 4971–4983. [CrossRef]
23. Mathews, R.; Richter, B.D. Application of the Indicators of Hydrologic Alteration Software in Environmental Flow Setting1: Tools for Environmental Flow Setting. *JAWRA J. Am. Water Resour. Assoc.* **2007**, *43*, 1400–1413. [CrossRef]
24. Barbalić, D.; Kuspilić, N. Trends of Indicators of Hydrological Alterations. *GRADEVINAR* **2014**, *66*, 613–624. [CrossRef]
25. Gao, Y.; Vogel, R.M.; Kroll, C.N.; Poff, N.L.; Olden, J.D. Development of Representative Indicators of Hydrologic Alteration. *J. Hydrol.* **2009**, *374*, 136–147. [CrossRef]
26. Chathuranika, I.M.; Gunathilake, M.B.; Azamathulla, H.M.; Rathnayake, U. Evaluation of Future Streamflow in the Upper Part of the Nilwala River Basin (Sri Lanka) under Climate Change. *Hydrology* **2022**, *9*, 48. [CrossRef]
27. Abeysingha, N.S. A Review of Recent Changes in Rainfall Trend in Sri Lanka. *Trop. Agric. Res. Ext.* **2022**, *25*, 1–13. [CrossRef]
28. Perera, A.; Ranasinghe, T.; Gunathilake, M.; Rathnayake, U. Comparison of Different Analyzing Techniques in Identifying Rainfall Trends for Colombo, Sri Lanka. *Adv. Meteorol.* **2020**, *2020*, 8844052. [CrossRef]
29. Alahacoon, N.; Edirisinghe, M. Spatial Variability of Rainfall Trends in Sri Lanka from 1989 to 2019 as an Indication of Climate Change. *Int. J. Geo-Inf.* **2021**, *10*, 84. [CrossRef]
30. Ruwangika, A.M.; Perera, A.; Rathnayake, U. Comparison of Statistical, Graphical, and Wavelet Transform Analyses for Rainfall Trends and Patterns in Badulu Oya Catchment, Sri Lanka. *Complexity* **2020**, *2020*, 7146593. [CrossRef]
31. Dhanapala, L.; Gunarathna, M.H.J.P.; Kumari, M.K.N.; Ranagalage, M.; Sakai, K.; Meegastenna, T.J. Towards Coupling of 1D and 2D Models for Flood Simulation—A Case Study of Nilwala River Basin, Sri Lanka. *Hydrology* **2022**, *9*, 17. [CrossRef]
32. Yatagai, A.; Arakawa, O.; Kamiguchi, K.; Kawamoto, H.; Nodzu, M.I.; Hamada, A. A 44-Year Daily Gridded Precipitation Dataset for Asia Based on a Dense Network of Rain Gauges. *SOLA* **2009**, *5*, 137–140. [CrossRef]
33. Yatagai, A.; Kamiguchi, K.; Arakawa, O.; Hamada, A.; Yasutomi, N.; Kitoh, A. APHRODITE: Constructing a Long-Term Daily Gridded Precipitation Dataset for Asia Based on a Dense Network of Rain Gauges. *Bull. Am. Meteorol. Soc.* **2012**, *93*, 1401–1415. [CrossRef]
34. Marie, M.; Yirga, F.; Haile, M.; Ehteshammajd, S.; Azadi, H.; Scheffran, J. Time-Series Trend Analysis and Farmer Perceptions of Rainfall and Temperature in Northwestern Ethiopia. *Environ. Dev. Sustain.* **2021**, *23*, 12904–12924. [CrossRef]
35. Mutiibwa, D.; Vavrus, S.J.; McAfee, S.A.; Albright, T.P. Recent Spatiotemporal Patterns in Temperature Extremes across Conterminous United States. *J. Geophys. Res. Atmos.* **2015**, *120*, 7378–7392. [CrossRef]
36. Mann, H.B. Nonparametric Tests Against Trend. *Econometrica* **1945**, *13*, 245–259. [CrossRef]
37. Kendall, M.G. *Rank Correlation Methods*, 4th ed.; Griffin: London, UK, 1970.
38. Yue, S.; Pilon, P.; Cavadias, G. Power of the Mann–Kendall and Spearman's Rho Tests for Detecting Monotonic Trends in Hydrological Series. *J. Hydrol.* **2002**, *259*, 254–271. [CrossRef]
39. Zhang, X.; Harvey, K.D.; Hogg, W.D.; Yuzyk, T.R. Trends in Canadian Streamflow. *Water Resour. Res.* **2001**, *37*, 987–998. [CrossRef]
40. Birsan, M.-V.; Molnar, P.; Burlando, P.; Pfaundler, M. Streamflow Trends in Switzerland. *J. Hydrol.* **2005**, *314*, 312–329. [CrossRef]

41. Kulkarni, A.; Storch, H.V. Monte Carlo Experiments on the Effect of Serial Correlation on the Mann-Kendall Test of Trend. *Meteorol. Z.* **1995**, *4*, 82–85. [CrossRef]
42. Hamed, K.H.; Rao, A.R. A Modified Mann-Kendall Trend Test for Autocorrelated Data. *J. Hydrol.* **1998**, *204*, 182–196. [CrossRef]
43. Sen, P.K. Estimates of the Regression Coefficient Based on Kendall's Tau. *J. Am. Stat. Assoc.* **1968**, *63*, 1379–1389. [CrossRef]
44. Pettitt, A.N. A Non-Parametric Approach to the Change-Point Problem. *Appl. Stat.* **1979**, *28*, 126. [CrossRef]
45. Zhang, X.; Yang, F. *RClimDex (1.0) User Manual*; Climate Research Branch Environment Canada: Downsview, ON, Canada, 2004.
46. Nisansala, W.D.S.; Abeysingha, N.S.; Islam, A.; Bandara, A.M.K.R. Recent Rainfall Trend over Sri Lanka (1987–2017). *Int. J. Climatol.* **2020**, *40*, 3417–3435. [CrossRef]
47. Wickramagamage, P. Spatial and Temporal Variation of Rainfall Trends of Sri Lanka. *Theor. Appl. Climatol.* **2016**, *125*, 427–438. [CrossRef]
48. Khaniya, B.; Jayanayaka, I.; Jayasanka, P.; Rathnayake, U. Rainfall Trend Analysis in Uma Oya Basin, Sri Lanka, and Future Water Scarcity Problems in Perspective of Climate Variability. *Adv. Meteorol.* **2019**, *2019*, 3636158. [CrossRef]
49. Pawar, U.; Rathnayake, U. Spatiotemporal Rainfall Variability and Trend Analysis over Mahaweli Basin, Sri Lanka. *Arab J. Geosci.* **2022**, *15*, 370. [CrossRef]
50. Mehta, D.; Yadav, S.M. An Analysis of Rainfall Variability and Drought over Barmer District of Rajasthan, Northwest India. *Water Supply* **2021**, *21*, 2505–2517. [CrossRef]
51. Dinethra, A.G.T.; Basnayake, D.M.L.A. Contribution of climate change and human activities on the runoff changes in the upper catchment of Nilwala basin, Sri Lanka. In Proceedings of the 20th IAHR APD Congress, Colombo, Sri Lanka, 28–31 August 2016.
52. Mekuriaw Bizuneh, A. *Climate Variability and Change in the Rift Valley and Blue Nile Basin, Ethiopia: Local Knowledge, Impacts, and Adaptation*; UAMR Studies on Development and Global Governance; Logos Verlag Berlin GmbH: Berlin, Germany, 2013.

remote sensing

Article

Drought Monitoring Using Landsat Derived Indices and Google Earth Engine Platform: A Case Study from Al-Lith Watershed, Kingdom of Saudi Arabia

Nuaman Ejaz [1], Jarbou Bahrawi [1], Khalid Mohammed Alghamdi [1], Khalil Ur Rahman [2] and Songhao Shang [2,*]

[1] Department of Hydrology and Water Resources Management, Faculty of Meteorology, Environment & Arid Land Agriculture, King Abdulaziz University, Jeddah 21589, Saudi Arabia
[2] State Key Laboratory of Hydro Science and Engineering, Department of Hydraulic Engineering, Tsinghua University, Beijing 100084, China
* Correspondence: shangsh@tsinghua.edu.cn; Tel.: +86-10-6279-6674

Abstract: Precise assessment of drought and its impact on the natural ecosystem is an arduous task in regions with limited climatic observations due to sparsely distributed in situ stations, especially in the hyper-arid region of Kingdom of Saudi Arabia (KSA). Therefore, this study investigates the application of remote sensing techniques to monitor drought and compare the remote sensing-retrieved drought indices (RSDIs) with the standardized meteorological drought index (Standardized Precipitation Evapotranspiration Index, SPEI) during 2001–2020. The computed RSDIs include Vegetation Condition Index (VCI), Temperature Condition Index (TCI), and Vegetation Health Index (VHI), which are derived using multi-temporal Landsat 7 ETM+, Landsat 8 OLI/TIRS satellites, and the Google Earth Engine (GEE) platform. Pearson correlation coefficient (CC) is used to find the extent of agreement between the SPEI and RSDIs. The comparison showed CC values of 0.74, 0.67, 0.57, and 0.47 observed for VHI/SPEI-12, VHI/SPEI-6, VHI/SPEI-3, and VHI/SPEI-1, respectively. Comparatively low agreement was observed between TCI and SPEI with CC values of 0.60, 0.61, 0.42, and 0.37 observed for TCI/SPEI-12, TCI/SPEI-6, TCI/SPEI-3, and TCI/SPEI-1. A lower correlation with CC values of 0.53, 0.45, 0.33 and 0.24 was observed for VCI/SPEI-12, VCI/SPEI-6, VCI/SPEI-3, and VCI/SPEI-1, respectively. Overall, the results suggest that VHI and SPEI are better correlated drought indices and are suitable for drought monitoring in the data-scarce hyper-arid regions. This research will help to improve our understanding of the relationships between meteorological and remote sensing drought indices.

Keywords: drought assessment; meteorological drought; remote sensing drought indices; standardized drought indices; Landsat; Google Earth Engine

Citation: Ejaz, N.; Bahrawi, J.; Alghamdi, K.M.; Rahman, K.U.; Shang, S. Drought Monitoring Using Landsat Derived Indices and Google Earth Engine Platform: A Case Study from Al-Lith Watershed, Kingdom of Saudi Arabia. *Remote Sens.* **2023**, *15*, 984. https://doi.org/10.3390/rs15040984

Academic Editor: Gabriel Senay

Received: 25 December 2022
Revised: 31 January 2023
Accepted: 4 February 2023
Published: 10 February 2023

1. Introduction

It is challenging to precisely monitor and evaluate the onset, intensity, frequency, persistence, and propagation of drought because of its complex nature, especially in hyper-arid regions characterized by data scarcity [1,2]. Drought is a frequently occurring hydrometeorological phenomenon [3], which is the direct cause of drought disasters and the second most detrimental natural hazard that causes social and economic instability after floods [4]. Drought events are categorized into four categories based on the affected sectors [5] including meteorological, hydrological, agricultural, and socio-economic droughts. Meteorological drought is characterized by an extended period of below-average precipitation (i.e., precipitation deficit) in relation to the region's average precipitation. In contrast, agricultural drought could be described as a drought resulting from soil moisture content below the level required for plant growth and development [6–9]. Hydrological drought refers to a decrease in the quantity of water both on surface and groundwater resources due to insufficient precipitation for an extended period [10]. Socio-economic drought focuses on the consequences of drought on water resources, agriculture, and industries [11,12].

Several approaches have been developed over the last few decades to monitor and statistically describe droughts, including the development of both standardized and unstandardized drought indices used in meteorology, hydrology, and agriculture [1–3]. In the past, drought monitoring methods were based on measurements taken at stations/gauges on the ground, such as the Palmer drought Severity Index (PDSI) [4], the Standardized Precipitation Index (SPI) [5], and the Standardized Precipitation Evapotranspiration Index (SPEI) [6]. Traditional approaches for assessing and monitoring drought depend on the in situ precipitation records, which are usually inaccurate and constrained both in time and space [7,8]. The sparse distribution of in situ weather stations per unit area and the associated uncertainties hinders the precise estimation of drought, which is most often for areas in arid and hyper-arid regions. Other natural impediments, such as mountains and dune fields, can also contribute to the said problem [9]. El Kenawy and McCabe [10] have confirmed these flaws in the meteorological network over Kingdom of Saudi Arabia (KSA). However, with the advancements in remote sensing and earth observation technologies (e.g., the launch of the National Aeronautics Space Administration (NASA) Landsat series in 1972) at the end of the 20th century, the way for drought monitoring has been changed [11]. In addition, there is an increasing curiosity and understanding regarding the climate change due to rising temperatures. This has led geospatial scientists to conclude that remote sensing must play a crucial role in delivering the data necessary to assess ecosystem conditions and monitor extreme climatic changes at both spatial and temporal scales [12–15].

Remote sensing (RS) products not only provide meteorological data but also monitor changes in the variables at the earth's surface such as the health of plants and the amount of available water, and provide a wide range of contextual data for monitoring drought [11]. RS and Geographic Information Systems (GIS) have made it easier for people to look at the world with sensors and see how it changes over time [16]. The main benefit of using RS and GIS techniques is the availability of continuous data over large areas in both space and time, which significantly contributed to the data scarcity issues as we might face in arid regions like KSA [17,18].

With the advancement of RS and GIS techniques, several remote-sensing-based drought indices are proposed and evaluated to monitor drought, including the Normalized Difference Vegetation Index (NDVI) [19], the Temperature Condition Index (TCI) [20,21], the Vegetation Condition Index (VCI) [22], and the Vegetation Health Index (VHI) [23,24]. TCI, VCI, and VHI are also characterized as vegetation indices since they describe the vegetation condition in a specific area, classify it into different drought classes, and are commonly employed as drought monitoring indices [25–27]. VCI is widely used to detect changes in vegetation from significantly worse to favorable conditions [20,28]. TCI detects vegetative stress induced by high temperatures and heavy moisture [29–31]. VHI, on the other hand, is the combination of TCI (temperature) and VCI (vegetation condition) that describes vegetation health [32,33].

Recently, the Google Earth Engine (GEE), a cloud-based geospatial data monitoring platform that calculates and presents raw and processed satellite-based datasets [33–35], is extensively used in various hydro-meteorological applications. Since its introduction in 2010, GEE capabilities have been tested in a variety of applications, including vegetation mapping and monitoring [36,37], land use/land cover change mapping [38,39], and flood mapping [40,41]. Furthermore, with a substantial volume of freely available satellite imageries and direct image processing, GEE has been proposed for time series analysis of drought in several studies [37,42,43]. Most of the researchers used the Moderate Resolution Imaging Spectroradiometer (MODIS) satellite dataset and GEE algorithms to monitor drought using the Remote Sensing-retrieved Drought Indices (RSDIs) [44–46]. Meanwhile, Pham and Tran [34] analyzed the temporal distribution of drought conditions in Vietnam using different Landsat-derived drought indices, which are calculated from Landsat-8 satellite data in the GEE platform. Benzougagh et al. [47] also used the GEE algorithm and a combination of Landsat-8 and Sentinel-2 datasets to monitor drought in Morocco. The above studies showed that Landsat-derived indices provide helpful spatial information for assessing drought conditions from the region to the country scales. The main theme of this research is to address the data scarcity issues in Lith watershed. There are four gauges in the watershed which are not sufficient to represent the variations in climate. Therefore, different remote sensing techniques should be applied to cope with data scarcity issues and thus we computed different remote sensing indices using the GEE.

Owing to the condition of data scarcity and minimal application of RS, GIS, and GEE applications in drought monitoring in KSA, this study aims: (1) to calculate the spatial extent of drought in the arid basin (Al-Lith watershed) using Landsat 7 and Landsat 8 datasets from 2001 to 2020; (2) to characterize the spatiotemporal pattern of drought conditions by SPEI, VCI, and TCI, and VHI using GEE; and (3) to compare the reliability of various drought indices (particularly SPEI with RSDIs) in drought monitoring by calculating the Pearson correlation coefficient (CC). The findings in this research will assist urban planners and environmental scientists in making decisions and implementing policies to mitigate drought in the Al-Lith watershed, particularly and other similar areas around the globe in general.

2. Study Area

The Al-Lith watershed is located between 20°00′N to 20°15′N longitude, and 40°10′E to 40°50′E latitude in the Makkah region of KSA, with elevation ranges from 0 to 2663 m above mean sea level (shown in Figure 1). The Al-Lith watershed has a total area of about 3262 km^2. The maximum temperature in Al-Lith is observed in July with an average of 41.9 °C, followed by June and August with the average temperatures of 41.3 °C and 41.2 °C, respectively [48]. A minimum of 20.0 °C temperature is observed in January.

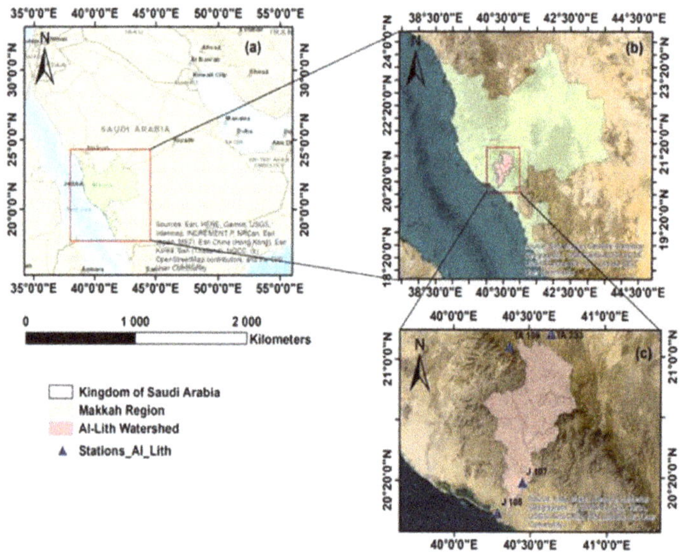

Figure 1. Geographical location of the Al-Lith Watershed: (**a**) Kingdom of Saudi Arabia; (**b**) Makkah Region; and (**c**) the Al-Lith Watershed.

The amount of precipitation in the Al-Lith Watershed varies with location and year. The average annual precipitation is 104.3 mm, with 145.2 mm, 167.7 mm, 56.7 mm, and 47.6 mm at stations TA-109, TA-233, J107, and J108, respectively. TA-109 and TA-233 are on the upstream side of the Al-Lith Watershed, whereas J107 and J108 are on the downstream side, as shown in Figure 1c. The rainy season from November to January contributed to 55% of annual precipitation, while the dry season from June to August contributed only 15%.

3. Data and Methods

The methodology used in this study is categorized into; (i) meteorological data acquisition, (ii) satellite data retrieval, (iii) meteorological drought monitoring using SPEI, (iv) drought monitoring using remotely sensed drought indices (i.e., VCI, TCI, and VHI), (v) temporal and geospatial analyses of the above indices using GEE, and (vi) correlation between SPEI and remote sensing drought indices. The workflow for this study is shown in Figure 2.

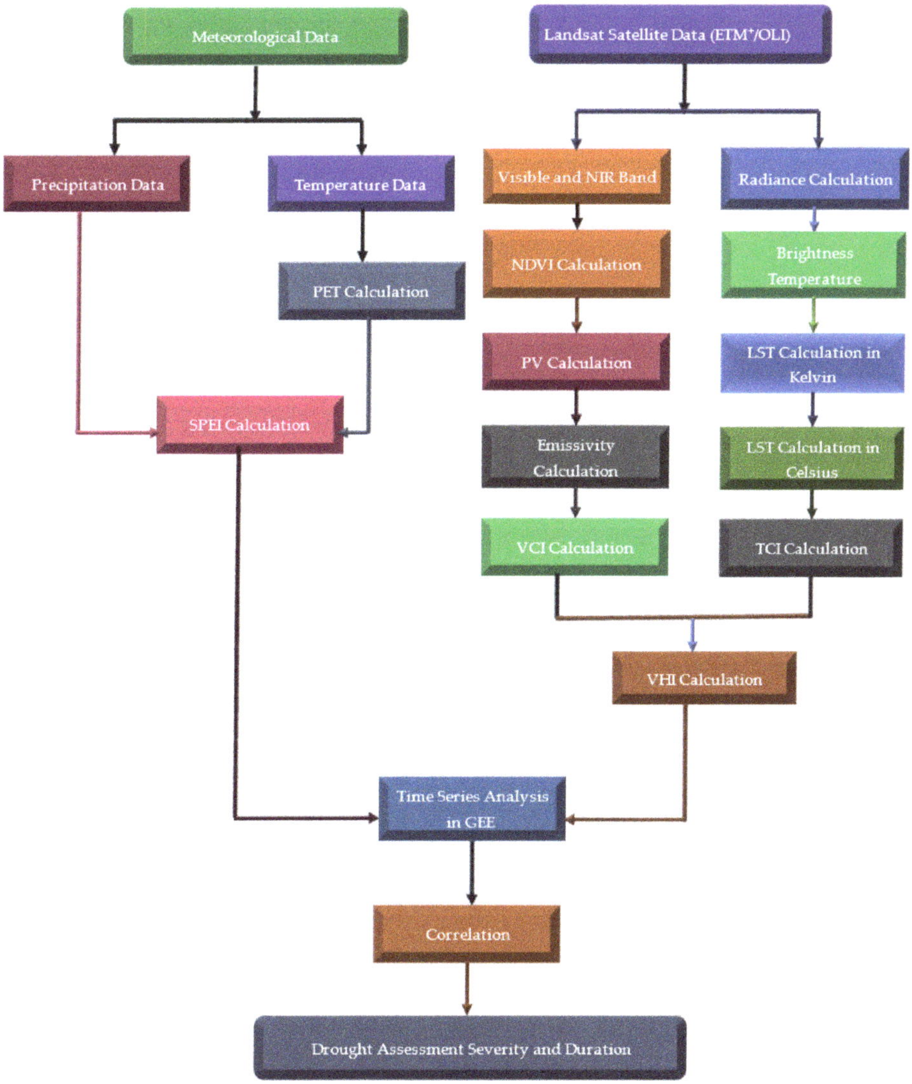

Figure 2. Methodological framework used in this study.

3.1. Meteorological Data Acquisition

This study utilized the daily precipitation and temperature data from the Ministry of Water Environment and Agriculture (MEWA). The data for four meteorological stations (J107, J108, TA109, and TA233, as shown in Figure 1) located in the Al-Lith watershed was collected spanning the period from 2001 to 2020.

3.2. Satellit Data Acquisitions

Landsat-7 and 8 Level 1 images at different dates were downloaded from the USGS Earth Explorer data portal between 2001 and 2020 for the study area. During the study period, Landsat-7 ETM+ data was acquired for the period of 2001–2012, while Landsat-8 OLI data was used for the period of 2013–2020. Images downloaded from the USGS Earth Explorer website were already georeferenced and projected in WGS UTM Zone 37 N for the Al-Lith watershed. However, no images were available for path 169 and row 046 on Earth Explorer, so the time series analysis for that specific location was done using Google Earth Engine. Table 1 shows the specifications, including sensors information, spatial and temporal resolutions, path, and rows for Landsat 7 and Landsat 8 satellites in Al-Lith watershed.

Table 1. Specification of Landsat Satellites over Al-Lith watershed.

Satellite	Sensor	Spatial Resolution	Temporal Resolution	Paths	Row	Years
Landsat 7	ETM+	30 m	16days	169	45	2001–2012
Landsat 8	OLI+	30 m	16days	169	46	2013–2020

3.3. Standard Precipitation Evapotranspiration Index (SPEI)

SPEI is the new and comprehensive drought index proposed by Vicentro et al. [6] and is used in this study to monitor meteorological drought. The SPEI is based on precipitation (P) and potential evapotranspiration (PET) and represents the fundamental calculation of the climate's water balance at various time scales. The PET can be calculated using many methods like Thornthwaite, Penman–Monteith, and Hargreaves. However, we used the Hargreaves equation in this study to calculate the PET because of less data requirement [49] suggested by many authors when solar radiation, relative humidity, and wind speed datasets are not available [50,51].

$$PET_i = 0.0135K_T(T + 17.78)(T_{max} - T_{min})^{0.5}R_a \quad (1)$$

where T, T_{max}, and T_{min} are the average, maximum, and minimum temperature in °C, respectively; R_a is the extraterrestrial radiation (mm/day); and K_T is an empirical coefficient (K_T = 0.162 for "interior" regions and K_T = 0.19 for coastal regions). Allen [52] also specified a criteria for calculating the observational coefficient as K_T = 0.17.

The monthly water balance (WB) equation is given by Equation (2), which is obtained by subtracting the calculated PET form monthly precipitation data:

$$WB_i = P_i - PET_i \quad (2)$$

where P is precipitation, and i shows the particular month.

Mostly, the SPEI is computed by first standardizing the differences in precipitation and PET values using the log-logistic probability distribution function. The log-logistic distribution function is given below by Equation (3):

$$f(x) = \frac{\beta}{\alpha}\left(\frac{x-\gamma}{\alpha}\right)^{\beta-1}\left[1+\left(\frac{x-\gamma}{\alpha}\right)^{\beta}\right]^{-2} \quad (3)$$

where α is the scale parameter, β is the shape parameter, γ is the beginning parameter, and x is the mean of the series of CWB values in each period. Then SPEI can be calculated from [6]

$$\mathrm{SPEI} = \mathrm{W} - \frac{\mathrm{C_o}+\mathrm{C_1W}+\mathrm{C_2W^2}}{1+\mathrm{d_1W}+\mathrm{d_2W^2}+\mathrm{d_3W^3}} \tag{4}$$

where $\mathrm{W} = \sqrt{-2\mathrm{Ln}(\mathrm{P})}$ for $p \leq 0.5$, and P is the probability of exceeding a determined WB value. The constants are C_0 = 2.515517, C_1 = 0.802853, C_2 = 0.010328, d_1 = 1.432788, d_2 = 0.189269, and d_3 = 0.001308 [6].

SPEI can be estimated at different time scales (1, 3, 6, and 12 months) at each station. SPEI-1 is calculated by taking the monthly precipitation and temperature. SPEI-3 is calculated by taking the mean of the three months (moving averaging of three-month precipitation and temperature inputs). Similarly, the remaining 6- and 12-month indices can be calculated. SPEI-1 is useful to study the short-term variations in drought frequency and severity, SPEI-3 and SPEI-6 are usually used to monitor the seasonal variations in drought, while SPEI-12 is useful to study the annual trend of drought. How severe a drought is can be estimated using the index, which compares actual precipitation to the amount of water lost through evaporation and transpiration, over a given period of time.

The severity of drought determined by the numeric values of SPEI, which is divided into different categories following McKee et al. [5] and Vicentro et al. [6] and are shown in Table 2.

Table 2. Division of drought severity based on SPEI values (after [6]).

SPEI	Categories
>2	Extremely wet
1.50 to 1.99	Severely wet
1.00 to1.49	Moderately wet
−0.99 to 0.99	Nearly Normal
−1.49 to −1.0	Moderately drought
−1.99 to −1.5	Severe drought
<−2	Extreme drought

3.4. Remote Sensing-Derived Indices

3.4.1. Vegetation Condition Index (VCI)

The Vegetation Condition Index (VCI) is considered a step forward in analyzing vegetation conditions, particularly in non-homogeneous environments [53]. VCI can extract the impact of weather on plants while removing the ecosystem signal from NDVI [54] and is defined as follows:

$$VCI_{ij}^{k} = 100 * \frac{NDVI_{ij}^{k} - \min\left(NDVI_{i}^{k}\right)}{\max\left(NDVI_{i}^{k}\right) - \min\left(NDVI_{i}^{k}\right)} \tag{5}$$

where VCI_{ij}^{k} and $NDVI_{ij}^{k}$ indicate VCI and NDVI values at pixel *k*, in *i*-th month, for the year *j*. The $\max\left(NDVI_{i}^{k}\right)$ and $\min\left(NDVI_{i}^{k}\right)$ shows the maximum and minimum values of NDVI in the period. The VCI values range from 0 to 100, where VCI values below 40 indicate drought conditions in the area (presented in Table 3) [45,55–57]. NDVI is calculated using the red and near-infrared (NIR) bands and is given by [19].

$$\mathrm{NDVI} = \frac{\mathrm{NIR} - \mathrm{RED}}{\mathrm{NIR} + \mathrm{RED}} \tag{6}$$

Table 3. Different drought categories for RSDIs (modified from [58]).

VHI/VCI/TCI Values	Drought Class
0 to 10	Extreme Drought
10 to 20	Severe Drought
20 to 30	Moderate Drought
30 to 40	Mild Drought
More than 40	No Drought

3.4.2. Temperature Condition Index (TCI)

The Temperature Condition Index (TCI) considers that a drought occurrence will reduce soil moisture and increase land surface thermal stress, i.e., there will be a higher land surface temperature (LST) in drought periods compared to normal ones. A high LST during the growing season of crops implies unfavorable or drought conditions, whereas a low land surface temperature suggests predominantly favorable conditions [31]. TCI is connected to the responsiveness of vegetation to any unfavorable changes in temperature. The following expression demonstrates the calculation of TCI [20]:

$$TCI_{ij}^{k} = 100 * \frac{\max\left(LST_{i}^{k}\right) - LST_{ij}^{k}}{\max\left(LST_{i}^{k}\right) - \min\left(LST_{i}^{k}\right)} \tag{7}$$

where TCI_{ij}^{k} and $LST\ _{ij}^{k}$ indicate TCI and LST values at *k* pixel, in *i*-th month, for the year *j*. The $\min\left(LST_{i}^{k}\right)$ and $\max\left(LST_{i}^{k}\right)$ shows the minimum and maximum values of LST in the period. TCI_{ij}^{k} values vary from 0 to 100, indicating stress to good thermal conditions of the vegetation, as shown in Table 3. To calculate LST the following steps are used [59].

$$L_{\lambda} = M_{L} \cdot Q_{CAL} + A_{L} \tag{8}$$

where L_{λ} is the Top of Atmosphere (TOA) spectral radiance $\left(Wm^{-2}sr^{-1}\ mm^{-1}\right)$, M_L is the band-specific multiplicative rescaling factor from the metadata, A_L is the band-specific additive rescaling factor from the metadata, and Q_{CAL} is the quantized and calibrated standard product pixel values (DN). All of these variables can be retrieved from the metadata file of Landsat 8 data.

TOA to brightness temperature can be calculated using the following Equation (9) [60,61]

$$BT = \left(\frac{K_2}{\left(\ln\left(\frac{K_1}{L}\right) + 1\right)}\right) - 273.15 \tag{9}$$

where K1 and K2 are band specific thermal conversion constants and can be found from metadata of the image.

The proportion of vegetation (Pv) is calculated by using the following Equation (10) [62]

$$Pv = \text{square}\left(\frac{NDVI - NDVI_{min}}{NDVI_{max} - NDVI_{min}}\right) \tag{10}$$

The emissivity is calculated using the following Equation (11) [63,64]

$$\varepsilon = 0.004 * P_v + 0.986 \tag{11}$$

Finally, the LST was derived using the following Equation (12) [65]

$$LST = \frac{BT}{(1 + [(\lambda BT/\rho)\ln\varepsilon\ ])} \tag{12}$$

where λ is the effective wavelength (10.9 mm for band 10 in Landsat 8 data) and ε is the emissivity.

$$\rho = h\frac{c}{\sigma} = 1.438 \times 10^{-2}\text{mK} \tag{13}$$

where σ is the Boltzmann constant (1.38×10^{-23} J/K), h is Plank's constant (6.626×10^{-34} Js), and c is the velocity of light in a vaccum (2.998×10^{8} m/s).

3.4.3. Vegetation Health Index (VHI)

The following equation is used to compute the VHI index, which is the weighted sum of VCI and TCI and is a useful source of information regarding the stress on vegetation by droughts. Gidey et al. [55] reported that the coefficient of the VHI equation was kept at 0.5 due to a lack of more precise information on the influence of VCI and TCI on the VHI.

$$\text{VHI} = a\text{VCI} + (1 - a)\text{TCI} \tag{14}$$

where $a = 0.5$ (the same contribution of VCI and TCI). Droughts based on VHI are classified into five categories, following to the recommendations of Kogan [53]. Table 3 shows different drought conditions based on VCI, TCI, and VHI values.

3.5. Pearson Corerlation Coefficient

The linear relationships between RSDIs and the meteorological drought index (SPEI) were evaluated on a 1-, 3-, 6-, and 12-month time scales in Wadi Al-Lith from 2001 to 2020 by utilizing the Pearson Correlation Coefficient (CC). The Standardized Anomaly Index (SAI) was used to discover anomalies in RSDIs by calculating a standardized deviation from the long-term mean. Then, the anomalies of RSDIs are compared with SPEI at different time scales to analyze the correlation between them. The following equation is used to calculate the SAI [66,67].

$$\text{SAI}_i = \frac{x_i - \overline{x}}{\sigma} \tag{15}$$

where $x_i, \overline{x}$, and σ represents the values of RSDIs at any month, the long-term mean, and the standard deviation, respectively.

4. Results

4.1. Evaluation of Drought Indices

The meteorological drought index (SPEI) at different time scale, i.e., 1-, 3-, 6-, and 12-months and RSDIs (i.e., VCI, TCI, and VHI) are used to analyze the drought severity/condition on both spatial and temporal scale from 2001 to 2020 in the Al-Lith watershed.

4.1.1. SPEI

Figure 3 shows the SPEI time series at different time scales (i.e., 1-, 3-, 6-, and 12-months) at stations J107, J108, TA-109, and TA-233 from 2001 to 2020 in the Al-Lith watershed. Figure 3a shows that more severe and extreme drought events are observed by SPEI-3, followed by SPEI-12 and SPEI-6, whereas SPEI-1 shows moderate to severe drought conditions. However, high fluctuations are observed in SPEI-1 and SPEI-3 time series compared to SPEI-6 and SPEI-12. Moreover, the time series analyses depict severe to extreme drought events in 2002, 2003, 2007, 2011, 2015, 2017, and 2019 (Figure 3a). Overall, the time series plot shown in Figure 3b demonstrates more severe to extreme drought events in 2003, 2007, 2009, 2012, 2016, 2019, and 2020. Further, drought at stations TA109 (Figure 3c) and TA233 (Figure 3d) are more severe at 3- and 12-month time scales followed by SPEI-6 and SPEI-1. Extreme and severe drought events at station TA109 are observed in 2002, 2007, and 2012. Likewise, severe to extreme drought events at station TA233 are observed in 2005, 2006, 2009, 2012, and 2019. Overall, the results (Figure 3a–d) demonstrate significant numbers of extreme and severe drought events in 2002, 2007, 2009, 2012, 2015, and 2019 in the Al-Lith watershed.

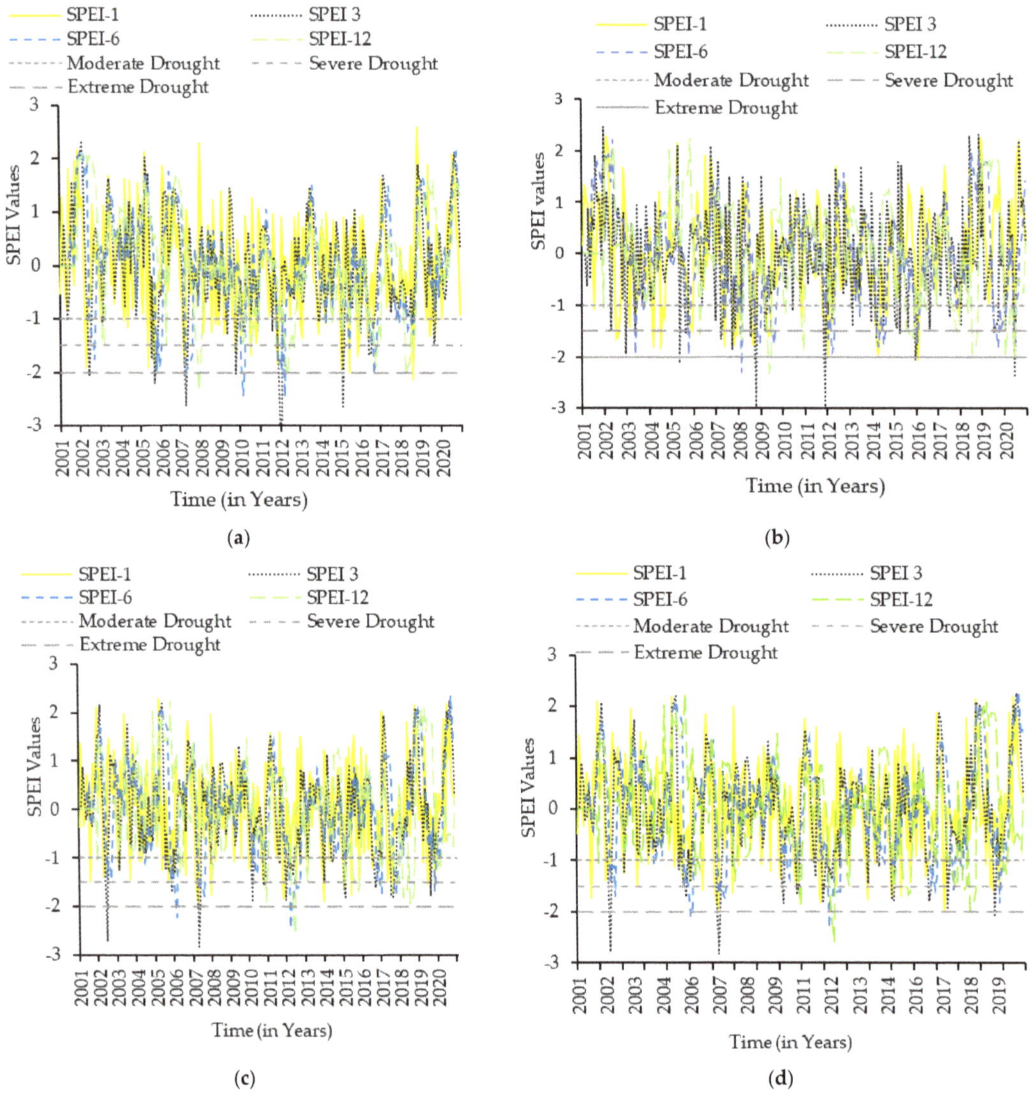

Figure 3. Time series analysis at four stations of the AlLith watershed on different time scales of SPEI (1-, 3-, 6-, and 12-month) at (**a**) J107, (**b**) J108, (**c**) TA109, and (**d**) TA233.

4.1.2. VCI

Figures 4 and 5 show the spatial and temporal distribution of VCI from 2001 to 2020 in the Al-Lith Watershed, which is calculated from the NDVI, shown in Figure A1 in Appendix A. The spatial distribution maps of VCI (Figure 4) indicate that 2001, 2002, 2004, 2007, 2008, 2010, and 2011 are the extreme drought years in the study period. In contrast, more wet conditions (relatively less drought) are observed in 2005, 2006, 2009, and 2016. Furthermore, it should be noted that drought is relatively more severe in northern areas of the Al-Lith watershed compared to southern areas.

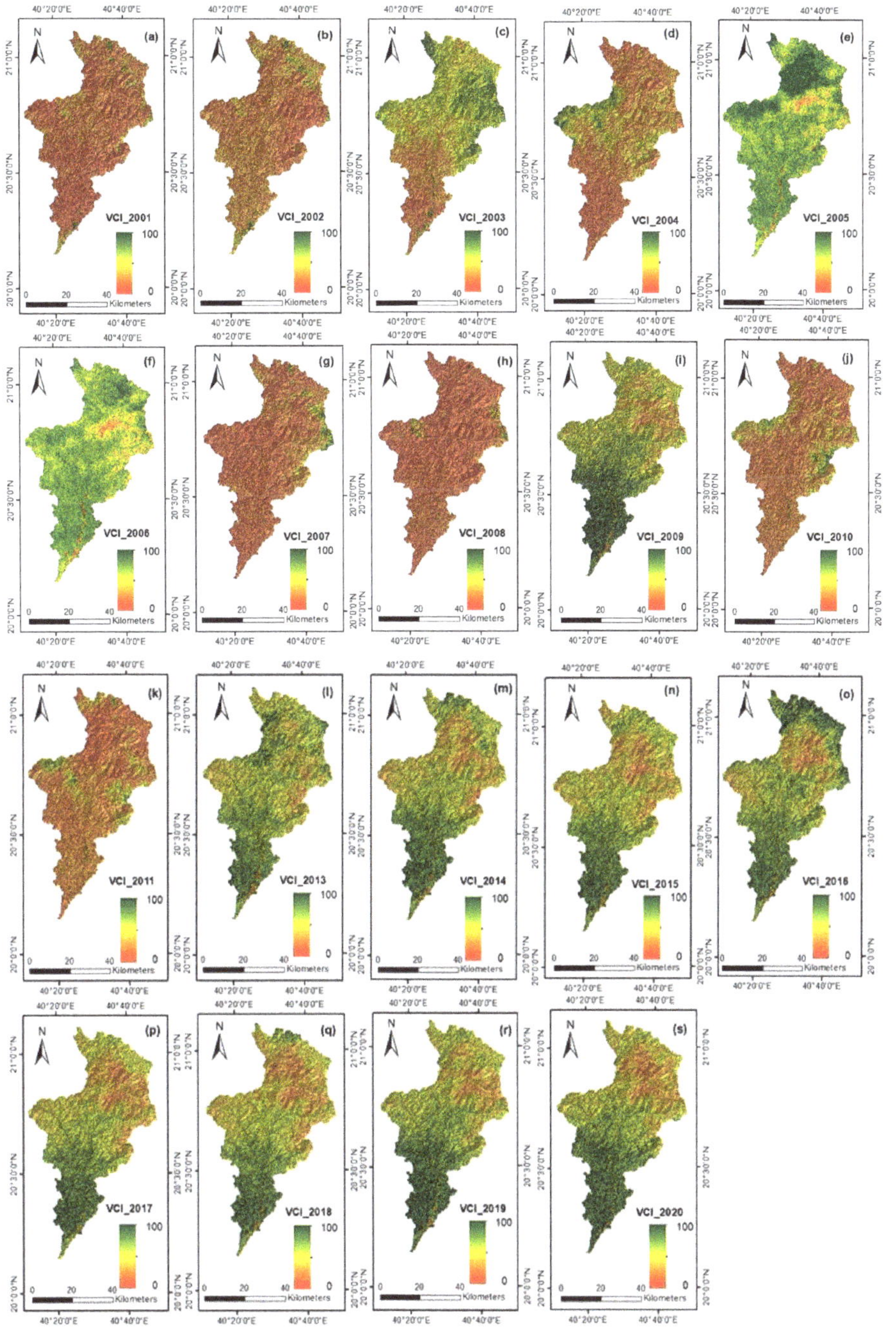

Figure 4. (**a–s**) Spatial distribution of VCI in Al-Lith Watershed retrieved from Landsat Satellites for the period of 2001–2020.

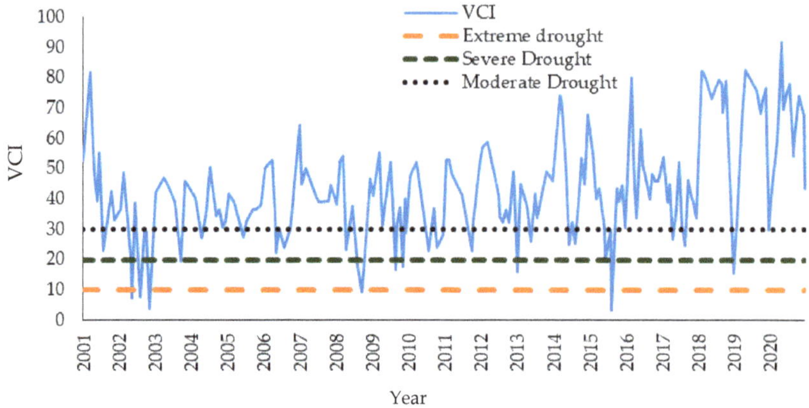

Figure 5. Time series plot of VCI in Al-Lith Watershed extracted using Google Earth Engine and Landsat Satellite datasets.

The time series of VCI extracted using the GEE (shown in Figure 5) illustrate that the mean VCI values range from 20.3 to 64.7 during the study period. According to the VCI time series plot, the Al-Lith watershed experienced minimum VCI values (severe and extreme drought events) in 2001, 2002, 2003, 2007, 2008, and 2010, in the first decade. Similarly, the minimum VCI values in the second decade are observed in 2013, 2015, 2016, and 2019. In other words, the VCI time series demonstrates that 2001, 2002, 2008, 2015, and 2018 are the extreme drought years.

4.1.3. TCI

Figures 6 and 7 show the spatial distribution and time series plot of TCI in the Al-Lith watershed spanning a period of 2001–2020 retrieved from Landsat satellites and GEE, respectively. As shown in Figure 6, the minimum TCI values are observed in 2001, 2003, 2005, 2006, 2009, 2011, 2013, 2015, 2017, 2018, and 2020. The temperature in the study area is increasing over time (Figure A2 in the Appendix A), therefore, the results show more severe TCI in the last decade compared with first decade. Since the Al-Lith watershed is located in a hyper-arid region, where the maximum temperature reaches above 40 °C and annual precipitation is usually less than 10 mm, the TCI will have a significant impact and contribution to VHI index. TCI has a similar spatial distribution trend to that of VCI, i.e., TCI is more severe in northern areas of Al-Lith compared with southern areas, except for a few years (2002, 2004, 2010, and 2014, as shown in Figure 6).

The spatial distributions of VCI and TCI are different in some periods, such as the year 2001 (Figures 4a and 6a), which is mainly caused by different views of these indices in describing the drought. VCI is calculated from NDIV and considers only the vegetation factor, while TCI is calculated from LST and is an integrated results of many factors, including vegetation, precipitation, topography, elevation, soil, and meteorology.

The time series plot of TCI for the Al-Lith watershed is shown in Figure 7. TCI values show that the Al-Lith watershed experienced frequent severe and extreme drought events in 2001, 2002, 2004, 2005, 2007, 2008, 2013, 2014, 2015, 2017, 2019, and 2020. Overall, the mean TCI values across different years ranges from 13.79 to 89.6 in the study area.

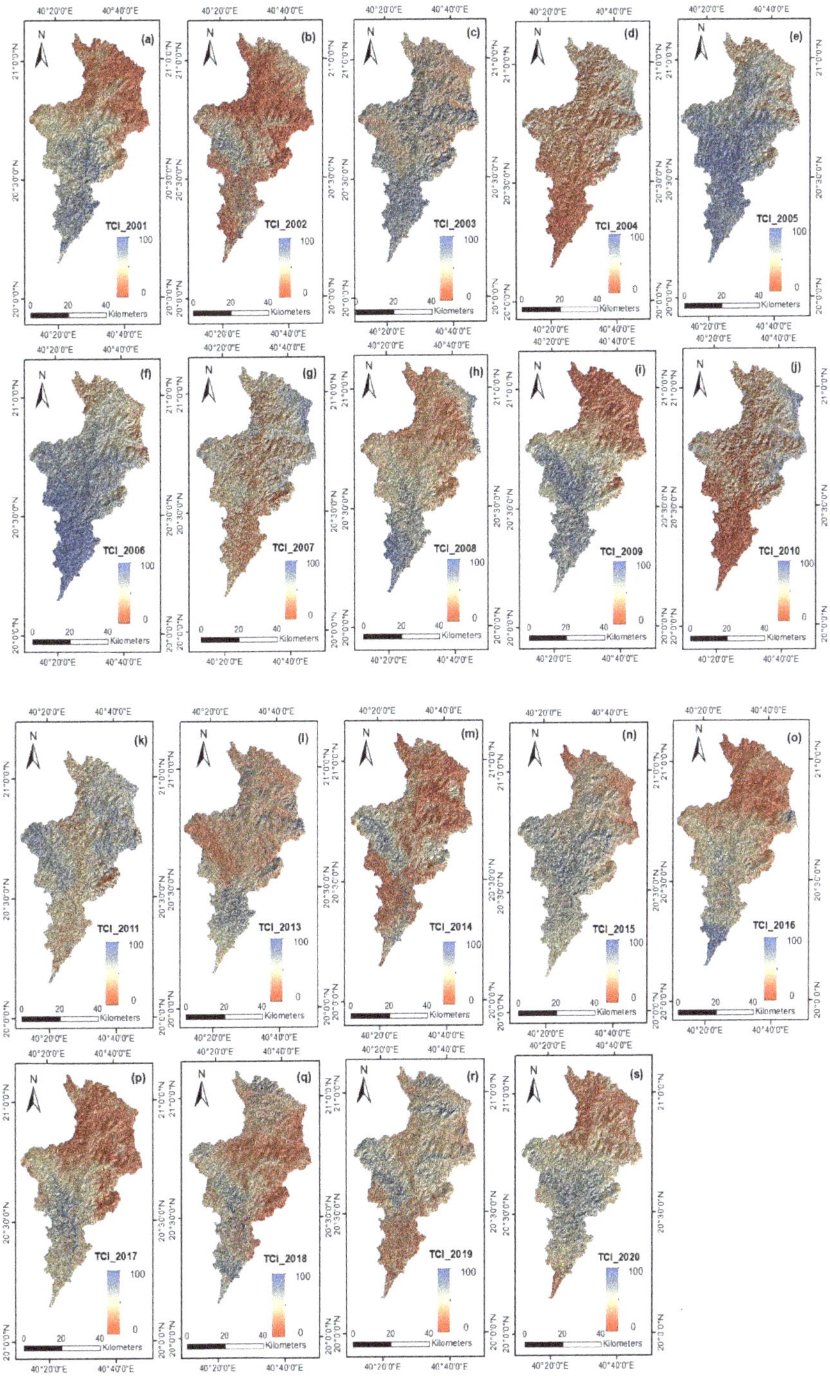

Figure 6. (**a–s**) Spatial distribution of TCI in Al-Lith Watershed retrieved from Landsat Satellites for the period of 2001–2020.

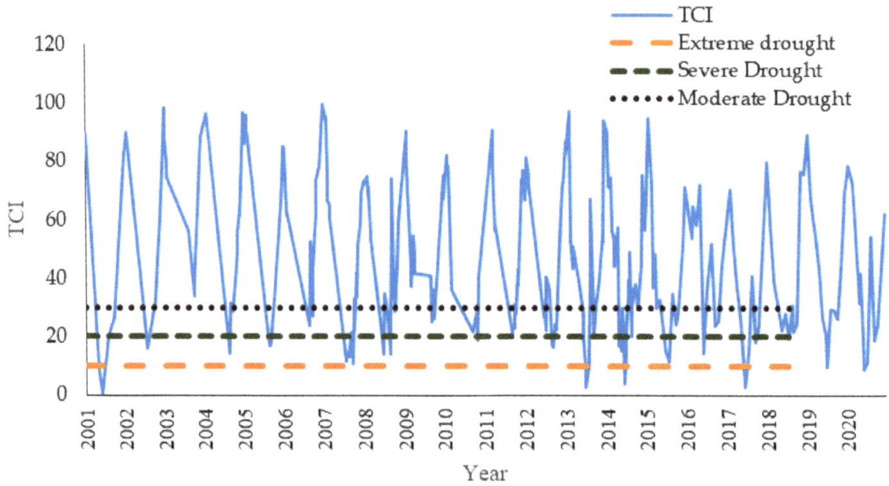

Figure 7. Time series plot of TCI in Al-Lith Watershed extracted using Google Earth Engine and Landsat Satellite datasets.

4.1.4. VHI

The spatial distribution and time series plot of VHI across the Al-Lith watershed extracted from Landsat satellite imageries and GEE, respectively, is shown in Figures 8 and 9. Figure 8 depicts that 2001, 2002, 2004, 2007, 2008, 2010, 2015, 2018, and 2019 are the severe drought years across the Al-Lith watershed. However, it should be noted that drought estimates using VHI depict that drought is more severe in the first decade than the second one. Furthermore, the figure also shows that drought is more severe in the downstream region rather than upstream region of the Al-Lith watershed. Since VHI is a more comprehensive drought index than VCI and TCI, the VHI shows that downstream areas of the Al-Lith watershed is more vulnerable to drought, and thus, it is advised to devise robust mitigation plans to encounter the adverse impacts of drought on available water reserves and agriculture in the downstream region of the Al-Lith watershed.

The time series plot of the VHI, extracted from GEE (shown in Figure 9), illustrates that the mean VHI values range from 21.03 to 61.05 during the entire study period. According to the VHI time series plot, the Al-Lith watershed experienced the minimum VHI values (severe and extreme drought events) in 2002, 2004, 2007, 2008, 2010, 2012, 2013, 2014, 2015, and 2017–2020. It is worth mentioning that intensity and frequency of severe drought events are significantly increased in the second decade, particularly after 2017.

The maximum, minimum, and average values of VCI, TCI, and VHI across the Al-Lith watershed on an annual scale from 2001 to 2020 are shown in Table 4. Table 4 shows extreme droughts in 2002, 2008, 2015 while moderate droughts in 2001, 2003, 2006, 2009, 2010, 2013, and 2018 for VCI. Similarly, the extreme (severe) drought events for TCI are observed in 2001, 2013, 2014, 2017, 2019, and 2020 (2002, 2004, 2005, 2007, 2008, 2010, 2012, 2015, 2016, and 2018). On the other hand, no extreme drought events are observed for VHI whereas severe drought is observed in 2004, 2007, 2012, 2013, and 2017–2020).

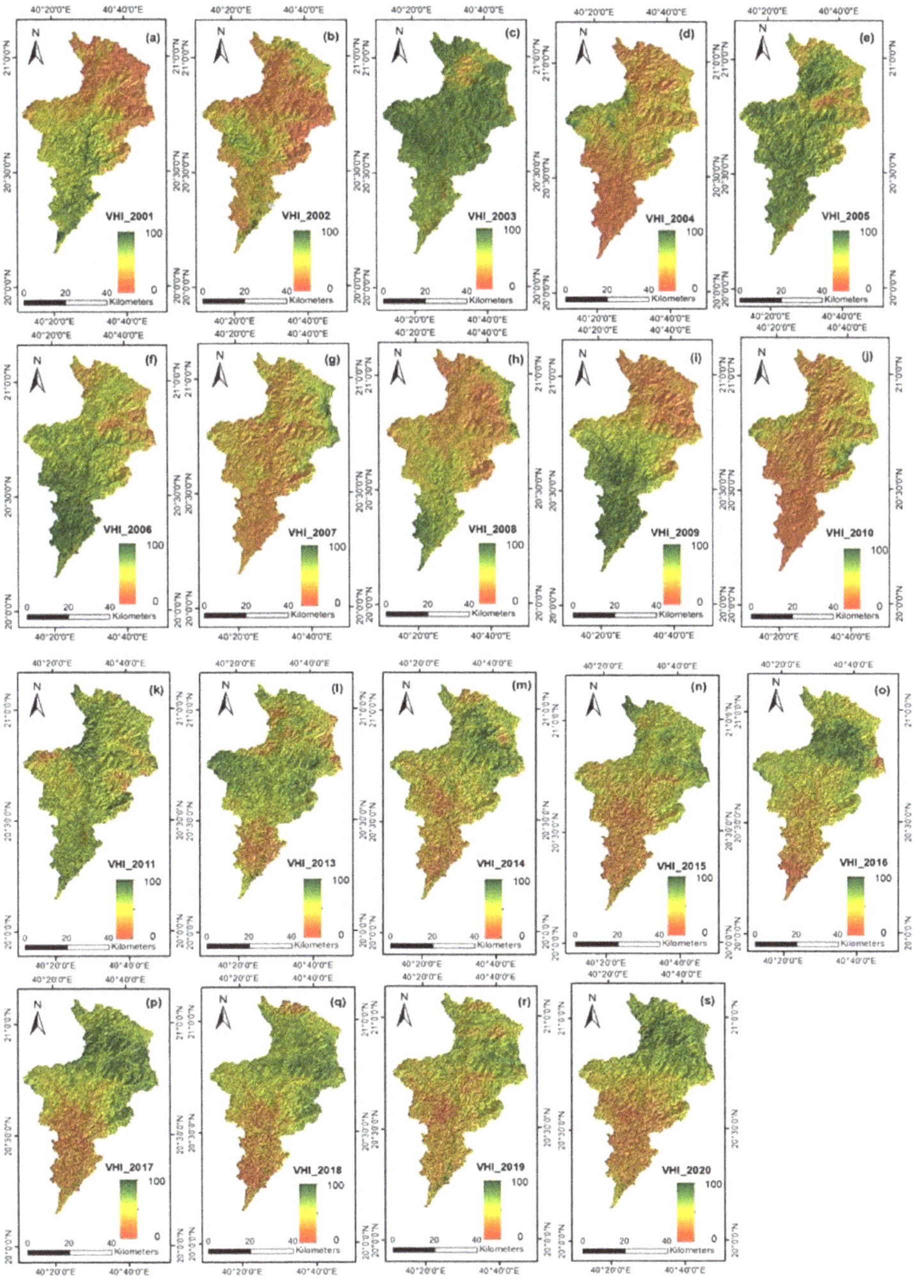

Figure 8. (**a–s**) Spatial distribution of VHI in Al-Lith Watershed retrieved from Landsat Satellites for the period of 2001–2020.

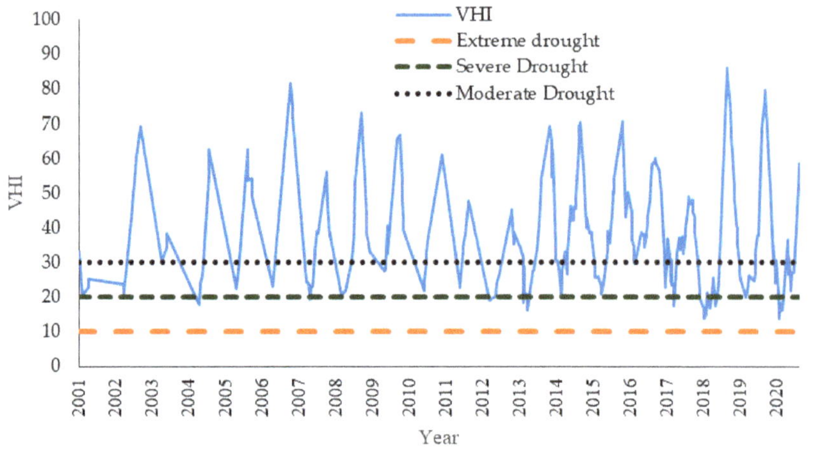

Figure 9. Time series plot of VHI in Al-Lith watershed extracted using Google Earth Engine and Landsat Satellite datasets.

Table 4. Minimum, Maximum, and Mean values of VCI, TCI, and VHI (2001–2020).

Year	VCI			TCI			VHI		
	Min	Max	Average	Min	Max	Average	Min	Max	Average
2001	19.68	93.97	56.83	0.52	89.56	45.04	20.82	33.25	27.03
2002	3.97	61.56	32.76	11.72	89.96	50.84	21.40	60.44	40.92
2003	19.66	53.78	36.72	33.92	98.52	66.22	30.33	69.27	49.80
2004	25.67	50.46	38.06	14.52	96.92	55.72	18.05	62.71	40.38
2005	25.43	46.15	35.79	16.69	95.98	56.33	22.65	62.72	42.68
2006	17.82	58.64	38.23	23.71	99.66	61.68	23.18	54.33	38.76
2007	26.00	64.52	45.26	10.79	95.35	53.07	19.54	81.81	50.68
2008	9.43	60.51	34.97	13.86	74.83	44.34	20.73	56.22	38.48
2009	16.68	55.61	36.15	25.00	90.24	57.62	27.42	73.19	50.30
2010	19.22	57.58	38.40	19.05	82.41	50.73	21.84	66.87	44.35
2011	23.02	52.94	37.98	20.87	90.94	55.91	22.77	61.23	42.00
2012	26.26	60.52	43.39	16.34	87.25	51.80	19.11	25.87	22.49
2013	16.03	49.03	32.53	2.58	97.19	49.88	16.45	53.90	35.17
2014	25.19	74.14	49.67	4.00	89.60	46.80	20.09	70.41	45.25
2015	3.48	54.69	29.09	11.43	94.88	53.15	21.03	58.64	39.83
2016	33.81	80.14	56.98	13.70	88.10	50.90	29.60	70.72	50.16
2017	22.56	54.46	38.51	2.49	71.84	37.17	17.72	60.16	38.94
2018	10.74	85.96	48.35	16.37	84.06	50.21	14.01	86.04	50.03
2019	28.06	84.65	56.35	9.40	96.63	53.02	19.99	75.30	47.65
2020	33.25	94.75	64.00	8.87	78.69	43.78	13.98	79.95	46.96
2001–2020	20.30	64.70	42.50	13.79	89.63	51.71	21.04	63.15	42.09

4.2. Correlation between SPEI, VCI, TCI, and VHI

The Pearson Correlation Coefficient (CC) between the meteorological drought index (SPEI) and RSDIs (VCI, TCI, and VHI) is calculated in the Al-Lith watershed from 2001 to 2020 and presented in Table 5. TCI and VHI have shown good agreement with SPEI in the study period, where a CC value of 0.64 is observed between the two indices, as illustrated in Table 5. VHI and VCI show reasonable agreement with each other where a CC value of 0.51 is observed between the two indices. However, the correlation drops significantly to a CC value of 0.39 between VCI and TCI.

Table 5. Correlation matrix between Remote sensing and meteorological drought indices based on GEE) and stations data over the Al-Lith watershed from 2001–2020.

	VHI	TCI	VCI	SPEI-1	SPEI-3	SPEI-6	SPEI-12
VHI	1.00	0.64	0.51	0.47	0.57	0.67	0.72
TCI	0.64	1.00	0.39	0.37	0.42	0.61	0.60
VCI	0.51	0.39	1.00	0.24	0.33	0.45	0.53

Further, the analyses are further extended and the correlation between SPEI and RSDIs is investigated at different time scales (1-, 3-, 6-, and 12-month). The results (Table 5) show that SPEI-12 strongly correlates with VHI, where a CC value of 0.72 is observed between VHI and SPEI-12. VHI also has a strong correlation with SPEI-6 (CC equals 0.67), which drops significantly to 0.47 with SPEI-1. In contrast to VHI, a lower correlation is observed between VCI and SPEI. The maximum and minimum CC is observed between VCI/SPEI-12 and VCI/SPEI-1, respectively. TCI shows a moderate correlation with SPEI, where the maximum correlation (with CC of 0.61) is observed between TCI/SPEI-6, followed by TCI/SPEI-6.

From Table 5, the correlation between VHI/VCI/TCI and SPEI increases with time scales of SPEI. In the calculations of VHI, TCI, and VCI, monthly data are used. However, the vegetation growth is related with wet/drought conditions of both the present month and previous months of vegetation growth, or even months before vegetation growth. Consequently, vegetation growth regime and vegetation index are more closely related with wet/drought conditions of longer time scales, and VHI, TCI, and VCI, closely related with vegetation index, all tend to have a stronger correlation with SPEI over a longer time frame.

5. Discussion

Drought catastrophes not only significantly affect the current agricultural productivity, human life, and economic development, but can potentially worsen climate conditions and exacerbate land desertification for a long period. Therefore, it is of utmost importance to investigate the causes of drought disasters, conduct in-depth analyses of their spatiotemporal characteristics, and employ scientific strategies for their prevention and management. The research on the disaster caused by drought is still in its infancy since researchers typically only look at one drought index at a time rather than the cumulative effects of numerous drought indices. Unlike previous studies, this study examined the spatial and temporal changes of various drought indices while taking precipitation, vegetation index, and surface temperature into account in the Arid region of KSA from 2001–2020.

The current study is an attempt to evaluate and correlate in situ meteorological drought (SPEI) and remote sensing-retrieved drought indices (VCI, TCI, and VHI) in the western region of KSA (the Al-Lith watershed) from 2001 to 2020. RSDIs are retrieved from Landsat 7 and Landsat 8 satellite datasets, and the GEE platform. The evaluation of in situ SPEI against the RSDIs is very time consuming in KSA because of the very limited availability of in situ climate data and the sparse distribution of weather stations. The climate data from weather stations is usually incomplete, with several missing observations and significant uncertainties [10,68,69]. The variations in temperature, precipitation, evaporation, etc., are hardly captured by the in situ data owing to the limited stations, its sparse distribution, and incomplete data [70]. Temperature in KSA are rising, droughts are getting worse, and nearly 70% of the country is affected by severe and frequent drought events [69,71]. For instance, the Al-Lith watershed has experienced severe drought events over the past 20 years, consistent with numerous regional and global assessments [25,44,69,72–75]. Therefore, it is of utmost importance to precisely estimate the increasing severity of drought using various indices, especially the RSDIs, to address the data scarcity issues and its contribution to the uncertainties in drought estimation.

GEE and the long-term continuous observations from the Landsat satellite datasets enabled us to compare the RSDIs (VCI, TCI, and VHI) with meteorological drought (SPEI). Many studies used GEE and Landsat satellite data to monitor droughts in different regions worldwide [63,72–74] by using VCI, TCI, and VHI. VHI is used to monitor agricultural drought and is suggested by many authors for drought monitoring [20,31,44,58,75]. According to Choi et al. [76], VHI reflects the anomalies in both vegetation cover and temperature. Table 5 illustrates that VHI showed good agreement with VCI, but the correlation is slightly weaker with TCI. VCI and TCI also play a critical role in drought monitoring because both indices depend on ecological conditions and weather-induced changes, respectively. VCI depends on the maximum and minimum NDVI, which reflects the regional vegetation health. On the other hand, TCI is based on the minimum and maximum of LST, i.e., more dependent on temperature. The results of VCI, TCI, and VHI indicate that the watershed experienced severe drought in 2001, 2003, 2007, 2008, 2013, 2014, 2017, and 2019, among other studied years.

Table 5 shows that SPEI and VHI show a strong correlation with each other in the study area, while the correlation is moderate to weak between SPEI/TCI and SPEI/VCI, respectively. Gidey et al. [44] found a very strong correlation between the meteorological drought index (SPI) and RSDIs (VCI/TCI/VHI) while studying the statistical relationship between meteorological and RSDIs in Northern Ethiopia. Del-Toro-Guerrero et al. [66] reported a very strong correlation between SPI and VHI annually while studying the surface reflectance derived indices for drought monitoring. Almeida-Ñauñay et al. [77] studied the impact of the meteorological and RSDIs in semi-arid Mediterranean Grass Land using station and satellite remote sensing data. They reported a good correlation between VHI and SPEI, which support the findings of this study.

Overall, this study utilized Landsat satellite data and generated time series analysis on GEE to investigate the spatial and temporal distribution of drought in the hyper-arid and data scarce regions of KSA. This study combined the meteorological and remote sensing indices for monitoring drought in an arid region. The results showed that VHI is a more robust drought index, showing a good correlation with station-based meteorological drought index (SPEI), and thus is more robust to represent drought over KSA. In other words, RSDIs have the capability to represent the drought condition comparatively well compared with station-based estimated drought. This research gives a clear picture of the drought assessment, which can be helpful for the policymakers, researchers, and government to take proper action for drought mitigation.

6. Conclusions

This study assessed the spatial and temporal distribution of droughts, including meteorological drought estimated through SPEI (SPEI-1, SPEI-3, SPEI-6, and SPEI-12) and remote sensing-retrieved drought indices (RSDIs, including VCI, TCI, and VHI). Moreover, the SPEI and RSDIs are correlated with each other at different time scales (i.e., 1-, 3-, 6-, and 12-month) using the Pearson correlation coefficient. The Standardized Anomaly Index (SAI) was used to calculate the anomalies for RSDIs and compared with SPEI at 1-, 3-, 6-, and 12-month time scales. The analyses are carried out over the hyper-arid region of KSA having limited in situ weather stations from 2001 to 2020. Our major findings are stated below.

(1) SPEI results showed that significant number of severe and extreme drought events are observed in 2002, 2007, 2009, 2012, 2015, and 2019. High fluctuations in drought severity are observed at smaller time scales (i.e., SPEI-1 and SPEI-3) compared with SPEI-6 and SPEI-12. However, interesting observations are observed at station J108 where drought is more severe at smaller time scales (SPEI-1) than larger times scales. Overall, the results showed significant regional variations in drought severity owing to regional changes in climate.

(2) VCI showed that northern areas of the Al-Lith watershed are prone to drought, particularly in 2001, 2002, 2004, 2007, 2008, 2010, and 2011. The spatial distribution of

VCI depicted extreme drought events in 2001, 2002, 2004, 2007, 2008, 2010, and 2011. VCI ranged from 20.3 to 64.7, where extreme (severe) drought events were observed in 2002, 2008, and 2018 (2001, 2003, 2006, 2009, 2010, 2013, and 2018).

(3) Based on the results of TCI, the Al-Lith watershed was prone to droughts in 2001, 2003, 2005, 2006, 2009, 2011, 2013, 2015, 2017, 2018, and 2020. Droughts were more intense in the second decade compared to the first decade. Time series plot of TCI showed that TCI values range from 10.60 to 91.34, where minimum TCI values are frequently observed in the last decade. Overall, the Al-Lith watershed was vulnerable to droughts in 2001, 2002, 2004, 2005, 2007, 2008, 2013, 2014, 2015, 2017, 2019, and 2020.

(4) VHI results depicted severe drought events in 2001, 2002, 2004, 2007, 2008, 2010, 2015, 2018, and 2019. The values of VHI ranged from a minimum of 19.47 to a maximum of 64.22. The VHI time series depicted severe and extreme drought events in 2002, 2004, 2007, 2008, 2010, 2012, 2013, 2014, 2015, and 2017–2020.

(5) The correlation analyses showed that VHI has a good correlation with VCI, with an average CC value of 0.51. On the other hand, a minimum correlation with CC value of 0.39 is observed between VCI and TCI. Highest correlation among RSDIs is observed between VHI and TCI, with a CC value of 0.64.

(6) The correlation between SPEI (at 1-, 3-, 6-, and 12-month) and RSDIs showed a good agreement between VHI/SPEI-12 and VHI/SPEI-6 with average CC values of 0.74 and 0.67, respectively. The correlation got weaker at smaller time scales, i.e., the average CC value between VHI/SPEI-3 and VHI/SPEI-1 are 0.42 and 0.37. A lower correlation of SPEI is observed with VCI, where the maximum CC value of 0.52 is estimated between VCI/SPEI-12 followed by 0.36 between VCI/SPEI-6. Further, a moderate correlation is observed between TCI and SPEI.

To conclude, droughts were monitored both from in situ data acquired from MEWA and remote sensing techniques. The comparison and correlation between in situ drought index (SPEI) and RSDIs indicated that RSDIs are accurate enough to represent drought conditions in hyper-arid regions like KSA. Contrasting results are observed for VCI and TCI, i.e., more severe droughts in the south than north as well as more severe droughts were observed in the first decade than the second because of interpolation and time series data, which is averaged for the whole Al-Lith watershed which does not represent the real scenario. Moreover, the remote sensing technology is found to be very useful in regions with limited weather stations and data availability. Therefore, the findings of this study are critical to our understanding of the nature of droughts in arid and hyper-arid regions. Prospective researchers will find useful information in the study's findings for resolving local and regional drought issues and in devising plans for drought mitigation.

Author Contributions: Conceptualization, J.B. and K.U.R.; methodology, J.B., K.U.R. and S.S.; software, N.E. and K.M.A.; validation, N.E. and K.M.A.; formal analysis, N.E. and K.U.R.; investigation, N.E. and K.M.A.; resources, N.E. and J.B.; data curation, N.E. and K.U.R.; writing—original draft preparation, N.E. and K.M.A.; writing—review and editing, J.B., K.U.R. and S.S.; visualization, N.E. and K.U.R.; supervision, J.B. and K.U.R.; project administration, J.B. and S.S.; funding acquisition, S.S. and K.U.R. All authors have read and agreed to the published version of the manuscript.

Funding: This research was funded by the Natural National Science Foundation of China (Grant Numbers 51839006 and 52250410336), China Postdoctoral Science Foundation (Grant number 2022M721872), and the Shuimu Scholar Program of Tsinghua University (Grant Number 2020SM072).

Data Availability Statement: Data will be available on request to the first two authors.

Acknowledgments: The authors acknowledge Ministry of Water Environment and Agriculture (MEWA) for providing daily meteorological data. Authors are also thankful to the developers of satellite products, which helped us to perform different analyses in the manuscript.

Conflicts of Interest: The authors declare no conflict of interest.

Appendix A

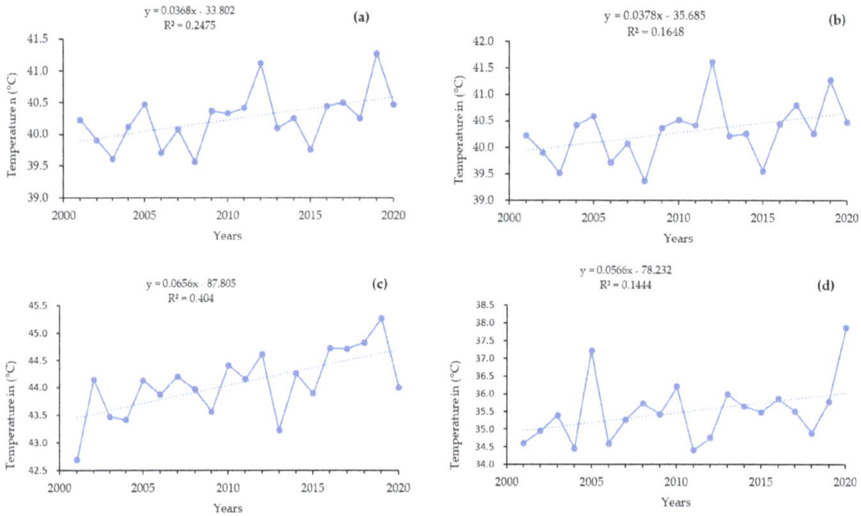

Figure A1. Variation trends in temperature from 2001–2020 at stations (**a**) TA 109, (**b**) TA 233, (**c**) J107, and (**d**) J108.

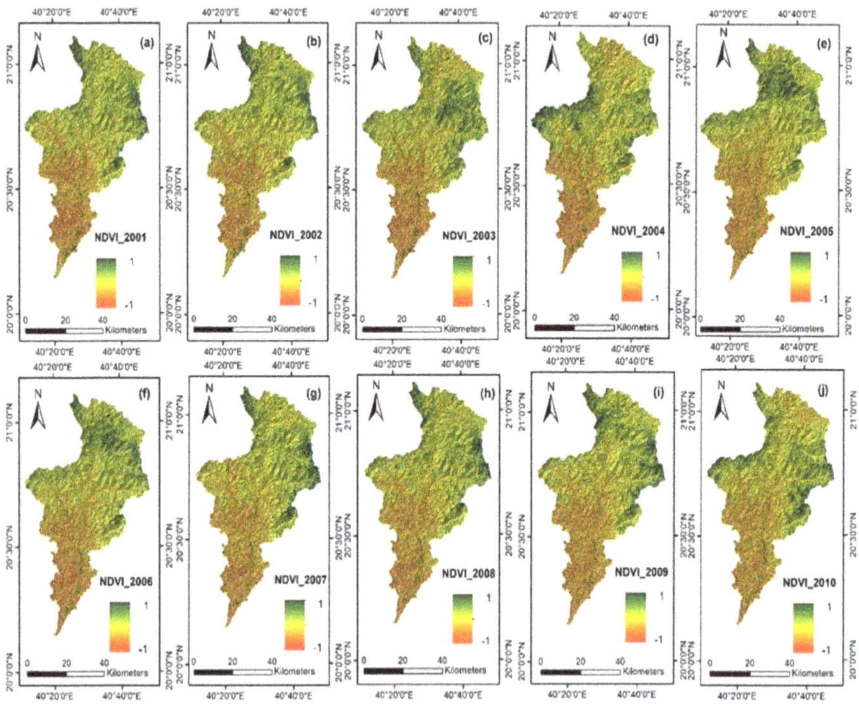

Figure A2. *Cont.*

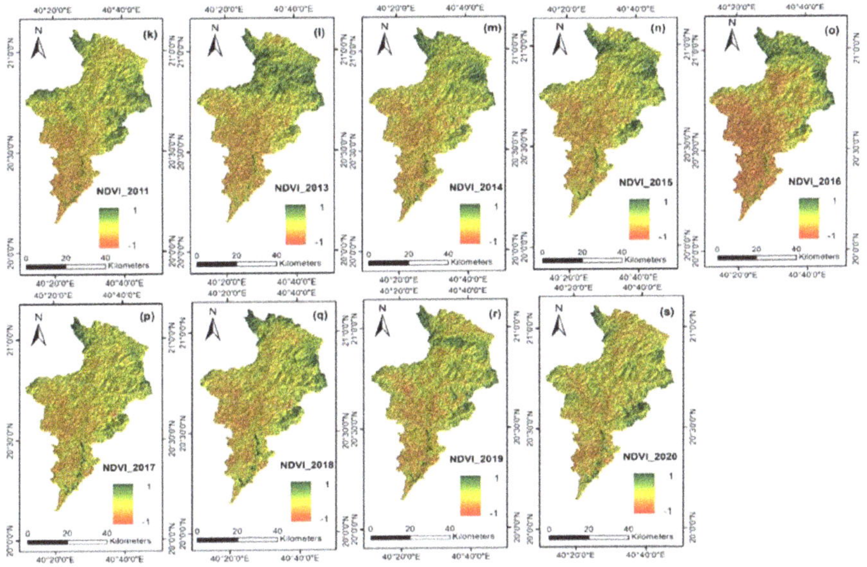

Figure A2. Spatial distribution of NDVI in Al-Lith Watershed retrieved from Landsat Satellites for the period of 2001–2020 (**a**–**s**).

References

1. Heim, R.R., Jr. A review of twentieth-century drought indices used in the United States. *Bull. Am. Meteorol. Soc.* **2002**, *83*, 1149–1166. [CrossRef]
2. Keyantash, J.; Dracup, J.A. The quantification of drought: An evaluation of drought indices. *Bull. Am. Meteorol. Soc.* **2002**, *83*, 1167–1180. [CrossRef]
3. Hao, Z.; Singh, V.P. Drought characterization from a multivariate perspective: A review. *J. Hydrol.* **2015**, *527*, 668–678. [CrossRef]
4. Palmer, W.C. *Meteorological Drought*; US Department of Commerce, Weather Bureau: Washington, DC, USA, 1965; Volume 30.
5. McKee, T.B.; Doesken, N.J.; Kleist, J. The relationship of drought frequency and duration to time scales. In Proceedings of the 8th Conference on Applied Climatology, Anaheim, CA, USA, 17–22 January 1993; pp. 179–183.
6. Vicente-Serrano, S.M.; Beguería, S.; López-Moreno, J.I. A multiscalar drought index sensitive to global warming: The standardized precipitation evapotranspiration index. *J. Clim.* **2010**, *23*, 1696–1718. [CrossRef]
7. Rahman, K.U.; Shang, S.; Shahid, M.; Wen, Y. Performance assessment of SM2RAIN-CCI and SM2RAIN-ASCAT precipitation products over Pakistan. *Remote Sens.* **2019**, *11*, 2040. [CrossRef]
8. Baez-Villanueva, O.M.; Zambrano-Bigiarini, M.; Ribbe, L.; Nauditt, A.; Giraldo-Osorio, J.D.; Thinh, N.X. Temporal and spatial evaluation of satellite rainfall estimates over different regions in Latin-America. *Atmos. Res.* **2018**, *213*, 34–50. [CrossRef]
9. Hereher, M.; El Kenawy, A.M. Exploring the potential of solar, tidal, and wind energy resources in Oman using an integrated climatic-socioeconomic approach. *Renew. Energy* **2020**, *161*, 662–675. [CrossRef]
10. El Kenawy, A.M.; McCabe, M.F. A multi-decadal assessment of the performance of gauge-and model-based rainfall products over Saudi Arabia: Climatology, anomalies and trends. *Int. J. Climatol.* **2016**, *36*, 656–674. [CrossRef]
11. West, H.; Quinn, N.; Horswell, M. Remote sensing for drought monitoring & impact assessment: Progress, past challenges and future opportunities. *Remote Sens. Environ.* **2019**, *232*, 111291.
12. Sarkar, A.; Kumar, V.; Jasrotia, A.S.; Taloor, A.K.; Kumar, R.; Sharma, R.; Khajuria, V.; Raina, G.; Kouser, B.; Roy, S. Spatial analysis and mapping of malaria risk in dehradun city India: A geospatial technology-based decision-making tool for planning and management. In *Geoecology of Landscape Dynamics*; Springer: Berlin/Heidelberg, Germany, 2020; pp. 207–221.
13. Smith, A.M.; Kolden, C.A.; Tinkham, W.T.; Talhelm, A.F.; Marshall, J.D.; Hudak, A.T.; Boschetti, L.; Falkowski, M.J.; Greenberg, J.A.; Anderson, J.W. Remote sensing the vulnerability of vegetation in natural terrestrial ecosystems. *Remote Sens. Environ.* **2014**, *154*, 322–337. [CrossRef]
14. Belal, A.-A.; El-Ramady, H.R.; Mohamed, E.S.; Saleh, A.M. Drought risk assessment using remote sensing and GIS techniques. *Arab. J. Geosci.* **2014**, *7*, 35–53. [CrossRef]
15. Smith, W.K.; Dannenberg, M.P.; Yan, D.; Herrmann, S.; Barnes, M.L.; Barron-Gafford, G.A.; Biederman, J.A.; Ferrenberg, S.; Fox, A.M.; Hudson, A. Remote sensing of dryland ecosystem structure and function: Progress, challenges, and opportunities. *Remote Sens. Environ.* **2019**, *233*, 111401. [CrossRef]

16. Taloor, A.K.; Manhas, D.S.; Kothyari, G.C. Retrieval of land surface temperature, normalized difference moisture index, normalized difference water index of the Ravi basin using Landsat data. *Appl. Comput. Geosci.* **2021**, *9*, 100051. [CrossRef]
17. Colesanti, C.; Wasowski, J. Investigating landslides with space-borne Synthetic Aperture Radar (SAR) interferometry. *Eng. Geol.* **2006**, *88*, 173–199. [CrossRef]
18. Casagli, N.; Frodella, W.; Morelli, S.; Tofani, V.; Ciampalini, A.; Intrieri, E.; Raspini, F.; Rossi, G.; Tanteri, L.; Lu, P. Spaceborne, UAV and ground-based remote sensing techniques for landslide mapping, monitoring and early warning. *Geoenvironmental Disasters* **2017**, *4*, 1–23. [CrossRef]
19. Tucker, C.J. Red and photographic infrared linear combinations for monitoring vegetation. *Remote Sens. Environ.* **1979**, *8*, 127–150. [CrossRef]
20. Kogan, F.N. Application of vegetation index and brightness temperature for drought detection. *Adv. Space Res.* **1995**, *15*, 91–100. [CrossRef]
21. Unganai, L.S.; Kogan, F.N. Drought monitoring and corn yield estimation in Southern Africa from AVHRR data. *Remote Sens. Environ.* **1998**, *63*, 219–232. [CrossRef]
22. Kogan, F.; Gitelson, A.; Zakarin, E.; Spivak, L.; Lebed, L. AVHRR-based spectral vegetation index for quantitative assessment of vegetation state and productivity. *Photogramm. Eng. Remote Sens.* **2003**, *69*, 899–906. [CrossRef]
23. Bento, V.A.; Gouveia, C.M.; DaCamara, C.C.; Libonati, R.; Trigo, I.F. The roles of NDVI and Land Surface Temperature when using the Vegetation Health Index over dry regions. *Glob. Planet. Chang.* **2020**, *190*, 103198. [CrossRef]
24. Karnieli, A.; Bayasgalan, M.; Bayarjargal, Y.; Agam, N.; Khudulmur, S.; Tucker, C. Comments on the use of the vegetation health index over Mongolia. *Int. J. Remote Sens.* **2006**, *27*, 2017–2024. [CrossRef]
25. Amalo, L.F.; Hidayat, R. Comparison between remote-sensing-based drought indices in East Java. In *IOP Conference Series: Earth and Environmental Science*; IOP Publishing: Bristol, UK, 2017.
26. Jiao, W.; Zhang, L.; Chang, Q.; Fu, D.; Cen, Y.; Tong, Q. Evaluating an enhanced vegetation condition index (VCI) based on VIUPD for drought monitoring in the continental United States. *Remote Sens.* **2016**, *8*, 224. [CrossRef]
27. Ghaleb, F.; Mario, M.; Sandra, A.N. Regional landsat-based drought monitoring from 1982 to 2014. *Climate* **2015**, *3*, 563–577. [CrossRef]
28. Dutta, D.; Kundu, A.; Patel, N.; Saha, S.; Siddiqui, A. Assessment of agricultural drought in Rajasthan (India) using remote sensing derived Vegetation Condition Index (VCI) and Standardized Precipitation Index (SPI). *Egypt. J. Remote Sens. Space Sci.* **2015**, *18*, 53–63. [CrossRef]
29. Kundu, A.; Dwivedi, S.; Dutta, D. Monitoring the vegetation health over India during contrasting monsoon years using satellite remote sensing indices. *Arab. J. Geosci.* **2016**, *9*, 1–15. [CrossRef]
30. Bhuiyan, C.; Saha, A.; Bandyopadhyay, N.; Kogan, F. Analyzing the impact of thermal stress on vegetation health and agricultural drought–a case study from Gujarat, India. *GIScience Remote Sens.* **2017**, *54*, 678–699. [CrossRef]
31. Singh, R.P.; Roy, S.; Kogan, F. Vegetation and temperature condition indices from NOAA AVHRR data for drought monitoring over India. *Int. J. Remote Sens.* **2003**, *24*, 4393–4402. [CrossRef]
32. Kirana, A.; Ariyanto, R.; Ririd, A.; Amalia, E. Agricultural drought monitoring based on vegetation health index in East Java Indonesia using MODIS Satellite Data. In *IOP Conference Series: Materials Science and Engineering*; IOP Publishing: Bristol, UK, 2020.
33. Alamdarloo, E.H.; Manesh, M.B.; Khosravi, H. Probability assessment of vegetation vulnerability to drought based on remote sensing data. *Environ. Monit. Assess.* **2018**, *190*, 1–11. [CrossRef]
34. Pham, H.T.T.; Tran, H.T. Application of Remote Sensing Imagery and Algorithms in Google Earth Engine platform for Drought Assessment. *J. Min. Earth Sci. Vol* **2020**, *62*, 53–67.
35. Gorelick, N.; Hancher, M.; Dixon, M.; Ilyushchenko, S.; Thau, D.; Moore, R. Google Earth Engine: Planetary-scale geospatial analysis for everyone. *Remote Sens. Environ.* **2017**, *202*, 18–27. [CrossRef]
36. Tamiminia, H.; Salehi, B.; Mahdianpari, M.; Quackenbush, L.; Adeli, S.; Brisco, B. Google Earth Engine for geo-big data applications: A meta-analysis and systematic review. *ISPRS J. Photogramm. Remote Sens.* **2020**, *164*, 152–170. [CrossRef]
37. Sazib, N.; Mladenova, I.; Bolten, J. Leveraging the Google Earth Engine for drought assessment using global soil moisture data. *Remote Sens.* **2018**, *10*, 1265. [CrossRef] [PubMed]
38. Midekisa, A.; Holl, F.; Savory, D.J.; Andrade-Pacheco, R.; Gething, P.W.; Bennett, A.; Sturrock, H.J. Mapping land cover change over continental Africa using Landsat and Google Earth Engine cloud computing. *PLoS ONE* **2017**, *12*, e0184926. [CrossRef] [PubMed]
39. Sidhu, N.; Pebesma, E.; Câmara, G. Using Google Earth Engine to detect land cover change: Singapore as a use case. *Eur. J. Remote Sens.* **2018**, *51*, 486–500. [CrossRef]
40. Sunar, A.; Yagmur, N.; Dervisoglu, A. Flood analysis with remote sensing data–A case study: Maritsa river, Edirne. *Int. Arch. Photogramm. Remote Sens. Spat. Inf. Sci.* **2019**, *42*, 497–502. [CrossRef]
41. DeVries, B.; Huang, C.; Armston, J.; Huang, W.; Jones, J.W.; Lang, M.W. Rapid and robust monitoring of flood events using Sentinel-1 and Landsat data on the Google Earth Engine. *Remote Sens. Environ.* **2020**, *240*, 111664. [CrossRef]
42. Aksoy, S.; Gorucu, O.; Sertel, E. Drought monitoring using MODIS derived indices and google earth engine platform. In Proceedings of the 2019 8th International Conference on Agro-Geoinformatics (Agro-Geoinformatics), Istanbul, Turkey, 16–19 July 2019; pp. 1–6.

43. Khan, R.; Gilani, H.; Iqbal, N.; Shahid, I. Satellite-based (2000–2015) drought hazard assessment with indices, mapping, and monitoring of Potohar plateau, Punjab, Pakistan. *Environ. Earth Sci.* **2020**, *79*, 1–18. [CrossRef]
44. Gidey, E.; Dikinya, O.; Sebego, R.; Segosebe, E.; Zenebe, A. Using drought indices to model the statistical relationships between meteorological and agricultural drought in Raya and its environs, Northern Ethiopia. *Earth Syst. Environ.* **2018**, *2*, 265–279. [CrossRef]
45. Kocaaslan, S.; Musaoğlu, N.; Karamzadeh, S. Evaluating Drought Events by Time-Frequency Analysis: A Case Study in Aegean Region of Turkey. *IEEE Access* **2021**, *9*, 125032–125041. [CrossRef]
46. Mehravar, S.; Amani, M.; Moghimi, A.; Javan, F.D.; Samadzadegan, F.; Ghorbanian, A.; Stein, A.; Mohammadzadeh, A.; Mirmazloumi, S.M. Temperature-Vegetation-soil Moisture-Precipitation Drought Index (TVMPDI); 21-year drought monitoring in Iran using satellite imagery within Google Earth Engine. *Adv. Space Res.* **2021**, *68*, 4573–4593. [CrossRef]
47. Benzougagh, B.; Meshram, S.G.; El Fellah, B.; Mastere, M.; Dridri, A.; Sadkaoui, D.; Mimich, K.; Khedher, K.M. Combined use of Sentinel-2 and Landsat-8 to monitor water surface area and evaluated drought risk severity using Google Earth Engine. *Earth Sci. Inform.* **2022**, *15*, 929–940. [CrossRef]
48. Ejaz, N.; Bahrawi, J. Assessment of Drought Severity and Their Spatio-Temporal Variations in the Hyper Arid Regions of Kingdom of Saudi Arabia: A Case Study from Al-Lith and Khafji Watersheds. *Atmosphere* **2022**, *13*, 1264. [CrossRef]
49. Hargreaves, G.H.; Samani, Z.A. Reference crop evapotranspiration from temperature. *Appl. Eng. Agric.* **1985**, *1*, 96–99. [CrossRef]
50. Dubrovsky, M.; Svoboda, M.D.; Trnka, M.; Hayes, M.J.; Wilhite, D.A.; Zalud, Z.; Hlavinka, P. Application of relative drought indices in assessing climate-change impacts on drought conditions in Czechia. *Theor. Appl. Climatol.* **2009**, *96*, 155–171. [CrossRef]
51. Abbasi, A.; Khalili, K.; Behmanesh, J.; Shirzad, A. Estimation of ARIMA model parameters for drought prediction using the genetic algorithm. *Arab. J. Geosci.* **2021**, *14*, 841. [CrossRef]
52. Allen, R.G. Assessing integrity of weather data for reference evapotranspiration estimation. *J. Irrig. Drain. Eng.* **1996**, *122*, 97–106. [CrossRef]
53. Kogan, F.N. Operational space technology for global vegetation assessment. *Bull. Am. Meteorol. Soc.* **2001**, *82*, 1949–1964. [CrossRef]
54. Yagci, A.L.; Di, L.; Deng, M. The effect of corn–soybean rotation on the NDVI-based drought indicators: A case study in Iowa, USA, using Vegetation Condition Index. *GIScience Remote Sens.* **2015**, *52*, 290–314. [CrossRef]
55. Gidey, E.; Dikinya, O.; Sebego, R.; Segosebe, E.; Zenebe, A. Analysis of the long-term agricultural drought onset, cessation, duration, frequency, severity and spatial extent using Vegetation Health Index (VHI) in Raya and its environs, Northern Ethiopia. *Environ. Syst. Res.* **2018**, *7*, 1–18. [CrossRef]
56. Bhuiyan, C.; Singh, R.; Kogan, F. Monitoring drought dynamics in the Aravalli region (India) using different indices based on ground and remote sensing data. *Int. J. Appl. Earth Obs. Geoinf.* **2006**, *8*, 289–302. [CrossRef]
57. Du, L.; Tian, Q.; Yu, T.; Meng, Q.; Jancso, T.; Udvardy, P.; Huang, Y. A comprehensive drought monitoring method integrating MODIS and TRMM data. *Int. J. Appl. Earth Obs. Geoinf.* **2013**, *23*, 245–253. [CrossRef]
58. Zanter, K. *LANDSAT 8 (L8) Data Users Handbook (LSDS-1574 Version 5.0)*; United States Geological Survey: Sioux Falls, SC, USA, 2019.
59. Wukelic, G.; Gibbons, D.; Martucci, L.; Foote, H. Radiometric calibration of Landsat Thematic Mapper thermal band. *Remote Sens. Environ.* **1989**, *28*, 339–347. [CrossRef]
60. Cao, L.; Li, P.; Zhang, L.; Chen, T. Remote sensing image-based analysis of the relationship between urban heat island and vegetation fraction. *Int. Arch. Photogramm. Remote Sens. Spat. Inf. Sci.* **2008**, *37*, 1379–1384.
61. Carlson, T.N.; Ripley, D.A. On the relation between NDVI, fractional vegetation cover, and leaf area index. *Remote Sens. Environ.* **1997**, *62*, 241–252. [CrossRef]
62. Sobrino, J.A.; Jiménez-Muñoz, J.C.; Paolini, L. Land surface temperature retrieval from Landsat TM 5. *Remote Sens. Environ.* **2004**, *90*, 434–440. [CrossRef]
63. Sobrino, J.; Raissouni, N.; Li, Z.-L. A comparative study of land surface emissivity retrieval from NOAA data. *Remote Sens. Environ.* **2001**, *75*, 256–266. [CrossRef]
64. Weng, Q.; Lu, D.; Schubring, J. Estimation of land surface temperature–vegetation abundance relationship for urban heat island studies. *Remote Sens. Environ.* **2004**, *89*, 467–483. [CrossRef]
65. Kogan, F.N. Global drought watch from space. *Bull. Am. Meteorol. Soc.* **1997**, *78*, 621–636. [CrossRef]
66. Del-Toro-Guerrero, F.J.; Daesslé, L.W.; Méndez-Alonzo, R.; Kretzschmar, T. Surface Reflectance–Derived Spectral Indices for Drought Detection: Application to the Guadalupe Valley Basin, Baja California, Mexico. *Land* **2022**, *11*, 783. [CrossRef]
67. Liou, Y.-A.; Mulualem, G.M. Spatio–temporal assessment of drought in Ethiopia and the impact of recent intense droughts. *Remote Sens.* **2019**, *11*, 1828. [CrossRef]
68. Almazroui, M.; Islam, M.N.; Jones, P.; Athar, H.; Rahman, M.A. Recent climate change in the Arabian Peninsula: Seasonal rainfall and temperature climatology of Saudi Arabia for 1979–2009. *Atmos. Res.* **2012**, *111*, 29–45. [CrossRef]
69. Syed, F.; Adnan, S.; Zamreeq, A.; Ghulam, A. Identification of droughts over Saudi Arabia and global teleconnections. *Nat. Hazards* **2022**, *112*, 2717–2737. [CrossRef]
70. Xu, K.; Yang, D.; Yang, H.; Li, Z.; Qin, Y.; Shen, Y. Spatio-temporal variation of drought in China during 1961–2012: A climatic perspective. *J. Hydrol.* **2015**, *526*, 253–264. [CrossRef]
71. Almazroui, M.; Nazrul Islam, M.; Athar, H.; Jones, P.; Rahman, M.A. Recent climate change in the Arabian Peninsula: Annual rainfall and temperature analysis of Saudi Arabia for 1978–2009. *Int. J. Climatol.* **2012**, *32*, 953–966. [CrossRef]

72. Alsubih, M.; Mallick, J.; Talukdar, S.; Salam, R.; AlQadhi, S.; Fattah, M.; Thanh, N.V. An investigation of the short-term meteorological drought variability over Asir Region of Saudi Arabia. *Theor. Appl. Climatol.* **2021**, *145*, 597–617. [CrossRef]
73. El Kenawy, A.M.; Al Buloshi, A.; Al-Awadhi, T.; Al Nasiri, N.; Navarro-Serrano, F.; Alhatrushi, S.; Robaa, S.; Domínguez-Castro, F.; McCabe, M.F.; Schuwerack, P.-M. Evidence for intensification of meteorological droughts in Oman over the past four decades. *Atmos. Res.* **2020**, *246*, 105126. [CrossRef]
74. Lotfirad, M.; Esmaeili-Gisavandani, H.; Adib, A. Drought monitoring and prediction using SPI, SPEI, and random forest model in various climates of Iran. *J. Water Clim. Chang.* **2022**, *13*, 383–406. [CrossRef]
75. Sultana, M.S.; Gazi, M.Y.; Mia, M.B. Multiple indices based agricultural drought assessment in the northwestern part of Bangladesh using geospatial techniques. *Environ. Chall.* **2021**, *4*, 100120. [CrossRef]
76. Choi, M.; Jacobs, J.M.; Anderson, M.C.; Bosch, D.D. Evaluation of drought indices via remotely sensed data with hydrological variables. *J. Hydrol.* **2013**, *476*, 265–273. [CrossRef]
77. Almeida-Ñauñay, A.F.; Villeta, M.; Quemada, M.; Tarquis, A.M. Assessment of Drought Indexes on Different Time Scales: A Case in Semiarid Mediterranean Grasslands. *Remote Sens.* **2022**, *14*, 565. [CrossRef]

sustainability

Article

Design and Numerical Simulation of the Headworks in the Shizuishan Section of the Yellow River

Mingyang Liu [1], Suiju Lv [1,*], Qiao Qiao [2] and Lulu Song [3]

[1] School of Civil Engineering, North Minzu University, Yinchuan 750021, China
[2] Department of Architecture, Lvliang University, Lvliang 033400, China
[3] School of Civil and Hydraulic Engineering, Ningxia University, Yinchuan 750021, China
* Correspondence: 2005060@nmu.edu.cn; Tel.: +86-137-095-853-70

Abstract: Irrigation water for agriculture in Ningxia during the summer is primarily sourced from the Yellow River self-flow irrigation region. However, the water conveyance system in this region is significantly influenced by hydrodynamic factors, morphological factors, human factors, and the infrastructure used for social purposes, all of which directly impact the irrigation water utilization coefficient. In order to improve the irrigation water utilization coefficient, reduce suspended sediment deposition in the water conveyance channels, and mitigate negative effects on the water supply system, this study implemented a sediment diversion system at the channel head. This is expected to increase water usage efficiency to a certain degree. Using actual data on hydrodynamic factors from the Shizuishan section of the Yellow River in Ningxia, a two-dimensional numerical simulation was performed, and a two-dimensional hydrodynamic model and sediment model of the Shizuishan section of the Yellow River in Ningxia were developed using MIKE 21. The water conveyance method at the channel head was simulated under two different operating conditions. Results indicated that compared to operating condition 1, operating condition 2 had a beneficial effect on diverting and reducing sediment at the fish mouth of the channel head: the sediment accumulation thickness of one day in operating condition 1 was 0.16 m, 0.003 m, 0.15 m, and 0.21 m under actual flow, scenario 1, scenario 2, and scenario 3, respectively; whereas in operating condition 2, the sediment accumulation thickness of one day was 0.11 m, 0.001 m, 0.09 m, and 0.12 m under the same conditions, respectively. Additionally, as the computation period lengthened, the sediment accumulation thickness of operating condition 2 was significantly smaller than that of operating condition 1. In conclusion, operating condition 2 is superior for the design of the channel head in the Yellow River self-flow irrigation region.

Keywords: the Yellow River Irrigation District; numerical simulation; river erosion; sediment movement; diversion channel

Citation: Liu, M.; Lv, S.; Qiao, Q.; Song, L. Design and Numerical Simulation of the Headworks in the Shizuishan Section of the Yellow River. *Sustainability* **2023**, *15*, 4564. https://doi.org/10.3390/su15054564

Academic Editors: Hamza Gabriel, Songhao Shang, Qianqian Zhang, Dongqin Yin and Magdy Mohssen

Received: 22 January 2023
Revised: 23 February 2023
Accepted: 24 February 2023
Published: 3 March 2023

1. Introduction

Agricultural modernization in Ningxia has led to the need for more diverse and stronger infrastructure in irrigation areas. As the focus of irrigation construction shifts, there is a need to improve quality and efficiency. One challenge in these areas is the low water level and high sand content of the Yellow River, which causes sediment accumulation in the channels of self-flow irrigation areas and decreases the irrigation water utilization coefficient. Researchers from within and outside China have studied the water and sediment movement in these sedimentary basins to develop models and management strategies that aim to improve the irrigation water utilization coefficient and enhance the ecological environment.

At the head of the diversion channel are critical hydraulic structures that serve several purposes. On one hand, they divert water from rivers into channels to fulfill the needs of water resource projects such as irrigation, hydropower generation, and domestic water supply [1,2]. On the other hand, they prevent the entry of coarser sediment particles into bends, thus reducing erosion and sedimentation, and ensuring normal bend operation. As a result, numerous scholars have conducted extensive research on the specific hydrodynamic and sediment transport characteristics of headworks bends, as well as the evolution patterns of the riverbed [3,4]. Sediment in channels is a significant factor that controls the morphological and hydraulic characteristics of the riverbed [5,6], and the grain size of the sediment significantly affects water flow resistance, sediment transport, and the intensity of riverbed erosion [7,8]. Furthermore, the sediment transport capacity of rivers can vary with changing downstream hydraulic conditions, and the difference in sediment transport capacity during flood events can cause changes in riverbed material. In turn, changes in riverbed material can result in irregular fluctuations in the bed surface and sediment transport, which are mechanisms of sediment transport in rivers [9–12].

As the mechanisms of river sediment transport continue to be studied by researchers, Bognold [13] developed equations for the calculation of suspended sediment and bedload sediment transport rates, which have significantly advanced our understanding of water and sediment dynamics. Chen et al. [14] conducted an analysis of annual runoff and sediment sequences, providing insight into the downstream impacts of water conservancy projects, including gradual weakening of runoff and sedimentation from upstream to downstream. An et al. [15] examined hydrodynamic data from the segment of the Yellow River Basin spanning from Ningxia to Inner Mongolia since the 1960s, determining that channel erosion and sedimentation are primarily the result of both mainstream and tributary processes. Krishnappan et al. [16] investigated the relationship between sediment particle size and settling velocity in the sedimentation process; smaller-grained sediments in rivers are more susceptible to deposition and accumulation due to their reduced volume, which results in a lower level of erosive force exerted by the water flow. The river current is composed of two water layers with varying velocities, with higher water speeds leading to smaller suspended particle sizes. As the particle size decreases, the likelihood of its being transported and subsequently deposited by the water flow increases. Therefore, smaller-grained sediment is more readily deposited and accumulated in rivers. Finally, Jin et al. [17] analyzed sediment initiation and water flow sediment carrying capacity, and investigated the advancement patterns of sediment in the Yellow River irrigation area.

With the advancement of computational capabilities and the integration of modern technologies, computer-aided visualization techniques have been employed to investigate the hydrodynamic behavior and sediment transport characteristics in irrigation channels [18–20]. In this context, Soulis [21] conducted a study on the Mornos drainage canal in Athens, utilizing numerical simulations to reproduce the mechanisms of channel damage in vulnerable areas. Oyarce [22] employed computational fluid dynamics (CFD) models to numerically evaluate the hydrodynamic characteristics of agricultural drainage channels with varying geometries, and analyzed the temporal variations in flow direction, velocity, relative soil moisture content and head pressure during the drainage process. Alomari [23] conducted physical experiments to investigate the effects of the channel angle on erosion and deposition in rivers at angles of 30°, 45°, 60°, 75°, and 90°. The results indicated that as the angle of the channel increased, the thickness of the deposited sediment decreased, with the smallest amount observed at 90°. However, it was observed that solely altering the angle of the channel had a limited impact on the hydrodynamic characteristics within the channel, and was not found to be an effective means of reducing sedimentation and erosion. The utilization of fish mouth diversion structures at the Dujiangyan irrigation hub leverages the principle of fish mouth diversion to optimize the distribution of water flows [24]. This study aims to investigate the impact of the bedform on sediment transport in the river channel and to adjust unfavorable flow conditions by changing the hydraulic characteristics of the river through the construction of fish mouth engineering. The Yellow River engineering channel serves as the subject of this study and the characteristics of fish mouth diversion and sedimentation at Dujiangyan are taken into consideration. Numerical simulation methods are employed to design various operating scenarios. The scouring and silting characteristics are analyzed under different conditions, reflecting different periods of water and sediment characteristics as well as incoming water and sediment conditions.

The integration of SMS software and MIKE21 software can enhance the efficiency and accuracy of river flow and hydrodynamic simulations [25]. In this study, the SMS software was utilized to partition the grid of the study region under various operational conditions. The generated grid files were imported into the MIKE 21 software, where the water dynamics and sediment transport modules were coupled to simulate the evolution of channel sediment. The sediment accumulation patterns were analyzed and the changes in channel sediment under different scenarios and operational conditions were compared in order to identify the optimal operational condition.

2. Materials and Methods

2.1. Introduction to the Study Area

The Yellow River Diversion Project is a significant engineering project aimed at ensuring the ecological water use of the people living along the route. The annual average flow of the Yellow River in Ningxia can reach 4 billion m3, with the majority of this water being used for agricultural irrigation. The irrigation period during the summer and autumn lasts approximately 40–50 days, while the winter period lasts approximately 10–15 days, resulting in a total irrigation period of approximately 50–60 days per year. The study area is located in the Shizuishan section of the Yellow River in Ningxia, with a length of 4 km (as depicted in Figure 1a). In order to measure the hydrodynamic elements of this region, acoustic Doppler profilers, GPS-RTK systems, and laser grain size distribution instruments were utilized on 20 October 2017, 20 October 2018, 20 October 2019, and 20 October 2020. The present study utilizes data collected on October 20, 2020, as an example, with the MIKE21 software being utilized to interpolate and generate a 3D view of the region based on the measurements taken at 10 different cross-sections (as shown in Figure 1b).

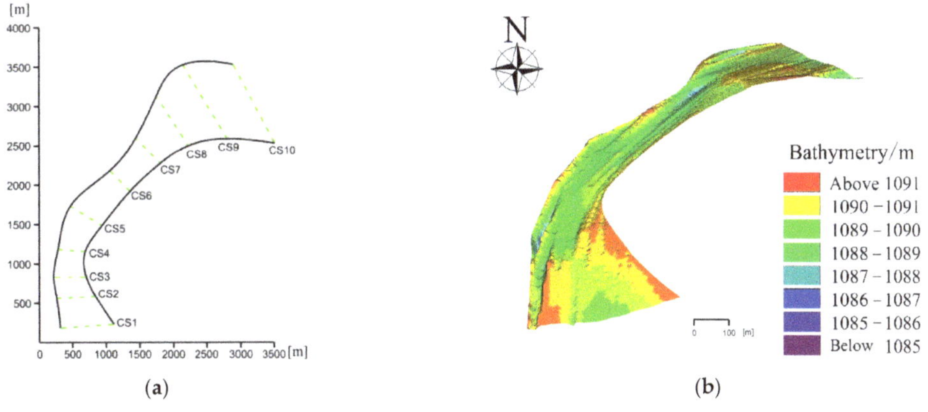

Figure 1. Diagram of measurement area, (**a**) Section division of measurement area; (**b**) 3D view of measurement area.

2.2. Mathematical Modelling

The present study employs a grid-based approach to evaluate the hydrodynamic characteristics of a water intake channel in different operating conditions. Surface Water Modeling System (SMS) software was utilized to partition the research region into grids, which were subsequently imported into MIKE 21 for analysis using the hydrodynamic module and sediment transport module. The sediment transport module is based on the calculations from the hydrodynamic module.

The present study employs a two-dimensional hydrodynamic model and a non-viscous mud–sediment coupling model to investigate the study region numerically. The model is founded upon the incompressible Reynolds-averaged Navier–Stokes equations, comprising the continuity equation, the x direction momentum equation, and the y direction momentum equation.

Hydrodynamic module principle

$$\frac{\partial h}{\partial t}+\frac{\partial h\overline{v}}{\partial x}+\frac{\partial h\overline{v}}{\partial y}=hS \tag{1}$$

The momentum equation in the X direction is

$$\begin{aligned}&\tfrac{\partial h\overline{u}}{\partial t}+\tfrac{\partial h\overline{u}^2}{\partial x}+\tfrac{\partial h\overline{u}\overline{v}}{\partial y}=f\overline{v}h-gh\tfrac{\partial \eta}{\partial x}-\tfrac{h}{\rho_0}\tfrac{\partial Pa}{\partial x}-\tfrac{gh^2}{2\rho_0}\tfrac{\partial \rho}{\partial x}+\tfrac{\tau_{sx}}{\rho_0}\\&-\tfrac{\tau_{bx}}{\rho_0}-\tfrac{1}{\rho_0}\left(\tfrac{\partial S_{xx}}{\partial x}+\tfrac{\partial S_{xy}}{\partial y}\right)+\tfrac{\partial}{\partial x}(hT_{XX})+\tfrac{\partial}{\partial y}(hT_{xy})+hU_sS\end{aligned} \tag{2}$$

The momentum equation in the Y direction is

$$\begin{aligned}&\tfrac{\partial h\overline{v}}{\partial t}+\tfrac{\partial h\overline{v}^2}{\partial x}+\tfrac{\partial h\overline{u}\overline{v}}{\partial y}=-f\overline{v}h-gh\tfrac{\partial \eta}{\partial y}-\tfrac{h}{\rho_0}\tfrac{\partial Pa}{\partial y}-\tfrac{gh^2}{2\rho_0}\tfrac{\partial \rho}{\partial y}+\tfrac{\tau_{sy}}{\rho_0}\\&-\tfrac{\tau_{bx}}{\rho_0}-\tfrac{1}{\rho_0}\left(\tfrac{\partial S_{yx}}{\partial x}+\tfrac{\partial S_{yy}}{\partial y}\right)+\tfrac{\partial}{\partial x}(hT_{XX})+\tfrac{\partial}{\partial y}(hT_{xy})+hV_sS\end{aligned} \tag{3}$$

where t represents the time variable, while h is the total water depth. The x and y variables represent the coordinates in a Cartesian coordinate system. The u and v variables represent the velocity components in the x and y directions, respectively. S refers to the source term, while f denotes the Coriolis force. The g represents the acceleration due to gravity. The variable d denotes the static water depth, while represents the density of water. S_{xx}, S_{xy}, and S_{yy} represent the components of the radiation stress, and Us and V_s represent the flow velocities of the water flow associated with the source term.

The non-cohesive sediment transport calculation is based on a two-dimensional hydrodynamic model that considers a single flow event. The calculation employs comprehensive sediment transport theory to determine the concentrations of suspended and bed load sediments.

Control equation of sediment transport model:

$$\frac{\partial \overline{c}}{\partial t}+u\frac{\partial \overline{c}}{\partial x}+v\frac{\partial \overline{c}}{\partial y}=\frac{1}{h}\frac{\partial}{\partial x}\left(hD_x\frac{\partial c}{\partial x}\right)+\frac{1}{h}\frac{\partial}{\partial y}\left(hD_y\frac{\partial c}{\partial y}\right)+Q_LC_L\frac{1}{h}-S \tag{4}$$

where $\overline{c}$ is the average sediment concentration in the water depth direction (kg/m^3), h is the water depth (m), D_x and D_y are the dispersion coefficients in the x and y directions, respectively (m^2/s), Q_L is the single wide source term flow in the horizontal direction (m^3/s/m^3); this source can be from a single location, such as a stream or river, or from multiple locations that contribute to the sediment transport in a similar manner; C_L is the source sediment concentration, g/m^3, and S is the scouring/silting item (representing any additional sources or sinks of sediment) kg/m^3/s.

2.3. Model Establishment and Parameter Calibration

Terrain data for the study region were processed using SMS software, with boundary attributes defined as open and land boundaries. The upstream open boundary extended 701 m in length and contained 48 nodes, while the downstream open boundary was 1004 m long with 78 nodes. The study region was divided into 15,665 unstructured triangular grid cells, with a grid resolution of 10 m and a minimum angle of 30°, as depicted in Figure 2. Elevation data were based on the 1985 Chinese elevation datum and the 1954 Beijing coordinate system was used for plane coordinates.

In the hydrodynamic module, the annual coefficient of the eddy flow is 0.28; the roughness coefficient of the river bed is calculated using the Manning coefficient, resulting in a value of 36 m$^{1/3}$/s. In the sediment module, flocculation settling is selected as the settling velocity for the suspended sediment, with a rate of 0.01 kg/m^3 and a density of 2650 kg/m^3 for the sediment.

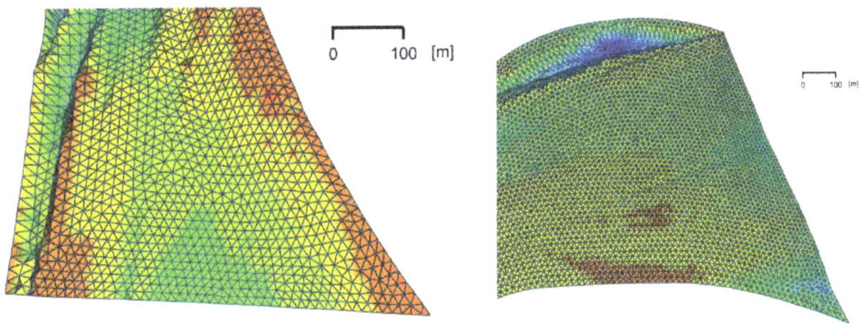

(**a**) (**b**)

Figure 2. Schematic diagram of the gridding of the measurement area, (**a**) Schematic of grid division in upstream study area; (**b**) Schematic of grid division in downstream study area.

2.4. Model Validation

The present study aims to assess the hydrodynamic and sedimentary characteristics of the Shizuishan section of the Yellow River using the hydraulic and sediment modules, respectively, in 2.2.2 and 2.2.3 as of 20 October 2020. The hydraulic data used in this study were obtained from in situ measurements carried out in the study area. The flow velocity and direction were determined through the use of acoustic Doppler profilers, the sediment concentration was estimated using laser particle size distribution instruments, and the water surface elevation was determined using GPS-RTK measurements. The upstream flow rate was set to 1861.5 m^3, based on the measured data, and the downstream elevation was set to 1092.3 m. For the sediment module, the upstream sediment concentration was 1.25 kg/m^3, and the median grain size was 0.15 mm.

The present study validated the hydrodynamic elements and sediment content of cross-sections CS4, CS5, CS6, and CS7, as the design of the water channel is situated between cross-sections CS5 and CS6. The results, shown in Figure 3a, demonstrate that the simulated flow velocity exhibits a similar trend to the measured flow velocity, with a maximum deviation of 0.37 m/s. Similarly, the sediment content in Figure 3b exhibits a maximum deviation of 0.21 kg/m^3. These results indicate that the model used in this study is reliable for simulation purposes, as the fitting error is minimal for both flow velocity and sediment content.

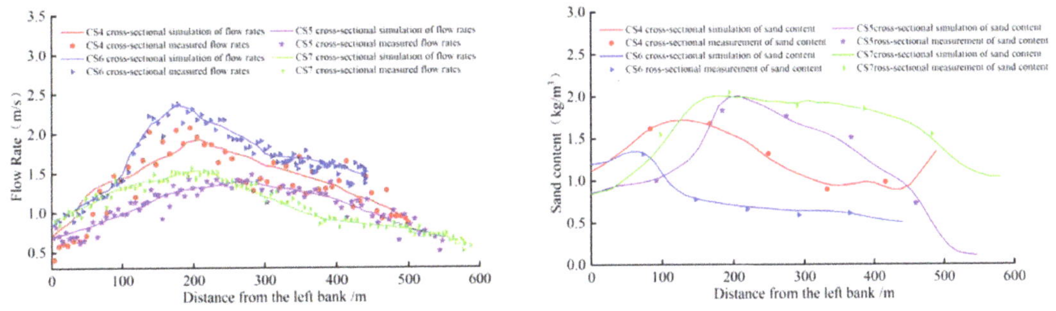

(**a**) (**b**)

Figure 3. Typical section simulation validation chart, (**a**) Typical section flow velocity validation chart; (**b**) Typical section sediment concentration validation chart.

3. Results

3.1. Experimental Design

3.1.1. Design for Working Conditions

To mitigate the negative impact of sediment on channel sedimentation in self-flow irrigation areas, two different operating conditions were designed based on the principle of fish mouth diversion and sediment reduction at the Dujiangyan water conservancy hub. A fish mouth is a specialized architectural structure designed to manage the velocity and orientation of fluidic currents (as shown in Figure 4). This study investigated the utilization of the fish mouth with a diminutive arch-shaped geometrical configuration, situated near the concave bank of the river channel, thus segmenting the river into an internal and external flow path. The fluidic flow entering the arch-shaped space through the upper portion of the fish mouth experienced a reduction in velocity, while the fluidic flow entering through the lower portion produced a robust fluidic current. Consequently, the fish mouth serves as a means to optimize fluidic distribution and effectively modulate fluidic velocity and orientation. Using MIKE 21 software, the erosion dynamics of the two operating conditions under similar upstream hydrodynamic conditions were simulated. Operating Condition 1 consisted of a direct water channel with a sediment blocking crest at the head (as depicted in Figure 4a), 100 m wide, and a channel 800 m long and 20 m wide with a longitudinal gradient of 1‰. The head of the channel employed a leaky bucket design to increase the drainage volume. Operating Condition 2 was similar to Condition 1, but with the addition of a fish mouth at the head of the channel. The sediment blocking crest connected to the natural river channel was 150 m wide and the outlet was 100 m wide. The fish mouth was located on the upper side of the bend of the leading river and was followed by a flying sand weir (as depicted in Figure 4b), 1093 m in elevation and 50 m wide. The fish mouth divided the flow into inner and outer sections, with the inner section being 150 m wide and the outer section being 100 m wide.

The design area was gridded using SMS software (as shown in Figure 5). In Condition 1 (as depicted in Figure 5a), the node spacing within the channel was set at 10 m, with a closer spacing of 5 m at the channel head to enhance the fidelity of this region. In Condition 2 (as shown in Figure 5b), the node division within the channel remained unchanged, but the node spacing in the vicinity of the fish mouth was decreased to 5 m in order to accurately reproduce the hydrodynamic and erosion dynamics in this area.

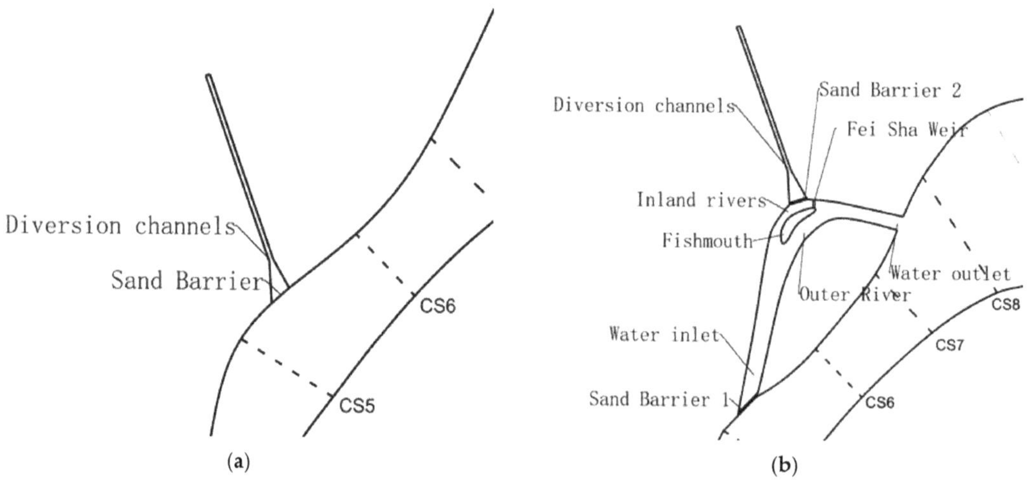

Figure 4. Design diagram of operating conditions, (**a**) Design schematic diagram of Operating condition 1; (**b**) Design schematic diagram of Operating condition 2.

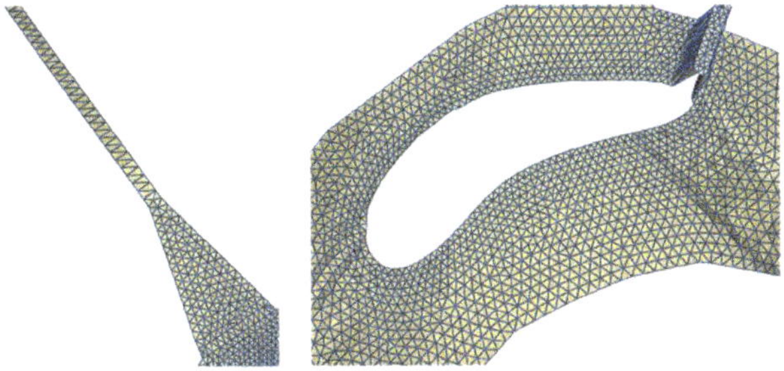

(**a**) (**b**)

Figure 5. Design area grid division map, (**a**) Grid division of water diversion channel; (**b**) Grid division of fish mouth.

3.1.2. Scenario Setting

According to the temporal distribution of irrigation in the Shizuishan section of the Yellow River in Ningxia, flow is typically concentrated from July to October and ranges from 1000 to 2000 m^3/s. However, runoff exhibits dynamic changes due to climatic factors. Therefore, this study establishes three scenarios based on variations in the runoff: Scenario 1 with a flow of 1000 m^3/s, Scenario 2 with a flow of 1500 m^3/s, and Scenario 3 with a flow of 2000 m^3/s.

3.1.3. Fish Mouth Flow Field Distribution

To investigate the spatial distribution of the water flow near a fish mouth, a simulation was conducted to examine the flow field at the fish mouth and the head of a channel under measured flow rates. As illustrated in the flow field vector distribution in Figure 6, the flow field in the outer river exhibits a lower intensity compared to that in the inner river, with flow velocities ranging from 0.2 to 0.48 m/s in the outer river and 0.8 to 0.9 m/s in the inner river, indicating that the outer river exhibits lower kinetic energy and weaker sediment-carrying capacity. At a discharge of 1861.5 m^3/s, the water surface elevation near the sand-ejecting weir is 1092.8 m and a vortex–like backflow is formed between the upstream and the head of the channel. Upon a further increase in the flow rate to the critical value, the water flow overflows the sand-ejecting weir and the sediment is entrained and transported downstream. Condition 2 reproduces the operation of the Dujiangyan water conservancy hub, including the functions of water diversion from the outer river and water discharge from the inner river in an artificial channel, which can be utilized for the purpose of simulation in this study.

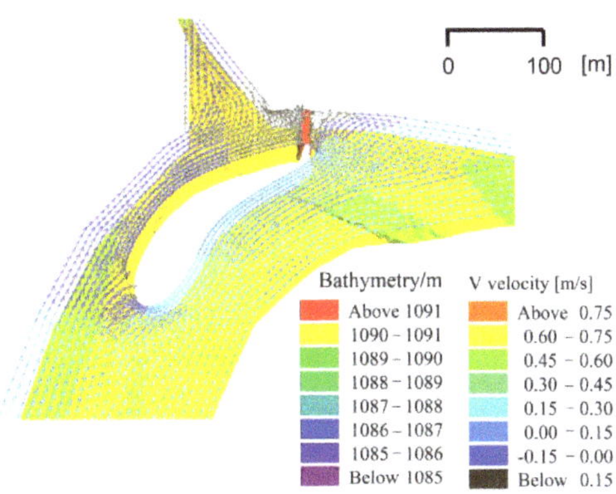

Figure 6. Vector map of flow distribution at the fish mouth.

3.2. Erosion of Diversion Channels at Different Times

The channel elevation and distribution of sediment concentration can effectively reflect the erosion behavior under different operating conditions during simulation. In this study, we simulated the changes in channel elevation and sediment concentration of the measured stream flow under different operating conditions for a one-day erosion period. As shown in Figure 7a, both operating conditions 1 and 2 are in a state of sedimentation, and the sediment thickness increases with the narrowing of the channel in the transitional zone. However, at the junction between the transitional zone and the 20 m rapid channel, the increase in hydrodynamic forces causes a sudden decrease in sediment thickness, followed by an increase. The flow then gradually penetrates deeper into the channel, leading to a decrease in sediment thickness. The trend of elevation change under operating conditions 1 and 2 is consistent, but the sediment thickness under operating condition 1 is consistently higher than under operating condition 2. As shown in Figure 7b, the distribution of sediment concentration in the channel under operating conditions 1 and 2 exhibits some variations, but the sediment concentration under operating condition 1 is consistently higher than under operating condition 2 throughout the channel.

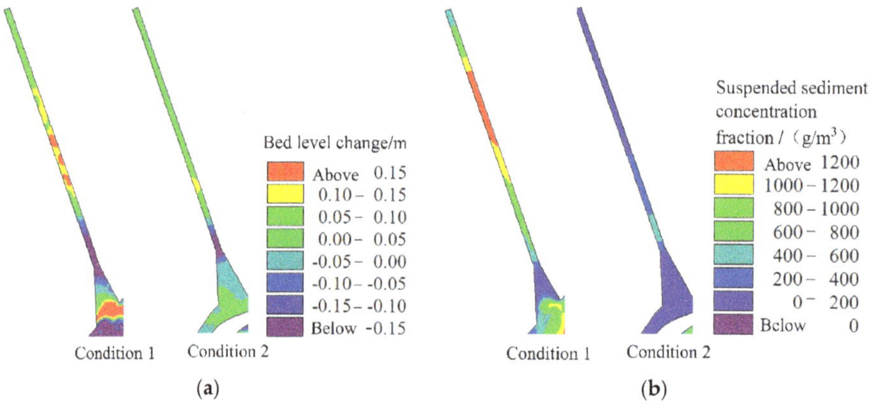

Figure 7. Erosion change map of rivers, (**a**) Elevation change map of channels; (**b**) Sediment content change map of channels.

This study conducted statistical analysis on the average height variation and sediment concentration along the channel within the study region, as illustrated in Figure 8. The maximum accumulation thickness under Condition 1 was 0.16 m, while the maximum accumulation thickness under Condition 2 was 0.115 m. The accumulation thickness was relatively similar in the area adjacent to the transitional zone and the linear segment, but the accumulation thickness in the broader region under Condition 2 was significantly lower compared to that under Condition 1. Additionally, the maximum sediment concentration under Condition 1 was 1.39 kg/m^3, which was significantly higher than the maximum sediment concentration of 0.477 kg/m^3 under Condition 2. Therefore, it can be inferred that Condition 2 effectively reduced the channel accumulation during a one-day calculation period when the flow rate was 1861.5 m^3/s.

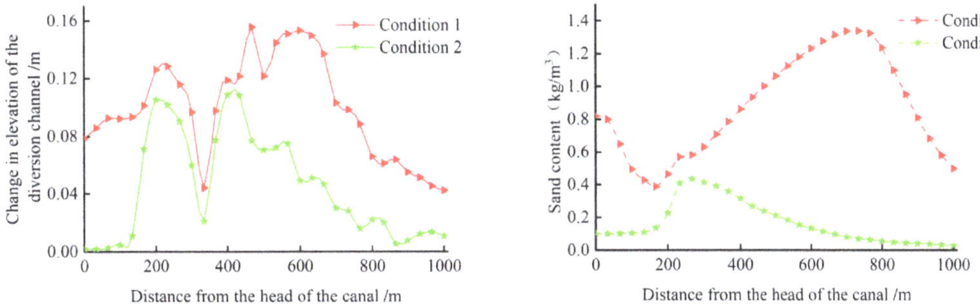

(a) (b)

Figure 8. Erosion change map of irrigation channels, (**a**) Elevation change map of irrigation channels; (**b**) Sediment content distribution map of irrigation channels.

3.3. Erosion of Diversion Channels for Different Scenarios

Based on long–term observations of water and sediment movement in the Yellow River in Ningxia, one–day erosion and sedimentation simulations were conducted for different flow scenarios (as illustrated in Figure 9). The results showed that the channel is in a state of sedimentation for all three scenarios (as illustrated in Figure 9a), with the sedimentation thickness first increasing and then decreasing. In addition, the sedimentation thickness of condition 1 is greater than that of condition 2 under the same scenario, and the sedimentation thickness of the channel increases with the increase in flow under the same condition. The maximum sedimentation thickness of condition 1 is 0.21 m in scenario 3, which is significantly larger than the maximum sedimentation thickness of 0.12 m in scenario 3 for condition 2. The minimum sedimentation thickness of condition 1 in scenario 1 is 0.003 m, but it is still larger than the minimum sedimentation thickness of 0.001 m in scenario 1 for condition 2. The sand content of the channel exhibits a similar trend to that of erosion and change (as illustrated in Figure 9b), with the maximum sand content for condition 1 being 1.74 kg/m^3 at the flow of scenario 3 and the minimum sand content being 0.38 kg/m^3 at the flow of scenario 1. The maximum sand content for condition 2 is 0.65 kg/m^3 at the flow of scenario 3, and the minimum sand content is 0.03 kg/m^3 at the flow of scenario 1. These findings suggest that condition 2 can effectively reduce the sedimentation thickness and sand content of the channel in the three scenarios to some extent.

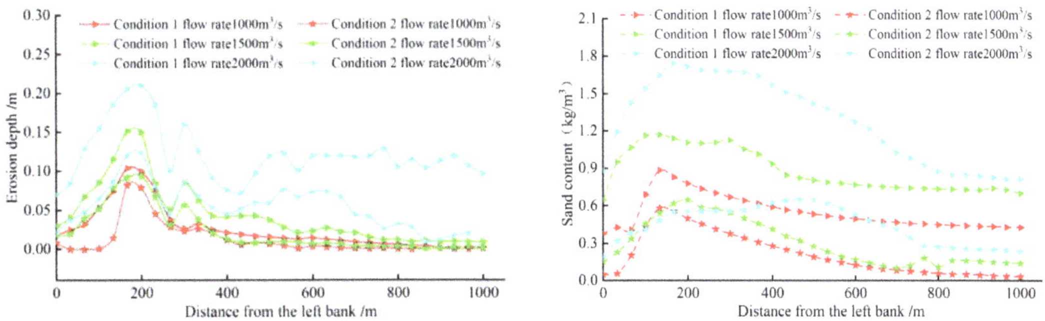

(a) (b)

Figure 9. Chart showing the changes in erosion caused by different channels in different scenarios, (**a**) Erosion depth of channels; (**b**) Graph of sediment concentration change in a channel.

3.4. Erosion of the Diversion Channel under Different Scenarios at the Same Time

In this study, the erosion characteristics of an irrigation area with long-term drainage were simulated under various operational conditions for a period of three days. The results showed that both operational conditions were in a state of sedimentation under different scenarios (as illustrated in Figure 10a). In the same scenario, the erosion thickness along the channel's straight segment exhibited a sinusoidal decline, with the sedimentation thickness of operational condition 1 being greater than that of operational condition 2 and exhibiting larger fluctuations. In scenario 3, the maximum erosion thickness and maximum erosion difference for operational conditions 1 and 2 reached their maximum values: the maximum erosion thickness for operational condition 1 was 0.37 m, with a maximum erosion difference of 0.25 m; the maximum erosion thickness for operational condition 2 was 0.28 m, with a maximum erosion difference of 0.23 m. In scenario 1, the maximum erosion thickness and maximum erosion difference for operational conditions 1 and 2 were at their minimum values: the maximum erosion thickness for operational condition 1 was 0.18 m, with a maximum erosion difference of 0.15 m; the maximum erosion thickness for operational condition 2 was 0.13 m, with a maximum erosion difference of 0.12 m. The sand content in the channel was found to vary as shown in Figure 10b, reaching a peak near the intersection of the gradient section and the straight section before decreasing steadily. The sand content was found to be similar to the value calculated for one day.

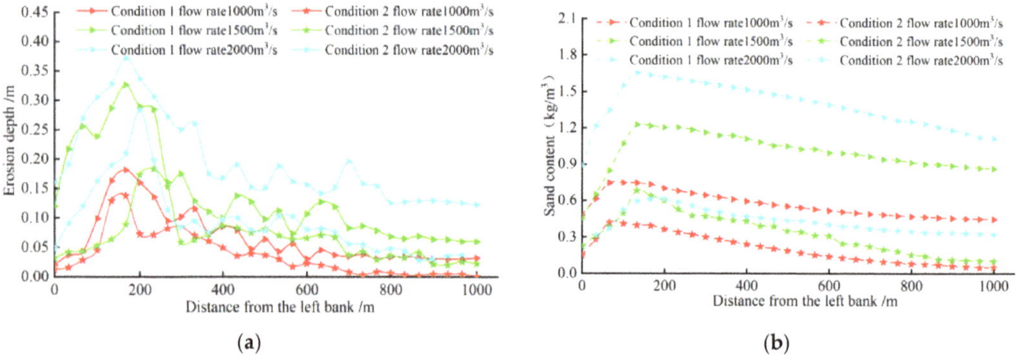

Figure 10. Chart showing the changes in erosion caused by different channels at different times and in different scenarios, (**a**) Erosion depth of channels; (**b**) Graph of sediment concentration change in a channel.

As the calculation time increases, the sediment accumulation thickness for both working conditions exhibits an upward trend, while the sediment content in the channel becomes increasingly stable.

4. Discussion

The present study performed a working condition design for the new water conveyance channel at the Shizuishan section of the Yellow River in Ningxia, China. Coupled simulation calculations were then conducted for various working conditions under various scenarios. The results showed that, compared to working condition 1, working condition 2 had three impacts on the self-irrigation area of the channel head: (1) the fishtail diverter divided and sieved the water, introducing outer river water with a lower sand content into the water supply system; (2) it decreased the flow rate of the channel head, weakening the water's sand-carrying capacity and thus reducing the sand content of the channel; and (3) it changed the hydrodynamic and erosion power changes, ultimately reducing channel sediment accumulation. The fishtail diverter in working condition 2 was found to be significantly more effective at preventing sediment accumulation and reducing sand compared to traditional Yellow River self-flow channel heads. Under scenarios 1, 2, and 3, the maximum expected thickness was reduced by 0.05, 0.14, and 0.09 m, respectively, over a period of three days of erosion. Therefore, it can be concluded that working condition 2 can effectively improve channel sediment accumulation in the irrigation area.

Fish mouths impede the flow of streamflow to some extent [26–29], and there are significant differences in the hydrodynamic elements of the inner and outer rivers of the canal. This is due to the fact that the fish mouth causes the outer riverbed of the canal to sink, while the inner river rises. According to the principle of hydrodynamic circulation, surface water containing a small amount of sediment flows towards the concave bank, that is, towards the inner river, and enters the canal via the sand-blocking weir. On the other hand, bottom water with a high concentration of sediment flows towards the convex bank, or the outer river, and the sediment is transported into the natural river along the outer river [30,31]. However, over time, sediment will also accumulate in the inner river due to canal irrigation. The water in the inner river flows towards the sand-blocking weir with a high impact force and, due to the top-support effect, forms a vortex near the sand-throwing weir. The sediment in the water is then ejected from the sand-throwing weir and discharged into the river through the outlet. When the water volume is high, the water level will surpass the sand-throwing weir and the sediment will flow over it, exiting the outlet at a fast speed. The Dujiangyan Irrigation System employs the hydraulic design of a fish mouth to effectively manage sediment transport, thereby preventing the siltation of the Inner River and maintaining balance. This system's design has demonstrated potential for scientific research and application in various fields [25]. In addition, contemporary irrigation systems in Africa [32] as well as ancient irrigation system in the Central Negev desert [33] employ the mechanism of fish mouth structures to achieve equilibrium between water and sediment.

Therefore, in this study, we apply the aforementioned bend circulation principle to the design of the head of the canal in the self-flow irrigation area of the Yellow River in Ningxia, in order to ensure the long-term operation of the water supply system.

The MIKE 21 software is capable of simulating the movement of water and sediment with high accuracy. The two-dimensional water-sediment coupling model can accurately depict the movement state of natural rivers through a mathematical model, simulating the intricate details of the movement of hydrodynamic elements in channels and thus enabling the visualization of the temporal variation of hydrodynamic elements in the canal under different hydrological conditions. However, this study also has some limitations. Firstly, the hydrodynamic elements of the river undergo changes after the addition of the canal, and field measurements are necessary to assess the changes in the movement of the natural river under different operating conditions. Secondly, the design in this study is based on the results of numerical simulation and lacks experimental validation from actual engineering projects. Therefore, in the forthcoming work it will be necessary to monitor the actual hydrodynamic elements after the construction of the canal head under the two operating conditions to ensure that the design of operating condition 2 is effective in reducing sediment in the canal.

5. Conclusions

The current study aims to simulate the construction of a new channel in the Yellow River irrigation area using MIKE 21 modeling. Based on observed hydrodynamic data from the Shizuishan section of the Yellow River in Ningxia, a mathematical model was developed and its accuracy was verified. The design of the channel's entrance, referred to as "Work Condition 1" and "Work Condition 2", was also carried out, and simulation calculations were performed on the flow under different scenarios. The following conclusions were obtained:

1. The fish mouth design utilizes the principle of bend circulation to divide the river into inner and outer channels. The inner channel exhibits higher flow velocity and higher sediment concentration compared to the outer channel, which has lower flow velocity

and lower sediment concentration. This design effectively reduces the hydrodynamic conditions and subsequently lowers the sediment-carrying capacity of the water flow.

2. Simulation calculations using a one-day time frame reveal that under different scenarios, the maximum sediment accumulation thickness for channel condition 1 is 0.21 m, while channel condition 2 exhibits a maximum sediment accumulation thickness of 0.12 m. These results demonstrate a significant reduction in sediment for channel condition 2.
3. As erosion time increases, both channel conditions exhibit sediment accumulation. However, when the calculation time and scenarios are held constant, the sediment accumulation thickness of channel condition 1 consistently exceeds that of channel condition 2.

Therefore, after the establishment and verification of the accuracy of the physical model for channel condition 2 at the head of the channel, it can potentially be applied in the construction of the head of the self-flowing irrigation channel in the Yellow River Irrigation District.

Author Contributions: Conceptualization, M.L. and S.L.; methodology, S.L.; software, Q.Q.; validation, M.L., L.S.; formal analysis, M.L.; investigation, Q.Q.; resources, S.L.; data curation, M.L.; writing—original draft preparation, M.L.; writing—review and editing, L.S.; visualization, L.S.; supervision, S.L.; project administration, S.L.; funding acquisition, S.L. All authors have read and agreed to the published version of the manuscript.

Funding: This research was funded by Ningxia Natural Science Foundation Project (grant number 2021AAC03173).

Institutional Review Board Statement: Not applicable.

Informed Consent Statement: Not applicable.

Data Availability Statement: The data presented in this study are available on request from the corresponding author.

Acknowledgments: We would like to thank the reviewer for their valuable comments and suggestions, to the editors for their serious and responsible attitude, to the publisher for this valuable opportunity. We also thank our other colleagues for their valuable comments and suggestions that helped to improve the manuscript.

Conflicts of Interest: The authors declare no conflict of interest.

References

1. Goel, A.; Pillai, N.N. A flowmeter for rectangular irrigation field channels. *Water Manag.* **2008**, *161*, 135–139. [CrossRef]
2. Khater, A.; Kitamura, Y.; Shimizu, K.; Abou El Hassan, W.; Fujimaki, H. Quantitative analysis of reusing agricultural water to compensate for water supply deficiencies in the Nile Delta irrigation network. *Paddy Water Environ.* **2015**, *13*, 367–378. [CrossRef]
3. Outeiro, J.C.; Umbrello, D.; M'saoubi, R. Experimental and numerical modelling of the residual stresses induced in orthogonal cutting of AISI 316L steel. *Int. J. Mach. Tools Manuf.* **2006**, *46*, 1786–1794. [CrossRef]
4. Qiao, Q.; Li, C.G.; Jing, H.F.; Huang, L.X.; Yang, C. Impact of an artificial chute cutoff on the river morphology and flow structure in Sipaikou area of the Upper Yellow River. *J. Mt. Sci.* **2021**, *18*, 16. [CrossRef]
5. Surian, N. Downstream variation in grain size along an Alpine river: Analysis of controls and processes. *Geomorphology* **2002**, *43*, 137–149. [CrossRef]
6. Moussavi-Harami, R.; Mahboubi, A.; Khanehbad, M. Analysis of controls on downstream fining along three gravel-bed rivers in the Band-e-Golestan drainage basin NE Iran. *Geomorphology* **2004**, *61*, 143–153. [CrossRef]
7. Bathurst, J.C.; Carling, P.A.; Reid, I.; Walling, D.E.; Webb, B. *Sediment Erosion, Transport, and Deposition*; Wiley: Hoboken, NJ, USA, 1997.
8. Petts, G.E.; Gurnell, A.M.; Gerrard, A.J.; Hannah, D.M.; Hansford, B.; Morrissey, I.; Edwards, P.J.; Kollmann, J.; Ward, J.V.; Tockner, K.; et al. Longitudinal variations in exposed riverine sediments: A context for the ecology of the Fiume Tagliamento, Italy. *Aquat. Conserv.* **2000**, *10*, 249–266. [CrossRef]
9. Surian, N. Fluvial Processes in Braided Rivers. In *Rivers–Physical, Fluvial and Environmental Processes*; Springer: Cham, Switzerland, 2015; pp. 403–425.
10. Mueller, E.R.; Pitlick, J. Sediment supply and channel mor phology in mountain river systems: 2. Single thread tobraided transitions. *J. Geophys. Res. Earth Surf.* **2014**, *119*, 1516–1541. [CrossRef]
11. Griffiths, G.A. Sediment translation waves in braided gravel-bed rivers. *J. Hydraul. Eng.* **1993**, *119*, 924–937. [CrossRef]

12. Ashmore, P.; Bertoldi, W.; Gardner, J.T. Active width of gravelbed braided rivers. *Earth Surf. Process. Landf.* **2011**, *36*, 1510–1521. [CrossRef]
13. Bagnold, R.A. Bed load transport by natural rivers. *Water Resour. Res.* **1977**, *13*, 303–312. [CrossRef]
14. Chen, G.S.; Wang, Y.M.; Bai, T.; Du, H.H. Diagnoses of runoff-sediment relationship based on variable diagnostic method-variable step length sliding correlation coefficient method in Ning-Meng reach. *Int. J. Hydrog. Energy* **2016**, *41*, 15909–15918. [CrossRef]
15. An, C.; Lu, J.; Qian, Y.; Wu, M.; Xiong, D. The scour-deposition characteristics of sediment fractions in desert aggrading rivers—Taking the upper reaches of the Yellow River as an example. *Quat. Int.* **2019**, *523*, 54–66. [CrossRef]
16. Krishnappan, B.G.; Lau, Y.L. Turbulence Modeling of Flood Plain Flows. *J. Hydraul. Eng.* **1986**, *112*, 251–266. [CrossRef]
17. Li, J.; Fei, L.; Chen, Z.; Sun, X. Jinshan. Particle size distribution and settling velocity of sediments in water diverted from the Yellow River during border-strip irrigation. *Tecnol. Cienc. Agua* **2017**, *8*, 31–41. [CrossRef]
18. Wang, L.; Nie, Z.; Liu, M.; Cao, L.; Zhu, P.; Yuan, Q. Rational Allocation of Water Resources in the Arid Area of Northwestern China Based on Numerical Simulations. *Sustainability* **2023**, *15*, 55. [CrossRef]
19. Kim, B.J.; Hwang, J.H.; Kim, B. FLOW-3D Model Development for the Analysis of the Flow Characteristics of Downstream Hydraulic Structures. *Sustainability* **2022**, *14*, 10493. [CrossRef]
20. Wang, P.; Li, J.; Wang, M.; Hu, J.; Zhang, F. Numerical Simulation of the Hydraulic Characteristics and Fish Habitat of a Natural Continuous Meandering River. *Sustainability* **2022**, *14*, 9798. [CrossRef]
21. Soulis, V.J. Numerical Investigation of the Failure Mechanism of Articulated Aqueduct. *J. Perform. Constr. Facil.* **2018**, *32*, 1–17. [CrossRef]
22. Oyarce, P.; Gurovich, L.; Calderón, I. Simulating Hydraulic Behavior of an Agricultural Drain Based on Experimental Data. *J. Irrig. Drain. Eng.* **2017**, *143*, 1–8. [CrossRef]
23. Alomari, N.K.; Yusuf, B.; Mohammad, T.A.; Ghazali, A.H. Experimental investigation of scour at a channel junctions of different diversion angles and bed width ratios. *Catena* **2018**, *166*, 10–20. [CrossRef]
24. Zheng, X.; Kazemi, E.; Gabreil, E.; Liu, X.; Chen, R. Sustainability of the Dujiangyan Irrigation System for over 2000 Years–A Numerical Investigation of the Water and Sediment Dynamic Diversions. *Sustainability* **2020**, *12*, 2431. [CrossRef]
25. Yan, J.; Chen, M.; Xu, L.; Liu, Q.; Shi, H.; He, N. Mike 21 Model Based Numerical Simulation of the Operation Optimization Scheme of Sedimentation Basin. *Coatings* **2022**, *12*, 478. [CrossRef]
26. Tabuchi, J.P.; Aboulouard, S.; Bernier, J.; Blanchet, B.; Guérin, S.; Saint-Germain, A.; Rocher, V. The Seine river flood in June 2016, implications for the operation of the sanitation system of the heart of Paris region. *Tech. —Sci. —Methodes* **2019**, *114*, 67–80.
27. Sun, J.; Xu, N.; Ding, L.; Ma, Y.; Liu, Z.; Huang, Z. Continuous Expansions of Yangtze River Islands after The Three Gorges Dam tracked by Landsat data based on Google Earth Engine. *IEEE Access* **2020**, *8*, 92731–92742. [CrossRef]
28. Itla, B.; Jcsc, D.; Mla, B. Architecture, sedimentary facies and chronology of a composite island: A model from the Upper Paraná River, Brazil—ScienceDirect. *Geomorphology* **2020**, *372*, 107457.
29. Shi, H.; Cao, Y.; Dong, C.; Xia, C.; Li, C. The Spatio-Temporal Evolution of River Island Based on Landsat Satellite Imagery, Hydrodynamic Numerical Simulation and Observed Data. *Remote Sens.* **2018**, *10*, 2046. [CrossRef]
30. Zhang, S.; Yi, Y.; Liu, Y.; Wang, X. Hydraulic Principles of the 2,268-Year-Old Dujiangyan Project in China. *J. Hydraul. Eng.* **2013**, *139*, 538–546. [CrossRef]
31. Fan, G.; Xia, J.; Song, J.; Sun, H.; Liang, D. Research on application of ecohydrology to disaster prevention and mitigation in China: A review. *Water Supply* **2022**, *22*, 2946–2958. [CrossRef]
32. Gebrehiwot, K.A.; Haile, A.M.; De Fraiture, C.; Chukalla, A.D.; Embaye, T.G. Optimizing flood and sediment management of spate irrigation in Aba'ala Plains. *Water Resour. Manag.* **2015**, *29*, 833–847. [CrossRef]
33. Shanan, L. Runoff, Erosion, and the Sustainability of Ancient Irrigation Systems in the Central Negev Desert. In *The Hydrology-Geomorphology Interface: Rainfall, Floods, Sedimentation, Land Use*; IAHS Press: Wallingford, UK, 2000; pp. 75–106.

remote sensing

MDPI

Article

Synergistic Change and Driving Mechanisms of Hydrological Processes and Ecosystem Quality in a Typical Arid and Semi-Arid Inland River Basin, China

Hongguang Chen [1,2,3], Fanhao Meng [1,2,3,*], Chula Sa [1,2,3], Min Luo [1,2,3], Huiting Zhang [1,2,3], Shanhu Bao [1,2,3], Guixiang Liu [1,4] and Yuhai Bao [1,2,3]

1 College of Geographical Science, Inner Mongolia Normal University, Hohhot 010022, China
2 Key Laboratory of Remote Sensing and Geographic Information System, Inner Mongolia Autonomous Region, Hohhot 010022, China
3 Key Laboratory of Disaster and Ecological Security on the Mongolia Plateau, Inner Mongolia Autonomous Region, Hohhot 010022, China
4 Grassland Research Institute, Chinese Academy of Agricultural Sciences, Hohhot 010010, China
* Correspondence: mfh320@imnu.edu.cn

Citation: Chen, H.; Meng, F.; Sa, C.; Luo, M.; Zhang, H.; Bao, S.; Liu, G.; Bao, Y. Synergistic Change and Driving Mechanisms of Hydrological Processes and Ecosystem Quality in a Typical Arid and Semi-Arid Inland River Basin, China. *Remote Sens.* **2023**, *15*, 1785. https://doi.org/10.3390/rs15071785

Academic Editor: Konstantinos X. Soulis

Received: 7 February 2023
Revised: 16 March 2023
Accepted: 23 March 2023
Published: 27 March 2023

Abstract: Global warming and human activities are complicating the spatial and temporal relationships between basin hydrologic processes and ecosystem quality (EQ), especially in arid and semi-arid regions. Knowledge of the synergy between hydrological processes and ecosystems in arid and semi-arid zones is an effective measure to achieve ecologically sustainable development. In this study, the inland river basin Ulagai River Basin (URB), a typical arid and semi-arid region in Northern China, was used as the study area; based on an improved hydrological model and remote-sensing and in situ measured data, this URB-focused study analyzed the spatial and temporal characteristics of hydrological process factors, such as precipitation, evapotranspiration (ET), surface runoff, lateral flow, groundwater recharge, and EQ and the synergistic relationships between them. It was found that, barring snowmelt, the hydrological process factors such as precipitation, ET, surface runoff, lateral flow, and groundwater recharge had a rising trend in the URB, since the 20th century. The rate of change was higher in the downstream areas when compared with what it was in the upstream and midstream areas. The multi-year average of EQ in the basin is 53.66, which is at a medium level and has an overall improving trend, accounting for 95.14% of the total area, mainly in the upstream, downstream southern, and downstream northern areas of the basin. The change in relationship between the hydrological process factors and EQ was found to have a highly synergistic effect. Temporally, EQ was consistent with the interannual trends of precipitation, surface runoff, lateral flow, and groundwater recharge. The correlation between the hydrological process factors and EQ was found to be higher than 0.7 during the study period. Spatially, the hydrological process factors had a synergistic relationship with EQ from strong to weak upstream, midstream, and downstream, respectively. In addition, ecosystem improvements were accelerated by government initiatives such as the policy of Returning Grazing Land to Grassland Project, which has played an important role in promoting soil and water conservation and EQ. This study provides theoretical support for understanding the relationship between hydrological processes and ecological evolution in arid and semi-arid regions, and it also provides new ideas for related research.

Keywords: hydrological processes; ecosystem quality; synergistic effect; arid and semi-arid regions; Ulagai River Basin

1. Introduction

Global climate change and human-activity-induced changes in water resources have had a tremendous consequence on the ecological environment and socioeconomics, thus attracting attention from the international community [1,2]. This is particularly obvious in arid and semi-arid regions with fragile regional ecosystems [3–5]. The evolution and relationship between hydrological processes and ecosystems in arid and semi-arid regions in a continuously changing environment is not clearly understood. The present research on hydrology–ecology mainly focuses on the relationship between a certain hydrological process factor and an ecological indicator, without focusing on the relationship between the linkage between the whole hydrological process factor and the ecosystem quality. Therefore, it is particularly important to understand the relationship between the whole hydrological processes and ecosystem quality (EQ) in arid and semi-arid regions and to design related policies. The results of the study can deepen the understanding of the relationship between hydrological processes and ecosystems in arid and semi-arid regions and provide a reference for the sustainable use of water resources and environmental protection in the region.

As arid and semi-arid regions are deeply inland and far from the ocean, not only is data about them deficient but the measurement data from meteorological and hydrological stations in regions such as the Ulagai River Basin and Xilin River Basin are lacking, [6]. Therefore, hydrological models and remote-sensing monitoring have become important monitoring tools. The Soil and Water Assent Tool (SWAT) model is a semi-distributed hydrological model developed by the United States Department of Agriculture and the Agricultural Research Service. The SWAT model can simulate the hydrological cycle of a basin and quantify the response of basin hydrological processes to changing environments [7]. This model is popular given its systematic and precise simulation and computational capabilities. Forecast simulations have been performed in several basins around the world, and they have achieved excellent results [6,8]. For example, Idrees et al. [9] used a modified SWAT model to simulate the extent of changes in hydrological process factors for different land-use types. The results of their study showed that the conversion of barren land to agricultural land had resulted in a decrease in surface runoff and water production, whereas the groundwater flow, lateral flow, and evapotranspiration (ET) had increased. Luan et al. [10] used the SWAT model to evaluate the impact of land use on hydrological processes (ET and streamflow) in the river suite irrigation area. They also evaluated dispersion and river flow, using the SWAT model.

Being a major data source for the study of several ecological and environmental issues, such as grassland degradation and soil sanding, remote-sensing data facilitate the quick identification of spatial and temporal changes as they relate to environmental quality [9]. Xiao et al. [11] used remote-sensing data concerning biomass and vegetation cover to study if and how the status of EQ in Inner Mongolia changed from 2000 to 2010 and explore whether and how it was affected by climate change and human activities. Wei et al. [12] explored the spatial and temporal characteristics of environmental evolution in inland river basins in the arid regions of Northwest China with the help of integrated remote-sensing-related indicators. However, previous studies have mostly focused on specific years and mostly taken into account aspects such as ecosystem service function, stability evaluation, ecosystem health evaluation, and ecosystem pattern. Only a few of these studies have carried out an integrated evaluation of EQ changes in arid and semi-arid regions over a long period. Recently, national and international researchers conducted several studies on the relationship between the hydrological cycle and EQ [13,14]. Zhang et al. [15] conducted a quantitative study of the relationships among soil, groundwater depth, and vegetation in terms of how these relationships have implications for EQ changes and found that the community cover, community height, leaf projection cover, and leaf area index (LAI) had all decreased significantly with increasing groundwater depth. Hao et al. [16] analyzed the different ecosystems of Inner Mongolia to which different grazing ban policies applied and found that the trend of positive influences of precipitation on these ecosystems had begun to weaken because of overgrazing. Liu et al. [17] analyzed the effect of different vegetation-change scenarios on ET in the Mongolian Plateau and discovered that ET increased with an improvement in vegetation. It was also found that changes in terrestrial ecosystem quality are strongly related to the relevant hydrological process factors such as precipitation, ET, and groundwater [18–20]. However, most of the existing studies have focused on the relationship between singular hydrological process factors and ecosystems. The research on the synergistic evolution of overall hydrological processes and ecosystems in arid and semi-arid regions is scarce.

Located in XilinGol League, Inner Mongolia, China, the Ulagai River Basin (URB) is a typical inland river basin in an arid and semi-arid region. As the second largest inland river basin in China, it is also an important livestock base and green ecological barrier [20]. The URB is subject to natural conditions and has a fragile ecological system, which is particularly sensitive to changing environmental conditions [21]. Due to climate change causing an increase in precipitation, temperature, and ET and the continuing influence of reclamation, irrigation, and grazing, a gradual increase has been observed in water shortage, river disruption, sanding of grasslands, and salinization in the URB. This poses a great threat to extant water resources and ecological balance in arid and semi-arid regions [21,22]. Moreover, some studies have shown that arid and semi-arid regions, such as the Mongolian Plateau, may experience more pronounced rises in temperature and more frequent droughts, leading to further water scarcity and deteriorating EQ [23,24]. Against the backdrop of global warming and the impact of human activities, the synergistic evolution of hydrological processes and EQ in the URB needed to be analyzed, for such an analysis could provide a basis for the conservation of water resources and sustainable ecological development of inland river basins in arid and semi-arid regions.

Although some progress has been made in the study of a certain hydrological process factor and ecosystem, relatively few studies have been conducted on the synergistic relationship between the whole hydrological cycle process and EQ. Specifically, this study aimed to achieve the following: characterize the evolution of the hydrological process factors in the inland river basins of arid and semi-arid regions from 2001 to 2019; comprehensively evaluate EQ of inland river basins in arid and semi-arid zones from 2001 to 2019; and explore the characteristics and differences in the synergistic evolution of the hydrological process factors and EQ in different river sections in the inland river basins of arid and semi-arid zones. The findings of this study can deepen the awareness of eco-hydrological processes and evolution patterns in semi-arid inland river basins, thereby providing an empirical basis for the sustainable use of water resources and ecological protection in semi-arid inland river basins.

2. Materials and Methods

2.1. Study Area

The URB is located at the junction of three leagues: XilinGol League, Xing'an League, and Tongliao City in Inner Mongolia in China. Its geographical location falls between 116°20′ and 119°59′E and 44°02′ and 46°42′N. The total basin area is 38,549.25 km^2, making it the largest inland river basin in Inner Mongolia and the second largest in China [25]. The annual average temperature here is about −0.9 °C, and the annual average precipitation is 250–400 mm, with the precipitation showing a decreasing trend from east to west. The URB terrain is at a higher elevation on its southern side than it is on its northern side, sloping from southeast to northwest, whereas the basin's average elevation is 990 m. The basin is composed of the Ulagai River, Gori Khan River, and Sayarchen Gorak River, and the multi-year average runoff from 2001 to 2019 was 1.28 m^3/s. The URB's vegetation type is mainly grassland, with meadow grassland in the northeast, typical grassland in the southwest, and a transition zone between the two parts [25–27]. To analyze the hydrological situation and EQ of the URB's different river sections, the river was divided into the following three sections: upstream (above the Ulagai reservoir), midstream (Ulagai reservoir to Hushao Temple), and downstream (Hushao Temple to Solinnur, Figure 1b). The 35 subbasins (sub) divided by the SWAT model are defined as upstream, midstream, and downstream by location (Figure 1d).

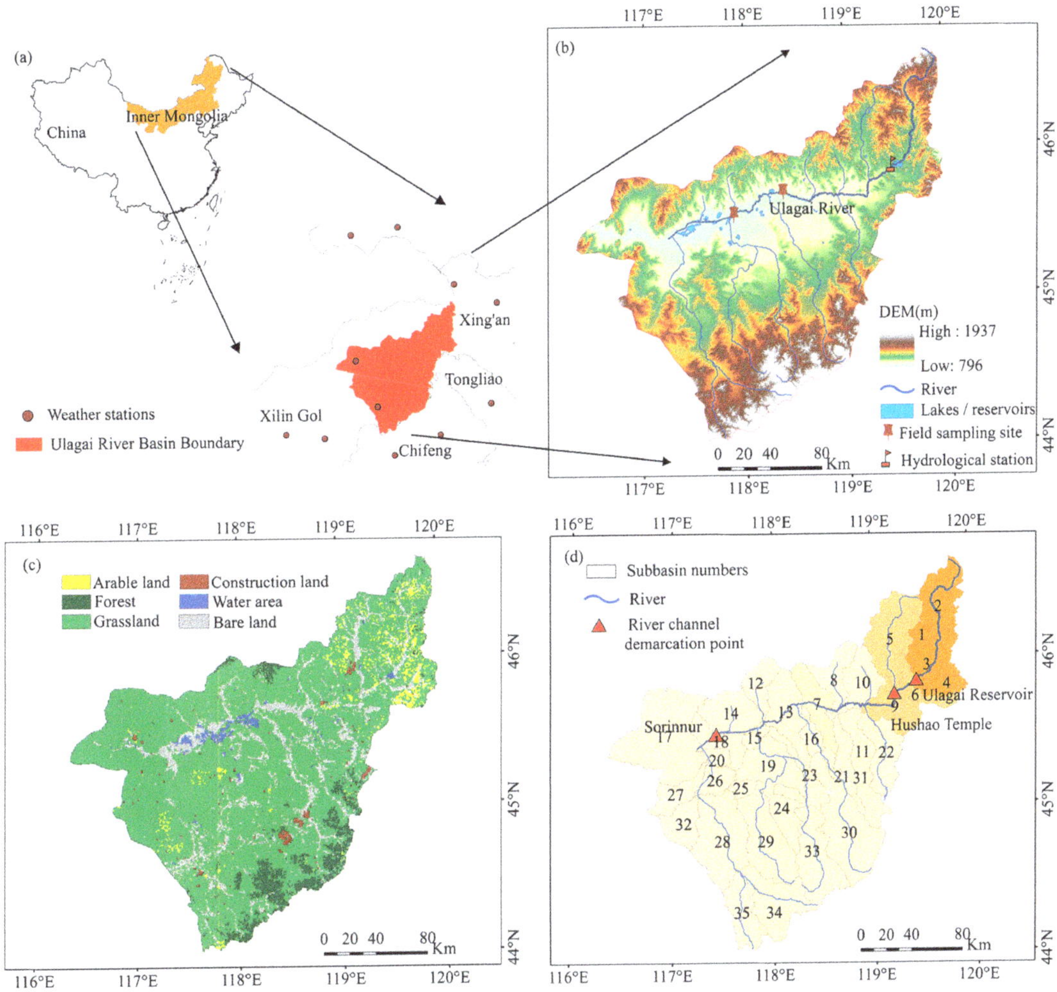

Figure 1. Overview map of (**a**) geographic location, (**b**) elevation, (**c**) land use/cover, and (**d**) subbasin delineation of the URB.

2.2. Data Sources

In this study, the Digital Elevation Model (DEM), land use/cover, soil, meteorology, and in situ measured runoff data were used to construct the SWAT hydrological model for the URB (Table 1). The DEM data were obtained from NASA with a spatial resolution of 30 m. Furthermore, to effectively characterize the surface cover changes—for instance, the degradation of grassland caused by overgrazing—the land-use/cover data were obtained via a secondary classification system for 2000, 2010, and 2020, with a spatial resolution of 1 km. The data were downloaded from the Environmental Resources and Data Center of the Chinese Academy of Sciences. Soil data were obtained from the Harmonized World Soil Database, with a spatial resolution of 1 km. Weather data were provided by the China Meteorological Data Network, with regard to daily precipitation, maximum/minimum temperatures, wind speeds, and relative humidity from 1981 to 2020 for the URB and 12 adjoining meteorological stations. The SWAT model employed multi-objective calibration, using measured runoff, field sampling, and Moderate Resolution Imaging Spectroradiometer (MODIS) snow cover and ET data. The measured runoff data were monthly runoff data, which were obtained by collating data between 1981 and 2000 from the Nunaimiao hydrological station and between 2004 and 2012 from the Ulagai reservoir inlet hydrological station. Field sampling data were converted into runoff based on the field measurements of water level and cross-section. The remote-sensing snow cover and ET data were selected from MODIS MOD10A1 and MOD16A2 products from 2001 to 2019, respectively, both having a spatial and temporal resolution of 8 days/500 m.

Table 1. Introduction to basic data sources.

Usage	Data Name	Data Type	Resolution	Source
For building SWAT model	ASTER DEM	Raster	30 m	NASA (http://www.nasa.gov (accessed on 15 February 2020))
	Soil types	Raster	1 km	HWSD (http://westdc.westgis.ac.cn/data/ (accessed on 15 February 2020))
	Land-use/cover	Raster	1 km	Environmental Resources and Data Center of Chinese Academy of Sciences (http://www.resdc.cn (accessed on 15 February 2020))
	Meteorology	Station	Daily scale	China Weather Data Network (http://data.cma.cn (accessed on 15 May 2021))
	Measured runoff	Station	Daily scale	Hydrological Yearbook of Inner Mongolia Autonomous Region
For evaluating ecosystem quality	Snow cover	Raster	500 m/8 days	MODIS MOD10A1
	ET	Raster	500 m/8 days	MODIS MOD16A2
	GPP	Raster	500 m/8 days	MODIS MOD17A2
	NDVI	Raster	1 km/30 days	MODIS MOD13A3
	LAI	Raster	500 m/8 days	MODIS MOD15A2
	LCT	Raster	500 m/year	MODIS MCD12Q1 (https://lpdaacsvc.cr.usgs.gov/appeears/ (accessed on 15 May 2021))
	Population Density	Raster	1 km	Demographic Data Network (https://www.worldpop.org/ (accessed on 25 May 2021))

Additionally, remote-sensing data, basic geographic information, and socioeconomic data were selected to evaluate the spatial and temporal variation characteristics of EQ in the URB from 2001 to 2019 (Table 1). Among these, remote-sensing data mainly included gross primary productivity (GPP), normalized difference vegetation index (NDVI), LAI, and land-cover type (LCT) data from 2001 to 2019. The MODIS MOD17A2, MOD13A3, MOD15A2, and MCD12Q1 products were selected and downloaded free of charge from the NASA Land Processes Distributed Data Archive Center, respectively. The spatiotemporal resolutions of MOD17A2, MOD13A3, and MOD15A2 products were 8 days/500 m, 30 days/1 km, and 8 days/500 m, respectively. The basic geographic-information data were vector boundary layers of the URB, and the socioeconomic data focused on population density, based on the population data network, with a spatial resolution of 1 km.

2.3. *Methods*

2.3.1. SWAT Model

The SWAT model is a long-time basin-distributed hydrological model [18]. It has been extensively used for the simulation and prediction of hydrological processes at the basin scale, with excellent simulation results [28]. In this study, the improved SWAT model by Meng [29] and Luo [30], including an improved snowmelt module and subbasin partitioning, was used to enhance the model simulation accuracy by adding cumulative temperature determination conditions to separate rainfall and snowfall types, while also adding land-use/cover-change nodes to the basin partitioning. The SWAT model automatically divides the URB into 35 subbasins and 76 hydrological response units. The model warm-up period for this study area was 1976–1980, the calibration period was 1981–2000, and the validation period was 2001–2012.

The SWAT model can be used to simulate hydrological processes at three times scales, such as daily, monthly, and annual, and the monthly scale is used in this study. The SWAT model determines the basin boundary and divides subbasins by analyzing the slope, slope direction, and elevation of the DEM (Digital Elevation Model) of the study area. On this basis, the hydrological processes under different land-use types and different soil types under climate change are simulated based on the input hydrometeorological data, land-use types, and soil data [10]. To improve the model accuracy, the SUFI-2 algorithm of SWAT-CUP software was used in this study for analyzing model parameter sensitivity and uncertainty [10]. The Nash–Sutcliffe efficiency (*NSE*), percentage bias (*PBIAS*), and coefficient of determination (R^2) were used to evaluate the model simulation accuracy. The evaluation index equations are as follows [10]:

$$NSE = 1 - \frac{\Sigma\left(Q_i^{obs} - Q_i^{sim}\right)^2}{\Sigma\left(Q_i^{obs} - Q^{mean}\right)^2} \tag{1}$$

$$PBIAS = \frac{\Sigma\left(Q_i^{obs} - Q_i^{sim}\right)}{\Sigma Q_i^{obs}} * 100 \tag{2}$$

$$R^2 = \frac{n\left(\Sigma Q_i^{obs} Q_i^{sim} - \Sigma Q_i^{obs} \Sigma Q_i^{sim}\right)^2}{\left[n\Sigma\left(Q_i^{obs}\right)^2 - \left(\Sigma Q_i^{obs}\right)^2\right]\left[n\Sigma\left(Q_i^{sim}\right)^2 - \left(\Sigma Q_i^{sim}\right)^2\right]} \tag{3}$$

where Q_i^{obs} is the measured value in m^3/s, Q_i^{sim} is the simulated value in m^3/s, Q^{mean} is the measured average value in m^3/s, and n is the measured data amount.

The simulation results were again validated with MODIS snow cover and ET data to meet multiple objectives. These results were also evaluated using *PBIAS* and R^2.

2.3.2. Ecosystem Quality Assessment

In this study, ecosystem quality (*EQ*) was comprehensively assessed through three components: ecosystem productivity index (*EPI*), ecosystem stability index (*ESI*), and ecosystem bearing capacity index (*EBCI*). The *EPI*, *ESI*, and *EBCI* were constructed based on multisource data, such as remote-sensing data and socioeconomic data from 2001 to 2019, and were later assigned weights of 0.40, 0.28, and 0.32, respectively, using the entropy value weighting method [31]. A comprehensive *EQ* evaluation model for the URB was established by the weighted summation method [32]. To better reflect the spatial and temporal variation characteristics of *EQ* caused by the hydrological process factors and external factors, *EQ* was normalized, and the range of values was delineated at (10,100). Its computation is in Equations (4)–(7), which are as follows:

$$EPI_{t,k} = \begin{cases} 10 & G_{t,k} \le G_{min} \\ 10 + \left(G_{t,k} - G_{min}\right) \times a & G_{min} < G_{t,k} < G_{max} \\ 100 & G_{t,k} \ge G_{max} \end{cases}$$
$$a = (100 - 10)/\left(G_{max} - G_{min}\right) \tag{4}$$

where $EPI_{t,k}$ is the dimensionless index of productive capacity in year *t* of image *k*, and its larger value indicates a higher level of ecosystem productive capacity; $G_{t,k}$ is the total GPP in year t of image k; G_{max} and G_{min} are the upper and lower limits of the GPP multi-year average, respectively; and a is the stretching constant.

$$ESI_{t,k} = \begin{cases} 10 & cv_{t,k} \geq cv_{max} \\ 10 + (cv_{max} - cv_{t,k}) \times a & cv_{min} < cv_{t,k} < cv_{max} \\ 100 & cv_{t,k} \leq cv_{min} \end{cases}$$
$$a = (100 - 10)/(cv_{max} - cv_{min}) \tag{5}$$

where $ESI_{t,k}$ is the stability index of the image element k in year t—the larger the value, the higher is the stability of the region; $cv_{t,k}$ denotes the coefficient of variation of the annual mean value of GPP in year t of the image element k; cv_{max} and cv_{min} are the upper and lower limits of the multi-year mean coefficient of variation of GPP, respectively; and a is the stretching constant.

$$EBCI_{t,k} = \begin{cases} 10 & EHI_{t,k} \leq EHI_{min} \\ 10 + (EHI_{t,k} - EHI_{min}) \times a & EHI_{min} < EHI_{t,k} < EHI_{max} \\ 100 & EHI_{t,k} \geq EHI_{max} \end{cases}$$
$$a = (100 - 10)/(EHI_{max} - EHI_{min}) \tag{6}$$

where $EBCI_{t,\,k}$ is the stability index of the image element k in year t—the larger the value, the higher is the stability of the region; $EHI_{t,k}$ denotes the coefficient of variation of the annual mean value of GPP in year t of the image element k; EHI_{max} and EHI_{min} are the upper and lower limits of the multi-year mean coefficient of variation of GPP, respectively; and a is the stretching constant.

$$EQ = \sum_{i=1}^{n} w_i \times y_i \tag{7}$$

where EQ is the EQ index; y_i represents EPI, ESI, and $EBCI$ indices; and w_i is the weight of each index.

The Natural Break clustering method was then used to classify EQ into the following five levels: (I) 0–45 as the lowest level, (II) 45–50 as the lower level, (III) 50–55 as the medium level, (IV) 55–65 as the higher level, and (V) 65–100 as the highest level, taking into account the URB's actual situation and the degree of influence of each index on the ecosystem.

2.3.3. Trend Analysis

This study employed the unidimensional linear regression method to analyze the spatial and temporal trends of the URB's hydrological processes and EQ from 2001 to 2019. The calculation equation [33] is shown below:

$$Slope = \frac{n\sum_{i=1}^{n}(iEQ_i) - \sum_{i=1}^{n} i \sum_{i=1}^{n} EQ_i}{n\sum_{i=1}^{n} i^2 - \left(\sum_{i=1}^{n} i\right)^2} \tag{8}$$

where n is the length of study, EQ_i is the mean value of EQ in year i, and *Slope* indicates the trend of change. If *Slope* > 0, the EQ is increasing; otherwise, it is decreasing. If *Slope* = 0, the EQ remains unchanged.

2.3.4. Analysis of Synergistic Effects

(a) Gray correlation analysis

The gray correlation analysis determines whether the relationship between sequences is close or not, mainly through the similarity of their curve geometries [34]. If the curves are more similar, the correlation between the series is greater, and if it is the other way around, the correlation is lower. This method is frequently used in the analysis of influence factors because it is less demanding and less computationally intensive than the mathematical and statistical methods. This method was used in this study to calculate the correlation between the hydrological process factors and EQ. The calculation formula is as follows:

$$\gamma_{oi} = \frac{1}{n}\sum_{k=1}^{n} \xi_{oi}(k) \quad (9)$$

$$\xi_{oi}(k) = \frac{\min_i \min_k \Delta_{oi}(k) + \rho \max_i \max_k \Delta_{oi}(k)}{\Delta_{oi}(k) + \rho \max_i \max_k \Delta_{oi}(k)}$$
$$\Delta_{oi} = |x_o'(k) - x_i'(k)|,\ i = 1, 2, \ldots, m;\ k = 1, 2, \ldots, n \quad (10)$$

where γ_{oi} is the gray correlation degree, ξ_{oi} is the gray correlation coefficient, and ρ is the discrimination coefficient.

(b) Pearson correlation analysis

The Pearson correlation analysis between the URB's *EQ* and hydrological process factors from 2001 to 2019 was conducted at the subbasin scale. Moreover, the correlation coefficients were tested for significance to reflect the degree of spatial and temporal correlation between *EQ* changes and hydrological process factors and identify the response of *EQ* to changes in the hydrological process factors. The calculation equation [35] is as follows:

$$r = \frac{\sum_{i=1}^{n}(EQ_i - \overline{EQ})(y_i - \overline{y})}{\sqrt{\sum_{i=1}^{n}(EQ_i - \overline{EQ})^2}\sqrt{\sum_{i=1}^{n}(y_i - \overline{y})^2}} \quad (11)$$

where r is the correlation coefficient between *EQ* and the hydrological process factors; EQ_i and y_i are the mean values of *EQ* and the hydrological process factors in year i, respectively; $\overline{EQ}$ and $\overline{y}$ are the mean values of *EQ* and the hydrological process factors in 19 years, respectively; and i represents the number of years. If $r > 0$, the *EQ* and the hydrological process factors are positively correlated and vice versa; the larger the r, the stronger the correlation between them.

3. Results

3.1. SWAT Model's Performance in the URB

As shown by analyzing the SWAT model simulation results, the evolutionary trends of runoff simulated values and measured values were generally consistent (Figure 2a). The model captured the seasonal variation characteristics of runoff, which are higher in summer and absent in winter, in the basin. The runoff from 1981 to 2000 was much higher than that from 2001 to 2012. The highest value, up to 134 m^3/s, was generated in 1998. The *NSE*, R^2, and percentage bias (*PBIAS*) for the calibration periods were 0.62, 0.62, and 18.8%, and for the validation periods, they were 0.71, 0.72, 8.5%, respectively. In this study, the validation period captured the peak runoff better compared to the calibration period. This may be due to the fact that the runoff variation in the study area during the calibration period is more drastic and the runoff process is more complex than that in the validation period, so the simulation results in the validation period are better than those in the calibration period. Moreover, the model-extracted snow area and ET matched the curve trend of MODIS snow and ET, and the correlation coefficients of both were higher than 0.8 (Figure 2b,c).

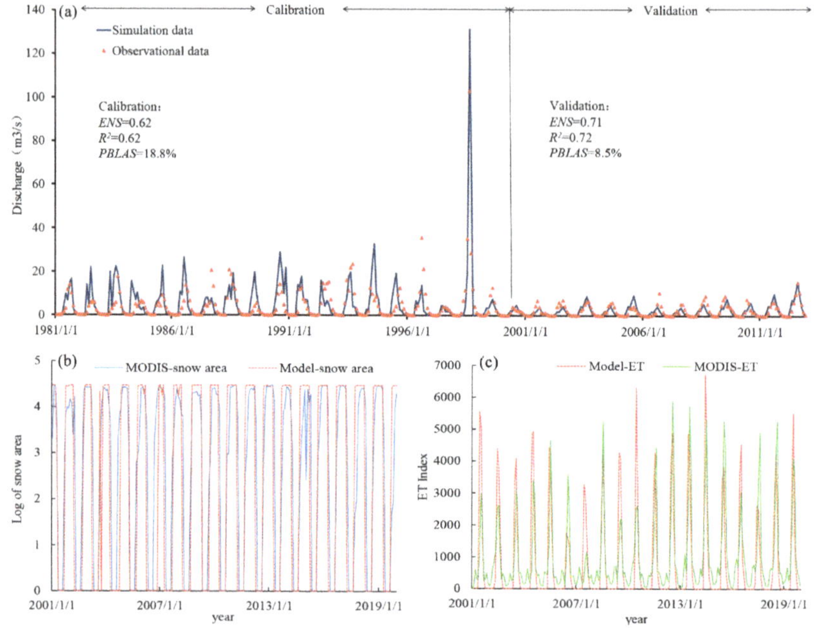

Figure 2. Comparison of SWAT model simulation results and measured discharged (**a**), MODIS snow cover (**b**), and evapotranspiration (**c**) in the URB.

3.2. Spatiotemporal Change Characteristics in the URB's Hydrological Process Factors

Upon analyzing the changes in the URB's hydrological process factors during 2001–2019, it was found that precipitation, ET, surface runoff, and lateral flow showed a non-significant increasing trend at the rates of 1.24 mm/yr, 1.66 mm/yr, 0.47 mm/yr, and 0.056 mm/yr, respectively (Figure 3a–d). Groundwater recharge showed a significant increasing trend at a rate of 0.18 mm/yr ($p < 0.1$), while snowmelt showed a non-significant decreasing trend at a rate of 0.01 mm/yr (Figure 3e). Precipitation, surface runoff, lateral flow, and groundwater recharge all peaked in 2012, while evaporation and snowmelt peaked in 2013.

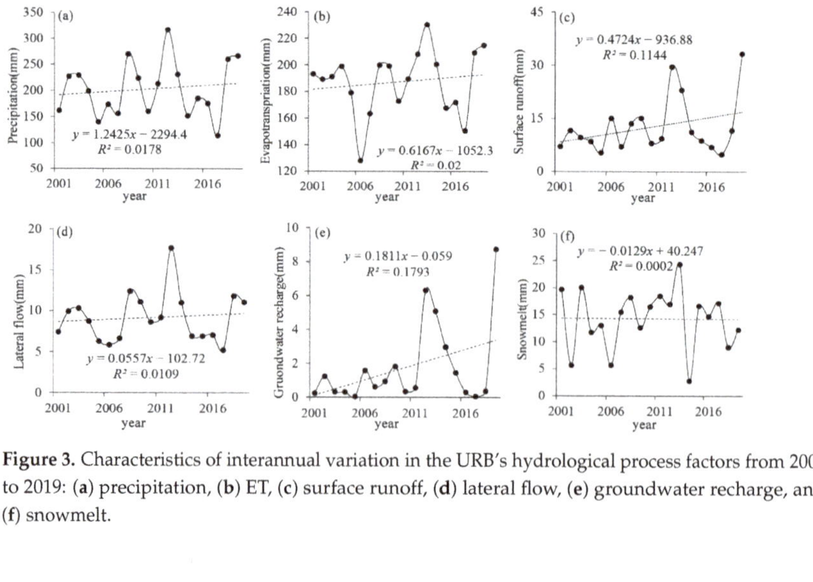

Figure 3. Characteristics of interannual variation in the URB's hydrological process factors from 2001 to 2019: (**a**) precipitation, (**b**) ET, (**c**) surface runoff, (**d**) lateral flow, (**e**) groundwater recharge, and (**f**) snowmelt.

Great spatial heterogeneity was observed in the spatial distribution and variation rates of the URB's different hydrological process factors (Figure 4). The annual total precipitation in the upstream region of the basin was 457.48 mm, and it gradually decreased from east to west; the overall precipitation variation rate was high, with the fastest variation reaching up to 2.033 mm/yr in the upstream, followed by that in the downstream (1.601 mm/yr; see Figure 4a). The ET in the southern part of URB reached 272.79 mm, while the ET in the northwestern part was lower at 261.74 mm, showing a decreasing trend from south to north. The rate of change was faster in the URB's upstream and southern parts and slower in its northern and downstream parts (Figure 4b). The runoff deep high value of the URB's runoff was mainly concentrated in the wetland area, decreasing from the wetland to the surrounding area. The rate of change varied similarly, with the downstream wetland area increasing at a rate of 8.23 mm/yr (Figure 4c). The lateral flow in the high-elevation area was higher than that in the low-elevation area; the lateral flow gradually increased from the URB's wetland to the surrounding area, and the change rate showed the same spatial distribution characteristics as the lateral flow did (Figure 4d). Groundwater recharge reached 3.12 mm in the URB's upper and western parts and was lower in the URB's southern and northern parts (1.5 mm); the rate of change was also faster in the higher groundwater recharge areas (Figure 4e). Snowmelt was observed more in the URB's upper and middle reaches, with the highest value being up to 16 mm. Snowmelt in the middle and south of the lower reaches was less than 6.27 mm. This shows a distribution pattern of gradual decrease from the URB's northeast to southwest, with the rate of decrease of snowmelt being higher in the middle (Figure 4f). Overall, the precipitation and snowmelt in the URB's upper and middle reaches were high and the runoff depth was small; for the downstream region, it was just the opposite, and the rate of change in the downstream was higher than that in the upstream.

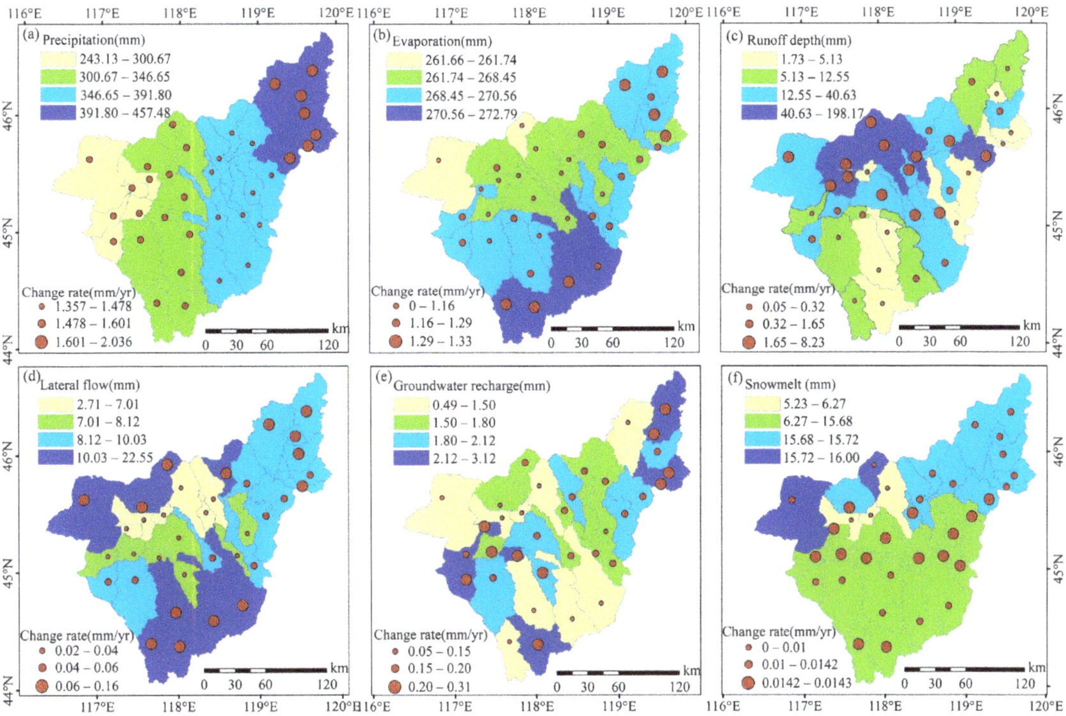

Figure 4. Spatial distribution and change rate of the URB's hydrological process factors, 2001–2019: (**a**) precipitation, (**b**) ET, (**c**) runoff depth, (**d**) lateral flow, (**e**) groundwater recharge, and (**f**) snowmelt.

3.3. Spatiotemporal Change Characteristics of the URB's EQ

The URB's EPI showed a significant increasing trend, at a rate of 0.43/yr from 2001 to 2019 ($p < 0.01$). EPI's highest and lowest values were 46.60 and 31.33, which appeared in 2012 and 2007, respectively (Figure 5a). The multi-year mean value was 41.01, indicating that the URB's overall EPI was low. The higher EPI areas were mainly distributed across the arable land dominant upstream and the downstream southern woodland coverage areas. Most areas in the middle and downstream areas had a lower EPI. The area of Level I accounted for 74.7% of the total area; Level II accounted for 11.51%; and Levels III, IV, and V accounted for 7.42%, 5.31%, and 1.08%, respectively (Figure 5b).

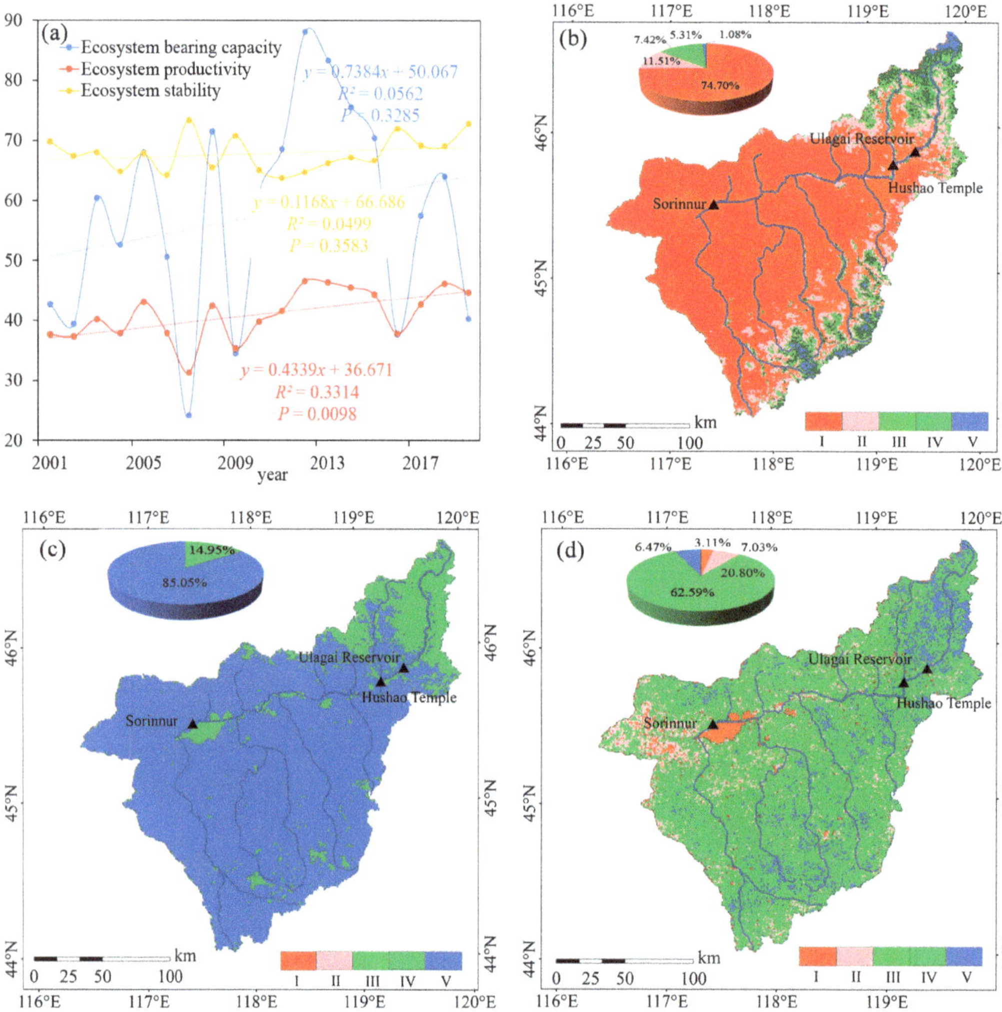

Figure 5. Interannual variation of the URB's EPI, ESI, and EBCI from 2001 to 2019 (**a**); the spatial distribution of EPI level (**b**); the spatial distribution of ESI level (**c**); and the spatial distribution of EBCI level (**d**).

The URB's ESI showed a non-significant increasing trend at a rate of 0.12/yr during the study period. The ESI was the highest in 2007, with a value of 73.43, and lowest in 2011, with a value of 63.79 (Figure 5a). The multi-year average ESI value reached 67.85, which is at a high level overall. As seen in Figure 5c, the URB's ESI from 2001 to 2019 had only two levels, Levels IV and V, with area ratios of 14.95% and 85.05%, respectively. Level V was mainly distributed in the downstream area, where grassland cover chiefly prevailed, whereas Level IV was mainly distributed in the upstream area, where arable land chiefly prevailed.

The URB's EBCI showed a non-significant increasing trend at a rate of 0.74/yr. Over the last 19 years, the highest value of EBCI was 88.09, which occurred in 2012, and the lowest value was 24.24 in 2007 (Figure 5a). The multi-year average value was 57.45, which is at a higher level. The higher EBCI areas were mainly spread over the upstream and downstream southern areas. On the other hand, the EBCI value in the downstream northwestern area was lower, which is similar to the spatial distribution characteristics of the EPI (Figure 5d).

As shown in Figure 6a, the EQ of the URB from 2001 to 2019 showed an overall distribution pattern of high in the northeast and southwest and low in the northwest. The best EQ levels were concentrated in the upstream and downstream southern areas of the URB, whereas the worst EQ areas lay in and around the wetlands of the downstream area. The highest area percentage at the medium level (III) was 51.97%, mainly in the central part of the URB downstream dominated by grass cover, which was followed by the higher level (IV) area percentage of 32.54, mainly in the upstream and midstream areas of the basin and the southern downstream parts.

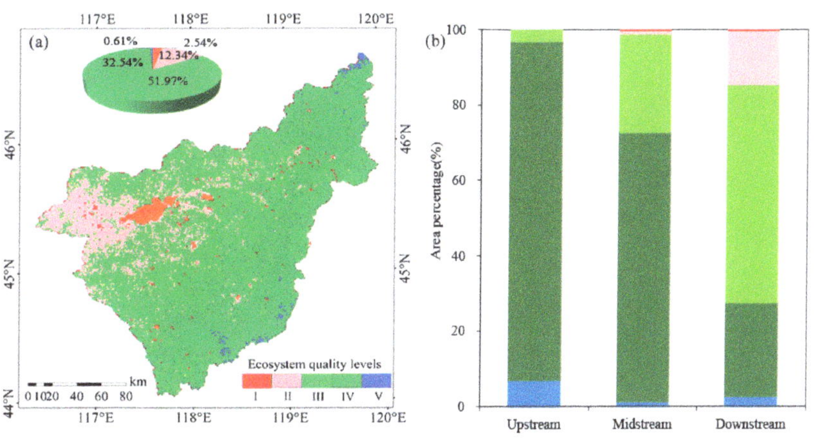

Figure 6. Ecosystem quality levels (**a**) and its area proportion in different river segments (**b**) in the URB from 2001 to 2019.

Statistics about area percentages at different EQ levels, across different reaches of the basin, revealed that levels III and IV have been dominant over the last 19 years. However, the difference in EQ among the reaches was relatively significant (Figure 6b). The areas of levels III and IV decreased gradually from the upstream to the downstream of the URB, whereas the areas of levels I and II increased subsequently. Moreover, the level IV area in the URB upstream had the highest proportion of 89.96%, whereas the level III area in the downstream region occupied the highest proportion of 54.80%. Thus, the upstream and midstream areas are at a high level, whereas the downstream area is at an intermediate level.

The URB's EQ showed a significant increasing trend at a rate of 0.39/yr from 2001 to 2019 ($p < 0.01$; see Figure 7a). EQ was the highest in the year 2012 at 64.91, fluctuating widely before and after. It was the lowest in 2007, at 40.99. The overall EQ was seen to be at a higher level between 2012 and 2017. Its multi-year average was 53.66, indicating that the URB's overall EQ is at a medium level. The URB's overall EQ showed an increasing trend (slope > 0) over the last 19 years, accounting for 95.14% of the total study area. Within this area, 19.75% was dominated by a significant increase ($p < 0.05$), mainly in the upstream and northern parts of the downstream. On the other hand, 4.86% of the URB showed a decreasing trend ($p < 0.05$), scattered across the southern and western parts of the downstream. On account of various river channels, EQ in the URB upstream increased significantly, whereas downstream EQ increased non-significantly, but a few downstream areas still showed decreasing trends (Figure 7b).

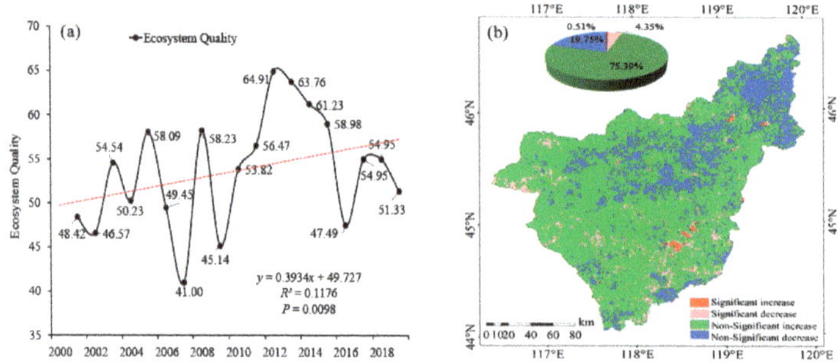

Figure 7. Interannual trends (**a**) and spatial distribution (**b**) of ecological quality in the URB from 2001 to 2019.

3.4. Synergy Relationships between the URB's Hydrological Process Factors and EQ

First, we analyzed the synergy between the hydrological process factors and EQ, using the gray correlation method. It was found that a high synergy existed between the URB's hydrological process factors and EQ during the study period, with correlations ranging between 0.7 and 0.95, along with large differences between different river sections (Figure 8a). The correlations of ET, precipitation, and lateral flow with EQ in the last 19 years were 0.95, 0.91, and 0.90, respectively. Taking 2012, the year with the largest EQ fluctuations, as a reference point, the synergy between EQ and the hydrological process factors was found to be high until 2012, with a mean value of 0.89, but decreased later, with a mean value of 0.83. The synergy between EQ and precipitation was found to be the highest in 2007 and the lowest in 2017 for ET. Moreover, surface runoff, lateral flow, and groundwater recharge mainly occurred during the early part of the study period; snowmelt was mainly concentrated toward the latter part (Figure 8b). The synergy between the hydrological process factors and EQ at the subbasin scale is in Figure 8c, with synergy coefficients higher than 0.8. The areas with lower synergy were mainly located west and south of the downstream regions. The synergy between the hydrological process factors and EQ decreased from the upstream to the downstream regions, according to varied river sections (Figure 8d).

Second, we analyzed the synergistic effects of the hydrological process factors and EQ through Pearson correlation analysis. Between 2001 and 2019, except for individual subbasins where ET was negatively correlated with EQ (e.g., Subbasins 11 and 13), all ET-EQ correlations were positive. Moreover, a large spatial heterogeneity was observed in the synergy (Figure 9). The synergistic relationships between precipitation, evapotranspiration, surface runoff, lateral flow, and groundwater recharge and EQ are stronger in the northwestern part of the downstream of the basin and in the southern part of the upstream—both around 0.5. However, the synergistic effect of the two is weaker in the southern part of the downstream. The synergistic coefficient between lateral flow and EQ in the lower reaches of the basin reached around 0.7. The synergy between snowmelt and EQ in the northwest region of the downstream basin was 0.37, while that in the upstream was only 0.1. In general, the synergy between the hydrological process factors and EQ was found to be stronger in the downstream basin and weaker in the upstream basin during 2001–2019.

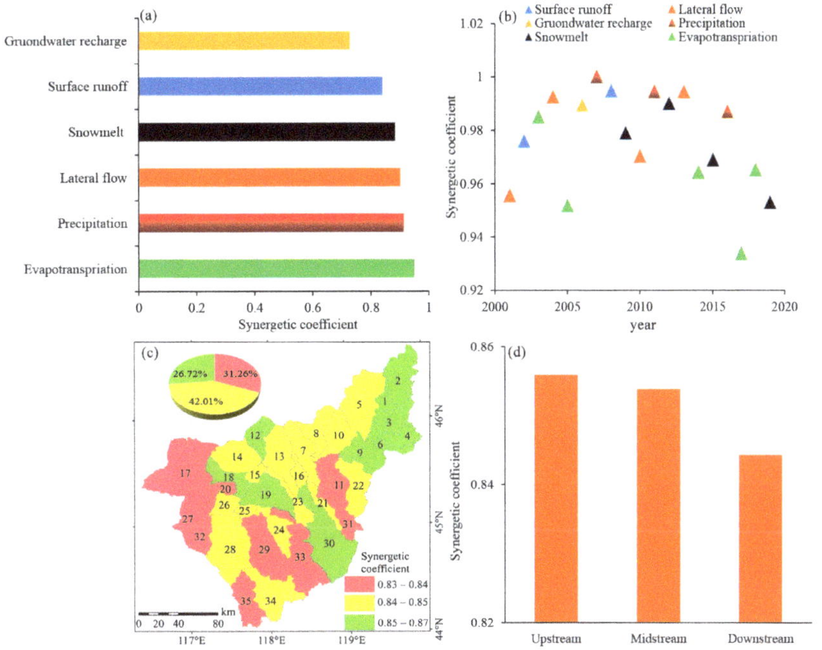

Figure 8. Synergy of hydrological process factors and EQ in URB (**a**), the most synergistic hydrological process factors (**b**), spatial distribution of average synergy between hydrological process factors and EQ in subbasins (**c**), and synergy between hydrological process factors and EQ system in different river sections (**d**).

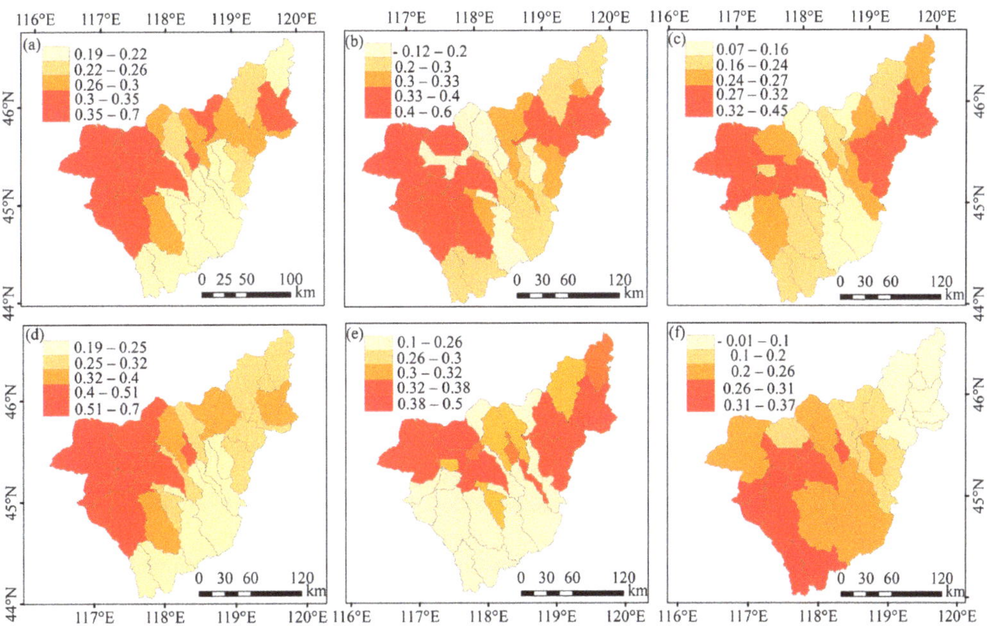

Figure 9. Synergy relationships between the URB's hydrological process factors and EQ: (**a**) precipitation, (**b**) ET, (**c**) runoff depth, (**d**) lateral flow, (**e**) groundwater recharge, and (**f**) snowmelt.

To gain deeper insights into the key hydrological process factors that account for EQ improvements in the URB, we further compared the synergistic coefficients of the hydrological process factors with EQ in each subbasin. As seen in Figure 10, EQ had the highest synergy with groundwater, lateral flow, ET, and snowmelt in the basin. The synergy between EQ and lateral flow is high in most parts of the basin, with the highest value of 0.626. It passed the significance test, accounting for 53.37% of the total basin area, mainly concentrated downstream. This was followed by groundwater, which accounts for 22.5% of the total basin area, and is mainly concentrated in the Southeastern URB. ET was most synergistic in the central part of the downstream area. The strongest synergy with snowmelt was mainly seen in the southern part of the downstream area. In general, the strongest synergistic hydrological process factors in the URB's upstream and midstream areas were relatively singular and mainly related to groundwater. However, the strongest synergistic factors in the URB downstream area were more diverse.

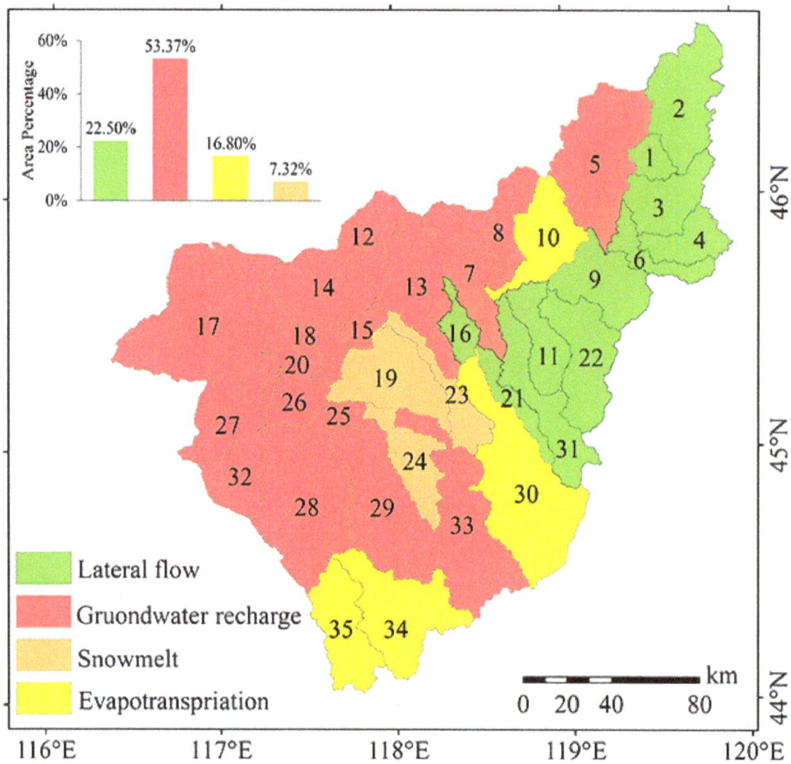

Figure 10. Spatial distribution of the URB's strongest synergistic hydrological process factors for EQ changes.

4. Discussion

4.1. Intrinsic Mechanisms of Change in the Synergistic Relationship between the URB's Hydrological Process Factors and EQ

Regional climatic changes and surface cover can effectively alter the distribution and balance of water and energy at the surface, thus affecting hydrological processes [3]. Precipitation, as the main source of hydration in arid and semi-arid regions, has a self-evident influence on hydrological processes [15]. The URB has experienced a warming and wetting trend from 2001 to 2019, with an increase in precipitation and that peaked in 2012 [36]. ET, surface runoff, lateral flow, and groundwater recharge also showed an increasing trend, and all of them peaked during the peak precipitation year or the year after.

The overall EQ trend in the URB is improving, showing an increasing trend from 2001 to 2012 and a decreasing trend from 2012 to 2019. Taking 2012, the year with large fluctuations in EQ, as the point, it was found that the synergy between EQ and hydrological process factors was high until 2012, with a mean value of 0.89, and then the synergy decreased, with a mean value of 0.83. As shown in the figure, from 2012 to 2019, the average correlations between ET, groundwater recharge, lateral flow, precipitation, and surface runoff and EQ have all decreased compared to 2001–2011. In particular, the average changes in lateral flow, precipitation, and surface runoff are large (Figure 11). This suggests that the climate and hydrology condition are not the only forces affecting EQ that were improved in this inland river basin and that there may be external drivers as well. Over the entire study period, EQ was consistent with the trends of the hydrological process factors such as precipitation, surface runoff, lateral flow, groundwater recharge, and snowmelt. In the arid and semi-arid regions, where water resources are scarce, increased precipitation results in a rise in the hydrological process factors such as surface runoff and groundwater recharge. This, in turn, provides more water for vegetation growth and can appropriately alleviate the stress caused earlier by water scarcity [35,37,38]. Additionally, the acceleration of the water circulation increases the activity of soil microorganisms to a certain extent, promotes vegetation nutrient absorption, and provides better growth conditions for vegetation [36]. In spatial terms, the URB's EQ is consistent with the spatial distribution characteristics of the precipitation changes there; thus, the precipitation in the region influences its EQ [36].

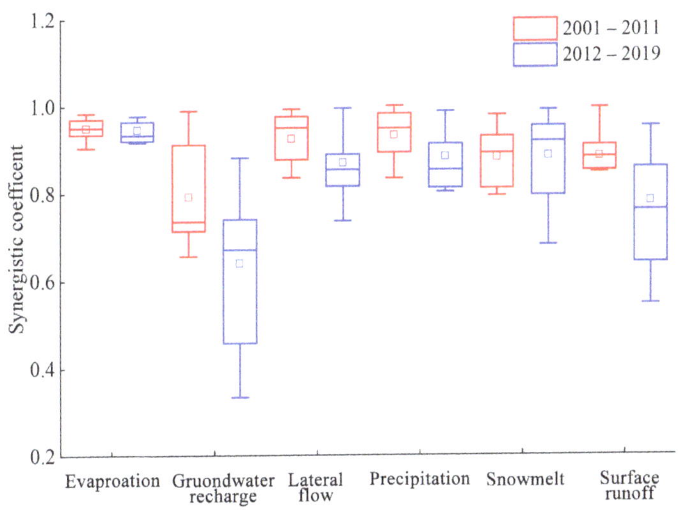

Figure 11. Interannual synergistic relationship between hydrological process factors and EQ.

The URB is covered mainly by grassland cover types, which are strongly dependent on hydrological processes [37]. The degree of synergy between the URB's hydrological process factors and EQ over the last 19 years was higher than 0.7. The synergy between the two was lower in the western and southern parts of downstream URB. This synergy coincides chiefly with EQ decline areas, indicating that EQ decline is influenced by external factors other than the hydrological process factors. The synergy between the hydrological process factors and EQ gradually decreased from upstream to downstream areas, indicating that the URB's EQ is affected more by external disturbances in the downstream areas than those in the upstream areas. The highest synergy was seen between the improvement of EQ and the groundwater, lateral flow, ET, and snowmelt. The increasing trend of ET, a water-consumptive state in the hydrological cycle, may harm vegetation [15,39]. However, the increase in precipitation, runoff, and groundwater recharge in the URB was larger than the increase in ET during the study period, thus offsetting the water loss while still increasing the amount of water available for vegetation [40,41].

4.2. Extrinsic Disturbances Affecting Changes in the Synergistic Relationship between the URB's Hydrological Processes and EQ

In the context of the URB warming and humidifying, EQ was dominated by a non-significant increase from 2001 to 2019. Even under supportive conditions for vegetation growth, EQ degradation still occurred in the southern and northwestern parts of the downstream region of the URB. Moreover, the results of this study showed that the correlation between the URB's hydrological process factors and EQ was relatively high until 2012 and decreases afterward. This suggests that climatic and hydrological conditions are not the only forces affecting improved EQ in this inland river basin and that external drivers may be behind it as well.

The rapid economic development and continuous population increase led to urban expansion and increased industrial and mineral land use, and road construction generates ecological problems such as increased water consumption and the destruction of grasslands [42]. This study collected data about population and socioeconomic indicators (primary, secondary, and tertiary industries) from the statistical annals of the URB, during 2001–2019, in an attempt to explore the link between the hydrological process factors and EQ (Figure 12) and found that the population and socioeconomic indicators in the URB showed an increasing trend. In particular, the secondary sector surged in 2010 and reached its peak in 2012. This might have been the main reason for the slight decrease in EQ and increase in the correlation with the URB's hydrological process factors after 2012. On the one hand, the construction of hydraulic reservoirs and industrial mines has contributed to the socioeconomic development of the region; on the other hand, the construction of factories and reservoirs has cut off the main streams of the river basin and occupied a large area of grassland, increasing the conflict between man and nature [15]. To achieve sustainable ecological development, the Inner Mongolia Autonomous Region has taken corresponding ecological restoration measures, such as the Beijing–Tianjin Sand Source Control Project, the Natural Forest Protection Project, and the policy of Returning Grazing Land to Grassland Project, mainly by reducing livestock and resting grazing [43,44]. As seen in Figures 12b and 13, the trends of gradual reduction of total livestock, reduction of arable land area, and subsequent increase of grassland, as well as forest land area, from 2001 to 2019 confirm the effective implementation of this series of measures. This has improved the carrying capacity of grasslands to some extent, by reducing the consumption of water resources and the burden on grasslands [45]. Ecological protection measures taken by the government have accelerated EQ improvement, as well as soil and water conservation, in the URB.

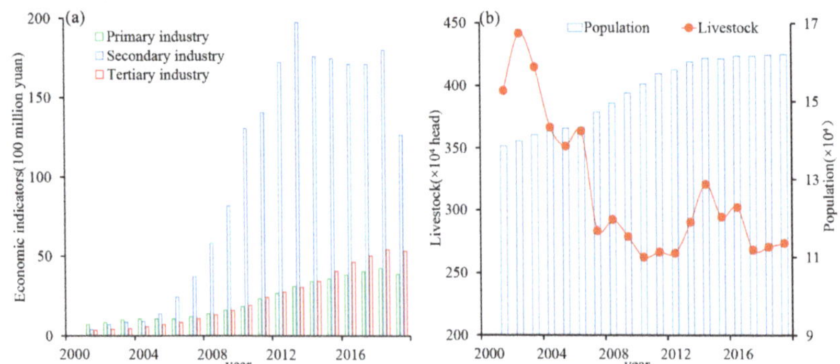

Figure 12. Economic indicators (**a**) and population (**b**) of the URB.

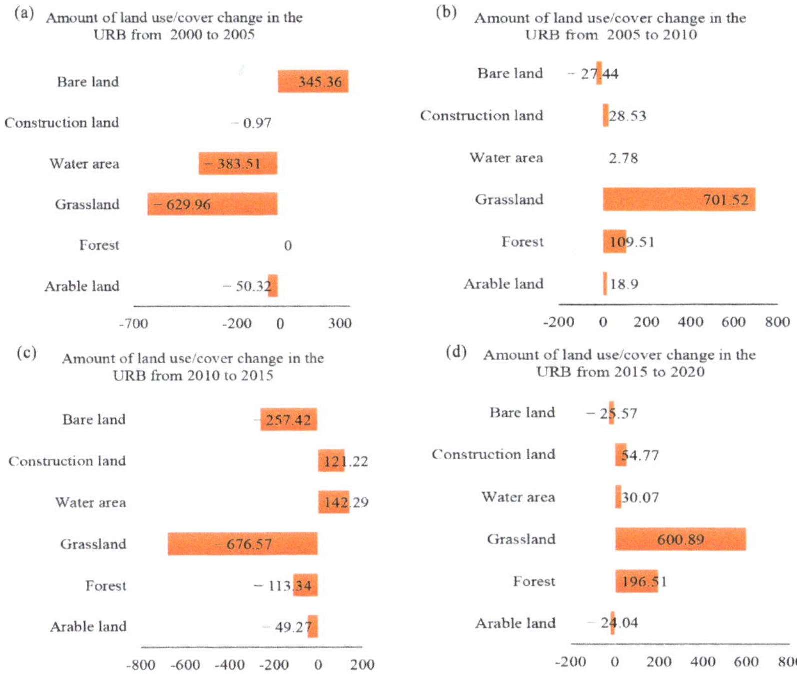

Figure 13. Land-use/cover net change in the URB in different periods: (**a**) 2000–2005, (**b**) 2005–2010, (**c**) 2010–2015, and (**d**) 2015–2020.

5. Conclusions

This study was based on the improved SWAT model and comprehensive EQ assessment model. It analyzed the spatial and temporal distribution and evolutionary characteristics of the URB's hydrological process factors and EQ and identified their synergy relationships and their internal and external change-driving mechanisms. The main conclusions of this study are as follows:

(1) Besides snowmelt, the URB's hydrological process factors, such as precipitation, ET, surface runoff, lateral flow, and groundwater recharge, are all on the rise; the change rate of the downstream hydrological process factors is higher than that of the upstream factors.

(2) During 2001–2019, the URB's EPI, ESI, and EBCI showed an increasing trend. The EPI and EBCI were higher in the upstream and southern part of the downstream, whereas the ESI of the downstream is higher than that of the upstream. The multi-year average value of the URB's EQ was 53.66, which is a medium level. The overall EQ trend improved, accounting for 95.14% of the total basin area, of which, of these, 19.75% were dominated by significant increases, mainly in the upstream and northern part of the downstream.

(3) The synergistic relationship between the hydrological process factors and EQ is strong. The degree of synergy between the URB's EQ and the hydrological process factors was higher than 0.7 during the study period. Moreover, this relationship showed obvious spatial heterogeneity, with a decreasing distribution pattern from upstream to downstream areas. The URB's EQ improved with an increase in precipitation, surface runoff, lateral flow, and groundwater recharge. Furthermore, the ecological protection measures proposed by the government have accelerated the improvement of the ecosystem, which plays an important role in promoting soil and water conservation

and EQ. Local governments should continue to strengthen the implementation of ecosystem protection strategies.

Overall, this study focused on the whole hydrological cycle process and evaluated EQ in an integrated way, with three aspects, i.e., EPI, ESI, and EBCI. On this basis, we analyzed the synergistic evolutionary relationship between hydrological processes and EQ in arid and semi-arid regions. This study contributes to theoretical support for understanding the synergistic relationship between ecological changes and hydrological processes in arid and semi-arid regions and provides new ideas for related studies. In this study, the external drivers of hydrologic processes and EQ were not discussed comprehensively. Therefore, the quantitative effects of industrial and mining exploitation, agricultural irrigation, and extreme weather on the relationship between hydrological processes and ecosystem quality should be considered in future studies.

Author Contributions: Conceptualization, methodology, and writing—original draft preparation, H.C.; conceptualization, supervision, inspection, review, and editing, F.M., C.S. and M.L.; investigation and validation, H.Z., S.B. and G.L.; investigation, Y.B. All authors have read and agreed to the published version of the manuscript.

Funding: This work was jointly supported by the National Natural Science Foundation of China (Nos. 42261079, 42101030, 41861014 and 41961058), Natural Science Foundation of Inner Mongolia Autonomous Region (Nos. 2022MS04004, 2020BS03042 and 2020BS04009), Key research, development and achievement transformation project of Inner Mongolia Autonomous Region (No. 2022YFDZ0061), Development Plan for Young Scientific and Technological Talents in Higher Education Institutions of Inner Mongolia Autonomous Region (Nos. NJYT23019 and NJYT22027), Fundamental Research Funds for the Inner Mongolia Normal University (Nos. 2022JBQN093 and 2022JBBJ014).

Data Availability Statement: Not applicable.

Acknowledgments: The authors would like to thank the editors and anonymous reviewers for their valuable comments and suggestions, which improved the quality of the manuscript.

Conflicts of Interest: The authors declare no conflict of interest.

References

1. Luo, M.; Liu, T.; Meng, F.; Duan, Y.; Bao, A.; Xing, W.; Feng, X.; De Maeyer, P.; Frankl, A. Identifying Climate Change Impacts on Water Resources in Xinjiang, China. *Sci. Total Environ.* **2019**, *676*, 613–626. [CrossRef] [PubMed]
2. Zhang, Q.; Xu, C.Y.; Zhang, Z.; Ren, G.; Chen, Y.D. Climate Change or Variability? The Case of Yellow River as Indicated by Extreme Maximum and Minimum Air Temperature During 1960–2004. *Theor. Appl. Climatol.* **2008**, *93*, 35–43. [CrossRef]
3. Yao, J.; Chen, Y.; Guan, X.; Zhao, Y.; Chen, J.; Mao, W. Recent Climate and Hydrological Changes in a Mountain–Basin System in Xinjiang, China. *Earth-Sci. Rev.* **2022**, *226*, 103957. [CrossRef]
4. Chen, Y.; Li, B.; Fan, Y.; Sun, C.; Fang, G. Hydrological and Water Cycle Processes of Inland River Basins in the Arid Region of Northwest China. *J. Arid Land.* **2019**, *11*, 161–179. [CrossRef]
5. Chen, J.; Brissette, F.P.; Poulin, A.; Leconte, R. Overall Uncertainty Study of the Hydrological Impacts of Climate Change for a Canadian Watershed. *Water Resour. Res.* **2011**, *47*, W12509. [CrossRef]
6. Meng, F.; Liu, T.; Huang, Y.; Luo, M.; Bao, A.; Hou, D. Quantitative Detection and Attribution of Runoff Variations in the Aksu River Basin. *Water* **2016**, *8*, 338. [CrossRef]
7. Li, C.; Qi, J.; Feng, Z.; Yin, R.; Zou, S.; Zhang, F. Parameters Optimization Based on the Combination of Localization and Auto-Calibration of SWAT Model in a Small Watershed in Chinese Loess Plateau. *Front. Earth Sci. China* **2010**, *4*, 296–310. [CrossRef]
8. Luo, Y.; Arnold, J.; Liu, S.; Wang, X.; Chen, X. Inclusion of Glacier Processes for Distributed Hydrological Modeling at Basin Scale with Application to a Watershed in Tianshan Mountains, Northwest China. *J. Hydrol.* **2013**, *477*, 72–85. [CrossRef]
9. Idrees, M.; Ahmad, S.; Khan, M.W.; Dahri, Z.H.; Ahmad, K.; Azmat, M.; Rana, I.A. Estimation of Water Balance for Anticipated Land Use in the Potohar Plateau of the Indus Basin Using SWAT. *Remote Sens.* **2022**, *21*, 5421. [CrossRef]
10. Luan, X.; Wu, P.; Sun, S.; Li, X.; Wang, Y.; Gao, X. Impact of Land Use Change on Hydrologic Processes in a Large Plain Irrigation District. *Water Resour. Manag.* **2018**, *32*, 3203–3217. [CrossRef]
11. Xiao, Y.; Ou Yang, Z.; Wang, L.; Rao, E.; Jiang, L. Spatial Characteristics of Ecosystem Quality and its Driving Forces in Inner Mongolia. *J. Ecol.* **2016**, *36*, 6019–6030. (In Chinese)
12. Wei, W.; Guo, Z.; Xie, B.; Zhou, J.; Li, C. Spatiotemporal Evolution of Environment Based on Integrated Remote Sensing Indexes in Arid Inland River Basin in Northwest China. *Environ. Sci. Pollut. Res.* **2019**, *26*, 13062–13084. [CrossRef] [PubMed]

13. Ma, L.; Bo, J.; Li, X.; Fang, F.; Cheng, W. Identifying Key Landscape Pattern Indices Influencing the Ecological Security of Inland River Basin: The Middle and Lower Reaches of Shule River Basin as an Example. *Sci. Total Env.* **2019**, *674*, 424–438. [CrossRef] [PubMed]
14. Huang, F.; Chunyu, X.; Zhang, D.; Chen, X.; Ochoa, C.G. A Framework to Assess the Impact of Ecological Water Conveyance on Groundwater-Dependent Terrestrial Ecosystems in Arid Inland River Basins. *Sci. Total Env.* **2020**, *709*, 136155. [CrossRef]
15. Zhang, Y.; Wang, Q.; Wang, Z.; Yang, Y.; Li, J. Impact of Human Activities and Climate Change on the Grassland Dynamics Under Different Regime Policies in the Mongolian Plateau. *Sci. Total Environ.* **2022**, *698*, 134304. [CrossRef]
16. Hao, L.; Sun, G.; Liu, Y.; Gao, Z.; He, J.; Shi, T.; Wu, B. Effects of Precipitation on Grassland Ecosystem Restoration Under Grazing Exclusion in Inner Mongolia, China. *Landsc. Ecol.* **2014**, *29*, 1657–1673. [CrossRef]
17. Liu, Y.; Zhuang, Q.; Chen, M.; Pan, Z.; Tchebakova, N.; Sokolov, A.; Kicklighter, D.; Melillo, J.; Sirin, A.; Zhou, G.; et al. Response of Evapotranspiration and Water Availability to Changing Climate and Land Cover on the Mongolian Plateau During the 21st Century. *Glob. Planet. Chang.* **2013**, *108*, 85–99. [CrossRef]
18. Woo, Y.H.; Min, J.P.; Jong, Y.P.; Geun, A.P.; Joon, K.S. The Spatial and Temporal Correlation Analysis Between MODIS NDVI and SWAT Predicted Soil Moisture during Forest NDVI Increasing and Decreasing Periods. *KSCE J. Civ. Eng.* **2010**, *14*, 931–939.
19. Zhang, X.; Yu, G.Q.; Li, Z.B.; Li, P. Experimental Study on Slope Runoff, Erosion and Sediment under Different Vegetation Types. *Water Resour. Manag.* **2014**, *28*, 2415–2433. [CrossRef]
20. Zhang, Y.; Liang, W.; Liao, Z.; Han, Z.; Xu, X.; Jiao, R.; Liu, H. Effects of Climate Change on Lake Area and Vegetation Cover over the Past 55 Years in Northeast Inner Mongolia Grassland, China. *Theor. Appl. Climatol.* **2019**, *138*, 13–25. [CrossRef]
21. Xue, L.; Wang, J.; Zhang, L.; Wei, G.; Zhu, B. Spatiotemporal Analysis of Ecological Vulnerability and Management in the Tarim River Basin, China. *Sci. Total Environ.* **2019**, *649*, 876–888. [CrossRef] [PubMed]
22. Su, B.D.; Yi, J.; Chen, J.; Bao, X.; Sa, R.N.S.Q. Analysis of Vegetation Degeneration Succession Trend in Middle and Lower Reaches of Wulagai Wetland of Inner Mongolia. *Chin. J. Grassl.* **2011**, *33*, 73–78.
23. Liu, J.; Luo, M.; Liu, T.; Bao, A.; De Maeyer, P.; Feng, X.; Chen, X. Local Climate Change and the Impacts on Hydrological Processes in an Arid Alpine Catchment in Karakoram. *Water* **2017**, *9*, 344. [CrossRef]
24. Gang, C.C.; Zhang, Y.Z.; Wang, Z.Q.; Chen, Y.Z.; Yang, Y.; Li, J.L.; Cheng, J.M.; Qi, J.G.; Odeh, I. Modeling the Dynamics of Distribution, Extent, And NPP of Global Terrestrial Ecosystems in Response to Future Climate Change. *Glob. Planet. Chang.* **2017**, *148*, 153–165. [CrossRef]
25. Chen, P. *Study of Plant Community Succession and Vegetation Dynamics in the Wulagai Basin in the Past 20 Years*; Inner Mongolia University: Inner Mongolia, China, 2019. (In Chinese)
26. Zhang, Y.; Liu, Z. Study on Water Resources Assessment and Reasonable use of Wulagai River and its Main Tributary. *J. Arid. Land Resour. Environ.* **1999**, *1*, 55–65. (In Chinese)
27. Nacun, B.; Nendel, C.; Hu, Y.; Lakes, T. Land-Use Change and Land Degradation on the Mongolian Plateau from 1975 to 2015—A Case Study from Xilingol, China. *Land Degrad. Dev.* **2018**, *29*, 1595–1606. (In Chinese)
28. Duan, Y.; Meng, F.; Liu, T.; Huang, Y.; Luo, M.; Xing, W.; De Maeyer, P. Sub-Daily Simulation of Mountain Flood Processes Based on the Modified Soil Water Assessment Tool (SWAT) Model. *Int. J. Environ. Res. Public Health* **2019**, *16*, 3118. [CrossRef] [PubMed]
29. Meng, F.; Liu, T.; Wang, H.; Luo, M.; Duan, Y.; Bao, A. An Alternative Approach to Overcome the Limitation of Hrus in Analyzing Hydrological Processes Based on Land Use/Cover Change. *Water* **2018**, *10*, 434. [CrossRef]
30. Luo, M.; Liu, T.; Meng, F.; Duan, Y.; Frankl, A.; Bao, A.; De Maeyer, P. Comparing Bias Correction Methods Used in Downscaling Precipitation and Temperature from Regional Climate Models: A Case Study from the Kaidu River Basin in Western China. *Water* **2018**, *10*, 1046. [CrossRef]
31. Wu, Y.; Li, Y.; Zhang, L.; Guo, L.; Li, H.; Xi, B.; Wang, L.; Li, C. Assessment of Lakes Ecosystem Health Based on Objective and Subjective Weighting Combined with Fuzzy Comprehensive Evaluation. *J. Lake Sci.* **2017**, *29*, 1091–1102. (In Chinese)
32. Chen, Q.; Chen, Y.; Wang, M.; Jiang, W.; Hou, P. Remote Sensing Comprehensive Evaluation and Change Analysis of Dongting Lake Ecosystem Quality from 2001 to 2010. *J. Ecol.* **2015**, *35*, 4347–4356. (In Chinese)
33. Huiting, Z.; Fanhao, M.; Chula, S.; Min, L.; Mulan, W.; Hao, L.C. Spatiotemporal Change and Cause Analysis of Ecosystem Quality in Mongolian Plateau during 2001 to 2019. *Chin. J. Ecol.* **2022**, *42*, 1–14. (In Chinese)
34. Li, M.; Weng, B.; Yan, D.; Bi, W.; Yang, Y.; Gong, X.; Wang, H. Spatiotemporal Characteristics of Surface Water Resources in the Tibetan Plateau Based on the Produce Water Coefficient Method Considering Snowmelt. *Sci. Total Environ.* **2022**, *851*, 158048. [CrossRef]
35. Yin, C.; Luo, M.; Meng, F.; Sa, C.; Yuan, Z.; Bao, Y. Contributions of Climatic and Anthropogenic Drivers to Net Primary Productivity of Vegetation in the Mongolian Plateau. *Remote Sens.* **2022**, *14*, 3383. [CrossRef]
36. Hu, S.; Ma, R.; Sun, Z.; Ge, M.; Zeng, L.; Huang, F.; Bu, J.; Wang, Z. Determination of the Optimal Ecological Water Conveyance Volume for Vegetation Restoration in an Arid Inland River Basin, Northwestern China. *Sci. Total Environ.* **2021**, *788*, 147775. [CrossRef]
37. Shen, Q.; Gao, G.; Lü, Y.; Wang, S.; Jiang, X.; Fu, B. River flow is Critical for Vegetation Dynamics: Lessons from Multi-Scale Analysis in a Hyper-Arid Endorheic Basin. *Sci. Total Environ.* **2017**, *603–604*, 290–298. [CrossRef]
38. Zhang, X.; Guan, T.; Zhou, J.; Cai, W.; Gao, N.; Du, H.; Jiang, L.; Lai, L.; Zheng, Y. Groundwater Depth and Soil Properties are Associated with Variation in Vegetation of a Desert Riparian Ecosystem in an Arid Area of China. *Forests* **2018**, *9*, 34. [CrossRef]

39. Guo, E.; Wang, Y.; Wang, C.; Sun, Z.; Bao, Y.; Mandula, N.; Jirigala, B.; Bao, Y.; Li, H. NDVI Indicates Long-Term Dynamics of Vegetation and its Driving Forces from Climatic and Anthropogenic Factors in Mongolian Plateau. *Remote Sens.* **2021**, *13*, 688. [CrossRef]
40. Luo, M.; Meng, F.; Sa, C.; Duan, Y.; Bao, Y.; Liu, T.; De Maeyer, P. Response of Vegetation Phenology to Soil Moisture Dynamics in the Mongolian Plateau. *Catena* **2021**, *206*, 105505. [CrossRef]
41. Chen, B.; Zhang, X.; Tao, J.; Wu, J.; Wang, J.; Shi, P.; Yangjian. The Impact of Climate Change and Anthropogenic Activities on Alpine Grassland Over the Qinghai-Tibet Plateau. *Agric. For. Meteorol.* **2014**, *189*, 11–18. [CrossRef]
42. Yang, Y.; Wang, Z.; Li, J.; Gang, C.; Zhang, Y.; Zhang, Y.; Odeh, I.; Qi, J. Comparative Assessment of Grassland Degradation Dynamics in Response to Climate Variation and Human Activities in China, Mongolia, Pakistan and Uzbekistan from 2000 to 2013. *J. Arid. Environ.* **2016**, *135*, 164–172. [CrossRef]
43. Ouyang, X.; Wang, J.; Chen, X.; Zhao, X.; Ye, H.; Watson, A.E.; Wang, S. Applying a Projection Pursuit Model for Evaluation of Ecological Quality in Jiangxi Province, China. *Ecol. Indic.* **2021**, *133*, 108414. [CrossRef]
44. Wang, J.; Brown, D.G.; Chen, J. Drivers of the Dynamics in Net Primary Productivity Across Ecological Zones on the Mongolian Plateau. *Landsc. Ecol.* **2013**, *28*, 725–739. [CrossRef]
45. Miao, L.; Sun, Z.; Ren, Y.; Schierhorn, F.; Müller, D. Grassland Greening on the Mongolian Plateau Despite Higher Grazing Intensity. *Land Degrad. Dev.* **2021**, *32*, 792–802. [CrossRef]

water

MDPI

Article

Evaluation of Different Methods on the Estimation of the Daily Crop Coefficient of Winter Wheat

Jingjing Fang [1], Yining Wang [1,2,*], Peng Jiang [1], Qin Ju [1], Chao Zhou [3], Yiran Lu [1], Pei Gao [1] and Bo Sun [1]

1 College of Hydrology and Water Resources, Hohai University, Nanjing 210024, China
2 Nanjing Hydraulic Research Institute, Nanjing 210029, China
3 Wudaogou Hydrologic Experimental Station, Water Resources Research Institute of Anhui Province (Huai Committee of the Ministry of Water Resources), Bengbu 233700, China
* Correspondence: ynwang@nhri.cn

Abstract: Various methods have been developed to estimate daily crop coefficients, but their performance varies. In this paper, a comprehensive evaluation was conducted to estimate the crop coefficient of winter wheat in four growth stages based on the observed data of weighing-type lysimeters and the high-precision automatic weather station in the Wudaogou Hydrological Experimental Station from 2018 to 2019. The three methods include the temperature effect method, the cumulative crop coefficient method, and the radiative soil temperature method. Our results suggest that the performance of these methods was different in each individual growth stage. The temperature effect method was better in the emergence-branching (RMSE = 0.06, r = 0.80) and heading-maturity stages (RMSE = 0.16, r = 0.94) because the temperature is suitable for crop growth during most of these two periods. The cumulative crop coefficient method was better in the greening-jointing (RMSE = 0.16, r = 0.88) and heading-maturity stages (RMSE = 0.20, r = 0.91) because this method is closely related to crop growth, which is vigorous during these two stages. The radiative soil temperature method was better in the emergence-branching (RMSE = 0.20, r = 0.35) and branch-overwintering stages (RMSE = 0.25, r = 0.52) as the energy balance can be ensured by the relatively high level of the effective energy during these periods. By comparing the estimation accuracy indices of the three methods, we found that the temperature effect method performed the best during the emergence-branching stage (RMSE = 0.06, MAE = 0.06, r = 0.80, d_{IA} = 0.88), branch-overwintering stage (RMSE = 0.13, MAE = 0.11, r = 0.44, d_{IA} = 0.55), and heading-maturity stage (RMSE = 0.16, MAE = 0.13, r = 0.94, d_{IA} = 0.97), while the cumulative crop coefficient method performed best during the greening-jointing stage (RMSE = 0.16, MAE = 0.13, r = 0.88, d_{IA} = 0.89). Based on this result, an integrated modelling procedure was proposed by applying the best method in each growth stage, which provides higher simulation precision than any single method. When the best method was adopted in each growth stage, the estimated accuracy of the whole growth process was RMSE = 0.13, MAE = 0.09, r = 0.98, d_{IA} = 0.99.

Keywords: crop coefficient; multiple methods; difference research; temperature effect method; cumulative crop coefficient method; radiative soil temperature method

Citation: Fang, J.; Wang, Y.; Jiang, P.; Ju, Q.; Zhou, C.; Lu, Y.; Gao, P.; Sun, B. Evaluation of Different Methods on the Estimation of the Daily Crop Coefficient of Winter Wheat. *Water* **2023**, *15*, 1395. https://doi.org/10.3390/w15071395

Academic Editors: Songhao Shang, Magdy Mohssen, Qianqian Zhang, Dongqin Yin and Hamza Gabriel

Received: 15 February 2023
Revised: 23 March 2023
Accepted: 30 March 2023
Published: 4 April 2023

1. Introduction

According to data released by the Food and Agriculture Organization (FAO), the FAO Food Price Index in 2021 reached a 10-year high, and the incidence of moderate or severe food insecurity in the world increased from 22.6% in 2014 to 30.4% in 2020. In 2020, an estimated 720 to 811 million people, or 9.5 to 10.7 percent of the global population, will face food insecurity [1]. Agricultural irrigation accounts for 87% of total water use [2], but water resources are increasingly in short supply under the pressure of climate change and population growth [3]. Accurate estimation of crop water demand is of great significance for rational allocation of water resources.

Wheat is one of the main crops grown and eaten worldwide [4]. From 2011 to 2020, the global average annual planting area reached 219 million hectares, with an annual yield of 733 million tons [5]. Wheat is the second main crop in China [6]; the wheat output of China accounts for approximately 18% of the world's output (ranking first in the world), and the sown area accounts for approximately 10% of that of the world [7]. Winter wheat is sown from October to November and harvested from May to June of the following year. Planting is performed in dry conditions with little rainfall. Accurate estimation of the daily evapotranspiration and crop coefficient of winter wheat is conducive to the formulation of accurate irrigation schemes to improve the yield of winter wheat and save water resources [8].

Evapotranspiration is an important link in the hydrological cycle and is involved in the surface energy balance and water balance [9]. The crop coefficient is the ratio of actual evapotranspiration and reference evapotranspiration of crops, reflecting the influence of soil, vegetation, and hydrometeorological conditions on evapotranspiration and is often used to calculate crop water requirements [10]. Finding a suitable crop coefficient estimation method is of great significance for further estimating actual evapotranspiration, making irrigation plans, and efficiently utilizing water resources.

Domestic and foreign scholars have performed much research on crop coefficient estimation methods. In the crop coefficient estimation method recommended by FAO, the basic crop coefficient constant is obtained by plotting tabulated values and drawing the crop coefficient curve with a simplified straight-line connection at each growth stage. Finally, the daily crop coefficient value is modified according to wind speed and humidity. This method has been widely adopted and is very convenient for practical application. Er-Raki et al. [11] directly used the basic crop coefficient provided by the FAO Irrigation and Drainage Paper No. 56 (FAO-56) to calculate the soil surface vegetation coverage. Vu et al. [12] compared the crop coefficient recommended by the FAO-56 with the field monitoring value and found that the applicability of the recommended value of FAO-56 was affected by crop variety and growth stage. Ali et al. [13] simulated the Kc curve using four crop coefficient estimation methods, including the FAO-56 recommendation method. Many scholars estimate the crop coefficient according to the crop growth character index and calculate the crop coefficient by measuring the leaf area, plant height, vegetation index, etc. The daily scale is usually obtained by linear interpolation. This method of estimation has strong physical significance and high accuracy. Spiliotopoulos et al. [14] estimated crop coefficients based on the vegetation index using mapping ET of high resolution and internalized calibration models. Zhang et al. [15] established two Ks regression models for crop coefficient inversion and found that the model established by TCARI/RDVI had a better correlation with the crop coefficient. Park et al. [8] estimated crop coefficients for cropland and mixed forest based on the normalized vegetation index, leaf area index, and soil moisture.

The above methods use the same method to estimate the crop coefficient at each growth stage, but the growth of crops at different growth stages and the hydrometeorological factors closely related to their growth are different. It is difficult to obtain satisfactory results with a single method, so it is necessary to select the best method to estimate the crop coefficient at different growth stages. Based on the temperature effect, Wang et al. [16] established a daily crop coefficient estimation model of winter wheat and summer maize with a high fitting degree and found that the estimation effect of each growth stage was different. Tang et al. [17] conducted a remote sensing study by UAV and found that there was a good linear correlation between biomass and the cumulative crop coefficient and cumulative transpiration in different treatment areas. Kool et al. [18] evaluated the dual source energy balance (TSEB) model using net radiation, soil heat flux, and surface temperature. Paulino et al. [19] proposed an empirical model based on linear multiple regression to estimate the number of fruits per plant of two sweet oranges and concluded that the estimates of the model varied greatly in the three growth stages. At present, most related studies from both China and abroad focus on crop coefficient estimation methods.

Few studies have compared the differences in crop coefficient estimation methods in different growth stages and provided the best estimation methods in each growth stage. In this paper, we attempt to select three methods, the temperature effect method, cumulative crop coefficient method, and radiative soil temperature method, to compare and analyse their estimation results to provide the best estimation method of crop coefficients in different growth stages and provide a basis for the accurate calculation of evapotranspiration in each growth stage.

2. Materials and Methods

2.1. Experimental Area Profile

The measured experimental data of the Wudaogou hydrological experimental station were used. The station is located in Guzhen County (33°09′ N, 117°21′ E), Bengbu City, Anhui Province, in the southern Huaibei Plain, covering an area of 27,000 square meters. It is affected by a subtropical humid monsoon climate and temperate subhumid monsoon climate, with rain and heat at the same time. The rainfall in the area where the experimental station is located varies greatly from year to year and is unevenly distributed within the year. A total of 61.8% of the annual rainfall is concentrated in the flood season (from June to September). The main soil types in the experimental area are sandy ginger black soil and yellow tidal soil, and the main crops are wheat, corn, and soybean. There are 62 sets of nonweighing ground lysimeters and 10 sets of large-scale weighing lysimeters in the station, which can record submersible evaporation, soil moisture, and evapotranspiration. A high-precision weather station is set up about 50 m south of the weighing lysimeters to automatically monitor net radiation, soil heat flux, air temperature, wind speed, and other hydrometeorological elements every 10 min.

According to the measured hydrometeorological data from 1986 to 2021, the average annual temperature is 15.2 °C, the average annual rainfall is 929.5 mm, the average annual flood season rainfall is 582.4 mm, the average annual evaporation is 931.9 mm, the average annual relative humidity is 79.28%, the average annual sunshine duration is 1723.6 h, the average annual wind speed is 1.6 m/s, the annual average surface temperature is 18.4 °C, the annual average soil temperature at 10 cm is 16.6 °C, the maximum temperature is 41.5 °C, and the minimum temperature is −22.7 °C.

2.2. Experimental Facilities and Data Selection

The difference in the three crop coefficient estimation methods at different growth stages and daily scales was studied by using a large-scale lysimeter. The winter wheat relies on rainfall. The lysimeter model is FR101A, the resolution is 0.025 mm, the soil column height is 4.0 m, the diameter area is 2.0 square meters, and the weight data are automatically collected every 10 min. The depth of the groundwater level in the shallow buried area of the Huaibei Plain is 1 to 3 m [20], and the root system of winter wheat is concentrated within 1 m [21]. Therefore, experimental data from a lysimeter with a burial depth of 1.0 m and measured data from high-precision weather stations were selected from 11 November 2018 to 4 June 2019.

2.3. Division of Growth Stages

According to the actual growth status of winter wheat, the whole growth process can be divided into four stages: emergence-branching stage, branch-overwintering stage, greening-jointing stage, and heading-maturity stage. The classification of the stages is based on the characteristics of the crops in each stage. The parameters of each stage are shown in Table 1.

Table 1. Growth stage division of winter wheat.

Stage of Growth	Emergence-Branching	Branch-Overwintering	Greening-Jointing	Heading-Maturity
Date	2018/11/11–2018/12/1	2018/12/2–2019/2/21	2019/2/22–2019/4/19	2019/4/20–2019/6/4
Number of days	21	82	57	46
ET_0 at this stage	21.80 mm	67.80 mm	216.48 mm	220.67 mm
Proportion of total ET_0	4.14%	12.87%	41.10%	41.89%
Average daily ET_0	1.04 mm	0.83 mm	3.80 mm	4.80 mm

2.4. Crop Coefficient, Actual Evapotranspiration, and Reference Evapotranspiration

The crop coefficient is divided into a single crop coefficient and a double crop coefficient. The single crop coefficient involves fewer factors and has higher estimation accuracy, so the single crop coefficient is selected. According to the definition of the crop coefficient, the calculation method is shown in Equation (1):

$$K_c = \frac{ET}{ET_0} \tag{1}$$

where K_c is the crop coefficient; ET is the actual evapotranspiration (mm); and ET_0 is the reference evapotranspiration (mm).

The actual evapotranspiration (ET) of winter wheat is automatically recorded by a large weighing lysimeter, and the difference between the total weights of two adjacent collections is the actual evapotranspiration of the stage. The basic principle is shown in Equation (2):

$$P + I + E_g = P_a + ET + R + \Delta S \tag{2}$$

where P is the rainfall (mm); I is the irrigation water volume (mm); E_g is the diving evaporation (mm); P_a is the deep leakage rate (mm); ET is the actual evapotranspiration (mm); R is the volume of runoff (mm); and ΔS is the soil storage variable (mm).

Xu et al. [22] studied East China and concluded that the FAO-56 Penman–Monteith formula was the best method to calculate the daily reference evapotranspiration. The reference evapotranspiration is calculated by Equation (3):

$$ET_0 = \frac{0.408\Delta(R_n - G) + \gamma\frac{900}{T+273}u_2(e_s - e_a)}{\Delta + \gamma(1 + 0.34u_2)} \tag{3}$$

where ET_0 is the reference evapotranspiration (mm × d^{-1}); R_n is the net surface radiation (MJ × m^{-2} × d^{-1}); G is the soil heat flux (MJ × m^{-2} × d^{-1}); T is the average daily temperature (°C); u_2 is the average wind speed at 2 m above the ground (m × s^{-1}); e_s is the saturated vapour pressure (kPa); e_a is the actual vapour pressure (kPa); Δ is the slope of the saturated vapour pressure and temperature curve (kPa × $°C^{-1}$); and γ is the dry and wet table constant (kPa × $°C^{-1}$). Data are from a high precision weather station.

The change curves of actual evapotranspiration and reference evapotranspiration are shown in Figure 1.

2.5. Crop Coefficient Estimation Method and Evaluation Indices

2.5.1. Temperature Effect Method

The temperature effect model proposed by Huang et al. [23] effectively simulated the dynamic process of crop growth and development. Wang et al. [16] used the model structure to construct the calculation formula of the crop coefficient considering the temperature of three basis points, as shown in Equation (4):

$$K_c = K_0 e^{-\left(\frac{T - T_0}{\beta}\right)^2} \tag{4}$$

where K_c is the crop coefficient; K_0 is the crop coefficient at the optimum temperature; T is the average temperature (°C); T_0 is the optimum temperature for physiological and ecological processes such as crop growth and photosynthesis (°C); and β is the parameter to be estimated.

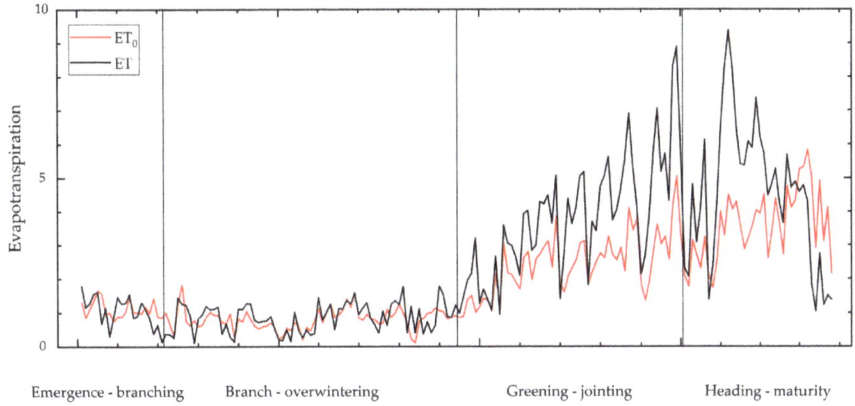

Figure 1. The variation of the actual evapotranspiration (ET) and reference evapotranspiration (ET_0).

Since an invalid temperature is not conducive to crop growth, the average temperature T is calculated using the eliminating invalid temperature method [24,25], where invalid temperature values above the upper limit and below the lower limit are discarded, as shown in Equation (5):

$$T = \frac{T'_x + T'_n}{2} \quad (5)$$

The specific calculation method of each parameter in Equation (5) is

$$T'_x = \max(T_x, T_{upper}),\ T'_n = \min(T_n, T_{upper}),\ T_x = \max(T_{max}, T_{base}),\ T_n = \max(T_{min}, T_{base}) \quad (6)$$

where T_{upper} is the upper limit temperature (°C), which is 30 °C and T_{base} indicates the lower limit temperature (°C), which is 3 °C [26–28].

The optimal values of unknown parameters K_0, T_0, and β were determined by SPSS software combined with the least squares method and sequential quadratic programming. First, the logarithm of Equation (4) can be obtained:

$$\ln(K_c) = \ln(K_0) - \frac{T_0^2}{\beta^2} + \frac{2T_0}{\beta^2} \times T - \frac{1}{\beta^2} \times T^2 \quad (7)$$

Setting $y = \ln(K_c)$, $x = T$, $a_1 = \ln(K_0) - \frac{T_0^2}{\beta^2}$, $a_2 = \frac{2T_0}{\beta^2}$, and $a_3 = -\frac{1}{\beta^2}$, Equation (7) is converted to

$$y = a_1 + a_2 \times x + a_3 \times x^2 \quad (8)$$

Let $Y = \begin{bmatrix} y_1 \\ \vdots \\ y_n \end{bmatrix}$, $A = \begin{bmatrix} a_1 \\ a_2 \\ a_3 \end{bmatrix}$, and $X = \begin{bmatrix} 1 & x_{11} & x_{21} \\ \vdots & \vdots & \vdots \\ 1 & x_{1n} & x_{2n} \end{bmatrix}$; if there are n samples in the whole growth process, then Equation (8) can be expressed as

$$Y = AX \quad (9)$$

The unknown parameter values in the whole growth process were estimated by combining the least square method:

$$\hat{A} = \left(X^T X\right)^{-1} X^T Y \quad (10)$$

The inverse solution is:

$$\widehat{K_0} = e^{\widehat{a_0} - \frac{\widehat{a_1}^2}{4 - \widehat{a_2}}},\ \widehat{T_0} = -\frac{\widehat{a_1}}{2 \times \widehat{a_2}},\ \widehat{\beta} = \sqrt{-\frac{1}{\widehat{a_2}}} \tag{11}$$

The obtained value of Equation (11) is set as the initial value, the sequential quadratic programming method is used to solve the optimal value of each parameter by SPSS software, and the objective function is set as

$$\min \sum_{i=1}^{m} \left(\hat{K}_c - K_{ci}\right)^2 \tag{12}$$

The constraint conditions are $T_{base} < T_0 < T_{upper}$, $0 < K_0 < 3$, and $\beta > 0$.

2.5.2. Cumulative Crop Coefficient Method

Many scholars have used cumulative evapotranspiration [29] and cumulative growth days [30] when estimating crop coefficients and evapotranspiration; in this study, we attempted to use cumulative crop coefficients. The cumulative crop coefficient refers to the cumulative value of the crop coefficient since the day of planting and only considers the days after seeding. This method does not require data on hydrometeorological elements and crop growth traits and is convenient for use in areas where observation conditions and crop growth data are scarce.

The curves of the cumulative crop coefficient and days after seeding of winter wheat are shown in Figure 2, showing an approximately elongated "S" shape. The daily growth value of the cumulative crop coefficient is that of the daily crop coefficient, which increases first and then decreases with time after seeding. The cumulative crop coefficient model was constructed as follows:

$$K_{c\ cumulative} = a + b \times \cos(c \times D) + d \times \sin(c \times D) \tag{13}$$

where $K_{c\ cumulative}$ is the cumulative crop coefficient value, D is the number of days after seeding, and a, b, c, and d are unknown parameters.

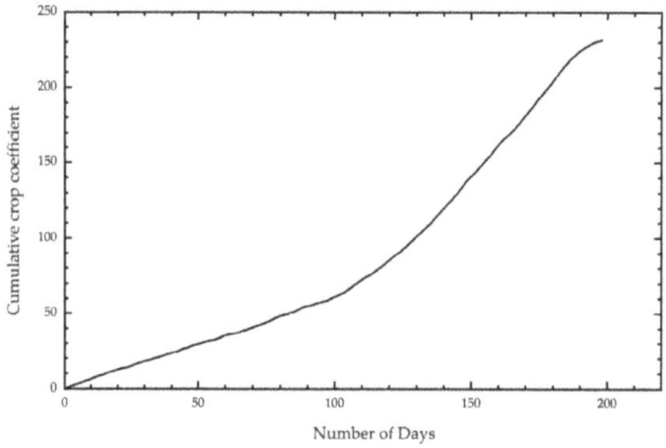

Figure 2. Correlation curve between cumulative crop coefficient and days after seeding.

For each growth stage, the least square method was used to estimate the values of unknown parameters a, b, c, and d. According to the definition of the function, its derivative is the daily value function of the crop coefficient, and the value of the crop coefficient on day D can be obtained by substituting D into it. The derivative of Equation (13) with respect to D can be obtained as follows:

$$K_c = -b \times c \times \sin(c \times D) + d \times c \times \cos(c \times D) \tag{14}$$

2.5.3. Radiative Soil Temperature Method

According to the study of Zhao et al. [31], there is a strong positive correlation between the crop coefficient and evapotranspiration and a significant positive correlation between evapotranspiration and effective energy. Effective energy (R_n-G) refers to the difference between net radiation and soil heat flux, showing a trend of being smaller at the emergence-branching and branch-overwintering stages and gradually increasing at the greening-jointing and heading-maturity stages (Figure 3). During the whole process of winter wheat growth, the correlation coefficient between the crop coefficient and effective energy reached 0.75, showing a strong correlation. The correlation coefficient between the crop coefficient and soil temperature from 0 to 160 cm decreased from 0.66 to 0.70 with increasing depth, as shown in Figure 4.

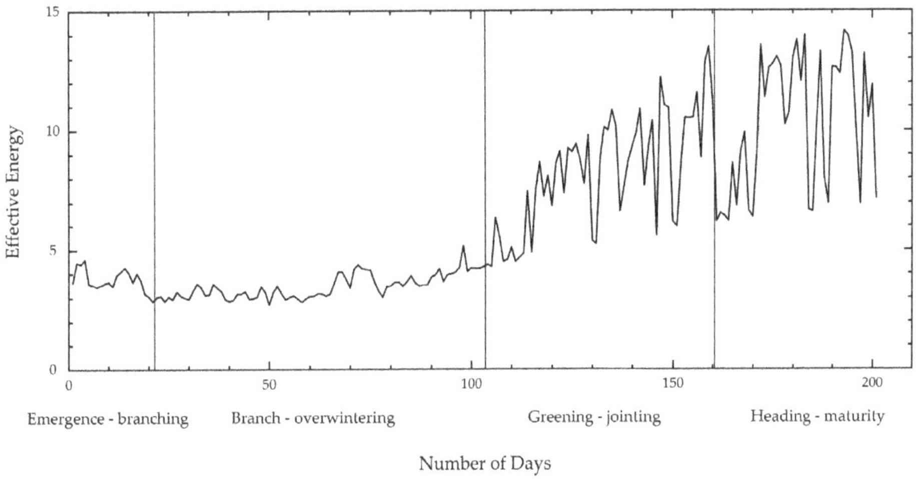

Figure 3. The variation of the effective energy (R_n-G).

Because FAO-56 notes that the penetration depth of temperature waves in soil is 0.1 to 0.2 m at a daily scale, soil temperatures at 0 and 10 cm were selected. The following mathematical model was constructed involving radiation and soil temperature:

$$K_c = me^{(R_n - G - n|\frac{D_0}{D_{10}}|)} \tag{15}$$

where D_0 is the surface temperature (°C); D_{10} is the soil temperature of 10 cm (°C); and m and n are unknown parameters. The meanings of the other symbols are the same as described previously.

Parameter calibration was realized by the particle swarm optimization algorithm. After finding the unknown parameter values corresponding to the daily scale data of each group in the whole growth process, the best unknown parameter of each growth stage was obtained by taking the average value. The particle swarm optimization algorithm is a random optimization method based on swarm intelligence proposed by Kennedy and Eberhart [32]. Each particle represents a candidate solution, and the problem is solved through assessment of the simple behaviour of individual particles and information interaction within the group.

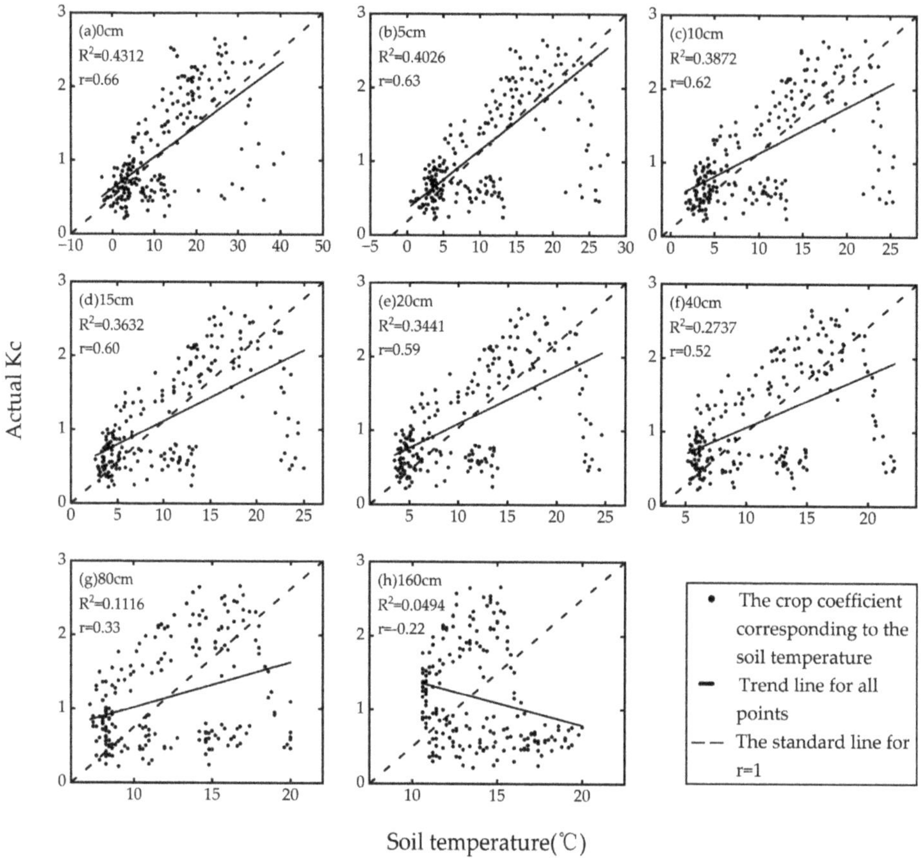

Figure 4. Correlation between crop coefficient and soil temperature from 0 to 160 cm(a to h in the figure represent the deepening of soil depth).

When solving the optimization problem, each particle has two state quantities, position and velocity, and the fitness value determined by the objective function. The flight process of the particle is the search process of the individual. In each iteration, individual particles record the best solution found as the current individual extreme value and share it with other particles. The best of all individual extreme values is the extreme value of the current group. All particles adjust their speed and position according to their current individual extreme value and the current group extreme value of the whole particle swarm [33]. Optimization was achieved using MATLAB software programming.

According to the properties of the model, in the two-dimensional solution search space, there is a particle swarm composed of 500 particles. The position of the particle swarm of the k-th iteration in the solution space is assumed to be expressed as Equation (16), and the flight velocity is expressed as Equation (17):

$$X_i^k = \left(X_{i1}^k, X_{i2}^k\right), i = 1, 2, \ldots, 500 \quad (16)$$

$$V_i^k = \left(V_{i1}^k, V_{i2}^k\right), i = 1, 2, \ldots, 500 \quad (17)$$

At this time, the best individual historical position of the jth (j < 500) particle is $\left(X_{mj1}^{k}, X_{mj2}^{k}\right)$, and the best historical position of the particle swarm is $\left(Y_{m1}^{k}, Y_{m2}^{k}\right)$. Then, at the k + 1 iteration, the velocity and position update formula of the particle is as follows:

$$\begin{cases} V_{j1}^{k+1} = w \times V_{j1}^{k} + c_1 \times r_1 \times \left(X_{mj1}^{k} - X_{j1}^{k}\right) + c_2 \times r_2 \times \left(Y_{m1}^{k} - X_{j1}^{k}\right) \\ V_{j2}^{k+1} = w \times V_{j2}^{k} + c_1 \times r_1 \times \left(X_{mj2}^{k} - X_{j2}^{k}\right) + c_2 \times r_2 \times \left(Y_{m2}^{k} - X_{j2}^{k}\right) \end{cases} \quad (18)$$

$$\begin{cases} X_{i1}^{k+1} = X_{i1}^{k} + V_{j1}^{k+1} \\ X_{i2}^{k+1} = X_{i2}^{k} + V_{j2}^{k+1} \end{cases} \quad (19)$$

$$w = w_{min} + (w_{max} - w_{min}) \times \frac{ger - times}{ger} \quad (20)$$

where w is the inertia weight, representing the influence of the current speed on the next movement; c_1 and c_2 are the learning factors, which measure the influence of the current individual historical optimal position and group historical optimal position on the next movement; r_1 and r_2 are random numbers from 0 to 1; w_{min} is the initial inertia weight with a typical value of 0.4; w_{max} is the inertia weight under the maximum number of iterations with a typical value of 0.9; ger is the maximum number of iterations; and times indicates the current iteration times.

2.5.4. Indices of Evaluation

The evaluation indices include the root mean square error (RMSE), mean absolute error (MAE), correlation coefficient (r), and consistency index (d_{IA}), which are used to evaluate the error and consistency between the estimated value and the measured value of each estimation method. See Equation (21) to Equation (24) for the calculation formula of each index.

$$RMSE = \sqrt{\frac{\sum_{i=1}^{n}(x_i - y_i)^2}{n}} \quad (21)$$

$$MAE = \frac{\sum_{i=1}^{n}|x_i - y_i|}{n} \quad (22)$$

$$r = \frac{\sum_{i=1}^{n}(x_i - \overline{x})(y_i - \overline{y})}{\sqrt{\sum_{i=1}^{n}(x_i - \overline{x})^2}\sqrt{\sum_{i=1}^{n}(y_i - \overline{y})^2}} \quad (23)$$

$$d_{IA} = 1 - \frac{\sum_{i=1}^{n}(x_i - y_i)^2}{\sum_{i=1}^{n}(|x_i - \overline{y}| + |y_i - \overline{y}|)^2} \quad (24)$$

where x_i is the estimated value of K_c; y_i is the actual value of K_c; i is the sample ordinal number, $i = 1, 2, \ldots, n$; $\overline{x}$ is the mean value of K_c's estimation; $\overline{y}$ is the mean value of K_c; and n is the number of samples of the estimated value.

Generally, the closer the root mean square error RMSE and mean absolute error MAE are to 0, the smaller the error and the greater the accuracy. The closer the correlation coefficient r and the consistency index d_{IA} are to 1, the closer the estimated value of the model is to the actual value, and the stronger its estimation ability is.

3. Results

Based on the observation data of the large weighing lysimeter and the data of the high-precision weather station from 2018 to 2019, the parameters of the three crop coefficient estimation models were calibrated. The unknown parameter values obtained by each method under different growth stages are shown in Table 2.

Table 2. Values of unknown parameters in different growth stages and methods of winter wheat.

Method	Temperature Effect			Cumulative Crop Coefficient				Radiative Soil Temperature	
Parameter	K_0	T_0	β	a	b	c	d	m	n
Emergence-branching stage	1.24	3.00	18.84	158.62	−7.17	0.07	6.54	4.19	5.87
Branch-overwintering stage	1.96	3.00	12.42	216.22	−79.90	0.01	60.00	0.26	3.21
Greening-jointing stage	2.14	20.95	6.61	180.23	−382.33	−0.01	26.25	0.41	5.62
Heading-maturity stage	2.39	20.37	5.16	227.91	0.02	−0.05	43.52	6.16	8.81

3.1. The Differences and Causes of Crop Coefficient Estimation by Different Methods

The results of the three methods for estimating winter wheat crop coefficients are shown in Figures 5 and 6. The results of the temperature effect method were better in the emergence-branching and heading-maturity stages, followed by the greening-jointing and branch-overwintering stages. The reason was that there were fewer ineffective temperatures below 3 °C and above 30 °C during the emergence-branching and heading-maturity stages, and crop growth was not inhibited by ineffective temperatures. The cumulative crop coefficient method was better in the greening-jointing and heading-maturity stages but worse in the emergence-branching and branch-overwintering stages. The reason was that the cumulative crop coefficient method is closely related to crop growth [17], the leaf index in the greening-jointing and heading-maturity stages was larger [34,35], and crop growth and development were vigorous. The radiative soil temperature method was better in the emergence-branching and branch-overwintering stages but worse in the greening-jointing and heading-maturity stages. The reason was that the effective energy cannot maintain a high level during the greening-jointing and heading-maturity stages, and the balance between energy and soil temperature was difficult to ensure.

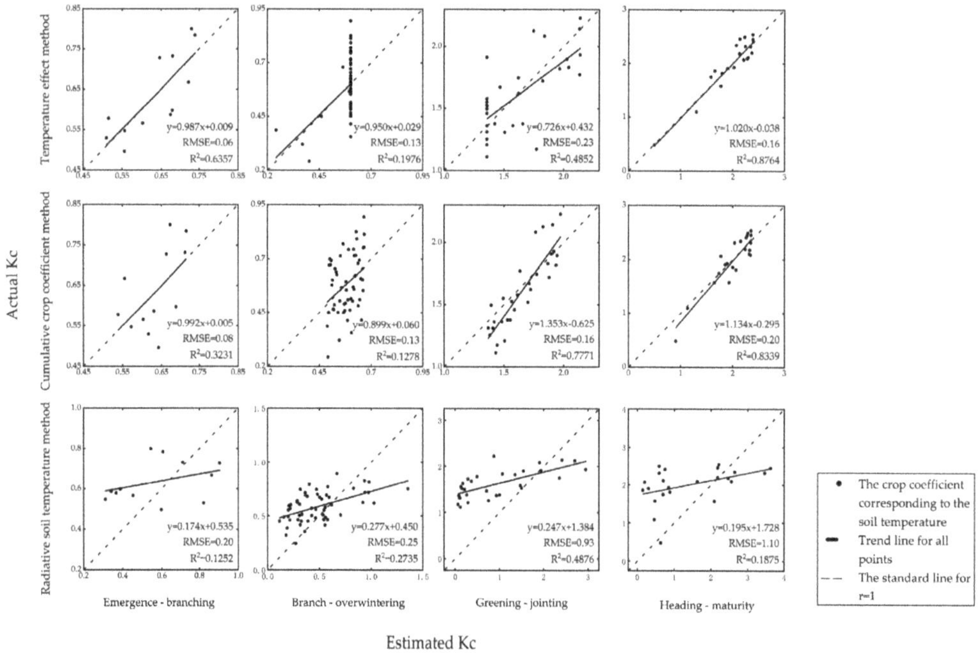

Figure 5. Results of three methods for estimating winter wheat crop coefficients.

3.2. Determination of The Best Estimation Method for Each Growth Stage

The accuracy of the results of the three methods in each growth stage are shown in Table 3 (in this table, "TE" means the temperature effect method, "CCC" means the cumulative crop coefficient method, and "RST" means the radiative soil temperature method).

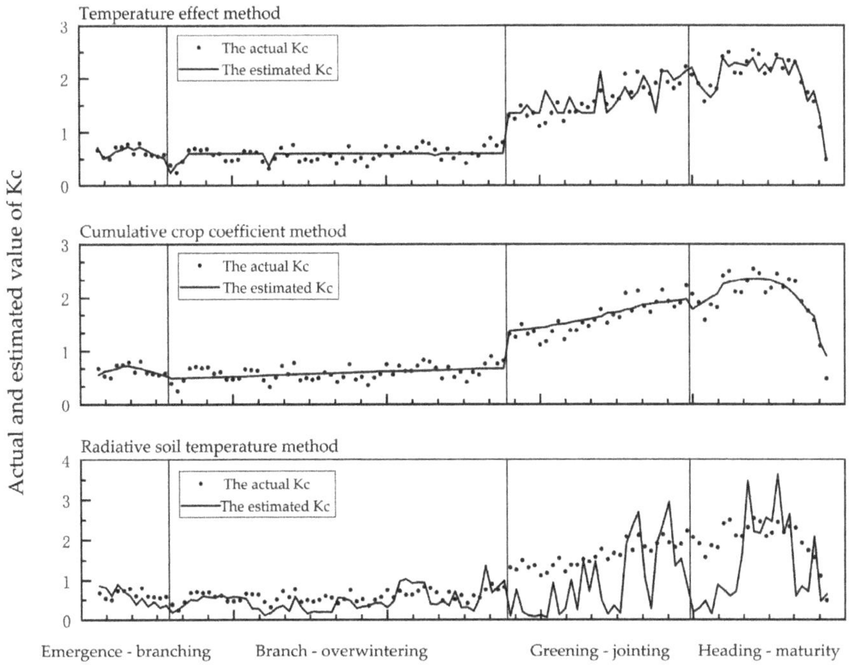

Figure 6. The estimated crop coefficient and actual crop coefficient during four growing stages.

Table 3. The estimated precision index value of the results of three methods in each growth stage.

Stage	Emergence-Branching Stage				Branch-Overwintering Stage				Greening-Jointing Stage				Heading-Maturity Stage			
Index	RMSE	MAE	r	dIA	RMSE	MAE	r	dIA	RMSE	MAE	r	dIA	RMSE	MAE	r	dIA
TE	0.06	0.06	0.80	0.88	0.13	0.11	0.44	0.55	0.23	0.18	0.70	0.83	0.16	0.13	0.94	0.97
CCC	0.08	0.07	0.57	0.69	0.13	0.12	0.36	0.51	0.16	0.13	0.88	0.89	0.20	0.16	0.91	0.94
RST	0.20	0.19	0.35	0.51	0.25	0.22	0.52	0.61	0.93	0.79	0.70	0.49	1.10	0.91	0.43	0.49

The estimation results were compared, and the best estimation method was selected according to the following steps. In the first step, the root mean square error, RMSE, and mean absolute error, MAE, were compared, and the method with the smallest error was selected. In the second step, when the difference in the root mean square error and mean absolute error between multiple methods are no more than 0.03, the correlation coefficient r and consistency index d_{IA} are compared in turn to select the method with the strongest estimation ability. In the third step, if the difference between the correlation coefficient and consistency index is less than 0.03, then the cumulative crop coefficient method should be directly adopted according to the principle of minimum required observations if it is still in the option list. Otherwise, the most appropriate method should be adopted in combination with the graph.

According to the above criteria, the temperature effect method during the emergence-branching stage was the best (RMSE = 0.06, MAE = 0.06, r = 0.80, d_{IA} = 0.88), followed by the cumulative crop coefficient method, and the radiative soil temperature method was inferior. The temperature effect method during the branch-overwintering stage was the best (RMSE = 0.13, MAE = 0.11, r = 0.44, d_{IA} = 0.55), followed by the cumulative crop coefficient method, and the radiative soil temperature method was inferior. The cumulative crop coefficient method during the greening-jointing stage was the best (RMSE = 0.16, MAE = 0.13, r = 0.88, d_{IA} = 0.89), followed by the temperature effect method, and the radiative soil temperature method was inferior. The temperature effect method during the heading-maturity stage was the best (RMSE = 0.16, MAE = 0.13, r = 0.94, d_{IA} = 0.97), followed by the cumulative crop coefficient method, and the radiative soil temperature method was inferior.

The estimated accuracy index of the whole growth process when a single method or the best method is used in each growth stage is shown in Table 4. When a single method was used in each growth stage, the cumulative crop coefficient method was the best method for estimating the whole growth process (RMSE = 0.25, MAE = 0.20, r = 0.93, d_{IA} = 0.96), followed by the temperature effect method, and the radiative soil temperature method was inferior. When the cumulative crop coefficient method and temperature effect method were used, the correlation coefficient and consistency index of the whole growth process estimation results were both greater than 0.80, which met the accuracy requirements of estimation and could be used for crop coefficient estimation. When the radiative soil temperature method was used, the correlation coefficient and consistency index of the whole growth process estimation results were only 0.50 and 0.69, respectively, which could not meet the requirement of estimation accuracy.

Table 4. The estimation precision index values of the whole growth process.

Method	Root Mean Square Error	Mean Absolute Error	Correlation Coefficient	Consistency Index
TE	0.34	0.25	0.87	0.93
CCC	0.25	0.20	0.93	0.96
RST	0.79	0.58	0.50	0.69
The best	0.13	0.09	0.98	0.99

When the best method was used in each growth stage, the four precision indices were better than when a single method was used; the root mean square error and mean absolute error were 0.13 and 0.09, and the correlation coefficient and consistency index reached 0.98 and 0.99, far greater than 0.80. The best method has higher estimation ability and accuracy than the single method and can be used for crop coefficient estimation.

4. Conclusions and Discussion

1. The results of the temperature effect method were better in the emergence-branching and heading-maturity stages, followed by the greening-jointing and branch-overwintering stages. The reason was that the ineffective temperature was lower during the emergence-branching and heading-maturity stages, and crop growth was not inhibited. The cumulative crop coefficient method was better in the greening-jointing and heading-maturity stages but worse in the emergence-branching and branch-overwintering stages. The reason was that the cumulative crop coefficient method is closely related to crop growth, and the crops grew vigorously during the greening-jointing and heading-maturity stages. The radiative soil temperature method was better in the emergence-branching and branch-overwintering stages but worse in the greening-jointing and heading-maturity stages. The reason was that the effective energy could not maintain a high level during the greening-jointing and heading-maturity stages, and the energy balance was difficult to ensure.

2. The temperature effect method during the emergence-branching stage was the best, followed by the cumulative crop coefficient method, and the radiative soil temperature method was inferior. The temperature effect method during the branch-overwintering stage was the best, followed by the cumulative crop coefficient method, and the radiative soil temperature method was inferior. The cumulative crop coefficient method during the greening-jointing stage was the best, followed by the temperature effect method, and the radiative soil temperature method was inferior. The temperature effect method during the heading-maturity stage was the best, followed by the cumulative crop coefficient method, and the radiative soil temperature method was inferior.

3. When a single method was used in each growth stage, the cumulative crop coefficient method was the best, followed by the temperature effect method, and the radiative soil temperature method was inferior. The cumulative crop coefficient method and temper-

ature effect method meet the accuracy requirements of estimation, but the radiative soil temperature method could not meet the accuracy requirements of estimation. The root mean square error, RMSE = 0.13; mean absolute error, MAE = 0.09; correlation coefficient, r = 0.98; and consistency index, dIA = 0.99 were all better than the single method when the best method was used in each growth stage. The best method had higher estimation ability and accuracy than the single method.

To improve upon the results presented in this paper, the adaptability of different crop coefficient estimation methods in other areas needs to be further explored. In addition, the daily crop coefficient estimation is only discussed under the condition of 1 m burial depth. The difference in the estimation effect of different methods under different burial depths needs to be further studied. In addition to crop growth traits, there is also the standard crop coefficient stage division table recommended by the FAO. Different division methods also have a certain impact on the accuracy of estimation. In the future, more crop coefficient estimation methods can be considered to provide more options for crop coefficient estimation.

Author Contributions: Conceptualization, J.F.; Formal analysis, J.F.; Investigation, J.F.; Resources, C.Z.; Data curation, C.Z., Y.L., P.G. and B.S.; Writing – original draft, J.F.; Writing – review & editing, Y.W., P.J. and Q.J.; Visualization, Y.L., P.G. and B.S.; Funding acquisition, Y.W. All authors have read and agreed to the published version of the manuscript.

Funding: This work was supported by the Youth Fund of National Natural Science Foundation of China (Grant No. 52109029), the National key research and development program funded project of China (Grant No. 2022YFC3202401), the Belt and Road Special Foundation of the State Key Laboratory of Hydrology-Water Resources and Hydraulic Engineering (Grant No. 2020490211; Grant No. 2020490208), the Water science and technology project of Jiangsu Province (Grant No. 2021047), the Independent research project of State Key Laboratory of Hydrology, Water Resources and Hydraulic Engineering (Grant No. 522012222), and the Basic research projects of central level public welfare research institutes (Grant No. Y522003).

Data Availability Statement: No new data were created or analyzed in this study. Data sharing is not applicable to this article.

Conflicts of Interest: The authors declare no conflict of interest.

Abbreviations

The following abbreviations are used in this manuscript:

FAO	Food and Agriculture Organization
FAO-56	FAO Irrigation and Drainage Paper No. 56
K_c	Crop coefficient
ET	Actual evapotranspiration
K_s	Water stress coefficient
TCARI	Transformed chlorophyll absorption in reflectance index
RDVI	Renormalized difference vegetation index
UAV	Unmanned aerial vehicle
TSEB	Two-source energy balance
ET_0	Reference evapotranspiration
SPSS	Statistical Product and Service Solutions
R_n-G	Effective energy
MATLAB	Matrix Laboratory
r	Correlation coefficient
d_{IA}	Consistency index
RMSE	Root mean square error
MAE	Mean absolute error
TE	Temperature effect method
CCC	Cumulative crop coefficient method
RST	Radiative soil temperature method

References

1. Daszkiewicz, T. Food Production in the Context of Global Developmental Challenges. *Agriculture* **2022**, *12*, 832. [CrossRef]
2. Wu, B.F.; Tian, F.Y.; Zhang, M.; Piao, S.L.; Zeng, H.W.; Zhu, W.W.; Liu, J.G.; Elnashar, A.; Lu, Y.M. Quantifying global agricultural water appropriation with data derived from earth observations. *J. Clean. Prod.* **2022**, *358*, 131891. [CrossRef]
3. Kabbadj, A.; Makoudi, B.; Mouradi, M.; Pauly, N.; Frendo, P.; Ghoulam, C. Physiological and biochemical responses involved in water deficit tolerance of nitrogen-fixing vicia faba. *PLoS ONE* **2017**, *12*, e0190284. [CrossRef] [PubMed]
4. Paul, A.S.; Panwar, N.L.; Salvi, B.L.; Jain, S.; Sharma, D. Experimental investigation on the production of bio-oil from wheat straw. *Energy Sources Part A Recovery Util. Environ. Eff.* **2020**, *6*, 1–16. [CrossRef]
5. Langridge, P.; Alaux, M.; Almeida, N.F.; Ammar, K.; Baum, M.; Bekkaoui, F.; Bentley, A.R.; Beres, B.L.; Berger, B.; Braun, H.J.; et al. Meeting the Challenges Facing Wheat Production: The Strategic Research Agenda of the Global Wheat Initiative. *Agronomy* **2022**, *12*, 2767. [CrossRef]
6. Xiang, C.; Huang, J.K. The role of exotic wheat germplasms in wheat breeding and their impact on wheat yield and production in china. *China Econ. Rev.* **2020**, *62*, 101239. [CrossRef]
7. Dong, J.; Fu, Y.Y.; Wang, J.J.; Tian, H.F.; Fu, S.; Niu, Z.; Han, W.; Zheng, Y.; Huang, J.X.; Yuan, W.P. Early-season mapping of winter wheat in china based on landsat and sentinel images. *Earth Syst. Sci. Data* **2020**, *12*, 3081–3095. [CrossRef]
8. Park, J.; Baik, J.; Choi, M. Satellite-based crop coefficient and evapotranspiration using surface soil moisture and vegetation indices in northeast asia. *Catena* **2017**, *156*, 305–314. [CrossRef]
9. Shan, N.; Zhang, Y.G.; Cheng, J.M.; Ju, W.M.; Migliavacca, M.; Penuelas, J.; Yang, X.; Zhang, Z.Y.; Nelson, J.A.; Goulas, Y. A model for estimating transpiration from remotely sensed solar-induced chlorophyll fluorescence. *Remote Sens. Environ.* **2021**, *252*, 112–134. [CrossRef]
10. Allen, R.G.; Pereira, L.S.; Raes, D.; Smith, M. *Crop Evapotranspiration-Guidelines for Computing Crop Water Requirements-FAO Irrigation and Drainage Paper 56*; FAO: Rome, Italy, 2008.
11. Er-Raki, S.; Chehbouni, A.; Guemouria, N.; Duchemin, B.; Ezzahar, J.; Hadria, R. Combining fao-56 model and ground-based remote sensing to estimate water consumptions of wheat crops in a semi-arid region. *Agric. Water Manag.* **2007**, *87*, 41–54. [CrossRef]
12. Vu, S.H.; Watanabe, H.; Takagi, K. Application of fao-56 for evaluating evapotranspiration in simulation of pollutant runoff from paddy rice field in japan. *Agric. Water Manag.* **2005**, *76*, 195–210. [CrossRef]
13. Mokhtari, A.; Noory, H.; Vazifedoust, M.; Bahrami, M. Estimating net irrigation requirement of winter wheat using model- and satellite-based single and basal crop coefficients. *Agric. Water Manag.* **2018**, *208*, 95–106. [CrossRef]
14. Spiliotopoulos, M.; Loukas, A. Hybrid Methodology for the Estimation of Crop Coefficients Based on Satellite Imagery and Ground-Based Measurements. *Water* **2019**, *11*, 1364. [CrossRef]
15. Zhang, Y.; Han, W.T.; Niu, X.T.; Li, G. Maize Crop Coefficient Estimated from UAV-Measured Multispectral Vegetation Indices. *Sensors* **2019**, *19*, 5250. [CrossRef] [PubMed]
16. Wang, Z.L.; Gu, N.; Lu, H.S.; Hu, Y.S.; Zhu, Y.H.; Yang, M. Calculation method of crop coefficient and evapotranspiration basedon temperature effect. *J. Hydraul. Eng.* **2019**, *50*, 242–251.
17. Tang, J.D. Estimation Method of Maize Water Use Efficiency and Biomass Based on UAV Remote Sensing. Master's Thesis, Northwest A F University, Xianyang, China, 2021.
18. Kool, D.; Kustas, W.P.; Ben-Gal, A.; Agam, N. Energy partitioning between plant canopy and soil, performance of the two-source energy balance model in a vineyard. *Agric. For. Meteorol.* **2021**, *300*, 108328. [CrossRef]
19. Paulino, S.E.P.; Mourao, F.D.A.; Maia, A.D.N.; Aviles, T.E.C.; Neto, D.D. Agrometeorological models for 'valencia' and 'hamlin' sweet oranges to estimate the number of fruits per plant. *Sci. Agric.* **2007**, *64*, 1–11. [CrossRef]
20. Ju, Q.; Liu, X.N.; Liu, Z.T.; Gao, H.B.; Zhao, W.J.; Nimazhaxi; Hao, W.L. Response and prediction of groundwater depth change to climate change in Huaibei Plain. *Trans. Chin. Soc. Agric. Eng.* **2022**, *38*, 136–145.
21. Zhu, Y.H.; Ren, L.L.; Horton, R.; Lu, H.S.; Wang, Z.L.; Yuan, F. Estimating the Contribution of Groundwater to the Root Zone of Winter Wheat Using Root Density Distribution Functions. *Vadose Zone J.* **2018**, *17*, 1–15. [CrossRef]
22. Xu, J.Z.; Peng, S.Z.; Ding, J.L.; Wei, Q.; Yu, Y.M. Evaluation and calibration of simple methods for daily reference evapotranspiration estimation in humid East China. *Arch. Agron. Soil Sci.* **2013**, *59*, 845–858. [CrossRef]
23. Huang, C.P.; Zhang, F.; Wang, A.H.; Hu, B.M.; Yang, J.H. Dynamic simulation of potato growth process. *J. Appl. Ecol.* **2004**, *15*, 1203–1206.
24. Qureshi, Z.A.; Neibling, H. Response of two-row malting spring barley to water cutoff under sprinkler irrigation. *Agric. Water Manag.* **2009**, *96*, 141–148. [CrossRef]
25. Rosa, H.T.; Walter, L.C.; Streck, N.A.; Alberto, C.M. Thermal time methods and sowing dates in phyllochron determination in wheat cultivars. *Pesqui. Agropecuária Bras.* **2009**, *44*, 1374–1382. [CrossRef]
26. Wechsung, F.; Ritter, M.; Wall, G.W. The upper homeostatic range for the temperature–yield response of irrigated us wheat down revised from a theoretical and experimental perspective. *Agric. For. Meteorol.* **2021**, *307*, 108478. [CrossRef]
27. Porter, J.R.; Gawith, M. Temperatures and the growth and development of wheat: A review. *Eur. J. Agron.* **1999**, *10*, 23–36. [CrossRef]
28. Qi, S.H.; Wang, B.J.; Wu, Z.S. The relationship between agricultural production and temperature. *J. Henan Inst. Sci. Technol. (Nat. Sci. Ed.)* **2007**, *35*, 20–23.

29. Su, L.J.; Wang, Q.J.; Bai, Y.G. An analysis of yearly trends in growing degree days and the relationship between growing degree day values and reference evapotranspiration in Turpan area, China. *Theor. Appl. Climatol.* **2013**, *113*, 711–724. [CrossRef]
30. Lewis, C.S.; Allen, L.N. Potential crop evapotranspiration and surface evaporation estimates via a gridded weather forcing dataset. *J. Hydrol.* **2017**, *546*, 450–463. [CrossRef]
31. Zhao, Y.; Mao, X.M.; Shukla, M.K.; Tian, F.; Hou, M.J.; Zhang, T.; Li, S.E. How does film mulching modify available energy, evapotranspiration, and crop coefficient during the seed–maize growing season in northwest china? *Agric. Water Manag.* **2021**, *245*, 106666. [CrossRef]
32. Weerasinghe, G.; Chi, H.M.; Cao, Y.Z. Particle Swarm Optimization Simulation via Optimal Halton Sequences. *Procedia Comput. Sci.* **2016**, *80*, 772–781. [CrossRef]
33. Benuwa, B.B.; Ghansah, B.; Wornyo, D.K.; Adabunu, S.A. A Comprehensive Review of Particle Swarm Optimization. *Int. J. Eng. Res. Afr.* **2016**, *23*, 141–161. [CrossRef]
34. Du, X.B.; Wang, Z.; Xi, M.; Wu, W.G.; Wei, Z.; Xu, Y.Z.; Zhou, Y.J.; Lei, W.X.; Kong, L.C. A novel planting pattern increases the grain yield of wheat after rice cultivation by improving radiation resource utilization. *Agric. For. Meteorol.* **2021**, *310*, 108625. [CrossRef]
35. Li, Y.X.; Liu, H.J.; Ma, J.C.; Zhang, L.X. Estimation of leaf area index for winter wheat at early stages based on convolutional neural networks. *Comput. Electron. Agric.* **2021**, *190*, 106480. [CrossRef]

hydrology

MDPI

Article

Soil–Water–Atmosphere Effects on Soil Crack Characteristics under Field Conditions in a Semiarid Climate

Jacques Carvalho Ribeiro Filho [1], Eunice Maia de Andrade [2], Maria Simas Guerreiro [3,*], Helba Araújo de Queiroz Palácio [4] and José Bandeira Brasil [1]

[1] Departamento de Engenharia Agrícola, Campus do Pici, Universidade Federal do Ceará, Fortaleza CEP 60455-760, Brazil
[2] Departamento de Conservação de Solo e Água, Universidade Federal Rural do Semi-Arido, Rua Francisco Mota, 572, Mossoró CEP 59625-900, Brazil
[3] I3ID, Universidade Fernando Pessoa, Praça 9 de Abril, 349, 4249-004 Porto, Portugal
[4] Instituto Federal de Educação, Ciência e Tecnologia do Ceará, Rodovia Iguatu-Várzea Alegre km, 5, Iguatu CEP 63503-790, Brazil
* Correspondence: mariajoao@ufp.edu.pt

Abstract: Soil's physical and hydrological properties influence the proper modeling, planning, and management of water resources and soil conservation. In areas of vertic soils subjected to wetting and drying cycles, the soil–water–atmosphere interaction is complex and understudied at the field scale, especially in dry tropical regions. This work quantifies and analyzes crack development under field conditions in an expansive soil in a semiarid region for both the dry and rainy seasons. Six 1 m^2 plots in an experimental 2.8 ha watershed were photographed and direct measurements were taken of the soil moisture and crack area, depth and volume once a week and after a rainfall event from July 2019 to June 2020. The rainfall was monitored for the entire period and showed a unimodal distribution from December to May after five months without precipitation. The cracks were first sealed in the plots with a predominance of sand and when the soil moisture was above 23% and had an accumulated precipitation of 102 mm. The other plots sealed their cracks when the soil moisture was above 32% and with an accumulated precipitation in the rainy season above 222 mm. The cracks redeveloped after sealing upon a reduction of 4% in the soil moisture. The depth of the cracks showed a better response to climatic variations (total precipitation, soil moisture and continuous dry and wet days). The higher clay content and the higher plasticity index plots developed more cracks with greater depth and volume.

Keywords: vertic soils; crack dynamics; tropical dry regions; semiarid

Citation: Ribeiro Filho, J.C.; Andrade, E.M.d.; Guerreiro, M.S.; Palácio, H.A.d.Q.; Brasil, J.B. Soil–Water–Atmosphere Effects on Soil Crack Characteristics under Field Conditions in a Semiarid Climate. *Hydrology* **2023**, *10*, 83. https://doi.org/10.3390/hydrology10040083

Academic Editors: Songhao Shang, Qianqian Zhang, Dongqin Yin, Hamza Gabriel and Magdy Mohssen

Received: 28 February 2023
Revised: 24 March 2023
Accepted: 2 April 2023
Published: 4 April 2023

1. Introduction

Arid and semiarid ecosystems account for approximately 29.8% of the Earth's surface [1]. Global warming and associated climate changes may increase the risk of extreme phenomena—droughts and heavy rains, which directly affect hydrological processes [1,2]. Forecasts of more extreme climate regimes will make arid and semiarid ecosystems become more vulnerable, with the possibility of increasing the total area of dry land globally [3].

Soil's physical, chemical and biological characteristics [1,4–7] influence runoff, infiltration and evapotranspiration [2,6,7], and some expansive clays (e.g., montmorillonite) swell and shrink during the wetting and drying processes [2,8], affecting the hydrologic processes.

Changes in the water content in expansive soils can significantly alter the hydromechanical behavior of a soil. These effects mainly include: (1) voluminous change due to swelling and shrinkage, and (2) variation in mechanical behavior, such as strength (or stress) and compression [8]. Soils and clay minerals absorb water and expand upon wetting, and they shrink and form desiccation cracks as they dry. Cracks modify the processes of infiltration, flow, evapotranspiration, and redistribution of water in the soil profile [4,6,9], and they also form preferential flow channels with faster transport of water and nutrients (solutes) that negatively affect irrigation and soil fertilization for agricultural use [5,10].

The expansion/contraction characteristic of these soils has implications for distinct fields—in geotechnical engineering, where the formation of crack networks can destroy the integrity of the soil structure, damaging road structures and other civil construction infrastructures [7,11]; and in agriculture, where soil cracks may impose limitations on crop production [12], promote physical damage to plant roots, encourage the vertical movement and leaching of dissolved nutrients beyond the root zone, provide extra surface for moisture loss, and even promote rill erosion.

Drought-induced cracks in the soil are usually complex network structures. The accurate acquisition of cracks' morphometric data is not only a prerequisite for obtaining the relevant crack networks' geometric parameters, but also a basis for better understanding the cracks' development mechanism and defining procedures to promote or minimize soil cracks [13,14].

Research on cracks in expansive soils has been mostly conducted in controlled environments for the quantification of the geometry of a single crack and the morphology of crack networks [1,2,5,15,16], and not so much in field-scale studies to understand the dynamics of the cracks through in situ observations of the soil surface [2,8,16]. Changes in the hydraulic properties of expansive soils (e.g., soil moisture) under field conditions may help explain the response of soil cracks' properties to climate dynamics in time [2,5].

The objectives of this study were to (a) assess and quantify the soil characteristics and dynamics that govern the crack formation and healing processes under natural conditions; and (b) quantify the soil moisture limits on the response of soil swelling and shrinking in the wet and dry seasons in a vertic soil in a semiarid region under natural conditions.

2. Materials and Methods

2.1. Study Area

The experimental area is a 2.8 ha first-order catchment with a 5.6% slope. The soil has a depth of 2.0 m and is classified as vertisol, with a predominance of expansive 2:1 clay minerals from the montmorillonite group [17]. It is located in a representative fragment of a seasonally dry tropical dry forest in northeastern Brazil (Figure 1) under vegetation regeneration after clearing, burning and planting pasture in 2010.

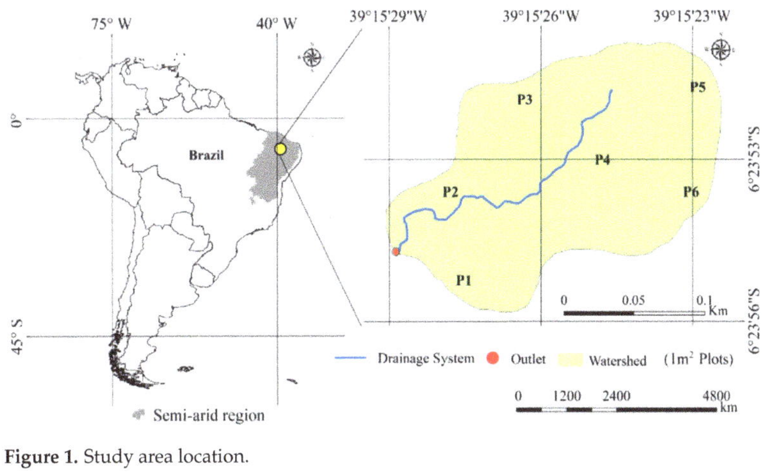

Figure 1. Study area location.

The climate is classified as a subtropical steppe (BSh—low-latitude semiarid or dry), according to Köppen's classification. With an aridity index of 0.48, it has a mean annual potential evapotranspiration of 2113 mm year^{-1} and a mean annual precipitation of

997 ± 300 mm. Some 89% percent of the annual rainfall is concentrated in the wet semester of December to May (Figure 2) [18,19].

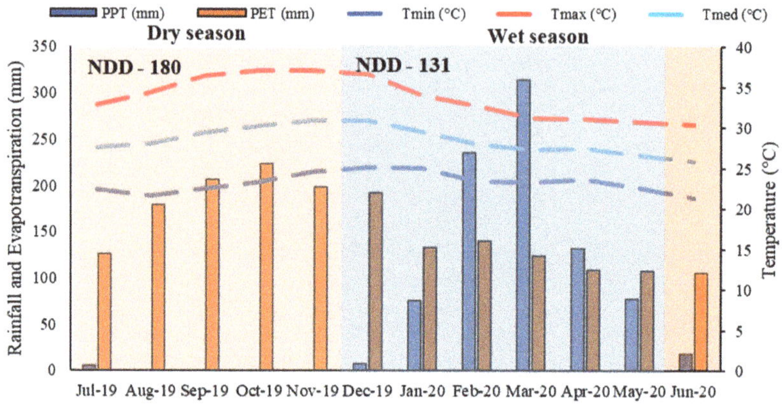

Figure 2. Mean rainfall; potential evapotranspiration; minimum, maximum and average temperatures; and average number of dry days for the wet and dry seasons in the study area from 1980 to 2020. PPT is the average monthly rainfall (mm); PET is the average monthly potential evapotranspiration (mm); Tmin, Tmax, and Tmed are the average monthly minimum, maximum, and mean temperatures (°C), respectively, for the study period; and NDD is the seasonal average number of dry days from 1980 to 2020.

Soil samples were collected with a sampling probe for textural classification and determination of the physical characteristics at each plot. A cluster analysis was performed on the soil characteristics to explore the naturally occurring groups using the software IBM SPSS Statistics 27.

2.2. Monitoring

Rainfall was assessed using a Ville de Paris rain gauge. The soil moisture content was determined weekly and after a rainfall event by means of the gravimetric method in triplicate. Soil samples were collected outside the border of each plot, as this border sampling pattern better estimates the soil moisture of the plot without compromising its soil structure.

The soil cracks were monitored at six 1 m^2 (1 m × 1 m) randomly located experimental plots (Figure 1) for one year—1 July 2019 to 30 June 2020 (Figure 3). All the vegetation inside and around the plots was removed before the field measurements (at least weekly), keeping the plots free of vegetation for the entire period of study to minimize the soil structure changes.

The crack monitoring consisted of the in situ location of the cracks and the measurement of their respective depth. Cracks were identified with the aid of a 0.05 m × 0.05 m net placed over the 1 m^2 experimental plots, totaling a mesh of 400 points (Figure 4). The crack depth was measured using a 4 mm diameter rod and a ruler at the intersection of each crack and the 0.05 m × 0.05 m net (Figure 4).

Figure 3. Monthly development of soil cracks in an experimental plot.

Figure 4. Crack monitoring in situ.

2.3. Crack Morphology and Dynamics

The recognition of individual cracks and the assessment of the respective surface area, average width, and length was performed using the software Crack Image Analysis System Version 2.32 (CIAS) [20] based on the in situ acquired images (Figure 5). A Sony® DSC-H9 camera placed on a tripod always at the same location collected the plot images. Photogra hic images were taken at the best light hours (between 12:00 and 1:00 p.m.), except on rainy days, when photos were taken after the end of the event or on the next morning.

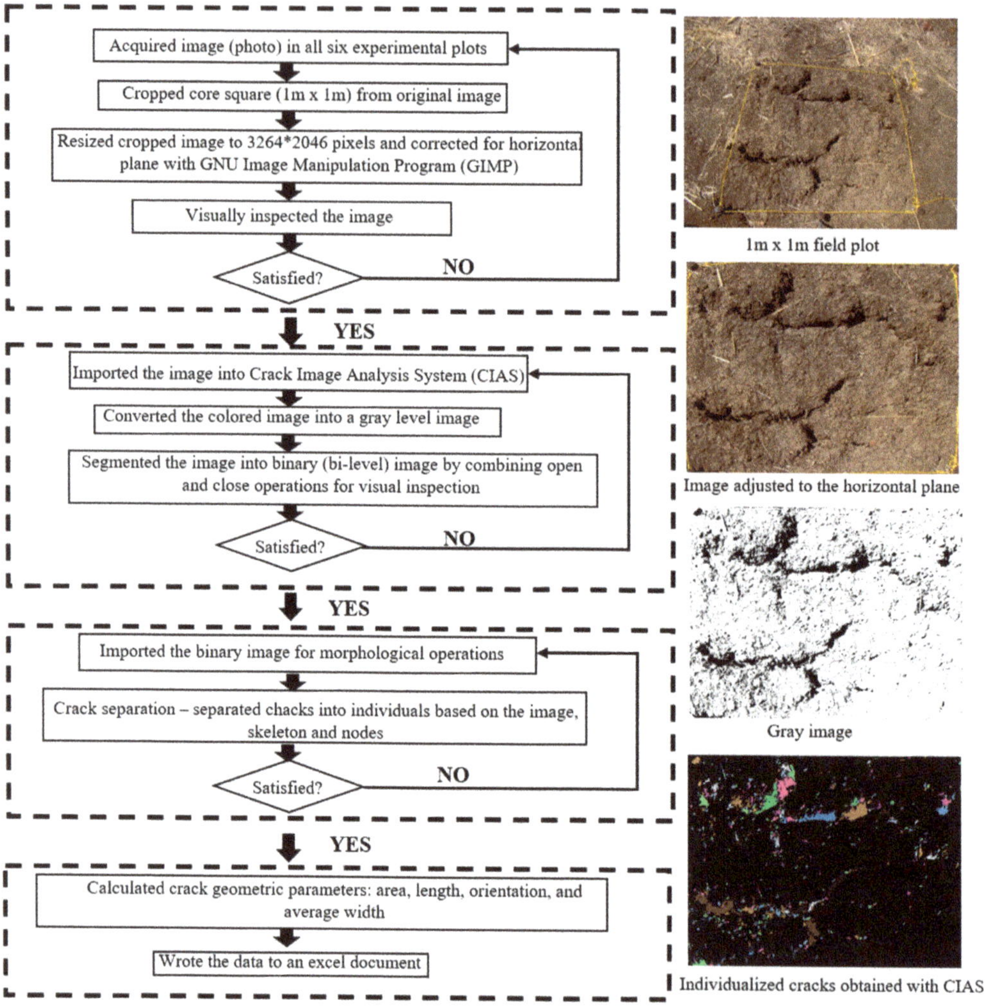

Figure 5. Methodology for image collection, correction and parameter calculation.

The volume of the cracks was computed using the crack average depth obtained via the in situ monitoring and the area was assessed using the CIAS software Version 2.32, as proposed by [21]. The soil crack area density (*Dc*) (Equation (1)) was evaluated using the method proposed also by [21]. The level of development (Table 1) was based on [16]:

$$Dc = \frac{a_c}{A_t} * 100 \quad (1)$$

where a_c is the crack area (m^2) and A_t is the total area (1 m^2).

The crack area velocity represents the rate of development of the crack area in both the swelling and the shrinking stages. The assessment of the temporal and spatial variability of the cracks' swelling and shrinking included the crack area formation velocity and associated correlations with climate factors, soil moisture and soil characteristics. We assumed the linear variation of the parameters (Dc, depth, and soil moisture) with time between field monitoring visits. All the analyses were based on the average depth, total area and total volume of the cracks at each plot.

Table 1. Soil cracks' level of development based on the crack area density.

Dc Intervals	Levels of Development
Dc ≤ 5%	Feeble
5 < Dc ≤ 10	Light
10 < Dc ≤ 22	Medium
22 < Dc ≤ 27	Intensive
Dc > 27	Extremely intensive

3. Results

3.1. Field Data

A total rainfall of 869 mm distributed in 54 events (Figure 6) was recorded during the study period (19 July–20 June). Only three rainfall events occurred in the dry season, with a total of 9 mm that led to no runoff. The wet season had seven runoff events in both February and March and one runoff event in May. February showed 23% of the rainfall of the wet season (12 events), and March concentrated 33% of that rainfall (17 events). The cracks were sealed for 16 days in March and opened after 2 consecutive days without precipitation.

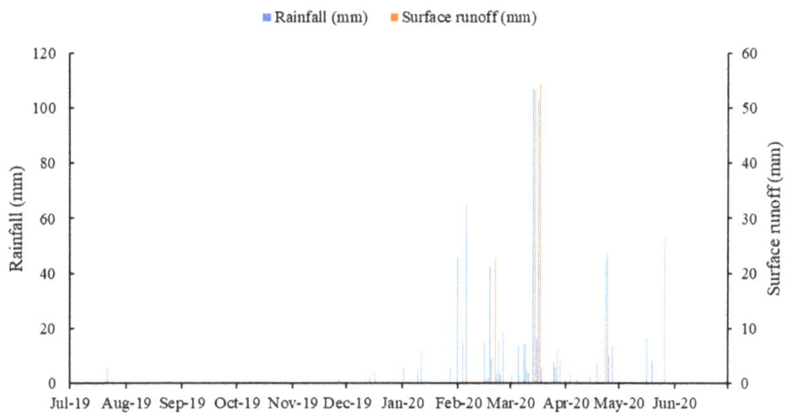

Figure 6. Rainfall and runoff events in the study period.

The total sealing of the cracks occurred on 18 February 2020, with an accumulated rainfall of 236 mm, of which 75% of the rainfall events were less than 10 mm and 52% less than 5 mm. During the first surface runoff event of the year (5 February 2020), the cracks were not totally sealed, although there were no surface cracks upon the occurrence of the other runoff events.

The spatial variability of the soils in the catchment was highlighted by the plots' physical soil properties (Table 2). Plots P3 and P6 stood out for the sand content (above 40%), classifying them as sandy clay loams—SCL (Table 2). The plasticity index (PI) and liquid limit (LL) in both plots showed the lowest values, as well as the base saturation (percentage of cation exchange capacity occupied by base cations) due to the sand content of these soils.

A cluster analysis of the soil properties of the experimental plots revealed two groups: one formed by the clay loam plots (CL) and the other formed by the sandy clay loam plots (SCL). This was confirmed by the significant difference (p-value < 0.05) in the physical parameters between the soils in plots P3 and P6 and the others (Figure 7). The P3 and P6 plots were more distant from the stream (Figure 1), suggesting that there was transport of finer particles to the lower zone (i.e., stream).

Table 2. Soil properties of the experimental plots.

Soil Properties	Experimental Plots					
	P1	P2	P3	P4	P5	P6
Grain size analysis						
Sand (%)	21	26	41	22	27	44
Silt (%)	43	46	33	41	41	36
Clay (%)	36	28	26	37	32	20
Organic matter (%)	0.8	1.5	1.3	1.4	1.9	2.2
Base saturation (%)	92	95	84	95	92	89
pH	6.6	7.1	6	6.8	6.6	6.7
Specific gravity	2.79	2.57	2.53	2.53	2.48	2.52
Liquid limit (%)	43	38	28	42	40	33
Plastic limit (%)	14	27	21	32	26	27
Plasticity index	30	12	7	10	14	7
Textural classification	Clay loam	Clay loam	Sandy clay loam	Clay loam	Clay loam	Sandy clay loam

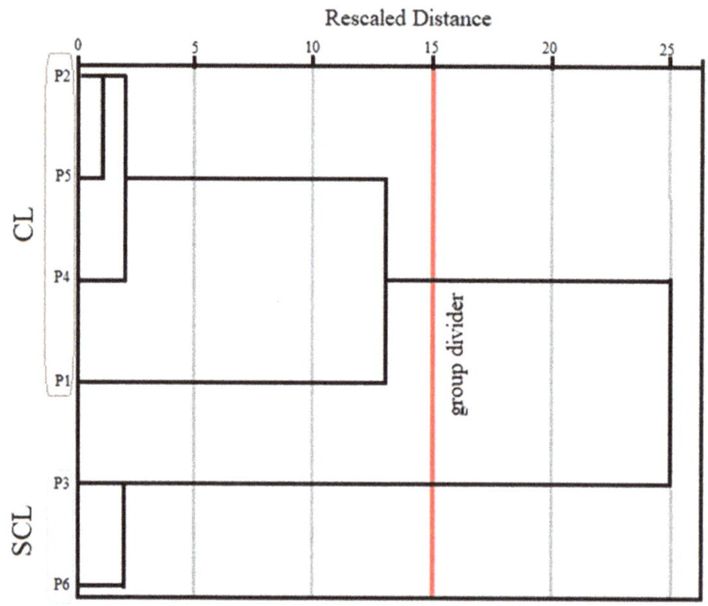

Figure 7. Cluster analysis results concerning the plots' soil characteristics. CL represents the clay loam cluster and SCL represents the sandy clay loam cluster.

3.2. *Crack Morphology*

The crack depth in the clay loam plots varied from zero (no surface cracks) to a maximum depth of 0.12 m to 0.22 m (Figures 8 and 9). The maximum values were observed in P1, closer to the outlet, where the soil showed greater plasticity (Table 1).

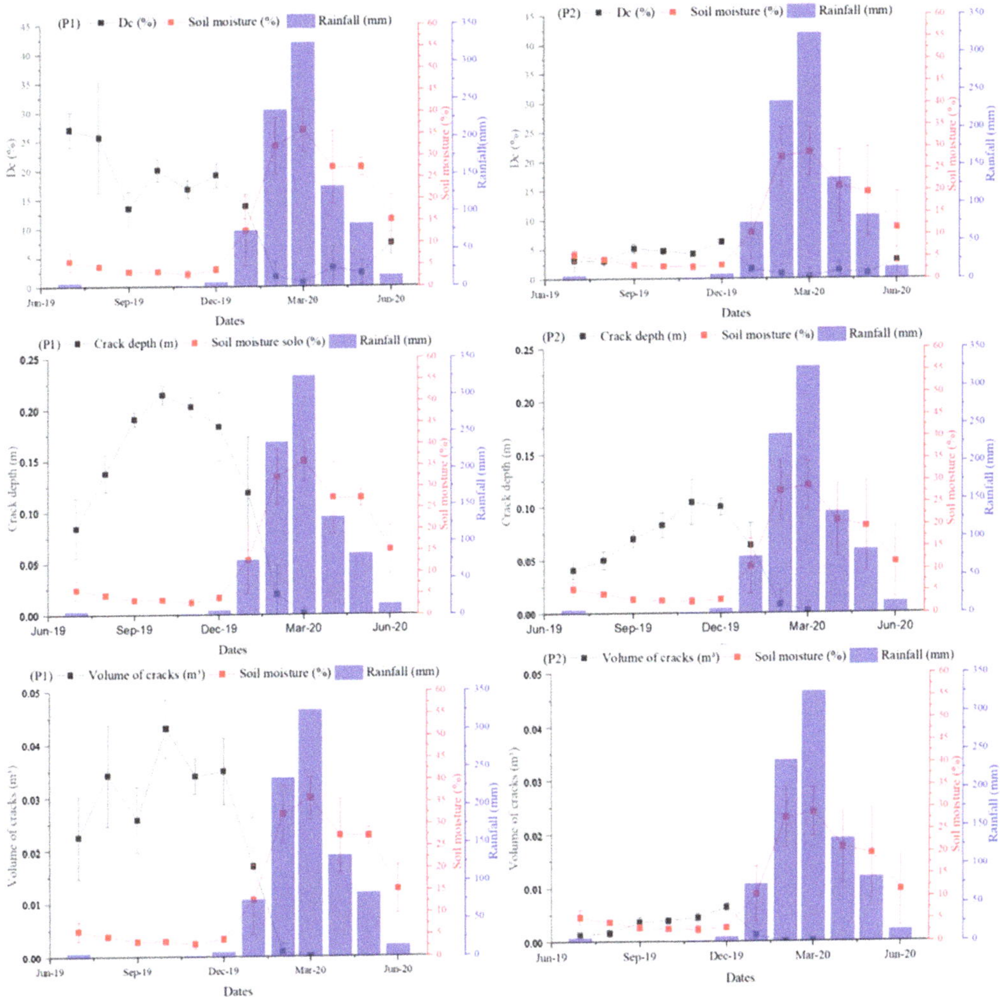

Figure 8. CL cluster plots P1 and P2—monthly Dc, area, depth, soil moisture and rainfall.

On the sandy clay loam plots, the cracks' depth varied from zero (no surface cracks) to 0.06 m (P3), as evidenced by Figure 10. The cracks showed an approximately constant value of 0.04 m after confirmation of the dry season in July–August and remained constant until the beginning of the rainy season in December. The plots with the least Dc and crack volume variability were the sandy clay loam plots—P3 and P6 (Figure 10), which also showed a lower fine particle percentage (<60%), a lower plasticity index (7.0), and a liquid limit below 33% (Table 2).

Based on the number of soil cracks and the respective morphometric characteristics, we observed a greater number, depth, and volume at the end of the dry season (Figures 8–10), as expected. The deepest cracks and higher Dc and crack volume were recorded in the CL group plots (P1—downstream and P5—upstream), as well as the greatest variability in these parameters. The smallest variability in the depth of the cracks, Dc and crack volume occurred in the sandy clay loam plots, with the lowest fine particle contents (P3 and P6). The lowest monthly mean values were recorded in the P6 plot—0.035 m and 0.0013 m^3 for the crack depth and crack volume, respectively (in the dry period).

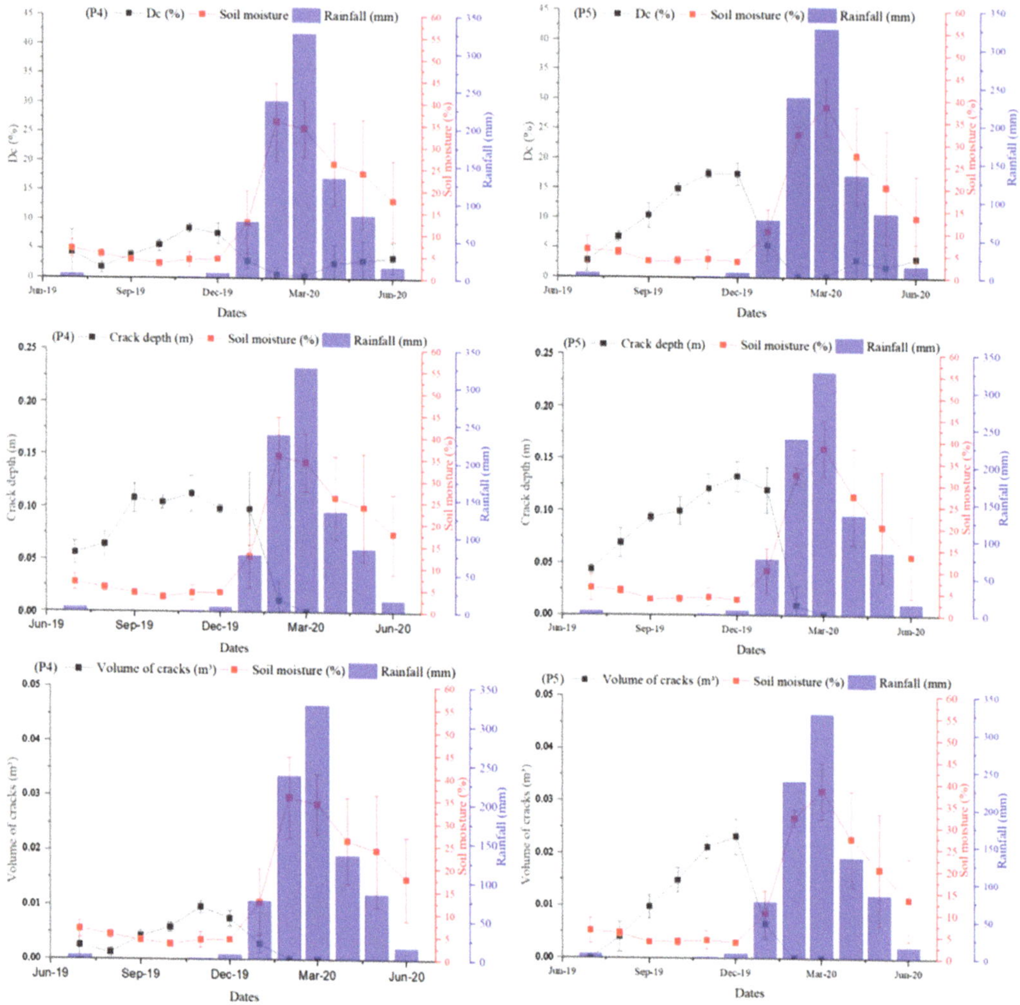

Figure 9. CL cluster plots P4 and P5—monthly Dc, area, depth, soil moisture and rainfall.

In the dry period, the crack depth varied from 0.034 m to 0.225 m (both in plot P1) in the clay loam plots (Figures 8 and 9), whereas in the sandy clay loam plots, the variation was smaller—0.026 m to 0.066 m (both in plot P6) (Figure 10). For a small variation in the soil moisture content, there was a quick response from the crack depth in the clay loam plots when compared to the sandy clay loam plots, which showed little response.

The rainfall accumulation until February was 87.1 mm, which was not enough to totally close the cracks—the crack depth was higher in the clay loam plots than in the sandy loam plots. Even though there was precipitation in the wet season after March, the wettest month of the year, the monthly rainfall decreased, with an increasing response from the Dc at all the plots (Figures 8–10).

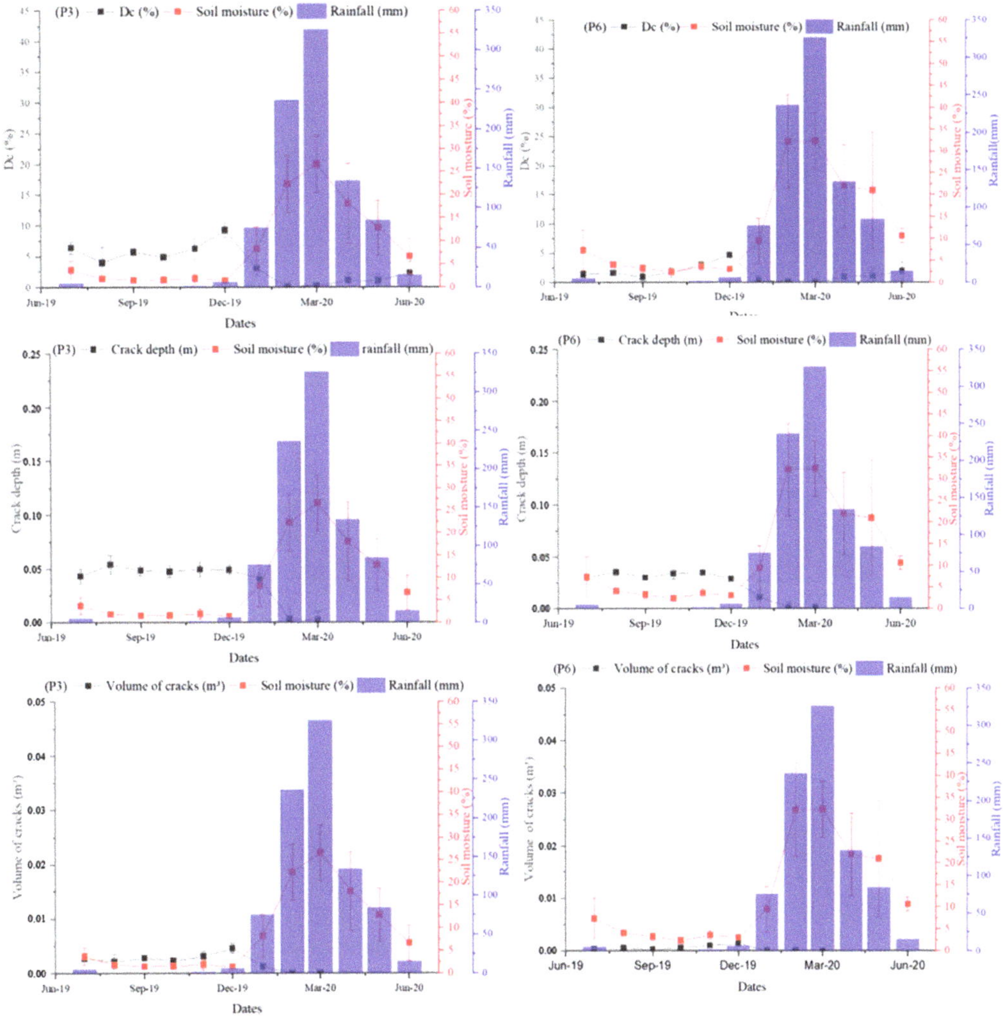

Figure 10. SCL cluster plots P3 and P6—monthly Dc, area, depth, soil moisture and rainfall.

3.3. Crack Dynamics

The Dc showed a significant logarithmic correlation with the soil moisture content ($p < 0.001$) in all the plots of the CL group (Figure 11). The SCL group (Figure 12) did not show a significant correlation ($p < 0.001$) in either plot.

The SCL cluster plots showed an initial crack sealing process (Dc = 0) 70 days after the beginning of the rainy season, with a cumulative rainfall of 102 mm and a soil moisture content above 23% (Figures 10 and 12). As for the CL group (Figures 8 and 9), the sealing only occurred 87 days after the beginning of the rainy season, with a total cumulative rainfall of 222 mm, a soil moisture content above 32% (Figure 11), and after a rainfall event of 52.4 mm. There seemed to be no pattern in the response of the soil crack area to a soil moisture content of up to 7% in the SCL and 13% in the CL plots, respectively.

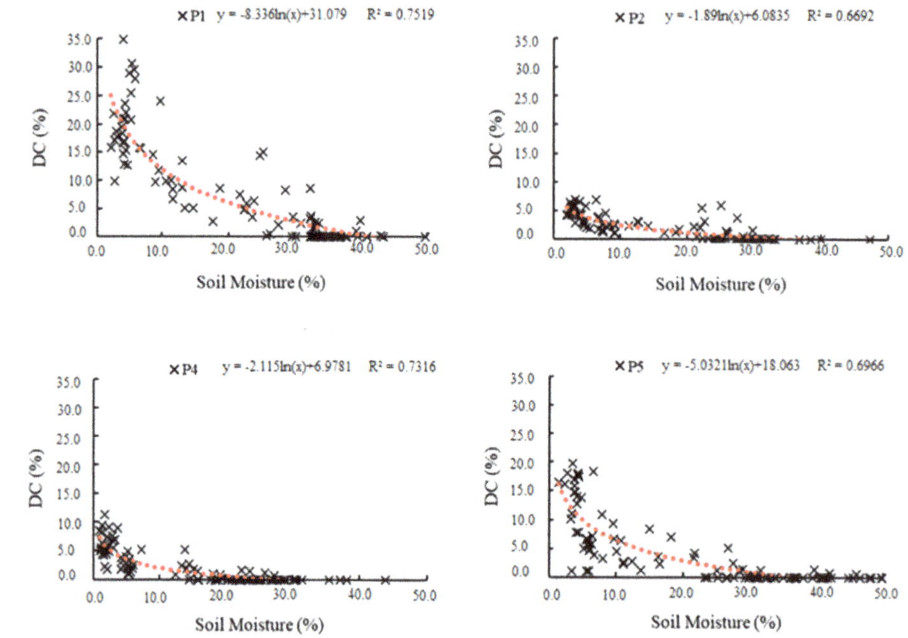

Figure 11. Soil moisture vs. Dc for the CL cluster plots P1, P2, P4 and P5.

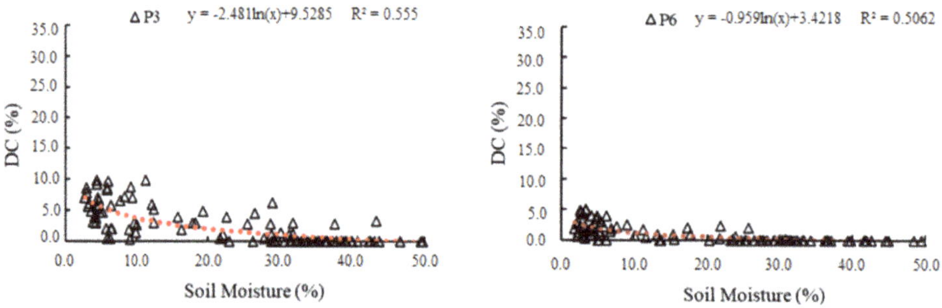

Figure 12. Soil moisture vs. Dc for the SCL cluster plots P3 and P6.

A closer look at the swelling and shrinking behavior of the soil highlighted a remarkable difference between the CL plots and the SCL plots—a maximum value of the soil crack area density of 0.37 m^2 in the CL plots against 0.10 m^2 in the SCL plots, and crack occurrence up to a soil moisture content of 43% in the CL against 27% in the SCL (Figure 13a,b).

The limit to the crack opening and sealing on the drying and wetting cycles showed a difference of 6% in the soil moisture content—when drying, the CL plots started opening cracks at a soil moisture below 43%, whereas as it reached a soil moisture content of 37% upon wetting, the cracks were sealed. Similar behavior was observed in the SCL plots, with a smaller difference of 3% in the soil moisture content, which began the crack opening process at a soil moisture content below 27% when drying and a soil moisture content below 24% when wetting (Figure 13c,d).

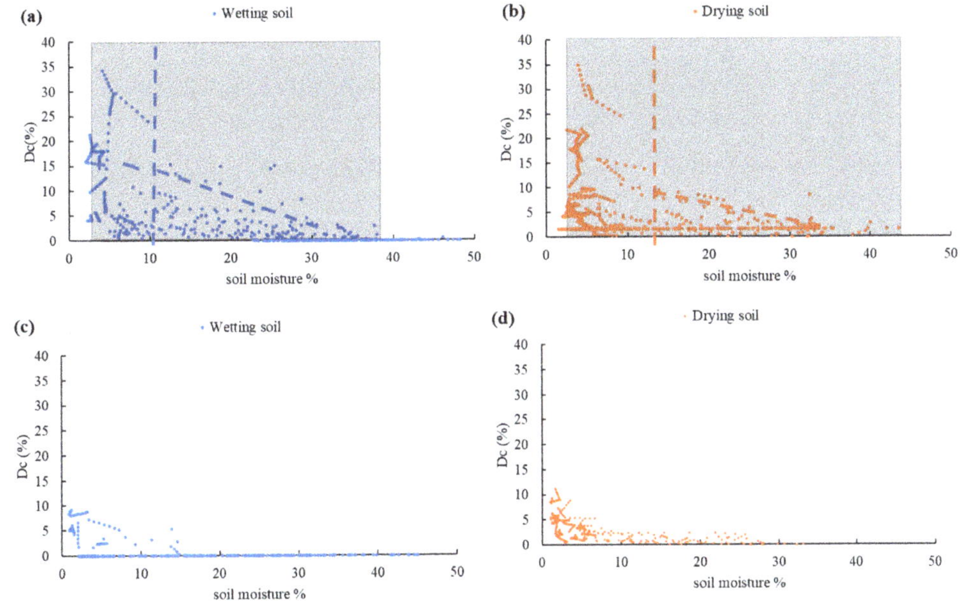

Figure 13. Crack area density: (**a**) for wetting soil—CL; (**b**) for drying soil—CL; (**c**) for wetting soil—SCL; and (**d**) for drying soil—SCL.

The crack area opening/closing response velocity (Figure 14) in the CL was below 0.1 m^2 day^{-1} and in the SCL below 0.06 m^2 day^{-1} (just above half the response velocity of the CL). The fastest response time of the crack opening/sealing to the soil moisture was below 45% in the CL and below 28% in the SCL. There was a remarkable difference in the crack response in the soils with finer particles above 70% (CL) relative to the soil moisture despite the clay content, both upon the drying and wetting cycles.

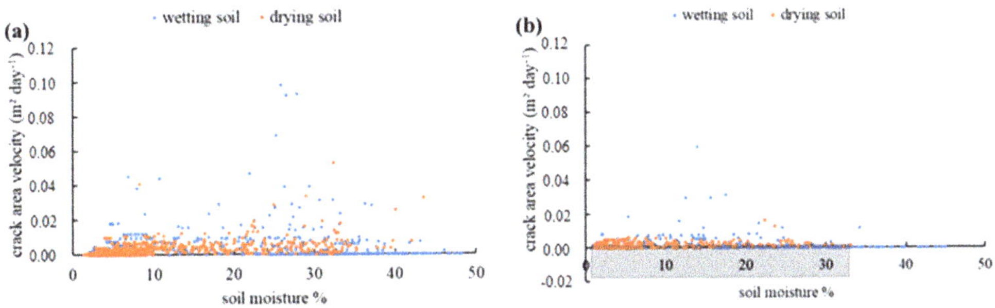

Figure 14. Velocity of crack opening/closure (m^2/day) in the (**a**) CL and (**b**) SCL plots.

4. Discussion

4.1. Crack Morphology

The swelling and shrinking processes in expansive clay soils showed sensitivity to the fine particle content. The limits of the soil moisture content on the soil crack formation and sealing were significantly different for the clay loams and sandy clay loams, as were the crack area density limits [22].

The shrinking of expansive clays during dry periods and swelling with the occurrence of wet days [23] promote an increase/decrease in the depth, intensity, and volume of cracks (Figures 6 and 8–10). After a rainfall event, both in the wet and dry seasons, the clay swelling process began in the surface layers as the hydrostatic forces increased [2], promoting the sealing of micro cracks even before the surface runoff process began. At the onset of the surface runoff, the cracks were not all sealed (Figures 6 and 8–10), and the deeper cracks acted as preferential flow paths [4,15], which influenced solute transport—the faster water and dissolved solutes flowed through the preferential paths with reduced soil sorption opportunities by the clay particles, promoting considerable water loss and groundwater pollution, as suggested by [24].

The smaller pores in the soils with higher clay contents promoted the development of high soil suction pressure [25] and cohesion [26], which led to crack formation. Water sorption by the clay particles at the deepest parts of the cracks promoted their closure in an upward movement and the reduction of the depth of the cracks throughout the rainy season (Figures 8–10).

The clay content in itself did not explain the different behavior of the crack area formation as a function of the soil moisture in this study—all the plots had a montmorillonite clay content above 30%. However, the behavior of the Dc as a function of the soil moisture (Figures 11 and 12) was different for both the sandy loam and the clay plots, which differed due to the fine particles content, liquid limit and plasticity. Such evidence resulted in the different volume of cracks between plot P1, with a higher volume and greater plasticity, and P4 and P5 (Figures 8–10), with lower volumes and lower plasticity indexes, although these plots, P1, P4 and P5, presented similar texture [2].

The process of crack sealing occurred gradually and inversely to the soil moisture content (Figures 8–10). From the beginning of the rainy season and until the total closure of the cracks, they generated preferential flow paths [4], depending mostly on the cumulative rainfall and depth of the event.

The expansion/contraction of the cracks in the soil was more sensitive in the wetting than in the drying process (Figure 13a,b). The SCL plots showed little response to crack formation for soil moisture contents between 5% and 25%, and no defined pattern relative to soil moisture contents below 5%. The CL plots showed a higher crack formation/sealing sensitivity for soil moisture contents between 13% and 42%, and no defined pattern relative to soil moisture contents below 13%. The expansion/shrinkage processes may respond differently to different climatic conditions due to the intrinsic soil characteristics, such as the granulometry and plasticity [2,27,28].

4.2. *Crack Dynamics*

The reduction of 4% in the soil moisture (Figures 8–10) was enough for the emergence of cracks. As cracks formed, soil water evaporated in two dimensions—vertically via the soil surface and horizontally through the walls of the cracks [2], provided there was available water. As the water availability decreased, the velocity of the crack formation remained constant for both the clay loams and sandy clay loams, although it was faster in the clay loam plots (Figure 14). The swelling and shrinking processes occurred at a faster pace in the clay loam plots [1,26], as stressed by the different soil sealing time (Figures 8–10), suggesting correlation with their higher plasticity index (Table 1).

The degree of development of the cracks in the CL plots recorded a medium level of development for the Dc (Table 1), evidencing that the reduction of moisture in soils with higher rates of plasticity resulted in initially mild and then pronounced deformations when compared with the less plastic soils (SCL) [2,8]. The lower fine particle content in plots P3 and P6 (Table 2) reduced the degree of self-healing, as governed by the soil plasticity, which determines the potential for soil expansion and contraction [26].

It is known that the Dc decreases with the addition of water to the soil. During the dry season in this region, the possible sources of water entry into the soil are the processes of capillary rise and/or condensation of water vapor during the early morning hours [29]. The process of capillary rise was discarded, since there was no reduction in the depth of the cracks (Figures 8–10), evidencing a surface phenomenon.

Although there were no records of isolated precipitation events during the dry season (July to December) (except for three events > 10 mm, Figure 6), in August and September there was a reduction in the value of the Dc (Figures 8–10). The increase in the soil surface moisture content was believed to be the result of dew formation in other semiarid regions [1,30,31]. The process of increasing the soil moisture via the condensation process in the months with the lowest minimum temperatures is supported by [29] in an area adjacent to the studied watershed, raising the soil moisture at night by as much as 5%, which may be responsible for the reduction of the Dc (Figures 8–10).

The CL group presented an extremely intensive level of development (Figures 8 and 9), suggesting a need for greater initial abstractions to seal the surface cracks (Figures 6 and 8–10). There seemed to be a greater risk of soil and aquifer contamination during the drying process as the total crack area and depth were greater, offering preferential flow paths and a reduced opportunity for the adsorption of fertilizers by plants [24].

The sealing and formation of cracks in the SCL plots occurred at soil moisture contents above 25% and 27% in the wetting and drying processes, respectively. In the CL plots, these values were 38% and 43%, respectively. These results highlighted the hysteresis of these processes in both soil types, which may have occurred due to the fact that there was no increase in the soil moisture in the drying process and an increase and decrease in the soil moisture in the wetting processes. The soil moisture losses were mainly due to evaporation that occurred in two dimensions: horizontally via the soil surface and vertically from the walls of the cracks [4] and the rainfall events (Figure 6).

5. Conclusions

The intensity of the occurrence, maximum depth and volume of cracks vary according to the texture, limits of plasticity and soil liquidity. A lower fine particle content reduces the cracks' healing process, which is governed mostly by the soil plasticity. Higher fine particle contents lead to larger occurrence, depth and volume despite the clay content if associated with greater plasticity.

Cracks form after two consecutive dry days even during the wet season. The expansion/contraction of cracks in the soil is more sensitive in the wetting than in the drying process. Clay loams and sandy clay loams show different limits on the soil moisture content to start the opening and the sealing processes of crack formation.

There is no pattern in the response of crack formation to soil moisture contents below 5% and 13% for sandy clay loams and clay loams, respectively. The process of capillary rise impact on crack healing may be discarded, evidencing it being a surface phenomenon. There is a greater risk of soil and aquifer contamination during the drying process.

Author Contributions: All the authors made a significant contribution to the final version of the manuscript. Conceptualization, J.C.R.F., M.S.G., E.M.d.A. and H.A.d.Q.P.; methodology, M.S.G., J.B.B. and J.C.R.F.; writing—original draft preparation, J.C.R.F., M.S.G. and E.M.d.A.; writing—review and editing, all authors; supervision, M.S.G. and E.M.d.A.; project administration, E.M.d.A. All authors have read and agreed to the published version of the manuscript.

Funding: This work was supported by CNPq—Conselho Nacional de Desenvolvimento Científico e Tecnológico, Brazil (grant number 558135/2009-9).

Data Availability Statement: Data sharing is not applicable to this article.

Acknowledgments: This study was carried out with the support of the Coordenação de Aperfeiçoamento de Pessoal de Nível Superior—Brasil (CAPES), the Conselho Nacional de Desenvolvimento Científico e Tecnológico (CNPq) and the Fundação Cearense de Apoio ao Desenvolvimento Científico e Tecnológico (FUNCAP).

Conflicts of Interest: The authors declare that they have no conflict of interest.

References

1. Wang, C.; Zhang, Z.Y.; Qi, W.; Fan, S.M. Morphological approach to quantifying soil cracks: Application to dynamic crack patterns during wetting-drying cycles. *Soil Sci. Soc. Am. J.* **2018**, *82*, 757–771. [CrossRef]
2. Tang, C.S.; Zhu, C.; Cheng, Q.; Zeng, H.; Xu, J.J.; Tian, B.G.; Shi, B. Desiccation cracking of soils: A review of investigation approaches, underlying mechanisms, and influencing factors. *Earth-Sci. Rev.* **2021**, *216*, 103586. [CrossRef]
3. Dai, A. Increasing drought under global warming in observations and models. *Nat. Clim. Chang.* **2013**, *3*, 52–58. [CrossRef]
4. Santos, J.C.N.; de Andrade, E.M.; Guerreiro, M.J.S.; Medeiros, P.H.A.; de Queiroz Palácio, H.A.; de Araújo Neto, J.R. Effect of dry spells and soil cracking on runoff generation in a semiarid micro watershed under land use change. *J. Hydrol.* **2016**, *541*, 1057–1066. [CrossRef]
5. Bordoloi, S.; Ni, J.; Ng, C.W.W. Soil desiccation cracking and its characterization in vegetated soil: A perspective review. *Sci. Total Environ.* **2020**, *729*, 138760. [CrossRef]
6. Sadeghi, S.H.; Kheirfam, H.; Darki, B.Z. Controlling runoff generation and soil loss from field experimental plots through inoculating cyanobacteria. *J. Hydrol.* **2020**, *585*, 124814. [CrossRef]
7. Cheng, Q.; Tang, C.S.; Xu, D.; Zeng, H.; Shi, B. Water infiltration in a cracked soil considering effect of drying-wetting cycles. *J. Hydrol.* **2021**, *593*, 125640. [CrossRef]
8. Tang, C.S.; Wang, D.Y.; Zhu, C.; Zhou, Q.Y.; Xu, S.K.; Shi, B. Characterizing drying-induced clayey soil desiccation cracking process using electrical resistivity method. *Appl. Clay Sci.* **2018**, *152*, 101–112. [CrossRef]
9. Tang, C.S.; Shi, B.; Liu, C.; Suo, W.B.; Gao, L. Experimental characterization of shrinkage and desiccation cracking in thin clay layer. *Appl. Clay Sci.* **2011**, *52*, 69–77. [CrossRef]
10. Ralaizafisoloarivony, N.; Degré, A.; Mercatoris, B.; Leonard, A.; Toye, D.; Charlier, R. Assessing Soil Crack Dynamics and Water Evaporation during Dryings of Agricultural Soil from Reduced Tillage and Conventional Tillage Fields. *Multidiscip. Digit. Publ. Inst. Proc.* **2020**, *30*, 59. [CrossRef]
11. Al-Jeznawi, D.; Sanchez, M.; Al-Taie, A.J. Using image analysis technique to study the effect of boundary and environment conditions on soil cracking mechanism. *Geotech. Geol. Eng.* **2021**, *39*, 25–36. [CrossRef]
12. Elias, E.A.; Salih, A.A.; Alaily, F. Cracking patterns in the Vertisols of the Sudan Gezira at the end of dry season. *Int. Agrophysics* **2021**, *15*, 151–155.
13. Wei, X.; Hattab, M.; Bompard, P.; Fleureau, J.M. Highlighting some mechanisms of crack formation and propagation in clays on drying path. *Géotechnique* **2016**, *66*, 287–300. [CrossRef]
14. Xu, J.J.; Zhang, H.; Tang, C.S.; Cheng, Q.; Liu, B.; Shi, B. Automatic soil desiccation crack recognition using deep learning. *Géotechnique* **2022**, *72*, 337–349. [CrossRef]
15. Dinka, T.M.; Morgan, C.L.; McInnes, K.J.; Kishné, A.S.; Harmel, R.D. Shrink–swell behavior of soil across a Vertisol catena. *J. Hydrol.* **2013**, *476*, 352–359. [CrossRef]
16. Xiong, D.; Yan, D.; Long, Y.; Lu, X.; Han, J.; Han, X.; Shi, L. Simulation of morphological development of soil cracks in Yuanmou Dry-hot Valley region, Southwest China. *Chin. Geogr. Sci.* **2010**, *20*, 112–122. [CrossRef]
17. Natural Resources Conservation Service. *Keys to Soil Taxonomy*, 12th ed.; Government Printing Office: Washington, DC, USA, 2014.
18. Campos, D.A.; de Andrade, E.M. Seasonal trend of climate variables in an area of the Caatinga phytogeographic domain. *Rev. Agro@mbiente On-Line* **2021**, *15*, 1–18. [CrossRef]
19. Guerreiro, M.S.; Maia de Andrade, E.; Palácio, H.A.D.Q.; Brasil, J.B.; Filho, J.C.R. Enhancing Ecosystem Services to Minimize Impact of Climate Variability in a Dry Tropical Forest with Vertisols. *Hydrology* **2021**, *8*, 46. [CrossRef]
20. Liu, C.; Tang, C.S.; Shi, B.; Suo, W.B. Automatic quantification of crack patterns by image processing. *Comput. Geosci.* **2013**, *57*, 77–80. [CrossRef]
21. Stewart, R.D.; Najm, M.R.A. Field measurements of soil cracks. *Soil Sci. Soci. Am. J.* **2020**, *84*, 1462–1476. [CrossRef]
22. Miller, W.L.; Kishné, A.S.; Morgan, C.L. Vertisol morphology, classification, and seasonal cracking patterns in the Texas Gulf coast prairie. *Soil Surv. Horiz.* **2010**, *51*, 10–16. [CrossRef]
23. Bullard, J.E.; Ockelford, A.; Strong, C.L.; Aubault, H. Impact of multi-day rainfall events on surface roughness and physical crusting of very fine soils. *Geoderma* **2018**, *313*, 181–192. [CrossRef]
24. Li, M.; Yao, J.; Yan, R.; Cheng, J. Effects of infiltration amounts on preferential flow characteristics and solute transport in the protection forest soil of southwestern China. *Water* **2021**, *13*, 1301. [CrossRef]
25. Ferreira, S.R.D.M.; Araújo, A.G.D.D.; Barbosa, F.A.S.; Silva, T.C.R.; Bezerra, I.M.D.L. Analysis of changes in volume and propagation of cracks in expansive soil due to changes in water content. *Revista Brasileira de Ciência do Solo* **2020**, *44*, 1–19. [CrossRef]
26. Rayhani, M.H.T.; Yanful, E.K.; Fakher, A. Physical modeling of desiccation cracking in plastic soils. *Eng. Geol.* **2008**, *97*, 25–31. [CrossRef]
27. Ribeiro Filho, J.C.; de Andrade, E.M.; Guerreiro, M.S.; de Queiroz Palácio, H.A.; Brasil, J.B. Climate Data to Predict Geometry of Cracks in Expansive Soils in a Tropical Semiarid Region. *Sustainability* **2022**, *14*, 675. [CrossRef]
28. Ribeiro Filho, J.C.; de Andrade, E.M.; de Sousa, M.M.M.; Brasil, J.B.; de Quairoz Palácio, H.A. Morphological characteristics of cracks in soil with and without vegetation cover. *Rev. Agro@mbiente On-Line* **2023**, *17*, 1–15. [CrossRef]
29. Guerreiro, M.S.; de Andrade, E.M.; de Sousa, M.M.M.; Brasil, J.B.; Filho, J.C.R.; de Queiroz Palácio, H.A. Contribution of non-rainfall water input to surface soil moisture in a tropical dry forest. *Hydrology* **2022**, *9*, 102. [CrossRef]

30. Dou, Y.; Quan, J.; Jia, X.; Wang, Q.; Liu, Y. Near-Surface Warming Reduces Dew Frequency in China. *Geophys. Res. Lett.* **2021**, *48*, e2020GL091923. [CrossRef]
31. Yu, B.; Liu, G.; Liu, Q.; Wang, X.; Feng, J.; Huang, C. Soil moisture variations at different topographic domains and land use types in the semi-arid Loess Plateau, China. *Catena* **2018**, *165*, 125–132. [CrossRef]

water

MDPI

Article

Spatial-Temporal Characteristics of Spring Maize Drought in Songnen Plain, Northeast China

Zhifang Pei [1,*] and Bin Wu [2]

[1] School of Architecture, Nanyang Institute of Technology, Nanyang 473004, China
[2] School of Marxism, Nanyang Institute of Technology, Nanyang 473004, China
* Correspondence: peizhifang@nyist.edu.cn

Abstract: With the intensification of global warming, food production will face serious drought risk. In view of the insufficient applicability of the existing crop drought index, a standardized crop water deficit index (SCWDI) was constructed based on the construction idea of the standardized precipitation evapotranspiration index (SPEI) and the crop water deficit index (CWDI) in this study. On this basis, the spatial and temporal characteristics of spring maize drought in Songnen Plain were explored by the slope trend analysis and Morlet wavelet analysis methods. The results show the following: (1) Compared with the existing drought index, the SCWDI shows obvious advantages in drought monitoring of spring maize. (2) In the whole growth stage of spring maize, the change trend of SCWDI was small in the temporal series (−0.012/10a). Spatially, the drought trend of spring maize was mainly decreasing (−0.14~0/10a). The drought frequency of spring maize in each growth stage was mainly light drought in most regions. (3) The three main drought cycles of spring maize in Songnen Plain were 29 years, 10 years, and 4 years. In the next few years, the drought of spring maize in Songnen Plain was controlled by the first main cycle, and the drought years may increase, which should be prevented. The research was expected to provide technical support for crop drought monitoring and agricultural disaster prevention.

Keywords: spring maize; drought; climate change; Songnen Plain; Northeast China

Citation: Pei, Z.; Wu, B. Spatial-Temporal Characteristics of Spring Maize Drought in Songnen Plain, Northeast China. *Water* **2023**, *15*, 1618. https://doi.org/10.3390/w15081618

Academic Editors: Songhao Shang, Magdy Mohssen, Qianqian Zhang, Dongqin Yin and Hamza Gabriel

Received: 31 March 2023
Revised: 14 April 2023
Accepted: 16 April 2023
Published: 21 April 2023

1. Introduction

Global warming is an obvious issue and the frequency and extent of extreme weather are also increasing [1–7]. Owing to its wide range and long duration, drought has become one of the most significant natural disasters facing humankind [8–10]. According to research and analysis, global economic losses caused by drought every year are incalculable, and the total number of people affected by drought has exceeded other natural disasters, affecting more than 120 countries and regions [11,12]. The United Nations report shows that the number and duration of global droughts have increased by 29% since 2000. The report estimates that, between 1998 and 2017 alone, global economic losses caused by drought amounted to USD 124 billion, and drought has affected at least 1.5 billion people [13]. Agriculture is more vulnerable to drought disasters because of its high sensitivity and vulnerability to drought [14]. According to statistics from the Food and Agriculture Organization of the United Nations, drought caused a huge loss of USD 29 billion in the global agricultural sector in 2005–2015 [15]. Some studies have shown that, if effective measures are not taken to deal with the risk of drought disasters, the area of global crop drought will increase significantly by the end of this century and food security will be seriously threatened [16,17]. At present, how to effectively reduce the loss of agricultural production caused by drought and further improve the ability of agriculture to resist risks have become key issues of urgent concern for countries and regions around the world [18,19].

In agricultural drought, the drought index plays an important role in reducing and preventing the adverse effects of drought on crops [20,21]. Owing to the long time and high precision of ground observation data, various types of drought indexes have been derived, which are widely used in agricultural drought monitoring, such as precipitation anomaly in percentage (Pa), relative moisture index (MI), standardized precipitation index (SPI), standardized precipitation evapotranspiration index (SPEI), crop water deficit index (CWDI), and crop water deficit abnormal index (CWDIa), among others [22]. However, owing to the different construction principles of each index, there are different applications in drought monitoring [23]. Pa and MI only consider the impact of drought on crops in the current period, but lack consideration of early precipitation, so accurate drought information may not be obtained in time, with poor sensitivity [24]. Under global warming, the rise in temperature has become one of the significant factors aggravating the drought process. However, Pa, MI, and SPI only consider precipitation, which may be inferior in some regional drought monitoring [25,26]. Although SPEI makes up for the disadvantages of SPI, it fails to take the crop coefficient into account when calculating the potential evapotranspiration and cannot accurately reflect the water deficit of crops [27–29]. CWDI and CWDIa consider the water demand characteristics of crops and can accurately evaluate the water status of crop growth in different growth stages in combination with the growth stage of crops [30,31]. However, the drought standards of CWDI and CWDIa in different regions are inconsistent, with poor universality [32]. In view of the shortcomings of the above drought indexes, how to modify it to improve its effectiveness and universality still needs further exploration.

Songnen Plain, located in Northeast China, is the most important grain-producing area in China. The cultivated land area of Songnen Plain accounts for more than 50% of the land area of Songnen Plain and about 8% of the total cultivated land area in China. It is a typical agricultural farming area [33,34]. The northeast region is one of the regions with the most significant temperature increases in China. Since the 21st century, the drought trend in the northeast region has become increasingly serious, posing a potential threat to the growth of maize, rice, soybean, and other major crops [35]. The main grain crops in Songnen Plain are spring maize, which accounts for 72.88% of the total grain output in the region [36,37]. The increase in drought frequency in the future will seriously threaten the stability of corn output in the region. Therefore, the prevention and mitigation of spring maize drought in Songnen Plain plays an important role in ensuring the sustainable stability of its agricultural economy and the national food security [38,39].

In view of the above analysis, the spatial and temporal characteristics of spring maize drought in Songnen Plain of Northeast China were analyzed on the basis of the revised crop drought index. The main objectives of this study were as follows: (1) to construct a new crop drought index and improve the applicability of crop drought index; (2) to compare the effect of the newly constructed drought index and other drought indexes in drought monitoring of spring maize; and (3) to investigate the spatio–temporal variations in spring maize drought in Songnen Plain based on the newly constructed drought index. It is expected that this study will provide ideas for the construction of crop drought indexes in the future, as well as provide a scientific basis and technical means for the prevention of crop drought in Songnen Plain of Northeast China.

2. Materials and Methods

2.1. Study Area

Songnen Plain (121°40′ E~128°30′ E, 42°50′ N~49°12′ N) is located in Northeast China [33]. It is a plain formed by the alluvial deposits of Songhua River and Nenjiang River. Together with Sanjiang Plain and Liaohe Plain, they form the three major plains in Northeast China. The total area of Songnen Plain is about 224,000 km^2, including mainly the southwest of Heilongjiang Province and the northwest of Jilin Province (Figure 1). Songnen Plain belongs to a typical temperate continental monsoon climate, with four distinct seasons. Its average annual temperature is between 2 °C and 6 °C, showing a gradually increasing trend from north to south; the annual precipitation of the whole region is between 400 mm and 600 mm, showing a decreasing trend from southeast to northwest [40]. Songnen Plain has a high concentration of precipitation, which is common in summer, accounting for about 70% of the annual precipitation. This distribution of precipitation means that drought and flood disasters can form very easily. The cultivation method of crops in Songnen Plain is generally one crop per year. It is an important grain production area and commodity grain production base in China, and plays a pivotal role in the social and economic development of Northeast China [41].

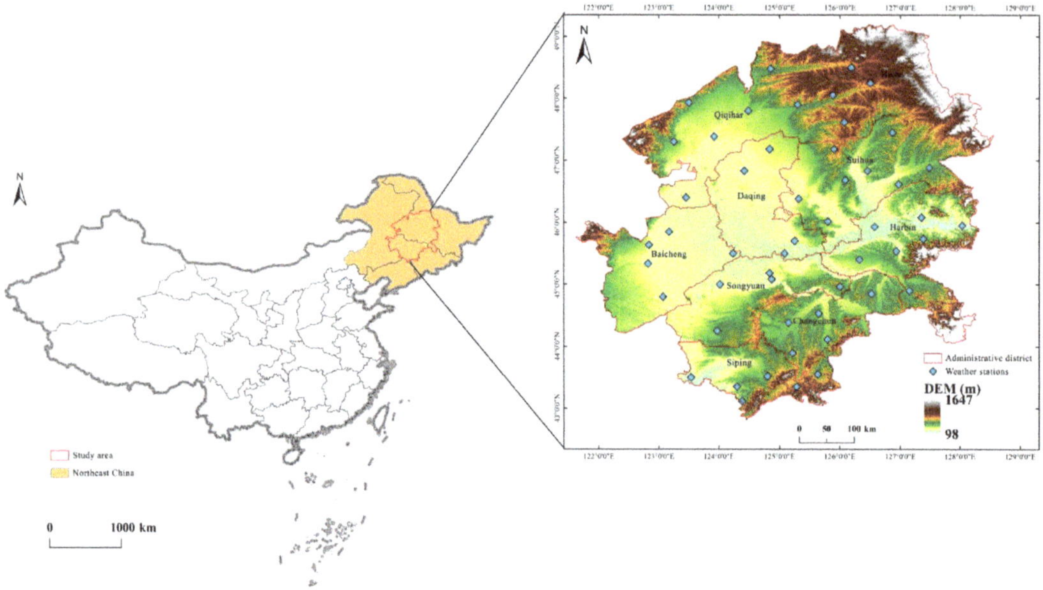

Figure 1. The location of Songnen Plain in Northeast China and distribution of meteorological stations.

2.2. *Data and Processing*

To calculate the drought index of spring maize, the daily meteorological data from 51 meteorological stations in Songnen Plain from 1981 to 2018 were obtained, and the data included average temperature, precipitation, average relative humidity, average pressure, sunshine hours, and average wind speed, among others. All data were from the China Meteorological Data Network (http://data.cma.cn/), accessed on 1 October 2022. The data have been repeatedly revised and supplemented by the China Meteorological Administration. For some missing data, linear interpolation is carried out in this study to improve the integrity and continuity of the data series to a certain extent [42].

In the study, the growth stage of spring maize was taken as the research period. Therefore, we obtained the growth stage data of spring maize from the agricultural meteorological station in the study area. According to the literature [43], five growth stages of spring maize were established, i.e., the sowing to seedling stage, the seedling to jointing stage, the jointing to heading stage, the heading to milk mature stage, and the milk mature to mature stage. For convenience of expression, we named the whole growth stage and the five growth stages A0, A1, A2, A3, A4, and A5, respectively. To verify the validity of the drought index, we also obtained the drought disaster data of spring maize in typical drought years in the study area. All data were also from the China Meteorological Data Network (http://data.cma.cn), accessed on 23 October 2022.

The normalized difference vegetation index (NDVI) can provide vegetation greenness information. When drought occurs, the value of NDVI will change, which better reflects the crop stress information. In the study, we obtained the MODIS NDVI data, which are provided by NASA (http://earthdata.nasa.gov), accessed on 11 November 2022 [44,45]. The vegetation condition index (VCI) was obtained through standardized processing, so as to test the validity of the drought index.

In addition, the administrative division data, DEM, and other basic data in the study were provided by the Data Center for Resources and Environmental Sciences, at the Chinese Academy of Sciences (http://www.resdc.cn/), accessed on 10 October 2022.

2.3. Methods

2.3.1. Construction of the SCWDI

Based on the construction principle of CWDI and SPEI, this study took the accumulated water deficit as the basic quantity and assumed that it follows the log-logistic probability function distribution of three parameters. Then, the accumulated water deficit was fitted and normalized and the standardized crop water deficit index (SCWDI) was constructed.

The specific calculation process was as follows:

Step 1: Calculate the water demand of crop in a growth stage. The formula is as follows:

$$ET_c = K_c \times ET_0 \tag{1}$$

where ET_c is crop water demand (mm); ET_0 is the reference evapotranspiration (mm) [46,47]; and K_c is the crop coefficient in a growth stage [48].

Step 2: Calculate the accumulated water deficit on a daily scale in a certain growth stage. The details are as follows:

$$I_{CWD,i} = a \times CWD_i + b \times CWD_{i-1} + c \times CWD_{i-2} + d \times CWD_{i-3} + e \times CWD_{i-4} \tag{2}$$

where $I_{CWD,i}$ is the accumulated water deficit on the i-th day in a growth stage (mm); CWD_i, CWD_{i-1}, CWD_{i-2}, CWD_{i-3}, and CWD_{i-4} are the water deficit (mm) of the i, $i-1$, $i-2$, $i-3$, and $i-4$ stages, respectively (i.e., the first 1–10 D, the first 11–20 D, the first 21–30 D, the first 31–40 D, and the first 41–50 D, respectively); and *a*, *b*, *c*, and *d* are weight coefficients, which are 0.3, 0.25, 0.2, 0.15, and 0.1, respectively. In the study, it was assumed that there is no irrigation in the study area and the influence of surface runoff and groundwater is ignored; that is, the crop water deficit is the difference between the accumulated water demand of crops and the accumulated precipitation.

Step 3: Establish the multi-year daily scale cumulative water deficit sequence of a certain growth stage, and set the cumulative water deficit (I_{CWD}) data sequence as follows: $x = \{x_1, x_2, \ldots, x_n\}$.

Step 4: Follow the principle of SPEI construction, use the log-logistic probability distribution function $F(x)$ with three parameters to fit the data series, and finally carry out normal normalization.

Set the cumulative probability such that the fitting distribution function is greater than any I_{CWD} value as $P = 1 - F(x)$, then the standardized crop water deficit index (SCWDI) corresponding to a certain growth stage can be obtained:

$$\text{SCWDI} = \begin{cases} W - \frac{C_0 - C_1 W - C_2 W^2}{1 + d_1 W + d_2 W^2 + d_3 W^3} & P \leq 0.5 \quad W = \sqrt{-2\ln P} \\ \frac{C_0 - C_1 W - C_2 W^2}{1 + d_1 W + d_2 W^2 + d_3 W^3} - W & P > 0.5 \quad W = \sqrt{-2\ln(1-P)} \end{cases} \tag{3}$$

where $C_0 = 2.515517$, $C_1 = 0.802853$, $C_2 = 0.010328$, $d_1 = 1.432788$, $d_2 = 0.189269$, and $d_3 = 0.001308$. According to the SPEI drought grading standard [28], SCWDI can be divided into five grades, i.e., no drought (SCWDI < 0.5), light drought (0.5 < SCWDI $\leq$ 1), moderate drought (1 < SCWDI $\leq$ 1.5), severe drought (1.5 < SCWDI $\leq$ 2), and extreme drought (SCWDI > 2).

In this study, the six commonly used drought indexes (Pa, MI, SPI, SPEI, CWDI, and CWDIa) were also calculated to compare and verify the effectiveness of SCWDI. The specific calculation methods of the six drought indexes can be obtained in the corresponding literature [49–54].

2.3.2. Morlet Wavelet Analysis

Wavelet analysis is based on various wavelet transforms. It is an analysis method with an adjustable time-domain window and frequency-domain window. In the analysis of climate change, wavelet analysis can fully show the variety of change information hidden in the time series and predict the future change trend [55]. In recent years, wavelet analysis has been widely used in the analysis and prediction of drought periodicity [56]. In the study, Morlet wavelet transform was used to analyze the periodicity of drought, which is mainly realized by matlab2016 software.

2.3.3. Other Methods

In the study, the ANUSPLIN interpolation method was used for the spatial interpolation of the drought indexes in the study area [57]. The Pearson correlation coefficient (R) was calculated to study the relationships among the drought indexes. The changes in drought in spring maize were analyzed by the slope trend analysis method [58–60].

3. Results

3.1. Applicability Analysis of the SCWDI

3.1.1. Feasibility of SCWDI Construction

The SCWDI was constructed on the premise that the cumulative crop water deficit follows the three-parameter log-logistic function distribution. Whether or not the cumulative crop water deficit followed the log-logistic function distribution needs to be tested.

In this study, the empirical distribution was used as a reference and the normal distribution was used as a comparison to test whether the accumulated crop water deficit follows the log-logistic distribution. Taking the jointing to heading stage of spring maize as an example, we calculated the time series of the accumulated water deficit of spring maize in this stage (A3) at typical stations (Keshan station, Qiqihar station, Nong'an station, and Lishu station) in Songnen Plain. As the density curve of the actual data series can intuitively represent the data distribution, it was taken as an empirical density function. Then, by calculating the key parameters of the normal distribution and log-logistic distribution of the actual data series, the respective density curves were drawn (Figure 2a). We can see that the three-parameter log-logistic density curve is closer to the empirical density curve than the normal distribution density curve.

(a)

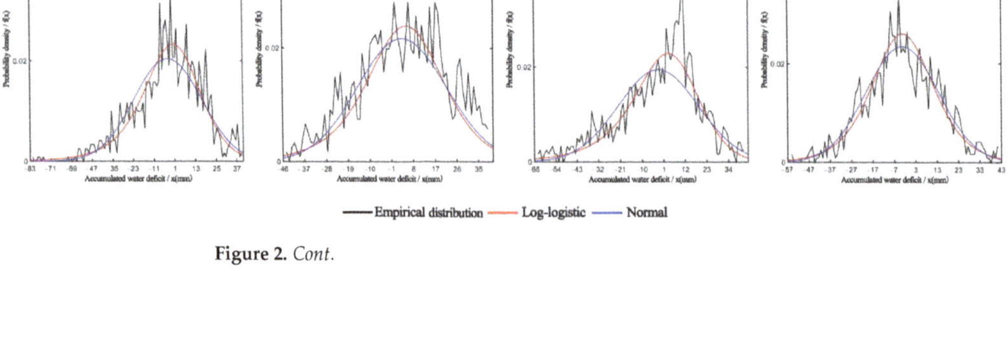

Figure 2. *Cont.*

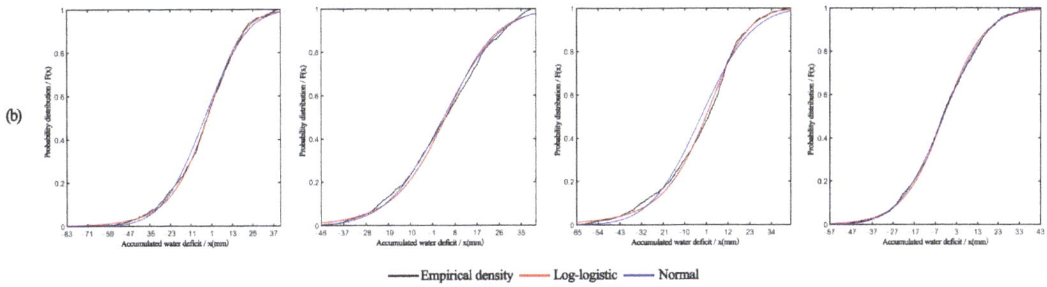

Figure 2. Comparison of three function fitting curves of crop water deficit sequence: (**a**) density curves and (**b**) distribution curve.

Similarly, from the distribution curve (Figure 2b), the coincidence degree of the three parameter Log-logistic distribution curve and the empirical distribution curve was relatively high. So the time series of cumulative crop water deficit of spring maize in Songnen Plain obeyed three parameter Log-logistic distribution, and it was feasible to fit and standardize it.

3.1.2. Comparison of Different Drought Indexes in Time Series

The commonly used drought indexes (Pa, MI, SPI, SPEI, CWDI, and CWDIa) in the study area from 1981 to 2018 were calculated to test the applicability of SCWDI. Among these indexes, SCWDI has the same meaning as CWDI and CWDIa. The larger the index value, the more serious the drought. However, it is opposite to the meaning of Pa, MI, SPI, and SPEI, that is, the smaller the index value, the more serious the drought.

From the perspective of time series (Figure 3), although there were differences in each drought index, it can reflect the dry and wet state of spring maize in different years at various growth stages. The drought period shown by SCWDI was consistent with other drought indexes and can reflect the typical drought years, such as 1982, 1989, 2001, 2004, 2007, 2017, and so on. However, there were differences in the monitoring results of each index. In 1982, the drought severity period of spring maize reflected by SCWDI was consistent with that reflected by SPI and CWDIa; in 1989, it was consistent with the drought period reflected by MI, CWDI, and CWDIa; in 2001, it was consistent with the drought period reflected by CWDI and CWDIa; in 2004, it was basically consistent with other drought indexes except that it was different from MI; in 2007, it was consistent with the drought period reflected by SPI, SPEI and CWDIa; and in 2017, it was consistent with the drought period reflected by SPI and CWDIa. In general, the drought periods identified by SCWDI in the drought monitoring of spring maize can be consistent with the commonly used drought index.

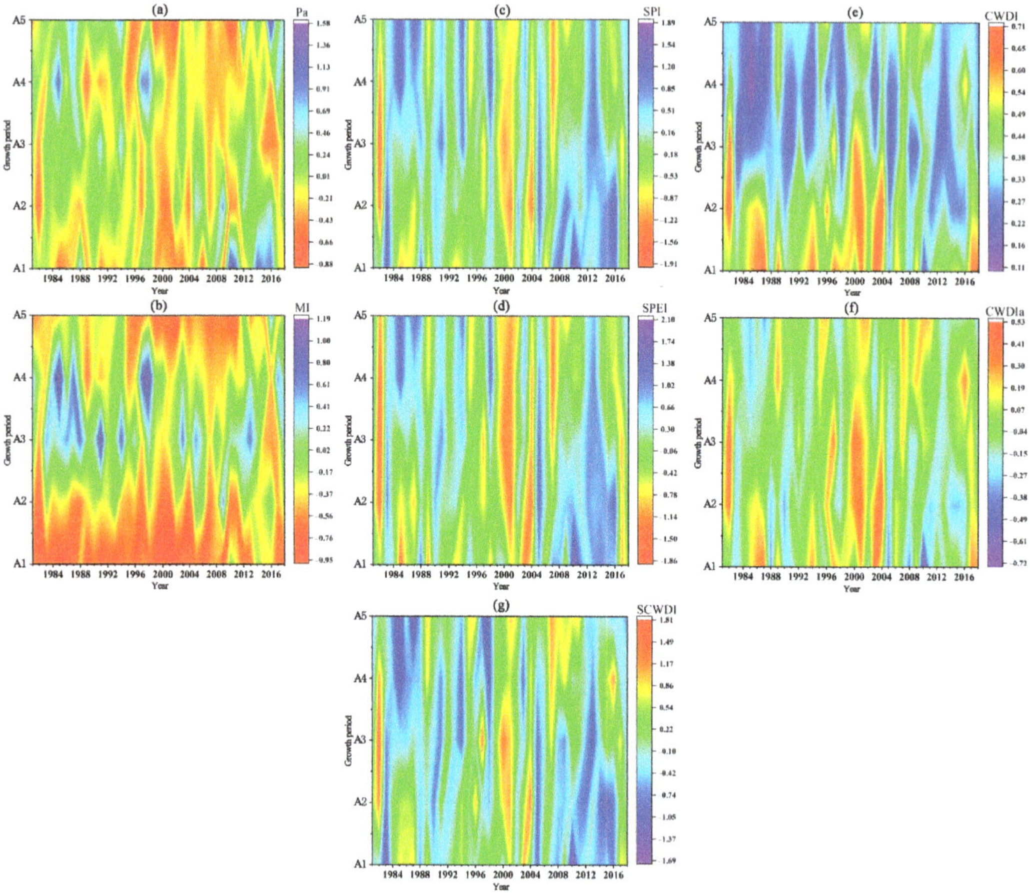

Figure 3. Comparison of different drought indexes in time series: (**a**) Pa; (**b**) MI; (**c**) SPI; (**d**) SPEI; (**e**) CWDI; (**f**) CWDIa; and (**g**) SCWDI.

3.1.3. Comparison of Different Drought Indexes in Space

Taking the typical drought year 2007 as an example, the applicability of SCWDI was analyzed in space, and the growth state of spring maize during drought was characterized by the vegetation state index (VCI) (Figure 4). In the A1 stage, the vegetation index of spring maize was generally low, and the VCI may be low. The VCI in the northwest of Songnen Plain was relatively low and may be subject to drought stress. The drought distribution identified by Pa, SPI, SPEI, and SCWDI was similar to that of VCI. In the A2 stage, the VCI in the West and north of Songnen Plain was low, and the drought distribution identified by CWDIa and SCWDI was similar to that of VCI. In the A3 stage, the VCI in the west and north of Songnen Plain was low, and the drought distribution identified by SPEI, CWDI, CWDIa, and SCWDI was similar to that of VCI. In the A4 stage, the VCI in the east and west of Songnen Plain was low, and the drought distribution identified by SPI, SPEI, and SCWDI was similar to that of VCI. In the A5 stage, the VCI in the east of Songnen Plain was low, and only the drought distribution identified by SCWDI was similar to that of VCI. In general, compared with other drought indexes, SCWDI had a greater advantage in drought monitoring of spring maize at various growth stages.

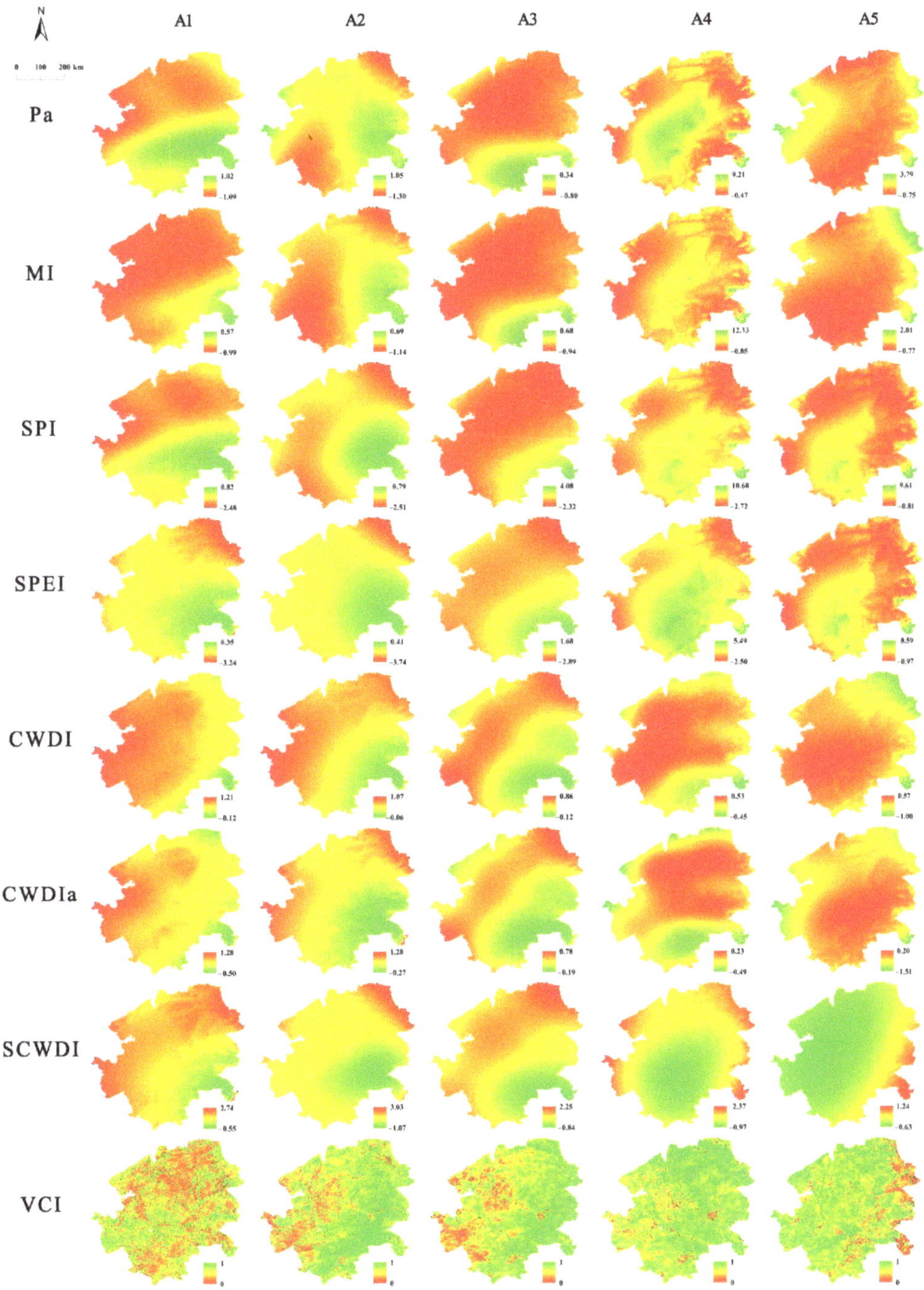

Figure 4. Comparison of different drought indexes in space.

3.2. Spatial and Temporal Variation in Spring Maize Drought in Songnen Plain

3.2.1. Temporal Variation in Spring Maize Drought

The temporal variation in spring maize drought at each growth stage in Songnen Plain from 1981 to 2018 was analyzed. In the A1 stage (Figure 5a), SCWDI also showed a decreasing trend (−0.026/10a), with a small range. The driest year was 2017. On the interdecadal scale, it showed a trend of first increasing, then decreasing, and then increasing. In the 2010s, there were many relatively dry years. In the A2 stage (Figure 5b), SCWDI showed a significant decreasing trend (−0.204/10a) ($p < 0.05$), and the driest year was 2001. On the interdecadal scale, SCWDI increased first and then decreased. The driest years were the mostly in the 1990s, and the drought has been relatively alleviated in recent years. In the A3 stage (Figure 5c), SCWDI showed a decreasing trend (−0.104/10a), and 1982 was the driest year; the change trend on the decadal scale was similar to that in the A2 stage. In the A4 and A5 stages (Figure 5d,e), the change trend of SCWDI was similar, showing an increasing trend (0.128/10a and 0.145/10a). In the A4 stage, the driest year was 1982 and, in the A5 stage, the driest year was 2007. On the interdecadal scale, SCWDI in both stages showed a trend of first increasing and then decreasing. The two stages in the 2000s showed relative drought, while the drought in recent years has been relatively alleviated. From 1981 to 2018, the drought index (SCWDI) of spring maize in the whole growth stage showed a decreasing trend with a small slope (−0.012/10a) (Figure 5f). The driest year was 1982. On the interdecadal scale, SCWDI showed a trend of first increasing and then decreasing. In the 2000s, it was relatively dry. In recent years, the drought of spring maize has been alleviated.

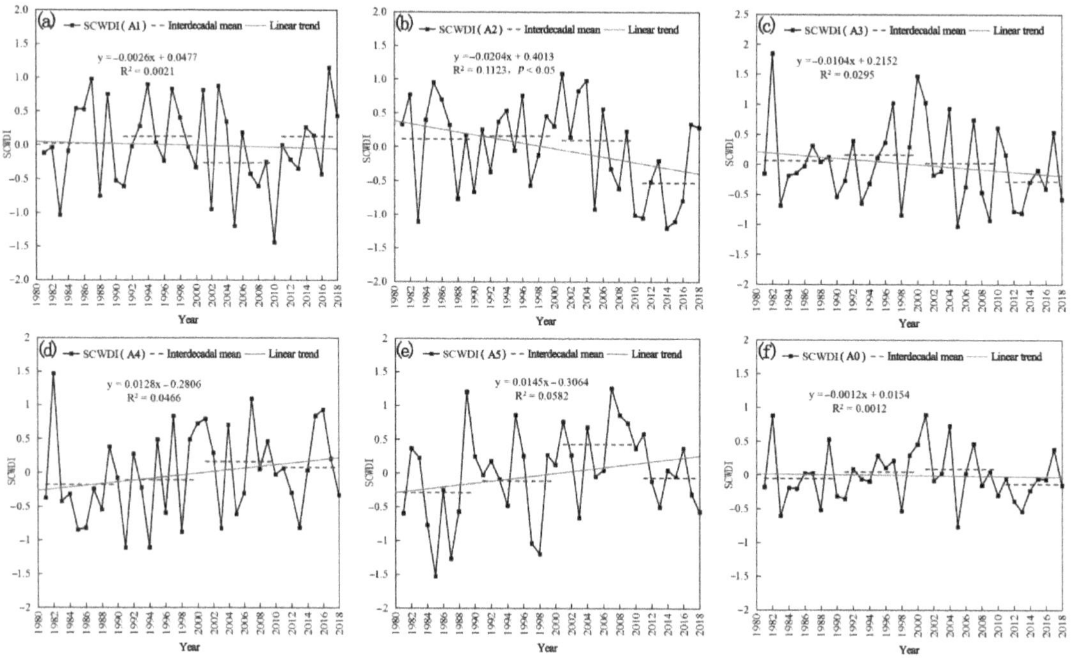

Figure 5. Temporal variation characteristics of SCWDI in different growth stages of spring maize: (**a**) the sowing to seedling stage; (**b**) the seedling to jointing stage; (**c**) the jointing to heading stage; (**d**) the heading to milk mature stage; (**e**) the milk mature to mature stage; and (**f**) the whole growth stage.

3.2.2. Spatial Variation in Spring Maize Drought

The drought characteristics of spring maize in Songnen Plain were analyzed in spatial. In the A1 stage (Figure 6a), the drought of spring maize in various regions of Songnen Plain showed an increasing trend (0~0.24/10a) or decreasing trend (−0.29~0/10a), of which the decreasing trend of drought was mainly distributed in Suihua City, Harbin City, and Changchun City, and the increasing trend of drought was mainly in other regions. In the A2 stage (Figure 6b), the drought of spring maize in the whole area of Songnen Plain showed a decreasing trend (−0.53~0/10a), and the significant areas were mainly distributed in the eastern and central parts of Songnen Plain. In the A3 stage (Figure 6c), the drought of spring maize in most areas of Songnen Plain showed a decreasing trend (−0.25~0/10a), and the increasing trend was mainly distributed in Siping City (0~0.18/10a). In the A4 stage (Figure 6d), the drought trend of spring maize in most areas of Songnen Plain mainly increased (0~0.45/10a), and a significant increase was distributed in the southwest. The drought of spring maize in the northwest of Suihua City showed a decreasing trend. In the A5 stage (Figure 6e), the drought trend of spring maize in most areas of Songnen Plain also increased (0~0.44/10a), and the drought reduction trend was sporadically distributed. In the whole growth stage (Figure 6f), the drought trend of spring maize in Songnen Plain was mainly decreasing (−0.14~0/10a), and the drought of spring maize in the south and southwest showed an increasing trend (0~0.21/10a).

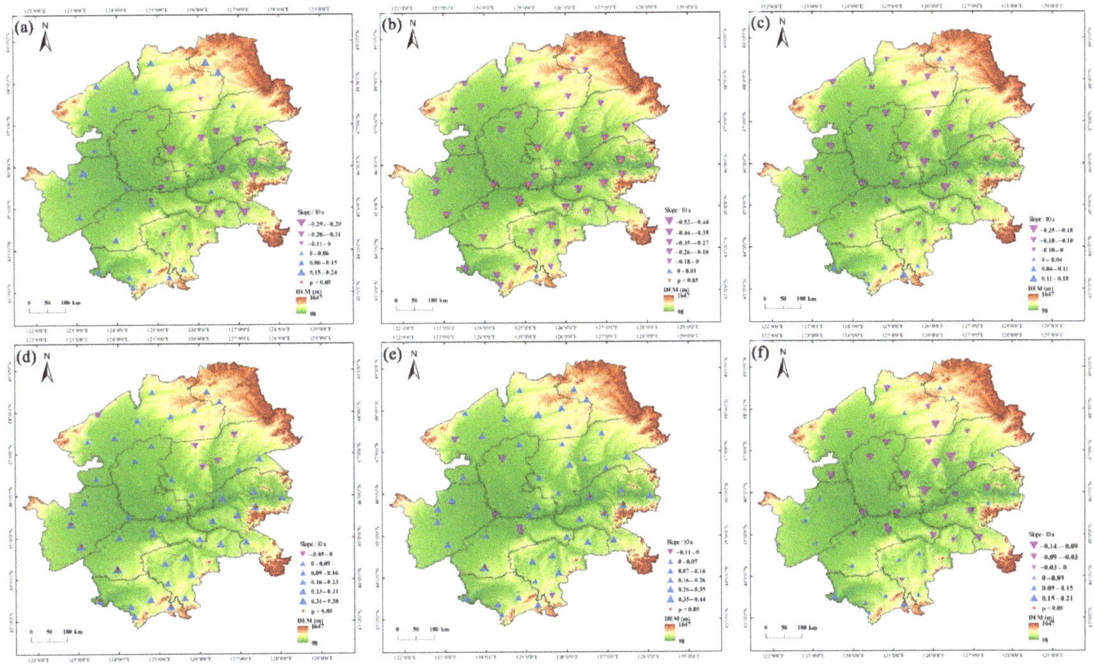

Figure 6. Spatial trend of SCWDI in different growth stages of spring maize: (**a**) the sowing to seedling stage; (**b**) the seedling to jointing stage; (**c**) the jointing to heading stage; (**d**) the heading to milk mature stage; (**e**) the milk mature to mature stage; and (**f**) the whole growth stage.

We also analyzed the spatial characteristics of drought frequency of spring maize in Songnen Plain. In the A1 stage (Figure 7a), the drought frequency range of spring maize in Songnen Plain was 16.3–41.8%, and the areas with high drought frequency were mainly distributed in the north and northwest. Most of the spring maize in Songnen Plain was dominated by light drought. The frequency of medium drought was relatively high in the central and southern regions, and the heavy drought was mainly distributed in the surrounding areas of Songnen Plain. In the A2 stage (Figure 7b), the drought frequency range of spring maize in Songnen Plain was 23.3–35.6%, and the drought frequency was high in most areas. The whole area of spring maize in Songnen Plain was dominated by light drought, while the frequency of medium drought was high in the central part and the heavy drought was mainly distributed in the west and east. In the A3 stage (Figure 7c), the drought frequency range of spring maize in Songnen Plain was 25.2–33.1%, and the areas with high drought frequency were mainly distributed in the northwest and east. In most areas of Songnen Plain, spring maize was mainly subject to light drought, and there were more areas with a relatively high frequency of medium drought and heavy drought. In the A4 stage (Figure 7d), the drought frequency range of spring maize in Songnen Plain was 17.3–48.8%, and the areas with high drought frequency were mainly distributed in the west. Songnen Plain was mainly characterized by light drought. There were many areas with relatively high moderate drought, and the frequency of heavy drought in the northwest and southeast was relatively high. In the A5 stage (Figure 7e), the drought frequency range of spring maize in Songnen Plain was 16.7–36.6%. The drought frequency in the whole region was relatively high, mainly light drought. The regions with a relatively high frequency of medium drought were mainly distributed in the middle, and the northern and southern regions had a relatively high frequency of heavy drought. In the whole growth stage (Figure 7f), the drought frequency range of spring maize in Songnen Plain was 18.7–35.8%, and the areas with a high drought frequency were mainly distributed in the northwest. Most areas of Songnen Plain were dominated by light drought. The frequency of moderate drought was relatively high in the central and southern regions and the frequency of severe drought was relatively high in the northwest and southeast.

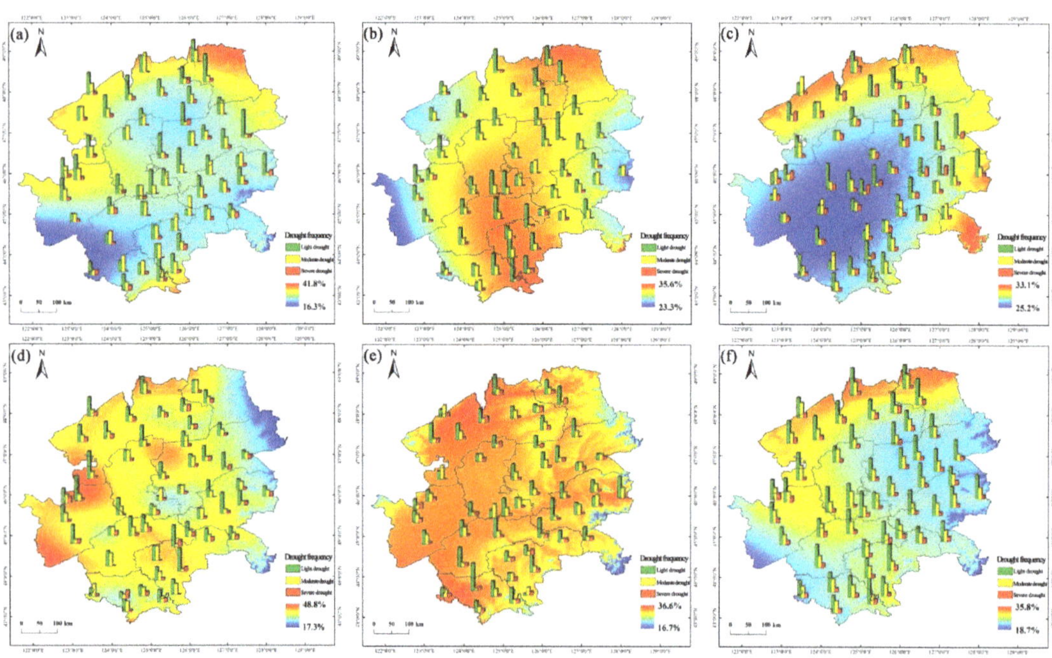

Figure 7. Spatial distribution of drought frequency in different growth stages of spring maize: (**a**) the sowing to seedling stage; (**b**) the seedling to jointing stage; (**c**) the jointing to heading stage; (**d**) the heading to milk mature stage; (**e**) the milk mature to mature stage; and (**f**) the whole growth stage.

3.3. Periodic Variation in Spring Maize Drought in Songnen Plain

Wavelet analysis can better show the time–frequency characteristics and periodicity of data series and is widely used in research. By analyzing the periodicity of spring maize drought, we can better understand the law of drought change and perform the corresponding disaster prevention and early warning measures. In this study, Morlet wavelet transform was used to analyze the drought of spring maize in the whole growth stage from 1981 to 2018 in Songnen plain to obtain the periodic characteristics of drought (Figure 8).

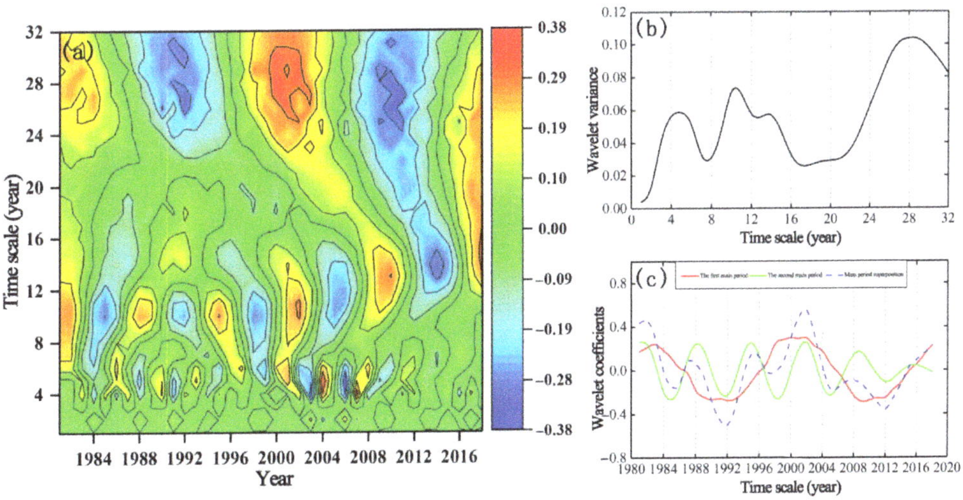

Figure 8. Periodicity characteristics of drought in the whole growth stage of spring maize in the Songnen Plain from 1981 to 2018: (**a**) the real part contour of wavelet coefficients; (**b**) wavelet variance; and (**c**) wavelet coefficients.

The strength of the drought signal was usually expressed by the size of the wavelet coefficient. In the contour of the wavelet coefficient, if the center was positive, the period was dry; if it was negative, the period was light. In the whole growth stage of spring maize, according to the contour map of the wavelet coefficient (Figure 8a), the drought of spring maize in Songnen Plain was obvious in three time scales of 4–6 years, 9–12 years, and 24–32 years. On the time scale of 24–32 years, some cycle centers were not completely closed, and the periodicity was obvious. Before 1986, in 1996–2006, and after 2015, the wavelet real part was positive, indicating that the drought of spring maize in Songnen Plain was more significant in this period. In 1987–1995 and 2007–2014, the wavelet real part was negative, indicating that the drought of spring maize in Songnen Plain was less significant in this period. On the time scale of 9–12 years, the periodic signal before 2012 was strong and the periodic signal after 2012 was weak. Before 2012, there were mainly five periods with more drought and four periods with less drought. On the time scale of 4–6 years, the periodic performance was unstable, with obvious performance during 1985–2012 and weak performance before 1985 and after 2012. The main time scale (i.e., the main period) was mainly determined from the maximum value of small wave variance. It can be seen that the three main time scales of drought in the whole growth period of spring maize in Songnen Plain were 29 years (the first main period), 10 years (the second main period), and 4 years (the third main period) (Figure 8b). In the study, the first two main cycles were often used to reflect the change in regional drought. By drawing the time–frequency diagram of the first main cycle and the second main cycle (Figure 8c), it can be seen that the first main cycle was periodic and stable, mainly experiencing two dry and wet cycles. After 2018, the third drought will be more significant than in previous years. The second main cycle was periodically stable before 2012 and then weakened. It experienced five dry and wet cycles in total. After 2018, there may be years with less drought. By superimposing the two main cycles, the two main cycles of 1981–1983, 2000–2003, and 2015–2017 fluctuated in positive phases, indicating that the drought was serious in these periods. The two main cycles of 1990–1993 and 2011–2014 fluctuated in negative phases, indicating that the drought was weak in these periods. From the drought cycle in recent years, the second main cycle of drought in the whole growth stage of spring maize in Songnen Plain was weakened, mainly controlled by the first main cycle, and the drought years may increase.

4. Discussion

The impact of drought on agricultural production is incalculable, and seriously restricts agricultural production and human life. Actively carrying out the monitoring and risk assessment of drought on agricultural production has important practical significance for reducing the loss from drought disasters and ensuring food security [39,41,52]. The commonly used drought index has played a great role in crop drought monitoring [22,24,28,31]. However, owing to the different construction principles, each index has different applicability in regional drought monitoring [32,52]. In this study, we can see that there were certain differences in analyzing the spatial and temporal changes in spring maize drought using different drought indexes. According to the previous analysis, the newly established index (SCWDI) was only consistent with some drought indexes in drought monitoring of spring maize. To further explore the relationship between SCWDI and common drought indexes (Pa, MI, SPI, SPEI, CWDI, and CWDIa), we calculated the correlation coefficient between SCWDI and common drought indexes in different growth stages of spring maize (Figure 9).

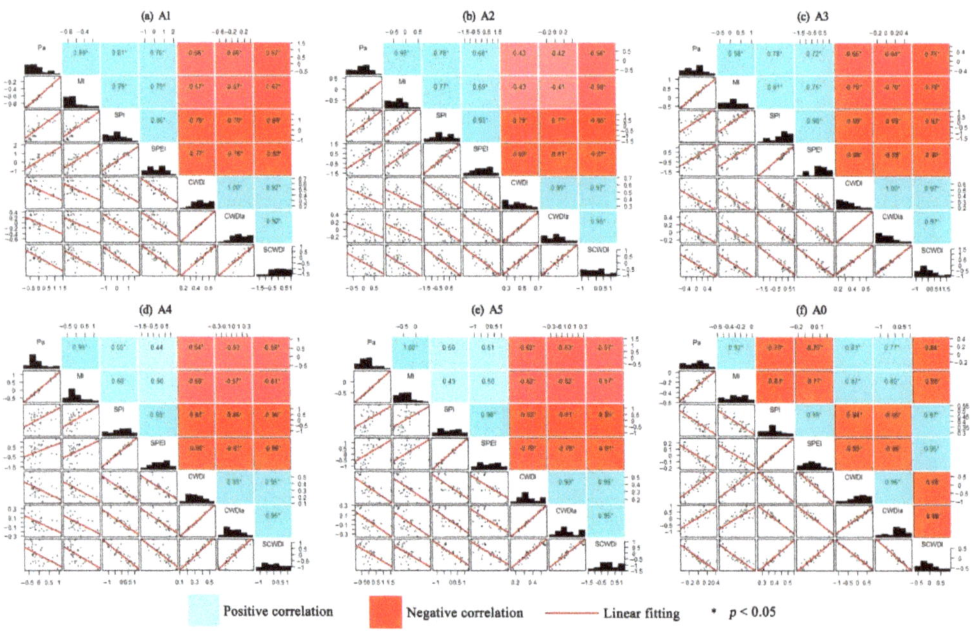

Figure 9. Correlation between SCWDI and the commonly used drought index in different growth stages of spring maize: (**a**) the sowing to seedling stage; (**b**) the seedling to jointing stage; (**c**) the jointing to heading stage; (**d**) the heading to milk mature stage; (**e**) the milk mature to mature stage; and (**f**) the whole growth stage.

In the A1 stage (Figure 9a), the correlation coefficient between SCWDI and each index was above 0.67 and significantly correlated, among which the correlation coefficient with CWDI and CWDIa was the highest, and the correlation coefficient was 0.92. In the A2 stage (Figure 9b), SCWDI was significantly correlated with each index, and the correlation coefficient with CWDI was 0.97. In the A3 stage (Figure 9c), the correlation between SCWDI and each index was significant, and the correlation coefficient was high, among which the correlation with CWDI and CWDIa was the best, and the correlation coefficient was 0.92. In the A4 stage (Figure 9d), the correlation between SCWDI and each index was also significant, among which the correlation with CWDI and CWDIa was the best, and the correlation coefficient was 0.95. In the A5 stage (Figure 9e), the correlation between SCWDI and each index was significant, among which the correlation with CWDI and CWDIa was the best, and the correlation coefficient was 0.95. In the whole growth stage (Figure 9f), the correlation between SCWDI and each index was significant, and the correlation coefficient was high, of which the correlation with CWDI was the best, and the correlation coefficient was 0.98. From the above analysis, it can be seen that SCWDI has a good correlation with the commonly used drought indexes (Pa, MI, SPI, SPEI, CWDI, and CWDIa), which can comprehensively contain some information of each drought index and has great advantages in drought monitoring.

To further reflect the applicability of SCWDI in drought monitoring, we collected actual drought events in a typical drought year (2007) to verify the feasibility of SCWDI. According to the actual drought event records in 2007 (Table 1), in the A1 stage, light drought occurred in Zhaoyuan and Lishu, moderate drought occurred in Songyuan and Changling, and drought mainly occurred in the south of Songnen Plain, which was consistent with the description of SCWDI. In the A2 stage, light drought occurred in Zhaoyuan and Lishu, moderate drought occurred in Changling, and drought mainly occurred in the south of Songnen Plain, which was also consistent with the description of SCWDI. In the A3 stage, Qinggang, Yushu, and Shuangyang suffered light drought and Songyuan, Changling, and Lishu suffered moderate drought, and these areas also suffered drought in the drought distribution described by SCWDI. In the A4 stage, moderate drought occurred in Wudalianchi, Qianguo, Qing'an, Longjiang, Wuchang, and Suiling and severe drought occurred in fufu, Tailai, Songyuan, Zhaoyuan, and Changling. SCWDI showed that drought occurred in these areas. In the A5 stage, Wuchang suffered from light drought; Songyuan suffered from moderate drought; and Zhaoyuan, Changling, and Fuyu suffered from severe drought. In the drought distribution indicated by SCWDI, these regions all suffered from different degrees of drought. In general, SCWDI has certain applicability in drought monitoring of spring maize in Songnen Plain.

Table 1. Drought disasters of spring maize in Songnen Plain in typical drought years.

Growth Stage	Drought Grade	Drought Range
A1	Light drought	Zhaoyuan, Lishu
	Moderate drought	Songyuan, Changling
A2	Light drought	Zhaoyuan, Lishu
	Moderate drought	Changling
A3	Light drought	Qinggang, Yushu, Shuangyang
	Moderate drought	Songyuan, Changling, Lishu
A4	Moderate drought	Wudalianchi, Qianguo, Qing'an, Longjiang, Wuchang, Suiling
	Severe drought	Fuyu, Tailai, Songyuan, Zhaoyuan, Changling
A5	Light drought	Wuchang
	Moderate drought	Songyuan
	Severe drought	Zhaoyuan, Changling, Fuyu

In practice, the process of crop drought is complex [4,48,52,53]. The drought index may have some shortcomings, but it can also play a certain role. Owing to the long time series and high accuracy of meteorological observation data, this study described more meteorological drought indexes, and soil relative humidity was also a commonly used agricultural drought monitoring index, which can also characterize the drought characteristics of crops [14]. Considering that the time and space of the soil relative humidity data in the existing area are not continuous, and it cannot well express the spatial and temporal characteristics of drought, it is rarely used in this study [61]. In future research, more applicable soil humidity data will be collected to further verify the applicability of the drought index constructed in this study.

5. Conclusions

In this study, a new crop drought index (SCWDI) was developed and its applicability was tested. On this basis, the spatial and temporal characteristics of spring maize drought in Songnen Plain were explored. The main findings were as follows:

(1) It was feasible to construct the standardized crop water deficit index (SCWDI) by combining the ideas of CWDI and SPEI. Compared with the commonly used drought indexes (Pa, MI, SPI, SPEI, CWDI, and CWDIa), SCWDI had great advantages in drought monitoring of spring maize.

(2) In the whole growth stage of spring maize, the change trend of SCWDI in Songnen Plain was small in the temporal series (−0.012/10a). Spatially, the drought trend of spring maize was mainly decreasing (−0.14~0/10a), while the drought of spring maize in the south and southwest showed an increasing trend (0~0.21/10a). The drought frequency of spring maize in each growth stage was mainly light drought in most regions. In the whole growth stage, the moderate drought frequency in the central and southern regions was relatively high and the severe drought frequency in the northwest and southeast was relatively high.

(3) In terms of periodicity, in the whole growth stage of spring maize, the three main drought cycles in Songnen Plain were 29 years, 10 years, and 4 years. In the next few years, the drought of spring maize in Songnen Plain was controlled by the first main cycle, and the drought years may increase.

The drought monitoring of crops was very complicated. Our research improved the applicability of the drought index on the basis of the existing drought monitoring and provided ideas for the construction of the drought index in the future. As SCWDI had only been verified in Songnen Plain, its universal applicability still needs to be studied, and future work will be carried out in other regions.

Author Contributions: Conceptualization, Z.P.; methodology, Z.P.; formal analysis, B.W.; investigation, B.W.; writing—original draft preparation, Z.P.; writing—review and editing, B.W.; visualization, Z.P.; supervision, B.W. All authors have read and agreed to the published version of the manuscript.

Funding: This research was supported by the Key Scientific Research Projects of Colleges and Universities of Henan Province (grant number 23A170021), the Interdisciplinary Sciences Project of Nanyang Institute of Technology (grant number NGJC-2022-16), and the Doctoral Research Initiation Fund Program of Nanyang Institute of Technology (grant number NGBJ-2022-36).

Institutional Review Board Statement: Not applicable.

Informed Consent Statement: Not applicable.

Data Availability Statement: The data presented in this study are available upon request from the corresponding author.

Acknowledgments: We would like to thank the National Climatic Centre of the China Meteorological Administration for providing the climate database, the NASA official website for providing the MODIS NDVI data, and the Data Center for Resources and Environmental Sciences and the Chinese Academy of Sciences for providing basic data used in this study.

Conflicts of Interest: The authors declare no conflict of interest.

References

1. IPCC. *Climate Change 2014: Impacts, Adaptation, and Vulnerability*; Cambridge University Press: Cambridge, UK, 2014.
2. Ju, J.; Wu, C.; Yeh, J.F.; Dai, H.; Hu, B.X. Global precipitation-related extremes at 1.5 °C and 2 °C of global warming targets: Projection and uncertainty assessment based on the CESM-LWR experiment. *Atmos. Res.* **2021**, *264*, 105868. [CrossRef]
3. Wu, J.; Han, Z.; Li, R.; Xu, Y.; Shi, Y. Changes of extreme climate events and related risk exposures in Huang-Huai-Hai river basin under 1.5–2 °C global warming targets based on high resolution combined dynamical and statistical downscaling dataset. *Int. J. Climatol.* **2021**, *41*, 1383–1401. [CrossRef]
4. Chen, Q.; Qu, Z.; Ma, G.; Wang, W.; Dai, J.; Zhang, M.; Wei, Z.; Liu, Z. Humic acid modulates growth, photosynthesis, hormone and osmolytes system of maize under drought conditions. *Agric. Water Manag.* **2022**, *263*, 107447. [CrossRef]
5. Xu, Y.; Zhao, Y.; Zhai, P. IPCC special report SRCCL's new cognition and Enlightenment on climate change and food security. *Res. Prog. Clim. Chang.* **2020**, *16*, 13. [CrossRef]
6. Liu, K.; Harrison, M.T.; Yan, H.; Liu, D.L.; Meinke, H.; Hoogenboom, G.; Wang, B.; Peng, B.; Guan, K.; Jaegermeyr, J.; et al. Silver lining to a climate crisis in multiple prospects for alleviating crop waterlogging under future climates. *Nat. Commun.* **2023**, *14*, 765. [CrossRef] [PubMed]
7. Yang, P.; Zhang, S.; Xia, J.; Zhan, C.; Cai, W.; Wang, W.; Luo, X.; Chen, N.; Li, J. Analysis of drought and flood alternation and its driving factors in the Yangtze River Basin under climate change. *Atmos. Res.* **2022**, *270*, 106087. [CrossRef]
8. Wan, W.; Zhao, J.; Popat, E.; Herbert, C.; Dll, P. Analyzing the Impact of Streamflow Drought on Hydroelectricity Production: A Global-Scale Study. *Water Resour. Res.* **2021**, *57*, e2020WR028087. [CrossRef]
9. Zhou, S.; Wang, Y.; Li, Z.; Chang, J.; Guo, A.; Zhou, K. Characterizing spatio-temporal patterns of multi-scalar drought risk in mainland China. *Ecol. Indic.* **2021**, *131*, 108189. [CrossRef]
10. Kamali, B.; Jahanbakhshi, F.; Dogaru, D.; Dietrich, J.; Nendel, C.; Aghakouchak, A. Probabilistic modeling of crop-yield loss risk under drought: A spatial showcase for sub-Saharan Africa. *Environ. Res. Lett.* **2022**, *17*, 024028. [CrossRef]
11. Tian, F.; Wu, J.; Liu, L.; Leng, S.; Shen, Q. Exceptional Drought across Southeastern Australia Caused by Extreme Lack of Precipitation and Its Impacts on NDVI and SIF in 2018. *Remote Sens.* **2019**, *12*, 54. [CrossRef]
12. Mishra, A.K.; Singh, V.P. A review of drought concepts. *J. Hydrol.* **2010**, *391*, 202–216. [CrossRef]
13. Schwartz, C.; Ellenburg, W.L.; Mishra, V.; Mayer, T.; Griffin, R.; Qamer, F.; Matin, M.; Tadesse, T. A statistical evaluation of Earth-observation-based composite drought indices for a localized assessment of agricultural drought in Pakistan. *Int. J. Appl. Earth Obs. Geoinf.* **2022**, *106*, 102646. [CrossRef]
14. Shang, Y.; Wang, J.; Wang, Z.; Hong, S.; Yun, S.U. Vulnerability Identification and Assessment of Agriculture Drought Disaster in China. *Adv. Earth Sci.* **2006**, *21*, 161–169. [CrossRef]
15. Ha, T.V.; Huth, J.; Bachofer, F.; Kuenzer, C. A Review of Earth Observation-Based Drought Studies in Southeast Asia. *Remote Sens.* **2022**, *14*, 3763. [CrossRef]
16. Chen, H.; Wang, J.; Huang, J. Policy support, social capital, and farmers' adaptation to drought in China. *Glob. Environ. Chang.* **2014**, *24*, 193–202. [CrossRef]
17. Kebede, A.; Kang, M.S.; Bekele, E. Advances in mechanisms of drought tolerance in crops, with emphasis on barley. *Adv. Agron.* **2019**, *156*, 265–314. [CrossRef]
18. Wang, Q.; Yan, D.H.; Weng, B.S.; Feng, J.; Shi, X. Response to drought disaster in North America and their experiences to China. *Arid Land Geogr.* **2012**, *35*, 332–338. [CrossRef]
19. Han, R.; Li, Z.; Li, Z.; Han, Y. Spatial–Temporal Assessment of Historical and Future Meteorological Droughts in China. *Atmosphere* **2021**, *12*, 787. [CrossRef]
20. Yu, M.; Liu, X.; Wei, L.; Li, Q.; Zhang, J.; Wang, G. Drought Assessment by a Short-/Long-Term Composited Drought Index in the Upper Huaihe River Basin, China. *Adv. Meteorol.* **2015**, *2016*, 7986568. [CrossRef]
21. Noureldeen, N.; Mao, K.; Mohmmed, A.; Yuan, Z.; Yang, Y. Spatiotemporal Drought Assessment over Sahelian Countries from 1985 to 2015. *J. Meteorol. Res.* **2020**, *34*, 104–118. [CrossRef]
22. Xie, W.; Sheng, W.; Tang, W.; Rong, W.; Dai, J. Comparative Analysis on the Applicability of Drought Indexes in the Huaihe River Basin. *J. Appl. Meteorol. Sci.* **2014**, *25*, 176–184. [CrossRef]
23. Sharafati, A.; Nabaei, S.; Shahid, S. Spatial assessment of meteorological drought features over different climate regions in Iran. *Int. J. Climatol.* **2020**, *40*, 1864–1884. [CrossRef]
24. Han, H.Q.; Bai, Y.M.; Zhang, Y.J.; Chen, M.L. Comparative Study on Applicability of Four Drought Indexes in Guizhou Province. *J. Qiannan Norm. Univ. Natl.* **2019**, *39*, 45–49.
25. Zhang, J.; Liu, Z.; Wang, J.; He, Y.; Luo, H. Construction and validation of comprehensive drought monitoring model in Southwest China. *Trans. Chin. Soc. Agric. Eng.* **2017**, *33*, 102–107. [CrossRef]
26. Kalisa, W.; Zhang, J.; Igbawua, T.; Ujoh, F.; Ebohon, O.J.; Namugize, J.N.; Yao, F. Spatio-temporal analysis of drought and return periods over the East African region using Standardized Precipitation Index from 1920 to 2016. *Agric. Water Manag.* **2020**, *237*, 106195. [CrossRef]

27. Begueria-Portugues, S.; Vicente-Serrano, S.M.; Angulo-Martínez, M.; López-Moreno, J.I.; Kenawy, A.E. The Standardized Precipitation-Evapotranspiration Index (SPEI): A multiscalar drought index. In Proceedings of the EMS Annual Meeting 2010, Zurich, Switzerland, 13–17 September 2010.
28. Li, W.G.; Yi, X.; Hou, M.T.; Chen, H.L.; Chen, Z.L. Standardized precipitation evapotranspiration index shows drought trend in China. *Chin. J. Eco-Agric.* **2012**, *5*, 643–649. [CrossRef]
29. Liu, X.; Wang, S.; Zhou, Y.; Wang, F.; Li, W.; Liu, W. Regionalization and Spatiotemporal Variation of Drought in China Based on Standardized Precipitation Evapotranspiration Index (1961–2013). *Adv. Meteorol.* **2015**, *2015*, 950262. [CrossRef]
30. Hu, Z.; Wu, Z.; Islam, A.R.M.T.; You, X.; Zhang, X. Spatiotemporal characteristics and risk assessment of agricultural drought disasters during the winter wheat-growing season on the Huang-Huai-Hai Plain, China. *Theor. Appl. Climatol.* **2021**, *143*, 1393–1407. [CrossRef]
31. Li, Y.; Li, H.; Wang, H.; Wang, Y. Spatiotemporal difference analysis of drought on wine grape in Ningxia based on crop water deficit index. *J. Nat. Disasters* **2014**, *23*, 203–211. [CrossRef]
32. Mu, J.; Qiu, M.; Gu, Y.; Ren, J.; Liu, Y. Applicability of five drought indices for agricultural drought evaluation in Jilin Province, China. *Ying Yong Sheng Tai Xue Bao = J. Appl. Ecol.* **2018**, *29*, 2624–2632. [CrossRef]
33. Yu, W.; Zhang, L.; Zhang, H.; Jiang, L.; Zhang, A.; Pan, T. Effect of farmland expansion on drought over the past century in Songnen Plain, Northeast China. *J. Geogr. Sci.* **2020**, *30*, 439–454. [CrossRef]
34. Wang, Y.; Shen, X.; Jiang, M.; Lu, X. Vegetation Change and Its Response to Climate Change between 2000 and 2016 in Marshes of the Songnen Plain, Northeast China. *Sustainability* **2020**, *12*, 3569. [CrossRef]
35. Zhou, Z.; Shi, H.; Fu, Q.; Li, T.; Liu, S. Assessing spatiotemporal characteristics of drought and its effects on climate-induced yield of maize in Northeast China. *J. Hydrol.* **2020**, *588*, 125097. [CrossRef]
36. Jian-Zhai, W.U.; Zhang, J.; Zhang-Ming, G.E.; Xing, L.W.; Kong, F.T. Impact of climate change on maize yield in China from 1979 to 2016. *J. Integr. Agric.* **2021**, *20*, 289–299.
37. Kang, L.; Zhang, H. Comprehensive study on agricultural drought situation in five major grain producing areas of China. *Chin. J. Ecol. Agric.* **2014**, *22*, 10. [CrossRef]
38. Zhu, X.; Xu, K.; Liu, Y.; Guo, R.; Chen, L. Assessing the vulnerability and risk of maize to drought in China based on the AquaCrop model. *Agric. Syst.* **2021**, *189*, 103040. [CrossRef]
39. Yin, X.; Olesen, J.E.; Wang, M.; Kersebaum, K.C.; Chen, H.; Baby, S.; Öztürk, I.; Chen, F. Adapting maize production to drought in the Northeast Farming Region of China. *Eur. J. Agron.* **2016**, *77*, 47–58. [CrossRef]
40. Kang, H.; Na, X.; Zang, S. Evaluation of ecological service function of Songnen Plain Wetland from 1980 to 2010. *Remote Sens. Land Resour.* **2017**, *29*, 8.
41. Qiao, S.; Zhang, L.; Yang, P.; Zhang, X. Analysis on the change of crop water content in the growing season of Songnen Plain from 2000 to 2012. *Prog. Meteorol. Sci. Technol.* **2015**, *5*, 66–69.
42. Tang, W.; Liu, S.; Kang, P.; Peng, X.; Li, Y.; Guo, R.; Jia, J.; Liu, M.; Zhu, L. Quantifying the lagged effects of climate factors on vegetation growth in 32 major cities of China. *Ecol. Indic.* **2021**, *132*, 108290. [CrossRef]
43. Ren, Z.; Liu, X.; Liu, J.; Chen, P. Study on the evolution of drought and flood trend of spring corn in Northeast China in recent 60 years. *Chin. J. Ecol. Agric.* **2020**, *28*, 12. [CrossRef]
44. Wang, L.; Zhu, H.; Lin, A.; Zo, L.; Qin, W. Evaluation of the Latest MODIS GPP Products across Multiple Biomes Using Global Eddy Covariance Flux Data. *Remote Sens.* **2017**, *9*, 418. [CrossRef]
45. Yang, Y.; Xiao, P.; Feng, X.; Li, H. Accuracy assessment of seven global land cover datasets over China. *ISPRS J. Photogramm. Remote Sens.* **2017**, *125*, 156–173. [CrossRef]
46. Carmona, F.; Rivas, R.; Kruse, E. Estimating daily net radiation in the FAO Penman–Monteith method. *Theor. Appl. Climatol.* **2017**, *129*, 89–95. [CrossRef]
47. Zhang, G.; Zhang, D.; Zhao, Y.; Tian, G.; An, W. Changes of surface moisture and dryness in Shanxi Province under the background of climate warming. *Geogr. Arid Area* **2020**, *43*, 9.
48. Wang, L.; Wang, T.; Li, Q.; Wu, D.; Hu, Z.; Lu, Y.; Zhang, Y. Spatial and temporal characteristics of winter wheat drought in Henan Province Based on crop water deficit index. *Jiangsu Agric. Sci.* **2019**, *47*, 6. [CrossRef]
49. Zarei, A.R.; Mahmoudi, M.R. Assessment of the effect of PET calculation method on the Standardized Precipitation Evapotranspiration Index (SPEI). *Arab. J. Geosci.* **2020**, *13*, 182. [CrossRef]
50. Wu, Y.; Li, W.; Wang, W.; Quan, Q.; Chen, X.; Yin, H.; Zhou, Q.; Xu, K. Drought characteristics in Inner Mongolia Based on precipitation anomaly percentage. *Study Arid Area* **2019**, *36*, 10. [CrossRef]
51. Wang, M.; Wang, X.; Huang, W.; Zhang, Y.; Ma, J. Spatial and temporal distribution characteristics of seasonal drought in Southwest China based on relative humidity index. *Trans. Chin. Soc. Agric. Eng.* **2012**, *28*, 9. [CrossRef]
52. Li, C.; You, S.; Wu, Y.; Wang, Y. Application of improved crop water deficit index to drought disaster monitoring of spring maize in Northeast China. *Trans. Chin. Soc. Agric. Eng.* **2019**, *35*, 11.
53. Pang, D.; Ren, C.; Wang, Y.; Cui, F.; Liu, H.; Tang, C. Construction and evaluation of maize water suitability index based on yield loss. *J. Nat. Disasters* **2020**, *29*, 10.
54. Musonda, B.; Jing, Y.; Iyakaremye, V.; Ojara, M. Analysis of Long-Term Variations of Drought Characteristics Using Standardized Precipitation Index over Zambia. *Atmosphere* **2020**, *11*, 1268. [CrossRef]

55. Nicolay, S.; Mabille, G.; Fettweis, X.; Erpicum, M. 30 and 43 months period cycles found in air temperature time series using the Morlet wavelet. *Clim. Dyn.* **2009**, *33*, 1117–1129. [CrossRef]
56. Islam, A.R.M.T.; Islam, H.M.T.; Shahid, S.; Khatun, M.K.; Almoajel, A.M. Spatiotemporal nexus between vegetation change and extreme climatic indices and their possible causes of change. *J. Environ. Manag.* **2021**, *289*, 112505. [CrossRef] [PubMed]
57. Guo, B.; Zhang, J.; Meng, X.; Xu, T.; Song, Y. Long-term spatio-temporal precipitation variations in China with precipitation surface interpolated by ANUSPLIN. *Sci. Rep.* **2020**, *10*, 81. [CrossRef] [PubMed]
58. Li, P.; Wang, J.; Liu, M.; Xue, Z.; Liu, M. Spatio-temporal variation characteristics of NDVI and its response to climate on the Loess Plateau from 1985 to 2015. *Catena* **2021**, *203*, 105331. [CrossRef]
59. Teshome, H.; Tesfaye, K.; Dechassa, N.; Tana, T.; Huber, M. Analysis of Past and Projected Trends of Rainfall and Temperature Parameters in Eastern and Western Hararghe Zones, Ethiopia. *Atmosphere* **2022**, *13*, 67. [CrossRef]
60. Mondol, M.; Zhu, X.; Dunkerley, D.; Henley, B.J. Observed meteorological drought trends in Bangladesh identified with the Effective Drought Index (EDI). *Agric. Water Manag.* **2021**, *255*, 107001. [CrossRef]
61. Řehoř, J.; Brázdil, R.; Trnka, M.; Fischer, M.; Balek, J.; Štěpánek, P.; Zahradníček, P.; Semerádová, D.; Bláhová, M. Effects of Climatic and Soil Data on Soil Drought Monitoring Based on Different Modelling Schemes. *Atmosphere* **2021**, *12*, 913. [CrossRef]

hydrology

MDPI

Article

Rainfall Partitioning in Amazon Forest: Implications of Reduced Impact Logging on Litter Water Conservation

Jeferson Alberto de Lima [1] and Kelly Cristina Tonello [2,*]

[1] Department of Environmental Engineering, Federal University of Rondônia, Ji-Paraná 76900-726, Brazil; jeferson.lima@unir.br
[2] Department of Environmental Sciences, Federal University of São Carlos, Sorocaba 18052-780, Brazil
* Correspondence: kellytonello@ufscar.br

Abstract: This study aimed to investigate how sustainable forest management can affect litter hydrological properties. We investigated the net precipitation, litter mass, water-holding capacity, effective water-holding and retention capacity, maximum water retention and water content in unlogged and logged forests over 13 months in the Amazon Forest, where reduced-impact logging is allowed. The mean litter mass was similar for unlogged and logged forests. The litter water-holding capacity was 220% for unlogged and 224% for logged forests, and for fractions followed: unstructured > leaves > seeds > branches for both forests. The effective water-holding capacity was 48.7% and 49.3% for unlogged and logged, respectively, and the effective water retention was 10.3 t·ha^{-1} for both forests. The effective water retention in the rainy and dry seasons accounted for 12.5 t ha^{-1} and 7.2 t ha^{-1} for unlogged and logged, respectively. The maximum water retention was slightly greater for logged forests (16.7 t ha^{-1}) than unlogged (16.3 t ha^{-1}). The litter water content had 40% less water in the dry season than in the rainy in both forests. In general, there were no significant differences in litter storage and hydrological properties between stands. This suggests that reduced-impact logging did not significantly affect the hydrological dynamics of the litter layer in the Amazonian forests studied.

Keywords: forest hydrology; logging concessions; reduced-impact logging; selective logging; stemflow; Jamari National Forest

Citation: de Lima, J.A.; Tonello, K.C. Rainfall Partitioning in Amazon Forest: Implications of Reduced Impact Logging on Litter Water Conservation. *Hydrology* **2023**, *10*, 97. https://doi.org/10.3390/hydrology10040097

Academic Editors: Songhao Shang, Qianqian Zhang, Dongqin Yin, Hamza Gabriel and Magdy Mohssen

Received: 22 March 2023
Revised: 6 April 2023
Accepted: 10 April 2023
Published: 21 April 2023

1. Introduction

Litter studies are mainly related to stocks, decomposition, and biogeochemical cycles [1–4]. However, it is important to highlight that litter also acts as an insulating layer that protects the water and soil loss [5,6], making it essential for regulating surface hydrological processes [7] as a stage in the forest hydrology dynamics. Although the role in mediating the impact of raindrops, reducing, delaying, or often eliminating surface runoff and soil erosion [8–10] is recognized, the relevance of litter is still often underestimated in forest hydrology [6,8,11], especially in tropical forests.

Rainfall interacting with the forest canopy can take multiple paths, with some being intercepted and evaporated, while others reach the soil through throughfall and stemflow, constituting the net precipitation. However, before the net precipitation can contribute to soil moisture, it must cross the intermediate layer of litter. The litter layer can either retain the net precipitation without allowing it to reach the soil, initiate surface runoff, or facilitate infiltration into the soil. For example, it can protect soil water from wind and radiation-induced evaporation, thereby minimizing soil evaporation by 26.1–53.8% and reducing the surface runoff by 71.1–90.5% in a *Pinus tabulaeformis* plantation [8]. The forest litter layer is thus a key factor for water conservation in forest ecosystems. The retention of water in the litter layer is essential for hydrological modeling, as it can alter the amounts of water available for infiltration and/or runoff [7,12,13]. However, it depends on the physical-chemical composition of the material [14], which is influenced by the forest's age and type, as well as climatic conditions. Understanding the structure and composition of the litter in the forest ecosystem is therefore crucial for determining the best way to manage the system, contributing to soil and water conservation. This is especially important in areas where exploration activities are carried out—like in Amazon forest—although the litter hydrological properties have not yet been quantitatively reported to the authors' knowledge.

Given the escalating rates of deforestation in tropical forests, particularly in the Amazon, it is crucial to conduct more studies on the ecosystem services provided by these forest environments. The Amazon Forest is globally significant, acknowledged for its environmental services, biological diversity, carbon sinks, and as a regulator of climate through its contribution to biogeochemical cycles [15,16]. However, the current situation is concerning, as the environmental impacts generated by human exploitation of the Amazon are imperiling its vast biodiversity reserves and globally important ecosystem services [17]. Deforestation in the Amazon hit a 15-year high in 2022, with a record 10,573 km^2 of deforestation between January and December [18]. In 2020, Brazil, which occupies 60% of the Amazon Forest [19], produced an estimated 29.2 million m^3 of tropical industrial roundwood, excluding plantations [20]. Given that current Brazilian law allows a maximum of 30 m^3 ha^{-1}, timber-harvesting activities affect a substantial area every year. To control the predatory exploitation of forests, the conservation potential of managed production tropical forests has promoted the implementation of timber harvesting practices generally referred to as 'low-impact' or 'reduced-impact' logging (RIL) [21]. Managers and certifiers widely recognize RIL as a sustainable and environmentally friendly approach to harvesting primary tropical forests [22]. This practice is a rational model of forest exploitation that aims to reduce environmental impacts by aligning natural resource conservation with forestry and the intensity of the impacts is related to the number and volume of trees removed from the natural ecosystems [23]. However, the impacts of RIL activities on tropical-forest ecosystem services, such as hydrological processes, are still understudied. To establish ecologically and economically productive harvesting cycles, a more detailed understanding of the effects of RIL on ecohydrological processes is also necessary.

This study was conducted in Jamari National Forest, a conservation unit whose primary objective is the sustainable multiple use of natural resources and scientific research [24], where sustainable forest management is allowed through RIL. As RIL is one of the most important sustainable economic activities in tropical forests, understanding these impacts is essential to evaluate RIL as a sustainable solution. Thus, to investigate how sustainable forest management actions can affect the hydrological processes of the forest, this study aims to test the following hypothesis: changes in vegetation structure by sustainable forest management activities will alter the composition of litter fractions, which in turn affects the water retention capacity of litter. To test this hypothesis, the study aims to (1) understand the dynamics and seasonality of litter stocks, (2) the water retention capacity of litter, and (3) evaluate the effective water retention and water content of litter in unlogged and logged (RIL) Amazonian forests. The goal of this study is to provide a scientific basis for understanding sustainable forest management actions in the Amazon and their impact on the hydrological dynamics of litter and water conservation.

2. Materials and Methods

The study site is located southwest of the Jamari National Forest (JNF) at the Forest Management Unit III (FMU-III). The physiognomic-ecological classification shows the predominance of Dense Tropical Rain Forest, with Open Rain Forest, characterized by spaced trees forming a canopy of 40 m in height [25]. The JNF area has a rainy tropical climate (Aw according to the Koppen classification) (Brazilian Forest Service 2019), precipitation from 1800~2200 mm y^{-1}. The weather in Amazon is well defined [26] with two distinct periods: rainy (from October to April) and dry (from June to August); May and September are transition months. The mean air temperature is 24~26 °C, and the relative humidity is 80~90%.

Open precipitation, net precipitation, and litter yield were studied for 13 months (from October 2019 to October 2020) on three plots of 20 × 20 m at unlogged and logged forest stands (total per stand: 1200 m^2) at FMU-III. The general characteristics of the stands are listed in Table 1, as also the methodology and the name of the tree species/family for each stand are in Tables S2 and S3, respectively. The unlogged forest (UF) (9°24′53.15″ S and 63°04′29.48″ W) represents vegetation that retains the characteristics of primary forest and logged forest (LF) (9°23′27.52″ S and 63°02′27.08″ W) in which the silvicultural system adopted is polycyclic with 25-year cutting cycles, a maximum cutting intensity of 25 m^3 ha^{-1} $year^{-1}$, and minimum cutting diameter equal to or greater than 50 cm, as determined by law no. 11.284/2006 [27,28]. The UF is located at UAP-18 and LF is located at UAP-11 which was exploited in 2018. UAP means the Units of Annual Production, i.e., subdivision of the Forest Management Area, designated for logging within one year [29].

Table 1. Forest structure in the studied stands (UF—unlogged and LF—logged forest), Jamari Na-tional Forest, Rondônia, Brazil.

Parameters	Forest Structure									
	D [cm]		Ht [m]		CA [m^2]		Vc [m^3]		Tree Density [trees ha^{-1}]	
	UF	LF	UF	LF	UF	LF	UF	LF	UF	LF
Min	12.7	12.7	12.0	9.0	7.2	0.4	8.0	1.3		
Max	111.5	140.8	28.0	36.0	199.7	284.8	795.4	1512.8		
Means	21.8	29.0	17.4	17.3	41.4	49.9	133.7	161.6	1625 *	1450 *
SD	18.0	24.5	3.5	4.9	37.5	58.8	154.5	284.0		

D = diameter at breast height; Ht = tree height; CA = crown area; Vc = crown volume. * Indicates differences between means ($p < 0.05$).

Rain gauges were installed outdoors, 50 m from each other in places free of treetops and other forest structures. Throughfall was monitored by 7 rain gauges randomly installed inside each plot, totaling 21 rain gauges per stand. To measure stemflow volume [L $tree^{-1}$], collectors were installed in 60 trees with a diameter of breast height [D] > 10 cm. Stemflow collars were constructed by wrapping individual tree stems with a polyurethane gutter, fixed at 1.3 m from the ground. Water running down the stem was captured by these gutters, then drained by a 16 mm hose (5/8 inch) connected to 20 L collection bins. Event stemflow volumes [L $tree^{-1}$] were calculated by each tree's projected canopy area [m^2 $tree^{-1}$]. Open precipitation, throughfall and stemflow samples were measured in the field in a measuring cylinder (1-L, at 0.5-L graduation) or in a graduated bucket (20-L, at 0.5-L graduation). Net precipitation [NP, mm] per plot was found by summing throughfall to stemflow and represents the portion of open precipitation that reaches the soil surface.

Litter samples were collected from a 100 × 100 cm litter square [30], which was partitioned into 4 quadrants. Collection was random and only materials in one quadrant (50 × 50 cm) were collected. A total of 390 litter bags per stand (3 plots × 10 random quadrants × 13 months) were collected. In the laboratory, soil was sieved and removed from the litter samples, which were classified into four fractions: branches, leaves, seeds and unstructured material. Monthly and annual litter yields were estimated by summing the fractions. The fresh mass [FM, g] for each fraction was determined on a suitably accurate scale [0.01 g] and rehydrated through immersion in water for 90 min. After this, the litter fractions were deposited on sieves and drained for 30 min for further humid litter mass [HM, g] determination. Subsequently, the amount of litter dried mass [DM, g] was determined by oven-drying samples at a forced circulation oven at 70 °C, until reaching constant mass. Finally, the equations applied to calculate the hydrological properties of litter can be found in Table 2.

Table 2. Hydrological properties of litter applied to unlogged and logged litter from Amazon Forest, Rondônia, Brazil.

Hydrological Properties	Equation	Notes
Water-holding capacity [WHC, %] [30–32]	$WHC = \frac{(HM-DM)}{DM} \times 100$	HM = humid litter mass, g; DM = dry mass, g
Effective water-holding capacity of the litter under ambient conditions [EWC, %]	$EWC = \left[\frac{FM-DM}{FM}\right] \times 100$	FM = fresh mass, g; DM = dry mass, g
Litter effective water-retention capacity [W_{eff}, t ha^{-1}]	$W_{eff} = \frac{(0.85*WHC-EWC)\times M}{100}$	WHC = water-holding capacity, %; EWC = effective water-holding capacity; %M = is the unit litter mass (t·ha^{-1})
Maximum water-retention capacity [W_{max}, t ha^{-1}]	$W_{max} = \frac{WHC\times M}{100}$	WHC = water-holding capacity, %; M = is the unit litter mass (t·ha^{-1})
Litter water content [WC, $g_{water}\ kg_{litter}^{-1}$]	$WC = \left[\frac{FM[g]-DM[g]}{FM[kg]}\right]$	FM = fresh mass, g; DM = dry mass, g

The differences in the litter stocks and hydrological properties were tested for homoscedasticity of variance using the Bartlett test and for normality using Lilliefors (Kolmogorov–Smirnov) test for statistical analysis. Analysis of variance was applied to normal data using the Student's *t*-test at a 5% probability level. Data that did not meet ANOVA assumptions were subjected to a non-parametric Mann-Whitney test. The analyses were performed at BioEstat 5.3 [33] and @Minitab 17. For further examination of seasonal changes, monthly groups were created from the data.

3. Results

3.1. Hydrometeorological Observations and Litter Accumulation

There was no significant difference in the mean open precipitation between UF (1868 mm) and LF (1771 mm) ($p = 0.316$). Similarly, there was no significant difference in the mean net precipitation between UF (2106 mm) and LF (1680 mm) ($p = 0.082$). The net precipitation in UF was 13% higher than the open precipitation, and 25% higher than LF (Figure 1). It is worth noting that there was no rainfall in July (see Supplementary Materials Table S3). Seasonally, the lowest accumulations were observed in the dry period (May to September), representing approximately 9.6% and 8.6% of the open precipitation recorded in UF and LF, respectively.

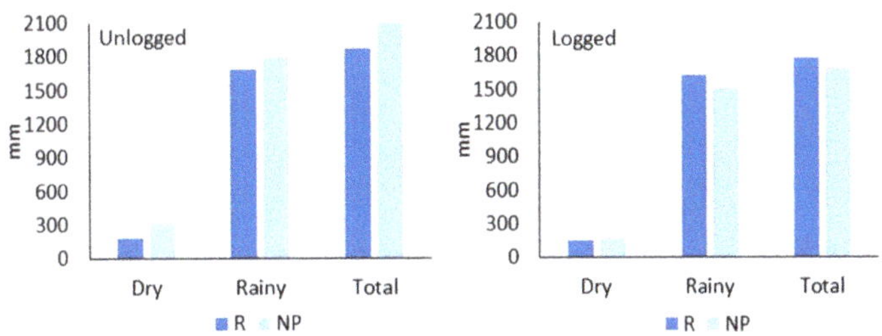

Figure 1. Open precipitation [R, mm] and net precipitation [NP, mm] partitioning in the dry and rainy season at unlogged and logged forests, Flona of Jamari, Rondônia—Brazil.

The monthly accumulated litter ranged from 3.8 to 12.2 Mg ha^{-1} in UF and from 4.9 to 13.3 Mg ha^{-1} in LF (Figure 2, Supplementary Materials Table S4), and no significant differences were observed between stands (p-value = 0.974). The number of fractions varied for both forests, with branches being the most abundant fraction (40% for UF and 38% for LF), followed by leaves (33% for UF and 37% for LF), unstructured (20% for UF and 19% for

LF) and seeds (3% and 2% for UF and LF, respectively). Despite no significant differences between fractions and seasons, it was observed that the highest litter mean accumulation was in the rainy season, with 69.1% (8.3 ± 2.3 t ha^{-1}) for UF and 67.3% (8.1 ± 2.9 t ha^{-1}) for LF. During the dry season, the accumulated litter decreased by 28% (6.0 ± 1.4 t ha^{-1}) and 22% (6.3 ± 1.3 t ha^{-1}) for UF and LF, respectively. The accumulation of all fractions was lower during the dry season.

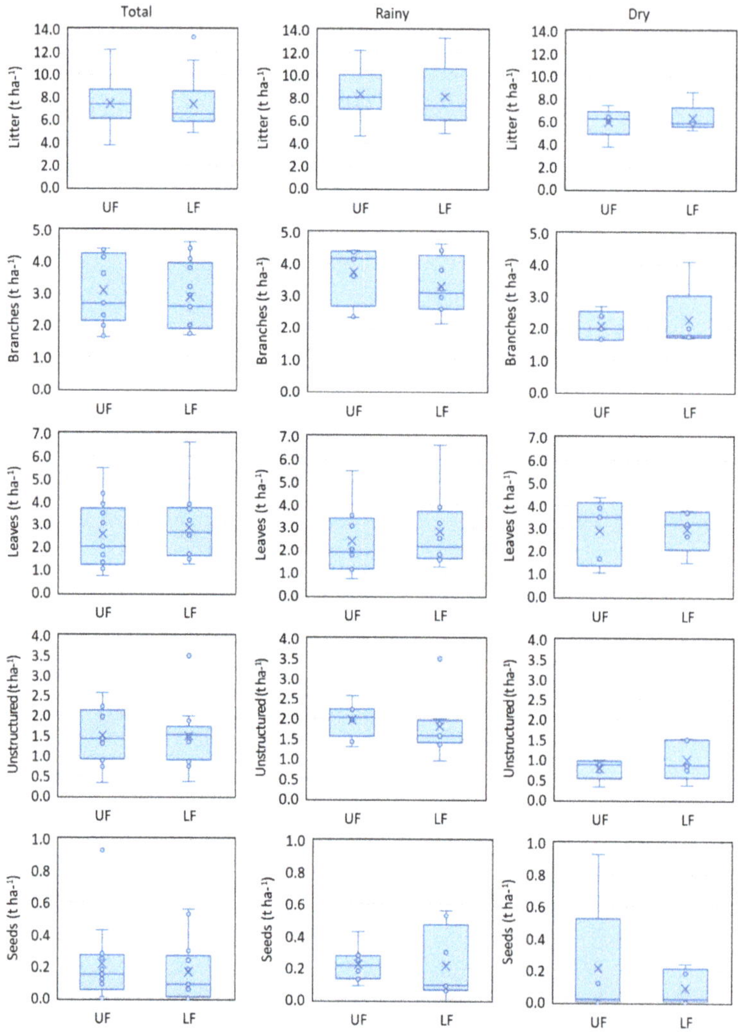

Figure 2. Monthly litter and fractions accumulation (dry mass) for total period, rainy and dry seasons at unlogged (UF) and logged forest (LF), Jamari National Forest, Rondônia—Brazil.

3.2. Litter-Water Interactions

Despite the lowest variation in water-holding capacity for UF (153–271%) than LF (140–332%), both stands showed similar properties (p-value = *0.856*), even when considering different fractions and seasons (Figure 3; Supplementary Materials Table S3). The mean water-holding capacity was 220 ± 43.7% for UF, distributed in unstructured material (320 ± 65.9%) > leaves (271 ± 79.6%) > seeds (222 ± 99.7%) > branches (136 ± 34.7%). Meanwhile, the mean WHC was slightly higher for LF at 224 ± 58.7%, and its fractions were ranked as follows: unstructured material (309 ± 67.8%) > leaves (283 ± 88.1%) > seeds (213 ± 99.5%) > branches (145 ± 43.3%). Regarding seasons, both stands showed the highest WHC during the rainy season, in which the litter from LF had the capacity to hold more water than UF, although only branches in the rainy season showed the highest WHC for LF.

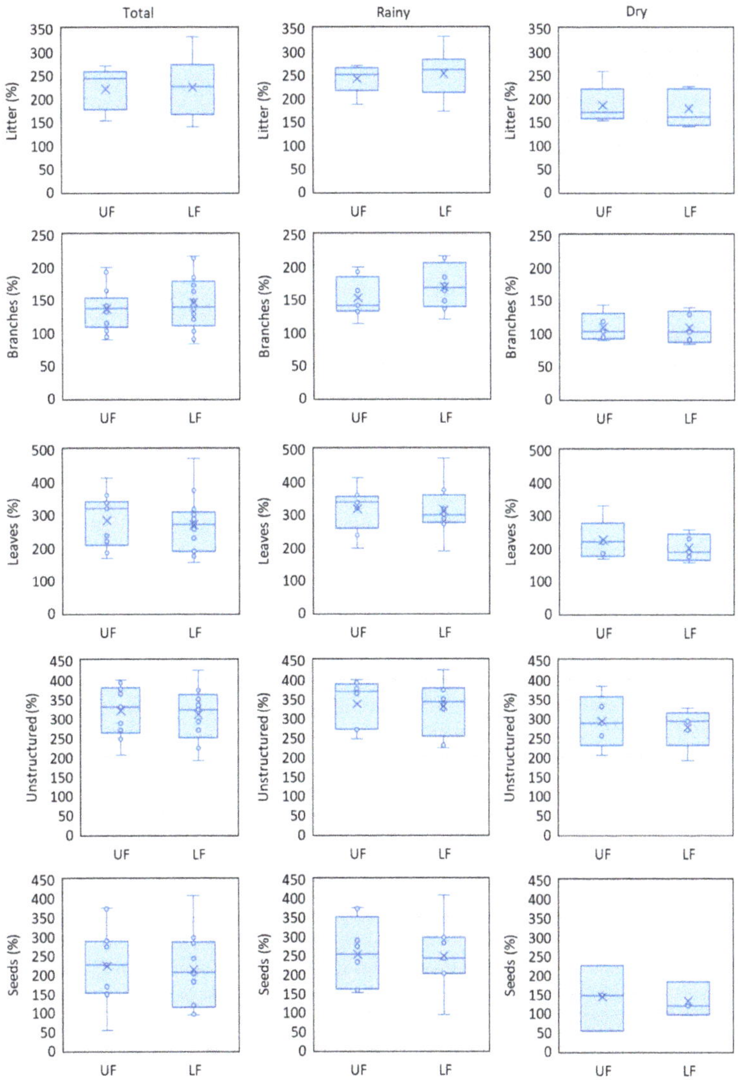

Figure 3. Monthly litter and fractions water-holding capacity for total period, rainy and dry seasons at unlogged (UF) and logged forest (LF), Jamari National Forest, Rondônia—Brazil.

The mean effective water-holding [EWC, %] was similar for both stands (p = 0.775) (UF = 48.7 ± 61.8% and LF = 49.3 ± 54.2%) (Figure 4). In general, both the annual and seasonal water content of litter did not differ significantly between stands but showed that litter contains more water during the rainy season. In the dry season, EWC was 41% and 35% lower than rainy for UF and LF, respectively. The W_{eff} ranged from 6.4

to 16.5 and 5.6 to 19.1 t·ha^{-1} for UF and LF, respectively (Supplementary Materials Table S5). The mean did not differ between the stands (p = 0.935) and represented 10.3 t·ha^{-1} month^{-1} for both. In the same way, for both stands, the rainy and dry seasons accounted 12.5 t ha^{-1} month^{-1} and 7.2 t ha^{-1} month^{-1}, respectively. The W_{max} varied from 9.7 to 25.7 t·ha^{-1} and from 8.2 to 28.7 t·ha^{-1} for UF and LF, respectively, and did not differ significantly between stands (p = 0.916). The mean W_{max} was slightly greater for LF (16.7 ± 1.5 t ha^{-1}) than UF (16.3 ± 1.0 t ha^{-1}). The rainy season was 45% (20.1 ± 4.5 t ha^{-1} month^{-1} and 20.6 ± 5.2 t ha^{-1} month^{-1} for UF and LF, respectively) higher than dry season (11.1 ± 1.1 t ha^{-1} month^{-1} and 11.2 ± 1.2 t ha^{-1} month^{-1} for UF and LF, respectively). During the study period, the litter WC ranged from 162 to 634 g_{water} kg_{litter}^{-1} for UF and from 176 to 696 g_{water} kg_{litter}^{-1} for LF and did not show significant differences (p = 0.809) between forests. The dry season had 40% less water than the rainy in both forests.

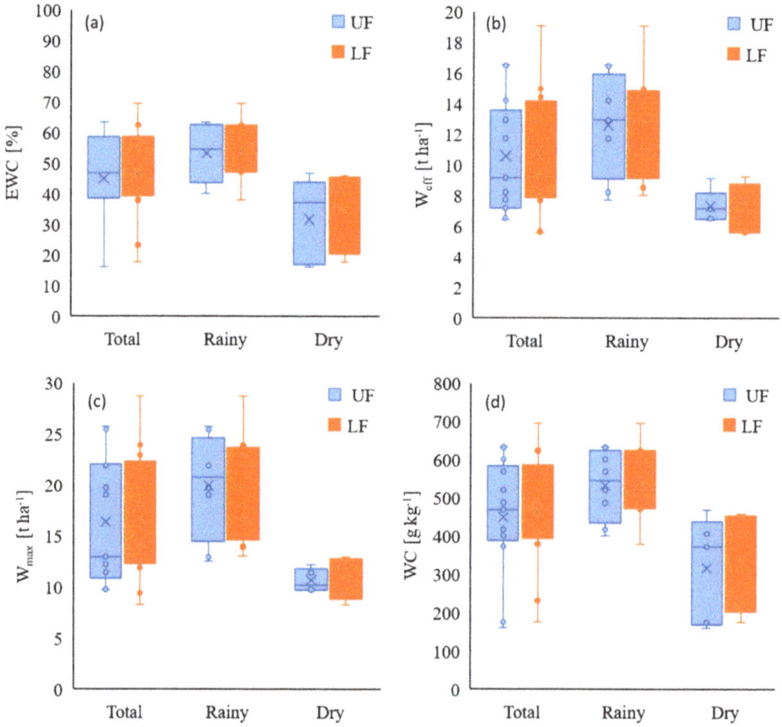

Figure 4. Monthly (**a**) effective water-holding capacity (EWC, %), (**b**) effective water-retention capacity (W_{eff}, t ha^{-1}), (**c**) maximum water-retention capacity (W_{max}, t ha^{-1}), (**d**) and litter water content (WC, g_{water} kg_{litter}^{-1}) in unlogged and logged forest. Jamari National Forest, Rondônia, Brazil.

4. Discussion

Net precipitation is an important variable to analyze the hydrological processes in forest ecosystems. This parameter corresponds to the total amount of open precipitation that reaches the ground after crossing various barriers. The results of this study showed that a larger volume of water—in addition to rainfall—reached the forest floor in the unlogged forest. This behavior has already been observed in the Amazon rainforest by other studies [34,35] and may correspond to the input of water from fog—which, being a horizontal movement, is not quantified by rain gauges. On the other hand, when fog condenses as it interacts with the surface of leaves, branches, and trunks, it contributes to the throughfall and stemflow, thus justifying the higher amount of water within the forest when compared to open precipitation. The study of net precipitation is important in hydrological research because it allows for the analysis of the amount of precipitation that reaches the soil surface and interacts with the ecosystem. This is especially critical in forest ecosystems, where various barriers, such as vegetation, topography, and fog, can alter the amount and distribution of water that reaches the soil. Understanding the dynamics of net precipitation can help us better comprehend hydrological processes in forest ecosystems and, consequently, promote their conservation and management more effectively. In this case, the reduced logging practices did not affect the distribution of net precipitation.

The mean annual litter deposition in the studied forest was similar to that reported for primary and successional stages of forests in the Brazilian Amazon [23,36,37]. The highest rates of litter deposition were observed during the rainy season and disagree with other studies [1,23,38] indicating that litterfall is not solely controlled by rainfall seasonality (rainy or dry season). Seasonal variation in litterfall production resulted in large variations in the amount of litter on the soil, with litter peaks occurring in several months of the year [39]. Factors such as forest structure, including age, species, soil nutrients, air temperature, luminosity, wind, and soil water, also play a role in litter deposition [3]. While leaves are typically reported as the predominant fraction in Amazon forests [1,40,41], the accumulation of branches was found to be more prevalent than leaves in the studied forest, which is consistent with observations in a Semideciduous Seasonal Forest [42,43]. The high amount of branches during the rainy season (October to April) may be attributed to the mechanical energy imposed by the wind during the rains, as well as the weight gain of the branches due to waterlogging, making them more susceptible to fall and consequently increasing their production at the onset of the rainy season.

The minor impacts of reduced-impact logging may minimize the effects on vegetation cover and, consequently, on litter deposition [23]. This is particularly important to maintain water conservation in sites under sustainable forest management and the results showed that reduced-impact logging did not affect the effectiveness per unit of mass litter in retaining water. Both stands showed higher water-holding capacity during the rainy season than the dry season, which is expected given the higher rainfall and humidity during the wet season. This similarity in hydrological properties between the stands could be explained by the similarity in the litter composition, such as the leaves, branches, seeds and unstructured materials. Despite the greater accumulation of branches in both sites, the highest rates of litter water-holding capacity were observed in the unstructured fraction. In fact, the water-holding capacity does not only depend on the amount of organic material, but also on the degree of decomposition of its fractions. The high-water retention rates observed in the amorphous fraction are due to the lower surface adhesion of this material [44,45]. This means that the greater the degree of litter decomposition, the greater the specific surface area, and consequently, the greater the potential for water retention compared to other more superficial and less decomposed fractions. The fact that the leaf fraction has a lower rate of water-holding capacity compared to the unstructured fraction can be understood as a function of the lower surface adhesion or leaf adsorption, which depends on aspects related to the leaf fraction itself, such as leaf area, structure, relief, shape, surface/weight ratio, and organic composition [45,46]. In relation to the branch fraction, which presented the lowest water retention rates, this is due to the nature of the woody material, which is more hydrophobic and presents less water absorption. These branches are mostly made of xylem with thick fibers and a relatively stable structure between cells, which is difficult for water absorption although the internal tubular structure is rich [47].

It is important to highlight that litter water-holding capacity reflects the water-holding condition in the laboratory, i.e., in the ideal state and the maximum capacity of water retention [30,48]. The effective water retention was also similar between the unlogged and logged stands, indicating that both stands were able to retain a significant amount of water and that the water retention capacity of litter was not significantly affected by RIL. As reported by [49,50], W_{eff} defines the effective interception of precipitation by litter, which is an important hydrological property that can be used to consistently evaluate the potential to absorb rainfall and reduce surface runoff [31,37]. The W_{eff} is also affected by the water content of litter, litter storage, and the nature of rainfall [27,40,45]. Our study shows that, regardless of the RIL management, the mean annual capacity of litter in the Amazon Forest to retain water was lower than that observed in *Eucalyptus mangium* and *Eucalyptus robusta* but greater than *Hevea brasiliensis* [32]. On the other hand, W_{max}, a measure of rainfall absorption, was higher than *Eucalyptus robusta* [32], and, especially in the rainy season, higher than *Hevea brasiliensis* [32]. If we consider that 1 mm of rainfall is equivalent to 1 $t \cdot ha^{-1}$ [32,49], the litter from unlogged or logged forests has the potential to intercept up to a mean of 12.5 mm and 7.2 mm of rainfall at the rainy and dry season, respectively, and a mean of 10.5 mm per year. In fact, the studied forest can retain in the litter the 3335 kg ha^{-1} and 3433 kg ha^{-1}—of water per month in unlogged and logged forests, respectively. Considering no statistical differences between unlogged and logged forests, the water content in the litter represented 3752 kg ha^{-1} and 2317 kg ha^{-1} per month during the rainy and dry seasons, respectively.

Studies investigating the patterns of litter and its fractions under field conditions to determine their effective capacity in retaining water are necessary, especially in complex and important environments undergoing accelerated changes, such as the Amazon biome. Although there are studies on litter production in different environments, there is a lack of knowledge on the water retention capacity of litter in field conditions. Research on the ecohydrological functions of litter in the water balance of forest ecosystems is essential [51]. The findings of this study have important implications for sustainable forest management in the Amazon. The results suggest that RIL can maintain the water retention capacity of litter, which is an important component of the hydrological cycle in tropical forests. This information can be used to develop more effective and sustainable forest management practices that maintain the ecological and economic values of tropical forests. It is important to note that while this study provides valuable insights into the impacts of RIL on the litter hydrological processes of tropical forests, further research is needed to fully understand the effects of RIL on other ecosystem services, such as carbon sequestration, biodiversity conservation, and soil erosion control. Additionally, more studies are needed to evaluate the long-term sustainability of RIL and its ability to maintain the ecological integrity of tropical forests.

5. Conclusions

The minor impacts of reduced-impact logging may minimize the effects on litter deposition, which is particularly important to maintain hydrological processes in sites under sustainable forest management. The litter water-holding capacity between unlogged and logged forests was similar, and the litter water dynamics kept the same properties. The results of the study indicate that both unlogged and logged forests have similar water-holding capacities, effective water holding and retention, maximum water retention and water content. Although there were some slight differences in the mean water-holding capacities of fractions, the overall differences between the two stands were not significant. The effective water retention was similar in both stands, and the W_{eff} and W_{max} did not differ significantly between the two stands. These findings suggest that sustainable forest management activities through reduced-impact logging did not significantly affect the hydrological dynamics of the litter layer in the Amazonian forests studied. More studies are needed to investigate the patterns of litter and its fractions under field conditions to determine its effective capacity to retain water, especially in complex and important environments such as the Amazonian biome. Understanding the ecohydrological functions of litter in the water balance of forest ecosystems is crucial for the management and conservation of these ecosystems.

Supplementary Materials: The following supporting information can be downloaded at: https://www.mdpi.com/article/10.3390/hydrology10040097/s1, Table S1. Dendrometry attributes applied for the characterization of the forest structure. Table S2. Families and forest species observed at unlogged (UL) and logged (L) plots at Jamari National Forest, Rondônia, Brazil. P1/P2/P3—plots at unlogged forest; P4/P5/P6—plots at logged forest. Table S3: Statistical analyses for open rainfall and net precipitation in the Unlogged (UL) and Logged (L) at Jamari National Forest, Rondônia, Brazil. Table S4: Litter fractions descriptive analyses for unlogged (UL) and logged (L) at Jamari National Forest, Rondônia, Brazil. Table S5: Effective litter water-holding capacity [EWC, %], water-holding capacity [WHC, %], litter effective water-retention capacity [W_{eff}, t ha^{-1}], maximum retention capacity [W_{max}, t ha^{-1}] and water content [WC, g_{water} $kg_{litter}{}^{-1}$] for unlogged (UL) and logged (L) at Jamari National Forest, Rondônia, Brazil.

Author Contributions: Conceptualization, K.C.T.; Data curation, J.A.d.L.; Formal analysis, J.A.d.L. and K.C.T.; Funding acquisition, J.A.d.L.; Investigation, J.A.d.L.; Methodology, K.C.T.; Supervision, K.C.T.; Writing—original draft, J.A.d.L.; Writing—review & editing, K.C.T. All authors have read and agreed to the published version of the manuscript.

Funding: This research was funded by Foundation to Support the Development of Scientific and Technological Actions and Research from the state of Rondônia (FAPERO).

Data Availability Statement: The data that support the findings of this study are available from the corresponding author upon reasonable request.

Acknowledgments: The forestry company AMATA S/A for providing the field databases and the Brazilian National Council for Scientific and Technological Development (CNPq).

Conflicts of Interest: The authors declare no conflict of interest.

References

1. Almeida, E.J.; Luizão, F.; Rodrigues, D.d.J. Produção de Serrapilheira Em Florestas Intactas e Exploradas Seletivamente No Sul Da Amazônia Em Função Da Área Basal Da Vegetação e Da Densidade de Plantas. *Acta Amazon.* **2015**, *45*, 157–166. [CrossRef]
2. Giweta, M. Role of Litter Production and Its Decomposition, and Factors Affecting the Processes in a Tropical Forest Ecosystem: A Review. *J. Ecol. Environ.* **2020**, *44*, 11. [CrossRef]
3. Holanda, A.C.; Feliciano, A.L.P.; Freire, F.J.; Sousa, F.Q.; Freire, S.R.d.O.; Alves, A.R. Aporte de Serapilheira e Nutrientes Em Uma Área de Caatinga. *Ciência Florest.* **2017**, *27*, 621–633. [CrossRef]
4. Seta, T.; Demissew, S.; Woldu, Z. Litterfall Dynamics in Boter-Becho Forest: Moist Evergreen Montane Forest of Southwestern Ethiopia. *J. Ecol. Nat. Environ.* **2018**, *10*, 13–21. [CrossRef]
5. Sayer, E.J. Using Experimental Manipulation to Assess the Roles of Leaf Litter in the Functioning of Forest Ecosystems. *Biol. Rev. Camb. Philos. Soc.* **2006**, *81*, 1–31. [CrossRef]
6. Xia, L.; Song, X.; Fu, N.; Cui, S.; Li, L.; Li, H.; Li, Y. Effects of Forest Litter Cover on Hydrological Response of Hillslopes in the Loess Plateau of China. *Catena* **2019**, *181*, 104076. [CrossRef]
7. Zagyvai-Kiss, K.A.; Kalicz, P.; Szilágyi, J.; Gribovszki, Z. On the Specific Water Holding Capacity of Litter for Three Forest Ecosystems in the Eastern Foothills of the Alps. *Agric. For. Meteorol.* **2019**, *278*, 107656. [CrossRef]
8. Cui, Y.; Pan, C.; Zhang, G.; Sun, Z.; Wang, F. Effects of Litter Mass on Throughfall Partitioning in a *Pinus tabulaeformis* Plantation on the Loess Plateau, China. *Agric. For. Meteorol.* **2022**, *318*, 108908. [CrossRef]
9. Giácomo, R.G.; Alves, M.C.; Camara, R.; Pereira, M.G.; de Arruda, O.G.; Souto, S.N.; de Moraes, M.L.T. Litterfall and Nutrient Input in a Degraded Area. *Floresta Ambiente* **2017**, *24*, 1–9. [CrossRef]
10. Gillerot, L.; Forrester, D.I.; Bottero, A.; Rigling, A.; Lévesque, M. Tree Neighbourhood Diversity Has Negligible Effects on Drought Resilience of European Beech, Silver Fir and Norway Spruce. *Ecosystems* **2021**, *24*, 20–36. [CrossRef]
11. Dunkerley, D. Percolation through Leaf Litter: What Happens during Rainfall Events of Varying Intensity? *J. Hydrol.* **2015**, *525*, 737–746. [CrossRef]
12. Winkler, J.P.; Negreiros, A.B.d. A Serrapilheira Como Bioindicador de Qualidade Ambiental Em Fragmentos de Eucalyptus. *Rev. Cont.* **2018**, *7*, 175–202.
13. Melissa, K.; Dickow, C.; Marques, R.; Pinto, C.B.; Höfer, H. Produção de Serapilheira Em Diferentes Fases Sucessionais de Uma Floresta Subtropical Secundária, Em Antonina, PR. *Cerne* **2011**, *18*, 75–86.
14. Santos, F.M.; Chaer, G.M.; Diniz, A.R.; Balieiro, F.d.C. Nutrient Cycling over Five Years of Mixed-Species Plantations of Eucalyptus and Acacia on a Sandy Tropical Soil. *For. Ecol. Manag.* **2017**, *384*, 110–121. [CrossRef]
15. Gatti, L.V.; Basso, L.S.; Miller, J.B.; Gloor, M.; Gatti Domingues, L.; Cassol, H.L.G.; Tejada, G.; Aragão, L.E.O.C.; Nobre, C.; Peters, W.; et al. Amazonia as a Carbon Source Linked to Deforestation and Climate Change. *Nature* **2021**, *595*, 388–393. [CrossRef] [PubMed]

16. Araujo, E.C.G.; Sanquetta, C.R.; Dalla Corte, A.P.; Pelissari, A.L.; Orso, G.A.; Silva, T.C. Global Review and State-of-the-Art of Biomass and Carbon Stock in the Amazon. *J. Environ. Manag.* **2023**, *331*, 117251. [CrossRef]
17. Albert, J.S.; Carnaval, A.C.; Flantua, S.G.A.; Lohmann, L.G.; Ribas, C.C.; Riff, D.; Carrillo, J.D.; Fan, Y.; Figueiredo, J.J.P.; Guayasamin, J.M.; et al. Human Impacts Outpace Natural Processes in the Amazon. *Science* **2023**, *379*, eabo5003. [CrossRef]
18. IMAZON. Amazônia Perdeu Quase 3 Mil Campos de Futebol Por Dia de Floresta Em 2022-Maior Desmatamento Em 15 Anos. Available online: https://imazon.org.br/imprensa/amazonia-perdeu-quase-3-mil-campos-d (accessed on 11 April 2023).
19. MMA Amazônia. Available online: https://www.gov.br/mma/pt-br/assuntos/ecossistemas-1/biomas/amazonia (accessed on 16 March 2023).
20. ITTO. *Biennial Review and Assessment of the World Timber Situation 2019–2020*; ITTO: Yokohama, Japan, 2021.
21. Sist, P.; Ferreira, F.N. Sustainability of Reduced-Impact Logging in the Eastern Amazon. *For. Ecol. Manag.* **2007**, *243*, 199–209. [CrossRef]
22. Wolff, S.; Schweinle, J. Effectiveness and Economic Viability of Forest Certification: A Systematic Review. *Forests* **2022**, *13*, 798. [CrossRef]
23. Barreiros, J.L.; Oliveira, N.S.; Cerboncini, R.A.S.; Junior, L.K. Does Selective Logging Affect Litter Deposition Rates in Central Brazilian Amazonia? *An. Acad. Bras. Ciências* **2022**, *94*, 1–10. [CrossRef]
24. Brasil. Lei nº 12.651, 25 de maio de 2012. Available online: https://www.planalto.gov.br/ccivil_03/_ato2011-2014/2012/lei/l12651.htm/ (accessed on 11 April 2023).
25. MMA. IBAMA Plano de Manejo Da Floresta Nacional Do Jamari; Brasília-DF, 2005. Available online: https://www.gov.br/agricultura/pt-br/assuntos/servico-florestal-brasileiro/concessao-florestal/concessoes-florestais-em-andamento-1/floresta-nacional-do-jamari-ro (accessed on 11 April 2023).
26. Carneiro, R.G.; Fisch, G. Observational Analysis of the Daily Cycle of the Planetary Boundary Layer in the Central Amazon during a Non-El Niño Year and El Niño Year (GoAmazon Project 2014/5). *Atmos. Chem. Phys.* **2020**, *20*, 5547–5558. [CrossRef]
27. SFB—Serviço Florestal Brasileiro. Especial: Concessões Florestais. In *O Manejo Florestal Sustentável Protege as Florestas*; SFB/MAPA: Santarém, Brazil, 2020. Available online: https://snif.florestal.gov.br/pt-br/ultimas-noticias/658-o-manejo-florestal-sustentavel-protege-as-florestas (accessed on 11 April 2023).
28. AMATA. *Plano de Manejo Florestal Sustentável: Categoria de PMFS Pleno 2007*; AMATA: Vila Olímpia, Brazil.
29. Brasil;Lei n° 11.284, 2 de Março de 2006. Available online: https://www.planalto.gov.br/ccivil_03/_Ato2004-2006/2006/Lei/L11284.htm (accessed on 11 April 2023).
30. Pereira, L.C.; Balbinot, L.; Lima, M.T.; Bramorski, J.; Tonello, K.C. Aspects of Forest Restoration and Hydrology: The Hydrological Function of Litter. *J. For. Res.* **2021**, *33*, 543–552. [CrossRef]
31. Bai, Y.; Zhou, Y.; Du, J.; Zhang, X.; Di, N. Effects of a Broadleaf-Oriented Transformation of Coniferous Plantations on the Hydrological Characteristics of Litter Layers in Subtropical China. *Glob. Ecol. Conserv.* **2021**, *25*, e01400. [CrossRef]
32. Tu, Z.; Chen, S.; Ruan, D.; Chen, Z.; Huang, Y.; Chen, J. Differential Hydrological Properties of Forest Litter Layers in Artificial Afforestation of Eroded Areas of Latosol in China. *Sustainability* **2022**, *14*, 14869. [CrossRef]
33. Ayres, M.; Ayres, M., Jr.; Ayres, D.L.; Santos, A.A. *Bioestat 5.0 Aplicações Estatísticas Nas Áreas Das Ciências Biológicas e Médicas*; IDSM: Belém, Brazil, 2007.
34. Andrade, A.M.D.; Moura, M.A. Condensação a Superfície na AMAZÔNIA em área de Pastagem Durante o Experimento LBA/SMOCC: Estudo de Casos. *Rev. Bras. Meteorol.* **2011**, *26*, 339–348. [CrossRef]
35. Bastable, H.G.; Shuttleworth, W.J.; Dallarosa, R.L.G.; Fisch, G.; Nobre, C.A. Observations of climate, albedo, and surface radiation over cleared and undisturbed Amazonian forest. *Int. J. Climatol.* **1993**, *13*, 783–796. [CrossRef]
36. Pereira, D.N.; Martins, W.B.R.; De Andrade, V.M.S.; De Assis Oliveira, F. Influence of Litter Removal in the Content of Phosphorus and Potassium in Eastern Amazon. *Rev. Bras. Cienc. Agrar.* **2017**, *12*, 380–385. [CrossRef]
37. Vasconcelos, S.S.; Zarin, D.J.; Araújo, M.M.; Rangel-Vasconcelos, L.G.T.; De Carvalho, C.J.R.; Staudhammer, C.L.; Oliveira, F.D.A. Effects of Seasonality, Litter Removal and Dry-Season Irrigation on Litterfall Quantity and Quality in Eastern Amazonian Forest Regrowth, Brazil. *J. Trop. Ecol.* **2008**, *24*, 27–38. [CrossRef]
38. Sanches, L.; Valentini, C.M.A.; Biudes, M.S.; De, J.; Nogueira, S. Dinâmica Sazonal da Produção e Decomposição de Serrapilheira em Floresta Tropical de Transição Seasonal Dynamics of the Litterfall Production and Decomposition in Tropical Transitional Forest. *Rev. Bras. Eng. Agrícola Ambient.* **2009**, *13*, 183–189. [CrossRef]
39. Zhang, H.; Yuan, W.; Dong, W.; Liu, S. Seasonal Patterns of Litterfall in Forest Ecosystem Worldwide. *Ecol. Complex.* **2014**, *20*, 240–247. [CrossRef]
40. Martius, C.; Höfer, H.; Garcia, M.V.B.; Römbke, J.; Hanagarth, W. Litter Fall, Litter Stocks and Decomposition Rates in Rainforest and Agroforestry Sites in Central Amazonia. *Nutr. Cycl. Agroecosyst.* **2004**, *68*, 137–154. [CrossRef]
41. Cianciaruso, M.V.; Pires, J.S.R.; Delitti, W.B.C.; Silva, É.F.L.P. Produção de Serapilheira e Decomposição Do Material Foliar Em Um Cerradão Na Estação Ecológica de Jataí, Município de Luiz Antônio, SP, Brasil. *Acta Bot. Bras.* **2006**, *20*, 49–56. [CrossRef]
42. Cunha Neto, F.V.; Leles, P.S.S.; Pereira, M.G.; Bellumath, V.G.H.; Alonso, J.M. Acúmulo e Decomposição Da Serapilheira Em Quatro Formações Florestais. *Ciência Florest.* **2013**, *23*, 379–387. [CrossRef]
43. Barbosa, V.; Barreto-Garcia, P.; Gama-Rodrigues, E.; de Paula, A. Biomassa, Carbono e Nitrogênio Na Serapilheira Acumulada de Florestas Plantadas e Nativa. *Floresta Ambiente* **2017**, *24*, 1–9. [CrossRef]

44. Santos, A.F.A.; Carneiro, A.C.P.; Martinez, D.T.; Caldeira, S.F. Capacidade de Retenção Hídrica Do Estoque de Serapilheira de Eucalipto. *Floresta Ambiente* **2017**, *24*, e20150303. [CrossRef]
45. Mateus, F.A.; do Couto Miranda, C.; Valcarcel, R.; Figueiredo, P.H.A. Estoque e Capacidade de Retenção Hídrica Da Serrapilheira Acumulada Na Restauração Florestal de Áreas Perturbadas Na Mata Atlântica. *Floresta Ambiente* **2013**, *20*, 336–343. [CrossRef]
46. Walsh, R.P.D.; Voigt, P.J. Vegetation Litter: An Underestimated Variable in Hydrology and Geomorphology. *J. Biogeogr.* **1977**, *4*, 253–274. [CrossRef]
47. Xie, J.; Su, D. Water-Holding Characteristics of Litter in Meadow Steppes with Different Years of Fencing in Inner Mongolia, China. *Water* **2020**, *12*, 2374. [CrossRef]
48. Li, W.; Li, Y.; Long, M.; Li, X. Study on Water-Holding Properties of Litters in Different Types of Forests of Yuntaishan Mountain Area in Shibing County, Guizhou Province. In *IOP Conference Series: Materials Science and Engineering*; Institute of Physics Publishing: Bristol, UK, 2018; Volume 394.
49. Tu, Z.; Chen, S.; Chen, Z.; Ruan, D.; Zhang, W.; Han, Y.; Han, L.; Wang, K.; Huang, Y.; Chen, J. Hydrological Properties of Soil and Litter Layers of Four Forest Types Restored in the Gully Erosion Area of Latosol in South China. *Forests* **2023**, *14*, 360. [CrossRef]
50. Li, Y.; Li, B.; Zhang, X.; Chen, J.J.; Zhan, F.D.; Guo, X.H.; Zu, Y.Q. Differential Water and Soil Conservation Capacity and Associated Processes in Four Forest Ecosystems in Dianchi Watershed, Yunnan Province, China. *J. Soil Water Conserv.* **2015**, *70*, 198–206. [CrossRef]
51. Zhou, Q.; Keith, D.M.; Zhou, X.; Cai, M.; Cui, X.; Wei, X.; Luo, Y. Comparing the Water-Holding Characteristics of Broadleaved, Coniferous, and Mixed Forest Litter Layers in a Karst Region. *Mt. Res. Dev.* **2018**, *38*, 220–229. [CrossRef]

sustainability

MDPI

Article

Content Variation and Potential Runoff Loss Risk of Nutrients in Surface Water of Saline-Alkali Paddy in Response to the Application of Different Nitrogen Fertilizer Types

Jiao Lyu [1,2,†], Xinyi Wang [2,3,4,†], Shengnan Hou [2,4], Anwar Zeb [2,5], Hui Zhu [2,4] and Yingying Xu [1,*]

1 Key Laboratory of Songliao Aquatic Environment, Ministry of Education, Jilin Jianzhu University, Changchun 130118, China
2 Key Laboratory of Wetland Ecology and Environment, Northeast Institute of Geography and Agroecology, Chinese Academy of Sciences, Changchun 130102, China
3 College of Resources and Environment, University of Chinese Academy of Sciences, Beijing 100049, China
4 Jilin Provincial Engineering Center of CWs Design in Cold Region & Beautiful Country Construction, Changchun 130102, China
5 School of Life Sciences, Northeast Normal University, Changchun 130021, China
* Correspondence: xuyingying.1019@aliyun.com
† These authors contributed equally to this work.

Abstract: As the saline-alkali paddy area continues to grow, the nutrient (e.g., nitrogen (N) and phosphorus (P)) runoff loss is becoming more serious in the world. The N-fertilizer application affects the nutrient runoff loss risk in paddy. Selecting suitable fertilizer types to reduce nutrient loss is beneficial to agricultural sustainability. However, the effects of N-fertilizer application in saline-alkali paddy are not clear. This study measured the N and P concentration of surface water in saline-alkali paddy, using various N—fertilizer treatments (i.e., urea (U), urea with urease—nitrification inhibitors (UI), organic–inorganic compound fertilizer (OCF), carbon—based slow—release fertilizer (CSF), and no N fertilization (CK)). Based on the structural equation model, both phosphate (PO_4^{3-}-P) and total−P (TP) concentrations had a positive influence on total-N (TN) concentration regardless of N−fertilizer types applied. Potential risks of ammonia—N (NH_4^+—N) and nitrate—N (NO_3^-—N) runoff losses were reduced in UI treatment, but the TN and TP losses were increased. At the panicle-initiation fertilizer stage, the NO_3^-−N, TN, and TP concentrations in CSF and OCF treatments were lower than U. The CSF application can control the TP runoff loss risk during the rice-growing season. UI should not be suggested for the control of nutrient runoff loss in saline-alkali paddy.

Keywords: saline-alkaline paddy; nitrogen fertilizer; nitrogen forms; runoff loss risk; phosphorus loss

Citation: Lyu, J.; Wang, X.; Hou, S.; Zeb, A.; Zhu, H.; Xu, Y. Content Variation and Potential Runoff Loss Risk of Nutrients in Surface Water of Saline-Alkali Paddy in Response to the Application of Different Nitrogen Fertilizer Types. *Sustainability* **2023**, *15*, 7040. https://doi.org/10.3390/su15097040

Academic Editors: Songhao Shang, Qianqian Zhang, Dongqin Yin, Hamza Gabriel and Magdy Mohssen

Received: 6 March 2023
Revised: 14 April 2023
Accepted: 20 April 2023
Published: 22 April 2023

1. Introduction

In recent years, the method of planting crops on saline-alkali lands to improve the property of saline-alkali soil has become increasingly common, but it has also brought about some agricultural pollution problems [1,2]. Especially in paddy fields, saline-alkali soil, typically with high salinity and/or high pH, can cause the nitrogen (N) content in the soil to be lower than that in non-saline-alkali soil, which requires more nutrient input to ensure a high yield. According to the statistics of the Food and Agriculture Organization of the United Nations (FAO) [3], the average seasonal N application rate adopted in China was 225 kg/ha, which was higher than the optimum N rate of 200 kg/ha suggested by Ju et al. (2009) [4], the optimum N rate calculated from the average of economic N rates from field experiments. A large amount of N has been lost due to the excessive and/or unreasonable application of N fertilizers, resulting in a series of environmental problems (e.g., nonpoint source pollution) [5,6]. The prevention and control of agricultural nonpoint source pollution have become an important environmental problem in the world [7,8]. More than 60% of surface water environmental problems are caused by agricultural activities in China [9].

An important source of agricultural nonpoint source pollution is the loss of N, phosphorus (P), and other nutrients from paddy fields [8,9]. Properties of saline-alkali soil (e.g., high salinity and pH) would cause N loss and phosphate recalcitrance [10]. Although P is not easy to lose from water, the improvement measures (e.g., drainage and salt washing after the reclamation of paddy fields) will aggravate the soil P loss via runoff in paddy fields [11]. At present, the area of saline-alkali paddy fields is expanding all over the world, and the degree of soil salinization is rising [8]. Therefore, the N and P losses from saline-alkali paddy fields have also attracted much attention.

Runoff loss is one of the main pathways of N and P losses in paddy fields [5,9]. Since the 1990s, N and P runoff losses from farmland have increased by 46% and 30%, respectively, resulting in an increasing nutrient (e.g., N and P) export to surrounding water bodies, thereby posing a threat to the aquatic ecosystems [12]. Fertilization intensity, fertilization types, and annual precipitation have effects on N and P runoff losses in paddy fields [5,11]. The application of fertilizer can increase the N and P concentration of saline-alkali paddy water; the potentiality of N and P runoff losses from the paddy soils owing to different fertilization types has not been investigated clearly. In addition, there is a coupling relationship between N and P [9]. The application of N fertilizers can stimulate the P release in the soil. In the process of topdressing, it will stir the topsoil of the paddy fields to release P in the soil, which will change the P concentration in the surface water. The concentrations of N and P in surface water can serve as an indirect indicator of potential nutrient runoff loss risk in paddy fields [9], as they are direct sources of nutrients in surface runoff [13]. Thus, it is of great significance to study the changes in N and P concentrations in surface water after fertilization for the prevention and control of N and P runoff losses in paddy fields.

Due to the complexity of N and P pollutants along the surface runoff, crop growth, and development and regionality, although the studies on the loss factors of N and P via surface runoff in paddy fields have been reported, the study on the control of field fertilization on the pollution runoff loads of N and P is still in the exploratory stage [5,8,9]. Zhao et al. [14] found that applications of organic and organic–inorganic compound fertilizers (OCF) reduced the N loss by 21.86% and 30.41%, respectively, compared with urea (U). Cui et al. [9] believed that the application of organic fertilizer could effectively reduce N loss but increase P loss. The studies on N and P losses via surface runoff are mainly based on non-saline-alkali paddy fields [5,14], while there are few studies on the saline-alkali paddy fields in Northeast China. Due to the unique properties of the saline-alkali soil, the response of different N—fertilizer types to N and P losses in saline-alkali paddy fields may be different from non-saline-alkali paddy fields. Therefore, this study set five treatments with different N—fertilizers to study the effect of nutrient runoff loss risk in saline-alkali paddy fields. The main objectives of this study were: (1) to explore the dynamic changes in various N and P forms in surface water of saline-alkali paddy fields under different N—fertilizer applications, and (2) to clarify the effect of N—fertilizer types on the potential risks of N and P runoff losses. This study will provide a theoretical reference for the sustainable development of the rice planting industry and provide an efficient strategy for formulating reasonable nonpoint source pollution control in saline-alkali paddy fields.

2. Materials and Methods

2.1. Experimental Design and Operation

Fifteen paddy mesocosms were established in a mobile intelligent canopy. The size of each mesocosm was 64 cm length × 49 cm width × 36 cm height, which was set up by the polyethylene material. This experiment was operated from 31 May to 15 October 2021, with a daily temperature of 2–30 °C and a relative humidity of 46–99%. The saline-alkali paddy soil used in this study was randomly collected from nine saline-alkali paddy fields (45°34′18–33″ N, 123°54′25–42″ E) in Baicheng City, Western Jilin Province, China. The physical and chemical properties of saline-alkali paddy soil are described in Table S1 of Supplementary Materials.

Five N−fertilizer treatments, with three replicates per treatment, were carried out in fifteen paddy mesocosms at the basal fertilizer (BF) stage. There were four treatments with different N fertilizers, i.e., U, urea with urease-nitrification inhibitors (i.e., 1% N-(N-butyl)thiophosphoric triamide (NBPT) and 1% 3,4-dimethylpyrazole phosphate (DMPP), refer to UI), OCF and carbon (C) based slow-release fertilizer (refer to CSF), and one treatment without additional N fertilizer as control (refer to CK). The organic–inorganic compound fertilizer (OCF) has 12% N and 3% K_2O, containing 20% organic matter, which comes from Chinese herbal materials. The carbon-based slow-release fertilizer has a proportion of N, P_2O_5, and K_2O with 24:8:10, containing 10% biochar. At the BF stage, these N−fertilizer types used were selected by the field investigation in Western Jilin, China. Urea was used at tillering fertilizer (TF) and panicle-initiation fertilizer (PIF) stages. The total N fertilizers were applied at 200 kg N/ha (i.e., 6.27 g N/mesocosm), which was consistent with the actual amount of N fertilizer applied by local farmers in all treatments (excluding CK); the ratio of three N—fertilizer application ratios with of BF, TF, and PIF was 5:3:2. Table 1 describes the specific application of different N fertilizers in each treatment. Along with the N fertilizer applied at the BF stage, ammonium phosphate (18% N and 46% P_2O_5) and potassium sulfate (50% K_2O) were employed as phosphate and potash fertilizers before rice transplanting, respectively. The total phosphate and potash fertilizers during the entire rice−growing season were 70 kg P_2O_5/ha (i.e., 2.20 g P_2O_5/mesocosm) and 90 kg K_2O/ha (i.e., 2.82 g K_2O/mesocosm), respectively. The selections of N, phosphate, and potash fertilizers were based on the field survey of saline-alkali paddy fields in Western Jilin Province, China. The BF was mixed completely with the collected saline-alkali paddy soil on 30 May 2021, and then each mesocosm was initiated flooded. Dongdao 4 (*Oryza sativa* L.), a saline-alkali-resistant rice variety, was transplanted into all paddy mesocosms on 31 May 2021. The same agricultural management, including irrigation, was performed in all paddy mesocosms. Each paddy mesocosm was regularly irrigated with the same amount of water, which was maintained at a water depth of 3–5 cm by intermittent irrigation before crop harvesting. In this study, no precipitation occurred, and insecticides and pesticides were not applied during the whole experiment period.

Table 1. Application of N fertilizer at different stages in each treatment.

Treatments	Basal Fertilizer Stage (BF)				Tillering Fertilizer Stage (TF)		Panicle-Initiation Fertilizer Stage (PIF)		Total N Amount (kg N/ha)
	N-Fertilizer Types	Amount (kg N/ha)	Other N Source	Amount (kg N/ha)	N-Fertilizer Types	Amount (kg N/ha)	N-Fertilizer Types	Amount (kg N/ha)	
CK	–	–		27.40	–	–	–	–	27.40
U	Urea (46%)	72.60		27.40	Urea	60	Urea	40	200
UI	Urea (46%) with 1% NBPT and 1% DMPP	72.60	Ammonium phosphate (18% N and 46% P_2O_5)	27.40	Urea	60	Urea	40	200
OCF	Organic–inorganic compound fertilizer (12% N and 3% K_2O)	72.60		27.40	Urea	60	Urea	40	200
CSF	C-based slow-release fertilizer (N:P_2O_5:K_2O = 24:8:10)	83.50		16.50	Urea	60	Urea	40	200

2.2. *Sampling and Chemical Analyses of Surface Water*

During the 137-day experiment, the sampling was conducted in a total of 34 days, including Day 0, 1, 3, 5, 7, and 10 at each initial fertilization and every 5–7 days thereafter. The fertilization dates of BF, TF, and PIF were 30 May, 13 June, and 10 August 2021, respectively. A 100 mL polyethylene sampling bottle was used to collect the surface water sample from each mesocosm on each sampling day. Surface water samples collected from all treatments were tested for electrical conductivity (EC) and pH using a quality analyzer (Bante ™, Shanghai, China). The concentrations of ammonia—N (NH_4^+—N), nitrite—N (NO_2^-—N), nitrate-N (NO_3^-—N), total—N (TN), phosphate (PO_4^{3-}—P), and total-P (TP) concentrations were analyzed via the automatic chemical analyzer (Mode Smartchem 200, Italy).

2.3. Statistical Analysis

All experimental data were graphically interpreted using Origin 2021 software (OriginLab Corporation, Northampton, MA, USA). Statistical analysis was performed using SPSS 22.0 software (IBM Corporation, New York, NY, USA). All experimental data were reported as means and standard deviations of three independent replicates (mean $\pm$ SD). The result of the one-way analysis of variance (ANOVA) was used to determine the significance of the difference between treatments. Levene's test was used to test the homogeneity of variances, and the least significant difference (LSD) was used to perform the multiple comparisons of mean values. A p-value less than or equal to 0.05 was considered significant in all analyses. The correlation between N and P in water parameters was described using Pearson correlation analysis. Principal component analysis (PCA) with the correlation matrix was also carried out with Origin 2021 software. The variables used in PCA were the values of pH and EC and the concentrations of NH_4^+—N, NO_2^-—N, NO_3^-—N, TN, PO_4^{3}—P, and TP. Amos 24.0 software (AMOS IBM, USA) was utilized to conduct a structural equation model (SEM).

3. Results

3.1. EC and pH of Surface Water in Saline-Alkali Paddy Fields

The EC and pH values in surface water of all treatments had a violent fluctuation trend during the entire rice-growing season (Figure 1). At the BF stage, the highest average EC value in surface water was observed in OCF treatment, followed by CSF, UI, U, and CK treatments. Compared to CK (1.07 $\pm$ 0.55 mS/cm), the EC values in UI (1.67 $\pm$ 0.98 mS/cm), OCF (1.89 $\pm$ 1.15 mS/cm), and CSF (1.70 $\pm$ 0.70 mS/cm) treatments demonstrated a statistically significant ($p < 0.05$) increase, respectively. The average EC values at the TF stage were CK < OCF < U < CSF < UI, and at PIF stage were CK < U < CSF < OCF < UI. The UI treatment at both TF and PIF stages had significantly ($p < 0.05$) higher average EC values than all the other four treatments. The average pH values at the BF stage were CK < CSF < UI < U < OCF, at the TF stage were U < CK < OCF < CSF < UI, and the UI treatment (8.57 $\pm$ 0.38) at the TF stage had a statistically significant ($p < 0.05$) higher pH value than U (8.24 $\pm$ 0.43) and CK (8.31 $\pm$ 0.25) treatments, respectively. At the PIF stage, the average pH values were U < CSF < CK < OCF < UI and the UI treatment had a significant ($p < 0.05$) higher average pH value than all the other four treatments. The average pH value in OCF treatment (8.53 $\pm$ 0.38) was significantly ($p < 0.05$) higher than U (8.16 $\pm$ 0.33) and CSF (8.18 $\pm$ 0.29), respectively.

3.2. Dynamic Changes in Different N Forms in Surface Water as Rice Grows

Regardless of the N—fertilizer types, the NH_4^+—N concentrations in surface water of all N—fertilizer treatments were higher than CK (Figure 2a). At the BF stage, the changing trend of NH_4^+—N concentration showed a gentle fluctuation, and the average NH_4^+—N concentrations were CK < UI < CSF < U < OCF. The OCF treatment was found to be significantly ($p < 0.05$) different from all the other four treatments in the average NH_4^+—N concentrations. The NH_4^+—N concentrations in surface water of all N—fertilizer treatments exhibited an increase–decrease trend after applying TF and PIF, respectively. At the TF stage, the peak values of all N—fertilizer treatments occurred from Day 3 to Day 5 after fertilizer application (i.e., from 16 to 18 June 2021). The average NH_4^+–N concentrations were CK < UI < U < OCF < CSF. The CSF, OCF, and U treatments had significant ($p < 0.05$) differences with UI and CK treatments, respectively. At the PIF stage, the peak values of all N—fertilizer treatments occurred from Day 1 to Day 5 after fertilizer application (i.e., from 10 to 14 August 2021), and the average NH_4^+—N concentrations were CK < UI < OCF< CSF < U. The difference in the average NH_4^+—N concentrations between CK and all N—fertilizer treatments was significant ($p < 0.05$) (Table 2).

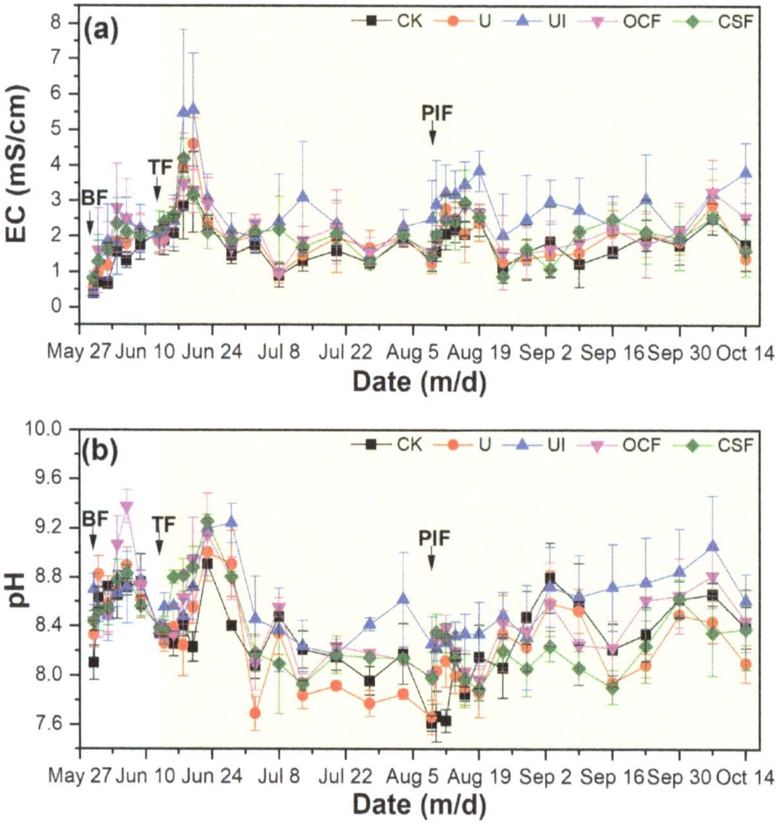

Figure 1. The EC (**a**) and pH (**b**) values in surface water of different N−fertilizer treatments during the rice-growing season. BF (blue background): basal fertilizer stage; TF (yellow background): tillering fertilizer stage; PIF (green background): panicle-initiation fertilizer stage. Data presented as mean ± standard deviation (n = 3).

For all N—fertilizer treatments, the concentrations of both NO_2^-—N and NO_3^-−N in surface water were decreased gradually over time in all N-fertilizer treatments at the BF stage (Figure 2b,c). At the BF stage, the average NO_2^-—N and NO_3^-—N concentrations in all N—fertilizer treatments were higher than CK. At TF and PIF stages, both NO_2^-—N and NO_3^-—N concentrations in surface water of most N—fertilizer treatments (except UI) had an increase–decrease trend, which was the same as NH_4^+—N. As shown in Table 2, the U, CSF, and OCF treatments at the TF stage had significantly ($p < 0.05$) higher average NO_2^-—N and NO_3^-—N concentrations than UI and CK treatments, respectively (Table 2). At the PIF stage, the average concentrations of both NO_2^-—N and NO_3^-—N followed the order of CK < UI < OCF < CSF < U, respectively. Therein, the U and CSF treatments of average NO_2^-—N and NO_3^-—N concentrations were significantly ($p < 0.05$) greater than UI and CK, respectively.

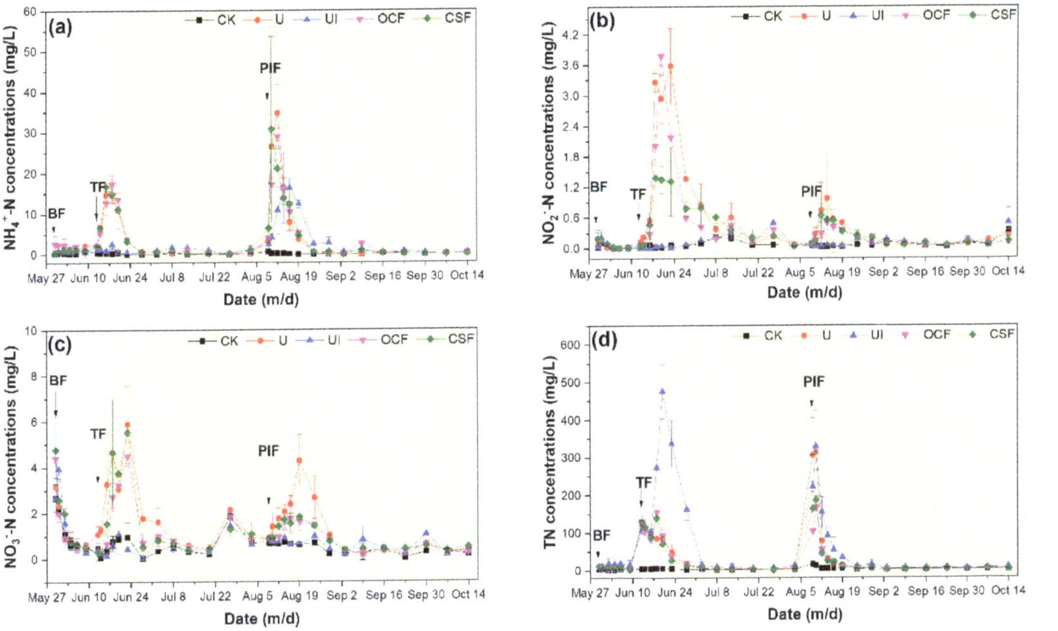

Figure 2. Concentrations of NH_4^+—N (**a**), NO_2^-—N (**b**), NO_3^-—N (**c**), and TN (**d**) in surface water of saline-alkali paddy fields with different N-fertilizer treatments during the rice-growing season. BF (blue background): basal fertilizer stage; TF (yellow background): tillering fertilizer stage; PIF (green background): panicle—initiation fertilizer stage. Data presented as mean ± standard deviation (n = 3).

Table 2. The average concentrations of various N forms in surface water of saline-alkali paddy fields with different N fertilizers.

Treatments	N Forms	BF Stage (mg/L)	TF Stage (mg/L)	PIF Stage (mg/L)
CK	NH_4^+—N	0.38 ± 0.31	0.21 ± 0.17	0.16 ± 0.22
	NO_2^-—N	0.03 ± 0.04	0.07 ± 0.07	0.07 ± 0.08
	NO_3^-—N	1.26 ± 0.95	0.58 ± 0.55	0.44 ± 0.26
	TN	3.62 ± 1.41	3.44 ± 1.80	4.51 ± 4.13
U	NH_4^+—N	1.09 ± 0.88	4.22 ± 5.60	6.29 ± 11.32
	NO_2^-—N	0.06 ± 0.07	1.10 ± 1.27	0.27 ± 0.44
	NO_3^-—N	1.44 ± 1.12	2.11 ± 1.87	1.26 ± 1.23
	TN	6.16 ± 2.95	45.58 ± 47.26	52.40 ± 109.58
UI	NH_4^+—N	0.60 ± 0.34	1.11 ± 0.83	4.73 ± 6.43
	NO_2^-—N	0.06 ± 0.10	0.12 ± 0.16	0.11 ± 0.14
	NO_3^-—N	1.62 ± 1.61	0.53 ± 0.52	0.62 ± 0.34
	TN	15.02 ± 12.20	125.89 ± 149.86	63.32 ± 98.96
OCF	NH_4^+—N	1.93 ± 1.71	4.55 ± 5.93	5.77 ± 8.42
	NO_2^-—N	0.08 ± 0.10	0.81 ± 1.10	0.19 ± 0.20
	NO_3^-—N	1.53 ± 1.56	1.37 ± 1.34	0.83 ± 0.60
	TN	6.63 ± 4.83	49.54 ± 54.56	28.31 ± 45.80
CSF	NH_4^+—N	0.89 ± 0.66	4.49 ± 5.88	6.15 ± 10.96
	NO_2^-—N	0.09 ± 0.10	0.59 ± 0.58	0.22 ± 0.22
	NO_3^-—N	1.88 ± 1.64	1.63 ± 1.94	0.91 ± 0.55
	TN	8.77 ± 4.73	46.85 ± 54.60	32.15 ± 71.15

The variation trend of TN concentrations in surface water of all treatments was consistent with NH_4^+—N during the entire rice—growing season (Figure 2d). The peak values of all N-fertilizer treatments occurred from Day 0 to Day 7 after TF application (i.e., from 13 to 20 June 2021) and on Day 1 after PIF application (i.e., 10 August 2021). At the BF stage, the highest average TN concentration was found in UI treatment in surface water, followed by CSF, OCF, U, and CK treatments. At the TF stage, the average TN concentrations were CK < U < CSF< OUF < UI (Table 2). The highest average TN concentrations at both BF and TF stages were observed in UI treatment, and the differences between UI and the other four treatments were significant ($p < 0.05$). At the PIF stage, the average TN concentrations were CK < OCF < CSF< U < UI; therein, UI treatment had higher average TN conventions than CK and OCF, respectively.

3.3. *Concentrations of PO_4^{3-}—P and TP in Surface Water*

The change trends of PO_4^{3-}—P concentrations in surface water of all N—fertilizer treatments are shown in Figure 3a. At the BF stage, the changes in PO_4^{3-}—P concentrations in UI and OCF treatments were greatly influenced compared with U and CSF, respectively. At both BF and TF stages, the average PO_4^{3-}—P concentrations were CK < CSF < U < OCF < UI, and the UI and OCF treatments had significant ($p < 0.05$) differences with U, CSF, and CK treatments, respectively. At the PIF stage, the UI treatment (0.19 ± 0.16 mg/L) had a significant ($p < 0.05$) higher average PO_4^{3-}—P concentration compared with OCF (0.11 ± 0.12 mg/L), U (0.07 ± 0.07 mg/L), and CK (0.07 ± 0.08 mg/L), respectively. During the entire rice-growing season, all N-fertilizer treatments had higher TP concentrations than CK in surface water (Figure 3b). At the BF stage, the highest average TP concentration was observed in UI treatment, followed by OCF, U, CSF, and CK treatments, and the UI treatment was found to be significantly different ($p < 0.05$) from all the other four treatments. At both TF and PIF stages, the average TP concentrations were all CK < CSF < OCF < U < UI. The average TP concentration in the surface water of UI treatment (0.98 ± 1.01 mg/L) at the TF stage was significantly ($p < 0.05$) increased with CSF (0.31 ± 0.23 mg/L), OCF (0.46 ± 0.49 mg/L) and CK (0.25 ± 0.22 mg/L), respectively.

3.4. *Correlation Analysis between N and P in Surface Water of Saline-Alkali Paddy Fields*

During the entire rice−growing season, the correlation coefficients between EC, pH, and the concentrations of various N and P forms in surface water were changed by applying different N fertilizers in saline-alkali paddy fields, and some directions were converted (Figure 4). Moreover, the positive correlation intensity between pH and EC in all N-fertilizer treatments was increased compared with CK. The negative correlation intensity between pH and TP was reduced. Compared with CK treatment, the negative correlation between EC and NH_4^+—N concentration was converted to a positive correlation by different N-fertilizer applications, with the intensities of UI < OCF < U < CSF. There was a positive correlation between NH_4^+—N and TN concentrations in all treatments, with UI < CK < U < OCF < CSF intensities. Compared with CK treatment, the negative correlation between NO_2^-—N and NO_3^-—N concentrations was turned into a positive correlation by applying different N fertilizers, with the intensities of UI < OCF < U < CSF. There was a positive correlation between NO_3^-—N and PO_4^{3-}—P concentrations with UI < CSF < CK < OCF <U intensities. The positive correlation intensities between PO_4^{3-}—P and TP concentrations were CK < U < CSF < OCF < UI.

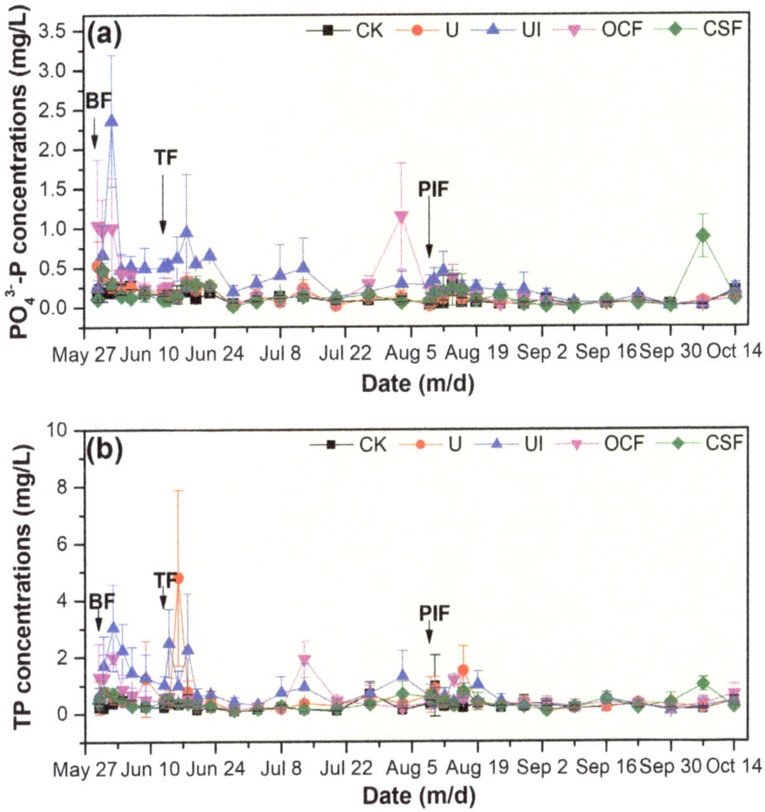

Figure 3. Concentrations of PO_4^{3-}—P (**a**) and TP (**b**) in surface water of different N—fertilizer treatments during the rice-growing season. BF (blue background): basal fertilizer stage; TF (yellow background): tillering fertilizer stage; PIF (green background): panicle—initiation fertilizer stage. Data presented as mean ± standard deviation (n = 3).

The results of PCA clearly showed the variations in EC, pH, and various N and P forms in surface water of saline-alkali paddy ecosystems with different N—fertilizer applications, and the first and second principal components jointly explained 41.9–52.8% (Figure 5). In the first and second principal components, for CK treatment (Figure 5a), there was a higher correlation between pH and NO_2^-—N compared with the other indices, while EC had a higher correlation with TN. For U treatment (Figure 5b), pH had a higher correlation with NO_2^-—N, NO_3—N, and PO_4^{3-}—P, while EC had a higher correlation with NH_4^+—N, TN, and TP, respectively. For UI treatment (Figure 5c), pH had no higher correlation with all indices, while EC had a higher correlation with NH_4^+—N and TN. For OCF treatment (Figure 5d), pH and EC had higher correlations with NH_4^+—N, NO_2—N, and TN, respectively. For CSF treatment (Figure 5e), pH had a higher correlation with NO_2^-—N and NO_3^-—N, while EC had a higher correlation with NH_4^+—N, TN, PO_4^3—P, and TP.

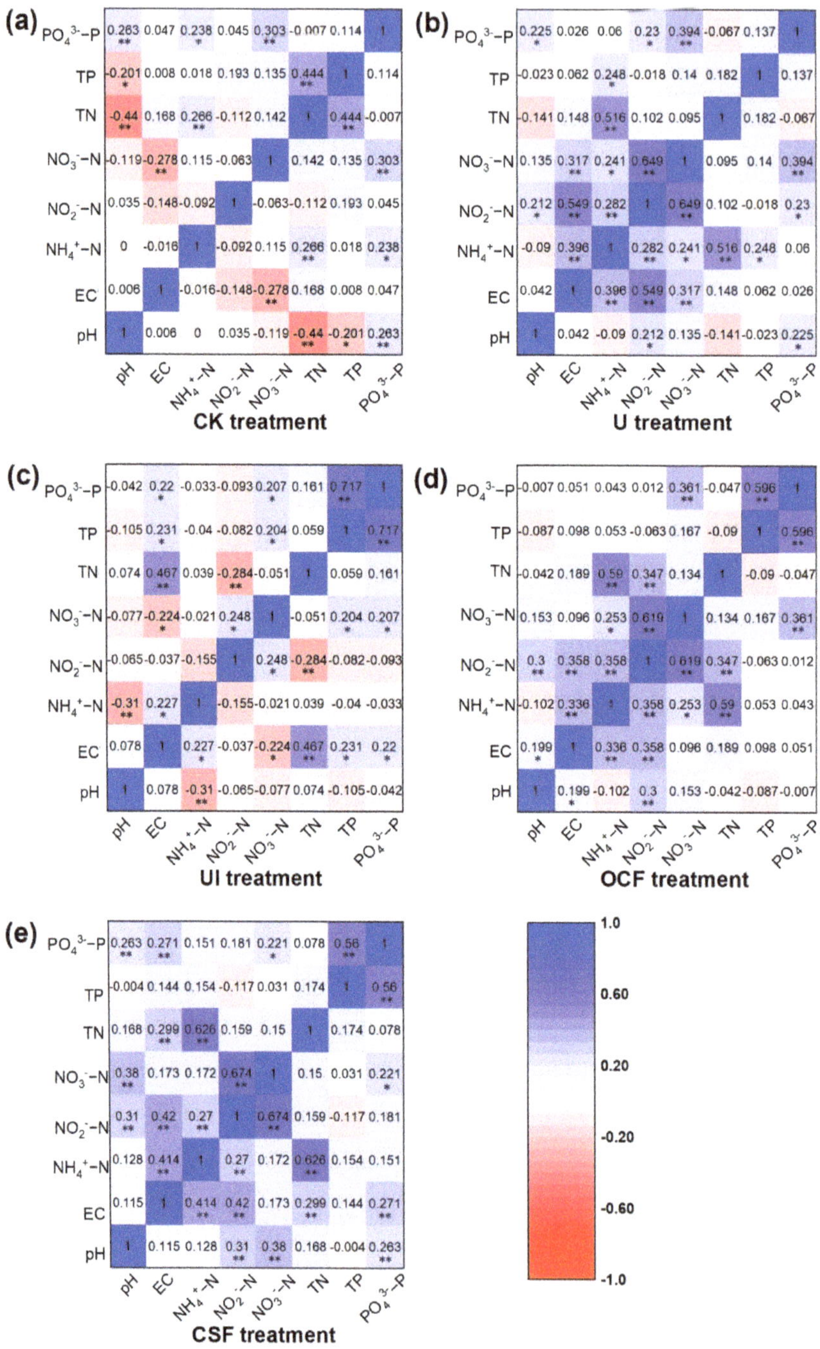

Figure 4. Correlations of pH, EC, and various forms of N and P in surface water of saline-alkali paddy fields with different N-fertilizer applications ((**a**): CK, (**b**): U, (**c**): UI, (**d**): OCF, and (**e**): CSF) during the entire rice—growing season (n = 102). * and ** represent significance at $p < 0.05$ and 0.01, respectively.

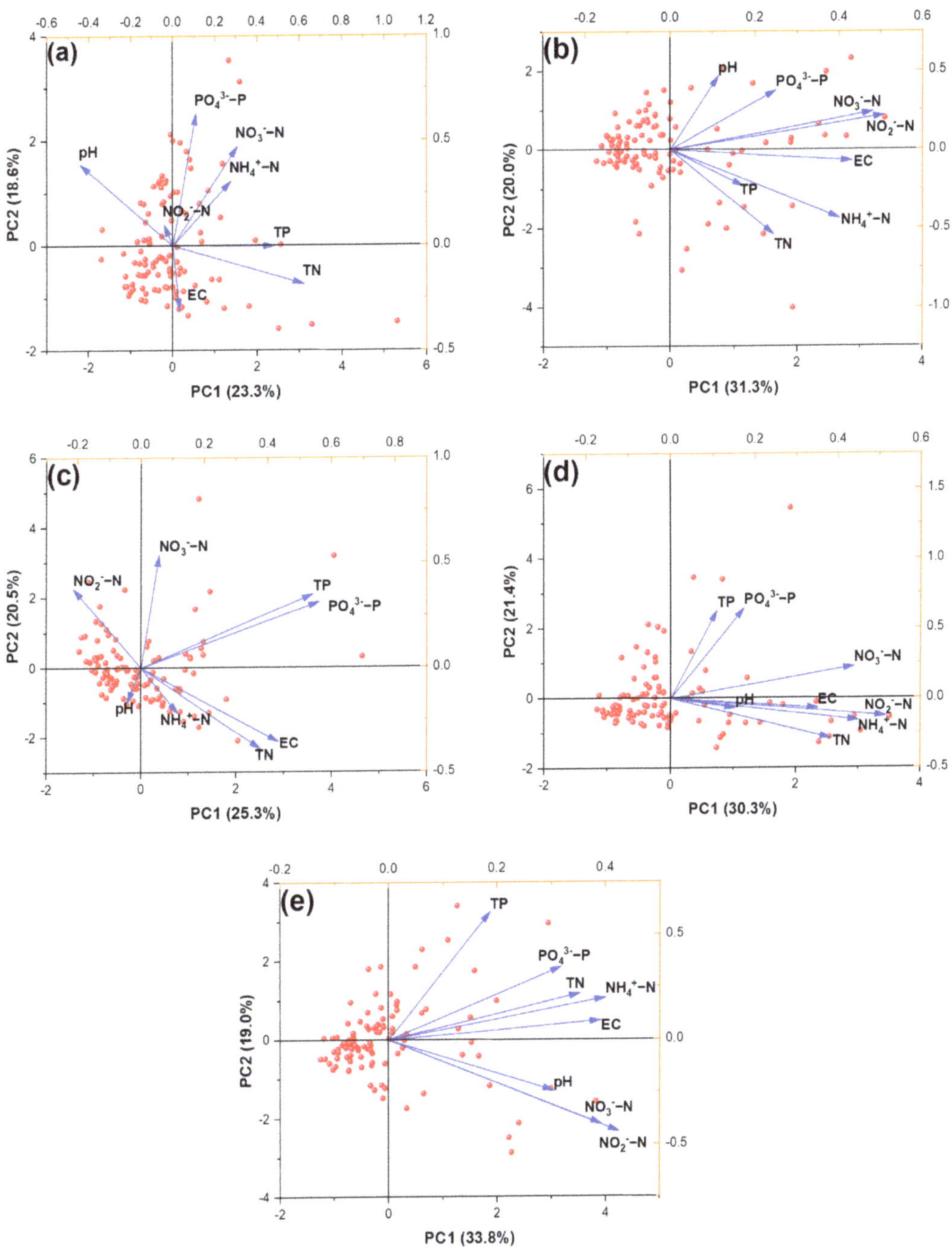

Figure 5. Principal component analysis (PCA) of N and P indices in surface water of saline-alkali paddy fields with different N-fertilizer application ((**a**): CK, (**b**): U, (**c**): UI, (**d**): OCF, and (**e**): CSF) during the entire rice—growing season (n = 102).

3.5. Multiple Interaction Pathways among NH_4^+-N, TN, and Physiochemical Parameters

To further understand causal relationships among NH_4^+—N, TN, and physiochemical parameters in surface water of saline-alkali paddy fields during the entire rice—growing season, regardless of the N—fertilizer types, the interaction model among detected factors was established (Figure 6). Based on the SEMs, pH presented a significant negative influence on NH_4^+—N ($\beta = -2.054$, $p < 0.01$) and TP ($\beta = -0.151$, $p < 0.05$), respectively. On the other hand, a significant positive influence was observed on NO_2^-—N ($\beta = 0.177$, $p < 0.01$) and PO_4^{3-}—P ($\beta = 0.084$, $p < 0.05$). For EC, a significant positive influence was observed on NH_4^+—N ($\beta = 1.652$, $p < 0.001$), NO_2^-—N ($\beta = 0.143$, $p < 0.001$), TN ($\beta = 19.172$, $p < 0.001$), PO_4^{3-}—P ($\beta = 0.079$, $p < 0.001$), and TP ($\beta = 0.087$, $p < 0.01$), respectively, while a significant negative influence ($\beta = -0.148$, $p < 0.001$) was observed on NO_3^-—N. For NH_4^+—N, there was a significant positive influence ($\beta = 3.366$, $p < 0.001$) on TN. The NO_2^-—N had a significant positive influence ($\beta = 2.144$ and 1.285, $p < 0.001$) on NH_4^+-N and NO_3^-—N, while a significant negative influence on PO_4^{3-}—P ($\beta = -0.128$, $p < 0.001$) and TP ($\beta = -0.141$, $p < 0.05$), respectively. For NO_3^--N, there was a significant positive pathway on NH_4^+—N ($\beta = 0.667$, $p < 0.05$), PO_4^{3-}—P ($\beta = 0.078$, $p < 0.001$), and TP ($\beta = 0.057$, $p < 0.05$), respectively. The PO_4^{3-}-P presented a significant positive influence ($\beta = 1.151$, $p < 0.001$) on TP. Meanwhile, TP had a significant positive ($\beta = 0.954$, $p < 0.05$) pathway on NH_4^+—N.

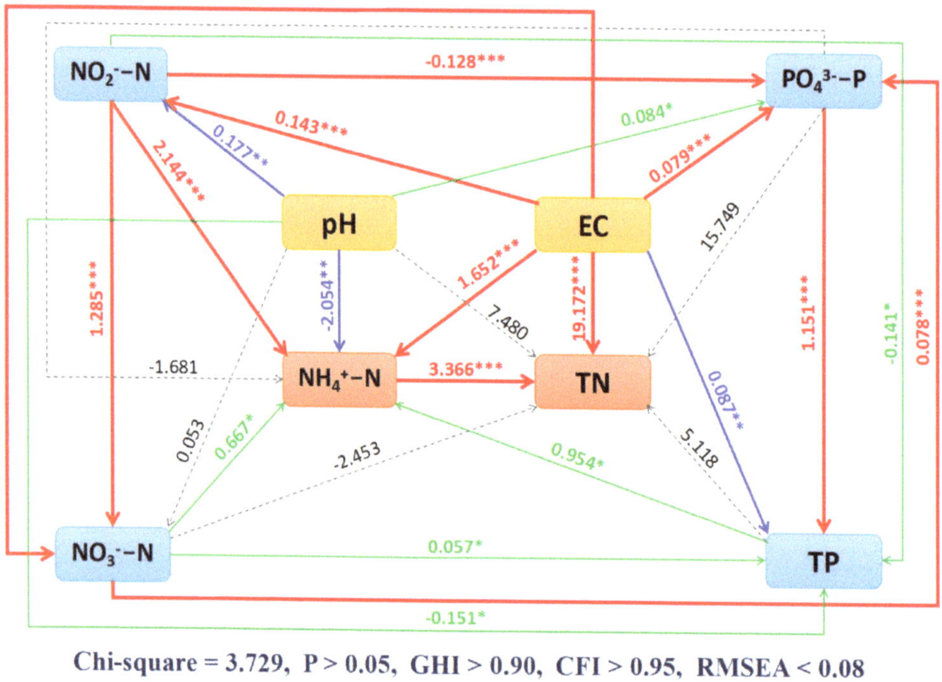

Figure 6. Simplified structural equation models (SEMs) representing hypothesized causal relationships among NH_4^+—N, TN, and physiochemical parameters in surface water of saline-alkali paddy fields during the entire rice-growing season. *, ** and *** represent the significant levels of $p < 0.05$, 0.01, and 0.001, respectively. Chi-Square: the difference between the expected covariance matrix and the covariance matrix of the data; P: the significance of fit index; GFI: the goodness of fit index; CFI: the Bentler's comparative fit index; RMSEA: the root mean square error of approximation. The numbers on the arrows indicate the strength of the relationships between variables and are used to analyze causal relationships.

4. Discussion

The runoff caused by drainage is one of the ways of nutrient (e.g., N and P) loss during the rice-growing season [15,16]. The main reason for agricultural nonpoint source pollution is N and P returning to the surface water with runoff [9]. The N and P concentrations in the surface water of paddy fields determine the nutrient supply level of rice growth [17,18]. The paddy field is in the state of soaking for a long time after fertilization. During rainfall and/or over−irrigation, the surface water in paddy fields is usually discharged randomly, which not only causes the runoff loss of nutrients but also increases the potential risk of agricultural nonpoint source pollution [5,19–21]. Thus, the study on the dynamic characteristics of N and P in surface water is significant for clarifying the law of N and P runoff losses in paddy fields to protect the surface water environment.

The dynamic changes in N and P concentrations in surface water of paddy fields can reflect the adsorption and/or fixed saturation of nutrients by the paddy soil [22]. Most N fertilizers are easily transported to adjacent water bodies with rainfall and/or artificial irrigation runoff due to their high solubility in flooded paddy fields. Based on statistical data, about 7% of the N fertilizer utilized within Chinese agricultural practices is lost via surface runoff and subsurface leaching [23]. In addition, the concentrations of TP and PO_4^{3-}−P in surface runoff were significantly correlated with the respective P forms in the field ponding water [24]. Therefore, the N and P concentrations in surface water can reflect the potential of nutrient loss from paddy fields. Fertilizers are the main source of N and P in the surface water of paddy fields. During the initial stage of fertilization, the N and P losses in paddy fields are significantly higher than those in unfertilized fields [9]. In this study, after just fertilization, regardless of the fertilization stages, the concentrations of various N and P forms in the surface water of paddy fields were increased compared with CK (Figures 2 and 3), indicating a potential risk of nutrient loss at the beginning of each fertilization stage. Application of suitable fertilizer types, reasonable control of fertilization amount, and improvement of the fertilizer utilization rate of crops are necessary measures to reduce nutrient loss via runoff in paddy fields [25–27]. Moreover, as the fertilizers with the largest amount, the rational selection of N−fertilizer types is the key to controlling the N loss via runoff in paddy fields [25]. Due to the physiological response of plants, the supply of N can increase the absorption and utilization of P by plants [28]. NH_4^+−N can promote the absorption of P by rice and the transport of P to the aboveground tissues, while the P absorbed by rice is mainly accumulated in the root system when NO_3^-−N is applied [28]. Thus, the selection of N fertilizer also plays an important role in improving the P absorption capacity of crops and avoiding the risk of P runoff loss in paddy fields. Compared with non−saline-alkali paddy fields, the soil physicochemical properties and biological processes of saline-alkali paddy fields are vulnerable to the negative effects of high salinity and pH, resulting in more nutrient loss via runoff [29,30]. The interaction and relationship between N and P in the surface water of saline-alkali paddy fields are also influenced by the N—fertilizer types (Figures 4 and 5). Therefore, selecting the appropriate N fertilizer for application is the key to controlling the N and P losses via runoff in saline-alkali paddy fields.

The concentrations of various N forms in the surface water of saline-alkali paddy fields were normally high in the initial stage of each fertilization stage and gradually decreased to a low value by the end of each fertilization stage, regardless of the types of N fertilizer applied (Figure 2). This result is almost similar to Xue et al. [21], who reported the highest N concentration mainly occurred from Day 0 to Day 10 after applying N fertilizers and then declined to a low value after 10 days. Both NH_4^+−N and NO_3^-−N are the main forms of N loss via runoff in paddy fields, which are the available N that can be directly used by rice [8,17]. The concentrations of both NH_4^+−N and NO_3^-−N in the surface water of paddy fields are affected by the N−fertilizer types, which can affect the N content of rice growth supplied by paddy soil. For the UI treatment of this study, the application of UI can inhibit the U hydrolysis and promote crop growth [31,32]; thus, the average NH_4^+—N concentration in surface water of UI treatment at each fertilization stage was observed to be lower than U treatment (Figure 2a). At both TF and PIF stages, the average NO_3^-—N concentrations in UI treatment were significantly ($p < 0.05$) reduced compared with U (Figure 2c). These results suggested that the potential risks of NH_4^+−N and NO_3^-−N runoff losses in saline-alkali paddy fields can be effectively controlled by the addition of inhibitors. Meanwhile, the concentrations of NO_3^-−N in OCF and CSF treatments at the initial stage of topdressing (i.e., on Day 0 to Day 20 of TF and PIF stages) were lower than U, and the average NO_3^-−N concentrations in OCF and CSF treatments at both TF and PIF stages were reduced compared with U (Figure 2c). Furthermore, both CSF and OCF applications significantly ($p < 0.05$) decreased the average NO_3^-−N concentrations in surface water at the PIF stage, which proved that both OCF and CSF have the potential to reduce the risk of NO_3^-−N runoff loss in saline-alkali paddy fields at the PIF stage.

TN is the sum of soluble−N and granular−N in the surface water of paddy fields, so the TN loss via runoff is the largest among different N forms [8]. Effectively controlling the TN runoff loss is one of the main tasks to effectively control the N loss in paddy fields. The TN concentration in the surface water of paddy fields is affected by N−fertilizer types, fertilization methods and fertilizing times, etc. [9,33]. In this study, regardless of N−fertilizer types, the average TN concentration at the BF stage by the deep placement of N fertilizer was lower than those at topdressing stages (i.e., TF and PIF stages) using the throwing method (Figure 2d). This result revealed that the deep placement of N fertilizers can effectively control TN loss via runoff in paddy fields, which was consistent with Min et al. [33]. Regardless of fertilization stages, the application of UI increased the average TN concentration compared with all the other treatments (Figure 2d). However, the average NH_4^+−N, NO_2—N, and NO_3^-−N concentrations in surface water of UI treatment were all lower than all the other N-fertilizer treatments at the topdressing stages (Figure 2a–c). These results indicated that the UI addition can increase the risk of TN runoff loss in saline-alkali paddy fields, and organic-N accounted for the main contribution. The reasons for these results may be as follows: (1) for UI treatment, the urease inhibitor (i.e., NBPT) can effectively inhibit the hydrolysis of U and control the speed of U conversion to NH_4^+—N, so the U as organic—N remains in the soil may be directly dissolved in surface water of paddy fields [31,32]; (2) after the nitrification process was inhibited by the nitrification inhibitor (i.e., DMPP) in UI treatment, U can remain in the paddy soil in the form of NH_4^+-N for a long time, avoiding the appearance of high NO_2^-−N and NO_3^-−N concentrations, and reducing the runoff losses of both NO_2^-−N and NO_3^-−N [34,35]; (3) the addition of inhibitors reduces the activity of relevant functional microorganisms (e.g., nitrifying bacteria) and enzymes (e.g., urease), and even leads to their death [35,36], which may cause the increase in organic−N in saline-alkali paddy fields. Therefore, although the UI application can effectively control the risk of inorganic-N (i.e., NH_4^+−N, NO_2—N, and NO_3^-—N) loss via runoff, there is a serious potential risk of TN runoff loss.

The loss of P, which is one of the necessary nutrients for rice growth, occurs mainly in the form of dissolved P via surface runoff in paddy fields [37]. The PO_4^{3-}−P and TP concentrations in the surface water of saline-alkali paddy fields showed remarkable variation among the different N-fertilizer treatments (Figure 3). Regardless of the fertilization stages, the average PO_4^{3-}−P and TP concentrations in the surface water of UI treatment were also higher than all the other N-fertilizer treatments (Figure 3), which was consistent with the result of TN (Figure 2d). Based on the SEMs, both PO_4^{3-}—P and TP concentrations in surface water had a positive influence on TN concentration regardless of N−fertilizer types applied at the BF stage (Figure 6). Meanwhile, the TN concentration in the surface water of UI treatment positively correlated with TP (Figure 4c). These results indicated that the UI application also has the potential risk of P loss via surface runoff in saline-alkali paddy fields. The PO_4^{3-}—P and TP concentrations in surface water can promote the increase in TN concentration, thus simultaneously causing the risk of TN runoff loss. For OCF treatment, the average PO_4^{3-}—P and TP concentrations were decreased with increasing fertilization times. Therein, the average TP concentration in OCF treatment at the BF stage was higher than U, while the average TP concentrations at TF and PIF stages were lower than U. These results indicated that the application of OCF can increase the P concentration in surface water of saline-alkali paddy fields in the early rice-growing season and generate a potential risk of P runoff loss, which are similar to the results of Zanon et al. [16] and Cui et al. [9]. For CSF treatment, the application of CSF can control the dissolution rate of nutrients so that the nutrient release rate of fertilizer is consistent with the nutrient absorption law of crops, thus improving crop nutrient use efficiency and reducing the risk of nutrient loss [38,39]. In this study, regardless of the fertilization stages, the average TP concentration in CSF treatment was reduced compared with U (Figure 3b), indicating that CSF can control the potential risk of P runoff loss in saline-alkali paddy fields. In summary, CSF is a better choice for avoiding the potential risk of P loss via runoff in saline-alkali paddy fields during the entire rice—growing season, but UI should not be suggested for the control of P runoff loss.

5. Conclusions

This study investigated the dynamic characteristics of different forms of N and P concentrations in surface water of saline-alkali paddy fields under different N−fertilizer applications and revealed their potential risk of nutrient loss via runoff. Based on the SEMs, there was a direct and/or indirect relationship among various forms of N and P in saline-alkali paddy fields. The N-fertilizer types can affect the interaction and relationship between N and P in the surface water of saline-alkali paddy fields, resulting in different potential risks of N and P losses via surface runoff. Comprehensively considering the average concentrations and variation laws of N and P in each fertilization stage, the application of UI can effectively control the potential risks of NH_4^+—N and NO_3^-—N losses via surface runoff, but increase the risks of TN and TP losses, indicating that UI is not suitable in saline-alkali paddy fields for controlling nutrient loss via runoff. The OCF application increased the N and P concentrations in surface water of saline-alkali paddy fields at the BF stage, thus enhancing the potential risk of nutrient loss via surface runoff compared with U. Meanwhile, OCF had a good potential to control N and P runoff losses at PIF stage. CSF is a good choice to control the risk of TP loss via runoff in saline-alkali paddy fields regardless of the fertilization stages and has an effective potential for controlling the risk of N runoff loss at the PIF stage.

Supplementary Materials: The following supporting information can be downloaded at https://www.mdpi.com/article/10.3390/su15097040/s1, Table S1: The physical and chemical properties of saline-alkali soil used in this study.

Author Contributions: Conceptualization, J.L., X.W. and Y.X.; methodology, J.L., X.W. and Y.X.; formal analysis, J.L. and X.W.; investigation, J.L., X.W. and S.H.; resources, H.Z. and Y.X.; writing—original draft preparation, J.L. and X.W.; writing—review and editing, Y.X. and A.Z.; supervision, H.Z.; project administration, Y.X.; funding acquisition, H.Z. All authors have read and agreed to the published version of the manuscript.

Funding: This study was supported by the Excellent Youth Foundation of Jilin Province, China (No. 20230101361JC), the Strategic Priority Research Program of the Chinese Academy of Sciences (No. XDA28040102), the CAS Interdisciplinary Innovation Team Project (No. JCTD-2020-14), and the Youth Innovation Promotion Association, CAS (No. 2017274 and No. Y2021068).

Institutional Review Board Statement: Not applicable.

Informed Consent Statement: Not applicable.

Data Availability Statement: Data is available on request.

Acknowledgments: We would like to thank the editors and reviewers for their helpful comments and suggestions.

Conflicts of Interest: The authors declare no conflict of interest.

References

1. Chi, C.M.; Zhao, C.W.; Sun, X.J.; Wang, Z.C. Reclamation of saline-sodic soil properties and improvement of rice (Oriza sativa L.) growth and yield using desulfurized gypsum in the west of Songnen Plain, northeast China. *Geoderma* **2012**, *187-188*, 24–30. [CrossRef]
2. Huang, L.H.; Liang, Z.W.; Suarez, D.L.; Wang, Z.C.; Wang, M.M.; Yang, H.Y.; Liu, M. Impact of cultivation year, nitrogen fertilization rate and irrigation water quality on soil salinity and soil nitrogen in saline-sodic paddy fields in Northeast China. *J. Agric. Sci.* **2015**, *154*, 632–646. [CrossRef]
3. FAO. FAO FAOSTAT. 2017. Available online: http://www.fao.org/faostat/en/#home (accessed on 7 February 2023).
4. Ju, X.T.; Xing, G.X.; Chen, X.P.; Zhang, S.L.; Zhang, L.J.; Liu, X.J.; Cui, Z.L.; Yin, B.; Christie, P.; Zhu, Z.L.; et al. Reducing environmental risk by improving N management in intensive Chinese agricultural systems. *Proc. Natl. Acad. Sci. USA* **2009**, *106*, 3041–3046. [CrossRef] [PubMed]
5. Zhang, S.; Zhang, G.; Wang, D.; Liu, Q.; Xu, M. Investigation into runoff nitrogen loss variations due to different crop residue retention modes and nitrogen fertilizer rates in rice-wheat cropping systems. *Agric. Water Manag.* **2021**, *247*, 106729. [CrossRef]
6. Zhao, Z.; Cao, L.; Sha, Z.; Deng, J.; Lv, W. Impacts of fertilization optimization on N loss from paddy fields: Observations and DNDC modeling case study in Shanghai, China. *Soil Tillage Res.* **2020**, *199*, 104587. [CrossRef]
7. Xiang, C.; Wang, Y.; Liu, H. A scientometrics review on nonpoint source pollution research. *Ecol. Eng.* **2017**, *99*, 400–408. [CrossRef]
8. Xu, Y.; Su, B.; Wang, H.; He, J.; Yang, Y. Analysis of the water balance and the nitrogen and phosphorus runoff pollution of a paddy field in situ in the Taihu Lake basin. *Paddy Water Environ.* **2020**, *18*, 385–398. [CrossRef]
9. Cui, N.; Cai, M.; Zhang, X.; Abdelhafez, A.A.; Zhou, L.; Sun, H.; Chen, G.; Zou, G.; Zhou, S. Runoff loss of nitrogen and phosphorus from a rice paddy field in the east of China: Effects of long-term chemical N fertilizer and organic manure applications. *Glob. Ecol. Conserv.* **2020**, *22*, e01011. [CrossRef]
10. Liu, Z.; Shang, H.; Han, F.; Zhang, M.; Li, Q.; Zhou, W. Improvement of nitrogen and phosphorus availability by Pseudoalteromonas sp. during salt-washing in saline-alkali soil. *Appl. Soil Ecol.* **2021**, *168*, 104117. [CrossRef]
11. Liu, J.; Ouyang, X.; Shen, J.; Li, Y.; Sun, W.; Jiang, W.; Wu, J. Nitrogen and phosphorus runoff losses were influenced by chemical fertilization but not by pesticide application in a double rice-cropping system in the subtropical hilly region of China. *Sci. Total Environ.* **2020**, *715*, 136852. [CrossRef]
12. Hou, X.; Zhou, F.; Leip, A.; Fu, B.; Yang, H.; Chen, Y.; Gao, S.; Shang, Z.; Ma, L. Spatial patterns of nitrogen runoff from Chinese paddy fields. *Agr. Ecosyst. Environ.* **2016**, *231*, 246–254. [CrossRef]
13. Ruan, S.; Zhuang, Y.; Zhang, L.; Li, S.; Chen, J.; Wen, W.; Zhai, L.; Liu, H.; Du, Y. Improved estimation of nitrogen dynamics in paddy surface water in China. *J. Environ. Manag.* **2022**, *312*, 114932. [CrossRef] [PubMed]
14. Zhao, Z.; Yue, Y.; Sha, Z.; Li, C.; Deng, J.; Zhang, H.; Gao, M.; Cao, L. Assessing impacts of alternative fertilizer manage-ment practices on both nitrogen loading and greenhouse gas emissions in rice cultivation. *Atmos. Environ.* **2015**, *119*, 393–401. [CrossRef]
15. Yu, Y.; Xu, J.; Zhang, P.; Meng, Y.; Xiong, Y. Controlled irrigation and drainage reduce rainfall runoff and nitrogen loss in paddy fields. *Int. J. Environ. Res. Public Health* **2021**, *18*, 3348. [CrossRef]
16. Zanon, J.A.; Favaretto, N.; Democh Goularte, G.; Dieckow, J.; Barth, G. Manure application at long-term in no-till: Effects on runoff, sediment and nutrients losses in high rainfall events. *Agric. Water Manag.* **2020**, *228*, 105908. [CrossRef]
17. Ishii, S.; Ikeda, S.; Minamisawa, K.; Senoo, K. Nitrogen cycling in rice paddy environments: Past achievements and future challenges. *Microbes Environ.* **2011**, *26*, 282–292. [CrossRef] [PubMed]
18. Wang, K.; Onodera, S.-i.; Saito, M.; Okuda, N.; Okubo, T. Estimation of phosphorus transport influenced by climate change in a rice paddy catchment using SWAT. *Int. J. Environ. Res.* **2021**, *15*, 759–772. [CrossRef]
19. Peng, S.; Yang, S.; Xu, J.; Gao, H. Field experiments on greenhouse gas emissions and nitrogen and phosphorus losses from rice paddy with efficient irrigation and drainage management. *Sci. China Technol. Sci.* **2011**, *54*, 1581–1587. [CrossRef]
20. Sun, C.; Chen, L.; Zhu, H.; Xie, H.; Qi, S.; Shen, Z. New framework for natural-artificial transport paths and hydrological connectivity analysis in an agriculture-intensive catchment. *Water Res.* **2021**, *196*, 117015. [CrossRef]
21. Xue, L.; Yu, Y.; Yang, L. Maintaining yields and reducing nitrogen loss in rice–wheat rotation system in Taihu Lake region with proper fertilizer management. *Environ. Res. Lett.* **2014**, *9*, 115010. [CrossRef]
22. Zhang, Q.; Fang, L.; Jian, Z.; Ling, M. Changes in concentrations of N and P in surface water of paddy field and its pollution assessment in Karst Hilly Areas. *Chin. Agric. Sci. Bull.* **2021**, *28*, 63–67.
23. Zhu, Z.; Sun, B. Study on the agricultural non-point pollution control in China. *Environ. Prot.* **2008**, *8*, 4–6.
24. Liu, J.; Zuo, Q.; Zhai, L.; Luo, C.; Liu, H.; Wang, H.; Liu, S.; Zou, G.; Ren, T. Phosphorus losses via surface runoff in rice-wheat cropping systems as impacted by rainfall regimes and fertilizer applications. *J. Integr. Agric.* **2016**, *15*, 667–677. [CrossRef]
25. Hou, P.; Jiang, Y.; Yan, L.; Petropoulos, E.; Wang, J.; Xue, L.; Yang, L.; Chen, D. Effect of fertilization on nitrogen losses through surface runoffs in Chinese farmlands: A meta-analysis. *Sci. Total. Environ.* **2021**, *793*, 148554. [CrossRef] [PubMed]
26. Husain, A.; Muneer, M.A.; Wu, F.; Yin, G.-F.; Shen, S.Z.; Wang, F.; Li, Y.; Zhang, K.-Q. Application of optimum N through different fertilizers alleviate NH_4^+-N, NO_3^--N and total nitrogen losses in the surface runoff and leached water and improve nitrogen use efficiency of rice crop in Erhai Lake Basin, China. *Commun. Soil Sci. Plan.* **2019**, *50*, 716–738. [CrossRef]

27. Shi, X.; Hu, K.; Batchelor, W.D.; Liang, H.; Wu, Y.; Wang, Q.; Fu, J.; Cui, X.; Zhou, F. Exploring optimal nitrogen management strategies to mitigate nitrogen losses from paddy soil in the middle reaches of the Yangtze river. *Agric. Water Manag.* **2020**, *228*, 105877. [CrossRef]
28. Li, B.; Wang, S.; Feng, H.; Xu, G. Effects of nitrogen forms on root morphology and phosphate uptake in rice. *Chin. J. Rice Sci.* **2008**, *22*, 665–668.
29. She, D.; Fei, Y.; Liu, Z.; Liu, D.; Shao, G. Soil erosion characteristics of ditch banks during reclamation of a saline/sodic soil in a coastal region of China: Field investigation and rainfall simulation. *Catena* **2014**, *121*, 176–185. [CrossRef]
30. Yuan, R.X.; Ye, J.Q.; Wang, Z.H.; Liu, J.S. Dynamic variation characteristics of phosphorus in paddy field runoff in saline land and its potential environmental effect. *Meteorol. Environ. Res.* **2011**, *2*, 72–75.
31. Cantarella, H.; Otto, R.; Soares, J.R.; Silva, A.G.B. Agronomic efficiency of NBPT as a urease inhibitor: A review. *J. Adv. Res.* **2018**, *13*, 19–27. [CrossRef]
32. Yang, G.; Ji, H.; Sheng, J.; Zhang, Y.; Feng, Y.; Guo, Z.; Chen, L. Combining Azolla and urease inhibitor to reduce ammonia volatilization and increase nitrogen use efficiency and grain yield of rice. *Sci. Total Environ.* **2020**, *743*, 140799. [CrossRef] [PubMed]
33. Min, J.; Sun, H.; Wang, Y.; Pan, Y.; Kronzucker, H.J.; Zhao, D.; Shi, W. Mechanical side-deep fertilization mitigates ammonia volatilization and nitrogen runoff and increases profitability in rice production independent of fertilizer type and split ratio. *J. Clean. Prod.* **2021**, *316*, 128370. [CrossRef]
34. Li, H.; Chen, X.; Liu, C.; Wang, S.; Guo, B.; Li, N.; Jin, Y.; Fu, Q.; Liang, X. Effect of various doses of 3,4-dimethylpyrazole phosphate on mineral nitrogen losses in two paddy soils. *J. Soils Sediments* **2020**, *20*, 3825–3834. [CrossRef]
35. Liu, C.; Zhang, Y.; Liu, H.; Liu, X.; Ren, D.; Wang, L.; Guan, D.; Li, Z.; Zhang, M. Fertilizer stabilizers reduce nitrous oxide emissions from agricultural soil by targeting microbial nitrogen transformations. *Sci. Total Environ.* **2022**, *806*, 151225. [CrossRef] [PubMed]
36. Florio, A.; Maienza, A.; Dell'Abate, M.T.; Stazi, S.R.; Benedetti, A. Changes in the activity and abundance of the soil microbial community in response to the nitrification inhibitor 3,4-dimethylpyrazole phosphate (DMPP). *J. Soils Sediments* **2016**, *16*, 2687–2697. [CrossRef]
37. Zhang, H.; Cao, Z.; Wang, G.; Zhang, H.; Wong, M.H. Winter runoff losses of phosphorus from paddy soils in the Taihu Lake Region of South China. *Chemosphere* **2003**, *52*, 1461–1466. [CrossRef] [PubMed]
38. Bakshi, S.; Banik, C.; Laird, D.A.; Smith, R.; Brown, R.C. Enhancing biochar as scaffolding for slow release of nitrogen fertilizer. *ACS Sustain. Chem. Eng.* **2021**, *9*, 8222–8231. [CrossRef]
39. Jien, S.-H.; Wang, C.-C.; Lee, C.-H.; Lee, T.-Y. Stabilization of organic matter by biochar application in compost-amended soils with contrasting pH values and textures. *Sustainability* **2015**, *7*, 13317–13333. [CrossRef]

hydrology

Article

Sensitivity of the Penman–Monteith Reference Evapotranspiration Equation to Meteorological Variables for Puerto Rico

Michelle Irizarry-Ortiz [1,*] and Eric W. Harmsen [2]

[1] U.S. Geological Survey, Caribbean-Florida Water Science Center, Orlando, FL 32826, USA
[2] Agricultural and Biosystems Engineering Department, University of Puerto Rico, Mayaguez, PR 00681, USA
* Correspondence: mirizarry-ortiz@usgs.gov; Tel.: +1-407-803-5533

Abstract: Spatiotemporal variations in reference evapotranspiration (ET_o) are sensitive to the meteorological data used in its estimation. The sensitivity of the ASCE standardized ET_o equation to meteorological variables from GOES-PRWEB dataset was evaluated for the island of Puerto Rico. Island wide, ET_o is most sensitive to daily mean relative humidity (RH_{mean}), followed by solar radiation, daily maximum (T_{max}) and minimum (T_{min}) air temperatures, and wind speed with average absolute relative sensitivity coefficients (SCs) of 0.98, 0.57, 0.50, 0.27, and 0.12, respectively. The derived SCs guided the prioritization of bias correction of meteorological data for ET_o estimation from two downscaled climate models (CNRM and CESM). The SCs were applied to evaluate how meteorological variables contribute to model errors and projected future changes in ET_o from 1985–2005 to 2040–2060 at irrigated farms in the south. Both models project a 5.6% average increase in annual ET_o due to projected increases in T_{max} and T_{min} and a decrease in RH_{mean}. Despite ET_o being most sensitive to relative changes in RH_{mean}, the contributions from RH_{mean}, T_{max}, and T_{min} to future changes in ET_o are similar. CESM projects increases in ET_o in March, November, and December, increasing the potential for crop water stress. Study limitations are discussed.

Keywords: evapotranspiration; evaporation; transpiration; water use; water balance; Puerto Rico; sensitivity analysis; climate change; WRF; GOES-PRWEB

Citation: Irizarry-Ortiz, M.; Harmsen, E.W. Sensitivity of the Penman–Monteith Reference Evapotranspiration Equation to Meteorological Variables for Puerto Rico. *Hydrology* **2023**, *10*, 101. https://doi.org/10.3390/hydrology10050101

Academic Editors: Songhao Shang, Qianqian Zhang, Dongqin Yin, Hamza Gabriel and Magdy Mohssen

Received: 9 February 2023
Revised: 10 March 2023
Accepted: 13 March 2023
Published: 25 April 2023

1. Introduction

Accurate quantification of evapotranspiration (ET) is critical to water resource management and planning, especially in the semi-arid southern part of the island of Puerto Rico where agriculture is dependent on irrigation. Currently, the island only produces 15% of the food consumed by residents [1]. Compared to most of the U.S., where agricultural irrigation was the primary freshwater withdrawal in 2015, irrigation withdrawals from surface water and groundwater sources in Puerto Rico accounted for only about 12% of the total freshwater withdrawals in 2015 [2]. Because agricultural production has declined since the 1960s and urbanization encroached into agricultural lands, an increasing proportion of the island's surface-water withdrawals have been for domestic water supply. In the south, canals that were built primarily to supply irrigation water to agricultural lands are now increasingly being used for public water supply. Increasing competition among water users has limited the capacity for recovery of the agricultural sector [3] by reducing the availability of water supplies for future agricultural production [4]. This may hinder the island's attempts to achieve food sovereignty which is also being threatened by climate change [5].

The island has historically been subject to periods of drought which have adversely affected the agricultural sector. Since the turn of the century, an extreme drought occurred during 2014–2016 and severe drought conditions occurred again in 2019–2022 [6]. A 2013–2016 Caribbean-wide drought was partly linked to 2015–2016 El Niño conditions, but analysis by Herrera et al. [7] suggests that climate change may have accounted for ~15–17% of the drought severity and/or ~7% of its spatial extent. Downscaled climate models project increased drought intensity and frequency in the future as a result of climate change [8,9]. Therefore, understanding potential future changes in rainfall and actual *ET* from crops is critical for water managers and planners to better prepare for the future. Actual *ET* depends on various factors, including atmospheric evaporative demand, which is quantified via reference evapotranspiration (ET_o), soil water availability, crop physiology, and crop management factors.

The American Society of Civil Engineers (ASCE) has defined a standardized grass-reference Penman–Monteith (P-M) evapotranspiration equation [10] to estimate ET_o on a daily timestep for short grass. When applied on a daily timestep, the ASCE standardized ET_o equation is the same as the Food and Agriculture Organization paper number 56 (FAO-56) P-M equation [11]. Understanding spatiotemporal variations in ET_o requires understanding the sensitivity of standardized equations to each meteorological variable used in its estimation. For this purpose, a sensitivity analysis of the ASCE standardized ET_o equation is performed for Puerto Rico. This involves the computation of non-dimensional relative sensitivity coefficients (SCs) of the ASCE standardized ET_o equation to each of the basic meteorological variables used in its estimation. When combined with a measure of the variability or measurement/estimation error in the basic meteorological variables, the SCs could be used to define monitoring priorities among the variables. ET_o may be sensitive to a particular variable but if that variable varies little in time and if it can be measured or estimated precisely, then the variable will not influence ET_o estimates significantly.

Gong et al. [12], and McKenney and Rosenberg [13] discuss difficulties in comparing results of ET_o sensitivity analysis from the literature, including the use of different ET_o models, parameterizations and meteorological variables, spatiotemporal scales, climatic settings, and SC definitions (i.e., absolute versus relative). The studies discussed hereafter are limited to those where the sensitivity of ET_o to various meteorological variables was determined based on the ASCE or the FAO-56 grass-reference P-M equations. For example, Irmak et al. [11] performed a sensitivity analysis of the ET_o equation at various U.S. locations computing absolute SCs numerically. They found that ET_o is most sensitive to vapor pressure deficit (VPD) at all the U.S. locations evaluated, but that sensitivity to solar radiation (R_s) dominates during the summer months at humid locations (Ft. Pierce, Florida with ~1200 mm of annual precipitation and at Rockport, Missouri with ~800 mm of annual precipitation). Debnath et al. [14] computed absolute SCs for ET_o at five stations in different agroecological regions of India where annual precipitation ranges from 680 to 1500 mm. They found ET_o to be most sensitive to either R_s or 2 m wind speed (u_2), and least sensitive to daily mean relative humidity (RH_{mean}) and daily minimum air temperature (T_{min}) with significant spatiotemporal variation in SCs.

Gong et al. [12] computed relative SCs for ET_o at meteorological stations in the Yangtze River Basin in China, where annual precipitation ranges from 400 to 1600 mm [15]. They found that RH_{mean} was the most sensitive variable, followed by R_s, daily mean air temperature (T_{mean}), and u_2. Seasonal and regional variations in sensitivity were observed. Furthermore, they found that the sensitivity of ET_o to a meteorological variable depended on other variables. For example, although the lower and middle regions of the Yangtze River Basins have similar RH_{mean} year round, the sensitivity of ET_o to changes in RH_{mean} was higher in the Lower basin region, where wind speeds are higher. Similarly, Liu et al. [16] derived relative SCs for ET_o at meteorological stations in the Yellow River Basin in China, which is more arid than the Yangtze River Basin and has annual precipitation of 372–671 mm. They found that R_s was the most sensitive variable in general, followed by RH_{mean}, T_{mean}, and u_2 at the basin scale. Biazar et al. [17] computed relative SCs for ET_o at meteorological stations in a humid region of Iran with annual precipitation of 1000–1850 mm and relative humidity exceeding 85% throughout the year. They found that the most sensitive parameter for ET_o was T_{max} and the least sensitive was T_{min}. Emeka et al. [18] evaluated the relative sensitivity of ET_o at seven distinct agroecological zones in Nigeria. They found that overall ET_o was most sensitive to RH_{mean}, followed by R_s, T_{max}, u_2, and T_{min}. However, there was significant spatiotemporal variation in sensitivity. In the south of Nigeria, which has a tropical rainforest climate with annual precipitation of 1200–3000 mm, ET_o was most sensitive to RH_{mean} and R_s and least sensitive to u_2. Meanwhile, in the north (annual precipitation of 400–1100 mm), the maximum sensitivity was for T_{max} and the minimum for T_{min}. For the very arid region of the Ejina Oasis in northwest China (annual precipitation less than 50 mm), Hou et al. [19] found that the relative sensitivity of ET_o is highest for R_s, followed by T_{mean}, u_2, and RH_{mean}.

The SCs can also be used to understand past historical and projected future changes in ET_o. Luo et al. [20] performed a sensitivity analysis of the ASCE standardized ET_o equation to understand meteorological drivers of historical trends (mainly increasing) in ET_o in the Yanhe River Basin in China. Liu et al. [16] performed a sensitivity analysis of ET_o in the Yellow River Basin in China to understand temporal trends in ET_o in different regions of the basin. They found that positive trends in the upper, middle, and whole Yellow River Basin resulted from a significant increasing trend in T_{mean} and a decreasing trend in RH_{mean}. Wang et al. [21] used sensitivity analyses to understand the causes of historical decreases in pan evaporation over China, also called the "pan evaporation paradox". The "pan evaporation paradox," first discovered by Peterson et al. [22], was observed for decades in the late 20th century in many areas of the globe and contrasts with the general expectation that atmospheric evaporative demand would increase under climate change as temperatures warm. Similar declining trends in reference and potential evapotranspiration were observed in many areas of the globe during the same period. Wang et al. [21] summarized the contribution of different meteorological variables to reductions in pan evaporation observed in many regions of the globe based on the attribution method proposed by Roderick et al. [23], which is essentially a sensitivity analysis of a pan evaporation equation. Wang et al. [21] found that changes in R_s, u_2, and RH_{mean} overcompensate for the positive contributions of increasing air temperatures on pan evaporation, which resulted in a net decrease in pan evaporation in China and many regions of the globe. However, an evaluation of output from general circulation models (GCMs) from the Coupled Model Intercomparison Phase 5 indicates that the evaporation paradox will not continue into the future, at least in China. McKenney and Rosenberg [13] show that future changes in potential evapotranspiration derived from GCM output using different methods may vary in magnitude and, in some cases, in sign depending on the estimation method used.

Our main objective in deriving SCs for Puerto Rico is to guide the prioritization of bias correction of meteorological output for ET_o estimation from dynamically downscaled climate projections for Puerto Rico by Bowden et al. [9] for the historical period 1985–2005 and the future period 2040–2060. The derived SCs are also applied to evaluate the drivers of potential future changes in ET_o in Puerto Rico. Of particular interest are potential future changes to agricultural water demand on irrigated farms and golf courses on the island, which are a function of future projected changes in precipitation and ET_o. Bias-corrected ET_o can be used to drive models (i.e., soil-moisture water-balance models [2]) to estimate potential future changes to agricultural irrigation requirements on the island.

2. Materials and Methods

2.1. Study Area

The study area consists of the main island of the Commonwealth of Puerto Rico, which has a wide range of climate types. Precipitation ranges from 800 mm/year in the subtropical dry forest on the south coast to over 4300 mm/year [24] in the subtropical rainforest of El Yunque, which is part of the Sierra de Luquillo on the northeast of the island (Figure 1). The spatial variability of rainfall results from easterly to northeasterly winds interacting with the local orography. Large rainfall amounts fall north [24] of the mountain range, which runs predominantly east–west over central portions of the island known as the Cordillera Central and the Sierra de Cayey, which runs southeastward on the eastern side of the island. The rainfall maximum occurs over the Sierra de Luquillo mountains in the northeast where the El Yunque rainforest is located. A westerly sea-breeze front develops on the western side of the island, which converges with the predominant easterly winds, resulting in strong convection and high rainfall amounts in the northwest part of the island. A rain shadow occurs over most of the southern coast of the island. Rainfall has a bimodal distribution with the highest rainfall in May–early June and late July–November. A mid-summer drought (MSD) develops in late June through early July and, based on climate model simulations, has been found to be the combined result of increased aerosol concentrations from Saharan dust events and changes in vertical wind shear [25].

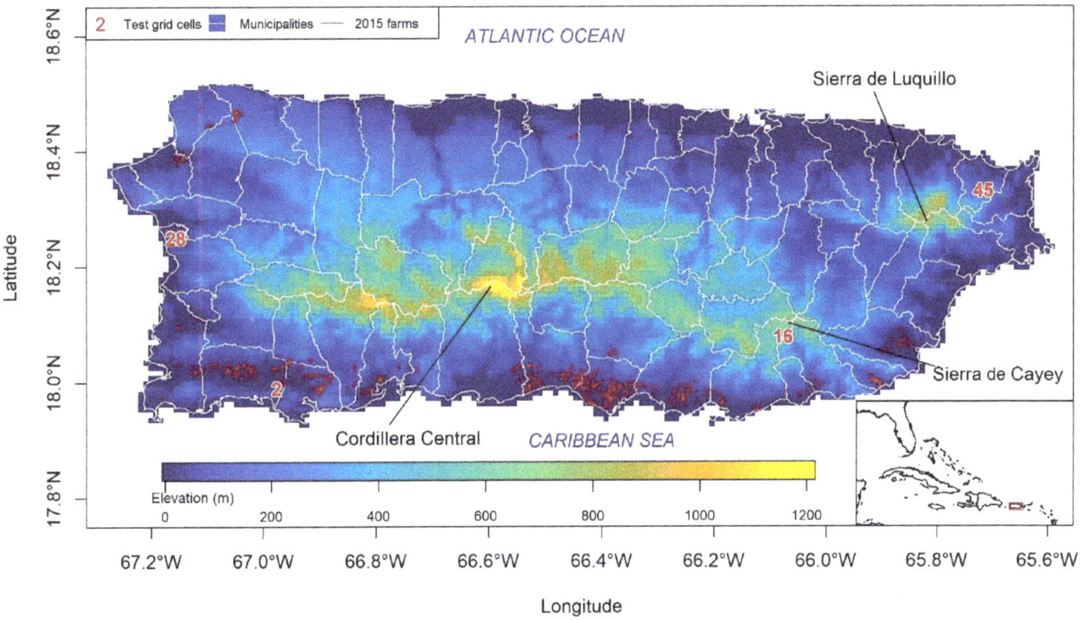

Figure 1. Study area consisting of the main island of Puerto Rico. Elevation in meters is shown. The white lines show the boundaries of the municipalities in Puerto Rico, and the red lines show the boundaries of irrigated farms in 2015. Numbers indicate the location of Geostationary Operational Environmental Satellite-Puerto Rico Water and Energy Balance (GOES-PRWEB) 1 km-scale grid cells at which timeseries of sensitivity coefficients will be presented in detail. Base for main map from 1:20,000 USGS Digital Line Graphs municipality boundaries. Base for inset map is from version 1.4.0 of Natural Earth data, country boundaries 1:10 m. Both are in geographic coordinate system and World Geodetic System 1984 datum.

Low relative humidity, high temperatures, high wind speed, and high incoming solar radiation on the south coast result in high atmospheric evaporative demand compared to rainfall. The south-coast region is characterized by alluvial floodplains with highly fertile soils (mollisols), which are rich in organic matter and minerals [26]. This has made the region suitable for agriculture, which, owing to the relatively low annual rainfall, is dependent on irrigation. Of particular interest to water managers are potential future changes in water availability and water use across the island [4]. There is increasing concern about potential future conflicts among domestic and agricultural water uses on the southern coast of the island [4]. Therefore, it is important to quantify potential future agricultural water demand on irrigated farms and golf courses on the island, which are predominantly located in this southern region (Figure 1; [27]). Agriculture also occurs in the interior mountainous parts of the island, which receive high amounts of rainfall during the year so as not to require irrigation, but these areas may also be subjected to reductions in rainfall in the future [9].

2.2. Data

2.2.1. Gridded Historical Data

The Geostationary Operational Environmental Satellite-Puerto Rico Water and Energy Balance (GOES-PRWEB) dataset [28] provides daily gridded data for water and energy budget components at a 1 km resolution over Puerto Rico for the period 2009–2020. GOES-PRWEB provides estimates of daily rainfall, actual evapotranspiration, reference evapotranspiration (ET_0) based on the FAO-56 (or ASCE) P-M equation (Equation (1) in Section 2.3), as well as data for five basic meteorological variables that are used in estimating ET_0: daily maximum (T_{max}) and minimum (T_{min}) air temperature at 2 meter height, daily mean relative humidity (RH_{mean}), daily mean incoming solar radiation at the land surface (R_s), and 2 m-height wind speed (u_2). Appendix S1 in Supplementary Materials summarizes the GOES-PRWEB computation of ET_0 from these five basic variables. GOES-PRWEB estimates actual ET based on the surface energy balance equation, as described by Harmsen et al. [28]. The GOES-PRWEB estimated ET_0 and its driving meteorological variables have been validated by Mecikalski and Harmsen [29] and Harmsen et al. [28,30] at a few stations on the island, supporting the use of GOES-PRWEB ET_0 in this study. For this study, GOES-PRWEB data for the period 2009–2017 is used.

The high-resolution regional statistical downscaled GPCC v7 for the Caribbean dataset (Herrera-Ault; [7,31]) provides estimates of monthly potential evapotranspiration based on the FAO-56 P-M equation [32] at an approximate resolution of 4 km. Although these data are referred to as potential evapotranspiration, Herrera and Ault [31] mention that this dataset refers to evapotranspiration from an idealized grass surface, that is, ET_0. ET_0 data from this dataset show higher values than GOES-PRWEB from December to April and lower values than GOES-PRWEB the rest of the year. Besides T_{max} and T_{min}, Herrera-Ault does not provide data for the remaining variables required for ET_0 estimation using the FAO-56 P-M equation. This lack of availability of all the meteorological data needed for ET_0 estimation makes it impossible to investigate the causes of the differences in seasonality between the two datasets.

2.2.2. WRF Dynamically Downscaled Climate Change Projections

Bowden et al. [9] used the Weather Research and Forecasting (WRF) model to dynamically downscale historical (1985–2005) and future climate projections (2040–2060) for Puerto Rico and the U.S. Virgin Islands under the greenhouse gas emission scenario RCP8.5 for two GCMs from the Coupled Model Intercomparison Project phase 5 (CMIP5): (1) the Community Climate System Model (CCSM4 or CESM) and (2) the Centre National de Recherches Météorologiques-CERFACS (CNRM). Hourly output from the innermost model domain (domain 3 at 2-km resolution) was downloaded from the Bowden et al. [33], converted from Greenwich Meridian Time to local time (Atlantic Standard Time), and the five basic daily meteorological variables used for ET_0 estimation in GOES-PRWEB were computed and bilinearly interpolated to the 1-km GOES-PRWEB grid. Appendix S2 in Supplementary Materials has a description of the computation of daily meteorological data from hourly data.

2.3. Methods

Actual ET depends on various factors, including atmospheric evaporative demand, soil water availability, crop physiology, and management practices. Absent actual ET data, crop potential evapotranspiration (ET_c) is often computed based on the potential evapotranspiration from a reference crop, typically grass or alfalfa, which is called reference evapotranspiration. Grass reference evapotranspiration is denoted as ET_0 and refers to a crop assumed to be 0.12 m in height, with an albedo of 0.23 and a daily surface resistance of 70 s/m [10]. Crop coefficients (K_c) are often used as multipliers to convert ET_0 to ET_c

(i.e., $ET_c = K_c\, ET_o$). ET_o accounts primarily for weather conditions (i.e., atmospheric evaporative demand), and K_c accounts for the characteristics of a specific crop with respect to the reference grass under fully watered conditions [32]. Therefore, the crop coefficients incorporate many factors that distinguish a particular crop from the reference grass, including planting date, plant growth stage, leaf area, albedo or reflectivity, canopy resistance, soil and climate conditions, evaporation from soil, and crop management practices, among others [32]. In conditions of insufficient rainfall or irrigation and a deficit in soil moisture, the actual combined soil evaporation and plant transpiration (actual *ET*) will be lower than ET_c. GOES-PRWEB estimates actual *ET* based on the surface energy balance equation, as described by Harmsen et al. [28]. In modeling applications, where actual *ET* cannot be estimated a priori but depends on the simulated hydrology, actual *ET* is often estimated from ET_o using modified crop coefficients, which not only account for crop-type differences but also for water availability. Water stress coefficients, which are often defined as a function of soil water content and/or depth to the water table, can be applied as multiplicative correction factors to the standard crop coefficients for well-watered conditions [32] to estimate actual *ET*.

The ASCE standardized grass-reference Penman–Monteith evapotranspiration equation [10] for short grass on a daily timestep is given by:

$$ET_o = \frac{0.408\,\Delta\,(R_n - G) + \gamma \frac{900}{T+273} u_2 (e_s - e_a)}{\Delta + \gamma(1 + 0.34 u_2)}, \quad (1)$$

where ET_o is the daily reference evapotranspiration (mm/day), R_n is the net radiation at the crop surface (MJ/m^2/day), G is the soil heat flux density at the soil surface (MJ/m^2/day), which is generally small compared to R_n beneath a fully vegetated reference surface, and hence, neglected for daily timesteps, T is the mean daily air temperature at 2 meters (°C), u_2 is the mean daily wind speed at 2 meters (m/s), e_s is the saturation vapor pressure at 2 meters (kPa), e_a is the mean actual vapor pressure at 2 meters (kPa), Δ is the slope of the saturation vapor pressure–temperature curve (kPa/°C), and γ is the psychrometric constant (kPa/°C). Appendix S1 summarizes how each of these variables are estimated from five basic meteorological variables defined hereafter and available in GOES-PRWEB. When applied on a daily timestep, the ASCE standardized ET_o equation is the same as the Food and Agriculture Organization paper number 56 (FAO-56) P-M equation [11].

A sensitivity analysis was performed for the ASCE ET_o equation to determine the sensitivity of ET_o to each of the five basic meteorological variables listed above. Neglecting higher-order terms, the sensitivity of ET_o to changes (or errors) in each of the individual basic meteorological variables is quantified by means of non-dimensional relative sensitivity coefficients [34]:

$$S_{V_i} = \frac{\partial ET_o}{\partial V_i} \times \frac{V_i}{ET_o}, \quad (2)$$

where V_i is one of five basic meteorological variables (T_{max} in °C, T_{min} in °C, RH_{mean} in %, R_s in MJ/m^2/day, or u_2 in m/s), i is the index for the variable (1–5), S_{V_i} is the relative SC for variable V_i, and $\frac{\partial ET_o}{\partial V_i}$ is the partial derivative of ET_o with respect to V_i. A positive (negative) SC for a variable indicates that ET_o will increase (decrease) as the variable increases and can be visualized as the slope of the tangent at the origin of the sensitivity curve. The derivatives, $\frac{\partial ET_o}{\partial V_i}$, were derived analytically (Appendix S3 of the Supplementary Materials) and also derived numerically, and their linearity assessed by computing and plotting the percentage change in ET_o due to +/− 5, 10, 15, 20, and 25% change in V_i alone. This is done by changing the value of the variable of interest at each particular location on each day by a given percentage while keeping the remaining variables fixed at their observed value and computing the resulting change in ET_o as a percentage. For RH_{mean}, only negative percent changes were evaluated in order to keep it from exceeding 100%. It is expected that the analytically derived SCs will be closest to those derived based on +/− 5% perturbations for a particular variable. The range of variation in a meteorological variable over which the slope of the sensitivity curve is linear is also assessed. The larger the range of linearity, the smaller the error incurred when applying the point SCs over large changes in the meteorological variable.

The SCs are estimated based on daily GOES-PRWEB gridded meteorological data for the period 2009–2017 and averaged for each day or month of the year to obtain the average annual cycle of the SCs for each variable of interest at each grid cell. Timeseries of average SCs by day of the year will be presented at four representative locations on the island (Figure 1) that capture most of the distinct patterns of variation in the SCs as well as maps of long-term monthly average SCs.

When combined with a measure of the variability or measurement/estimation error in the basic meteorological variables, the SCs could be used to define monitoring priorities among the variables. The SCs could also be used to guide the prioritization of bias correction of meteorological output for ET_o estimation from climate models. When combined with projections of the percent change in the basic meteorological variables, the SCs can aid in understanding past trends and future projected changes in ET_o derived from observations and climate model output. Based on the chain rule, which helps differentiate composite functions, the combined contributions of fractional changes or errors in the basic meteorological variables to fractional changes or errors in ET_o can be approximated by:

$$\frac{dET_o}{ET_o} = \sum_i \left(S_{V_i} \times \frac{dV_i}{V_i} \right) + error = \sum_i \left(\frac{\partial ET_o}{\partial V_i} \times \frac{V_i}{ET_o} \times \frac{dV_i}{V_i} \right) + error, \quad (3)$$

where *error* is the approximation error.

An example application of the SCs and Equation (3) to understanding biases and potential future changes in ET_o from 1985–2005 to 2040–2060 based on the dynamically downscaled climate projections for Puerto Rico by Bowden et al. [9] will be presented in this paper. Of particular interest are potential future changes to agricultural water demand on irrigated farms on the island, which are a function of future projected changes in precipitation and evaporative demand (ET_o). As of the year 2015, most irrigated farms are located in the fertile semi-arid southern coast of Puerto Rico (Figure 1; [27]). The average annual cycle of the SCs and the contribution of projected changes in each of the basic meteorological variables to projected changes in ET_o were calculated for all model grid cells whose center is located within a 2015 irrigated farm. No comprehensive gridded meteorological dataset including daily data for all the basic meteorological variables used in estimating ET_o has been located for the simulated historical period 1985–2005; therefore, GOES-PRWEB data for 2009–2017 is used to examine potential biases in ET_o calculated from downscaled model output and the contribution of individual meteorological variables to those biases.

Finally, the Pearson correlation coefficient (r) between long-term monthly average maps of SCs, the five basic meteorological variables, the aerodynamic and energetic components of ET_o and their fractions ($ET_{o,energetic}$, $ET_{o,aerodynamic}$, $ET_{o,energetic\ fraction}$, $ET_{o,aerodynamic\ fraction}$, respectively, given by Equations (S1.1)–(S1.5) in the Supplementary Materials), rainfall, and elevation will be presented. The Pearson correlation coefficient is a measure of the strength of the linear association between two sets of data and will be used here to better understand how different variables affect or are correlated with ET_o. Because the correlation coefficient is calculated between all the monthly maps, it captures both temporal and spatial correlations between the different fields. In the discussion, the correlation (or anti-correlation) between variables is considered moderate for absolute values of r of 0.50–0.69, high for absolute values of r of 0.70–0.89, and very high for absolute values of r of 0.90–1.00.

3. Results

Spatiotemporal variations in rainfall, ET, and other components of the hydrologic cycle in Puerto Rico will be presented as maps and seasonal and annual plots. Results of the sensitivity analysis will be presented as long-term (2009–2017) monthly average maps of SCs and timeseries of SCs at various locations. Results from the application of

SCs to understand model biases and projected future changes in ET_0 based on dynamically downscaled climate model output will also be presented.

3.1. Spatiotemporal Variations in Rainfall and ET

On the basis of GOES-PRWEB data [28], it was found that annual rainfall (Figure 2a) and annual actual *ET* (Figure 2b) are similar in magnitude across large portions of the island. About 65% of rainfall on average island wide, and upwards of 90% in the south coast, returns to the atmosphere as evapotranspiration on an average year (Figure 2c). Only a small fraction of annual rainfall recharges the shallow aquifer (Figure 2e), while the fraction of annual rainfall that generates runoff is higher in areas with higher rainfall, such as on the central northwest part of the island and on the Sierra de Luquillo in the northeast (Figure 2f). Annual atmospheric evaporative demand, as quantified by ET_0, is more than twice as high as annual rainfall over the south coast of Puerto Rico (Figure 2d), where agriculture has to rely on irrigation to meet the atmospheric evaporative demand. Low relative humidity, high temperatures, high wind speed, and high incoming solar radiation on the south coast result in high ET_0 compared to rainfall.

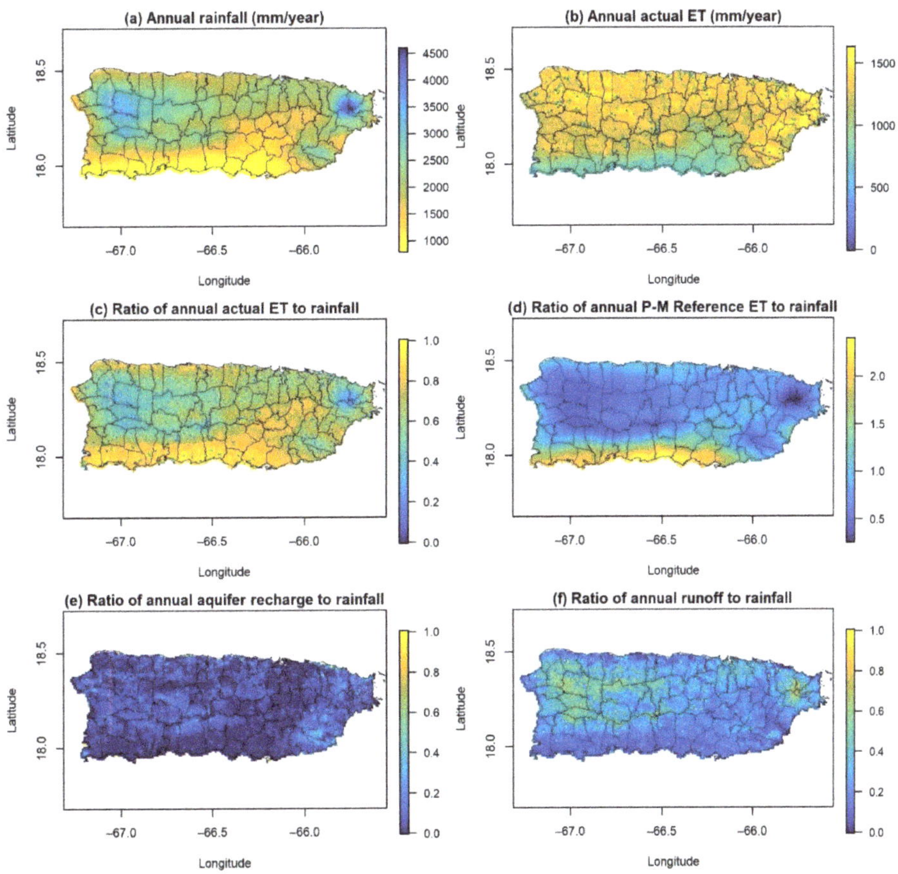

Figure 2. (**a**) Annual rainfall (mm/year); (**b**) annual actual *ET* (mm/year); (**c**) ratio of annual actual *ET* to rainfall; (**d**) ratio of annual Penman–Monteith ET_0 to rainfall; (**e**) ratio of annual aquifer recharge to rainfall; and (**f**) ratio of annual runoff to rainfall for Puerto Rico based on Geostationary Operational Environmental Satellite-Puerto Rico Water and Energy Balance (GOES-PRWEB) data for 2009–2017. The black lines show the location of the municipalities in Puerto Rico.

Figure 3a,b illustrate the annual cycle of monthly rainfall and ET_o and Figure 3c illustrates the interannual variability of ET_o at the 2015 irrigated farm locations. The bimodal distribution of rainfall is evident in Figure 3a, with the highest rainfall in May–early June and late July–November. The mid-summer drought (MSD) that develops in late June through early July is also evident. The annual cycle of monthly $ET_{o.}$ (Figure 3b) from GOES-PRWEB with peak values in July appears to match station-based estimates of ET_o in Harmsen et al. [35] better than the Herrera-Ault ET_o which peaks in March. This justifies our use of GOES-PRWEB ET_o in this study.

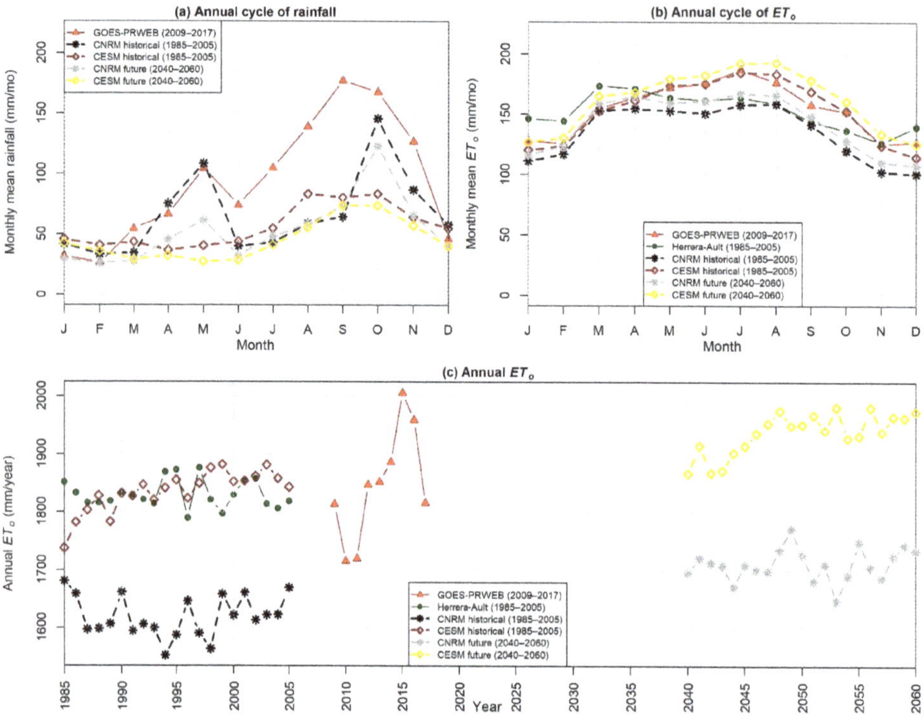

Figure 3. (**a**) Annual cycle of rainfall (mm/mo.); (**b**) annual cycle of ET_o (mm/mo.); (**c**) annual ET_o (mm/year) from Geostationary Operational Environmental Satellite-Puerto Rico Water and Energy Balance (GOES-PRWEB), Herrera-Ault dataset, and WRF dynamically downscaled climate models at the 2015 irrigated farm locations.

3.2. Sensitivity Coefficients

Sensitivity coefficients for ET_o, with respect to each of the five basic meteorological variables, were computed analytically and numerically at each GOES-PRWEB grid cell for each day in the period 2009–2017. Maps of the long-term (2009–2017) monthly average sensitivity coefficients derived analytically are shown in Figures 4–8. The maps derived numerically for a +/− 5% change in each meteorological variable (not shown) are very similar, corroborating the analytical calculations. The SC for RH_{mean} is negative because an increase in atmospheric water content reduces ET_o. To make the SC for RH_{mean} of comparable magnitude to the SC of the other meteorological variables, it is multiplied by −1 in Figure 4. Island wide, ET_o is most sensitive to RH_{mean}, followed by R_s, T_{max}, T_{min}, and u_2, with average absolute relative SCs of 0.98, 0.57, 0.50, 0.27, and 0.12, respectively. This overall ranking of sensitivities is similar to that found by Emeka et al. [18] based on relative sensitivities for the tropical rainforest climate of southern Nigeria, where precipitation ranges from 1200 to 3000 mm per year, similar to the range of annual precipitation in Puerto

Rico (Figure 2a). The only exception is that in southern Nigeria, ET_o was found to be most sensitive to R_s followed by RH_{mean} in the non-monsoon season. The ranking of sensitivities also agrees well with that obtained by Gong et al. [12] for the Yangtze River Basin in China, especially during winter when the sensitivity to RH_{mean} is higher than that of R_s and T_{mean}. Annual precipitation ranges from 400 in the upper Yangtze River Basin to 1600 mm in the lower basin [15]. Only the middle and lower portions of the Yangtze River Basin have annual precipitation comparable to that of some regions in Puerto Rico (Figure 2a).

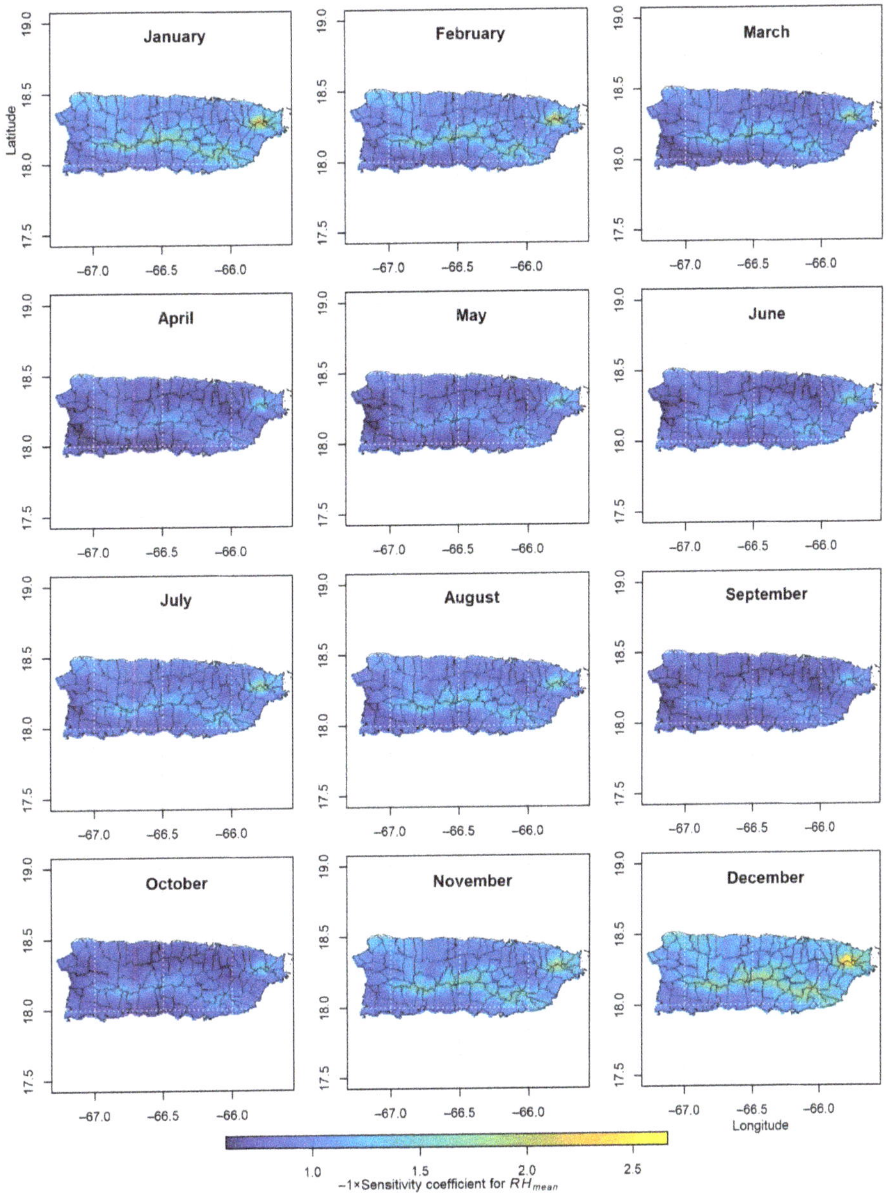

Figure 4. Long-term (2009–2017) monthly average sensitivity coefficients, multiplied by −1, for mean relative humidity (RH_{mean}). The black lines show the location of the municipalities in Puerto Rico.

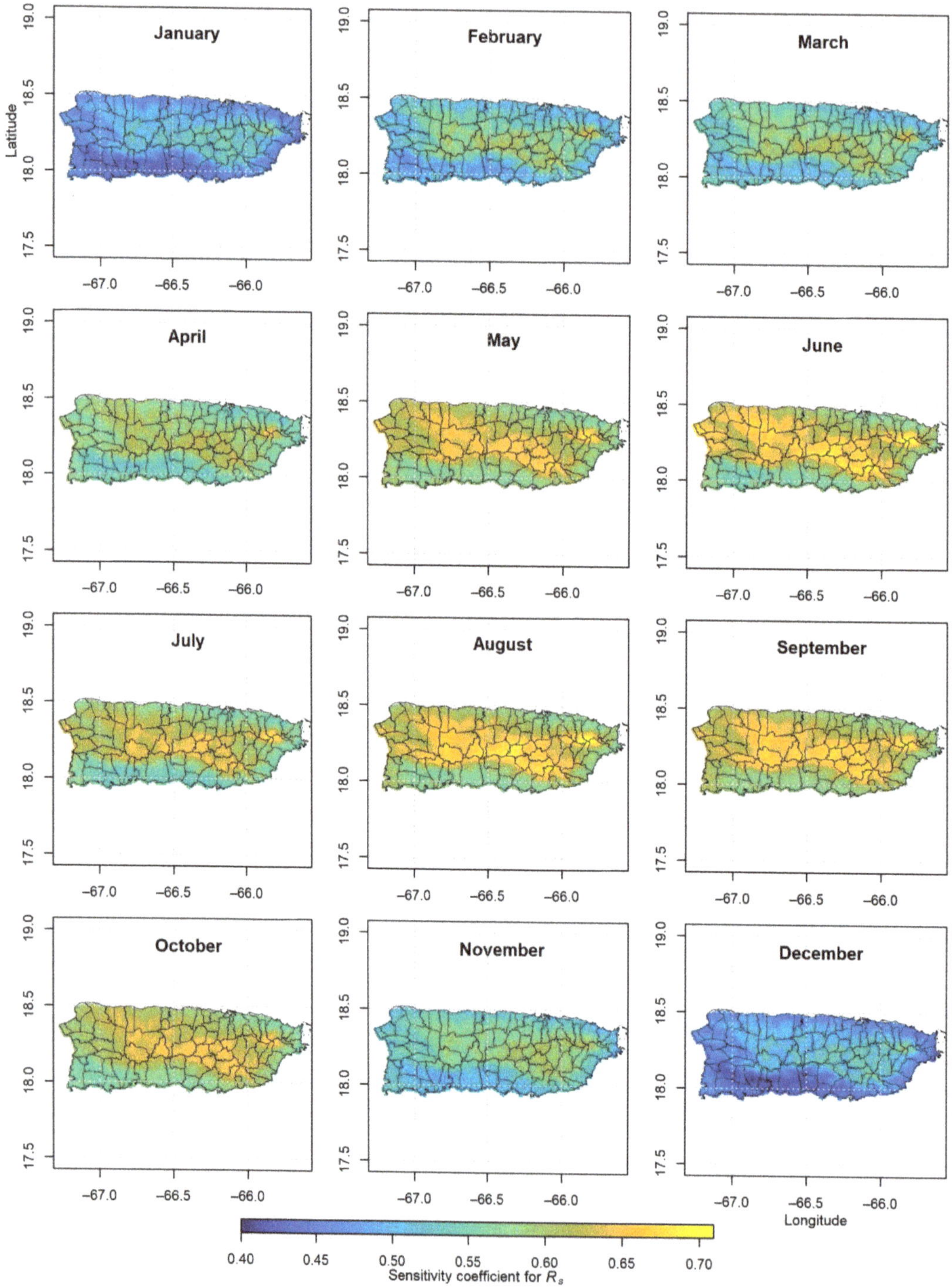

Figure 5. Long-term (2009–2017) monthly average sensitivity coefficients for solar radiation (R_s). The black lines show the location of the municipalities in Puerto Rico.

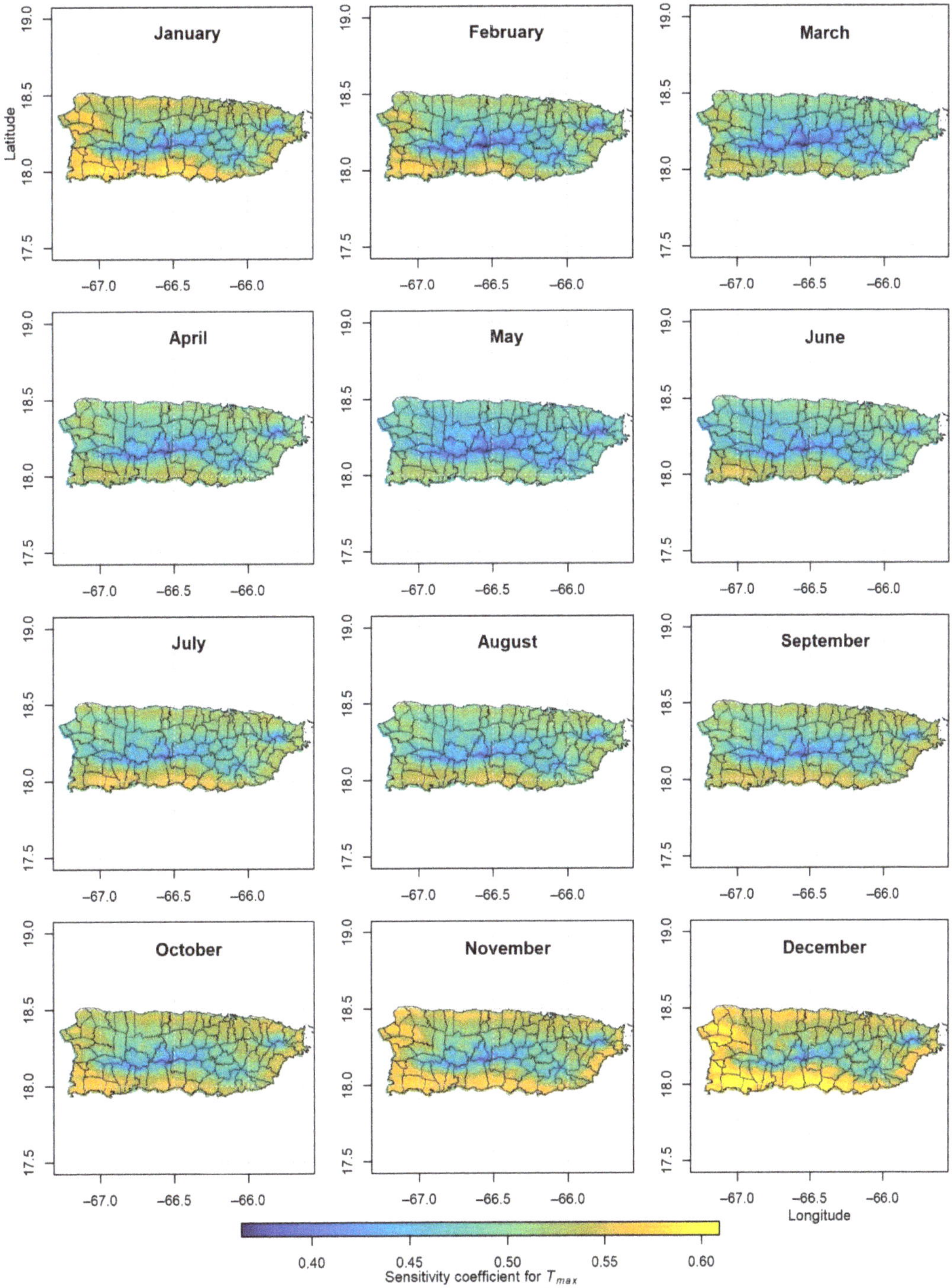

Figure 6. Long-term (2009–2017) monthly average sensitivity coefficients for daily maximum air temperature (T_{max}). The black lines show the location of the municipalities in Puerto Rico.

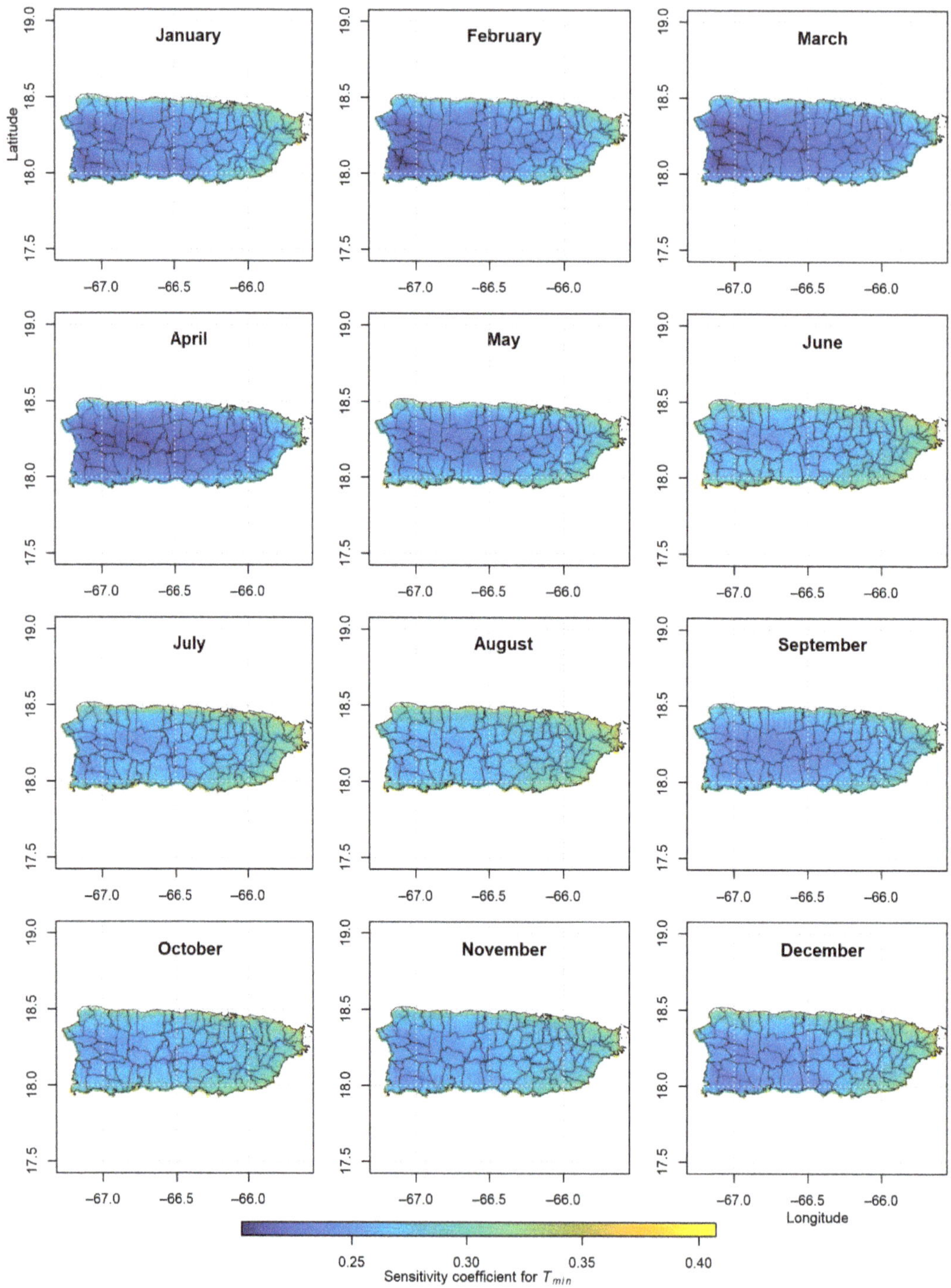

Figure 7. Long-term (2009–2017) monthly average sensitivity coefficients for daily minimum air temperature (T_{min}). The black lines show the location of the municipalities in Puerto Rico.

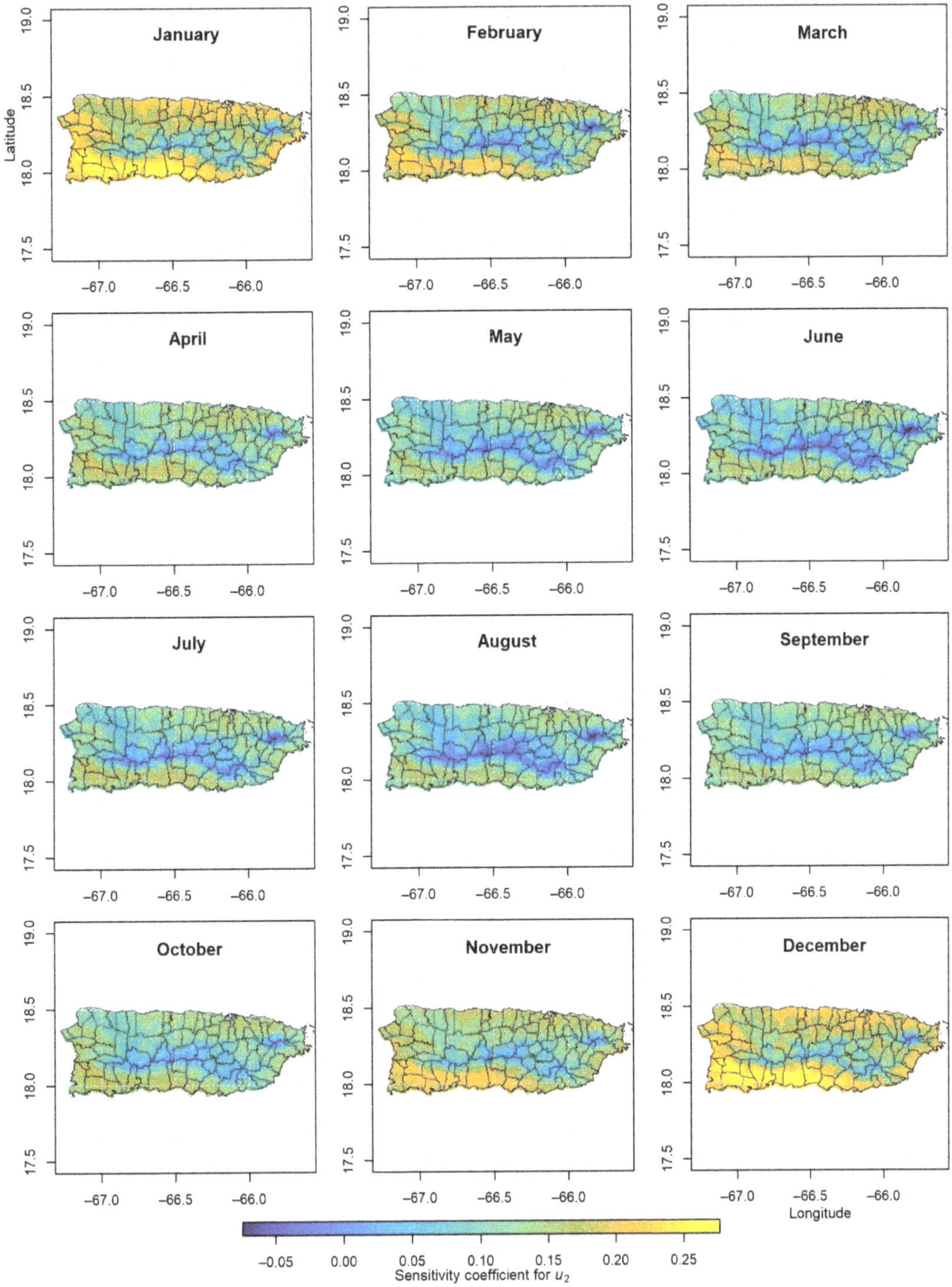

Figure 8. Long-term (2009–2017) monthly average sensitivity coefficients for 2 m wind speed (u_2). The black lines show the location of the municipalities in Puerto Rico.

Timeseries of SCs by day of the year averaged over the period 2009–2017 at the chosen four GOES-PRWEB grid cell locations (cells 2, 16, 28, 45; Figure 1) are shown in Figure 9. In addition, the average timeseries of SCs for the location of the farms as of 2015 shown in Figure 1 is also shown in Figure 9. The SC for RH_{mean} (Figures 4 and 9) dominates the sensitivity of ET_o throughout most of the island, except for the western side of the island (cell 28 in Figures 1 and 9), where the SC for R_s is as or more dominant during the summer. The SC for RH_{mean} has a large amplitude annual cycle, especially along the Cordillera Central mountains and the Sierra de Luquillo with larger values and less day-to-day variability in the dry season than in the wet season. Coastal areas including the 2015 irrigated farm locations have a lower relative sensitivity to RH_{mean}, with a less marked annual cycle than interior areas.

Higher SCs for R_s are apparent (Figure 5) along an interior band that goes from the northwest to the southeast on the island and generally coincides with more humid regions (Figure 2a), consistent with the findings of Tabari and Talaee [36] for Iran and Emeka et al. [18] for Nigeria. ET_o is relatively less sensitive to R_s along the northeast coast and especially along the south coast. Although a clear seasonal cycle of higher relative sensitivity to R_s during the summer months is evident (Figures 5 and 9), its amplitude is much smaller than that of RH_{mean}. Although, overall, the SCs for R_s and T_{max} are quite close on an island-wide basis (0.57 and 0.50, respectively), their spatial and temporal patterns are generally opposite (Figures 5, 6 and 9). ET_o is relatively less sensitive to T_{max} along interior mountainous regions of the island and during the summer months, and more sensitive along coastal areas, especially in the southern and western coasts during the dry season months. Seasonal variations in the relative sensitivity to T_{max} are relatively minor, especially in mountainous areas (cell 45 in Figures 1 and 9). Minimal spatial and temporal variation in the SCs for T_{min} was found, with only slightly lower values in the western interior areas of the island from January to April and slightly higher values in coastal areas to the east (Figure 7).

Many similarities in the spatial and temporal patterns of SCs to u_2 (Figure 8) and the spatial and temporal patterns of SCs to T_{max} (Figure 6) were found. However, ET_o is much less sensitive to relative changes in u_2 than in T_{max} (Figure 9). The relative sensitivity of ET_o to changes in T_{max} and u_2 is lower along humid and warm interior mountainous regions of the island and during the rainy summer months, and higher along coastal areas, especially in the southern coast during the dry season months. These findings are consistent with those of Emeka et al. [18] for Nigeria and Tabari and Talaee [36] for Iran. The lower sensitivity of ET_o to changes in wind speed in humid and warm environments compared to its higher sensitivity in hot and dry environments is explained by Allen et al. [32]. In humid and warm conditions, such as in the interior mountains, the wind (no matter how intense) can only replace saturated air at the vegetation's surface with air from above that is only slightly less saturated, thereby limiting potential increases in evapotranspiration. Increases in wind speed may, in fact, increase sensible heating more than evapotranspiration in these humid and warm environments, resulting in slightly negative SCs for u_2 at times (Figure 8 and cell 16 in Figure 9). In contrast, in hot and dry environments, such as the island's south coast, increases in wind speed can increase the evapotranspiration rate more since the atmosphere is thirstier and has more energy available.

Figure 10 shows the percent change in ET_o due to $+/-$ 5, 10, 15, 20, and 25% changes in basic meteorological variables at selected GOES-PRWEB grid cell locations shown in Figure 1 and the average at the 2015 irrigated farm locations. The larger relative sensitivity of ET_o to changes in RH_{mean} compared to other variables is evident in the larger slope for the RH_{mean} line. Linearity in the SCs is observed over the tested range of percentage changes in the basic meteorological variables at all grid cells, which means that the application of the SC values at the origin over a relatively wide range of changes in the meteorological variables is appropriate.

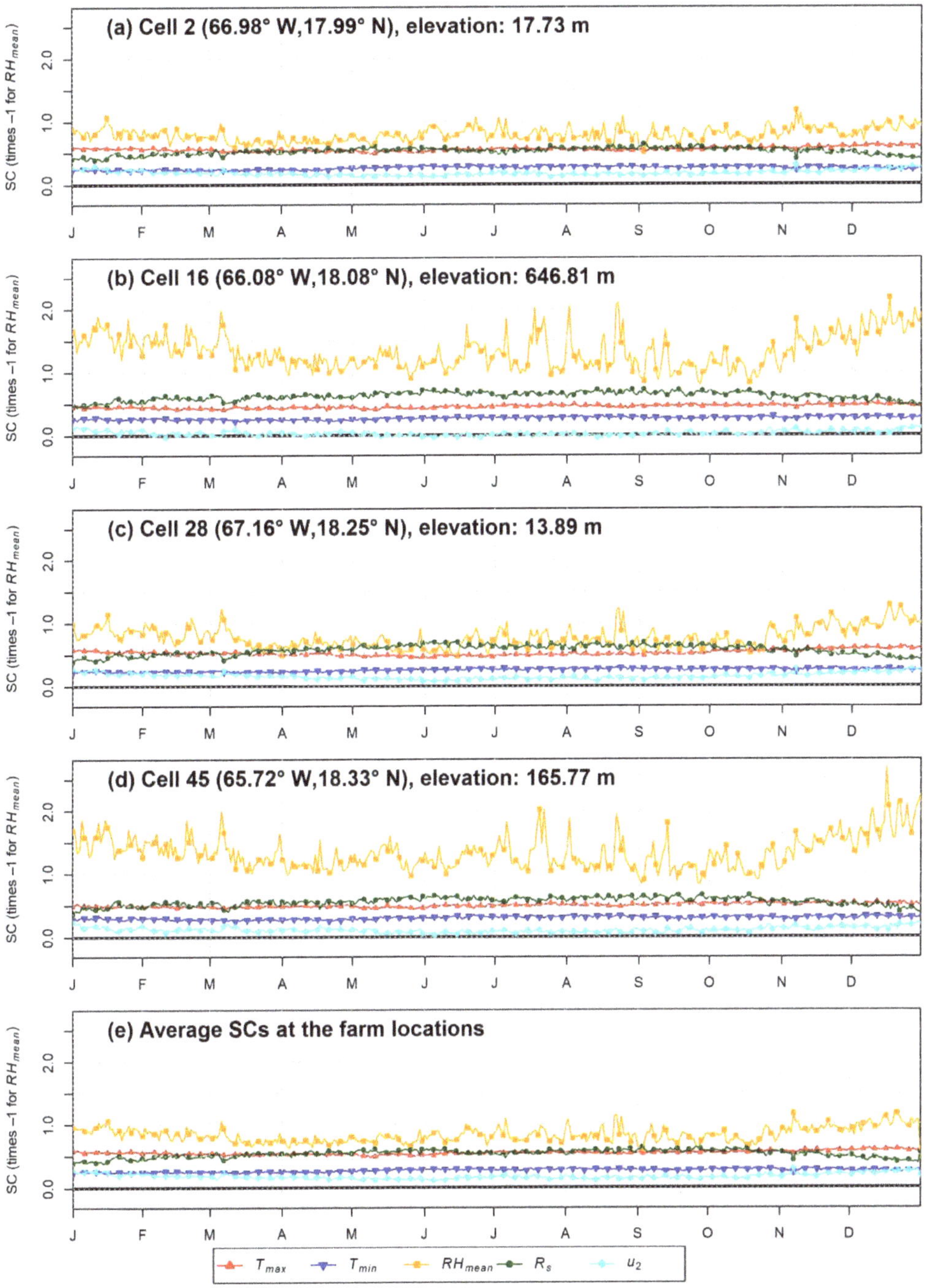

Figure 9. Daily timeseries of sensitivity coefficients (SCs) at: (**a**–**d**) representative GOES-PRWEB grid cell locations (Figure 1) and (**e**) average at grid cells associated with the 2015 irrigated farm locations. Markers are shown every 5 days. Cell 2 is on the southwest coast; cell 16 is located near the Sierra de Cayey; cell 28 is on the west coast; cell 45 is located near the Sierra de Luquillo.

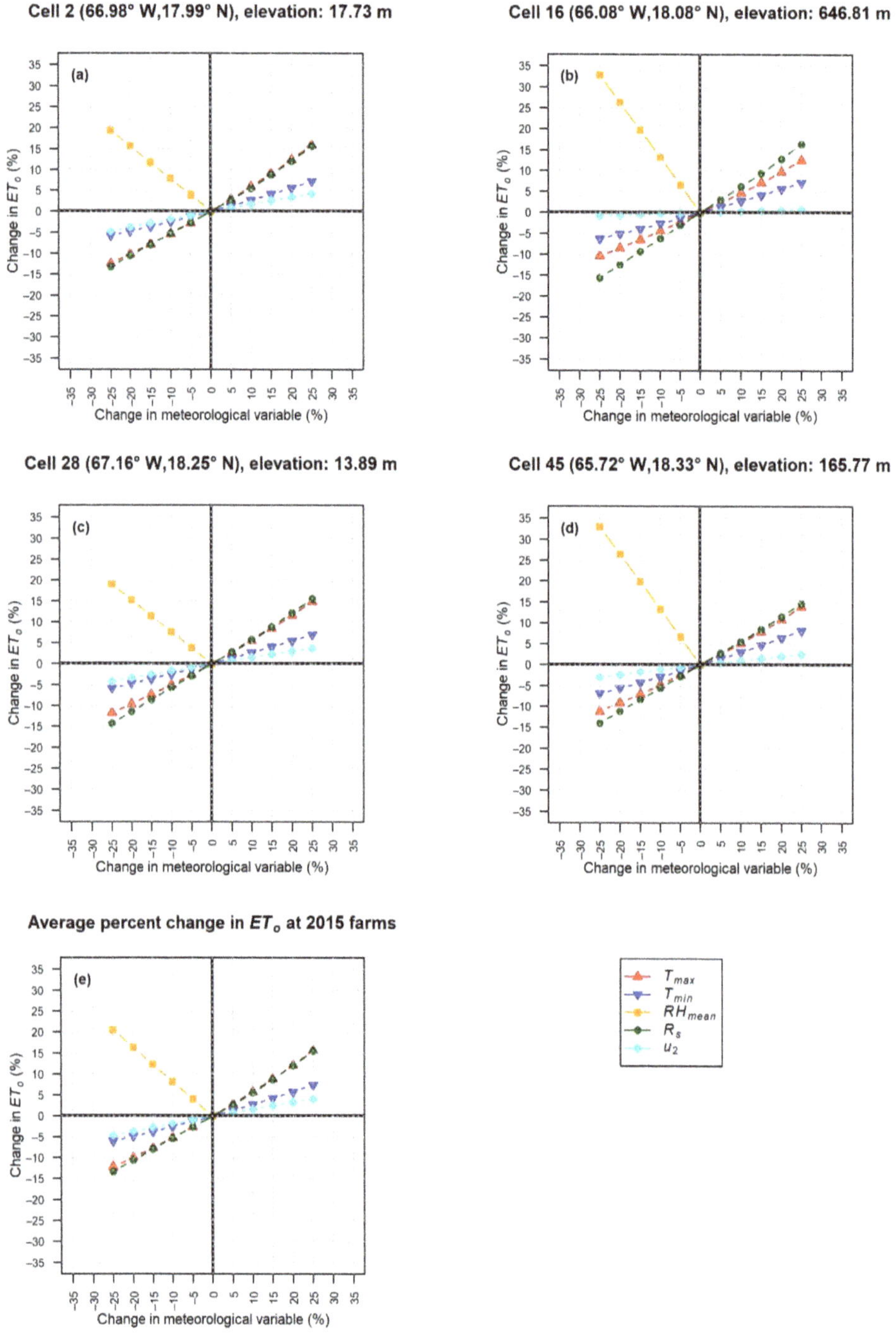

Figure 10. Percentage change in ET_0 due to +/− 5, 10, 15, 20, and 25% change in basic meteorological variables at: (**a**–**d**) representative GOES-PRWEB grid cell locations (Figure 1) and (**e**) average at grid cells associated with the 2015 irrigated farm locations.

The Pearson correlation matrix is presented between long-term monthly maps of basic meteorological variables, their SCs, components of ET_0, rainfall, and elevation (Table 1). The SC for RH_{mean} has not been multiplied by -1 when calculating correlations. To aid in the discussion, Figures S1–S11 show maps of the long-term monthly average for the basic meteorological variables, rainfall, and components of ET_0. The SC for RH_{mean} (Figure 4 shows the SC for RH_{mean} multiplied by -1) is highly anti-correlated to RH_{mean} (Figure S1), specifically to the spatial pattern in RH_{mean}. In other words, ET_o is most sensitive to changes in RH_{mean} in areas where RH_{mean} already tends to be high and limiting, such as in the high-elevation areas of the Cordillera Central and the Sierra de Luquillo, where it exceeds about 82% on average (Figure S1). In these areas, lower R_s (Figure S2) also results in a lower contribution from $ET_{o,energetic}$ (Figure S7). The absolute value of the SC for RH_{mean} is also highest in winter when aerodynamic effects become more important due to lower solar radiation. RH_{mean} is highly anti-correlated to $ET_{o,aerodynamic}$ and $ET_{o,aerodynamic\,fraction}$, as expected, since an increase in RH_{mean} decreases the VPD and reduces these terms. RH_{mean} is also highly correlated to elevation (Figure 1) and highly anti-correlated to T_{max} reflecting the higher RH_{mean} and the lower T_{max} in the interior mountainous areas, and the opposite trends in coastal areas. RH_{mean} is also highly anti-correlated to the SCs for T_{max} and for u_2. This is expected since these three variables act synergistically to increase $ET_{o,aerodynamic}$, especially during winter.

The SC for R_s (Figure 5) is generally higher along the higher-elevation interior areas and moderately correlated to R_s (Figure S2). They are highly correlated in time (i.e., months with lowest R_s are also the months with lowest SC for R_s) but somewhat anti-correlated in space (i.e., areas with the lowest R_s in the interior mountains have the highest SC for R_s). The SC for R_s is moderately correlated to $ET_{o,energetic}$ (Figure S7), especially in time. The sensitivity to R_s is highest from northwest to southeast interior areas, especially in summer when R_s and $ET_{o,energetic}$ are at their highest and tend to dominate over aerodynamic effects (Figures S9 and S10). In winter, the aerodynamic term dominates more, so that R_s is not as important in driving ET_o on the southwest coast for example (Figure 5). As expected, $ET_{o,energetic}$ (Figure S7) is very highly correlated to R_s and to elevation since R_s is moderately anti-correlated to elevation (Figure 1) due to the common presence of clouds at higher elevations. R_s is moderately correlated to T_{max} and T_{min}. This is most evident in the interior mountainous areas where R_s, T_{max}, and T_{min} tend to be the lowest. R_s is moderately correlated to the SC for RH_{mean} (Figure 4) since R_s tends to be higher along coastal areas where the SC for RH_{mean} is higher (less negative).

Rainfall from GOES-PRWEB was only found to be moderately correlated to the SC for R_s and $ET_{o,energetic\,fraction}$. The correlation between rainfall and elevation was found to be quite low at only 0.02. Figure 2a shows two distinct areas of high rainfall—in the Sierra de Luquillo on the eastern side of the island, and in the northwest central part of the island (Figure 1). The spatial variability of rainfall results from easterly to northeasterly winds interacting with the local orography. However, rainfall appears to occur predominantly on the windward side of the mountains and not exactly at the locations with highest elevations. Even in the Sierra de Luquillo, the expectation of higher rainfall at higher elevations may not hold. Based on a relatively large set of rain gauges, Murphy et al. [37] found that the long-held assumption of precipitation increasing consistently with elevation within the Sierra de Luquillo does not hold. They found that leeward (western) watersheds in the mountains generally receive lower mean annual precipitation than windward (eastern) watersheds.

Table 1. Pearson correlation coefficient (r) matrix between long-term monthly maps of basic meteorological variables, their sensitivity coefficients (SCs), components of ET_o, rainfall, and elevation. ET_o is the reference evapotranspiration, RH_{mean} is the daily mean relative humidity, R_s is the daily incoming solar radiation, T_{max} is the daily maximum air temperature, T_{min} is the daily minimum air temperature, and u_2 is the 2 m wind speed. ET_o *energ. (frac.)* and ET_o *aero. (frac.)* are the energetic and aerodynamic ET_o components (and their fractions), respectively.

Variable	RH_{mean}	R_s	T_{max}	T_{min}	u_2	$SC\ RH_{mean}$ *	$SC\ R_s$	$SC\ T_{max}$	$SC\ T_{min}$	SCu_2
RH_{mean}	1	−0.29	−0.72	−0.49	−0.02	−0.73	0.47	−0.69	0.07	−0.80
R_s	−0.29	1	0.54	0.57	0.30	0.56	0.55	−0.07	0.25	−0.26
T_{max}	−0.72	0.54	1	0.88	0	0.76	−0.01	0.67	0.31	0.51
T_{min}	−0.49	0.57	0.88	1	0.22	0.51	0.06	0.49	0.70	0.33
u_2	−0.02	0.30	0	0.22	1	−0.39	−0.25	0.08	0.55	−0.01
$SC\ RH_{mean}$ *	−0.73	0.56	0.76	0.51	−0.39	1	0.18	0.33	−0.17	0.34
$SC\ R_s$	0.47	0.55	−0.01	0.06	−0.25	0.18	1	−0.64	−0.01	−0.83
$SC\ T_{max}$	−0.69	−0.07	0.67	0.49	0.08	0.33	−0.64	1	0.20	0.86
$SC\ T_{min}$	0.07	0.25	0.31	0.70	0.55	−0.17	−0.01	0.20	1	0.04
$SC\ u_2$	−0.80	−0.26	0.51	0.33	−0.01	0.34	−0.83	0.86	0.04	1
ET_o *energ.*	−0.30	0.96	0.65	0.65	0.10	0.66	0.62	−0.05	0.23	−0.23
ET_o *aero.*	−0.86	0.48	0.72	0.67	0.49	0.49	−0.44	0.67	0.31	0.66
ET_o *energ. frac.*	0.68	0.24	−0.30	−0.23	−0.39	−0.04	0.94	−0.77	−0.16	−0.90
ET_o *aero. frac.*	−0.68	−0.24	0.30	0.23	0.39	0.04	−0.94	0.77	0.16	0.90
Rainfall	0.27	0.17	0.17	0.19	−0.32	0.15	0.60	−0.27	0.05	−0.38
Elevation	0.78	−0.65	−0.92	−0.92	−0.10	−0.66	0.68	−0.85	−0.47	−0.84

Variable	ET_o *energ.*	ET_o *aero.*	ET_o *energ. frac.*	ET_o *aero. frac.*	*Rainfall*	*Elevation*
RH_{mean}	−0.30	−0.86	0.68	−0.68	0.27	0.78
R_s	0.96	0.48	0.24	−0.24	0.17	−0.65
T_{max}	0.65	0.72	−0.30	0.30	0.17	−0.92
T_{min}	0.65	0.67	−0.23	0.23	0.19	−0.92
u_2	0.10	0.49	−0.39	0.39	−0.32	−0.10
$SC\ RH_{mean}$ *	0.66	0.49	−0.04	0.04	0.15	−0.66
$SC\ R_s$	0.62	−0.44	0.94	−0.94	0.60	0.68
$SC\ T_{max}$	−0.05	0.67	−0.77	0.77	−0.27	−0.85
$SC\ T_{min}$	0.23	0.31	−0.16	0.16	0.05	−0.47
$SC\ u_2$	−0.23	0.66	−0.90	0.90	−0.38	−0.84
ET_o *energ.*	1	0.42	0.32	−0.32	0.33	−0.88
ET_o *aero.*	0.42	1	−0.72	0.72	−0.32	−0.80
ET_o *energ. frac.*	0.32	−0.72	1	−1	0.56	0.77
ET_o *aero. frac.*	−0.32	0.72	−1	1	−0.56	−0.77
Rainfall	0.33	−0.32	0.56	−0.56	1	0.02
Elevation	−0.88	−0.80	0.77	−0.77	0.02	1

* The SC for RH_{mean} has not been multiplied by −1 when calculating correlations.

The SC for T_{max} (Figure 6) is higher along the low-elevation coastal areas, especially along the south coast, and it is moderately correlated to T_{max} (Figure S3). They are highly correlated spatially; however, they appear anti-correlated in time. That is, the cooler winter months with lower T_{max} show a higher sensitivity to T_{max} as the energetic term decreases and the aerodynamic term's importance increases, especially along the southwest coast (Figures S9 and S10). T_{max} (Figure S3), T_{min} (Figure S4), and elevation (Figure 1) are very highly correlated due to lower temperatures in the interior mountain regions. T_{max} stays relatively high in the southwest coast year-long, whereas T_{min} is reduced during winter and early spring in the area, resulting in a higher daily temperature range in the area in winter and early spring. T_{max} is moderately to highly correlated to both $ET_{o,energetic}$ (Figure S7) and $ET_{o,aerodynamic}$ (Figure S8) because it affects both terms; however, it appears to affect ET_o and especially the aerodynamic component more in the winter when it acts synergistically with u_2 and RH_{mean} to increase it, especially along coastal areas. T_{max} is also highly correlated to the SCs for RH_{mean} (Figure 4) and has a moderate correlation to the SC for u_2 (Figure 8). This is expected since these three variables act synergistically to increase $ET_{o,aerodynamic}$, especially during winter. The SC for T_{min} (Figure 7) is highly correlated to T_{min} (Figure S4). The correlation is stronger in time than in space. T_{min} is moderately correlated to both $ET_{o,energetic}$ (Figure S7) and $ET_{o,aerodynamic}$ (Figure S8) since it affects both terms.

The SC for u_2 (Figure 8) is generally higher along coastal areas and has no correlation to the spatial pattern in u_2 (Figure S5). As expected, u_2 is moderately correlated to $ET_{o,aerodynamic}$ (Figure S8); u_2 is also moderately correlated to the sensitivity coefficient for T_{min} (Figure 7). The SCs for R_s (Figure 5) and for T_{max} (Figure 6) are moderately anti-correlated. This is especially true in time (i.e., in winter months, the SC for R_s decreases while the SC for T_{max} increases, especially in coastal areas), but also to a lesser extent in space (i.e., interior areas have the highest SCs for R_s and the lowest SC for T_{max}). The SCs for R_s (Figure 5) and for u_2 (Figure 8) are highly anti-correlated. This is especially true in time (i.e., in winter months, the SC for R_s decreases while the SC for u_2 increases, especially in coastal areas), but also to a lesser extent in space (i.e., interior areas have the highest SCs for R_s and the lowest SC for u_2). The SCs for T_{max} (Figure 6) and for u_2 are highly correlated in both space and time. The SC for RH_{mean} (Figure 4) is moderately correlated to $ET_{o,energetic}$ (Figure S7) and has a moderate correlation to $ET_{o,aerodynamic}$ (Figure S8). The SC for R_s (Figure 5) is moderately correlated to $ET_{o,energetic}$ (Figure S7) and very highly correlated to $ET_{o,energetic\,fraction}$ (Figure S9), as expected.

3.3. *Application*

Timeseries of SCs are presented by day of the year (Figure 9e) at the GOES-PRWEB grid cells at the 2015 irrigated farms, most of which are located along the south coast (Figure 1). Compared to interior areas, the areas including the 2015 irrigated farm locations show a lower relative sensitivity to RH_{mean} (Figure 4) with a less marked annual cycle. ET_o is also relatively less sensitive to R_s (Figure 5) in these areas, especially during winter months when ET_o becomes more sensitive to relative changes in T_{max} (Figure 6). The relative sensitivity of ET_o to changes in u_2 is lower than for other variables but higher along the southern coast than in other areas, especially during the dry season. In hot and dry environments, such as the south coast, increases in wind speed can increase the evapotranspiration rate more since the atmosphere has a lower moisture content and more energy available.

The lowest $ET_{o,energetic\,fraction}$ (Figure S9) and the highest $ET_{o,aerodynamic\,fraction}$ (Figure S10) occur in the south coast of Puerto Rico. This is due to this region having the highest T_{max} (Figure S3) and the lowest RH_{mean} (Figure S1) (which results in a high VPD) combined with moderate wind speeds in this area (Figure S5). These factors result in comparatively high $ET_{o,aerodynamic}$ (Figure S8) and $ET_{o,aerodynamic\,fraction}$ (Figure S10) compared to other areas. The low RH_{mean} and relatively low cloudiness on the south coast also traps less longwave radiation in the atmosphere, reducing the net radiation, which would tend to reduce the energetic term. However, this effect appears to be counteracted by higher R_s (Figure S2) due to the lower cloudiness for a net result of higher incoming net radiation and higher $ET_{o,energetic}$ (Figure S7) compared to other areas.

3.3.1. Causes of Errors in Reference ET Estimated from Model Output for 1985–2005

Combining the average SCs at the 2015 irrigated farm locations on a monthly basis (Figure 11a) with the percent error in the simulated basic meteorological variables for the period 1985–2005 with respect to GOES-PRWEB for 2009–2017 according to Equation (3) gives the total contribution of each variable to errors in estimated ET_o for the two downscaled GCMs (Figure 11b,c). The individual lines in these plots show the product of the SC for a particular basic meteorological variable times the percent error in the variable, which gives the contribution of that variable to the total error in ET_o as a percentage. The brown line shows the total computed error in ET_o ($\frac{dET_o}{ET_o}$ in Equation (3), as a percentage), and the black line shows the sum of the contributions from all the five basic meteorological variables to the total error in ET_o ($\sum_i \left(\frac{\partial ET_o}{\partial V_i} \times \frac{V_i}{ET_o} \times \frac{dV_i}{V_i} \right)$ in Equation (3)). The difference between the black and brown lines is the approximation error (*error* in Equation (3)). The individual error components add up to large negative monthly biases in ET_o of up to −21.1% for CNRM and −12.3% annually. The largest and most consistent contributors to those negative biases appear to be the well-documented [9] cold model biases for both T_{max} and T_{min}. However, biases in all five meteorological variables act synergistically to result in a large underestimation of ET_o during October–December in CNRM. Despite ET_o being most sensitive to relative errors in RH_{mean} than in the other variables, smaller relative biases in RH_{mean} in CNRM compared to other variables resulted in a smaller contribution of RH_{mean} to errors in ET_o.

In CESM, the cold biases for T_{max} and T_{min} are compensated by positive contributions to ET_o bias from other meteorological variables during certain parts of the year, resulting in overall lower annual bias (−0.5%) than CNRM, and monthly biases ranging from −9.8% to 7.2%. Therefore, only bias correcting model output for precipitation and temperature, as is commonly done due to the widespread availability of observational gridded datasets for these two variables, is not always appropriate. Here, it would decrease the biases in ET_o for CNRM but increase them for CESM. Although the existence of a cold bias in the downscaled models has been identified in other studies, it is important to note that some of the apparent errors in the other variables may be due to the different periods being compared between the model historical simulation and the observations.

3.3.2. Causes of Changes in Reference ET Estimated from Model Output for 2040–2060

The monthly SCs at the 2015 irrigated farm locations (Figure 11a) were also combined with the percent change in the simulated basic meteorological variables from 1985–2005 to 2040–2060 to obtain the total contribution of each variable to projected changes in ET_o for the two downscaled GCMs (Figure 11d,e). The individual lines in these plots show the product of the SC for a meteorological variable times the percent change in the variable from 1985–2005 to 2040–2060, which gives the contribution of that variable to the total change in ET_o as a percentage. The brown line shows the total computed change in ET_o ($\frac{dET_o}{ET_o}$ in Equation (3), as a percentage), and the black line shows the sum of the contributions from all five basic meteorological variables to the total change in ET_o ($\sum_i \left(\frac{\partial ET_o}{\partial V_i} \times \frac{V_i}{ET_o} \times \frac{dV_i}{V_i} \right)$ in Equation (3)). The difference between the black and brown lines is the approximation error (*error* in Equation (3)). Both models project an overall 5.6% increase in annual ET_o mainly due to projected increases in T_{max} and T_{min} and a decrease in RH_{mean}; however, the seasonality of the changes is different between the two models, as shown in Figure 11d,e. The projected decline in RH_{mean} indicates a sub-Clausius–Clapeyron relationship (i.e., humidity increasing at rates lower than the 7% predicted by the Clausius–Clapeyron relationship; [38]) for the region which, together with the projected reductions in rainfall, may be a result of decreased moisture convergence into the region during the period 2040–2060. Despite ET_o being most sensitive to relative changes in RH_{mean} than in the other variables, smaller relative changes in RH_{mean} in both models compared to T_{max} and T_{min}, result in similar contributions to changes in ET_o from the three variables. The contribution of increases in u_2 to increases in ET_o is also positive but minor. CNRM shows small changes in R_s during most months, while CESM shows a small decrease in R_s during most of the year. CESM projects larger increases in ET_0 during March, November, and December, increasing the potential for crop water stress during those times.

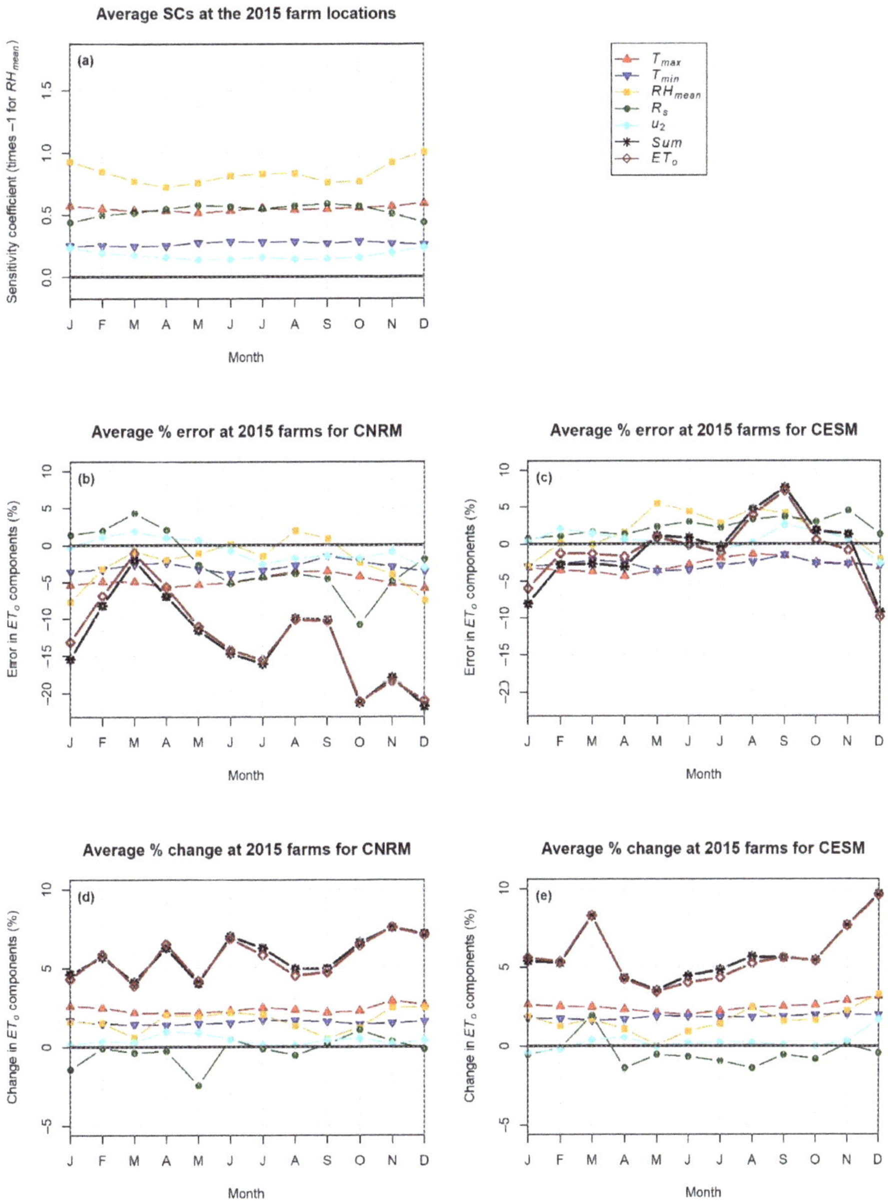

Figure 11. (**a**) Timeseries of SCs at the 2015 irrigated farm locations. (**b**,**c**) Average percent error in ET_0 estimated for the period 1985–2005, contributions from the basic meteorological variables, and their sum for CNRM and CESM, respectively. (**d**,**e**) Average percent change in ET_0 from 1985–2005 to 2040–2060, contributions from the basic meteorological variables, and their sum for CNRM and CESM, respectively.

3.4. Limitations

Some key factors that were not considered in the application of the derived SCs to understanding potential future changes in ET_o (and ET_c) include increases in stomatal resistance and increases in photosynthesis in C3 plants and in C4 plants under drought [39] due to projected increases in atmospheric CO_2 concentrations (CO_2). These changes in plant physiology may counteract some of the expected increases in ET_c from increased evaporative demand (ET_o) due to higher temperatures [40]. Changes in stomatal resistance due to increases in CO_2 can be informed by free-air carbon dioxide enrichment (FACE) experiments [39] and could be applied as modifications to the crop potential transpiration term to obtain estimates of ET_c under future conditions, as in Kruijt et al. [41] and Scarpare et al. [42]. However, crop growth models are needed to better understand potential future changes in ET due to changes in plant biomass from increased photosynthesis and other factors, such as changes in stomatal resistance due to meteorological changes and changes in soil moisture, changes in planting dates and speed of plant growth, and feedbacks between these factors. For example, Scarpare et al. [42] used the CropSyst model to investigate potential future changes in ET_c for five major irrigated crops in the Columbia Basin for 15 bias-corrected and statistically downscaled GCMs under the CMIP5 RCP8.5 scenario. Full irrigation was assumed to avoid crop water stress. For the 2090s, they found that changes in ET_c ranged from −12.5% to +14.3% depending on the crop and the interplay of different meteorological, crop, and management factors. The application of a crop growth model to estimate changes in ET_c for the main crops in Puerto Rico would help clarify potential future changes in agricultural irrigation requirements and aid water management planning efforts.

Additional sources of uncertainty include the effect of neglecting changes in soil heat flux density at the soil surface and changes in water heat flux density over areas with ponding water, which may be important during advective conditions such as cold fronts. In addition, there is an assumption that GOES-PRWEB meteorological and ET_o data are reasonably accurate compared to observations, which may not necessarily be the case given that some of the data are derived from climate models, as described in Harmsen et al. [28]. As previously mentioned, the GOES-PRWEB estimated ET_o, and its driving meteorological variables have been validated by Mecikalski and Harmsen [29] and Harmsen et al. [28,30] at a few stations on the island. The GOES-PRWEB precipitation and T_{max} and T_{min} fields have also been validated at stations throughout the island. However, efforts to validate the GOES-PRWEB dataset in a more comprehensive manner have been limited in particular by the lack of homogeneous, high-quality humidity, solar radiation, and wind speed data at a large number of stations on the island. For solar radiation in particular, some of the existing station data can show sensor issues, such as change points and drifts. In addition, at stations along the interior of the island, the upper envelope of daily solar radiation values is often well below a reasonable clear-sky radiation curve for the time of year, indicating that the station may be shaded by the local canopy or nearby obstructions (buildings, topographic features) for the majority of the year.

Similar concerns occur for wind speed data and are exacerbated by the lack of documented (changes in) sensor heights at meteorological stations. The sensitivity of ET_o to changes in RH_{mean} has been found to increase with increasing wind speed [12]. This is consistent with our finding of moderate autocorrelation between SC for RH_{mean} and u_2 (r of −0.39 in Table 1) which means that as u_2 increases, the SC for RH_{mean} becomes more negative (i.e., the absolute of the SC for RH_{mean} increases). Therefore, errors in u_2 from GOES-PRWEB may affect the relative ranking of SCs across meteorological variables. In particular, an overestimation of u_2 in GOES-PRWEB would result in an overestimation of the SC for RH_{mean}, which was found to be the meteorological variable with the highest relative sensitivity. For this reason, in Appendix S5, wind-speed records at stations throughout the island were compared with GOES-PRWEB wind speeds at the grid cell closest to each station (Table S1 and Figure S12). When sensor height information is available at a station (Table S1), it was used to convert the wind speeds to 2 m heights using the relationship in [10] for consistency with u_2 from GOES-PRWEB. Generally, wind sensors are installed at meteorological stations at 2 to 10 m heights, for which conversion factors to 2 m (based on [10]) range from 1 to 0.75, respectively. Therefore, when sensor height information is not available, a discrepancy of up to about 25% between station wind speeds and u_2 from GOES-PRWEB can be expected on average.

Figure S12b shows that the percent difference between GOES-PRWEB u_2 and sensor wind speed (converted to 2 m when possible) ranges between −23% and 225%. Notably, 16 out of 22 stations have percent differences between +/− 35%, which appear reasonable. However, high percent differences are observed at some stations along the southwest and southeast coasts and along the interior of the island. Furthermore, some of these stations with the higher percent differences are relatively close to other stations with lower percent differences, such as at the Yabucoa stations for example. These large discrepancies in percent differences within short distances may be due to various reasons, including incorrect measurement units, coordinates, or sensor height reported for the station, station sheltering, comparison of 1 km wind speeds from GOES-PRWEB versus point values, and large gradients occurring within small areas. The Global Wind Atlas [43] is a web-based application that provides estimates of wind speed for wind power generation worldwide. It uses the WRF model to downscale reanalysis data in the ERA5 dataset from the European Centre for Medium-Range Forecasts (ECMWF; [44]). The resulting 3 km wind speeds are then generalized and applied to a microscale modeling system to determine local wind climates at 250 m resolution at heights ranging from 10 to 200 m. The microscale model considers orography, roughness and roughness-change effects, including speed-up effects as the wind moves up a mountain or hill. The Global Wind Atlas shows a large gradient in the annual average 10 m wind speed from coastal to offshore areas of Puerto Rico and high variability in wind speeds over mountainous areas resulting from changes in roughness and orography. The 1 km GOES-PRWEB dataset is not expected to capture these localized changes. This likely explains the large changes in wind speed performance over small distances and the overestimation of wind speeds in GOES-PRWEB at interior stations.

Due to the uncertainty in wind speeds and the potential for its overestimation in GOES-PRWEB at some station locations, a sensitivity analysis was performed on the SCs where GOES-PRWEB u_2 was multiplied by factors equal to 0.67 and 0.5 to counteract possible overestimation of wind speeds in GOES-PRWEB by 50% and 100%, respectively. This sensitivity analysis represents the worst-case scenario, where it is assumed that the GOES-PRWEB u_2 is overestimated island wide, when Figure S12 shows that is clearly not the case. Results from this sensitivity analysis are shown in Appendix S5, where Figure 9 is recreated for the two factors (Figures S13 and S14; Table S2). As the wind speed is reduced, the SC for RH_{mean} decreases and the SC for R_s increases. At some locations, such as cells 16 and 45 (Figure 1), the SC for RH_{mean} remains mostly larger than the SC for R_s throughout the year when 0.67 u_2 is used (Figure S13), but only in the winter months when 0.5 u_2 is used (Figure S14). In the case when 0.67 u_2 is used, the SC for RH_{mean} becomes comparable to the SC for R_s at many locations throughout the island, whereas when 0.5 u_2 is used, it becomes smaller than the SC for R_s, as reflected in the island-wide average SCs (Table S2).

As observed in Figure 3c, it was also found that the standard deviation of annual ET_o at the location of 2015 irrigated farms and golf courses is much higher for GOES-PRWEB (97.0 mm/year for 2009–2017) than for the two downscaled climate models (35.7 and 35.6 mm/year for CESM, and CNRM, respectively, in the period 1985–2005) and the Herrera-Ault dataset (24.6 mm/year for 1985–2005). The larger interannual variability in GOES-PRWEB ET_o may be due to the switch of datasets used in its derivation, as described by Harmsen et al. [28]. As previously discussed, the annual cycle of station-based estimates of ET_o appears to match that of GOES-PRWEB better than that of Herrera-Ault (Figure 3b), with peak values in July as in GOES-PRWEB rather than in March as in Herrera-Ault. Since the exact magnitude of interannual variability of ET_o in the study area is unknown, and the downscaled climate models show interannual variability within the range of the two observational datasets, it may be sensible to bias correct model-based ET_o using GOES-PRWEB on monthly timescales only (i.e., no corrections for interannual variability).

4. Discussion

It was found that the SC for RH_{mean} dominates the sensitivity of ET_0 throughout most of the island, even more so than the SCs for R_s and T_{max}. This is because the high relative humidity over most of the island currently limits ET_0 and there is plenty of energy (R_s and T_{max}) available for ET_0 in most areas during most of the year. The SC for RH_{mean} has a large-amplitude annual cycle, especially along the Cordillera Central mountains and the Sierra de Luquillo with larger values and less day-to-day variability in the dry season than in the wet season. Coastal areas, including the 2015 irrigated farm locations, show a lower relative sensitivity to RH_{mean}, with a less marked annual cycle than interior areas. The overall ranking of sensitivities found in our study (SC for RH_{mean} > SC for R_s > SC for T_{max} > SC for T_{min} > SC for u_2) is similar to that obtained for the tropical rainforest climate of southern Nigeria [18] and for the Yangtze River Basin in China [12]. However, the SCs are not very meaningful on their own unless combined with a measure of the variability measurement/estimation error or changes in the basic meteorological variables with time.

Our main objective in deriving SCs for Puerto Rico was to guide the prioritization of bias correction of meteorological output for ET_o estimation from dynamically downscaled climate projections, which can then be used to drive a soil-moisture water-balance model [2] for estimating future agricultural irrigation requirements on the island. This objective was met by applying the SCs to understand the sources of error and potential future changes in ET_o estimated from downscaled model output at the 2015 irrigated farm locations. The results illustrate how the ranking of variable importance may change when the SCs are combined with estimates of error or change in meteorological variables. For example, despite ET_0 being most sensitive to relative errors in RH_{mean}, relative biases in RH_{mean} in the CNRM model are small compared to other variables. This results in a smaller contribution of RH_{mean} to errors in CNRM-estimated ET_o. Similarly, despite ET_0 being most sensitive to relative changes in RH_{mean} than in the other variables, smaller relative future changes in RH_{mean} compared to T_{max} and T_{min} in both climate models result in similar contributions to future changes in ET_o from the three variables.

For the period 2040–2060, it was found that both climate models project an overall 5.6% increase in annual ET_0 over the 2015 irrigated farm locations with respect to the historical period 1985–2005. This increase in ET_0 is primarily a result of projected increases in T_{max} and T_{min} and a decrease in RH_{mean}, with differences in the seasonality of changes between the two models. Absent compensatory increases in rainfall and/or a compensatory decrease in ET_0 due to stomatal closure as CO_2 increases, the future increase in ET_c resulting from the 5.6% increase in ET_o may have to be met by an additional volume of irrigation. Future research could attempt to estimate the additional agricultural irrigation requirements and their sensitivity to various bias-correction choices.

Various limitations of the current study are highlighted in the results. They include uncertainties in the stomatal and plant-growth responses to projected increases in atmospheric CO_2, uncertainties in crop suitability under climate change, the effect of neglecting changes in soil and water heat fluxes in the ET_0 equation, and the lack of sufficient meteorological data at weather stations to corroborate meteorological data and ET_0 estimates and the derived sensitivity coefficients. The development of reliable homogeneous spatially distributed long-term observational datasets of meteorological and hydrological variables is imperative for improving the understanding of drivers of hydrological processes, evaluating model performance, and for bias correction of model output.

Supplementary Materials: The following supporting information can be downloaded at: https://www.mdpi.com/article/10.3390/hydrology10050101/s1, Appendix S1: Aerodynamic and energetic components of the ASCE standardized reference evapotranspiration (ET_o) equation; Appendix S2: Computation of daily meteorological variables from hourly WRF output; Appendix S3: Analytical sensitivity coefficient equations; Appendix S4: Maps of long-term (2009–2017) monthly average meteorological variables and ET_o components; Appendix S5: Performance of GOES-PRWEB wind speed at 2 meters and sensitivity to errors in wind speed; Figure S1: Long-term (2009–2017) monthly average mean relative humidity (RH_{mean}); Figure S2: Long-term (2009–2017) monthly average incoming solar radiation (R_s); Figure S3: Long-term (2009–2017) monthly average maximum air temperature (T_{max}); Figure S4: Long-term (2009–2017) monthly average minimum air temperature (T_{min}); Figure S5: Long-term (2009–2017) monthly average wind speed (u_2); Figure S6: Long-term (2009–2017) monthly average daily rainfall rate; Figure S7: Long-term (2009–2017) monthly average energetic component or reference evapotranspiration ($ET_{o,energetic}$); Figure S8: Long-term (2009–2017) monthly average aerodynamic component of reference evapotranspiration ($ET_{o,aerodynamic}$); Figure S9: Long-term (2009–2017) monthly average energetic fraction of reference evapotranspiration ($ET_{o,energetic\ fraction}$); Figure S10: Long-term (2009–2017) monthly average aerodynamic fraction of reference evapotranspiration ($ET_{o,aerodynamic\ fraction}$); Figure S11: Long-term (2009–2017) monthly average daily evapotranspiration (ET_o); Figure S12: (a) Location of stations with homogeneous wind speed data in Puerto Rico; (b) Percent difference, and (c) coefficient of determination (R^2), between GOES-PRWEB and station wind speeds converted to 2-m height according to [10] when sensor elevation data are available; Figure S13: Daily timeseries of sensitivity coefficients (SCs) with wind speed multiplied by 0.67; Figure S14: Daily timeseries of sensitivity coefficients (SCs) with wind speed multiplied by 0.5; Table S1: Stations with homogeneous wind speed data in Puerto Rico; Table S2: Island-wide average absolute relative SCs for case with original GOES-PRWEB 2 m wind speeds and wind speeds multiplied by factors equal to 0.5 and 0.67. References [45–50] are cited in the supplementary materials.

Author Contributions: Conceptualization, M.I.-O.; methodology, M.I.-O.; software, M.I.-O.; validation, M.I.-O.; formal analysis, M.I.-O.; investigation, M.I.-O.; resources, M.I.-O.; data curation, M.I.-O. and E.W.H.; writing—original draft preparation, M.I.-O.; writing—review and editing, M.I.-O. and E.W.H.; visualization, M.I.-O.; supervision, M.I.-O. and E.W.H.; project administration, M.I.-O.; funding acquisition, M.I.-O. All authors have read and agreed to the published version of the manuscript.

Funding: This research was partially funded by the USGS, USDA-NIFA (W4128), and NSF (grant 1832576). Findings, opinions, or conclusions expressed in this paper do not necessarily reflect those of the USDA or NSF.

Data Availability Statement: GOES-PRWEB data can be found here: https://pragwater.com (accessed on 31 August 2021). Weather Research and Forecasting (WRF): Puerto Rico and US Virgin Islands Dynamical Downscaled Climate Change Projections are publicly available at https://doi.org/10.5066/F7GB23BW (accessed on 23 March 2021).

Conflicts of Interest: The authors declare no conflict of interest. The funders had no role in the design of the study; in the collection, analyses, or interpretation of data; in the writing of the manuscript; or in the decision to publish the results.

References

1. Puerto Rico Institute of Statistics. Food Security in Puerto Rico, Calendar Year 2015. 2019. Available online: https://estadisticas.pr/files/Comunicados/Seguridad%20Alimentaria%20en%20Puerto%20Rico%20-%20Final%20%28300519%29.pdf (accessed on 20 July 2022).
2. Molina-Rivera, W.L.; Irizarry-Ortiz, M.M. Estimated water withdrawals and use in Puerto Rico. *U.S. Geol. Surv. Open-File Rep.* **2021**, *2021–1060*, 1–38.
3. Puerto Rico Planning Board. Memorial of the Land Use Plan, Guide for the Classification of Territories. 2015. Available online: https://www.sciencebase.gov/catalog/item/576bfe89e4b07657d1a26ee5 (accessed on 8 February 2023).
4. Puerto Rico Department of Natural and Environmental Resources. Integrated Water Resources Plan for Puerto Rico. 2016. Available online: https://drna.pr.gov/wp-content/uploads/formidable/PIRA-2016.pdf (accessed on 8 February 2023).
5. Harmsen, E.W.; Howard-Harmsen, R. Agricultural Water Management and Puerto Rico's Food Insecurity. *Ethos Gubernamental.* Special Edition. September 2019. Available online: https://academic.uprm.edu/hdc/HarmsenPapers/Harmsen_and_Harmsen_Ethos.pdf (accessed on 18 July 2022).

6. NOAA (National Oceanographic and Atmospheric Administration). National Integrated Drought Information System. Available online: https://www.drought.gov/states/puerto-rico (accessed on 20 July 2022).
7. Herrera, D.A.; Ault, T.R.; Fasullo, J.T.; Coats, S.J.; Carrillo, C.M.; Cook, B.I.; Park Williams, A. Exacerbation of the 2013–2016 pan-Caribbean drought by anthropogenic warming. *J. Geophys. Res.* **2018**, *45*, 10619–10626. [CrossRef] [PubMed]
8. Khalyani, A.H.; Gould, W.A.; Harmsen, E.; Terando, A.; Quinones, M.; Collazo, J.A. Climate change implications for tropical islands: Interpolating and interpreting statistically downscaled GCM projections for management and planning. *J. Appl. Meteorol. Climatol.* **2016**, *55*, 265–282. [CrossRef]
9. Bowden, J.; Terando, A.J.; Misra, V.; Wootten, A.; Bhardwaj, A.; Boyles, R.; Gould, W.; Collazo, J.A.; Spero, T.L. High-resolution dynamically downscaled rainfall and temperature projections for ecological life zones within Puerto Rico and for the U.S. Virgin Islands. *Int. J. Climatol.* **2021**, *41*, 1305–1327. [CrossRef]
10. ASCE (American Society of Civil Engineers); EWRI (Environmental and Water Resources Institute). Task Committee on Standardization of Reference Evapotranspiration. The ASCE Standardized Reference Evapotranspiration Equation. 2005. Available online: https://ascelibrary.org/doi/book/10.1061/9780784408056 (accessed on 20 April 2023).
11. Irmak, S.; Payero, J.O.; Martin, D.L.; Irmak, A.; Howell, T.A. Sensitivity analyses and sensitivity coefficients of standardized daily ASCE Penman–Monteith equation. *J. Irrig. Drain Eng.* **2006**, *132*, 564–578. [CrossRef]
12. Gong, L.; Xu, C.-y.; Chen, D.; Halldin, S.; Chen, Y.D. Sensitivity of the Penman–Monteith reference evapotranspiration to key climatic variables in the Changjiang (Yangtze River) basin. *J. Hydrol.* **2006**, *329*, 620–629. [CrossRef]
13. McKenney, M.S.; Rosenberg, N.J. Sensitivity of some potential evapotranspiration estimation methods to climate change. *Agric. For. Meteorol.* **1993**, *64*, 81–110. [CrossRef]
14. Debnath, S.; Adamala, S.; Raghuwanshi, N.S. Sensitivity analysis of FAO-56 Penman–Monteith method for different agro-ecological regions of India. *Environ. Process.* **2015**, *2*, 689–704. [CrossRef]
15. Cao, L.; Zhang, Y.; Shi, Y. Climate change effect on hydrological processes over the Yangtze River basin. *Quat. Int.* **2011**, *244*, 202–210. [CrossRef]
16. Liu, Q.; Yang, Z.; Cui, B.; Sun, T. The temporal trends of reference evapotranspiration and its sensitivity to key meteorological variables in the Yellow River Basin, China. *Hydrol. Process.* **2010**, *24*, 2171–2181. [CrossRef]
17. Biazar, S.M.; Dinpashoh, Y.; Singh, V.P. Sensitivity analysis of the reference crop evapotranspiration in a humid region. *Environ. Sci. Pollut. Res.* **2019**, *26*, 32517–32544. [CrossRef] [PubMed]
18. Emeka, N.; Ikenna, O.; Okechukwu, M.; Chinenye, A.; Emmanuel, E. Sensitivity of FAO Penman–Monteith reference evapotranspiration (ETo) to climatic variables under different climate types in Nigeria. *J. Water Clim. Chang.* **2021**, *12*, 858–878. [CrossRef]
19. Hou, L.-g.; Zou, S.-b.; Xiao, H.-l.; Yang, Y.-g. Sensitivity of the reference evapotranspiration to key climatic variables during the growing season in the Ejina oasis northwest China. *SpringerPlus* **2013**, *2* (Suppl. S1), S4. [CrossRef]
20. Luo, Y.; Gao, P.; Mu, X. Influence of meteorological factors on potential evapotranspiration in Yanhe River Basin, China. *Water* **2021**, *13*, 1222. [CrossRef]
21. Wang, T.; Zhang, J.; Sun, F.; Liu, W. Pan evaporation paradox and evaporative demand from the past to the future over China: A review. *WIREs Water* **2017**, *4*, e1207. [CrossRef]
22. Peterson, T.; Golubev, V.S.; Groisman, P.Y. Evaporation losing its strength. *Nature* **1995**, *377*, 687–688. [CrossRef]
23. Roderick, M.L.; Rotstayn, L.D.; Farquhar, G.D.; Hobbins, M.T. On the attribution of changing pan evaporation. *Geophys. Res. Lett.* **2007**, *34*, 251–270. [CrossRef]
24. USGS (U.S. Geological Survey). Climate of Puerto Rico. 2016. Available online: https://www.usgs.gov/centers/caribbean-florida-water-science-center-%28cfwsc%29/science/climate-puerto-rico (accessed on 20 July 2022).
25. Angeles, M.E.; González, J.E.; Ramírez-Beltrán, N.D.; Tepley, C.A.; Comarazamy, D.E. Origins of the Caribbean rainfall bimodal behavior. *J. Geophys. Res.* **2010**, *115*, D11106. [CrossRef]
26. Muñoz, M.M.; Lugo, W.I.; Santiago, C.; Matos, M.; Ríos, S.; Lugo, J. Taxonomic Classification of the Soils of Puerto Rico, 2017. University of Puerto Rico, Mayagüez Campus, College of Agricultural Sciences, Agricultural Experiment Station Bulletin 313. January 2018. Available online: https://www.uprm.edu/tamuk/wp-content/uploads/sites/299/2019/06/Taxonomic_classification_soils_PR_2018_reduced.pdf (accessed on 20 July 2022).
27. Dixon, J.F.; Molina-Rivera, W.L.; Irizarry-Ortiz, M.M.; Christesson, K.R. Spatial and Tabular Datasets of Water Withdrawals and Use in Puerto Rico, 2015. U.S. Geological Survey Data Release 2021. Available online: https://www.sciencebase.gov/catalog/item/600aec34d34e162231fb26d9 (accessed on 9 August 2022).
28. Harmsen, E.W.; Mecikalski, J.R.; Reventos, V.J.; Alvarez Perez, E.; Uwakweh, S.S.; Adorno Garcia, C. Water and energy balance model GOES-PRWEB: Development and Validation. *Hydrology* **2021**, *8*, 113. [CrossRef]
29. Mecikalski, J.R.; Harmsen, E.W. The use of visible geostationary operational meteorological satellite imagery in mapping the water balance over Puerto Rico for water resource management. In *Satellite Information Classification and Interpretation;* Rustamov, R.B., Ed.; IntechOpen: London, UK, 2019.
30. Harmsen, E.W.; Tosado Cruz, P.; Mecikalski, J.R. Calibration of selected pyranometers and satellite derived solar radiation in Puerto Rico. *Int. J. Renew. Energy Technol.* **2014**, *5*, 43–54. [CrossRef]
31. Herrera, D.; Ault, T. Insights from a new high-resolution drought Atlas for the Caribbean spanning 1950–2016. *J. Clim.* **2017**, *30*, 7801–7825. [CrossRef]

32. Allen, R.G.; Pereira, L.S.; Raes, D.; Smith, M. Crop evapotranspiration: Guidelines for computing crop water requirements. *FAO Irrig. Drain. Paper* **1998**, *56*. Available online: https://www.fao.org/3/x0490e/x0490e00.htm (accessed on 18 July 2022).
33. Bowden, J.; Wootten, A.; Terando, A.; Boyles, R. Weather Research and Forecasting (WRF): Puerto Rico and US Virgin Islands Dynamical Downscaled Climate Change Projections. U.S. Geological Survey Data Release 2018. Available online: https://www.sciencebase.gov/catalog/item/5977b95de4b0ec1a4888e493 (accessed on 23 March 2021).
34. McCuen, R.H. A sensitivity and error analysis of procedures for estimating evaporation. *Water Resour. Bull.* **1974**, *10*, 486–498. [CrossRef]
35. Harmsen, E.W.; Goyal, M.R.; Torres-Justiniano, S. Estimating evapotranspiration in Puerto Rico. *J. Agric. Univ. Puerto Rico* **2002**, *86*, 35–54. [CrossRef]
36. Tabari, H.; Talaee, P.H. Sensitivity of evapotranspiration to climatic change in different climates. *Glob. Planet. Chang.* **2014**, *115*, 16–23. [CrossRef]
37. Murphy, S.; Stallard, R.F.; Scholl, M.A.; González, G.; Torres-Sánchez, A.J. Reassessing rainfall in the Luquillo Mountains, Puerto Rico: Local and global ecohydrological implications. *PLoS ONE* **2017**, *12*, e0180987. [CrossRef] [PubMed]
38. Koutsoyiannis, D. Revisiting the global hydrological cycle: Is it intensifying? *Hydrol. Earth Syst. Sci.* **2020**, *24*, 3899–3932. [CrossRef]
39. Ainsworth, E.A.; Long, S.P. 30 years of free-air carbon dioxide enrichment (FACE): What have we learned about future crop productivity and its potential for adaptation? *Glob. Chang. Biol.* **2020**, *27*, 27–49. [CrossRef]
40. Islam, A.; Ahuja, L.R.; Garcia, L.A.; Ma, L.; Sassendran, A.S. Modeling the effect of elevated CO_2 and climate change on reference evapotranspiration in the semi-arid Central Great Plains. *Trans. ASABE* **2012**, *55*, 2135–2146. [CrossRef]
41. Kruijt, B.; Witte, J.-P.M.; Jacobs, C.; Kroon, T. Effects of rising atmospheric CO_2 on evapotranspiration and soil moisture: A practical approach for the Netherlands. *J. Hydrol.* **2008**, *349*, 257–267. [CrossRef]
42. Scarpare, F.V.; Rajagopalan, K.; Liu, M.; Nelson, R.L.; Stöckel, C.O. Evapotranspiration of irrigated crops under warming and elevated atmospheric CO_2: What is the direction of change? *Atmosphere* **2022**, *13*, 163. [CrossRef]
43. Global Wind Atlas Version 3.0. Available online: https://globalwindatlas.info/en/area/Puerto%20Rico/ (accessed on 30 January 2023).
44. Hersbach, H.; Bell, B.; Berrisford, P.; Hirahara, S.; Horányi, A.; Muñoz-Sabater, J.; Nicolas, J.; Peubey, C.; Radu, R.; Schepers, D.; et al. The ERA5 global reanalysis. *Q. J. R. Meteorolog. Soc.* **2020**, *146*, 1999–2049. [CrossRef]
45. Wallace, J.M.; Hobbs, P.V. *Atmospheric Science—An Introductory Survey*, 2nd ed.; Elsevier: New York, NY, USA, 2006; p. 483.
46. American Meteorological Society. Glossary of Meteorology. 2012. Available online: https://glossary.ametsoc.org/wiki/Mixing_ratio (accessed on 8 February 2023).
47. CariCOOS (Caribbean Coastal Ocean Observing System). Weather Flow Mesonet Meteorological Station Aggregation Data. Available online: http://52.55.122.42/thredds/caricoos_mesonet_agg.html (accessed on 13 September 2021).
48. IOOS (Integrated Ocean Observing System). Environmental Sensor Map. Available online: https://sensors.ioos.us/#map (accessed on 18 November 2021).
49. NOAA (National Oceanic and Atmospheric Administration); NCEI (National Centers for Environmental Information). Global Summary of the Day—GSOD. User Engagement and Services Branch. DOC/NOAA/NESDIS/NCDC > National Climatic Data Center, NESDIS, NOAA, U.S. Department of Commerce. Available online: https://www.ncei.noaa.gov/cdo-web/ (accessed on 16 November 2021).
50. WRCC (Western Regional Climate Center). National Interagency Fire Center Remote Automatic Weather Stations (RAWS). Available online: https://raws.dri.edu/prF.html (accessed on 10 November 2021).

water

MDPI

Article

Effect of Soil Texture on Water and Salt Transport in Freeze—Thaw Soil in the Shallow Groundwater Area

Erqing Wang [1], Junfeng Chen [1,*], Lei Liu [1], Lihong Cui [1], Jing Xue [1], Jiameng Ren [1] and Qi Du [2]

[1] College of Water Resources and Engineering, Taiyuan University of Technology, Taiyuan 030024, China; wangerqing2021@163.com (E.W.); liulei4525@163.com (L.L.); cuilihong@tyut.edu.cn (L.C.); xuejing@tyut.edu.cn (J.X.); renjiameng0619@163.com (J.R.)

[2] Taigu Water Balance Experimental Field, Hydrology and Water Resources Survey Station of Shanxi Province, Jinzhong 030800, China; duqi0416@163.com

* Correspondence: chenjunfeng@tyut.edu.cn; Tel.: +86-139-3424-7805

Abstract: Research on the variation in soil water, heat, and salt in unsaturated zones during the freeze–thaw process has great significance in efficiently utilizing water resources and preventing soil salinization. The freeze–thaw field experiment was carried out with the lysimeter as the test equipment to analyze characteristics of the soil freeze–thaw process, profile water content, main ion content, and salt content of three textured soils with the groundwater table depth of 0.5 m. The results showed that the soil temperature gradient and freezing depth were greater as the average soil particle size increased. The increment of water content at the depth of 0 to 30 cm in sandy loam and loamy sand decreased by 40.20~93.10% and 28.14~65.52% compared with that in sandy soil, and the average increment of salt content at the depth of 0 to 30 cm decreased as the average soil particle size increased during the freeze–thaw period. The average content of Ca^{2+}, Na^{+}, Cl^{-}, and $SO_4{}^{2-}$ in loamy sand and sandy soil decreased by 4.37~45.50% and 22.60~70.42% compared with that in sandy loam at the end of the freeze–thaw period, and the correlation between soil salt content and water content decreased with the increase in the average soil particle size. The research results can provide a theoretical basis for soil salinization prevention and crop production in shallow groundwater areas.

Keywords: freeze–thaw period; ion content; soil salt content; soil texture

Citation: Wang, E.; Chen, J.; Liu, L.; Cui, L.; Xue, J.; Ren, J.; Du, Q. Effect of Soil Texture on Water and Salt Transport in Freeze—Thaw Soil in the Shallow Groundwater Area. *Water* **2023**, *15*, 2587. https://doi.org/10.3390/w15142587

Academic Editors: Songhao Shang, Qianqian Zhang, Dongqin Yin, Hamza Gabriel and Magdy Mohssen

Received: 9 June 2023
Revised: 8 July 2023
Accepted: 14 July 2023
Published: 16 July 2023

1. Introduction

Most of the northern regions of China are seasonally frozen soil areas, where the transformation of phreatic water in the shallow groundwater zone is unusually strong [1]. The salt in the groundwater is transported to the unsaturated zone as the phreatic water migrates upward under the action of the soil water potential gradient during the freezing process [2]. It is easy to form soil salinization due to drought and intense evaporation in winter and spring [3,4]. The redistribution of water and salt in shallow groundwater areas caused by freezing and thawing processes is one of the main causes of soil salinization [5,6]. Salinized land is widely distributed in the world, and there are many types of it due to different soil textures and hydrological characteristics between regions [7]. Soil salinization not only affects the balance of the ecological environment but also restricts the virtuous cycle of the natural environment [8] and limits crop growth and production [9]. Therefore, the in-depth study of the migration law of water and salt in the unsaturated zone of soils with different textures in shallow groundwater areas under freeze–thaw action can provide a theoretical basis for the prevention of soil salinization in different regions, which has great significance for the efficient utilization of water resource and agricultural production in seasonally frozen soil areas.

The migration of soil salt under freeze–thaw action is the result of the combined action of water convection, concentration gradient, and temperature gradient [5]; soil salt moves convectively along with soil water under the action of the matrix potential gradient and temperature gradient, and it diffuses under the action of the concentration potential gradient. Scholars have explored the transport law of soil salt in the unsaturated zone during the freeze–thaw period through experimental research [10,11] and numerical simulation [12–14]. It was found that surface mulching [15,16], autumn irrigation [13], and barrier measures [17] could inhibit the accumulation of soil salt on the surface, while soil bulk density [18], initial soil salinization degree [19], the freeze–thaw cycle [20], particle gradation [21], and initial soil water content [22] also affected the variation in salt content in the soil profile during the freeze–thaw period.

Soil texture affects soil hydraulic properties [23] and thermal conductivity [24], which relate to the distribution of soil water, heat, and salt. In non-freeze–thaw periods, a lot of research has been carried out on the effect of soil texture on soil salt transport under different experimental conditions [25–28], under the condition of salt water irrigation, the soil texture affected the process of salt stress [29], and soil salt content was significantly affected by clay content [30]. The thickness of the interlayer [31], the constitution of the soil layer [26], and the soil texture under the intermittent irrigation condition [32] also affected the migration and distribution of water and salt in the soil. In addition, soil texture affected the degree of retardation of salt on water migration, and the inhibition effect of salt on evaporation was stronger in homogeneous soil than that in heterogeneous soil [33]. However, the transformation of phreatic water and the corresponding salt transport become extremely complex in the freeze–thaw period; quantitative analyses of the influence of soil texture on soil salt content and the correlation between soil water and salt at different depths have yet to be performed, but some research has been conducted on the transformation of phreatic water [34,35]. The inadequate understanding of the law of salt transport in different textured soils during the freeze–thaw period prompted this study.

The freeze–thaw cycle of soil is a complex process, accompanied by physical, chemical, and mechanical effects such as heat conduction, water phase change, solute transport, etc. The distribution of soil pollutants [36–41] and the climate types in different regions and different periods in the same region are complex and variable, and therefore, outdoor experiments and numerical simulations are conducted more frequently to monitor and simulate the transport process of water and solute in freeze–thaw soils under conditions of natural environmental change. Multiple sets of outdoor experiments were carried out during a freeze–thaw period in this paper. Only by comprehensively understanding the migration path and evolution law of soil salt in unsaturated zones of different textured soils can we grasp the dynamics of soil salt migration in different regions and make reasonable use of and improvements in salinized land.

An experimental study on the variation in soil water, heat, and salt in the unsaturated zone of shallow groundwater areas under freeze–thaw action was carried out; the study discussed the spatial and temporal variation in soil temperature and soil water content, analyzed the effect of soil texture on the migration of soil anions, cations, and salt during the freeze–thaw period, and explored the correlation between soil salt content and water content at different depths via Pearson correlation analysis, all of which revealed the characteristics of variation in water content and salt content in different textured soils and provided a basis for the rational planning and utilization of salinized land.

2. Materials and Methods

2.1. Experimental Conditions

The field experiment was carried out from November 2020 to March 2021 at the Taigu Water Balance Experimental Field of the Hydrology and Water Resources Survey Station in the Jinzhong Basin, Shanxi Province (Figure 1), geographically located at 112°30′32.58″ E and 37°26′11.74″ N, with an altitude of 777.0 m and a ground slope of 0.3%. The area is characterized by a continental semi-arid climate, with an annual average temperature of 9.95 °C, annual precipitation of 415 mm, mainly concentrated in June to September, and an annual average water surface evaporation of 1627.9 mm. The annual average freezing depth is 34~70 cm, the historical maximum freezing depth is 92 cm, the annual average relative humidity is 74%, the annual average wind speed is 0.9 m/s, and the annual average frost-free period is 200 days. The variation in daily average air temperature, relative humidity, precipitation, wind speed, and solar radiation during the experimental period is shown in Figure 2.

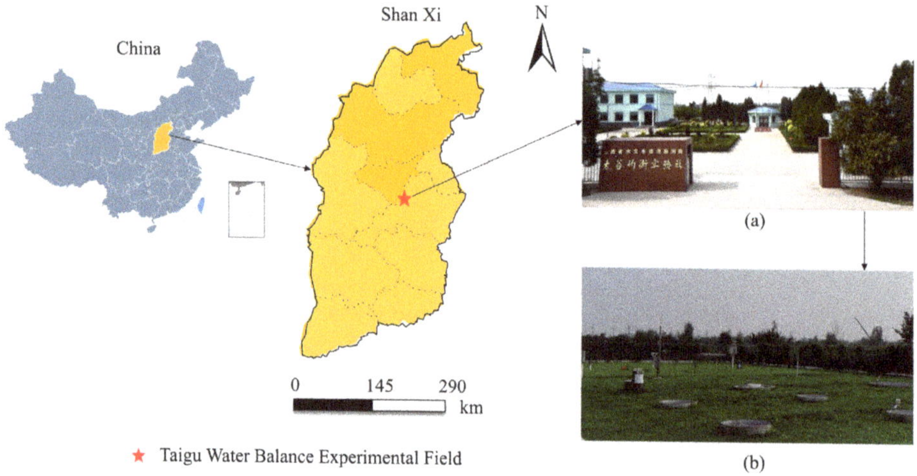

Figure 1. Location of Taigu Water Balance Experimental Field. (**a**) Field station and (**b**) automatic weather station.

2.2. Experimental Methods

The experimental equipment consisted of lysimeters (Figure 3), which were cylindrical in cross-section with a cross-sectional area of 0.5 m^2, and the interval between different lysimeters was 0.5 m. The lysimeter was filled with three representative homogeneous soils in Taiyuan Basin, which were classified as sandy loam, loamy sand, and sandy soil according to the USDA classification system. The physical parameters of three textured soils are shown in Table 1, and the soil surface was exposed to the atmosphere. Mariotte bottles were used to control the groundwater table depth to maintain 0.5 m in the lysimeter system and were recharged with local groundwater.

Table 1. Main physical parameters of soils with different textures.

Soil Texture	Mass Percentage of Different Textured Soils (%)			Maximum Capillary Height (cm)	Specific Yield (m^3/m^3)	Bulk Density ($g{\cdot}cm^{-3}$)
	Clay	Silt	Sand			
Sandy Loam	16.4	27.5	56.1	185	0.08	1.45
Loamy Sand	7.3	7.5	85.2	77	0.18	1.55
Sandy Soil	4.1	5.9	90	60	0.21	1.59

(1) The monitoring of soil temperature profile

The soil temperature was monitored by thermistors embedded before winter at depths of 0 cm, 5 cm, 10 cm, 20 cm, 30 cm, 40 cm, and 50 cm, respectively, and the soil temperature was monitored at intervals of 5~7 days at around 9 a.m. from 1 November 2020 to 6 March 2021. The relationship between the thermistor value and soil temperature is shown in Equation (1).

$$T = \frac{B}{\ln\frac{R_x}{R_{25}} + \frac{B}{298.15}} - 273.15, \quad (1)$$

where T is the soil temperature, °C; R_x is the measured thermistor value, Ω; R_{25} is the thermistor value at a temperature of 25 °C, Ω; and B is a constant, generally taken to be 3950.

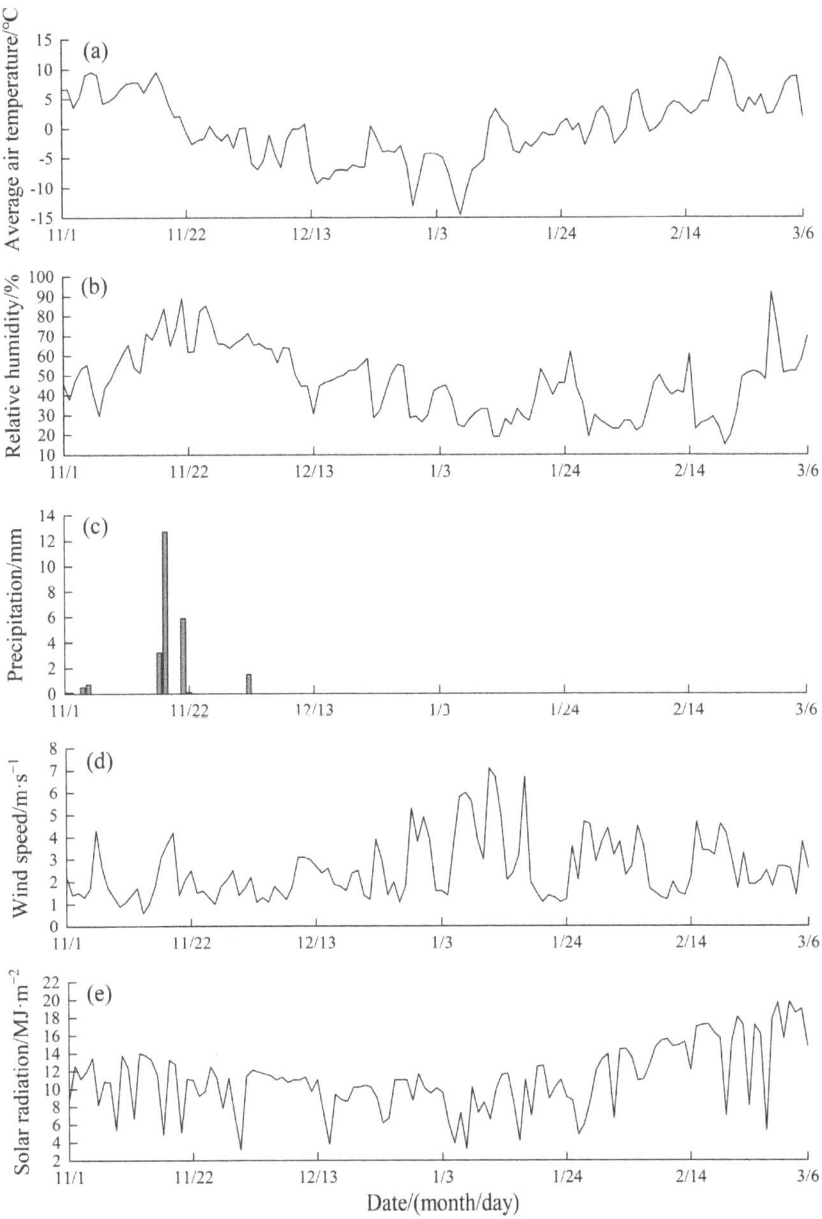

Figure 2. Meteorological conditions during the experimental period. Variations in (**a**) average air temperature, (**b**) relative humidity, (**c**) precipitation, (**d**) wind speed, and (**e**) solar radiation from November 2020 to March 2021.

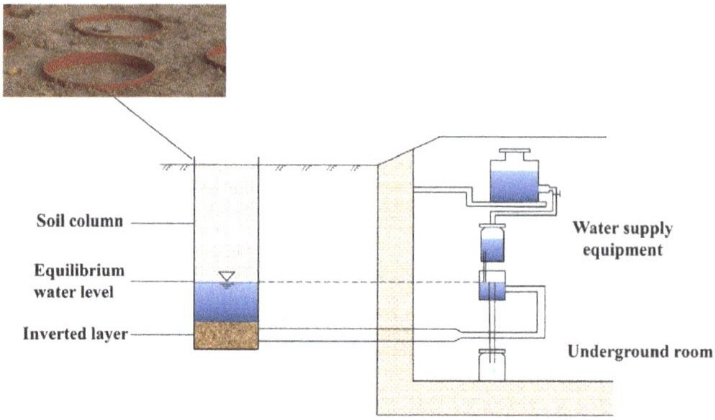

Figure 3. Schematic diagram of the lysimeter.

(2) The monitoring of soil water content and salt content

The monitoring depth of soil water content, anion and cation content, and salt content was the same as that of the soil temperature profile. Soil samples were obtained at different depths using a soil drill and then sealed in aluminum boxes; each soil sample was placed in an oven and then dried at a constant temperature of 105 °C for 8 h to obtain the soil water content. The soil solution was configured using deionized water according to a soil-to-water ratio of 1:5 after drying, and the clarified soil extract obtained after shaking and filtering was used to measure the soil ion content, among which the content of K^+, Ca^{2+}, Na^+, and Mg^{2+} was determined using an atomic spectrophotometer (TAS-990AFG) (Beijing Purkinje GENERAL Instrument Co., Ltd., Beijing, China) under the following conditions: wavelength ranged from 190 to 900 nm; grating scribing was 1200 or 1800; and the accuracy and repeatability of wavelength was ±0.25 nm and 0.15 nm, respectively. The content of Cl^-, NO^{3-}, and $SO_4{}^{2-}$ was determined via ion chromatograph (883 Basic IC plus) (Metrohm China, Hong Kong, China), which consists of iCulumn, iPump, and iDetectror: the pump speed of iPump ranged from 0.001 to 20 mL/min, the measurement range and accuracy of the iDetectror was 0~15,000 µs/cm and 0.0047 ns/cm, respectively. The content of HCO^{3-} and $CO_3{}^{2-}$ was determined by titration. The soil salt content was the sum of each ion content. The absorbance of potassium, sodium, calcium, and magnesium ions was measured at wavelengths of 766.4 nm, 589.0 nm, 422.7 nm, and 285.2 mm, respectively, to obtain the standard curve. Then, a 10.00 mL sample was taken into the colorimetric tube, and 0.50 mL of cesium nitrate solution (China National Pharmaceutical Group Shanxi Co., Ltd., Taiyuan, China) (K^+, Na^+) or 0.2 mL of lanthanum nitrate solution (China National Pharmaceutical Group Shanxi Co., Ltd., Taiyuan, China) (Ca^{2+}, Mg^{2+}) was added to measure the absorbance of each ion, and the ion concentration was obtained by comparing the standard curve. The determination range of K^+, Na^+, Ca^{2+}, and Mg^{2+} was 0.01~4.0, 0.01~4.0, 0.1~6.0, and 0.01~0.6 mg/L, respectively, and that of Cl^-, NO^{3-}, and $SO_4{}^{2-}$ was 0.1~12, 0.02~10, and 0.2~12 mg/L, respectively.

2.3. Grey Relation Analysis

The correlation analysis in grey system theory is a new method of factor analysis, which mainly analyzes the degree of correlation between various factors in the system by comparing the geometric relationship of system data sequences [42,43].

(1) Original data transformation. Because each sequence has different dimensions and orders of magnitude, in order to obtain correct analysis results, the original data should be dimensionless to facilitate the analysis and calculation, and the standardized transformation is used in this paper; that is, the average value and standard deviation of each sequence are first calculated, and the average value of each original data is subtracted

from the original data and then divided by the standard deviation. The new data sequences $x_0(k)$ and $x_i(k)$ obtained are standardized sequences.

(2) Correlation coefficient calculation. The calculation formula is as follows (Equation (2)):

$$\delta_i(k) = r(x_0(k), x_i(k)) = \frac{\underset{i\in m}{\min}\underset{k\in n}{\min}|x_0(k) - x_i(k)| + \rho\underset{i\in m}{\max}\underset{k\in n}{\max}|x_0(k) - x_i(k)|}{|x_0(k) - x_i(k)| + \rho\underset{i\in m}{\max}\underset{k\in n}{\max}|x_0(k) - x_i(k)|}, \quad (2)$$

where $\delta_i(k)$ is the relative difference between the comparison curve x_i and the reference curve x_0 at the kth moment, that is, the correlation coefficient of the data series of x_i and x_0 at the kth moment. The value of ρ, which is the resolution coefficient, is generally between 0 and 1, where $\rho = 0.5$ in the article.

(3) Correlation degree calculation. With the calculation formula for the correlation coefficient, according to the grey correlation space, the calculation formula for correlation degree is shown in Equation (3).

$$r_i = \frac{1}{n}\sum_{k=1}^{n}\delta_i(k), \quad (3)$$

where r_i is the correlation degree between two sequences, and N is the number of each subsequence.

2.4. Pearson Correlation Analysis

The Pearson correlation coefficient is used to reflect the degree of linear correlation between two random variables and describe the trend of two sets of linear data changing together. The value range is between −1 and 1. Given two random variables, the Pearson correlation coefficient is equal to the covariance of the two variables divided by the standard deviation of the two variables (Equation (4)). SPSS 19.0 is used to calculate the Pearson correlation coefficient in this article.

$$\rho_{X,Y} = \frac{cov(X,Y)}{\sigma_X\sigma_Y} = \frac{E[(X-\mu_X)(Y-\mu_Y)}{\sigma_X\sigma_Y}, \quad (4)$$

where $Cov(X,Y)$ is the covariance between X and Y; and σ_X and σ_Y are the standard deviation.

3. Results

3.1. Freeze–Thaw Processes of Different Textured Soils

Soil freezing and thawing processes significantly affect the migration and distribution of water and salt in the soil. According to the freeze–thaw characteristic of soil (Figure 4), the freeze–thaw process was divided into three stages [44,45]: unstable freezing stage, stable freezing stage, and thawing stage, as shown in Table 2. The representative 1 day of three freeze–thaw stages, which was 18 November 2020 (D1), 9 January 2021 (D2), and 13 February 2021 (D3), respectively, was selected to analyze various characteristics of water content and salt content in the soil profile during the freeze–thaw period.

Table 2. Division of soil freeze–thaw period.

Soil Freezing and Thawing Stage	Soil Texture		
	Sandy Loam	Loamy Sand	Sandy Soil
Unstable Freezing Stage	From 18 November to 5 December 2020	From 18 November to 3 December 2020	From 18 November to 3 December 2020
Stable Freezing Stage	From 6 December 2020 to 12 January 2021	From 4 December 2020 to 9 January 2021	From 4 December 2020 to 10 January 2021
Thawing Stage	From 13 January to 13 February 2021	From 10 January to 14 February 2021	From 11 January to 15 February 2021

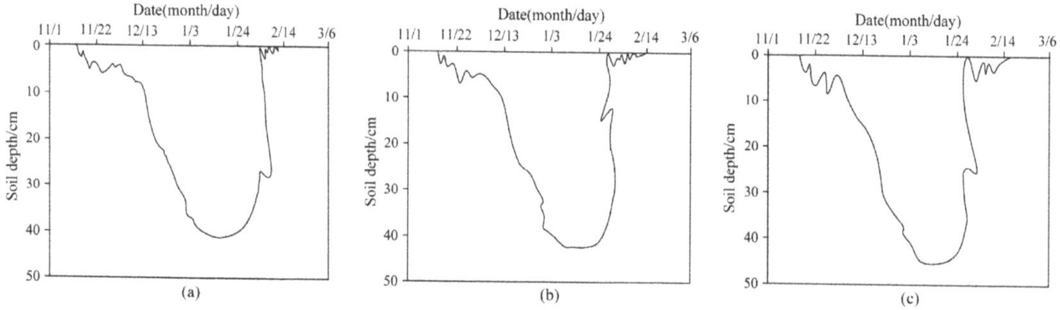

Figure 4. Soil freezing and thawing processes in (**a**) sandy loam, (**b**) loamy sand, and (**c**) sandy soil.

As the average soil particle size became larger, the porosity was smaller, and the thermal conductivity increased [21], which made the soil temperature drop quickly and led to a greater freezing depth. With the decrease in air temperature and the accumulation of negative surface temperature, sandy loam, loamy sand, and sandy soil began to freeze and entered the unstable freezing stage on November 18. The average freezing rate was 0.36~0.56 cm·d^{-1} with a short-term freeze–thaw cycle in the surface soil, and the maximum freezing depth of three kinds of soils reached 6.5 cm, 7.5 cm, and 9 cm in the unstable freezing stage. When entering the stable freezing stage in December, the average rate of frozen layer development downward was 0.98 cm·d^{-1}, 1.01 cm·d^{-1}, and 1.11 cm·d^{-1}, respectively, and the maximum freezing depths of three textured soils in the freezing period reached 41.3 cm, 42.5 cm, and 46 cm on 12 January, 9 January, and 10 January, respectively. The soil entered the thawing stage with the increase in air temperature, part of the thawing water recharged the phreatic water downward under the action of gravity potential, while another part migrated to the surface under the action of evaporation. Sandy loam, loamy sand, and sandy soil thawed completely on 13 February, 14 February, and 15 February, with the average thawing rate of 1.18~1.29 cm·d^{-1}.

During the freezing and thawing process, the maximum temperature gradients in the profile of three kinds of soils were observed when the freezing depth reached the maximum, which was 0.167 °C/cm, 0.172 °C/cm, and 0.178 °C/cm, respectively, indicating that the temperature gradient was greater, and the driving force of water and salt transport became stronger as the average soil particle size increased.

3.2. *Characteristics of Water Transport in Different Textured Soils*

The maximum rising height of capillary water in sandy loam, loamy sand, and sandy soil was 187 cm, 77 cm, and 60 cm, respectively, and the capillary water could reach the frozen layer. The water content of sandy loam increased at the depth of 10 to 50 cm, but decreased slightly at the depth of 0 to 10 cm under the effect of soil evaporation [46] in the process of D1 to D2 (Figure 5). On D2, the capillary water gathered at the freezing front under the blocking effect of the frozen layer [47], resulting in the greatest increment of the water content at a depth of 40 cm in the sandy loam. During the process of D2 to D3, the soil water mainly moved to the depth of 0 to 20 cm as the frozen layer thawed in both directions. The soil water content at the depth of 20 to 50 cm on D3 was 1.01~5.74% lower than that on D2, with the largest decrease in soil water content at the depth of 30 cm.

The water content of loamy sand increased by 0.37~33.33% at the depth of 5 to 50 cm in the process of D1 to D2 and increased by 0.0067~0.093% in the soil profile from D2 to D3. On D2 and D3, the soil water content increased with the increase in depth, and the loamy sand was almost saturated at the depth of 40 cm, while the soil water content was only 0.006% at the depth of 0 cm on D2, and the surface was seriously dry.

The water content of sandy soil increased at the depth of 0 to 50 cm during the process of D1 to D2 and D2 to D3 due to the continuous hydraulic link and the transformation of phreatic water throughout the whole freeze–thaw process [34], with the greatest increase at the depth of 5 cm. The water content in the sandy soil increased with the increase in depth during the freeze–thaw process and was greatest at the end of the thawing period, with the average soil water content in the sandy soil increasing by 18.46% and 0.19% compared with that in sandy loam and loamy sand, respectively.

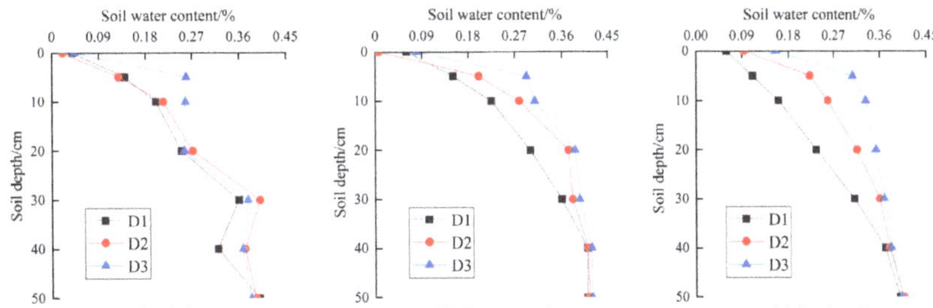

Figure 5. Vertical distribution characteristics of water in different textured soils during the freeze–thaw period.

3.3. Characteristics of Ion Content Variation in Different Textured Soils

The phreatic water continuously migrated upwards under the action of the soil water potential gradient, and the soil in the unsaturated zone continued to freeze from D1 to D2 in the shallow groundwater area. The lower edge of the frozen layer in sandy loam, loamy sand, and sandy soil was within the maximum rising height of the capillary water, so variation characteristics of soil ions content were closely related to the migration of groundwater and soil ions were redistributed in the profile.

3.3.1. Correlation Analysis between Soil Salt Content and Ions

The monitoring result of ion content in soil extract showed that the contents of soluble K^+, Mg^{2+}, NO_3^-, and CO_3^{2-} in the soil were very low, making the measurement error larger; therefore, only variations in Na^+, Ca^{2+}, Cl^-, SO_4^{2-}, and HCO_3^- were analyzed in the article. The correlation between salt content and ions in three textured soils was analyzed by Grey Relation Analysis.

Na^+ and HCO_3^- had high relativity with soil salt content in sandy loam (Table 3), with correlation coefficients of 0.9 and 0.909, respectively, indicating that Na^+ and HCO_3^- were the most abundant ions in groundwater. However, the correlation coefficient between Ca^{2+}, SO_4^{2-} and Cl^- and salt content was 0.73~0.817, showing that the content of Ca^{2+}, SO_4^{2-}, and Cl^- in groundwater was second only to Na^+ and HCO_3^-. Na^+ had a good correlation with Cl^-, SO_4^{2-}, and HCO_3^-. Ca^{2+} had a good correlation with SO_4^{2-} and HCO_3^- because Ca^{2+} was easy to precipitate with SO_4^{2-} and HCO_3^-. It can be seen that the main salt composition in sandy loam was Na_2SO_4, NaCl, $NaHCO_3$, $CaSO_4$, and $Ca(HCO_3)_2$.

Table 3. Correlation coefficients between salt content and ions in sandy loam.

Ionic Species	Soil Salt Content	Na^+	Ca^{2+}	Cl^-	SO_4^{2-}	HCO_3^-
Soil Salt Content	1					
Na^+	0.9	1				
Ca^{2+}	0.817	0.694	1			
Cl^-	0.73	0.749	0.749	1		
SO_4^{2-}	0.806	0.798	0.833	0.847	1	
HCO_3^-	0.909	0.772	0.82	0.442	0.516	1

Soil salt content was highly correlated to Na^+, HCO_3^-, and SO_4^{2-} in loamy sand (Table 4) with correlation coefficients of 0.99, 0.995, and 0.873, respectively. The correlation between Na^+ and HCO_3^- was the highest followed by the correlation between Ca^{2+} and Cl^- and HCO_3^-. It can be seen that the main salt composition in loamy sand was $NaHCO_3$, Na_2SO_4, $CaCl_2$, and $Ca(HCO_3)_2$.

Table 4. Correlation coefficients between salt content and ions in loamy sand.

Ionic Species	Soil Salt Content	Na^+	Ca^{2+}	Cl^-	SO_4^{2-}	HCO_3^-
Soil Salt Content	1					
Na^+	0.99	1				
Ca^{2+}	0.694	0.591	1			
Cl^-	0.531	0.477	0.729	1		
SO_4^{2-}	0.873	0.872	0.616	0.49	1	
HCO_3^-	0.995	0.986	0.712	0.484	0.823	1

The salt content in sandy soil was highly correlated to Na^+, HCO_3^-, and SO_4^{2-} (Table 5) with correlation coefficients of 0.990, 0.989, and 0.874, respectively. The correlation between Na^+ and HCO_3^- was the highest with a correlation coefficient of 0.988, followed by the correlation between Ca^{2+} and SO_4^{2-}. It was shown that the main salt component in the sandy soil was $NaHCO_3$, Na_2SO_4, and $CaSO_4$.

Table 5. Correlation coefficients between salt content and ions in sandy soil.

Ionic Species	Soil Salt Content	Na^+	Ca^{2+}	Cl^-	SO_4^{2-}	HCO_3^-
Soil Salt Content	1					
Na^+	0.99	1				
Ca^{2+}	0.608	0.493	1			
Cl^-	0.719	0.692	0.644	1		
SO_4^{2-}	0.874	0.884	0.752	0.659	1	
HCO_3^-	0.989	0.988	0.536	0.651	0.593	1

According to the correlation analysis between salt content and ions in three different textured soils, it was shown that the correlation coefficient between soil salt content and Na^+ and HCO_3^- in sandy loam, loamy sand, and sandy soil was higher than 0.9, and the correlation between Na^+ and HCO_3^- was better, so $NaHCO_3$ was the main component of salt in three kinds of soils. Na^+, Cl^-, and HCO_3^- migrate easily with water, while Ca^{2+} and SO_4^{2-} do not. As the average soil particle size increased, the pore diameter in the soil increased, the suction of soil particles to soil salt decreased, the migration ability of Ca^{2+} and SO_4^{2-} increased, and the main salt composition in the soil changed, with calcium salts changing from $CaSO_4$ and $Ca(HCO_3)_2$ to $CaSO_4$, and sodium salts changing from Na_2SO_4, NaCl, and $NaHCO_3$ to $NaHCO_3$ and Na_2SO_4. Since Na^+ and HCO_3^- were the main ions in the soil, the most abundant sodium salt in sandy soil was still $NaHCO_3$, although the content of SO_4^{2-} increased.

3.3.2. Characteristics of Ca^{2+} Content Variation

In the process of D1 to D2, the variation in Ca^{2+} gradually decreased with the increase in soil particle size (Figure 6), which in loamy sand and sandy soil was 12.36~67.38% and 53.88~97.47% lower than that in sandy loam at the depth of 0 to 50 cm. On D2, the distribution characteristic of Ca^{2+} in the sandy loam was similar to that in the sandy soil, and the content of Ca^{2+} decreased continuously at the depth of 0 to 30 cm and increased at the depth of 30 to 50 cm. The content of Ca^{2+} in loamy sand increased with the increase in depth from 5 to 30 cm and tended to be stable at the depth of 30 to 50 cm.

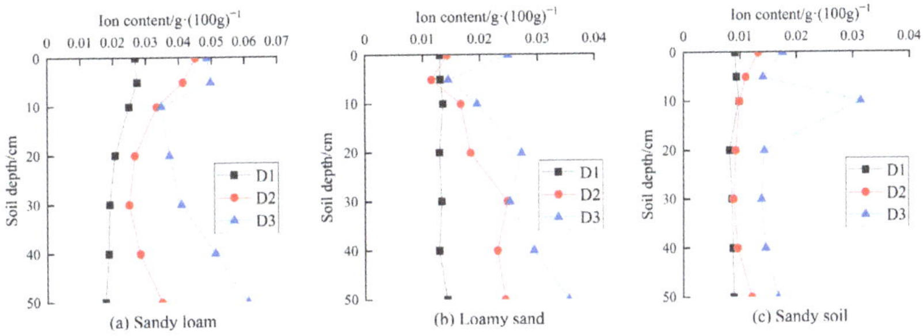

Figure 6. Vertical distribution characteristics of Ca^{2+} in different textured soils during the freeze–thaw period.

From D2 to D3, the content of Ca^{2+} in three textured soils increased in the vertical profile. The content of Ca^{2+} in sandy loam increased by a smaller amount of 4.53~20.58% at the depth of 0 to 10 cm, and the increase of Ca2+ content increased with depth within the range of 10 to 50 cm, with the increase at the depth of 10 cm being 6% of that at the depth of 50 cm. The content of Ca^{2+} in loamy sand increased by 1.84~74.04%, and the amplitude on the surface was the largest. The content of Ca^{2+} in sandy soil increased by 27.83~218.92% with the greatest increment at the depth of 10 cm. At the end of the thawing stage (D3), the content of Ca^{2+} in sandy loam was 0.0347~0.0611 g·(100 g)$^{-1}$, with a decrease of 41.90~58.21% and 48.61~60.23% in the loamy sand and sandy soil, respectively. The content of Ca^{2+} in different textured soils was less than that of other ions, the reason was that soluble $SO_4{}^{2-}$ was easy to combine with Ca^{2+} to form sulfate precipitate $CaSO_4$.

3.3.3. Characteristics of Na^+ Content Variation

From D1 to D2, Na^+ mainly accumulated at the depth of 10 to 30 cm in sandy loam (Figure 7), while at the depth of 0 to 5 cm in loamy sand, and it migrated from the depth of 5 to 30 cm to both sides in sandy soil, the average content of Na^+ at depths of 0 to 5 cm and 30 to 50 cm increased by 24.13% and 15.82%, respectively. On D2, the content of Na^+ was the highest at the depth of 20 cm in sandy loam, and the Na^+ in the loamy sand and sandy soil accumulated on the surface.

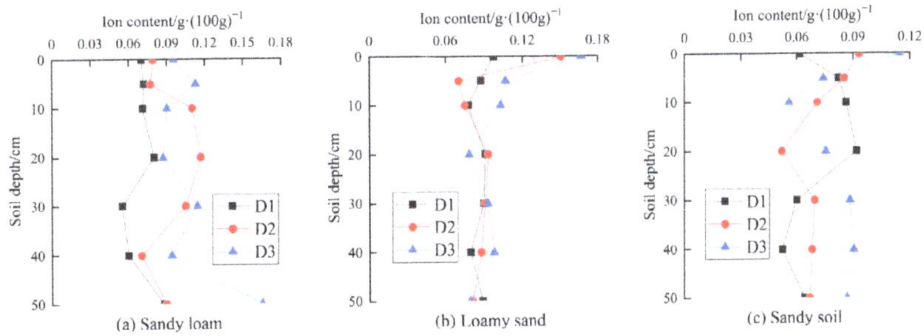

Figure 7. Vertical distribution characteristics of Na^+ in different textured soils during the freeze–thaw period.

From D2 to D3, Na^+ migrated with water from the thawing front to the non-frozen layer on both sides with the thawing of the frozen layer. Na^+ in sandy loam, loamy sand, and sandy soil migrated from depths of 8 to 30 cm, 15 to 30 cm, and 5 to 15 cm to both sides, respectively. On D3, the content of Na^+ was the highest at a depth of 50 cm in sandy loam, and Na^+ accumulated on the surface of loamy sand and sandy soil. The average content of Na^+ at the depth of 0 to 50 cm in sandy soil was the lowest at 0.084 g·(100 g)$^{-1}$, which was 22.60% and 19.07% lower than that in sandy loam and loamy sand, respectively.

Excessive Na^+ content can lead to the dispersion of soil particles and aggregates, resulting in a decrease in soil pore diameter and permeability.

3.3.4. Characteristics of Cl^- Content Variation

In the process of D1 to D2, the increment of Cl^- content at the depth of 10 to 30 cm was the most in sandy loam (Figure 8), which was 51.56~61.44%; however, Cl^- was mainly accumulated at the depth of 30 to 50 cm on D2. The Cl^- content in loamy sand increased throughout the vertical profile, with the largest increment and the highest Cl^- content on D2 at the depth of 0 cm. Cl^- in sandy soil migrated to the depth of 0 to 35 cm from D1 to D2; however, the average content of Cl^- at the depth of 30 to 50 cm increased by 39.51% compared with that at the depth of 0 to 30 cm on D2.

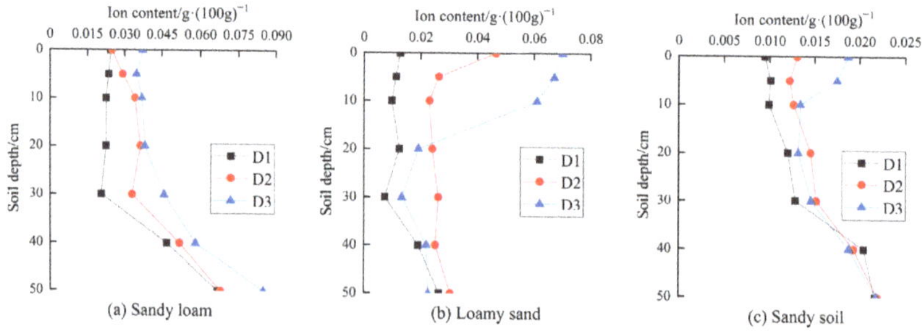

Figure 8. Vertical distribution characteristics of Cl^- in different textured soils during the freeze--thaw period.

In the process of D2 to D3, Cl^- in sandy loam mainly migrated to depths of 0 to 10 cm and 20 to 50 cm. Cl^- in loamy sand mainly accumulated at the depth of 0 to 20 cm, which made the average content of Cl^- in this depth increase by 39.51%. Cl^- in sandy soil mainly migrated at the depth of 0 to 10 cm, and Cl^- content at the depth of 10 to 50 cm decreased by 0.65~9.53%. On D3, the content of Cl^- in sandy loam increased with the increase in depth from 5 to 50 cm. Cl^- was mainly distributed at the depth of 0 to 20 cm in loamy sand, and was gathered at depths of 0 to 10 cm and 30 to 50 cm in sandy soil. The average content of Cl^- in sandy loam was the highest, with a decrease of 17.96% and 64.96% in loamy sand and sandy soil, respectively.

3.3.5. Characteristics of SO_4^{2-} Content Variation

In the D1 to D2 stage, SO_4^{2-} in sandy loam mainly accumulated in the frozen layer at the depth of 5 to 40 cm, the average content of SO_4^{2-} in this depth was 12.37% and 12.42% higher than that at depths of 0 to 5 cm and 40 to 50 cm at the end of the freezing stage (Figure 9). The SO_4^{2-} in loamy sand mainly aggregated at the depth of 0 to 20 cm, and the content of SO_4^{2-} was basically unchanged at the depth of 20 to 50 cm. The content of SO_4^{2-} in sandy soil increased by 16.74~105.60% at the depth of 5 to 50 cm, with the highest SO_4^{2-} content of 0.046 g·(100 g)$^{-1}$ at a depth of 50 cm on D2.

In the process of D2 to D3, SO_4^{2-} in sandy loam mainly accumulated at depths of 0 to 10 cm and 30 to 50 cm with the thawing of the frozen layer. The content of SO_4^{2-} increased in the profile of loamy sand. SO_4^{2-} in sandy soil mainly migrated to the depth of 22 to 50 cm, and the maximum increment of SO_4^{2-} was 38.75% at a depth of 30 cm. On D3, the content of SO_4^{2-} was the highest at a depth of 50 cm in sandy loam and sandy soil, and SO_4^{2-} was mainly concentrated at the depth of 0 to 10 cm in loamy sand. The content of SO_4^{2-} in the profile of sandy soil was lower than that of sandy loam. The average content of SO_4^{2-} in sandy loam was the highest, which was 0.115 g·(100 g)$^{-1}$, while that in loamy sand and sandy soil decreased by 23.55% and 70.42%.

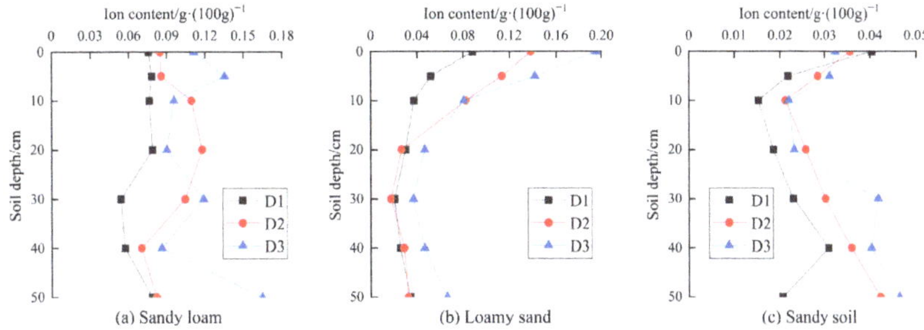

Figure 9. Vertical distribution characteristics of SO_4^{2-} in different textured soils during the freeze–thaw period.

3.3.6. Characteristics of HCO_3^- Content Variation

During the process of D1 to D2, the solubility of $Ca(HCO_3)_2$ increased with the decrease in soil temperature, and the content of HCO_3^- at the freezing front was higher, resulting in the movement of HCO_3^- from the area of high concentration to the area of low concentration, and the HCO_3^- content in the frozen layer decreased. HCO_3^- in sandy loam mainly migrated from the depth of 5~40 cm to both sides from D1 to D2 (Figure 10), the average content of HCO_3^- at the depth of 10 to 30 cm was the lowest, which was 36.72% and 18.34% lower than that at depths of 0 to 10 cm and 30 to 50 cm on D2. HCO_3^- in loamy sand mainly migrated from the depth of 0~20 cm to the bottom, and the content of HCO_3^- at the depth of 20 to 50 cm increased by 2.51~20.43%. HCO_3^- in sandy soil mainly migrated from the depth of 5~30 cm to both sides, and its average content decreased by 27.70%.

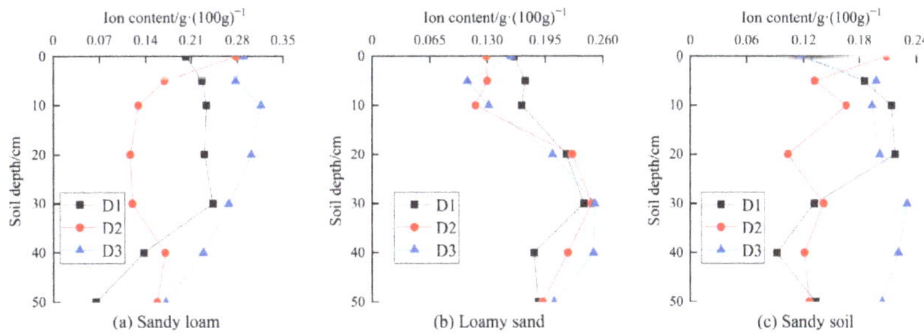

Figure 10. Vertical distribution characteristics of HCO_3^- in different textured soils during the freeze–thaw period.

In the process of D2 to D3, the solubility of $Ca(HCO_3)_2$ decreased with the increase in soil temperature, and some HCO_3^- precipitated in the form of crystals. The content of HCO_3^- increased in the soil profile of sandy loam, with the greatest increase of 146.21% at a depth of 10 cm, the content of HCO_3^- in loamy sand remained basically unchanged, and HCO_3^- in sandy soil accumulated at the depth of 5 to 50 cm, with an increase in the content of 17.20~95.45%. At the end of the thawing stage, the content of HCO_3^- in sandy loam decreased with the increase in depth from 10 to 50 cm. The HCO_3^- in loamy sand was gathered at the depth of 20 to 50 cm, accounting for 69.63% of the ion content in the whole soil profile. The content of HCO_3^- was the least on the surface of sandy soil and remained basically unchanged with increasing depth from 5 to 50 cm. The average content of HCO_3^- in loamy sand was the smallest, which was 29.50% and 4.65% lower than that in sandy loam and sandy soil, respectively.

3.4. Characteristics of Soil Salt Content Variation

From D1 to D2, the depth of the frozen layer increased, and the salt in three textured soils gathered to the surface and the soil at deeper depths, which mainly migrated from the middle frozen layer at the depth of 5~30 cm to both sides in the sandy loam (Figure 11); accumulated at depths of 0 to 5 cm and 20 to 50 cm in the loamy sand, with an increase in the soil salt content of only 2.08~6.11% at the depth of 5 to 20 cm; and migrated from the depth of 3~30 cm to both sides in sandy soil. On D2, the salt content on the surface was largest in the sandy loam, loamy sand, and sandy soil, which was 0.36~0.51 g·(100 g)$^{-1}$. The salt content in the profile of sandy soil was 0.20~0.36 g·(100 g)$^{-1}$, which increased by 41.67~95% and 33.33~55% in sandy loam and loamy sand, respectively.

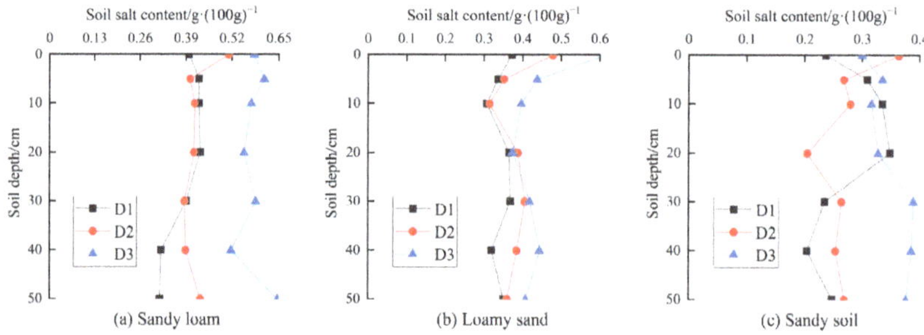

Figure 11. Vertical distribution characteristics of soil salt in different textured soils during the freeze–thaw period.

From D2 to D3, the salt content in the sandy loam at the depth of 0 to 50 cm increased by 13.69~51.71% with the thawing of the frozen layer and strong soil evaporation; the salt in the loamy sand accumulated at depths of 0 to 20 cm and 30 to 50 cm; differently from sandy loam and loamy sand, the salt content on the surface of sandy soil decreased, and the soil salt mainly gathered at the depth of 5 to 50 cm. On D3, the salt content in the soil profile of sandy loam was 0.52~0.65 g·(100 g)$^{-1}$, with the greatest soil salt content at a depth of 50cm. The average salt content at the depth of 10 to 20 cm was the smallest in the loamy sand, which was 20.16% and 6.37% lower than that at depths of 0 to 10 cm and 20 to 50 cm. The average salt content of the sandy soil at the depth of 30 to 50 cm increased by 15.01% compared with it at the depth of 0 to 30 cm.

At the end of the thawing period, the salt content of three textured soils at the depth of 5 to 50 cm gradually decreased as the average soil particle size increased, but the surface salt content in the loamy sand was the largest, the average soil salt content of sandy loam, loamy sand, and sandy soil was 0.58, 0.44, and 0.35 g·(100 g)$^{-1}$, respectively, and the degree of soil salinization gradually decreased. Among them, both sandy loam and loamy sand were severely salinized soil, while sandy soil was moderately salinized soil.

4. Discussion

Crop production plays a vital role in ensuring food security, providing livelihoods and supporting economic development, and many researchers have discussed different aspects of crop productions [48–51]. However, soil salinity poses a significant threat to crop production. Salinity negatively impacts plant growth and productivity [52,53]. High salt levels hinder water uptake by plant roots, leading to water stress and reduced crop yields. Excess salts can cause toxicity, damaging plant tissues and disrupting important physiological processes such as nutrient uptake and photosynthesis. Additionally, salinity alters soil structure, reducing its fertility and nutrient availability, further impairing crop growth. Managing and mitigating soil salinity is essential to safeguard crop production, maintain food supplies, and sustain agricultural livelihoods.

4.1. Correlation Analysis between Soil Water and Salt

In the seasonally frozen soil area, the migration of phreatic water and soil water driven by gravity potential, matrix potential, and temperature potential [2,54], makes a concomitant migration of salts dissolved in water, resulting in the redistribution of soil salt and the change in solute potential in the soil profile. There was a positive correlation between soil water content and electrical conductivity [55], and the transport of water and salt in the soil was not synchronized due to convection and diffusion [56]. The correlation between soil water and salt varies in different textured soils. Pearson correlation analysis was performed on the linear relationship between water and salt in the soil at different depths based on the monitoring data of soil water content and soil salt content (Figure 12).

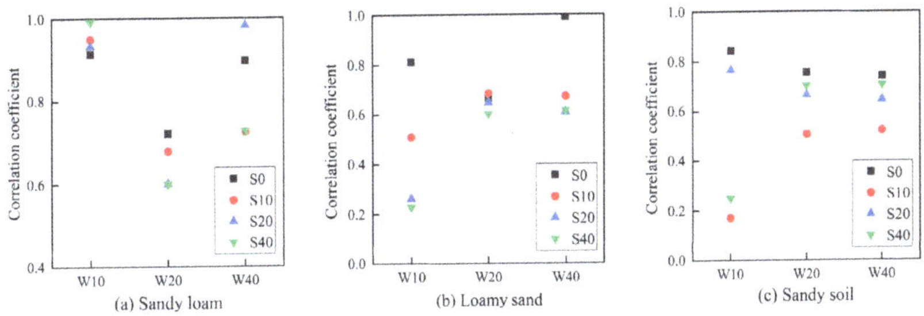

Figure 12. Correlation coefficients between salt content and water content at different depths of (**a**) sandy loam, (**b**) loamy sand, and (**c**) sandy soil. W10, W20, and W40 are the soil water content at the depth of 10 cm, 20 cm, and 40 cm, respectively, and S0, S10, S20, and S40 are the soil salt content at the depth of 0 cm, 10 cm, 20 cm, and 40 cm, respectively.

The correlation coefficient between soil water content at depths of 10 cm, 20 cm, and 40 cm and soil salt content at depths of 0 cm, 10 cm, 20 cm, and 40 cm was greater than 0.6 in sandy loam, which indicated that the soil salt content increased with the increase in soil water content, and soil salt mainly migrated with soil water convectively, which was consistent with previous research results [57,58]. However, soil salt content and water content did not increase synchronously, and there was a depth difference between their corresponding depth when the correlation was higher; for example, correlations between soil salt content at a depth of 0 cm and soil water content at a depth of 10 cm, and between soil salt content at a depth of 20 cm and soil water content at a depth of 40 cm, were higher. The distribution of soil salt depends on the salt migration caused by convection and diffusion [59].

The correlation coefficient between soil salt content at depths of 10 cm, 20 cm, and 40 cm and soil water content at a depth of 10 cm in loamy sand was less than 0.6, and the correlation was significantly lower than that in sandy loam, due to the greater water migration in loamy sand than that in sandy loam, and the soil water content at a depth of 10 cm was greatly affected by soil evaporation. The correlation coefficient between soil salt content at depths of 0 cm, 10 cm, 20 cm, and 40 cm and soil water content at depths of 20 cm and 40 cm was greater than 0.6.

The correlation coefficient between soil water content at depths of 10 cm, 20 cm, and 40 cm and soil salt content at a depth of 10 cm in sandy soil was less than 0.6, and the correlation was further weakened compared with that in loamy sand. In summary, the correlation between soil salt content and water content gradually weakened with the increase in average soil particle size. The soil with larger particles has stronger water and salt transport capacity [60]. In the thawing stage, soil water and salt migrated to the surface. Under the action of soil evaporation, the soil water content decreased, and the salt accumulated on the surface, so the correlation between soil water content and soil salt content was weaker.

4.2. *Effect of Freeze–Thaw Process on Water and Salt Transport in Different Textured Soils*

The distribution of soil water has an important influence on salinization prevention. The influence of management system [61], initial water content [62], and loess interlayer [63] on water migration, and the correlation between soil texture and soil water content in non-freeze–thaw periods [64] were researched. In the freeze–thaw period, the interlayer with fine particle inhibited the water migration in the soil [65]. For homogeneous soil, average particle size, and inhomogeneity coefficient of the sandy soil affected the soil moisture [34].

At the end of the thawing period, the water content in three textured soils increased compared with that at the begin of the freeze–thaw period at the depth of 0 to 40 cm, among which the increment of water content at the depth of 0 to 30 cm in sandy loam and loamy sand decreased by 40.20~93.10% and 28.14~65.52% compared with that in sandy soil (Table 6). Soil texture is closely associated with soil hydraulic properties [66] and temperature gradient [67]. Since phreatic water and soil water in the deeper depth are transported to shallow soil through the capillary, the variation in water content in the soil profile is affected by both the rising height and transport capacity of capillary water. Soil with a smaller average particle size has a stronger water holding capacity [68], but its hydraulic conductivity is smaller, making the soil water conductivity and permeability weaker, and the transport capacity of capillary water smaller. Among three textured soils, sandy loam has the largest rising height of capillary water; however, the maximum rising heights of capillary water in three kinds of soils are greater than 50 cm, the water transported upwards in sandy loam, loamy sand, and sandy soil can reach the frozen layer, and continuous hydraulic links can be formed. Therefore, the migration of soil water and phreatic water during the freeze–thaw period is mainly affected by the transport capacity of capillary water and temperature gradient. With the increase in groundwater table depth, the rising height of capillary water becomes the main factor affecting the phreatic evaporation and soil water content [34], and the temperature gradient is the principal factor affecting water and salt transfer in red silty clay [5]. The porosity of sandy soil is smaller, and the thermal conductivity is larger [69], which makes the temperature gradient larger, further leading to a greater freezing depth and soil water potential gradient, as well as a stronger driving force for water migration, but the maximum temperature gradient of three kinds of soils is 0.167~0.178 °C/cm, and the difference in temperature gradient is small. The transport capacity of capillary water is the main factor influencing the water transport. The particle size of sandy loam is smaller and the hydraulic conductivity is weaker, the amount of water transport is smaller under the same conditions, but the increment of water content in sandy loam at a depth of 40 cm was the largest, because the water holding capacity of sandy loam is stronger, and the specific yield is smaller [70]. When the thawing water in the frozen layer migrates downward under the action of gravity, it is easier to remain in the soil, and the gravity water is more easily lost from soil with a larger particle size [71]. The water content of the three textured soils at a depth of 50 cm was basically unchanged, and all fluctuated around the saturated water content.

Table 6. Variations in soil water content and salt content during the freeze–thaw period.

Depth (cm)	Soil Water Content (%)			Soil Salt Content (g·(100 g)$^{-1}$)		
	Sandy Loam	Loamy Sand	Sandy Soil	Sandy Loam	Loamy Sand	Sandy Soil
0	0.004	0.020	0.094	0.184	0.242	0.063
5	0.119	0.143	0.199	0.182	0.100	0.026
10	0.058	0.083	0.173	0.146	0.088	−0.019
20	0.006	0.086	0.120	0.122	0.009	−0.021
30	0.018	0.035	0.058	0.195	0.050	0.154
40	0.048	0.009	0.012	0.198	0.124	0.181
50	−0.016	0.010	0.004	0.330	0.057	0.128

During the freezing process, the phreatic water migrated upward under the action of the soil water potential gradient, so that the salt in the phreatic water and soil at a deeper depth was brought to the frozen layer [72]. The salt content in the sandy loam was larger, which reduced the freezing point of the soil and prolonged the freezing time of the soil [73]. The soil salt content increased in the freezing stage [74]; however, the one-way freezing experiment in the column with red silty clay found that salt content in the frozen soil changed little, while the soil water content increased more [5]. At the end of the thawing stage, the salt content of sandy loam and loamy sand in the soil profile increased compared with that at the beginning of the freeze–thaw period, while the salt content of sandy soil decreased at the depth of 10 to 20 cm. Under the same water content, the water migration ability of the sandy soil is stronger for the smaller matrix suction [75] and the larger specific yield [34]. In the thawing stage, the thawing water of the sandy soil migrated downward quickly and resulted in a decrease in soil salt content at the depth of 10–20 cm. The thickness and texture of the interlayer could affect the change in soil salt content [76,77], the finer soil formed a thinner salt shell and had a stronger evaporation capacity [25], and the bulk density of paddy and natural land changed in the freeze–thaw period [18]. Soil deformation led to the change in soil hydrothermal properties [78]; however, the frost heave was not obvious in this experiment. The influence of soil texture on soil salt migration is mainly controlled by the soil water migration. As the average soil particle size increases, the number of macropores increases, the amount of soil water migration becomes greater [79], and the migration ability of soil salt becomes stronger. At the same time, the capacity of adsorbing soil ions varies with the size of soil particles [80], the increase in fine particles weakens the permeability of soil, making the downward migration of water and salt decrease, and therefore, the capacity of adsorbing soil ions is smaller with a larger soil particle size. The variation in salt content in the sandy loam at the depth of 5 to 50 cm was the greatest, and the variation in loamy sand was larger than that in sandy soil at the depth of 5 to 10 cm. However, the variation in sandy soil was much larger than that in loamy sand at the depth of 20 to 50 cm due to the larger pores and better conductivity in sandy soil [81], which is conducive to the formation of preferential flow; the salt dissolved in soil water is gradually brought to the soil at deeper depth with the downward migration of thawing water. In contrast, the complex porous geometric structure formed by fine pores and the adsorption on charged aggregates will hinder ion migration in loamy sand. The average increment of salt content for sandy loam, loamy sand, and sandy soil in the tillage layer with the depth of 0 to 30 cm was 0.166, 0.098, and 0.041 $g \cdot (100\ g)^{-1}$, respectively, and the freeze–thaw process had the least effect on salinization in sandy soil. The effect of freezing and thawing processes on salt transport can be weakened by changing the hydraulic characteristics of frozen soil, such as surface coverage or plowing before freezing [82].

5. Conclusions

The effect of soil texture on the distribution characteristics of water, heat, and salt in the soil profile during the freeze–thaw period was quantitatively analyzed through a freeze–thaw field experiment. The main conclusions were as follows:

The soil temperature decreased quicker, and the freezing depth was greater as the average soil particle size increased. The increment of soil water content increased with the increasing average soil particle size at the depth of 0 to 30 cm during the freeze–thaw period, and the average water content in sandy soil at the depth of 0 to 50 cm was 18.46% and 0.19% higher than that in sandy loam and loamy sand at the end of the thawing stage, respectively.

$NaHCO_3$ was the main salt composition in the soil. The average content of Ca^{2+}, Na^+, Cl^-, and SO_4^{2-} in loamy sand and sandy soil decreased by 4.37~45.50% and 22.60~70.42% compared with that in sandy loam, respectively, and the average soil salt content decreased with the increase in soil particle size at the end of the thawing stage. The average increment of salt content at the depth of 0 to 30 cm in sandy loam and loamy sand increased by

304.88% and 139.02% compared with that in sandy soil during the freeze–thaw period; that is, the freeze–thaw process had the least effect on soil salinization in sandy soil. The correlation between soil salt content and water content decreased with the increase in soil particle size.

The effects of different treatments on water and salt transport in different textured soils during freeze–thaw periods need to be further studied, which can provide a basis for the management and efficient utilization of salinized land.

Author Contributions: Writing—original draft preparation, E.W.; resources, J.C.; data curation, L.L.; writing—review and editing, E.W., J.C., L.L., L.C., J.X. and J.R.; supervision, Q.D. All authors have read and agreed to the published version of the manuscript.

Funding: This research was funded by the National Natural Science Foundation of China (Grant No. 41572239); the National Natural Youth Science Foundation of China (Grant No. 41502243); the Natural Science Foundation of Shanxi Province, China (Grant No. 202203021221048); and the Natural Science Youth Foundation of Shanxi Province, China (Grant No. 202203021212195).

Data Availability Statement: The data that support the findings of this study are available upon reasonable request from the corresponding author.

Acknowledgments: The authors are grateful for the assistance with the research from Taigu Water Balance Experimental Field, Hydrology and Water Resources Survey Station of Shanxi Province.

Conflicts of Interest: The authors declare no conflict of interest.

References

1. Chen, S.; Ouyang, W.; Hao, F.; Zhao, X. Combined impacts of freeze-thaw processes on paddy land and dry land in Northeast China. *Sci. Total Environ.* **2013**, *456–457*, 24–33. [CrossRef] [PubMed]
2. Ireson, A.M.; van der Kamp, G.; Ferguson, G.; Nachshon, U.; Wheater, H.S. Hydrogeological processes in seasonally frozen northern latitudes: Understanding, gaps and challenges. *Hydrogeol. J.* **2012**, *21*, 53–66. [CrossRef]
3. Bechtold, M.; Haber-Pohlmeier, S.; Vanderborght, J.; Pohlmeier, A.; Ferré, T.P.A.; Vereecken, H. Near-surface solute redistribution during evaporation. *Geophys. Res. Lett.* **2011**, *38*, L17404. [CrossRef]
4. Li, Z.; Wang, Z. Experimental study on the relation between the fractal characteristics and solute transport parameters of sandy soil. *J. Soils Sediments* **2020**, *20*, 3181–3191. [CrossRef]
5. Bing, H.; He, P.; Zhang, Y. Cyclic freeze–thaw as a mechanism for water and salt migration in soil. *Environ. Earth Sci.* **2015**, *74*, 675–681. [CrossRef]
6. Han, Y.; Wang, Q.; Xia, W.; Liu, J.; Wang, J.; Chen, Y.; Shen, J. Experimental study on the hydraulic conductivity of unsaturated dispersive soil with different salinities subjected to freeze-thaw. *J. Hydrol.* **2020**, *583*, 124297. [CrossRef]
7. Yang, J.; Zhao, J.; Zhu, G.; Wen, Y.; Wang, Y.; Liu, J.; Yang, Z. Effects of Ecological Water Conveyance on Soil Salinization in the Shiyang River Basin's Terminal Lake—Qingtu Lake—Area. *Sustainability* **2022**, *14*, 311. [CrossRef]
8. Ma, Y.; Tashpolat, N. Current Status and Development Trend of Soil Salinity Monitoring Research in China. *Sustainability* **2023**, *15*, 5874. [CrossRef]
9. Mirbakhsh, M.; Sedeh, S.S. Effect of short and long period of salinity stress on physiological responses and biochemical markers of *Aloe vera* L. *Purdue Univ. Alzahra Univ.* **2022**, *7*, 178–187. [CrossRef]
10. Wu, M.; Huang, J.; Tan, X.; Wu, J. Water, Salt and Heat Influences on Carbon and Nitrogen Dynamics in Seasonally Frozen Soils in Hetao Irrigation District, Inner Mongolia, China. *Pedosphere* **2019**, *29*, 632–641. [CrossRef]
11. Wu, D.; Zhou, X.; Jiang, X. Water and Salt Migration with Phase Change in Saline Soil during Freezing and Thawing Processes. *Ground Water* **2018**, *56*, 742–752. [CrossRef] [PubMed]
12. Wan, H.; Bian, J.; Zhang, H.; Li, Y. Assessment of future climate change impacts on water-heat-salt migration in unsaturated frozen soil using CoupModel. *Front. Environ. Sci. Eng.* **2020**, *15*, 10. [CrossRef]
13. Lu, X.; Li, R.; Shi, H.; Liang, J.; Miao, Q.; Fan, L. Successive simulations of soil water-heat-salt transport in one whole year of agriculture after different mulching treatments and autumn irrigation. *Geoderma* **2019**, *344*, 99–107. [CrossRef]
14. Xun, Y.; Xiao, X.; Sun, C.; Meng, H.; Gao, Y.; Huang, G.; Xu, X. Modeling heat-water-salt transport, crop growth and water use in arid seasonally frozen regions with an improved coupled SPAC model. *J. Hydrol.* **2022**, *615*, 128703. [CrossRef]
15. Hou, R.; Qi, Z.; Li, T.; Fu, Q.; Meng, F.; Liu, D.; Li, Q.; Zhao, H.; Yu, P. Mechanism of snowmelt infiltration coupled with salt transport in soil amended with carbon-based materials in seasonally frozen areas. *Geoderma* **2022**, *420*, 115882. [CrossRef]
16. Tan, X.; Wu, J.; Wu, M.; Huang, J.; Tan, B.; Li, L. Effects of ice cover on soil water, heat, and solute movement: An experimental study. *Geoderma* **2021**, *403*, 115209. [CrossRef]
17. You, Z.; Wen, W.; Lai, Y.; Zhang, M.; Zhang, J. Model tests of the barrier measures on moisture and salt migration in soils subjected to freeze-thaw cycles. *Cold Reg. Sci. Technol.* **2022**, *201*, 103607. [CrossRef]

18. Liu, S.; Huang, Q.; Zhang, W.; Ren, D.; Xu, X.; Xiong, Y.; Huang, G. An improved estimation of soil water and salt dynamics by considering soil bulk density changes under freeze/thaw conditions in arid areas with shallow groundwater tables. *Sci. Total Environ.* **2023**, *859*, 160342. [CrossRef]
19. Li, L.; Liu, H.; Gong, P.; Lin, E.; Bai, Z.; Li, P.; Wang, C.; Li, J. Multi-objective optimization of winter irrigation for cotton fields in salinized freeze-thaw areas. *Eur. J. Agron.* **2023**, *143*, 126715. [CrossRef]
20. Wan, X.; Gong, F.; Qu, M.; Qiu, E.; Zhong, C. Experimental Study of the Salt Transfer in a Cold Sodium Sulfate Soil. *KSCE J. Civ. Eng.* **2019**, *23*, 1573–1585. [CrossRef]
21. Huang, X.; Zhang, Z.; Hao, R.; Guo, Z. The Effects of Particle Gradation on Salinized Soil in Arid and Cold Regions. *Water* **2022**, *14*, 236. [CrossRef]
22. Hou, R.-j.; Li, T.-x.; Fu, Q.; Liu, D.; Li, M.; Zhou, Z.-q.; Yan, J.-w.; Zhang, S. Research on the distribution of soil water, heat, salt and their response mechanisms under freezing conditions. *Soil Tillage Res.* **2020**, *196*, 104486. [CrossRef]
23. Mohawesh, O.; Janssen, M.; Maaitah, O.; Lennartz, B. Assessment the effect of homogenized soil on soil hydraulic properties and soil water transport. *Eurasian Soil Sci.* **2017**, *50*, 1077–1085. [CrossRef]
24. Zhang, N.; Yu, X.; Pradhan, A.; Puppala, A.J. Effects of Particle Size and Fines Content on Thermal Conductivity of Quartz Sands. *Transp. Res. Rec.* **2015**, *2510*, 36–43. [CrossRef]
25. Li, X.; Guo, M.; Wang, H. Impact of soil texture and salt type on salt precipitation and evaporation under different hydraulic conditions. *Hydrol. Process.* **2022**, *36*, e14763. [CrossRef]
26. Li, X.; Chang, S.X.; Salifu, K.F. Soil texture and layering effects on water and salt dynamics in the presence of a water table: A review. *Environ. Rev.* **2014**, *22*, 41–50. [CrossRef]
27. Wang, J.-j.; Huang, Y.-f.; Long, H.-y. Water and salt movement in different soil textures under various negative irrigating pressures. *J. Integr. Agric.* **2016**, *15*, 1874–1882. [CrossRef]
28. Wang, J.; Huang, Y.; Long, H.; Hou, S.; Xing, A.; Sun, Z. Simulations of water movement and solute transport through different soil texture configurations under negative-pressure irrigation. *Hydrol. Process.* **2017**, *31*, 2599–2612. [CrossRef]
29. Panta, S.; Flowers, T.; Doyle, R.; Lane, P.; Haros, G.; Shabala, S. Growth responses of Atriplex lentiformis and Medicago arborea in three soil types treated with saline water irrigation. *Environ. Exp. Bot.* **2016**, *128*, 39–50. [CrossRef]
30. Liu, D.; She, D.; Mu, X. Water flow and salt transport in bare saline-sodic soils subjected to evaporation and intermittent irrigation with saline/distilled water. *Land Degrad. Dev.* **2019**, *30*, 1204–1218. [CrossRef]
31. Liu, Q.; Liu, Y.; Jin, M.; He, J.; Ferré, P.A.T. Impacts of an Internal Finer-Textured Layer on Soil Evaporation and Salt Distribution. *Transp. Porous Media* **2021**, *140*, 603–620. [CrossRef]
32. Xu, C.; Zeng, W.-Z.; Wu, J.-W.; Huang, J.-S. Effects of Different Irrigation Strategies on Soil Water, Salt, and Nitrate Nitrogen Transport. *Ecol. Chem. Eng. S* **2015**, *22*, 589–609. [CrossRef]
33. Nachshon, U.; Weisbrod, N.; Dragila, M.I.; Grader, A. Combined evaporation and salt precipitation in homogeneous and heterogeneous porous media. *Water Resour. Res.* **2011**, *47*, W03513. [CrossRef]
34. Chen, J.; Wang, E.; Xue, J.; Cui, L.; Zheng, X.; Du, Q. Effects of soil particle size and gradation on the transformation between shallow phreatic water and soil water under laboratory freezing-thawing action. *J. Hydrol.* **2023**, *619*, 129323. [CrossRef]
35. Miao, C.; Chen, J.; Zheng, X.; Zhang, Y.; Xu, Y.; Du, Q. Soil Water and Phreatic Evaporation in Shallow Groundwater during a Freeze–Thaw Period. *Water* **2017**, *9*, 396. [CrossRef]
36. Ambade, B.; Sethi, S.S.; Giri, B.; Biswas, J.K.; Bauddh, K. Characterization, Behavior, and Risk Assessment of Polycyclic Aromatic Hydrocarbons (PAHs) in the Estuary Sediments. *Bull Environ. Contam. Toxicol.* **2022**, *108*, 243–252. [CrossRef] [PubMed]
37. Ambade, B.; Sethi, S.S.; Chintalacheruvu, M.R. Distribution, risk assessment, and source apportionment of polycyclic aromatic hydrocarbons (PAHs) using positive matrix factorization (PMF) in urban soils of East India. *Environ. Geochem. Health* **2023**, *45*, 491–505. [CrossRef]
38. Kumar, A.; Ambade, B.; Sankar, T.K.; Sethi, S.S.; Kurwadkar, S. Source identification and health risk assessment of atmospheric PM2.5-bound polycyclic aromatic hydrocarbons in Jamshedpur, India. *Sustain. Cities Soc.* **2020**, *52*, 101801. [CrossRef]
39. Ambade, B.; Sethi, S.S.; Kurwadkar, S.; Mishra, P.; Tripathee, L. Accumulation of polycyclic aromatic hydrocarbons (PAHs) in surface sediment residues of Mahanadi River Estuary: Abundance, source, and risk assessment. *Mar. Pollut. Bull* **2022**, *183*, 114073. [CrossRef]
40. Sankar, T.K.; Kumar, A.; Mahto, D.K.; Das, K.C.; Narayan, P.; Fukate, M.; Awachat, P.; Padghan, D.; Mohammad, F.; Al-Lohedan, H.A.; et al. The Health Risk and Source Assessment of Polycyclic Aromatic Hydrocarbons (PAHs) in the Soil of Industrial Cities in India. *Toxics* **2023**, *11*, 515. [CrossRef]
41. Ambade, B.; Rao, C.M. Assessment of groundwater quality with a special emphasis on fluoride contamination in Rajnandgaon district of Chhattisgarh state in central India. *Int. J. Environ. Sci.* **2012**, *3*, 851–858. [CrossRef]
42. Li, S.; Lu, L.; Gao, Y.; Zhang, Y.; Shen, D. An Analysis on the Characteristics and Influence Factors of Soil Salinity in the Wasteland of the Kashgar River Basin. *Sustainability* **2022**, *14*, 3500. [CrossRef]
43. Zhang, Z.; Abuduwaili, J.; Jiang, F. Relationship of Heavy Metals and Soil N, P, K and Total Salts in Tianshan Mountains, Central Asia. *Asian J. Chem.* **2013**, *25*, 8971–8975. [CrossRef]
44. Chen, J.; Xie, X.; Zheng, X.; Xue, J.; Miao, C.; Du, Q.; Xu, Y. Effects of sand-mulch thickness on soil evaporation during the freeze–thaw period. *Hydrol. Process.* **2020**, *34*, 2830–2842. [CrossRef]

45. Chen, J.; Gao, X.; Zheng, X.; Miao, C.; Liu, P.; Du, Q.; Xu, Y. Transformation between Phreatic Water and Soil Water during Freeze–Thaw Periods. *Water* **2018**, *10*, 376. [CrossRef]
46. Li, Y.; Huo, S.; Guo, J.; Sun, J.; Pan, J.; Wang, D.; Tan, Q.; Pei, B. Using hydrogen and oxygen stable isotopes to estimate soil water evaporation loss under continuous evaporation conditions. *Hydrol. Process.* **2023**, *37*, e14885. [CrossRef]
47. Wu, M.; Huang, J.; Wu, J.; Tan, X.; Jansson, P.-E. Experimental study on evaporation from seasonally frozen soils under various water, solute and groundwater conditions in Inner Mongolia, China. *J. Hydrol.* **2016**, *535*, 46–53. [CrossRef]
48. Jamshidi, S.; Zand-Parsa, S.; Kamgar-Haghighi, A.A.; Shahsavar, A.R.; Niyogi, D. Evapotranspiration, crop coefficients, and physiological responses of citrus trees in semi-arid climatic conditions. *Agric. Water Manag.* **2020**, *227*, 105838. [CrossRef]
49. Boini, A.; Manfrini, L.; Bortolotti, G.; Corelli-Grappadelli, L.; Morandi, B. Monitoring fruit daily growth indicates the onset of mild drought stress in apple. *Sci. Hortic.* **2019**, *256*, 108520. [CrossRef]
50. Fathian, M.; Bazrafshan, O.; Jamshidi, S.; Jafari, L. Impacts of climate change on water footprint components of rainfed and irrigated wheat in a semi-arid environment. *Environ. Monit. Assess.* **2023**, *195*, 324. [CrossRef]
51. Wan, W.; Liu, Z.; Li, K.; Wang, G.; Wu, H.; Wang, Q. Drought monitoring of the maize planting areas in Northeast and North China Plain. *Agric. Water Manag.* **2021**, *245*, 106636. [CrossRef]
52. Xiao, L.; Yuan, G.D.; Feng, L.R.; Shah, G.M.; Wei, J. Biochar to Reduce Fertilizer Use and Soil Salinity for Crop Production in the Yellow River Delta. *J. Soil Sci. Plant Nutr.* **2022**, *22*, 1478–1489. [CrossRef]
53. Stojšin, M.M.; Petrović, S.; Banjac, B.; Zečević, V.; Nikolić, S.R.; Majstorović, H.; Đorđević, R.; Knežević, D. Assessment of Genotype Stress Tolerance as an Effective Way to Sustain Wheat Production under Salinity Stress Conditions. *Sustainability* **2022**, *14*, 6973. [CrossRef]
54. Lu, Y.; Liu, S.; Zhang, Y.; Wang, L.; Li, Z. Hydraulic conductivity of gravelly soils with various coarse particle contents subjected to freeze–thaw cycles. *J. Hydrol.* **2021**, *598*, 126302. [CrossRef]
55. Ma, D.; He, Z.; Wang, L.; Zhao, W.; Chen, L.; Lin, P.; Zhao, P.; Wang, W.; Gao, Y.; Li, J. Soil water and salt migration in oasis farmland during crop growing season. *J. Soils Sediments* **2022**, *23*, 355–367. [CrossRef]
56. Liu, J.; Yang, P.; Yang, Z. Water and salt migration mechanisms of saturated chloride clay during freeze-thaw in an open system. *Cold Reg. Sci. Technol.* **2021**, *186*, 103277. [CrossRef]
57. Qin, Y.; Bai, Y.; Chen, G.; Liang, Y.; Li, X.; Wen, B.; Lu, X.; Li, X. The effects of soil freeze-thaw processes on water and salt migrations in the western Songnen Plain, China. *Sci. Rep.* **2021**, *11*, 3888. [CrossRef] [PubMed]
58. Cao, Q.; Yang, B.; Li, J.; Wang, R.; Liu, T.; Xiao, H. Characteristics of soil water and salt associated with Tamarix ramosissima communities during normal and dry periods in a semi-arid saline environment. *Catena* **2020**, *193*, 104661. [CrossRef]
59. Wang, M.; Zhu, Y.; Zhao, T.; Cui, L.; Mao, W.; Ye, M.; Wu, J.; Yang, J. Chemical characteristics of salt migration in frozen soils during the freezing-thawing period. *J. Hydrol.* **2022**, *606*, 127403. [CrossRef]
60. Zhang, R.; Sun, Z.; Li, G.; Wang, H.; Cheng, J.; Hao, M. Influences of water chemical property on infiltration into mixed soil consisting of feldspathic sandstone and aeolian sandy soil. *Sci. Rep.* **2020**, *10*, 19497. [CrossRef]
61. Zhang, S.X.; Zhang, Z.Y.; Jiang, S.W.; Yu, J. Effect of Different Management Systems on Soil Water Content in the Black Soil of Northeast China. *Adv. Mater. Res.* **2012**, *610–613*, 2912–2915. [CrossRef]
62. Guo, K.; Liu, X. Effect of initial soil water content and bulk density on the infiltration and desalination of melting saline ice water in coastal saline soil. *Eur. J. Soil Sci.* **2019**, *70*, 1249–1266. [CrossRef]
63. Wu, C.; Bi, Y.; Christie, P. Influence of Loess Interlayer Thickness on Water Transport in Sand. *Pol. J. Environ. Stud.* **2022**, *31*, 3317–3325. [CrossRef]
64. Li, T.; Hao, X.; Kang, S. Spatiotemporal Variability of Soil Moisture as Affected by Soil Properties during Irrigation Cycles. *Soil Sci. Soc. Am. J.* **2014**, *78*, 598–608. [CrossRef]
65. Chen, J.; Ren, J.; Wang, E.; Cui, L.; Xue, J.; Zheng, X.; Du, Q. Effect of the interlayer on soil temperature and the transformation between phreatic water and soil water under laboratory freeze-thaw action. *Front. Earth Sci.* **2023**, *11*, 677. [CrossRef]
66. Wösten, J.H.M.; Pachepsky, Y.A.; Rawls, W.J. Pedotransfer functions: Bridging the gap between available basic soil data and missing soil hydraulic characteristics. *J. Hydrol.* **2001**, *251*, 123–150. [CrossRef]
67. Zhang, N.; Yu, X.; Pradhan, A.; Puppala, A.J. A new generalized soil thermal conductivity model for sand–kaolin clay mixtures using thermo-time domain reflectometry probe test. *Acta Geotech.* **2016**, *12*, 739–752. [CrossRef]
68. Libohova, Z.; Seybold, C.; Wysocki, D.; Wills, S.; Schoeneberger, P.; Williams, C.; Lindbo, D.; Stott, D.; Owens, P.R. Reevaluating the effects of soil organic matter and other properties on available water-holding capacity using the National Cooperative Soil Survey Characterization Database. *J. Soil Water Conserv.* **2018**, *73*, 411–421. [CrossRef]
69. Zhao, Y.; Si, B.; Zhang, Z.; Li, M.; He, H.; Hill, R.L. A new thermal conductivity model for sandy and peat soils. *Agric. For. Meteorol.* **2019**, *274*, 95–105. [CrossRef]
70. Tan, S.B.K.; Shuy, E.B.; Chua, L.H.C.; Mzila, N. Estimation of areal specific yield in sands using the central limit theorem. *Hydrol. Process.* **2006**, *20*, 3975–3987. [CrossRef]
71. Shen, J.-H.; Hu, M.-J.; Wang, X.; Zhang, C.-Y.; Xu, D.-S. SWCC of Calcareous Silty Sand Under Different Fines Contents and dry Densities. *Front. Environ. Sci.* **2021**, *9*, 682907. [CrossRef]
72. Zhang, X.; Shu, C.; Wu, Y.; Ye, P.; Du, D. Advances of coupled water-heat-salt theory and test techniques for soils in cold and arid regions: A review. *Geoderma* **2023**, *432*, 116378. [CrossRef]

73. Hao, G.; Zhang, Z.; Guan, C.; Liu, G.; Hao, Y.; Fu, T. Influence of Soil Salinization on Active Layer Thickness of Frozen Soil. *Atmosphere* **2023**, *14*, 296. [CrossRef]
74. Zhang, X.; Wang, Q.; Wang, G.; Wang, W.; Chen, H.; Zhang, Z. A Study on the Coupled Model of Hydrothermal-Salt for Saturated Freezing Salinized Soil. *Math. Probl. Eng.* **2017**, *2017*, 4918461. [CrossRef]
75. Yang, C.; Wu, J.; Li, P.; Wang, Y.; Yang, N. Evaluation of Soil-Water Characteristic Curves for Different Textural Soils Using Fractal Analysis. *Water* **2023**, *15*, 772. [CrossRef]
76. Chen, S.; Mao, X.; Shukla, M.K. Evaluating the effects of layered soils on water flow, solute transport, and crop growth with a coupled agro-eco-hydrological model. *J. Soils Sediments* **2020**, *20*, 3442–3458. [CrossRef]
77. Liu, B.; Wang, S.; Kong, X.; Liu, X. Soil matric potential and salt transport in response to different irrigated lands and soil heterogeneity in the North China Plain. *J. Soils Sediments* **2019**, *19*, 3982–3993. [CrossRef]
78. Liu, S.; Huang, Q.; Zhang, W.; Ren, D.; Huang, G. Improving soil hydrological simulation under freeze–thaw conditions by considering soil deformation and its impact on soil hydrothermal properties. *J. Hydrol.* **2023**, *619*, 129336. [CrossRef]
79. Jia, J.; Zhang, P.; Yang, X.; Zhang, X.; Lupwayi, N. Feldspathic sandstone addition and its impact on hydraulic properties of sandy soil. *Can. J. Soil Sci.* **2018**, *98*, 399–406. [CrossRef]
80. El-Hefnawy, M.E.; Selim, E.M.; Assaad, F.F.; Ismail, A.I. The effect of chloride and sulfate ions on the adsorption of Cd2+ on clay and sandy loam Egyptian soils. *Sci. World J.* **2014**, *2014*, 806252. [CrossRef]
81. Pei, Y.; Huang, L.; Li, D.; Shao, M.A. Characteristics and controls of solute transport under different conditions of soil texture and vegetation type in the water-wind erosion crisscross region of China's Loess Plateau. *Chemosphere* **2021**, *273*, 129651. [CrossRef] [PubMed]
82. Luo, J.M.; Yang, F.; Wang, Y.J.; Ya, Y.J.; Deng, W.; Zhang, X.P.; Liu, Z. Mechanism of Soil Sodification at the Local Scale in Songnen Plain, Northeast China, as Affected by Shallow Groundwater Table. *Arid. Land Res. Manag.* **2011**, *25*, 234–256. [CrossRef]

hydrology

Article

Hydrological Properties of Litter in Different Vegetation Types: Implications for Ecosystem Functioning

Lara Castagnolli [1], Fernando Santos Boggiani [1], Jeferson Alberto de Lima [2], Marcelle Teodoro Lima [1] and Kelly Cristina Tonello [1,*]

[1] Department of Environmental Sciences, Federal University of São Carlos, Sorocaba 18052-780, Brazil; lara.castagnolli@estudante.ufscar.br (L.C.); fsboggiani@estudante.ufscar.br (F.S.B.); marcellelima@ufscar.br (M.T.L.)

[2] Department of Environmental Engineering, Federal University of Rondônia, Ji-Paraná 76900-726, Brazil; jeferson.lima@unir.br

* Correspondence: kellytonello@ufscar.br

Abstract: This study investigated the hydrological properties of litter in different vegetation cover types, including *Eucalyptus* sp. plantation, Agroforestry, and Restoration Forest. The research focused on evaluating litter accumulation, composition, water holding capacity, and effective water retention. The results revealed variations in litter accumulation among the stands, and especially *Eucalyptus* sp., which had a higher proportion of branches compared to leaves. The water holding capacity of the litter differed among the stands. Agroforest and Restoration Forest showed higher litter water capacities than *Eucalyptus* sp. The composition and decomposition stage of the litter fractions influenced their water retention capabilities, with leaves exhibiting superior water retention. In contrast, branches had lower water absorption due to their hydrophobic nature. Despite these differences, the effective water retention, which indicates the ability of litter to intercept precipitation, was similar among the stands. The findings highlight the importance of considering litter composition and species-specific characteristics in understanding the hydrological functions of litter. This knowledge contributes to effective conservation and management strategies for sustainable land use practices and water resource management. Further research is recommended to expand the study's scope to include a wider range of forest types and natural field conditions, providing a more comprehensive understanding of litter hydrological functions and their implications for ecosystem processes.

Keywords: forest hydrology; ecohydrology; litter water conservation; soil and water conservation; soil management; soil moisture

Citation: Castagnolli, L.; Boggiani, F.S.; Lima, J.A.d.; Lima, M.T.; Tonello, K.C. Hydrological Properties of Litter in Different Vegetation Types: Implications for Ecosystem Functioning. *Hydrology* **2023**, *10*, 165. https://doi.org/10.3390/hydrology10080165

Academic Editors: Songhao Shang, Qianqian Zhang, Dongqin Yin, Hamza Gabriel and Magdy Mohssen

Received: 20 July 2023
Revised: 4 August 2023
Accepted: 8 August 2023
Published: 10 August 2023

1. Introduction

Litter plays a vital role in nutrient cycling and organic matter decomposition, with notable relevance for global biogeochemical cycles. It refers to the layer of organic or decomposing material present on the forest floor, consisting of a diverse range of components such as leaves, branches, flowers, fruit, seeds, and animal residue. The composition and characteristics of litter are influenced by various factors. For instance, in different types of ecosystems, such as tropical rainforests [1–5] or temperate forests [6–9], the dominant tree species and their specific leaf traits can significantly affect litter composition and decomposition rates [2,10]. Additionally, forest disturbances, such as logging or fire events, can alter litter dynamics and nutrient cycling processes [11–13]. As litter decomposes, it releases nutrients into the soil, which are then taken up by plants, contributing to their growth and overall ecosystem productivity [14–18]. Furthermore, litter provides habitat and food sources for a wide range of organisms [19,20], playing a crucial role in supporting biodiversity within the ecosystem.

From a hydrological perspective, the litter acts as sponge layer, separating precipitation from mineral soil, mediating rainfall splash impact, and reducing or often eliminating overland flow and soil erosion from the forest floor [21,22]. This has implications for infiltration during and after rainfall [2,23]. Understanding the interactions between rainfall, vegetation, and litter on the soil surface is of great importance to forest hydrologists and researchers interested in the partitioning and disposition of rainfall [6,24,25]. The water balance of an ecosystem involves many processes, including the interception of rainfall by vegetation, the portion of rainfall that passes through gaps in the canopy (throughfall), and the water that flows down the trunks or stems of trees (stemflow) [26–29]. The latter two represent the portion of rainfall that reaches the soil and interacts with a boundary layer between the soil surface and its interior [4,5], known as litter [2,23]. Additionally, the presence of a litter layer serves to minimize water evaporation from the soil, aiding the preservation of moisture within the topsoil [30]. However, it is important to note that water retention in the litter layer is an essential hydrological indicator in forest hydrology, and a significant portion of the precipitation may be retained by the litter layer, influencing both evaporation and soil–moisture dynamics [31–34].

Soil water conservation capacity involves various natural processes, land management practices, and the characteristics of the landscape that influence water infiltration, retention, and storage in the soil. This conservation capacity is crucial for sustaining ecological balance, agricultural productivity, and mitigating the negative impacts of water-related issues such as erosion, flooding, and drought. Conserving water and soil resources is essential for sustainable land management, agriculture, and overall environmental health. Implementing practices that enhance water and soil conservation capacity helps protect ecosystems, support biodiversity, and secure the availability of clean water for human and ecological needs. However, it is not possible to generalize the hydrological properties of litter independently of vegetation type (e.g., forest or agriculture crop) or structure. The dynamics of litter water interception vary depending on factors such as stand species composition, thickness, storage, water holding capacity, and degree of decomposition, which contribute to different levels of rainfall interception [4,6,35]. As litter has close contact with soil, the water and soil conservation capacity of a forest is also influenced by factors such as forest type, management practices, soil bulk density, and porosity [4,5,36]. Therefore, in the practice of forestry production on agricultural land, in addition to the state of litter coverage, we should also consider the differences in litter caused by the presence of different tree species. While numerous studies have investigated the impact of litter on hydrological processes, insufficient attention has been paid to the hydrological properties of litter under natural conditions, and especially different vegetation types or management.

In this study, our objective was to assess the water holding characteristics of litter and the physical properties of soil in three vegetation types: *Eucalyptus* sp. plantation, Agroforestry, and Restoration Forest. We aimed to investigate the impact of vegetation cover on litter water conservation. Our hypotheses were as follows: (1) litter hydrological properties exhibit variations among different vegetation types and (2) soil water content, bulk density, and resistance to penetration are influenced by litter hydrological properties. To the best of the authors' knowledge, this is the first study to compare different vegetation types, incorporating various management techniques. A comprehensive understanding of the role of litter in ecosystem water balance and nutrient cycling is crucial for evaluating ecosystem functioning, carbon sequestration, and sustainable land management practices. Further research in this field can offer valuable insights into the intricate relationships between precipitation, vegetation, and ecosystem processes, thereby contributing to the development of effective conservation and management strategies.

2. Materials and Methods

2.1. Study Sites

The experimental site is located at the Federal University of São Carlos—Sorocaba, Brazil (23°35′07″ S, 47°31′03″ W, Figure 1), representing three different soil coverings or vegetation types: EU: a stand of *Eucalyptus* sp. (6 years old); AF: a biodiverse successional agroforestry system including trees from the Atlantic Forest, *Musa* spp. (Banana), and some leguminous species such as *Cajanus cajan* (Feijão guandú), with exotic species (*Brachiaria* sp.) controlled through weeding (established 6 years ago); and RF: Restoration Forest including tree species from the Atlantic Forest and Cerrado (10 years old). Table 1 presents information on the stands. The plots were located 100 m from each other. The mean altitude is 580 m AMSL, and the climate is classified as Cwa or temperate, with dry and hot summers [37,38]. Mean annual temperature is 22 °C and mean annual rainfall is 1311 mm [39]. This site is in a transition region between the Atlantic Plateau and the Paulista Peripheral Depression. It comprises the following soil types: Red Yellow Dystrophic Argisol and Red Dystrophic Latosol [40]. Biotic and abiotic variables were collected simultaneously, allowing a reliable comparison between stands. Three 10 × 10 m plots were established in each stand, and data on throughfall, litter production, hydrological properties, soil bulk density, resistance penetration, and water content were monitored for 8 months, from December 2020 to July 2021.

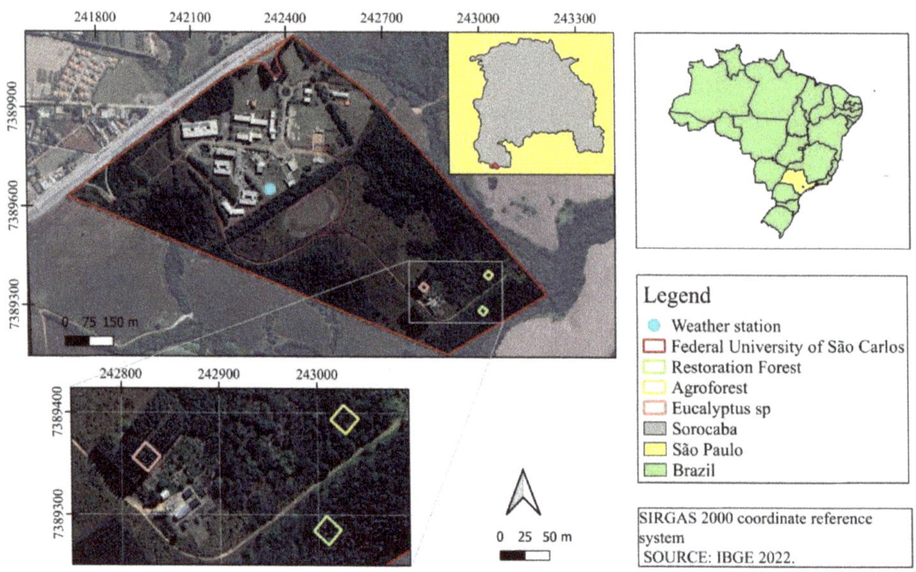

Figure 1. Studied site location at the Federal University of São Carlos—experimental site. Sorocaba, Brazil.

Table 1. Basic information on the sample sites.

Information	*Eucalyptus* sp. (EU)	Agroforest (AF)	Restoration Forest (RF)
Density [trees ha^{-1}]	1667	1250	1667
Management	3 × 2 m planting system. Established in 2014.	Trees on 4 × 2 m, with *Musa* spp. (Banana) and leguminous species between rows. Exotic species controlled through weeding. Established in 2014.	3 × 2 m planting system. Established in 2010.
Diameter at breast height [m]	13.5 ± 0.1	14.6 ± 1.5	13.8 ± 2.6
Height [m]	21.5 ± 0.3	8.2 ± 0.2	10.3 ± 0.1
Litter thickness [cm]	1.3 ± 0.1	2.0 ± 0.1	1.5 ± 0.1
Slope aspect	N-NW	N-NW	N-NW
Slope (°)	10.0%	10.3%	10.2%

2.2. Rainfall and Throughfall

The total rainfall data [R, mm] during the experiment were collected from the weather station located at the experimental site. Throughfall data were obtained using 6 rain gauge collectors positioned 1.20 m above the forest floor in each plot. The collectors were placed under the trees, along the center of the planted row, with a 3 m distance between each other. Throughfall values were obtained following [4,27,41]. Each sampling consisted of one or more consecutive rainfall events. An event is defined as rainfall of at least 1 mm in depth preceded by a dry period of a minimum of 12 h. Sample collections were performed as soon as possible after the end of rainfall.

2.3. Determination of Litter Hydrological Properties

Litter samples were collected from a 100 × 100 cm litter square divided into 4 quadrants. The collection process followed a random sampling method [4,5], where one quadrant measuring 50 × 50 cm was randomly chosen for material collection on the forest floor. The collected samples were then immediately placed in plastic bags and transported to the laboratory. Each stand yielded a total of 240 litter bags (3 plots × 10 random quadrants × 8 months). In the laboratory, the litter samples were sieved to remove soil (mesh 6—3 mm, approximately) and then sorted into four fractions: branches, leaves, seeds, and unstructured material. The monthly and annual litter yield was estimated by summing the fractions. The fresh mass [FM, g] of each fraction was determined using an accurate scale [0.01 g] and rehydrated through immersion in water for 90 min. Subsequently, the litter fractions were placed on sieves and drained for 30 min to determine the humid litter mass [HM, g]. The dried mass [DM, g] of the litter was then determined by oven-drying the samples at 70 °C until a constant mass was achieved. Finally, the litter hydrological properties were calculated as shown in Table 2.

Table 2. Litter hydrological properties studied.

Indicator	Description	Equation
Water holding capacity [WHC, %]	is the amount of water that can be preserved in litter	$WHC = \left[\frac{HM-DM}{DM}\right] \times 100$
Effective water holding capacity [EWC, %]	is the water holding capacity of litter under ambient conditions [5,9]	$EWC = \left[\frac{FM-DM}{FM}\right] \times 100$
Effective water retention capacity [W_{eff}, t ha^{-1}]	is the maximum amount of rainwater that can be retained by the litter layer in the forest in the natural field environment. Is numerically smaller than water retention capacity [42]	$W_{eff} = \frac{(0.85 \times WHC - EWC) \times M}{100}$ M = is the unit litter mass, t ha^{-1}
Maximum water retention capacity [W_{max}, t ha^{-1}]	is the maximum amount of water that can be retained after removing the amount of water contained in the litter under normal conditions [42]	$W_{max} = \frac{WHC \times M}{100}$ M = is the unit litter mass, t ha^{-1}

2.4. Soil Physical Properties

Soil bulk density and soil water content were determined by collecting three random samples from the 0 to 20 cm soil profiles in each plot (three plots per stand). Undisturbed samples were obtained using 100 cm^3 metallic volumetric rings. The weight of the samples was measured using a precision scale. The determination of these attributes was carried out using the thermogravimetric method, which involves weighing the freshly collected samples and then drying them in a forced circulation oven at 105 °C for 24 h. Soil bulk density (BD, g cm^{-3}) was calculated as the ratio of the dry soil mass (Ms, g) to the ring volume (V, m^3). Soil water content (SWC, %) was measured gravimetrically and expressed as a percentage of the weight of soil water to the weight of dry soil (g). Soil resistance to penetration (SRP, MPa) was assessed using Digital Falker PLG1020 Penetrograph, with three repetitions per plot, resulting in a total of nine measurements per stand. BD, SWC, and SRP were measured monthly, simultaneously with the collection of litter samples.

2.5. Statistical Analysis

To assess the homoscedasticity of variance in litter stocks and hydrological properties, the Bartlett test was employed. Additionally, the normality of the data was evaluated using the Lilliefors (Kolmogorov–Smirnov) test for statistical analysis. For normally distributed data, analysis of variance (ANOVA) with a significance level of 5% (Student's *t*-test) was conducted. The non-parametric Mann–Whitney test was used for data that did not meet the assumptions of ANOVA. The statistical analyses were performed using BioEstat 5.3 [43] and @Minitab 17.

3. Results

3.1. Weather Conditions

Accumulated rainfall throughout the research period was 667 mm; that is, approximately 55% of the annual precipitation. Mean temperature was 21.6 ± 3.5 °C, 5% higher than the normal recorded temperature (20.6 °C). The highest mean temperature was recorded in January (24.1 °C) and the lowest in July (15.1 °C) (Figure 2). The mean air humidity was 80.3 ± 1.4%. Throughfall (TF) in the *Eucalyptus* sp. (EU), Agroforestry (AF), and Restoration Forest (RF) areas was 395, 412, and 471 mm, respectively.

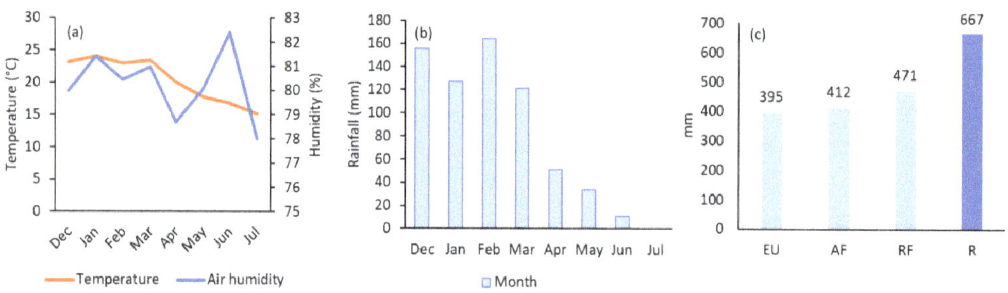

Figure 2. Mean air temperature and humidity (**a**), rainfall (**b**), throughfall in *Eucalyptus* sp. (EU), Agroforest (AF), Restoration Forest (RF), and total rainfall (R) during period of study (**c**).

3.2. Litter Accumulation and Composition

The mean litter accumulation ranged from 5.4 to 8.3 t ha^{-1} in the EU, 4.2 to 11 t ha^{-1} in the Agroforestry (AF) area, and 3.5 to 8.0 t ha^{-1} in the Restoration Forest (RF) area. Among the different components, branches accounted for 39% of the total litter in the EU, while leaves were the predominant fraction in the AF area (44%) and the RF area (47%). Seeds represented the lowest fraction in all stands, with percentages of 3.4% in the EU, 9.3% in the AF, and 3.2% in the RF. Although the litter accumulation was similar among the stands, there were significant differences in the mass of leaf and branch fractions (Figure 3).

3.3. Hydrological Properties of Litter

The total litter water holding capacity (WHC) differed between the Eucalyptus (EU) area and the Agroforestry (AF) and Restoration Forest (RF) areas (Table 3). The order of WHC was EU < RF < AF, with AF and RF having 1.33 and 1.30 times the water holding capacity of EU, respectively. The WHC for the unstructured fraction was similar among the stands, and it increased in the following order: RF < AF < EU. EU had the lowest values for leaves, branches, and seeds. The water holding capacity of leaves and seeds was similar between AF and RF, while branches showed significant variation among the three stands ($p < 0.05$). Although the mean effective water holding capacity (EWC, %) was similar among the stands ($p > 0.05$), the EWC was consistently higher for RF, followed by AF > EU, for total litter as well as for the unstructured fraction, leaves, branches, and seeds.

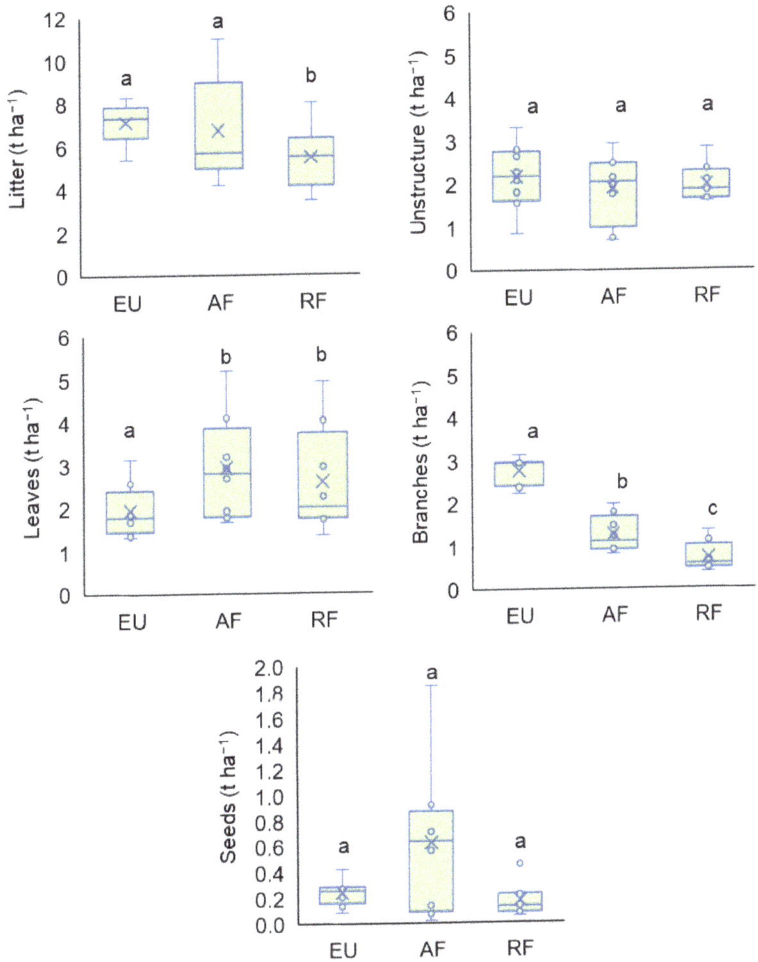

Figure 3. Mean litter accumulation, unstructured, leaves, branches, and seeds in *Eucalyptus* sp. (EU), Agroforest (AF), and Restoration Forest (RF). Different lowercase letters mean significant differences ($p < 0.05$).

Table 3. Water holding capacity [WHC, %], effective water holding capacity [EWC, %], effective water retention [W_{eff}, t ha^{-1}], and maximum retention capacity [W_{max}, t ha^{-1}].

Stand	Total	Unstructured	Leaves	Branches	Seeds
		Water holding capacity [WHC, %]			
Eucalyptus sp.	164 ± 6.8 a	228 ±16 a	193 ± 28 a	94 ± 3.8 a	141 ± 8 a
Agroforest	218 ± 12 b	222 ± 20 a	272 ± 32 b	200 ± 19 b	179 ± 15 b
Restoration	212 ±14 b	220 ± 18 a	265 ± 15 b	160 ± 24 c	204 ± 39 b
		Effective water holding capacity [EWC, %]			
Eucalyptus sp.	23 ± 3 a	25 ± 4 a	26 ± 2 a	17 ± 2 a	24 ± 2 a
Agroforest	25 ± 3 ab	25 ± 3 a	28 ± 3 a	28 ± 2 b	27 ± 2 a
Restoration	34 ± 4 b	34 ± 5 a	33 ± 3 a	35 ± 3 b	36 ± 3 a

Different lowercase letters in the same column mean significant differences ($p < 0.05$).

The maximum water holding capacity (W_{max}) was significantly different only between EU and AF, ranging from 10.3 t ha^{-1} (AF) to 31.7 t ha^{-1} (RF) (Figure 4). In EU and RF, the W_{max} was 1.6 and 1.4 times that of AF, respectively. The W_{max} of the unstructured layer varied among the stand types, with AF (4.3 ± 0.7 t ha^{-1}) < RF (12.1 ± 1.5 t ha^{-1}) < EU (16.3 ± 1.2 t ha^{-1}). There was a significant difference in the W_{max} of leaves between EU and AF ($p < 0.05$), with the order being EU (4.1 ± 0.36 t ha^{-1}) < RF (5.5 ± 1.0 t ha^{-1}) < AF (8.2 ± 1.3 t ha^{-1}). RF had the highest W_{max} for branches, followed by AF and EU. Seeds showed the order of W_{max} as AF < RF < EU, with significant differences between AF and RF for branches and seeds.

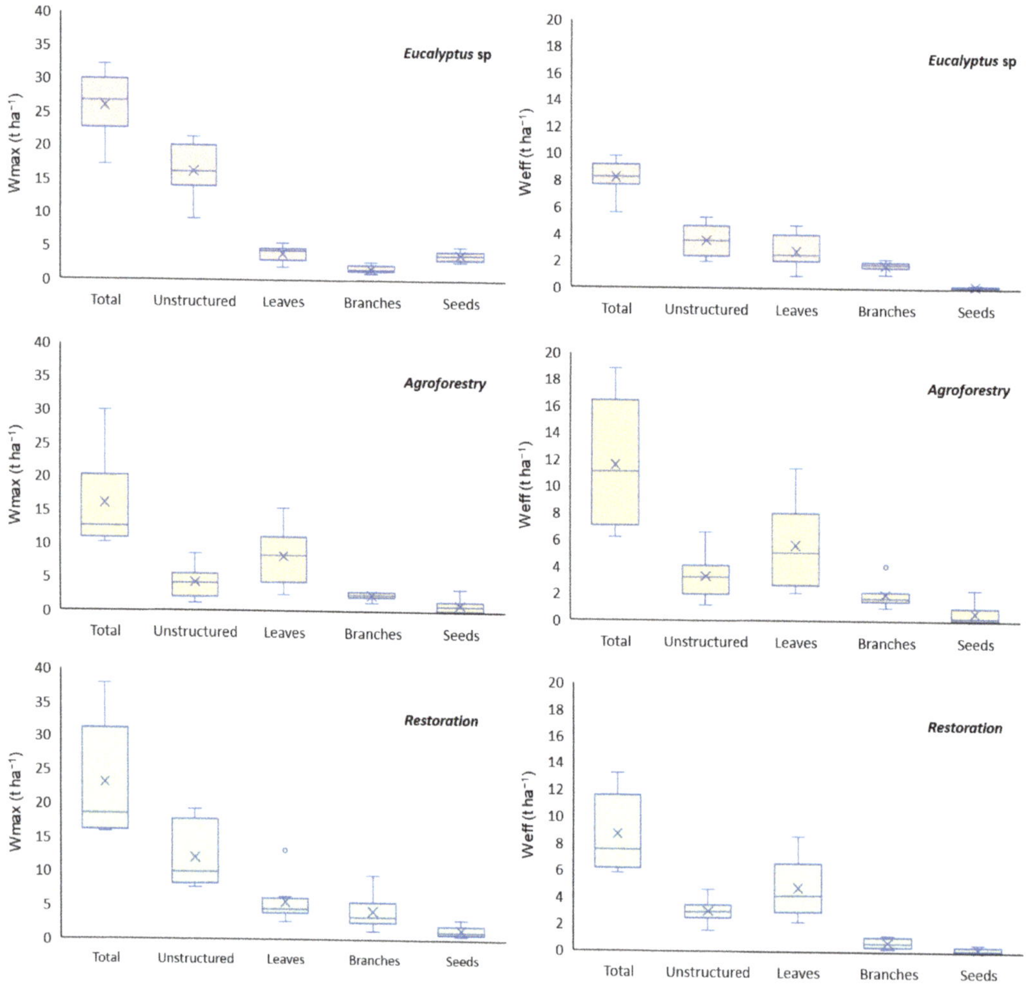

Figure 4. Litter maximum water retention capacity [W_{max}, t ha^{-1}] and effective water retention [W_{eff}, t ha^{-1}].

The effective water retention capacity (W_{eff}) of litter did not differ among stands and ranged from 5.6 t ha^{-1} for EU to 18.3 t ha^{-1} for AF (Figure 3). The mean W_{eff} followed the order: EU (8.3 ± 0.5 t ha^{-1}) < RF (8.7 ± 1.0 t ha^{-1}) < AF (11.6 ± 1.7 t ha^{-1}). For all stands, leaves had the highest W_{eff}, followed by the unstructured fraction, branches, and seeds. The W_{eff} for litter was similar among stands. The W_{eff} for the unstructured fraction ranged from 2.0 t ha^{-1} to 5.3 t ha^{-1} for EU, from 1.7 t ha^{-1} to 6.6 t ha^{-1} for AF, and from 1.6 t ha^{-1}

to 4.6 t ha^{-1} for RF. The mean W_{eff} for the unstructured fraction was similar among stands, following the order: RF < AF < EU. Significant differences were observed between leaves for EU × AF and RF. The mean W_{eff} for leaves followed the order: EU (2.8 ± 0.4 t ha^{-1}) < RF (4.8 ± 0.8 t ha^{-1}) < AF (5.7 ± 1.2 t ha^{-1}). For branches, W_{eff} was similar between EU and AF, and followed the order: RF (0.7 ± 0.1 t ha^{-1}) < EU (1.7 ± 0.1 t ha^{-1}) < AF (2.0 ± 0.3 t ha^{-1}). W_{eff} for seeds was similar among stands, increasing in the order of: EU (0.2 ± 0.02 t ha^{-1}) < RF (0.3 ± 0.07 t ha^{-1}) < AF (0.6 ± 0.3 t ha^{-1}).

3.4. Soil Physical Properties

There were no significant differences in the mean values of soil bulk density (BD), soil retention potential (SRP), and soil water content (SWC) among the stands (Figure 5). The BD was 1.0 ± 0.05 g cm^{-3} for all three stands. EU had the highest SRP (5.1 ± 1.5 MPa), followed by RF (4.9 ± 1.3 MPa) and AF (4.6 ± 1.2 MPa). The mean SWC was highest in RF (19.0 ± 3.7%), followed by AF (16.9 ± 3.5%) and EU (15.1 ± 3.1%). Soil water content ranged from 2.4% to 34.0% in EU, from 1.6% to 36.6% in AF, and from 2.5% to 40.3% in RF.

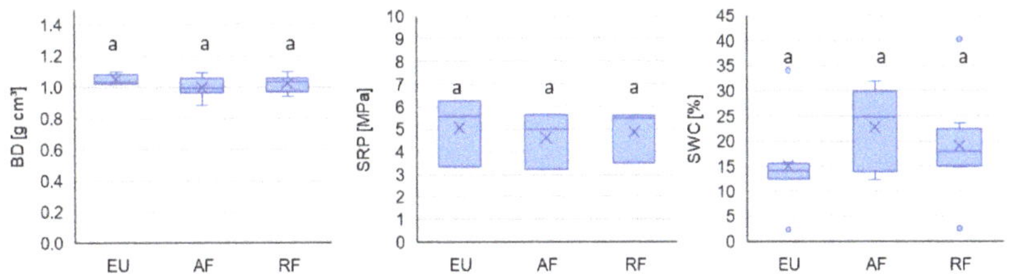

Figure 5. Soil bulk density [BD, g cm^3], soil resistance of penetration [SRP, MPa], and soil water content [SWC, %] in *Eucalyptus* sp. (EU), Agroforest (AF), and Restoration Forest (RF). Different lowercase letters in the same column mean significant differences ($p < 0.05$).

4. Discussion

4.1. Hydrological Properties and Water Retention Capacity of Litter

Litter serves as a crucial component in the hydrological cycle within ecosystems [8,44,45] and it also plays a significant role in various soil management practices and vegetation covers [45]. In our study, the litter mass of *Eucalyptus* sp. was above the average of other studies, although it varied within the range reported in the literature. For example, it exceeded the average deposition of 6.33 Mg ha^{-1} at 14 years of age [46] but was lower than the 16.6 t ha^{-1} at 7 years of age [47]. Compared with other forest types, the average litter mass of *Eucalyptus* sp. and Agroforest was higher than that of approximately 46-year-old Cerrado (5.5 t ha^{-1}) [4], while the latter was similar to the litter mass of Restoration Forest. It is worth noting that the composition and dynamics of litter formation can vary depending on the species present in the study area. In this case, the *Eucalyptus* sp. planting consisted of trees of the same species, while the Agroforest and Restoration Forest were composed of a diversity of plants species. This difference in species composition likely contributed to the higher proportion of branches in the litter of *Eucalyptus* sp. compared with other studies where the leaf fraction was more dominant [10,48].

The results demonstrate variations in litter hydrological properties among the different vegetation types. Understanding the hydrological properties of litter fractions is essential, as it influences the overall water holding capacity of the litter layer. The water holding capacity of litter depends not only on the quantity of organic material deposited but also on the composition and degree of decomposition of its fractions. Litter layers with higher decomposition levels have a larger specific surface area, enhancing their water retention potential [5,49,50]. It is important to note that water holding capacity reflects the ideal water holding condition and may not fully represent litter interception under field conditions [4,5,7]. The water holding capacity of the litter was found to differ significantly between the stands, with Agroforestry and Restoration Forest showing higher capacity compared with *Eucalyptus* sp. This variation can be attributed to differences in litter composition, particularly the proportion of leaves and branches. The water holding capacity of the unstructured fraction of *Eucalyptus* sp. was relatively high due to its lower surface adhesion [51,52], resulting in greater water retention rates. In contrast, in Agroforest and Restoration Forest, the leaf fraction exhibited the highest water holding capacity due to its higher surface adhesion. The composition and decomposition stage of the litter fractions, particularly the higher proportion of branches in *Eucalyptus* sp., influenced the water holding capacity. The hydrophobic nature of branches limited their water absorption capacity, while leaves showed higher surface adhesion and superior water retention. Stems or branches, mainly composed of xylem with thick fibers and a relatively stable structure between cells, have limited water absorption capacity despite the presence of an internal tubular structure [24]. Additionally, the hydrophobic nature of branches is a well-known characteristic that can also be attributed to the presence of lignin in their composition [53,54]. Lignin, being a complex aromatic polymer, contributes to the structural integrity of plant cell walls and imparts rigidity to woody tissues. As a hydrophobic substance, lignin naturally repels water [53–57]. It is important to note that the lignin content can vary not only between different tree species [53] but also between different parts of the same tree, such as leaves, stems, and branches. Moreover, the proportion of branches in the litter layer can vary depending on the tree species composition and forest management practices. This variability can influence the overall water holding capacity of the litter and its hydrological functions in different ecosystems. Overall, for all stands, the water holding capacity was lower than that observed for Amazon Forest [5] and Cerrado under various stages of regeneration [4]. Notably, the water retention rates of *Eucalyptus* sp. in this study were lower than those reported in other *Eucalyptus* sites [48].

Despite differences in water holding capacity, the effective water retention capacity (W_{eff}) of the litter was found to be similar among the stands. The similarity in W_{eff} indicates that *Eucalyptus* sp., Agroforest, and Restoration Forest have the ability to effectively retain water, contributing to enhanced rainwater storage within the ecosystem. This fact may explain the similarity in soil bulk density, soil resistance to penetration, and soil water content among the studied sites, although the water retention capacity was higher for Agroforest. This suggests that the hydrological properties of litter did not have a direct impact on soil physical properties in the studied stands. However, it is important to note that in this study, the soil physical properties were investigated at the topsoil level (0–20 cm for soil bulk and water content and 0–60 cm for resistance of penetration), and these properties can vary according to the soil profile [58–60]. Nevertheless, as reported by [36,61], W_{eff} defines the effective interception of precipitation by litter, which is an important hydrological property that can be used to consistently evaluate the potential to absorb rainfall and reduce surface runoff [9,42,61]. Effective water retention is influenced by factors such as litter water content, storage capacity, and rainfall characteristics [5]. Our study shows that the mean annual capacity of litter for both stands to retain water was higher for both sites than that observed for acacia–grass forest, eucalyptus–grass forest, and bamboo–grass forest [61]. However, *Eucalyptus* sp., Agroforest, and Restoration Forest were lower than *Acacia mangium* and higher than *Hevea brasiliensis* [9]. Moreover, Agroforest was W_{eff} slightly higher than Amazon Forest [5]. However, the W_{max}, which measures the rainfall absorption capacity, was higher in *Eucalyptus* sp., Agroforest, and Restoration Forest compared with previous studies on acacia–grass forest, eucalyptus–grass forest, and bamboo–grass forest [61]. It was also higher than *Eucalyptus robusta* [9] but lower than *Acacia mangium* [9]. Moreover, W_{max} for *Eucalyptus* sp. and Restoration Forest was higher than Amazon Forest [5]. Considering that 1 mm of rainfall is equivalent to 1 t ha^{-1} [9,61], the litter in *Eucalyptus* sp., Agroforest, and Restoration Forest could intercept an average of 26 mm, 16 mm, and 23 mm of rainfall, respectively, during the studied period. Considering the effective water retention for the studied period, *Eucalyptus* sp., Agroforest, and Restoration Forest intercepted a mean of 8.3, 11.6, and 8.7 mm of rainfall, respectively.

4.2. Implications for Ecosystem Functioning

It is important to highlight that a previous study conducted on *Pinus tabulaeformis* plantations revealed that around half of the throughfall was retained in the soil without any litter mass. In the litter-covered treatments, this proportion ranged from 77.0% to 87.9% [44], indicating that the presence of litter significantly enhanced rainwater storage [62]. These findings highlight the importance of considering litter composition and species-specific characteristics when assessing the hydrological functions of litter. The variations in litter hydrological properties observed in this study have important ecological implications for ecosystem functioning and water resource management. The higher water holding capacity of Agroforestry and Restoration Forest litter implies that these stands can retain more water, which has implications for water availability within the ecosystem. Increased water retention can lead to enhanced soil moisture, providing a vital water resource for plants and microorganisms, especially during dry periods. Restoration Forest, which exhibited higher effective water retention capacity, can serve as a valuable model for ecosystem restoration efforts. Restoring degraded areas with diverse native tree species can improve litter hydrological properties, leading to enhanced ecosystem services and ecological functionality. Agroforestry systems, with their capacity for higher water retention, can be integrated into water-sensitive agricultural practices to improve water availability for crops and reduce water-related risks, such as soil erosion and flooding. This improved water availability can contribute to the overall productivity and resilience of the ecosystem. Furthermore, litter acts as a natural barrier against soil erosion by reducing the impact of rainfall and slowing down or preventing surface runoff. The ability of Agroforestry and Restoration Forest to retain more water in their litter layers can be particularly beneficial in mitigating soil erosion and preserving soil health. Regarding biodiversity support, litter provides a habitat and food source for various organisms, supporting biodiversity within the ecosystem. The differences in litter composition and hydrological properties among the stands can influence the diversity and abundance of soil-dwelling organisms, contributing to overall ecosystem biodiversity.

Finally, the results obtained in this study provide valuable insights into the hydrological properties of litter in the specific stands investigated. However, further research is needed to expand the scope of the study and examine a wider range of forest types and species to obtain a more comprehensive understanding of litter hydrological functions. Additionally, field conditions and natural variability should be considered to better assess the actual litter interception and water retention capacities of different ecosystems. Such knowledge can contribute to improved water resource management and the development of sustainable land use practices that consider the hydrological role of litter in ecosystem functioning.

5. Conclusions

This study provides valuable insights into the hydrological properties of litter in different vegetation cover types, highlighting variations in water holding capacity and effective water retention among *Eucalyptus* sp., Agroforestry, and Restoration Forest. The results revealed variations in litter accumulation, composition, water holding capacity, and effective water retention among the stands. The litter accumulation in *Eucalyptus* sp. had a higher proportion of branches compared to leaves. The water holding capacity of the litter varied among the stands, with Agroforest and Restoration Forest exhibiting higher capacities than *Eucalyptus* sp. Leaves showed higher surface adhesion and superior water retention, while branches exhibited lower water absorption due to their hydrophobic nature. Despite differences in water holding capacity, the effective water retention, which defines the effective interception of precipitation by litter, was similar among the stands. Moreover, litter hydrological properties did not affect soil bulk density, soil retention potential, and soil water content. These findings emphasize the importance of considering litter composition and species-specific characteristics when assessing the hydrological functions of litter. The results contribute to our understanding of the intricate relationship between vegetation cover, litter properties, and water balance in ecosystems. This knowledge can aid in the development of effective conservation and management strategies for sustainable land use practices, carbon sequestration, and water resource management. Further research is recommended to broaden the scope of the study, encompassing a wider range of forest types and species, as well as considering field conditions and natural variability. This would provide a more comprehensive understanding of litter hydrological functions and their implications for ecosystem processes.

Author Contributions: All authors contributed equally. All authors have read and agreed to the published version of the manuscript.

Funding: This research was funded by Brazilian National Council for Scientific and Technological Development (CNPq).

Data Availability Statement: The data that support the findings of this study are available from the corresponding author upon reasonable request.

Conflicts of Interest: The authors declare no conflict of interest.

References

1. Vasconcelos, S.S.; Zarin, D.J.; Araújo, M.M.; Rangel-Vasconcelos, L.G.T.; Carvalho, C.J.R.; Staudhammer, C.L.; Oliveira, F. de A. Effects of Seasonality, Litter Removal and Dry-Season Irrigation on Litterfall Quantity and Quality in Eastern Amazonian Forest Regrowth, Brazil. *J. Trop. Ecol.* **2008**, *24*, 27–38. [CrossRef]
2. Giácomo, R.G.; Alves, M.C.; Camara, R.; Pereira, M.G.; de Arruda, O.G.; Souto, S.N.; de Moraes, M.L.T. Litterfall and Nutrient Input in a Degraded Area. *Floresta E Ambiente* **2017**, *24*, e20160028. [CrossRef]
3. Martius, C.; Höfer, H.; Garcia, M.V.B.; Römbke, J.; Hanagarth, W. Litter Fall, Litter Stocks and Decomposition Rates in Rainforest and Agroforestry Sites in Central Amazonia. *Nutr. Cycl. Agroecosyst.* **2004**, *68*, 137–154. [CrossRef]
4. Pereira, L.C.; Balbinot, L.; Lima, M.T.; Bramorski, J.; Tonello, K.C. Aspects of Forest Restoration and Hydrology: The Hydrological Function of Litter. *J. For. Res.* **2021**, *33*, 543–552. [CrossRef]
5. Lima, J.A.; Tonello, K.C. Rainfall Partitioning in Amazon Forest: Implications of Reduced Impact Logging on Litter Water Conservation. *Hydrology* **2023**, *10*, 97. [CrossRef]
6. Su, S.; Liu, X. The Water Storage Function of Litters and Soil in Five Typical Plantations in the Northern and Southern Mountains of Lanzhou, Northwest China. *Sustainability* **2022**, *14*, 8231. [CrossRef]
7. Li, W.; Li, Y.; Long, M.; Li, X. Study on Water-Holding Properties of Litters in Different Types of Forests of Yuntaishan Mountain Area in Shibing County, Guizhou Province. In *IOP Conference Series: Materials Science and Engineering*; Institute of Physics Publishing: Bristol, UK, 2018; Volume 394.
8. Xia, L.; Song, X.; Fu, N.; Cui, S.; Li, L.; Li, H.; Li, Y. Effects of Forest Litter Cover on Hydrological Response of Hillslopes in the Loess Plateau of China. *Catena* **2019**, *181*, 104076. [CrossRef]
9. Tu, Z.; Chen, S.; Ruan, D.; Chen, Z.; Huang, Y.; Chen, J. Differential Hydrological Properties of Forest Litter Layers in Artificial Afforestation of Eroded Areas of Latosol in China. *Sustainability* **2022**, *14*, 14869. [CrossRef]
10. Santos, A.F.A.; Carneiro, A.C.P.; Martinez, D.T.; Caldeira, S.F. Capacidade de Retenção Hídrica Do Estoque de Serapilheira de Eucalipto. *Floresta E Ambiente* **2017**, *24*, e20150303. [CrossRef]
11. Throop, H.L.; Abu Salem, M.; Whitford, W.G. Fire Enhances Litter Decomposition and Reduces Vegetation Cover Influences on Decomposition in a Dry Woodland. *Plant Ecol.* **2017**, *218*, 799–811. [CrossRef]
12. Bryanin, S.; Kondratova, A.; Abramova, E. Litter Decomposition and Nutrient Dynamics in Fire-Affected Larch Forests in the Russian Far East. *Forests* **2020**, *11*, 882. [CrossRef]
13. Bomfim, B.; Silva, L.C.R.; Pereira, R.S.; Gatto, A.; Emmert, F.; Higuchi, N. Litter and Soil Biogeochemical Parameters as Indicators of Sustainable Logging in Central Amazonia. *Sci. Total Environ.* **2020**, *714*, 136780. [CrossRef] [PubMed]
14. Watanabe, T.; Fukuzawa, K.; Shibata, H. Temporal Changes in Litterfall, Litter Decomposition and Their Chemical Composition in Sasa Dwarf Bamboo in a Natural Forest Ecosystem of Northern Japan. *J. For. Res.* **2013**, *18*, 129–138. [CrossRef]
15. Zhu, X.; Liu, W.; Chen, H.; Deng, Y.; Chen, C.; Zeng, H. Effects of Forest Transition on Litterfall, Standing Litter and Related Nutrient Returns: Implications for Forest Management in Tropical China. *Geoderma* **2019**, *333*, 123–134. [CrossRef]
16. Horodecki, P.; Nowiński, M.; Jagodziński, A.M. Advantages of Mixed Tree Stands in Restoration of Upper Soil Layers on Postmining Sites: A Five-Year Leaf Litter Decomposition Experiment. *Land Degrad. Dev.* **2019**, *30*, 3–13. [CrossRef]
17. Ge, X.; Zeng, L.; Xiao, W.; Huang, Z.; Geng, X.; Tan, B. Effect of Litter Substrate Quality and Soil Nutrients on Forest Litter Decomposition: A Review. *Acta Ecol. Sin.* **2013**, *33*, 102–108. [CrossRef]
18. Tonello, K.C.; Pereira, L.C.; Balbinot, L.; Nnadi, E.O.; Mosleh, M.H.; Bramorski, J. Patterns of Litter and Nutrient Return to the Soil during Passive Restoration in Cerrado, Brazil. *Biologia* **2023**, *78*, 399–414. [CrossRef]

19. Peng, Y.; Vesterdal, L.; Peñuelas, J.; Peguero, G.; Wu, Q.; Heděnec, P.; Yue, K.; Wu, F. Soil Fauna Effects on Litter Decomposition Are Better Predicted by Fauna Communities within Litterbags than by Ambient Soil Fauna Communities. *Plant Soil* **2023**, *487*, 49–59. [CrossRef]
20. Yue, K.; De Frenne, P.; Van Meerbeek, K.; Ferreira, V.; Fornara, D.A.; Wu, Q.; Ni, X.; Peng, Y.; Wang, D.; Heděnec, P.; et al. Litter Quality and Stream Physicochemical Properties Drive Global Invertebrate Effects on Instream Litter Decomposition. *Biol. Rev.* **2022**, *97*, 2023–2038. [CrossRef]
21. Neris, J.; Tejedor, M.; Rodríguez, M.; Fuentes, J.; Jiménez, C. Effect of Forest Floor Characteristics on Water Repellency, Infiltration, Runoff and Soil Loss in Andisols of Tenerife (Canary Islands, Spain). *Catena* **2013**, *108*, 50–57. [CrossRef]
22. Chen, S.; Cao, T.; Tanaka, N.; Gao, T.; Zhu, L.; Zou, C.B. Hydrological Properties of Litter Layers in Mixed Forests in Mt. Qinling, China. *IForest* **2018**, *11*, 243–250. [CrossRef]
23. Cheng, J.D.; Lin, J.P.; Lu, S.Y.; Huang, L.S.; Wu, H.L. Hydrological Characteristics of Betel Nut Plantations on Slopelands in Central Taiwan. *Hydrol. Sci. J.* **2008**, *53*, 1208–1220. [CrossRef]
24. Xie, J.; Su, D. Water-Holding Characteristics of Litter in Meadow Steppes with Different Years of Fencing in Inner Mongolia, China. *Water* **2020**, *12*, 2374. [CrossRef]
25. Bai, Y.; Zhou, Y.; Du, J.; Zhang, X.; Di, N. Effects of a Broadleaf-Oriented Transformation of Coniferous Plantations on the Hydrological Characteristics of Litter Layers in Subtropical China. *Glob. Ecol. Conserv.* **2021**, *25*, e01400. [CrossRef]
26. Tonello, K.C.; Van Stan, J.T.; Rosa, A.G.; Balbinot, L.; Pereira, L.C.; Bramorski, J. Stemflow Variability across Tree Stem and Canopy Traits in the Brazilian Cerrado. *Agric. For. Meteorol.* **2021**, *308–309*, 108551. [CrossRef]
27. Tonello, K.C.; Rosa, A.G.; Pereira, L.C.; Matus, G.N.; Guandique, M.E.G.; Navarrete, A.A. Rainfall Partitioning in the Cerrado and Its Influence on Net Rainfall Nutrient Fluxes. *Agric. For. Meteorol.* **2021**, *303*, 108372. [CrossRef]
28. Van Stan, J.T.; Gutmann, E.; Friesen, J. *Precipitation Partitioning by Vegetation: A Global Synthesis*; Springer International Publishing: Berlin/Heidelberg, Germany, 2020; ISBN 9783030297022.
29. Lima, J.; Tonello, K.C. Rainfall Partitioning in Amazon Forest: Implications of Reduced Impact Logging for Hydrological Processes. *Agric. For. Meteorol.* **2023**, *327*, 109505. [CrossRef]
30. Van Stan, J.T.; Coenders-Gerrits, M.; Dibble, M.; Bogeholz, P.; Norman, Z. Effects of Phenology and Meteorological Disturbance on Litter Rainfall Interception for a Pinus Elliottii Stand in the Southeastern United States. *Hydrol. Process.* **2017**, *31*, 3719–3728. [CrossRef]
31. Acharya, B.S.; Stebler, E.; Zou, C.B. Monitoring Litter Interception of Rainfall Using Leaf Wetness Sensor under Controlled and Field Conditions. *Hydrol. Process.* **2017**, *31*, 240–249. [CrossRef]
32. Melissa, K.; Dickow, C.; Marques, R.; Pinto, C.B.; Höfer, H. Produção de Serapilheira Em Diferentes Fases Sucessionais de Uma Floresta Subtropical Secundária, Em Antonina, PR. *Cerne* **2011**, *18*, 75–86.
33. Winkler, J.P.; Negreiros, A.B. A Serrapilheira Como Bioindicador de Qualidade Ambiental Em Fragmentos de Eucalyptus. *Rev. Cont. (UFRRJ)* **2018**, *7*, 175–202. [CrossRef]
34. Floriancic, M.G.; Allen, S.T.; Meier, R.; Truniger, L.; Kirchner, J.W.; Molnar, P. Potential for Significant Precipitation Cycling by Forest-Floor Litter and Deadwood. *Ecohydrology* **2022**, *16*, e2493. [CrossRef]
35. Zagyvai-Kiss, K.A.; Kalicz, P.; Szilágyi, J.; Gribovszki, Z. On the Specific Water Holding Capacity of Litter for Three Forest Ecosystems in the Eastern Foothills of the Alps. *Agric. For. Meteorol.* **2019**, *278*, 107656. [CrossRef]
36. Li, Y.; Li, B.; Zhang, X.; Chen, J.J.; Zhan, F.D.; Guo, X.H.; Zu, Y.Q. Differential Water and Soil Conservation Capacity and Associated Processes in Four Forest Ecosystems in Dianchi Watershed, Yunnan Province, China. *J. Soil Water Conserv.* **2015**, *70*, 198–206. [CrossRef]
37. Dubreuil, V.; Fante, K.P.; Planchon, O.; Sant'Anna Neto, J.L. Climate Change Evidence in Brazil from Köppen's Climate Annual Types Frequency. *Int. J. Climatol.* **2019**, *39*, 1446–1456. [CrossRef]
38. Pinheiro, R.C.; Tonello, K.C.; Valente, R.O.A.; Mingoti, R.; Santos, I.P. Occupation and Hydrologic Characterization of Ipaneminha Watershed, Sorocaba-SP. *IRRIGA* **2011**, *16*, 234. [CrossRef]
39. Lima, M.T.; Urso-Guimarães, M.V.; Van Stan, J.T.; Tonello, K.C. Stemflow Metazoan Transport from Common Urban Tree Species (São Paulo, Brazil). *Ecohydrology* **2023**, *16*, e2517. [CrossRef]
40. Oliveira, J.B.; Camargo, M.N.; Rossi, M.; Calderano Filho, B. *Solos Do Estado de São: Descrição Das Classes Registradas No Mapa Pedológico*; Embrapa Solos: Rio de Janeiro, Brazil; Instituto Agronômico: Campinas, Brazil, 1999.
41. Lloyd, C.R.; Marques, A.O. Spatial Variability of Throughfall and Stemflow Measurements in Amazonian Rainforest. *Agric. For. Meteorol.* **1988**, *42*, 73. [CrossRef]
42. Zhou, Q.; Keith, D.M.; Zhou, X.; Cai, M.; Cui, X.; Wei, X.; Luo, Y. Comparing the Water-Holding Characteristics of Broadleaved, Coniferous, and Mixed Forest Litter Layers in a Karst Region. *Mt. Res. Dev.* **2018**, *38*, 220–229. [CrossRef]
43. Ayres, M.; Ayres, M., Jr.; Ayres, D.L.; Santos, A.A. *Bioestat 5.0 Aplicações Estatísticas Nas Áreas Das Ciências Biológicas e Médicas*; IDSM: Belém, Brazil, 2007.
44. Cui, Y.; Pan, C.; Zhang, G.; Sun, Z.; Wang, F. Effects of Litter Mass on Throughfall Partitioning in a Pinus Tabulaeformis Plantation on the Loess Plateau, China. *Agric. For. Meteorol.* **2022**, *318*, 108908. [CrossRef]
45. Zhu, H.; Wang, G.; Yinglan, A.; Liu, T. Ecohydrological Effects of Litter Cover on the Hillslope-Scale Infiltration-Runoff Patterns for Layered Soil in Forest Ecosystem. *Ecol. Eng.* **2020**, *155*, 105930. [CrossRef]

46. Souza, J.A.; Davide, A.C. Decomposição de Serapilheira e Nutrientes Em Uma Mata Não Minerada e Em Plantações de Bracatinga (Mimosa Scabrella) e de Eucalipto (Eucalyptus Saligna) Em Áreas de Mineração de Bauxita. *Cerne* **2001**, *7*, 101–113.
47. Balieiro, F.D.C.; Franco, A.A.; Pereira, M.G.; Campello, E.F.C.; Dias, L.E.; de Faria, S.M.; Alves, B.J.R. Dinâmica Da Serapilheira e Transferência de Nitrogênio Ao Solo, Em Plantios de Pseudosamanea Guachapele e Eucalyptus Grandis. *Pesqui. Agropecuária Bras.* **2004**, *39*, 597–601. [CrossRef]
48. Melos, A.R.; Sato, A.M.; Coelho Netto, A.L. Produção, Estoque e Retenção Hídrica Da Serrapilheira Em Encosta Sob Plantio de Híbridos de Eucalyptus Urophylla e Eucalyptus Grandis: Médio Vale Do Rio Paraíba Do Sul. *Anuário Do Inst. De Geociências* **2010**, *33*, 66–73. [CrossRef]
49. Pereira, M.G.; Menezes, L.F.T.; Schultz, N. Aporte e Decomposição Da Serapilheira Na Floresta Atlântica, Ilha Da Marambaia, Mangaratiba, RJ. *Ciência Florest.* **2008**, *18*, 443–454. [CrossRef]
50. Pires, L.A.; Miranda De Britez, R.; Martel, G.; Pagano, S.N. Produção, Acúmulo e Decomposição Da Serapilheira Em Uma Restinga Da Ilha Do Mel, Paranaguá, PR, Brasil. *Acta Bot. Bras.* **2006**, *20*, 173–184. [CrossRef]
51. Santos, F.M.; Chaer, G.M.; Diniz, A.R.; Balieiro, F.C. Nutrient Cycling over Five Years of Mixed-Species Plantations of Eucalyptus and Acacia on a Sandy Tropical Soil. *For. Ecol. Manag.* **2017**, *384*, 110–121. [CrossRef]
52. Mateus, F.A.; Couto Miranda, C.; Valcarcel, R.; Figueiredo, P.H.A. Estoque e Capacidade de Retenção Hídrica Da Serrapilheira Acumulada Na Restauração Florestal de Áreas Perturbadas Na Mata Atlântica. *Floresta E Ambiente* **2013**, *20*, 336–343. [CrossRef]
53. Rahman, M.M.; Tsukamoto, J.; Rahman, M.M.; Yoneyama, A.; Mostafa, K.M. Lignin and Its Effects on Litter Decomposition in Forest Ecosystems. *Chem. Ecol.* **2013**, *29*, 540–553. [CrossRef]
54. Vogelmann, E.S.; Prevedello, J.; Reichert, J.M. Origem Dos Compostos Hidrofóbicos e Seus Efeitos Em Florestas de Pinus e Eucalyptus. *Ciência Florest.* **2015**, *25*, 1067–1079. [CrossRef]
55. Tonello, K.C.; Campos, S.D.; de Menezes, A.J.; Bramorski, J.; Mathias, S.L.; Lima, M.T. How Is Bark Absorbability and Wettability Related to Stemflow Yield? Observations From Isolated Trees in the Brazilian Cerrado. *Front. For. Glob. Chang.* **2021**, *4*, 650665. [CrossRef]
56. Cunha, G.M.; Costa, G.S.; Gama-Rodrigues, A.C. Litter Stock and Quality in Eucalyptus Grandis in Northern Rio de Janeiro State, Brazil. *Floresta E Ambiente* **2020**, *27*, e20180129. [CrossRef]
57. Ma, X.; Chen, J.; Zhu, J.; Yan, N. Lignin-Based Polyurethane: Recent Advances and Future Perspectives. *Macromol. Rapid Commun.* **2021**, *42*, 2000492. [CrossRef] [PubMed]
58. Meena, R.S.; Lal, R.; Yadav, G.S. Long-Term Impacts of Topsoil Depth and Amendments on Soil Physical and Hydrological Properties of an Alfisol in Central Ohio, USA. *Geoderma* **2020**, *363*, 114164. [CrossRef]
59. Haruna, S.I.; Anderson, S.H.; Udawatta, R.P.; Gantzer, C.J.; Phillips, N.C.; Cui, S.; Gao, Y. Improving Soil Physical Properties through the Use of Cover Crops: A Review. *Agrosystems Geosci. Environ.* **2020**, *3*, e20105. [CrossRef]
60. Wei, Y.; Wu, X.; Xia, J.; Miller, G.A.; Cai, C.; Guo, Z.; Arash, H. The Effect of Water Content on the Shear Strength Characteristics of Granitic Soils in South China. *Soil Tillage Res.* **2019**, *187*, 50–59. [CrossRef]
61. Tu, Z.; Chen, S.; Chen, Z.; Ruan, D.; Zhang, W.; Han, Y.; Han, L.; Wang, K.; Huang, Y.; Chen, J. Hydrological Properties of Soil and Litter Layers of Four Forest Types Restored in the Gully Erosion Area of Latosol in South China. *Forests* **2023**, *14*, 360. [CrossRef]
62. Biao, Z.; Wenhua, L.; Gaodi, X.; Yu, X. Water Conservation of Forest Ecosystem in Beijing and Its Value. *Ecol. Econ.* **2010**, *69*, 1416–1426. [CrossRef]

hydrology

MDPI

Article

Development and Automation of a Photovoltaic-Powered Soil Moisture Sensor for Water Management

Denilson Alves de Melo [1], Patrícia Costa Silva [1], Adriana Rodolfo da Costa [1], Josué Gomes Delmond [1], Ana Flávia Alves Ferreira [1], Johnny Alves de Souza [2], José Francisco de Oliveira-Júnior [3], Jhon Lennon Bezerra da Silva [4], Alexandre Maniçoba da Rosa Ferraz Jardim [5,6], Pedro Rogerio Giongo [1], Maria Beatriz Ferreira [7], Abelardo Antônio de Assunção Montenegro [5], Henrique Fonseca Elias de Oliveira [8], Thieres George Freire da Silva [5] and Marcos Vinícius da Silva [5,*]

1 Departamento de Engenharia Agrícola, Câmpus Suodoeste Unidade Universitária de Santa Helena de Goiás, Universidade Estadual de Goiás, Via Protestato Joaquim Bueno 945 Santa Helena de Goiás, Goiás 75920-000, GO, Brazil; denilsonmello@hotmail.com.br (D.A.d.M.); patricia.costa@ueg.br (P.C.S.); adriana.costa@ueg.br (A.R.d.C.); josue.delmond@ueg.br (J.G.D.); ferreiraanaflavia27@gmail.com (A.F.A.F.); pedro.giongo@ueg.br (P.R.G.)

2 Faculdade de Direito, Universidade de Rio Verde, Avenida Universitária, Qd.07, Lt2, Residencial Tocantins, Rio Verde, Goiás 75901-970, GO, Brazil; direitojohnny@gmail.com

3 Institute of Atmospheric Sciences (ICAT), Federal University of Alagoas (UFAL), Maceió 57072-260, AL, Brazil; jose.junior@icat.ufal.br

4 Center for Information Management and Popularization of Science, National Institute of the Semiarid Region (INSA), Av. Francisco Lopes de Almeida, s/n-Serrotão, Campina Grande 58434-700, PB, Brazil; jhon.lennon@insa.gov.br

5 Department of Agricultural Engineering, Federal Rural University of Pernambuco, Dom Manoel de Medeiros Avenue, s/n, Dois Irmãos, Recife 52171-900, PE, Brazil; alexandre.jardim@ufrpe.br (A.M.d.R.F.J.); abelardo.montenegro@ufrpe.br (A.A.d.A.M.); thieres.silva@ufrpe.br (T.G.F.d.S.)

6 Department of Biodiversity, Institute of Bioscience, São Paulo State University—UNESP, Av. 24A, 1515, Rio Claro 13506-900, SP, Brazil

7 Department of Forest Science, Federal Rural University of Pernambuco (UFRPE), Recife 52171-900, PE, Brazil; beatriz.ferreira2@ufrpe.br

8 Cerrado Irrigation Graduate Program, Goiano Federal Institute, Ceres 76300-000, GO, Brazil; henrique.fonseca@ifgoiano.edu.br

* Correspondence: marcos.viniciussilva@ufrpe.br

Citation: de Melo, D.A.; Silva, P.C.; da Costa, A.R.; Delmond, J.G.; Ferreira, A.F.A.; de Souza, J.A.; de Oliveira-Júnior, J.F.; da Silva, J.L.B.; da Rosa Ferraz Jardim, A.M.; Giongo, P.R.; et al. Development and Automation of a Photovoltaic-Powered Soil Moisture Sensor for Water Management. *Hydrology* **2023**, *10*, 166. https://doi.org/10.3390/hydrology10080166

Academic Editors: Songhao Shang, Qianqian Zhang, Dongqin Yin, Hamza Gabriel and Magdy Mohssen

Received: 13 July 2023
Revised: 4 August 2023
Accepted: 8 August 2023
Published: 10 August 2023

Abstract: The objective of this study was to develop and calibrate a photovoltaic-powered soil moisture sensor (SMS) for irrigation management. Soil moisture readings obtained from the sensor were compared with gravimetric measurements. An automated SMS was used in two trials: (i) okra crop (*Abelmoschus esculentus*) and (ii) chili pepper (*Capsicum frutescens*). All sensors were calibrated and automated using an Arduino Mega board with C++. The soil moisture data were subjected to descriptive statistical analysis. The data recorded by the equipment was correlated with the gravimetric method. The determination coefficient (R^2), Pearson correlation (r), and root mean square error (RMSE) were adopted as criteria for equipment validation. The results show that our SMS achieved an R^2 value of 0.70 and an r value of 0.84. Notably, there was a striking similarity observed between SMS and gravimetric data, with RMSE values of 3.95 and 4.01, respectively. The global model developed exhibited highly efficient outcomes with R^2 (0.98) and r (0.99) values. The applicability of the developed SMS facilitates irrigation management with accuracy and real-time monitoring using digital data. The automation of the SMS emerges as a real-time and precise alternative for performing irrigation at the right moment and in the correct amount, thus avoiding water losses.

Keywords: tensiometer; soil moisture; solar energy; automation

1. Introduction

Water resource management and increasing freshwater productivity are among the most effective options for conserving water resources, especially in irrigated agriculture [1–5]. In terms of water use, irrigated agriculture is the largest consumer of freshwater [6,7]. Given these aspects and issues, improving water use efficiency and irrigation water savings will enhance the effects and factors related to irrigation management. Technologies in the research field are essential in contributing to water management and soil conservation [8,9].

The expansion and diffusion of new technologies, along with the growth of the technology market, have led to dependence on and increased demand for technologies in the agricultural sector. They are essential in managing inputs used in the agricultural and livestock markets, such as controlling the amount and timing of water to be applied to the soil [10–12]. The increasing demand for water use in irrigated crops over the last three decades has constantly raised awareness about the rational use of water resources [13–16].

Therefore, effective irrigation management is crucial and aims to apply the exact amount of water that the plant needs at the right time [17,18]. However, it is important to know and monitor variables such as soil moisture. The methods used to determine soil moisture can be classified as direct and indirect [19–23]. Direct methods are those that quantify soil moisture by weighing (e.g., gravimetric), and indirect methods, through reflectance, neutron moderation, and soil stress (e.g., frequency domain reflectometry—FDR, time domain reflectometry—TDR, and tensiometers) [24,25]. Furthermore, methods that measure the tension generated by water retained in soil particles have evolved and adapted to technological advancements [26–29].

Among the methods that offer ease of application and indirect measurement, the tensiometric method stands out [30–32]. In irrigation management, a moisture sensor can be used, which is designated as the primary device for measuring the matric potential of water in the soil, which can be converted into current soil water content [33–35].

The development of tensiometers has emerged, intending to expand their operating range and, most importantly, meet the technological advancements in irrigation management [31,36,37]. The creation of digital reading systems that allow precise, effective, and rapid measurement of the matric potential of water in the soil has become essential [28,31]. Currently, developed sensors are installed directly in the soil and, with the aid of reading equipment, enable the collection of soil moisture data, analyzing the interrelationships between soil and water, thus allowing greater control, precision, and savings during cultivation.

With the advancement of digital agriculture 4.0, there has also been a greater diffusion of automation, especially in irrigated systems, due to the search for technological alternatives that help achieve greater control and productivity, generate higher profitability and sustainability, and reduce labor costs [38,39]. The integration of embedded sensors for data collection automation has become an accessible and viable alternative for advancing agriculture, which, along with other technologies, assists in real-time data collection, processing, analysis, and transfer of crop status, resulting in quick and cautious decision-making [10,40,41]. In addition, the use of the Internet of Things (IoT), data analytics, sensor nodes, and solar energy contribute to technological development and are growing in the rural environment. These elements lead to savings and income, reduce energy consumption from conventional sources, and significantly contribute to the sustainability of productive rural activities.

Considering the above, the applicability of digital agriculture 4.0 in crop management further contributes to meeting the water needs of crops, optimizing water use, and advancing technological advancements in irrigated systems, assisting in quick and precise irrigation management decision-making.

Therefore, the objective was to develop a soil water tension sensor for moisture determination, powered by solar energy, with automation using Arduino programming techniques calibrated by determining soil moisture through the oven drying method for different crops.

2. Materials and Methods

2.1. Study Dynamics and Characterization

An automated soil water tension sensor (using a pressure sensor) was developed to estimate soil moisture. Temperature and air humidity sensors were also integrated into the system, which was powered by a photovoltaic module. The negative pressure is related to the operating principle of the pressure sensor. The sensor assumes that the soil has a pressure of −100 kPa, and when the sensor is inserted into the soil, it calculates the pressure differential, allowing for the collection of data on water retention in the soil, like the operating principle of manual tensiometers. The design of the moisture sensor was like the one developed by Livingston [42], with precision technologies [43,44].

The soil moisture readings estimated by the sensor were compared with measurements obtained through the gravimetric method [45]. For calibration purposes, the sensor was used in two experiments in a protected environment. The first experiment involved okra [*Abelmoschus esculentus* (L.) Moench] subjected to 5 irrigation levels (50%, 75%, 100%, 125%, and 150%) determined based on crop evapotranspiration (ET_c), with a surface drip irrigation system employed. The second experiment involved chili pepper [*Capsicum frutescens* (L.)] subjected to 4 irrigation levels: 50%, 75%, 100%, and 125% of ET_c, with two drip irrigation systems: surface and subsurface. Both experiments were conducted in 15-L pots filled with clayey Red Latosol soil, which was sieved and homogenized. A localized drip irrigation system with 90% water application efficiency and a pressure of 8 m water column with a flow rate of 1.40 L h^{-1} was used for irrigation. It is worth mentioning that the moisture and temperature sensors were calibrated using meteorological data obtained from the thermohygrometer in the greenhouse of the State University of Goiás—UEG, Santa Helena University Unit.

2.2. Assembly of the Automated Moisture Sensor

The moisture sensor has a structure similar to a conventional tensiometer [42]. It consists of a polyvinyl chloride (PVC) pipe with dimensions of 60 cm × 5 cm, as shown in Figure 1. The components include:

- A set of photovoltaic modules (12 V each);
- A temperature and humidity sensor (DHT11);
- A pressure sensor (BMP280);
- A LCD display with 16 × 2 blue backlight (2-lines × 16-characters);
- An Arduino Mega board;
- Rechargeable battery with a voltage of 9 V and 250 mAh.

A set of 12 V (3 Watts) solar panels was necessary to power all the components and charge the battery in the system. All the data generated by the sensors were displayed on the LCD screen (Figure 1), located on the surface of the moisture sensor and directly connected to a breadboard. The LCD has 16 columns by 2 rows, a blue backlight, and white writing.

The temperature sensor element is an NTC transmitter, and the humidity sensor is of the DHT11 type. The internal circuit reads the sensors and communicates with an 8-bit microcontroller through a one-way serial signal, both connected to a controller. The protocol used to transfer data between the microcontroller and DHT11 involves a single-wire bus (ELETROGATE, 201-A). The ambient temperature and humidity sensor (Figure 1) was positioned near the LCD screen, requiring direct contact with the environment, and should not be enclosed.

According to Table 1, the equipment used for assembling and programming the moisture sensor, the number of necessary units, the unit price, and the total price to produce the equipment are presented.

The connections with the Arduino board were made through the Serial Data Line and SCL (Serial Clock Line) pins, allowing various ports on the Arduino board to easily connect to the other sensors used (Figure 2). The connection was made using the I2C module, linking the I2C screen to the Arduino Mega board.

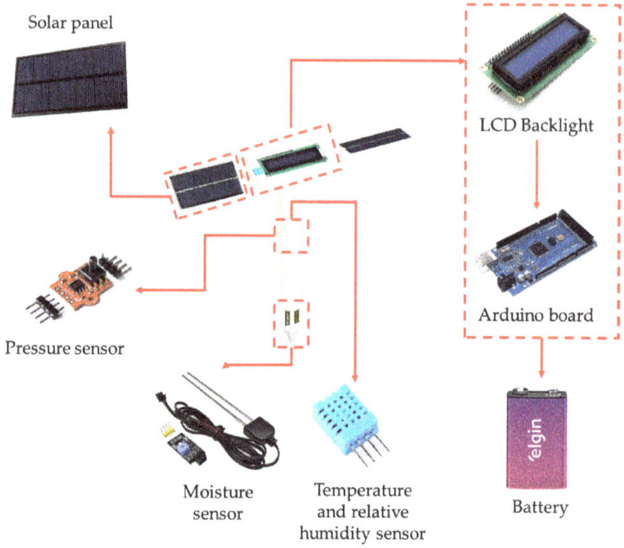

Figure 1. A graphical representation of the solar-powered soil moisture sensor and its main components.

Table 1. Parts used to assemble the humidity sensor, followed by price and technical specifications.

Quantity	Description	Unit Amount (USD)	Total Amount (USD)	Specifications
1	Arduino maker kit	70.64	70.64	Includes 136 pieces.
1	Pressure and Temperature Sensor (BMP280)	3.10	3.10	Operating voltage: 3 V; Current consumption: 2.7 μA; Interfaces: I2C and SPI; Pressure measurement range: 300–1100 hPa (equivalent +9000 to −500 m above/below sea level); Accuracy: ±0.12 hPa (±1 m equivalent); Temperature range: −40 to 85 °C; Temperature accuracy: ±1.0 °C.
1	Corrosion Resistant Soil Moisture Sensor, Arduino, Model S12	9.35	9.35	Operating voltage: 3.3 to 12 V DC input; Current: less than 20 mA; less than 30 mA (output); Output: Digital and analogue; Probe dimensions: 60 × 19 × 9 mm; Module dimensions: 36 × 15 × 7 mm; Probe cable length: 1 m.
1	Room temperature and humidity sensor (DHT11)	2.68	2.68	Power 3.0 to 5.0 VDC (5.5 VDC maximum); Humidity measurement range: 20 to 95% RH; Temperature measurement range: 0° to 50 °C; Humidity measurement accuracy: ±5.0% RH; Temperature measurement accuracy: ±2.0 °C.
1	Hikari Power-30 Soldering Iron	6.84	6.84	-
1	Transparent Organizer Box	5.17	5.17	-
1	Telijia 31-Piece Precision Wrench Kit (TE-6036)	4.14	4.14	-
4	Solar Panel System (12 V-3 W)	15.57	62.26	12 V-3 W-250 mA Photovoltaic Solar Energy Board Panel Cell, with 20 cm soldered wire, dimensions 145 × 145 mm
1	Elgin 12 V Rechargeable Battery	37.41	37.41	Blister with 1 rechargeable battery 12 V 250 mAh.
3	Tin Solder Wire Cobix Tube (1 mm, 22 g)	3.26	9.79	-
1	I2C Serial Module for 16 × 2 Blue Backlight LCD Display for Arduino	8.21	8.21	The I2C module operates with a minimum supply voltage of 5 V.

Table 1. *Cont.*

Quantity	Description	Unit Amount (USD)	Total Amount (USD)	Specifications
1	Ethernet Shield W5100	24.94	24.94	Supply Voltage: 3 to 5 VDC; Communication: SPI; Operating temperature: −40 to 85 °C; Indicators: TX, RX, COL, FEX, SPD, LNK; Current: 100 mA; Support: Full-duplex and half-duplex, Auto MDI/MDIX, ADSL connection; Works directly with the official Arduino library; TX/RX RAM Buffer: 16 kBytes; Dimensions: 55.8 × 68.58 × 1.6 mm; Datasheet: W5100 Ethernet Shield Module.
1	Fiberglass Structure	124.69	124.69	-
1	Lenovo Ideapad 330 laptop	519.55	519.55	-
Total amount		835.55	888.77	

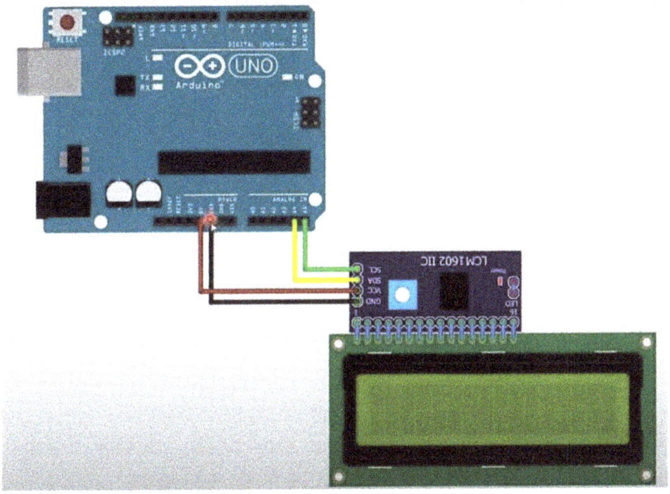

Figure 2. Connection of the LCD screen to the Arduino Mega board.

2.2.1. Arduino Board

All the sensors were calibrated and controlled using an Arduino Mega board (Figure 3), and the programming language used was C++. The software used was the Arduino IDE. The modules directly connected to the board are the soil moisture sensor, air humidity sensor, air temperature sensor, and pressure sensor.

After the testing phases, the Arduino Mega board was permanently connected to the soil moisture sensor structure, containing the programming for all the sensors and the memory for intelligent joint operation (Figure 4). All the information was displayed on the LCD screen (Figure 1).

2.2.2. Soil Moisture Sensor

The soil moisture sensor was calibrated using a potentiometer for dry and wet soil conditions. The data readings were performed in Siemens, the standard unit of electrical conductivity in the International System of Units (SI). The threshold between dry and wet soil conditions was compared and adjusted using the potentiometer present in the sensor, regulating the digital output D0.

Figure 3. Arduino Mega 2560 board.

Figure 4. Soil collection flowchart, construction, and programming of the soil moisture sensor.

The soil moisture sensor was connected to a digital port of the Arduino board, providing information between high and low states, i.e., dry and wet soil, respectively. The verification limits could be adjusted through a potentiometer located on the sensor body (ELETROGATE,

201-C). The moisture sensor was positioned at the bottom in direct contact with the soil, and through electrical conductivity and the connection with the voltage sensors, it was possible to quantify the water content in the soil and assist in irrigation management.

The accuracy of the soil moisture sensor was assessed based on the gravimetric method, with readings and calibration performed using the standard oven-drying method through linear regressions.

2.2.3. BMP280 Pressure Sensor

The pressure sensor was programmed to perform negative readings to verify the pressure difference generated by water loss in the soil. The BMP280 sensor is factory-configured to read pressure data in hectopascals (hPa), so the conversion of the data is necessary since the conventional tensiometer reads in kilopascals (kPa). The sensor has a reading range of 0 to −1100 hPa.

The pressure sensor was positioned on the surface of the moisture sensor to measure air pressure data, with its lower probe in contact with the soil to measure the pressure generated by water retention in the soil and obtain a pressure result based on the difference between the two. The pressure sensor reads the pressure difference between the environment and the force of water retention in the soil. The soil moisture sensor was calibrated through analyses conducted with soil samples in the laboratory using the standard oven-drying method [46].

After converting the units of measurement, it was necessary to configure the measurement range of the pressure sensor. The Arduino Mega board was used to program the pressure sensor to work in conjunction with the moisture sensor. Both sensors were configured using a numerical scale, where a pressure of 0 kPa indicates saturated soil and a scale reaching −100 kPa means the soil is very dry. All this information was outputted by the programmed system and displayed on the LCD screen (Figure 1).

The pressure sensor was programmed to work together with the moisture sensor so that the readings would provide soil moisture information under different moisture conditions. The pressure sensor was integrated inside the PVC tube to read pressure differences between the soil and air. The soil moisture and air temperature sensors are precise sensors with low power consumption. The sensor came pre-programmed and pre-configured from the factory. Additionally, the sensor was directly connected to the Arduino Mega board to be powered by the same source as all the other sensors.

2.2.4. Photovoltaic Modules

The photovoltaic modules were connected using standard power supply cables, with connections made to the GND (power ground) and VCC (positive power supply) terminals. The cables were connected to the battery and integrated into the moisture sensor structure, allowing excess energy generated by the solar panel to be stored for future use during periods of low solar radiation. This reduced the need for battery replacement.

2.2.5. DHT11 Ambient Relative Humidity and Temperature Sensor

The DHT11 (Figure 1) consists of two sensors: a temperature sensor (NTC thermistor) and a humidity sensor (HR202). The internal circuitry reads the sensors and communicates with a microcontroller via a one-way serial signal. Its temperature readings range from 0 to 50 °C, and its humidity readings range from 20 to 90%. The DHT11 sensor for relative humidity and air temperature has a simple 3-pin connection, facilitating programming and connectivity with other sensors. It has two power supply pins and one pin for data decoding between the sensor and the Arduino board.

2.3. Statistical Modeling and Validation of Moisture Sensor

2.3.1. Descriptive Statistics

The soil moisture data from the moisture sensor and the gravimetric method were subjected to descriptive statistical analysis to obtain the mean, median, minimum, maximum, standard deviation (SD), and coefficient of variation (CV, %). The percentage value of CV was categorized as low (CV < 12%), medium (if CV = 12–24%), and high (when CV > 24%) [47]. The normality test using the Kolmogorov–Smirnov test was applied to the studied variables, with a significance level (alpha) of 0.01. Descriptive statistics were performed using R software version 4.0.3 [48].

2.3.2. Regression Analysis

To validate the developed moisture sensor, the recorded data from the device were correlated with the gravimetric method to estimate soil moisture. The coefficient of determination (R^2), Pearson correlation (r), and root mean square error (RMSE) were adopted as criteria for equipment validation. Finally, an analysis of variance (ANOVA) of the established models for pepper and okra crops was conducted, with a significant F-value at a 1% probability and a p-value less than 0.01 ($p < 0.01$) for the validation of the established model and moisture sensor. The statistical modeling was performed using R software version 4.0.3 [48].

3. Results and Discussion

Based on greenhouse measurements, linear regressions were established to validate the solar-powered moisture sensor against the gravimetric method for estimating soil moisture in the pepper crop (Figure 5). Figure 5a shows the regression for Moisture Sensor vs. Gravimetric validation, and Figure 5b shows the regression for Gravimetric vs. Moisture Sensor validation. Based on the validation components, the coefficient of determination (R^2) and the Pearson correlation coefficient (r) did not change regardless of the order of the X and Y factors. They are interpreted as the proportion of variation in Y that is explained by the variable X and vice versa, being inversely proportional and unchangeable components, as indicated by studies [9,49].

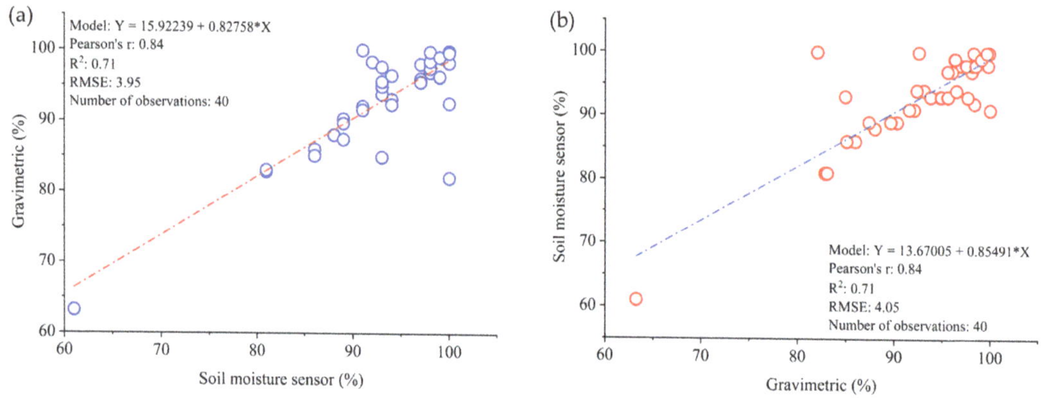

Figure 5. Regression models for Soil Moisture Sensor vs. Gravimetric validation and (**a**) Gravimetric vs. Soil Moisture Sensor validation (**b**) in the pepper crop, accompanied by their respective coefficients of determination (R^2), Pearson correlation coefficient (r), and root mean square error (RMSE).

Regarding R^2, it showed a satisfactory fit, with a value around 0.70, indicating that the accuracy of the solar-powered moisture sensor represents approximately 70.75% of the gravimetric method. On the other hand, r showed a fit of 0.84, reinforcing the accuracy of the developed moisture sensor and the reliability of its applicability in the field and the consumer market. Supporting the results of the present study, Thalheimer [50], who developed a low-cost solar-powered system for measuring soil water potential, obtained an R^2 of 1, recommending the applicability of the equipment in the field.

Furthermore, the values for the root mean square error (RMSE) were low and similar for Moisture Sensor vs. Gravimetric (Figure 5a) and Gravimetric vs. Moisture Sensor

(Figure 5b), with values around 3.95 and 4.01, respectively. Consistent with the results of this study, Sanches et al. [51], who developed and calibrated a low-cost, high-efficiency automated moisture sensor for irrigation control based on real-time monitoring, observed maximum errors of around 2.84 for the performed analyses. RMSE values are crucial for assessing the accuracy of a model, regardless of its r and R^2, as errors have a significant influence on the spatial variability of data precision.

Table 2 presents the analysis of variance (ANOVA) for the validation of the solar-powered moisture sensor against the gravimetric method in pepper crops under irrigation depths of 50, 75, 100, and 125% of crop evapotranspiration (ET_c). The generated model's F-value was found to be significant at a 1% probability level, indicating the precision and effectiveness of the developed moisture sensor and thus recommending its applicability in pepper cultivation. As for the p-value, it showed a satisfactory fit ($p < 0.01$). Silva et al. [8] emphasize the importance of exploring the components of a regression model's ANOVA (F-value and p-value) for validation purposes.

Table 2. Analysis of variance (ANOVA) for regression models validating the soil moisture sensor in pepper cultivation.

	[1] DF	[2] SS	[3] MS	F Value	p-Value
Model	1	1435.53	1435.53	91.92	<0.0001
Error	38	593.46	15.62		
Total	39	2028.99			

[1] DF—Degree of freedom; [2] SS—Sum of squares; [3] MS—Mean square.

To assess the spatial distribution of soil moisture data in the treatments of 50, 75, 100, and 125% of ET_c for the solar-powered moisture sensor and the gravimetric method, descriptive statistics were performed on the collected data, obtaining the mean, median, minimum, maximum, standard deviation (SD), and coefficient of variation (CV) (Table 3). It can be observed that the mean and median values for all treatments in both soil moisture estimation methods were close, indicating data normality, as also evidenced by the Kolmogorov–Smirnov test at a 1% probability level for the entire dataset. Supporting the results of this study, Silva et al. [49], through conventional statistical tests and geostatistical modeling, stated in their study that close mean and median values are indicative of data normality, as supported by the Kolmogorov–Smirnov test at a 1% probability level.

According to the criterion of Warrick and Nielsen [47], the CV was consistently low (<12%) for all treatments, except for the 100% ET_c treatment in the solar-powered moisture sensor. Based on the observed results, it is possible to affirm the distribution efficiency and uniformity of the subsurface drip system in pepper cultivation, which provides low spatial variability of moisture, as confirmed by the SD, which was low for all treatments. Furthermore, the values were close to the CV, substantiating the efficiency of the adopted irrigation system [52]. Supporting the results of this study, Colak [53], who evaluated leaf water potential in drip-irrigated bell pepper under various deficit irrigation strategies using surface and subsurface irrigation, highlights that the subsurface irrigation system exhibits a low CV, indicating its efficiency in water distribution and uniformity.

Figure 6a shows the regression for validating the soil moisture sensor vs. gravimetric method, and Figure 6b presents the regression for validating the gravimetric vs. soil moisture sensor method. Based on the pepper crop analyses, the observed results of R^2 and r for the okra crop were higher, with values around 0.98 and 0.99, respectively. These results indicate a greater sensitivity of the gravimetric method and the solar-powered moisture sensor in quantifying soil moisture. However, local abiotic conditions (e.g., temperature, relative humidity, wind speed, and incident solar radiation), physical and biological soil conditions, and the greater water demand sensitivity of the okra crop may have influenced the results. Supporting the observed results in this study, Aliku et al. [54], who estimated okra crop evapotranspiration using drainage lysimeters under dry season conditions, state that okra is one of the vegetables with the highest water demand.

Table 3. Descriptive statistics of the solar-powered soil moisture sensor and gravimetric method for treatments with 50, 75, 100, and 125% of crop evapotranspiration (ET_c) in pepper cultivation.

Variable	Mean	Median	Minimum	Maximum	[1] SD	[2] CV
			50%			
Soil moisture sensor	91.00	91.50	81.00	100.00	6.06	6.65
Gravimetric	91.19	90.05	82.82	99.93	6.68	7.33
			75%			
Soil moisture sensor	93.70	95.50	81.00	99.00	5.79	6.18
Gravimetric	94.22	96.17	83.04	100.02	4.94	5.24
			100%			
Soil moisture sensor	91.50	95.00	61.00	100.00	11.37	12.43
Gravimetric	91.67	95.25	63.24	98.29	10.66	11.63
			125%			
Soil moisture sensor	96.90	98.00	93.00	100.00	3.04	3.13
Gravimetric	95.38	97.07	82.04	99.82	5.40	5.67

[1] SD—Standard deviation; [2] CV—Coefficient of variation.

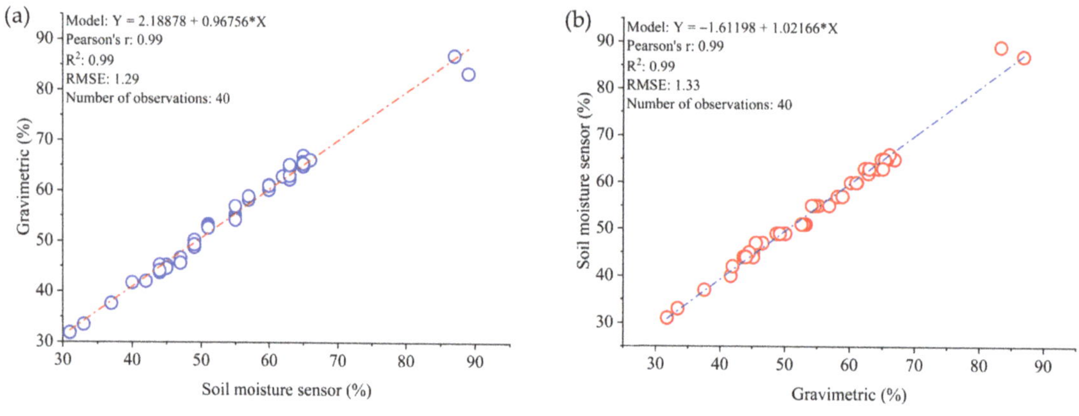

Figure 6. Regression models for validating the soil moisture sensor vs. gravimetric method and (**a**) gravimetric vs. soil moisture sensor method (**b**) in the okra crop are preceded by their respective coefficients of determination (R^2), Pearson correlation coefficient (r), and root mean square error (RMSE).

The RMSE values were lower compared to the measurements in the pepper crop. This reinforces the higher efficiency of the moisture sensor in recording soil moisture in the okra crop, making it recommended for irrigation management and accurate compared to the gravimetric method. Additionally, we emphasize that the developed moisture sensor's recording efficiency for both crop management (pepper and okra) is efficient and accurate for soil moisture between 30 and 100%.

Table 4 presents the ANOVA for validating the moisture sensor against the gravimetric method in the okra crop under irrigation depths of 50, 75, 100, 125, and 150% of ET_c. The F-value of the generated models was significant at a 1% probability level, with a value around 3269.20, indicating the precision and effectiveness of the developed moisture sensor and recommending its applicability to the okra crop. As for the *p*-value, it showed a satisfactory fit ($p < 0.01$).

Table 4. Analysis of variance (ANOVA) of regression models for validating the soil moisture sensor to okra crop.

	[1] DF	[2] SS	[3] MS	F Value	*p*-Value
Model	1	5445.47	5445.47	3269.20	<0.0001
Error	38	63.30	1.67		
Total	39	5508.76			

[1] DF—Degrees of freedom; [2] SS—Sum of squares; [3] MS—Mean square.

From the descriptive statistics (Table 5), it can be observed that the mean and median values are close, which, as discussed earlier, is indicative of data normality, as evidenced by the Kolmogorov–Smirnov test at a 1% probability level for the entire dataset. According to the criterion of Warrick and Nielsen [47], the CV ranged from moderate (CV = 12–24%) to high (CV > 24%), reinforcing the high water sensitivity of the okra crop and resulting in greater soil moisture variability.

Table 5. Descriptive statistics of the solar-powered soil moisture sensor and the gravimetric method for treatments with 50, 75, 100, 125, and 150% of crop evapotranspiration (ET_c) for the okra crop.

Variable	Mean	Median	Minimum	Maximum	[1] SD	[2] CV
			50%			
Soil moisture sensor	64.50	63.00	49.00	89.00	15.92	24.68
Gravimetric	64.52	63.37	48.74	86.92	14.40	22.31
			75%			
Soil moisture sensor	55.00	56.00	45.00	65.00	7.35	13.36
Gravimetric	55.41	56.73	45.15	65.76	7.69	13.88
			100%			
Soil moisture sensor	55.50	55.00	40.00	66.00	8.14	14.67
Gravimetric	56.31	55.79	41.69	66.15	7.65	13.58
			125%			
Soil moisture sensor	49.25	49.00	31.00	65.00	13.01	26.42
Gravimetric	49.77	49.10	31.78	65.41	13.01	26.15
			150%			
Soil moisture sensor	47.38	44.00	37.00	63.00	9.10	19.21
Gravimetric	47.75	43.99	37.58	65.13	9.79	20.50

[1] SD—Standard deviation; [2] CV—Coefficient of variation.

Regarding the minimum and maximum values, it can be observed that the okra crop requires a greater amount of water compared to the pepper crop. In the okra crop, the lowest observed soil moisture value was 31%, while the maximum was 89% (Table 5). In the pepper crop, these moisture values were higher, in the range of 61% to 100%, for the lowest and highest values, respectively. Therefore, it is evident that the okra crop requires more water than pepper.

To establish a global model for soil moisture estimation and test the sensitivity of moisture sensor estimates in both crops, Figure 7 presents the validation established for the global model. The global model proved to be more efficient than the model and validation established for the pepper crop (Figure 5), with an R^2 of 0.98 and an r of 0.99, making it the most satisfactory validation for the explored dataset. Based on these results, the use of the moisture sensor in the field is recommended for both pepper and okra crops.

Based on the RMSE, it was found to be low, indicating a satisfactory fit of the moisture sensor with the gravimetric method, with a value of around 3.00762%. With a low margin of error, good coefficient adjustments (R^2 and r), and finally, a significant F-value and *p*-value at a 1% probability level, as observed in Table 6, the use of the solar-powered moisture sensor for characterizing soil moisture in pepper and okra crops is recommended.

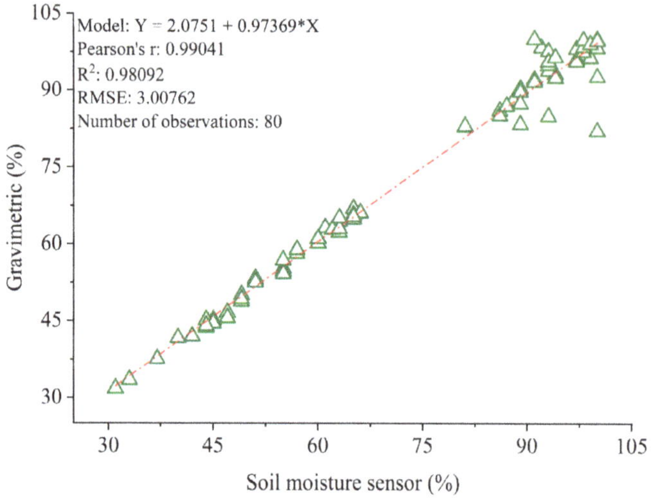

Figure 7. Global regression model for validating the soil moisture sensor vs. gravimetric method, preceded by their respective coefficients of determination (R^2), Pearson correlation coefficient (r), and root mean square error (RMSE).

Table 6. Analysis of variance (ANOVA) of the global regression model.

	[1] DF	[2] SS	[3] MS	F Value	*p*-Value
Model	1	36,268.22	36,268.22	4009.39	0.001
Error	79	705.57	9.04		
Total	80	36,973.79			

[1] DF—Degrees of freedom; [2] SS—Sum of squares; [3] MS—Mean square.

4. Conclusions

The solar-powered moisture sensor developed proved to be effective in characterizing soil moisture and was properly validated against the gravimetric method for soil moisture estimation. The parameters of coefficient of determination, Pearson correlation, and root mean square error were satisfactory for both pepper and okra crops, as well as for the global model.

The applicability of the developed moisture sensor will facilitate precise irrigation management by providing real-time and digital data, as most commonly used methods require time for moisture estimation and/or method calibration.

The automation of the soil moisture sensor emerges as a real-time alternative for irrigating at the right moment and in the right amount, thus avoiding water waste.

The solar powered soil moisture sensor is efficient and accurate. However, the present equipment has some limitations, the main one being the need for calibration, when using it in a soil with physical-chemical characteristics different from those used in this study. Therefore, it is recommended to calibrate the soil moisture photovoltaic sensor, depending on whether soils with different characteristics are used in this study.

Author Contributions: Conceptualization, D.A.d.M., P.C.S., J.G.D., J.A.d.S., T.G.F.d.S. and M.V.d.S.; methodology, D.A.d.M., P.C.S., A.R.d.C., J.A.d.S. and M.V.d.S.; software, D.A.d.M., P.C.S., M.B.F. and H.F.E.d.O.; validation, D.A.d.M., M.B.F. and M.V.d.S.; formal analysis, D.A.d.M., P.C.S., J.G.D., H.F.E.d.O. and P.R.G.; investigation, D.A.d.M., P.C.S., A.R.d.C., A.F.A.F., J.A.d.S. and A.A.d.A.M.; resources, P.C.S., J.F.d.O.-J., J.L.B.d.S., A.A.d.A.M. and M.V.d.S.; data curation, A.M.d.R.F.J., T.G.F.d.S. and M.V.d.S.; writing—original draft preparation, D.A.d.M., P.C.S., A.R.d.C., J.G.D., A.F.A.F. and J.F.d.O.-J.; writing—review and editing, J.F.d.O.-J., J.L.B.d.S., A.M.d.R.F.J., M.B.F., A.A.d.A.M., H.F.E.d.O., P.R.G., T.G.F.d.S. and M.V.d.S.; visualization, D.A.d.M., P.C.S., J.G.D., A.F.A.F., J.L.B.d.S., A.M.d.R.F.J., M.B.F., T.G.F.d.S. and M.V.d.S.; supervision, P.C.S. and J.L.B.d.S.; project administration, P.C.S. and A.R.d.C.; funding acquisition, J.F.d.O.-J., J.L.B.d.S., A.A.d.A.M., P.R.G. and M.V.d.S. All authors have read and agreed to the published version of the manuscript.

Funding: This research received no external funding.

Institutional Review Board Statement: Not applicable.

Informed Consent Statement: Not applicable.

Data Availability Statement: Not applicable.

Acknowledgments: Thanks to the undergraduate program in Agricultural Engineering at the State University of Goiás (UEG)—Santa Helena de Goiás Campus for the support in the development of this research. Thanks to the Pro-rectorate of Research and Graduate Studies at UEG, the Coordination for the Improvement of Higher Education Personnel (CAPES-Finance Code 001) and São Paulo Research Foundation (FAPESP, 2023/05323-4). We also thank two anonymous reviewers for their insightful comments, which led to substantial improvements to this paper.

Conflicts of Interest: The authors declare no conflict of interest.

References

1. Haldar, K.; Kujawa-Roeleveld, K.; Dey, P.; Bosu, S.; Datta, D.K.; Rijnaarts, H.H.M. Spatio-Temporal Variations in Chemical-Physical Water Quality Parameters Influencing Water Reuse for Irrigated Agriculture in Tropical Urbanized Deltas. *Sci. Total Environ.* **2020**, *708*, 134559. [CrossRef] [PubMed]
2. Muzammil, M.; Zahid, A.; Breuer, L. Water Resources Management Strategies for Irrigated Agriculture in the Indus Basin of Pakistan. *Water* **2020**, *12*, 1429. [CrossRef]
3. Singh, A. Assessment of Different Strategies for Managing the Water Resources Problems of Irrigated Agriculture. *Agric. Water Manag.* **2018**, *208*, 187–192. [CrossRef]
4. Singh, A. Hydrological Problems of Water Resources in Irrigated Agriculture: A Management Perspective. *J. Hydrol.* **2016**, *541*, 1430–1440. [CrossRef]
5. Silva, J.L.B.; Bezerra, A.C.; Moura, G.B.d.A.; Jardim, A.M.d.R.F.; Batista, P.H.D.; de Jesus, F.L.F.; Sanches, A.C.; da Silva, M.V. Spatiotemporal Dynamics of Agricultural Areas with Central Pivot Using Orbital Remote Sensing in the Brazilian Semiarid. *Smart Agric. Technol.* **2022**, *2*, 100052. [CrossRef]
6. Kamienski, C.; Soininen, J.P.; Taumberger, M.; Dantas, R.; Toscano, A.; Cinotti, T.S.; Maia, R.F.; Neto, A.T. Smart Water Management Platform: IoT-Based Precision Irrigation for Agriculture. *Sensors* **2019**, *19*, 276. [CrossRef]
7. Ngxumeshe, A.M.; Ratsaka, M.; Mtileni, B.; Nephawe, K. Sustainable Application of Livestock Water Footprints in Different Beef Production Systems of South Africa. *Sustainability* **2020**, *12*, 9921. [CrossRef]
8. da Silva, M.V.; Pandorfi, H.; de Almeida, G.L.P.; de Lima, R.P.; dos Santos, A.; Jardim, A.M.d.R.F.; Rolim, M.M.; da Silva, J.L.B.; Batista, P.H.D.; da Silva, R.A.B.; et al. Spatio-Temporal Monitoring of Soil and Plant Indicators under Forage Cactus Cultivation by Geoprocessing in Brazilian Semi-Arid Region. *J. South Am. Earth Sci.* **2021**, *107*, 103155. [CrossRef]
9. Silva, P.C.; Ferreira, A.F.A.; Araújo, E.S.; Bessa Neto, J.V.; da Costa, A.R.; Fernandes, L.d.S.; Martins, A.A.S.; Cândido, R.d.S.; Jardim, A.M.d.R.F.; Pandorfi, H.; et al. Cherry Tomato Crop Management Under Irrigation Levels: Morphometric Characteristics and Their Relationship with Fruit Production and Quality. *Gesunde Pflanz.* **2022**, *75*, 1277–1288. [CrossRef]
10. Pramanik, M.; Khanna, M.; Singh, M.; Singh, D.K.; Sudhishri, S.; Bhatia, A.; Ranjan, R. Automation of Soil Moisture Sensor-Based Basin Irrigation System. *Smart Agric. Technol.* **2022**, *2*, 100032. [CrossRef]
11. Gutierrez, J.; Villa-Medina, J.F.; Nieto-Garibay, A.; Porta-Gandara, M.A. Automated Irrigation System Using a Wireless Sensor Network and GPRS Module. *IEEE Trans. Instrum. Meas.* **2014**, *63*, 166–176. [CrossRef]
12. Freire, A.G.; de Alencar, T.L.; Chaves, A.F.; Nascimento, V.D.; de Assis, R.N.; Lier, Q.d.J.v.; Mota, J.C.A. Comparison of Devices for Measuring Soil Matric Potential and Effects on Soil Hydraulic Functions and Related Parameters. *Agric. Water Manag.* **2018**, *209*, 134–141. [CrossRef]
13. Pathak, S.; Adusumilli, N.C.; Wang, H.; Almas, L.K. Irrigation Water Demand and Elasticities: A Case Study of the High Plains Aquifer. *Irrig. Sci.* **2022**, *40*, 941–954. [CrossRef]
14. Li, Y. Water Saving Irrigation in China. *Irrig. Drain.* **2006**, *55*, 327–336. [CrossRef]

15. Sauer, T.; Havlík, P.; Schneider, U.A.; Schmid, E.; Kindermann, G.; Obersteiner, M. Agriculture and Resource Availability in a Changing World: The Role of Irrigation. *Water Resour. Res.* **2010**, *46*, 6503. [CrossRef]
16. Gbode, I.E.; Diro, G.T.; Intsiful, J.D.; Dudhia, J. Current Conditions and Projected Changes in Crop Water Demand, Irrigation Requirement, and Water Availability over West Africa. *Atmosphere* **2022**, *13*, 1155. [CrossRef]
17. Bwambale, E.; Abagale, F.K.; Anornu, G.K. Smart Irrigation Monitoring and Control Strategies for Improving Water Use Efficiency in Precision Agriculture: A Review. *Agric. Water Manag.* **2022**, *260*, 107324. [CrossRef]
18. Oliveira, R.A.; Ramos, M.M.; Aquino, L.A. Chapter 8—Irrigation Management. In *Agricultural Production, Bioenergy and Ethanol;* Academic Press: Cambridge, MA, USA, 2015; ISBN 9780128022399.
19. Lesiak, P.; Wolínski, T.; Jaroszewicz, L.; Majcher, J.; Kafarski, M.; Szypłowska, A.; Wilczek, A.; Lewandowski, A.; Gał, L.; Skierucha, W. Point Measurements of Soil Moisture Using TDR Technique—Comparison of Probes. *Eng. Proc.* **2022**, *21*, 8. [CrossRef]
20. Serrano, D.; Ávila, E.; Barrios, M.; Darghan, A.; Lobo, D. Surface Soil Moisture Monitoring with Near-Ground Sensors: Performance Assessment of a Matric Potential-Based Method. *Measurement* **2020**, *155*, 107542. [CrossRef]
21. Bertocco, M.; Parrino, S.; Peruzzi, G.; Pozzebon, A. Estimating Volumetric Water Content in Soil for IoUT Contexts by Exploiting RSSI-Based Augmented Sensors via Machine Learning. *Sensors* **2023**, *23*, 2033. [CrossRef]
22. Radi; Murtiningrum; Ngadisih; Muzdrikah, F.S.; Nuha, M.S.; Rizqi, F.A. Calibration of Capacitive Soil Moisture Sensor (SKU:SEN0193). In Proceedings of the 2018 4th International Conference on Science and Technology, ICST 2018, Yogyakarta, Indonesia, 7–8 August 2018; Institute of Electrical and Electronics Engineers Inc.: Yogyakarta, Indonesia, 2018.
23. Dobriyal, P.; Qureshi, A.; Badola, R.; Hussain, S.A. A Review of the Methods Available for Estimating Soil Moisture and Its Implications for Water Resource Management. *J. Hydrol.* **2012**, *458–459*, 110–117. [CrossRef]
24. Hardie, M. Review of Novel and Emerging Proximal Soil Moisture Sensors for Use in Agriculture. *Sensors* **2020**, *20*, 6934. [CrossRef] [PubMed]
25. Rasheed, M.W.; Tang, J.; Sarwar, A.; Shah, S.; Saddique, N.; Khan, M.U.; Imran Khan, M.; Nawaz, S.; Shamshiri, R.R.; Aziz, M.; et al. Soil Moisture Measuring Techniques and Factors Affecting the Moisture Dynamics: A Comprehensive Review. *Sustainability* **2022**, *14*, 11538. [CrossRef]
26. Mohanty, B.P.; Cosh, M.H.; Lakshmi, V.; Montzka, C. Soil Moisture Remote Sensing: State-of-the-Science. *Vadose Zone J.* **2017**, *16*, 1–9. [CrossRef]
27. Brocca, L.; Ciabatta, L.; Massari, C.; Camici, S.; Tarpanelli, A. Soil Moisture for Hydrological Applications: Open Questions and New Opportunities. *Water* **2017**, *9*, 140. [CrossRef]
28. Salam, A.; Vuran, M.C.; Irmak, S. Di-Sense: In Situ Real-Time Permittivity Estimation and Soil Moisture Sensing Using Wireless Underground Communications. *Comput. Netw.* **2019**, *151*, 31–41. [CrossRef]
29. Sanuade, O.A.; Hassan, A.M.; Akanji, A.O.; Olaojo, A.A.; Oladunjoye, M.A.; Abdulraheem, A. New Empirical Equation to Estimate the Soil Moisture Content Based on Thermal Properties Using Machine Learning Techniques. *Arab. J. Geosci.* **2020**, *13*, 377. [CrossRef]
30. Bagheri, M.; Rezania, M. Effect of Soil Moisture Evaporation Rate on Dynamic Measurement of Water Retention Curve with High-Capacity Tensiometer. *Int. J. Geomech.* **2022**, *22*, 04021301. [CrossRef]
31. Abdelmoneim, A.A.; Khadra, R.; Derardja, B.; Dragonetti, G. Internet of Things (IoT) for Soil Moisture Tensiometer Automation. *Micromachines* **2023**, *14*, 263. [CrossRef]
32. Patwa, D.; Bharat, T.V. Influence of Hydraulic Response Time of Tensiometer in Hydraulic Characteristics Estimation for Riverbank Sand. *Geotech. Geol. Eng.* **2022**, *41*, 413–427. [CrossRef]
33. Mesquita, M.; Machado, A.L.P.; Dos Santos, A.P.; Da Silva, M.V.; De Oliveira, H.F.E.; Battisti, R.; Nascimento, A.R. Assessing the Effects of Deficit Irrigation Techniques on Yield and Water Productivity of Processing Tomato. *Chem. Eng. Trans.* **2019**, *75*, 181–186. [CrossRef]
34. de Oliveira, H.F.E.; Campos, H.d.M.; Mesquita, M.; Machado, R.L.; Vale, L.S.R.; Siqueira, A.P.S.; Ferrarezi, R.S. Horticultural Performance of Greenhouse Cherry Tomatoes Irrigated Automatically Based on Soil Moisture Sensor Readings. *Water* **2021**, *13*, 2662. [CrossRef]
35. Mesquita, M.; Dos Santos, A.P.; Machado, A.L.P.; De Oliveira, H.F.E.; Casaroli, D.; Junior, J.A. Qualitative Characteristics of Processing Tomato Cultivated Under Water Deficit Induced in the Vegetative Growth Stage. *Chem. Eng. Trans.* **2019**, *75*, 175–180. [CrossRef]
36. Vatta, K.; Sidhu, R.S.; Lall, U.; Birthal, P.S.; Taneja, G.; Kaur, B.; Devineni, N.; MacAlister, C. Assessing the Economic Impact of a Low-Cost Water-Saving Irrigation Technology in Indian Punjab: The Tensiometer. *Water Int.* **2018**, *43*, 305–321. [CrossRef]
37. Pardossi, A.; Incrocci, L.; Incrocci, G.; Malorgio, F.; Battista, P.; Bacci, L.; Rapi, B.; Marzialetti, P.; Hemming, J.; Balendonck, J. Root Zone Sensors for Irrigation Management in Intensive Agriculture. *Sensors* **2009**, *9*, 2809–2835. [CrossRef]
38. Araújo, S.O.; Peres, R.S.; Barata, J.; Lidon, F.; Ramalho, J.C. Characterising the Agriculture 4.0 Landscape—Emerging Trends, Challenges and Opportunities. *Agronomy* **2021**, *11*, 667. [CrossRef]
39. Monteleone, S.; de Moraes, E.A.; de Faria, B.T.; Aquino Junior, P.T.; Maia, R.F.; Neto, A.T.; Toscano, A. Exploring the Adoption of Precision Agriculture for Irrigation in the Context of Agriculture 4.0: The Key Role of Internet of Things. *Sensors* **2020**, *20*, 7091. [CrossRef] [PubMed]
40. Amiri, Z.; Gheysari, M.; Mosaddeghi, M.R.; Amiri, S.; Tabatabaei, M.S. An Attempt to Find a Suitable Place for Soil Moisture Sensor in a Drip Irrigation System. *Inf. Process. Agric.* **2022**, *9*, 254–265. [CrossRef]

41. Domínguez-Niño, J.M.; Oliver-Manera, J.; Girona, J.; Casadesús, J. Differential Irrigation Scheduling by an Automated Algorithm of Water Balance Tuned by Capacitance-Type Soil Moisture Sensors. *Agric. Water Manag.* **2020**, *228*, 105880. [CrossRef]
42. Livingston, B.E. A Method for Controlling Plant Moisture. *Plant World* **1908**, *11*, 39–40.
43. Pereira, R.M.; Sandri, D.; Rios, G.F.A. Tensiômetros Eletrônicos Integrados a Placa Microcontroladora Arduino No Manejo da Irrigação de Alface em Diferentes Potenciais Matriciais Críticos e Tipos de Solo. *Irriga* **2022**, *27*, 311–327. [CrossRef]
44. Teixeira, A.S.; Coelho, S.L. Desenvolvimento e Calibração de Um Tensiômetro Eletrônico de Leitura Automática. *Eng. Agrícola* **2005**, *25*, 367–376. [CrossRef]
45. Richards, L.A. Methods of Measuring Soil Moisture Tension. *Soil Sci.* **1949**, *68*, 95. [CrossRef]
46. Wang, Z.; Ren, X.; Sun, X. Research on Calibration Method of Mechanical Belt Tension Meter. *Meas. Sens.* **2021**, *18*, 100226. [CrossRef]
47. Warrick, A.W.; Nielsen, D.R. Spatial Variability of Soil Physical Properties in the Field. In *Applications of Soil Physics*; Hillel, D., Ed.; Academic Press: New York, NY, USA, 1980; pp. 319–344.
48. R Core Team. R: A Language and Environment for Statistical Computing. Available online: https://www.r-project.org/ (accessed on 7 February 2023).
49. da Silva, M.V.; Pandorfi, H.; de Almeida, G.L.P.; Jardim, A.M.d.R.F.; Batista, P.H.D.; da Silva, R.A.B.; Lopes, I.; de Oliveira, M.E.G.; da Silva, J.L.B.; Moraes, A.S. Spatial Variability and Exploratory Inference of Abiotic Factors in Barn Compost Confinement for Cattle in the Semiarid. *J. Therm Biol.* **2020**, *94*, 102782. [CrossRef] [PubMed]
50. Thalheimer, M. A Low-Cost Electronic Tensiometer System for Continuous Monitoring of Soil Water Potential. *J. Agric. Eng.* **2013**, *44*, e16. [CrossRef]
51. Sanches, A.C.; Alves, C.d.O.; de Jesus, F.L.F.; Theodoro, F.L.; da Cruz, T.A.C.; Gomes, E.P. Low-Cost and High-Efficiency Automated Tensiometer for Real-Time Irrigation Monitoring. *Rev. Bras. Eng. Agrícola Ambient.* **2022**, *26*, 390–395. [CrossRef]
52. Jardim, A.M.d.R.F.; Silva, J.R.I.; da Silva, M.V.; de Souza, L.S.B.; Araújo Júnior, G.d.N.; Alves, H.K.M.N.; Mesquita, M.; Souza, P.J.d.O.P.d.; Teixeira, A.H.d.C.; da Silva, T.G.F. Modelling the Darcy–Weisbach Friction Factor and the Energy Gradient of the Lateral Line. *Irrig. Drain.* **2022**, *71*, 320–332. [CrossRef]
53. Colak, Y.B. Leaf Water Potential for Surface and Subsurface Drip Irrigated Bell Pepper under Various Deficit Irrigation Strategies. *Chil. J. Agric. Res.* **2021**, *81*, 491–506. [CrossRef]
54. Aliku, O.O.; Oshunsanya, S.O.; Aiyelari, E.O.A. Estimation of Crop Evapotranspiration of Okra Using Drainage Lysimeters under Dry Season Conditions. *Sci. Afr.* **2022**, *16*, e01189. [CrossRef]

hydrology

MDPI

Article

Improvements and Evaluation of the Agro-Hydrologic VegET Model for Large-Area Water Budget Analysis and Drought Monitoring

Gabriel B. Senay [1,*], Stefanie Kagone [2], Gabriel E. L. Parrish [3], Kul Khand [2], Olena Boiko [4] and Naga M. Velpuri [5]

1 U.S. Geological Survey Earth Resources Observation and Science (EROS) Center, North Central Climate Adaptation Science Center, Fort Collins, CO 80523, USA
2 ASRC Federal Data Solutions, Contractor to USGS EROS Center, Sioux Falls, SD 57198, USA; skagone@contractor.usgs.gov (S.K.); kkhand@contractor.usgs.gov (K.K.)
3 Innovate! Inc., Contractor to USGS EROS Center, Sioux Falls, SD 57198, USA; gparrish@contractor.usgs.gov
4 U-Spatial, Research Computing, University of Minnesota, Minneapolis, MN 55455, USA; oboiko@umn.edu
5 International Water Management Institute, Colombo P.O. Box 2075, Sri Lanka; n.velpuri@cgiar.org
* Correspondence: senay@usgs.gov

Abstract: We enhanced the agro-hydrologic VegET model to include snow accumulation and melt processes and the separation of runoff into surface runoff and deep drainage. Driven by global weather datasets and parameterized by land surface phenology (LSP), the enhanced VegET model was implemented in the cloud to simulate daily soil moisture (SM), actual evapotranspiration (ETa), and runoff (R) for the conterminous United States (CONUS) and the Greater Horn of Africa (GHA). Evaluation of the VegET model with independent data showed satisfactory performance, capturing the temporal variability of SM (Pearson correlation r: 0.22–0.97), snowpack (r: 0.86–0.88), ETa (r: 0.41–0.97), and spatial variability of R (r: 0.81–0.90). Absolute magnitudes showed some biases, indicating the need of calibrating the model for water budget analysis. The seasonal Landscape Water Requirement Satisfaction Index (L-WRSI) for CONUS and GHA showed realistic depictions of drought hazard extent and severity, indicating the usefulness of the L-WRSI for the convergence of an evidence toolkit used by the Famine Early Warning System Network to monitor potential food insecurity conditions in different parts of the world. Using projected weather datasets and landcover-based LSP, the VegET model can be used not only for global monitoring of drought conditions, but also for evaluating scenarios on the effect of a changing climate and land cover on agriculture and water resources.

Keywords: VegET model; soil moisture; actual evapotranspiration; runoff; land surface phenology; drought; water budget

Citation: Senay, G.B.; Kagone, S.; Parrish, G.E.L.; Khand, K.; Boiko, O.; Velpuri, N.M. Improvements and Evaluation of the Agro-Hydrologic VegET Model for Large-Area Water Budget Analysis and Drought Monitoring. *Hydrology* **2023**, *10*, 168. https://doi.org/10.3390/hydrology10080168

Academic Editor: Hamza Farooq Gabriel

Received: 1 July 2023
Revised: 6 August 2023
Accepted: 8 August 2023
Published: 10 August 2023

1. Introduction

Large-area modeling of rainfall–runoff processes has been an important component of many environmental assessments, particularly for flood early warning [1,2], drought monitoring and impact assessment [3,4], water accounting [5], and hydrologic studies [6–10]. Although hydrologic models vary in their degree of complexity in terms of model components and parameters, the fundamental principle remains the same in that all models are designed to conserve mass through water budget accounting at all time scales over a defined volume. By its nature, large-area rainfall–runoff modeling benefits from spatially distributed inputs and parameters. As the primary driver for all hydrologic models is precipitation, the availability of global satellite-based gridded precipitation data allows model implementation over large basins, continents, and the globe [11–13]. Similarly, model-assimilated gridded datasets are available for potential evapotranspiration [14–17], another key input to hydrologic models. Based on the purpose and desired accuracy, hydrologic models are parameterized to account for water storage and flux quantities over the landscape. Almost all hydrologic models define the soil moisture storage capacity using parameters such as water holding capacity (WHC) derived from soil texture properties [18,19]. More comprehensive models also include snow and canopy interception storage terms [7,8,20]. In addition to storage terms, flux-controlling parameters (surface runoff, drainage, and evapotranspiration) are mainly tied to land cover, soil properties, and climatic factors.

The purpose and availability of data may determine the complexity of the model from short time interval (minutes) flood prediction models to monthly water balance models such as the U.S. Geological Survey (USGS) water balance model [21]. Simple models with one-dimensional (vertical) accounting of fluxes are well suited for drought monitoring and basin-scale water budget studies at longer time scales. For agricultural drought monitoring purposes, simple bucket models that only account for the root-zone water balance status have been used by various modeling groups [3,4,22,23] with numerous simplifying assumptions. One of the early models is the Water Requirement Satisfaction Index (WRSI) by the Food and Agriculture Organization (FAO) [4] that parameterizes the seasonality of crop water use (actual evapotranspiration, ETa) using published crop coefficients (Kc) [24]. The WRSI model is based on the ratio of ETa (as a result of precipitation over a season) to an ideal water requirement of a well-watered crop, which is defined by the potential ET (atmospheric demand) and the seasonally prescribed Kc. Because of the difficulty in defining Kc values over large areas due to unknown crops and/or unreliability of published values outside of their experimental region/continent, the VegET model incorporated land surface phenology (LSP) derived from remotely sensed Normalized Difference Vegetation Index (NDVI) [3]. Evaluation and application of the VegET model showed good performance for ETa [25] and runoff estimation [26].

The original version of the VegET model [3] did not include snow accumulation and snowmelt processes, which limited its representation in snow-influenced landscapes for simulating soil moisture and runoff using the principle of saturation excess [27,28]. Furthermore, the previous version of the VegET model did not partition runoff into surface runoff and deep drainage. The main objectives of this study are to (1) describe the updated components and parameterizations to the VegET model, (2) evaluate the performance of the VegET model using independent data for soil moisture, snowpack, ETa, and runoff, and (3) demonstrate the applications of the updated VegET model for drought monitoring and early warning.

2. Materials and Methods

2.1. Study Area and Data Sources

The updated VegET v2.0 model [29] was implemented over the conterminous United States (CONUS) and the Greater Horn of Africa (GHA) (Figure 1), making use of Open Source Python libraries and leveraging a combination of cloud computing and local servers at the USGS Earth Resources Observation and Science (EROS) Center.

The model uses different input datasets including precipitation, reference ET, air temperature, and soil properties. The data for the CONUS are described in Table 1.

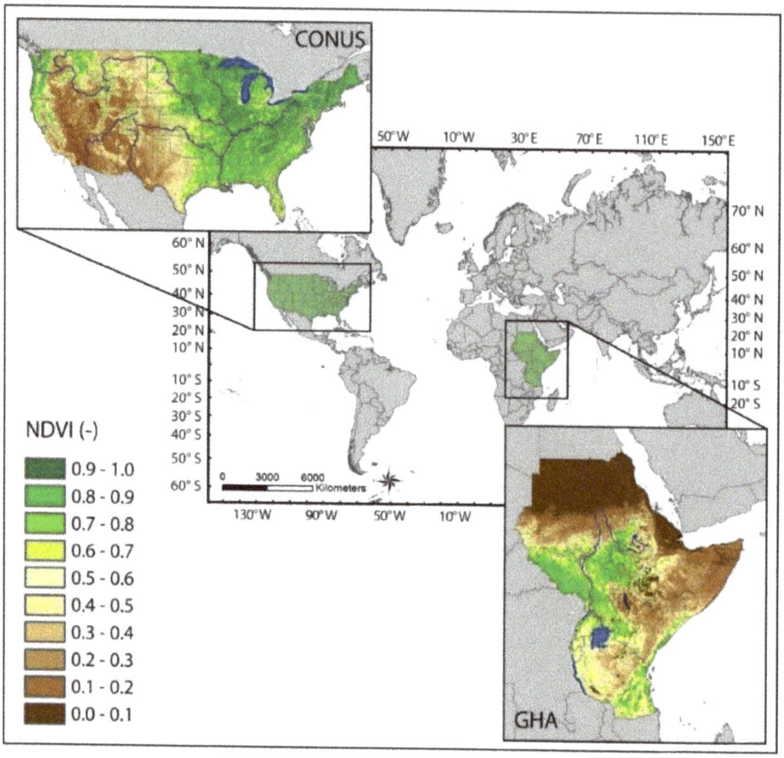

Figure 1. Study areas including the conterminous United States (CONUS) and the Greater Horn of Africa (GHA) using Normalized Difference Vegetation Index (NDVI) to capture the spatial distribution of relative vegetation productivity for July 2018.

Table 1. Characteristics of model inputs and parameters for the conterminous United States (CONUS).

Parameters	Spatial Resolution	Temporal Resolution	Source
Precipitation	4000 m	Daily, 1980–current	gridMET [30]
Land Surface Phenology	1000 m	16 days (Terra), 2003–2017 *	MODIS NDVI [31] (MOD13A2.061)
Reference Evapotranspiration	4000 m	Daily, 1981–2010 *	gridMET [30]
Air Temperature	4000 m	Daily, 1984–2017 *	gridMET [30]
Soil Properties	90 m	Static	gNATSGO [32,33]
Interception	250 m	Static	MODIS VCF [34] (MOD44B.061)

* Median climatology generated from the specified time period.

The precipitation, reference evapotranspiration (ETo), and air temperature (Ta) were downloaded from the Gridded Surface Meteorological (gridMET) website [30] and converted from the native netcdf format to geotiff. The air temperature data (daily minimum, maximum, and average) also were converted from Kelvin (K) to degree Celsius (°C) and a median climatology was created from 1984 to 2017. The land surface phenology (LSP) is based on the Moderate-Resolution Imaging Spectroradiometer (MODIS) NDVI provided by National Aeronautics and Space Administration (NASA) Land Processes Distributed Active Archive Center (LP DAAC). A daily median climatology NDVI for 2001–2019 (19 years) was established with linear interpolation from the 16-day dataset. The soil properties included WHC (also referred to as available water holding capacity, AWC), field capacity

(FC), and soil porosity (POR). The WHC represents the difference between FC and the wilting point (WP). Detailed information on the soil data can be found in [33]. The schematic representation of the WHC and associated parameters are shown in Figure 2.

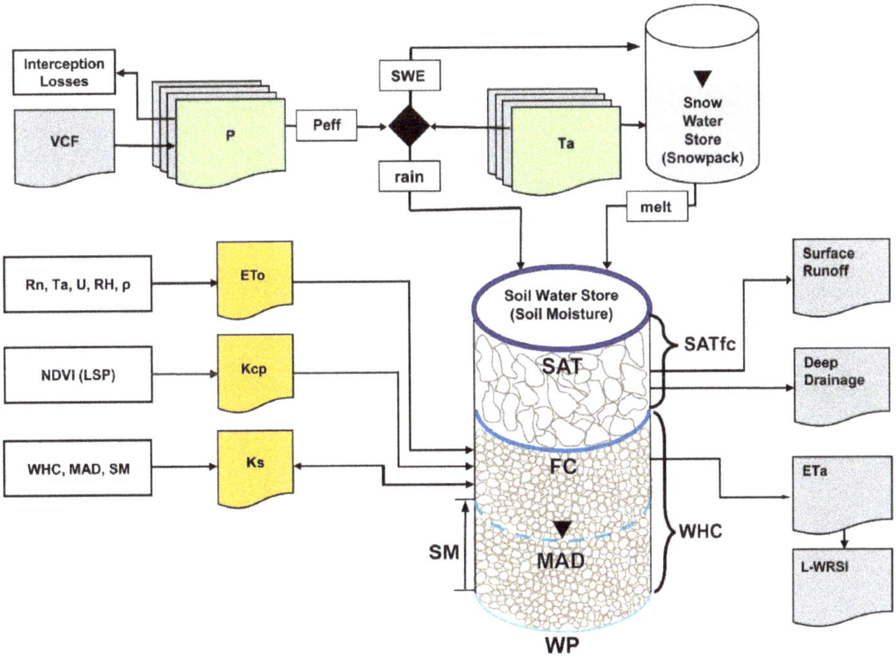

Figure 2. Schematic representation of the updated VegET model (v2.0) [29]. The Soil Water Store is defined by two major sections: "gravity water" is filled once the soil moisture is above field capacity (FC) and "plant available water" is the section between FC and permanent wilting point (WP). VCF: vegetation continuous field, P: precipitation, Peff: effective precipitation, SWE: snow water equivalent, Ta: air temperature; ETo: reference ET, Rn: net radiation, U: wind speed, RH: relative humidity, ρ = air pressure, Kcp: landscape water use coefficient, LSP: land surface phenology, Ks: soil stress coefficient, WHC: water holding capacity, MAD: maximum allowable depletion, SM: soil moisture, SAT, soil saturation, SATfc: volume between SAT and FC, L-WRSI: landscape Water Requirement Satisfaction Index.

The interception fraction layer was created from the MODIS Vegetation Continuous Fields (VCF: [35]) in proportion to a pixel's percentage of tree, herbaceous, and bare coverage using Equation (1) [3]:

$$\text{Interception} = 0.15 \times T_{cover} + 0.1 \times H_{cover} + 0.0 \times B_{cover} \quad (1)$$

where T_{cover} is the tree cover layer (%) with a maximum interception of 15%; H_{cover} is the herbaceous cover layer (%) with a maximum interception of 10%; and B_{cover} is the bare ground cover layer (%) with no interception.

Input datasets for GHA are described in Table 2. The precipitation data source is Climate Hazards Group InfraRed Precipitation with Stations (CHIRPS) [36], which uses remote sensing data in combination with station information to create a dataset suitable for crop monitoring and hydrologic applications. The LSP was created using MODIS NDVI (Aqua and Terra) to generate a 15-year median climatology (2003–2017) and then interpolated linearly from the 8-day time step to daily. The reference evapotranspiration (ETo) was obtained from NOAA [37]. Air temperature (minimum, maximum, and mean) was sourced from Climatologies at High resolution for the Earth's Land Surface Areas (CHELSA); the monthly climatology from 1981 to 2010 was interpolated linearly from monthly to daily and converted from K to °C units. The soil property raster data were provided by the International Soil Reference and Information Centre (ISRIC—World Soil Information) through their Soil Data Hub [38]. A list of soil parameters was extracted to generate WHC, FC, and POR: AWCh3_M_Sl6_250m_II.tif, (WWP_M_sl6_250m_II.tif and AWCh3_M_Sl6_250m_II.tif), and AWCtS_M_sl6_250m_ll.tif. POR layer was used to represent the soil saturation (SAT) level. To convert the raster data from volumetric percent representation (m^3/m^3) to depth (mm) per meter root-zone, they were multiplied by a unit conversion factor of 10 ((soil raster × 1000)/100). Additionally, the SAT value was capped to not be lower than FC. Interception was determined using Equation (1).

Table 2. Characteristics of model inputs and parameters for the Greater Horn of Africa (GHA).

Parameters	Spatial Resolution	Temporal Resolution	Reference
Precipitation	0.05°	Daily; 1981—current	CHIRPS [36]
Land Surface Phenology	1000 m	Every 8 days (Aqua and Terra); 2003–2017 *	MODIS NDVI [31]
Reference Evapotranspiration	0.625° × 0.5°	daily; 1981–2010 *	NOAA ETo [37]
Air Temperature	1000 m	Monthly; 1981–2010 *	CHELSA [39]
Soil Properties	250 m	Static	ISRIC [38]
Interception	250 m	Static	MODIS VCF [34]

* Median climatology generated from the specified time period.

2.2. Model Formulation

2.2.1. Original Model Setup

The original VegET model by [3] was developed to timely process and integrate readily available global weather and remote sensing datasets using water balance modeling techniques for drought monitoring purposes. ETo, a soil stress coefficient (Ks), and a phenology-based crop coefficient (Kcp) are used to determine daily soil moisture, runoff (R), and ETa using the root-zone as the control volume (Figure 2). The soil water level is determined using a daily soil water balance using Equation (2).

$$SM_i = SM_{i-1} + Peff_i - ETa_i \tag{2}$$

where *SM* is soil moisture (mm), *Peff* is effective precipitation (mm), *ETa* is simulated actual evapotranspiration (mm) and *i* represents the current day and *i*–1 represents the previous day. *ETa* is calculated using Equation (3) as follows:

$$ETa = Kcp \times Ks \times ET_o \tag{3}$$

where Kcp is the LSP-derived landscape "crop" coefficient (-); *Ks* is soil water stress coefficient (-), and ET_o is the grass reference ET (mm).

The innovation in the VegET model is on the calculation of Kcp, which is comparable to the Kc that is widely used by agronomists [40]. The key difference between the two parameters is that Kcp is derived from remotely sensed data as opposed to region-specific field experiments for Kc. Kcp represents both the spatial and temporal dynamics of the landscape water-use pattern on a grid basis. LSPs are characterized and converted into Kcp parameter functions for each modeling grid from NDVI climatology datasets with the assumption that the LSP climatology represents the target vegetation condition of the landscape where water requirement is met by precipitation. Thus, *ETa* is calculated using the modified version of the classical crop coefficient approach [24] using the LSP-derived crop coefficient.

Ks is determined from a soil water balance model such as the one developed by [22] for USGS Famine Early Warning Systems Network (FEWS NET) applications using Equations (4) and (5). The dimensionless K_s coefficient varies from 0 to 1 depending on the soil water level in the root zone and is calculated as:

$$Ks = \frac{SM_i}{MAD}; \quad SM_i < MAD \tag{4}$$

$$Ks = 1.0; \quad SM_i \geq MAD \tag{5}$$

where SM_i is the soil water of current time step in depth unit (mm); *MAD* (mm) is the maximum allowable depletion level of soil water in the root zone below which the vegetation ETa is less than "potential" and will be constrained by the availability of soil water.

Although *MAD* varies by crop/vegetation type, a nominal value of 50% of the WHC can be used for most generalized crops, such as cereals and natural vegetation. Thus, *MAD* was estimated as 50% of the WHC (i.e., 0.5 × WHC). More discussion on the setup and application of the soil water balance model for operational crop monitoring is available in [22].

The model estimates a combined surface runoff and deep drainage based on the principle of saturation excess where soil water in excess of the WHC is considered to be unavailable for plant use in the root zone; thus, *SMi* is set to a maximum of WHC and a minimum of 0 during the modeling time step.

$$R = SM_i - WHC; \quad SM_i > WHC \tag{6}$$

$$R = 0; \quad SM_i \leq WHC \tag{7}$$

where *R* is total runoff (surface runoff and deep drainage); *WHC* is soil water holding capacity (mm), i.e., the difference between FC and WP (Figure 2).

2.2.2. Model Updates

The original VegET model has been updated with improved parameterization to be more inclusive of hydrologic processes and for computing efficiency. The new modifications to the model include the incorporation of snowpack and snowmelt processes and the separation of runoff into surface runoff and deep drainage. Furthermore, parameterization of the LSP has been simplified to use a set of linear equations without the need to specify the minimum and maximum *Kc* that was part of the original formulation.

Figure 2 shows the schematic representation of the updated VegET v2.0 model [29]. Interception losses are first estimated to determine effective precipitation using the MODIS VCF (Equation (1)). Effective precipitation (Equation (2)) is split into rain and SWE (snow water equivalent) to enter the Soil Water Store (soil moisture) or Snow Water Store (snowpack) based on a temperature-index (Equations (10)–(13)) approach [41]. MODIS NDVI is used to create the LSP for the Kcp function (Equations (8) and (9)). Rn (net radiation), Ta (air temperature), U (wind speed), RH (relative humidity), and ρ (atmospheric pressure) are parameters used to estimate ETo (reference ET). Kcp and *Ks* are critical parameters to calculate outfluxes: surface runoff, deep drainage, and ETa. The Soil Water Store is defined by the soil-texture properties. Saturation (SAT), defined in the model, is equivalent to soil porosity (POR) from the soils database [38]; FC defines the maximum amount of water retention by the soil matrix that is available to plants; and WP (permanent wilting point) represents the water retention level at which point plants are unable to access moisture. WHC (difference between FC and WP) is the readily available water for plant access, but plant stress occurs in proportion to the remaining soil moisture (SM) once SM reduces below the MAD limit (Equations (4) and (5)). Runoff (R) is generated once SM is in excess of WHC (Equations (6) and (7)). All SM in excess of SAT will be surface runoff, but SM that is within SATfc (between SAT and FC) will be split into surface runoff and deep drainage (Equations (17)–(20)). L-WRSI (Landscape-Water Requirement Satisfaction Index) is determined using ETa and landscape water requirement (ETc) (Equations (21) and (22)). The VegET model is initialized with empty (0) amounts for SM and snowpack with a one complete year spin-up period.

Land Surface Phenology and Landscape Coefficients (Kcp)

The crop coefficient (Kc) determines the ideal (water unlimited condition) demand of the crop based on the type and stage of the crop [24]. In VegET, the crop water requirement (demand) is replaced with the landscape water requirement. The traditional tabular Kc by Allen et al. [24] is replaced by the phenology-based Kc known as Kcp. The main assumption is that the NDVI-derived Kcp represents the landscape "crop" water requirement in regions where a major land cover change does not occur for a large area. For example, the Land Change Monitoring, Assessment, and Projection (LCMAP) group indicates a less than 1% land cover change per year on average over the CONUS [42]. For drought monitoring purposes where the VegET is applied, the main goal is to determine if the precipitation amount and distribution meet the average demand of the landscape. The use of a climatology NDVI creates smoother and more realistic seasonal water use patterns compared to Kc, but it may underestimate the demand during years of vigorous vegetation activity. However, its effectiveness for drought monitoring would not be affected under such favorable wet conditions. Figure 3 illustrates the development and seasonal progression of LSP-based Kcp derived from climatology NDVI and its schematic Kc equivalent.

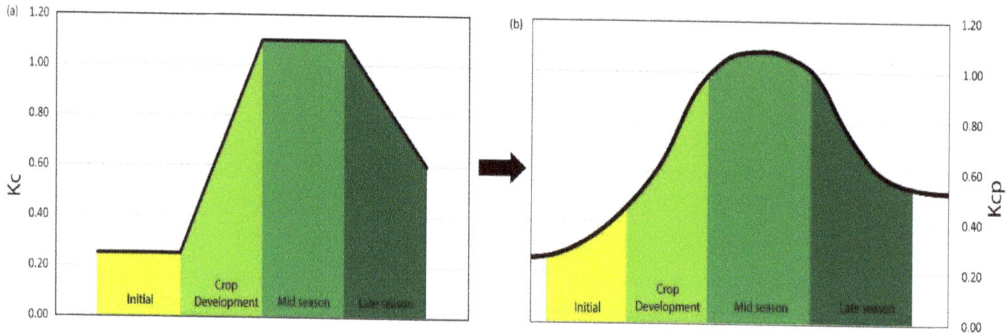

Figure 3. Schematic representation of (**a**) the traditional crop coefficient (Kc) and (**b**) the land surface phenology (LSP) (Kcp) coefficient.

The Kcp in the VegET model is estimated as:

$$Kcp = 1.25 \times NDVI + 0.20; \quad NDVI > 0.4 \tag{8}$$

$$Kcp = 1.25 \times NDVI; \quad NDVI \leq 0.4 \tag{9}$$

where the NDVI threshold of 0.4 is based on vegetation sparsity classification by [43] and a similar application in [3].

Equation (8) is similar to the one proposed by [44] when Kc is designed to be used in combination with grass reference ETo. The conditional elimination of the 0.2 intercept in Equation (9) is based on observations of overestimation of ETa over sparsely vegetated surfaces [3].

Snowpack and Snowmelt

The updated VegET model accounts for snowpack and snowmelt processes using air temperature-based empirical equations by [41]. A given day's precipitation is split into rainfall and snow water equivalent based on air temperature thresholds.

$$rain_{frac} = 1.0; \quad T_{avg} > 6.0\ ^{\circ}\mathrm{C} \tag{10}$$

$$rain_{frac} = 0.0; \quad T_{avg} < 0.0\ ^{\circ}\mathrm{C} \tag{11}$$

$$rain_{frac} = \frac{1}{12}(T_{avg} - 0.0); \quad 0.0 <= T_{avg} <= 6.0\ ^{\circ}\mathrm{C} \tag{12}$$

where $rain_{frac}$ is the rain fraction of precipitation that falls as rain (as opposed to snow) based on daily average air temperature T_{avg} for that day. If T_{avg} for a given day is below 0 °C, all precipitation is assumed to fall as snow $\left(rain_{frac} = 0\right)$; if T_{avg} is above 6 °C all precipitation is assumed to be rainfall $\left(rain_{frac} = 1\right)$; if T_{avg} is between 0 and 6 °C, the rain fraction is interpolated using Equation (12).

The rainfall and snow (snow water equivalent, SWE) components are then partitioned as follows:

$$SWE = \left(1 - rain_{frac}\right) \times Peff \tag{13}$$

where SWE is the snow water equivalent (mm) and $Peff$ (mm) is the effective precipitation (precipitation minus canopy interception losses), determined using the interception parameters from Equation (1). The rainfall component is simply a product of the $rain_{frac}$ and $Peff$ while SWE is the difference between $Peff$ and the rainfall (rain) component (Equations (10)–(12)).

The consideration of the timing of accumulation and melting of snow is useful for regions where snowpack (Snow Water Store) retains the precipitation instead of immediately releasing it as runoff during a cold season. The snowpack accumulates and melts based on the addition of new SWE and melting of snowpack using a daily snowpack ($Snow_{pack}$) balance. The daily snowmelt is calculated based on the *melt* (mm) rate as:

$$melt = 0.06\left(T_{max}^2 - T_{max} \times T_{min}\right); \quad Snow_{pack} \geq melt \tag{14}$$

$$melt = Snow_{pack}; \quad Snow_{pack} < melt \tag{15}$$

where 0.06 is the melt factor (mm/°C ^{2}), T_{max} is the daily maximum air temperature (°C), and T_{min} is the daily minimum air temperature (°C). The equation was adapted from [41]. All snow related parameters such as snowpack and melt are expressed in SWE forms.

$$Snow_{pack,\ i} = Snow_{pack,\ i-1} + SWE_i - melt_{\ i} \tag{16}$$

where $Snow_{pack,\ i}$ is the current snowpack in SWE unit (mm); $Snow_{pack,\ i-1}$ is the previous day's snowpack (mm); SWE_i is the additional fresh snow in SWE unit (mm) and $melt_{\ i}$ is today's snowmelt in SWE unit (mm) on the current day (i).

Deep Drainage and Surface Runoff Partitioning

The original VegET model estimates total runoff without the separation of quick flow (surface runoff) and deep percolation (deep drainage). A simple approximation coefficient is used to differentiate the quick flow (part of the total runoff that joins the stream network as overland flow) from the deep drainage (part of the flow that may combine interflow and deep percolation to groundwater). Although the separation of surface runoff and deep drainage does not affect the soil moisture and ETa estimation, the potential application of the VegET runoff in flood and streamflow simulation could benefit from this separation. VegET does not have a flow routing routine; therefore, runoff from one pixel does not affect soil moisture and evapotranspiration (ET) on nearby pixels. It is important to note that the VegET model is more optimized to simulate ET; thus, its use for hydrologic applications would benefit from more investigation, evaluation, and refinement.

In the updated VegET model, the deep drainage (*dd*) amount is estimated as the difference between total runoff (R) (Equations (6) and (7)) and surface runoff (srf) as follows:

$$dd = R - srf \tag{17}$$

where R is determined as daily SM in excess of the soil water holding capacity; srf is estimated based on the daily soil water, a quick-flow (qc)/drainage (dc) coefficient, soil saturation (SAT), and field capacity (FC) parameters (Figure 2):

$$SAT_{fc} = SAT - FC \tag{18}$$

$$srf = qc \times R; \quad R \leq SAT_{fc} \tag{19}$$

$$srf = R - SAT_{fc} + qc \times SAT_{fc}; \quad R > SAT_{fc} \tag{20}$$

where SAT_{fc} is the difference between SAT and FC; qc is the quick flow coefficient, which is a complement to the drainage coefficient (dc) as $qc = 1 - dc$. In this study, a uniform value of 0.35 is used for qc as a first approximation; however, this partitioning coefficient is expected to vary by soil type and topography, and thus a calibration procedure is required to estimate this coefficient more accurately.

2.2.3. Evaluation Data

The VegET model output parameters were evaluated using limited illustrative data from the Soil Climate Analysis Network (SCAN) [45] measurements for soil moisture, snow measurements from SNOpack TELemetry (SNOTEL) [46], AmeriFlux Network and FluxNet 2015 [47,48] measurements for ETa, and runoff data from the USGS [49] (Figure 4).

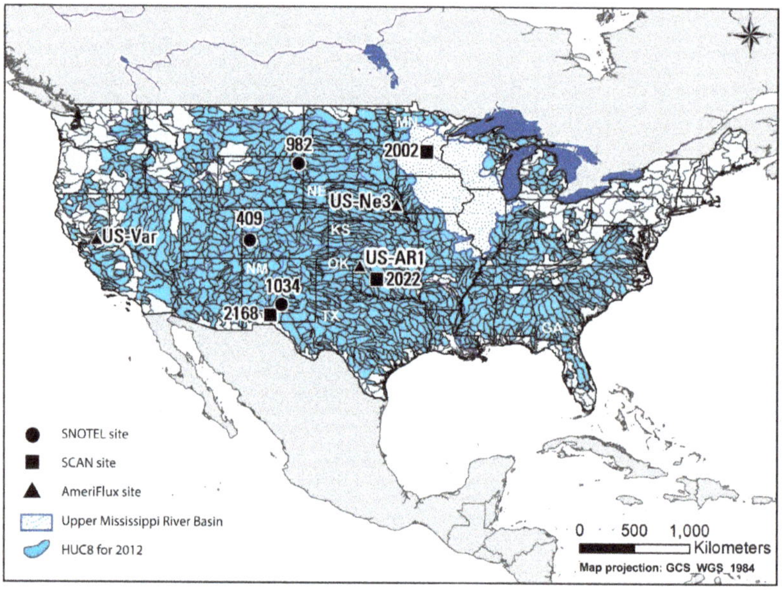

Figure 4. Location map for Soil Climate Analysis Network (SCAN) sites, SNOTEL (SNOpack TELemetry) sites, AmeriFlux Tower sites, and eight-digit Hydrologic Unit Code (HUC8) in the conterminous United States for water year 2012 evaluation. The map shows the Upper Mississippi River Basin used for the detailed water budget analysis. GA: Georgia, KS: Kansas, MN: Minnesota, NE: Nebraska, NM: New Mexico, OK: Oklahoma, TX: Texas.

Evaluation for Soil Moisture

One of the outputs of the VegET model is the daily soil moisture for a 1 m (39.4 inch) root zone. VegET SM was evaluated at three SCAN sites (Table 3) administered by the Natural Resources Conservation Service (NRCS) of the U.S. Department of Agriculture (USDA) [45]. Data for the growing season of May to September 2019 were used. The daily soil moisture measurements for five different depths (2 inches, 4 inches, 8 inches,

20 inches, and 40 inches) were averaged and converted from volumetric water content (m^3/m^3) percentage to depth of water per meter depth (mm/m) with a unit conversion factor of 10.

Table 3. Summary of the Soil Climate Analysis Network (SCAN) soil moisture sites [45] used for evaluation.

Site ID	Name	State	Location (Latitude, Longitude in Degrees)	Time Period
2002	Crescent Lake #1	Minnesota	45.42°, −93.95°	October 1993 to current
2022	Fort Reno #1	Nebraska	35.33°, −98.02°	November 1998 to current
2168	Jornada Exp Range	New Mexico	32.56°, −106.70°	October 2009 to current

Evaluation for Snow Water Equivalent

SNOTEL site measurements [46] were used to evaluate the simulated SWE by comparing the model output with in situ observations listed in Table 4. The time period used for evaluation was 2015–2020. SNOTEL data of SWE were converted from inches to mm prior to analysis and the comparison.

Table 4. Summary of the SNOpack TELemetry (SNOTEL) snow sites [46] used for evaluation.

Site ID	Site Name	Elevation (m)	Location (Latitude, Longitude in Degrees)	Time Period
982	Cole Canyon	5910	44.48°, −104.42°	2000 to current
409	Columbine Pass	9171	38.42°, −108.39°	1985 to current
1034	Sierra Blanca	10268	33.40°, −105.80°	2002 to current

Evaluation for Actual Evapotranspiration

The ETa results from the VegET model were evaluated using eddy covariance (EC) flux tower data from the AmeriFlux network [47]. For this evaluation, three EC towers were selected (Table 5) across the CONUS for availability of data to represent rainfed systems simulated by VegET. The locations of the towers are shown in Figure 4. Monthly data were obtained from the FLUXNET2015 dataset [48].

Table 5. Summary of AmeriFlux EC sites [47] used for actual evapotranspiration (ETa) evaluation.

Site ID	Name Name	Landcover	Location (Latitude, Longitude in Degrees)	Time Period Available
US-AR1	ARM USDA	Grassland	36.43, −99.42	2003–2021
US-Ne3	Mead	Rainfed crop	41.12, −96.44	2001–2020
US-Var	Vaira Ranch–Ione	Grassland	38.41, −120.95	2000–2014

Evaluation for Runoff

The VegET runoff (R) was evaluated against independent runoff obtained from the USGS WaterWatch [49] at 8-digit hydrologic unit code (HUC8) scale [50] across the CONUS (Table 6). The runoff data are generated from historical flow observations at the USGS streamgage locations, drainage basin boundaries of the streamgages, and the HUC8 boundaries [51]. The daily VegET runoff were summed by water year and the pixel values were spatially averaged within the HUC8 boundaries to obtain a single value and compared with the runoff (non-spatial single value) for water years 2012 (dry year), 2016 (wet year), and 2018 (average year). The HUC8s with high runoff values from the USGS WaterWatch were excluded from the comparison. For example, runoff more than 40% of precipitation (R/P > 40%) with potential regional groundwater flow contributions and possibility of watershed water balance closure issues [52,53].

Table 6. Summary of runoff data source [49], spatial and temporal resolution, study years, and number of HUC8 watersheds applied for evaluation.

Spatial Resolution	Temporal Resolution	Study Years	Number of HUC8s (R/P * $\leq$ 0.40)
HUC8 scale	Water year (October 1–September 30)	2012 (dry), 2016 (wet), 2018 (average)	1762 (1441) 1762 (1432) 1762 (1396)

*: R/P refer to the ratio of runoff (R) to precipitation filter for number of HUC8 watersheds in bracket.

3. Results and Discussion

3.1. Water Balance Components

The VegET model produces several parameters, fluxes, state variables, and indices including interception losses, rainfall, snow water equivalent, snowpack, snow melt, soil moisture, surface runoff, deep drainage, actual evapotranspiration, landscape water requirement, and the drought monitoring product L-WRSI [54]. Figure 5 provides an illustrative overview of the model inputs, outputs, and parameters for a flux tower location in Minnesota.

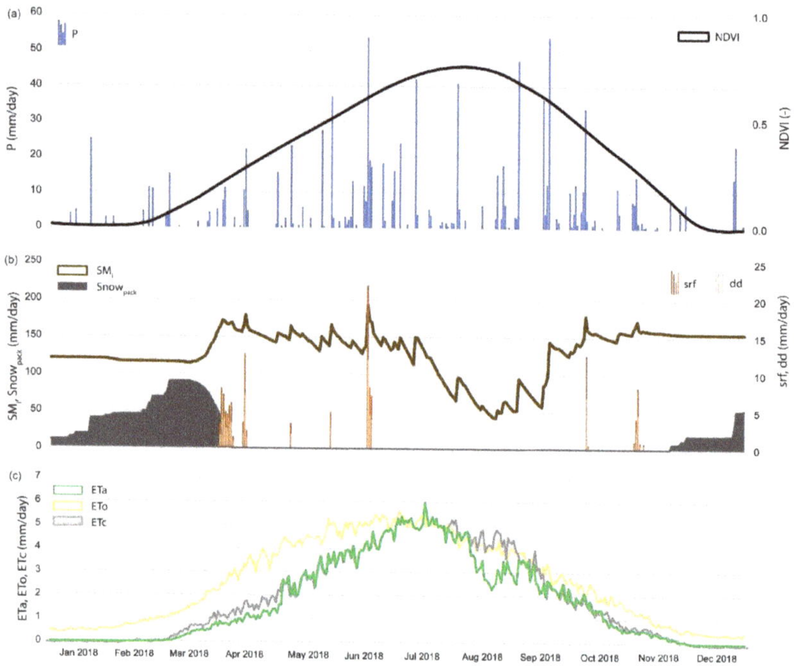

Figure 5. VegET model input and output parameters for a pixel at the AmeriFlux station in Minnesota (US-Ro1) for 2018. (**a**) Normalized Difference Vegetation Index (NDVI) and precipitation; (**b**) soil moisture (SM), snowpack ($Snow_{pack}$), surface runoff (srf), and deep drainage (dd); (**c**) actual ET (ETa), reference ET (ETo), and landscape water requirement (ETc).

Figure 5a shows precipitation and NDVI as a reference for providing an overview of the water supply and vegetation demand over the year. Precipitation is the main input and NDVI is an important rate controlling parameter representing the state of vegetation and its water use phenology. Figure 5b includes soil moisture level, snowpack, deep drainage, and surface runoff. The soil moisture (SM) shows a steady increase once snowpack decreases due to melt and additional rainfall and remains high for much of the spring with small variability around 150 mm, which is close to WHC = 160 mm. We note that no runoff component is generated during the summer, which requires SM to exceed the WHC. SM shows a substantial reduction in mid-August due to reduced precipitation events in frequency and magnitude, which leads to a reduction in ETa (Figure 5c). When SM reduces below the MAD level (half of WHC), ETa (green line, Figure 5c) will be lower than landscape water requirement (ETc) (Figure 5c), which leads to a deficit. The ETc is created as the product of ETo (Figure 5c) and Kcp.

VegET takes spatially explicit inputs and parameters and produces spatially explicit outputs, making it useful to create a continuous surface for agro-hydrologic applications. The annual ETa maps for CONUS are shown in Figure 6, in which water years 2012, 2016, and 2018 represent a drier year, a wetter year, and an average year, respectively. The drier landscape responses (ETa < 400 mm/yr) are noticeable for 2012 in large parts of Nebraska, Kansas, and Texas where drought conditions were reported by the U.S. Drought Monitor [55].

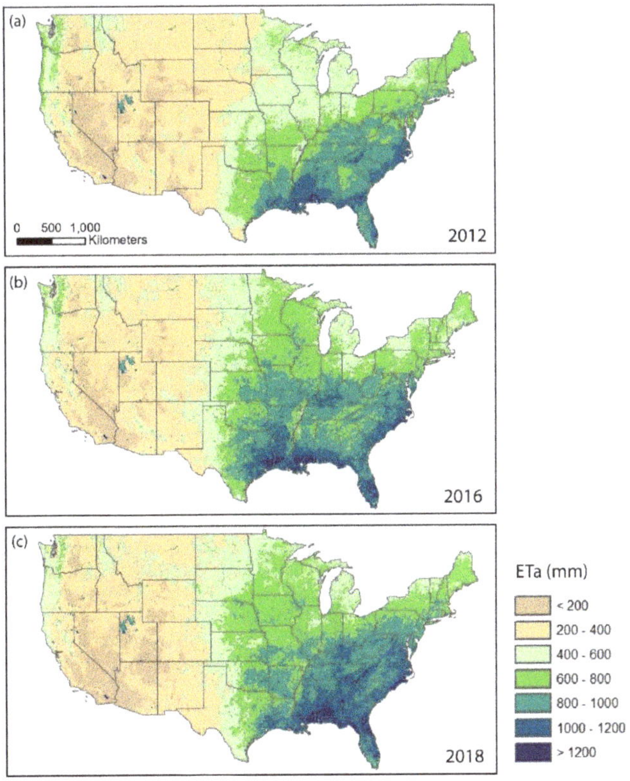

Figure 6. VegET annual actual evapotranspiration (ETa) for the years (**a**) 2012, (**b**) 2016, and (**c**) 2018. Brown colors with low ETa dominate low-precipitation and sparsely vegetation regions in contrast with green and blue tones on well-vegetated and precipitation-rich regions.

The calculation of ETa depends heavily on SM because reduction from ETa begins when SM falls below the MAD level. Due to differences in soil texture, maps of normalized SM as percentage of the WHC across the CONUS is shown in Figure 7 for selected days (1 October, 1 January, 1 April, and 1 July) during water year 2018. In this example, the relative SM in the soil started out low in October after the end of the growing season and increased over the next months until April. In July, SM showed a general reduction, especially in the southwestern CONUS.

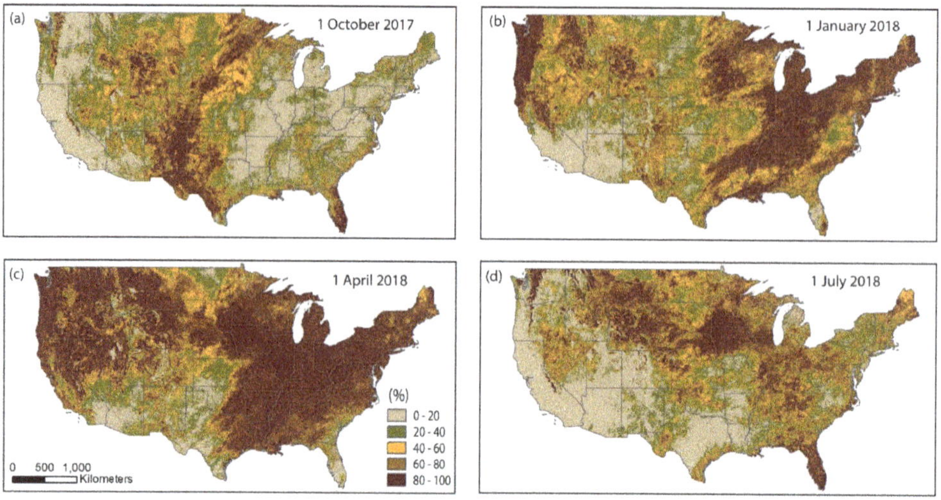

Figure 7. VegET relative soil moisture as percentage of water holding capacity (WHC) for (**a**) 1 October 2017, (**b**) 1 January 2018, (**c**) 1 April 2018, and (**d**) 1 July 2018.

The major improvement in VegET v2.0 model [29] is the inclusion of the snowpack and snowmelt processes. Figure 8 shows the state of snowpack based on the simple temperature-index algorithm to accumulate and melt the snow. As expected, the largest coverage of snowpack was observed on 1 January (Figure 8b) and the least snow was on 1 October after the summer (Figure 8a). Because of the simplicity of the model, only the relative magnitudes are reliable, which is sufficient for drought monitoring purposes. These maps are useful when comparing relative snowpack build-up and timing of melt across regions and years.

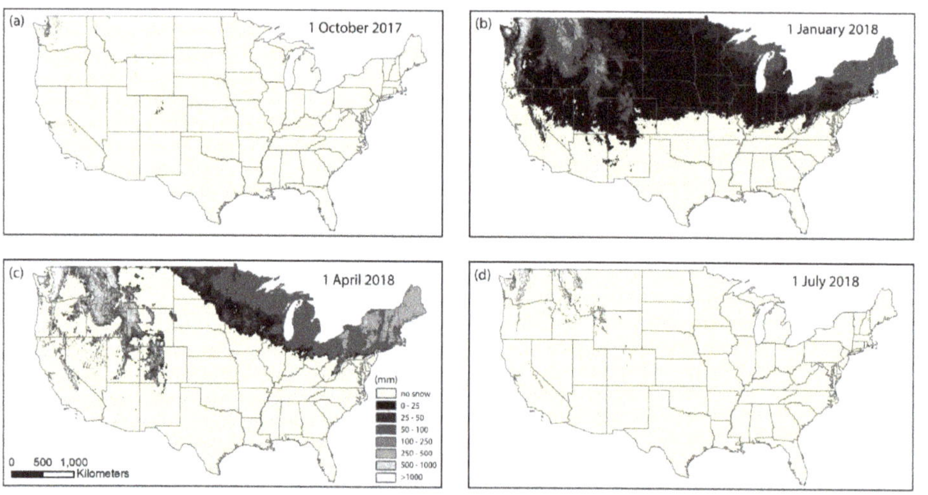

Figure 8. VegET snowpack as snow water equivalent (SWE) (mm) for (**a**) 1 October 2017, (**b**) 1 January 2018, (**c**) 1 April 2018, and (**d**) 1 July 2018. October shows the least amount of snow spatial coverage with January showing the largest areal extent.

3.2. Evaluation

Although VegET products are not calibrated with independently measured datasets and are not expected to be highly accurate in magnitude, the relative distribution in time and space can be evaluated. Illustrative comparisons with in situ observations for soil moisture, snowpack, ETa, and runoff are presented below.

3.2.1. Soil Moisture (SM)

VegET estimates SM for the entire 1 m (39 inch) depth root zone; SM represents the readily available water for plants, i.e., with a soil suction pressure between FC and WP. Comparisons with observed measurements from three different sites in the SCAN network are illustrated in Figure 9 for the growing season from 1 May to 30 September 2019. The sites located from north to south have different characteristics for soil properties and precipitation. Crescent Lake #1 (Figure 9a), Minnesota, received about 900 mm of precipitation in 2019 with a WHC = 70 mm. The temporal patterns of observed and simulated SM show a strong agreement (Pearson correlation r = 0.82). However, the observed SM shows a much higher magnitude. This can be attributed to the fact that the simulated SM only accounts for the amount of water between FC and WP with any moisture above FC (gravity water) that is assumed to be excess and would drain from the root-zone as runoff. The Fort Reno site (WHC = 190 mm) in Oklahoma (Figure 9b) portrays similar strong temporal agreement (r = 0.97), with observed SM showing much higher magnitudes and comparable decreasing rates during the growing season with limited precipitation events and amount. At the drier (~400 mm annual precipitation) New Mexico site (WHC = 140 mm), not only the temporal agreement is weak (r = 0.22), in contrast to the other two sites, the observed SM is lower than the simulated SM. This is probably due to overestimated WHC (140 mm) data used in the model, causing even greater estimation than the Minnesota (Figure 9a) site (WHC = 70 mm). This highlights the importance of acquiring accurate quality soils data for hydrologic modeling.

Despite the differences in absolute magnitude, the simulated SM shows satisfactory performance in terms of capturing the temporal variability, which is key for ETa estimation and drought monitoring applications.

3.2.2. Snow Water Equivalent (SWE)

Snow water equivalent (SWE) of snowpacks at three SNOTEL sites over six years (2015–2020) was used for evaluation. Generally, agreement is good (r: 0.86–0.88) on the timing and duration of SWE accumulation at the three sites (Figure 10). Although the magnitude is reasonable at the Cole Canyon site (Figure 10a), bias is large at Columbine Pass (Figure 10b). The Sierra Blanca (Figure 10c) site shows a good agreement on timing and mixed results on bias in water-year 2016, with a reasonable agreement during 2017–2020. The difference in magnitude can be partially attributed to errors in gridMET dataset used in VegET, which underestimated precipitation by as much as 200 mm for a calendar year at Columbine Pass. Furthermore, any differences between actual air temperature and simulated air temperature could cause discrepancy in the timing of melt and magnitude of snowpack. In winter, the average temperature input in the VegET model exceeded the average temperature recorded by SNOTEL by 1.5 °C at Cole Canyon, whereas Columbine Pass temperature input into VegET was warmer by 7.7 °C on average, which explains some of the differences in snow accumulation between the two sites.

Although large biases in snowpack SWE magnitude exist at a few sites such as Columbine Pass (Figure 10b), the consistent performance of VegET for timing and duration makes it useful for monitoring water availability in areas of the world with limited in situ observations. Moreover, relative variations in SWE are more important than actual magnitudes for predicting relative changes in river flows for irrigation. The simplified snow module in VegET can be used to provide valuable and timely insight into yearly changes and trends in snow accumulation and melt over watersheds and regions as well as to generate future scenarios with projected climate datasets.

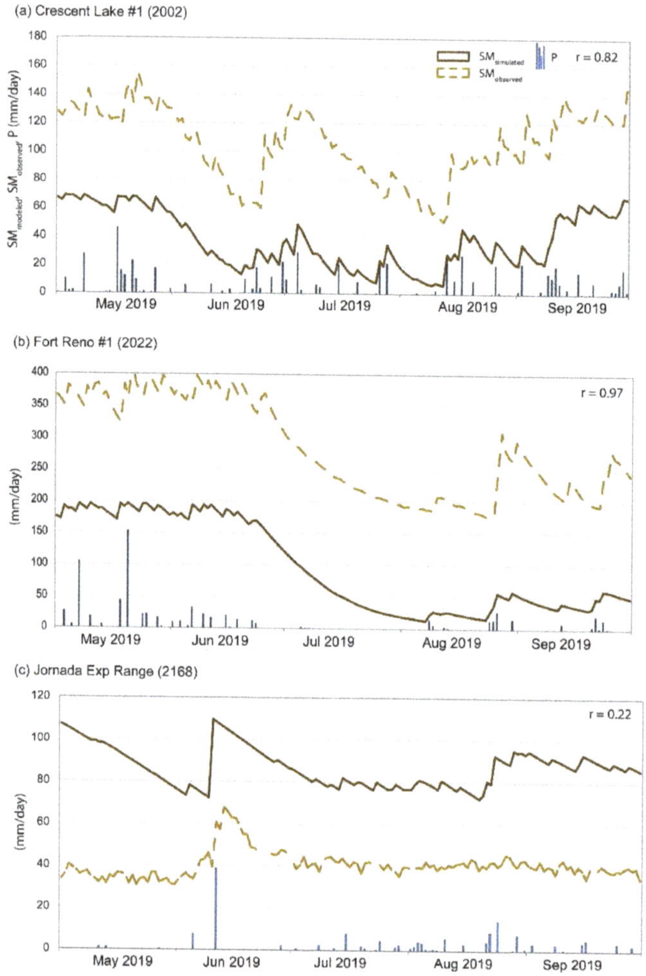

Figure 9. Three Soil Climate Analysis Network (SCAN) sites showing daily simulated (VegET model) and observed soil moisture (SM) [45] along with precipitation for the growing season (May–September) in 2019. (**a**) Crescent Lake # 1, water holding capacity (WHC) = 70 mm; (**b**) Fort Reno #1, WHC = 190 mm; and (**c**) Jornada Exp Range, WHC = 140 mm. The maximum magnitude of the simulated SM corresponds to field capacity (FC) of the soil. The observed SM is not confined to a maximum of FC and thus could include gravity water between soil saturation (SAT) and FC. Temporal-pattern comparison is more meaningful than absolute magnitudes.

3.2.3. Actual Evapotranspiration (ETa)

VegET ETa was compared to EC ETa over several years (Figure 11) using monthly ETa data obtained from the FLUXNET2015 dataset [48] for the following sites: US-AR1 (2009–2012), US-Ne3 (2009–2012), and US-Var (2009–2012). Because of the strong connection between ETa and biomass/yield, the performance of VegET ETa is crucial for accurate biomass estimation and drought monitoring.

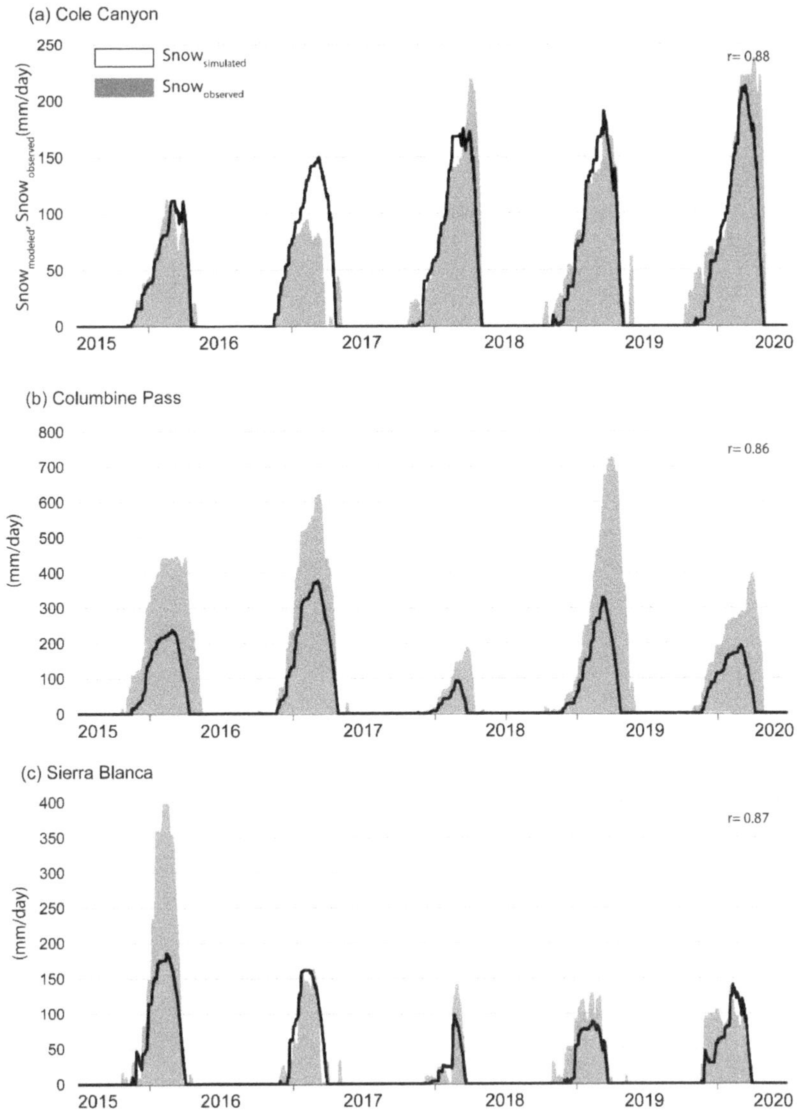

Figure 10. SNOpack TELemetry (SNOTEL) sites showing daily observed [46] and simulated (VegET model) snowpack as snow water equivalent (SWE) (mm) for 2015/2016 to 2019/2020 at (**a**) Cole Canyon, (**b**) Columbine Pass, and (**c**) Sierra Blanca.

Although the general seasonal pattern of VegET shows good agreement with the observed ETa, there are some seasonal inconsistencies in the two sites (US-AR1 and US-Ne3). VegET ETa captures well the winter and spring ETa at all sites but tends to show a relatively dry condition compared to the observed ETa in the summer during reduced precipitation periods at US-Ne3 and US-AR1. One explanation is that the footprint of the EC tower may include ETa from landscapes that have access to additional sources of water such as groundwater by deep-rooted trees or from nearby irrigated fields, especially for the Nebraska site (Figure 11b). The Oklahoma grassland site (US-AR1, Figure 11a) shows reasonable agreement in 2009 and 2010 but showed an out-of-phase behavior in 2011 and

2012. There was a reduction in precipitation during the summer of 2012, which is reflected in VegET ETa, but the EC tower shows a high ETa, contrary to the expected drought-year response. Furthermore, the EC data for the winter months of 2009 (January–March) show unrealistically high values compared to other years, casting doubt on the accuracy of the EC data from this site.

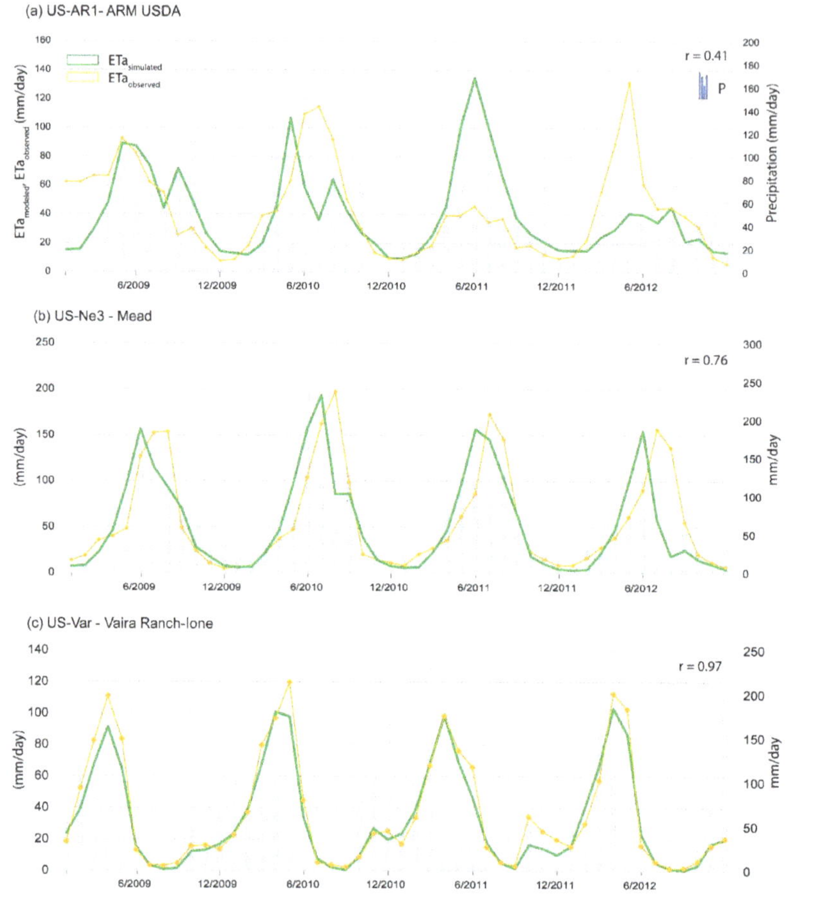

Figure 11. Monthly traces of observed [48] and simulated (VegET) ETa time series at three eddy covariance (EC) flux sites during 2009 to 2012: (**a**) US-AR1 (grassland), (**b**) US-Ne3 (rainfed crop), and (**c**) US-Var (grassland). Daily data were aggregated to monthly for clarity of display and interpretation.

The simulated ETa from VegET corresponds consistently to precipitation, usually exhibiting increases in ETa with a month lag, after peak precipitation in the summer (Figure 11a–c). Except for US-Var (Figure 11c), the peak seasonal ETa from VegET generally lags behind the peak ETa recorded by the EC tower. This is reflected in the lower r statistic in the comparison of US-AR1 and US-Ne3 (r = 0.41 and r = 0.76, respectively) (Figure 11a,b) relative to US-Var (r = 0.97) (Figure 11c), where it performs favorably in both pattern and magnitude. A further investigation with more sites would be useful to help understand and characterize the spatiotemporal dynamics of the performance of the simulated ETa.

The monthly temporal patterns of simulated ETa at point locations and the annual ETa maps over CONUS are consistent with seasonal and regional patterns of vegetation and precipitation in the CONUS. This reinforces the proposed application of the VegET model for quantifying green-water ETa (from precipitation and soil moisture), which is an important parameter in the determination of net irrigation water use (blue water) as the difference between total ETa from energy balance models and VegET ETa as suggested by [25].

3.2.4. HUC8 Runoff

Although the main purpose of the VegET model is to estimate precipitation-driven landscape ETa to develop an integrated drought monitoring product L-WRSI, one byproduct of VegET is runoff, which can be evaluated with independent data sources. The annual total runoff from VegET was compared with the model-assimilated observed runoff from the USGS WaterWatch [49] to evaluate the performance of VegET runoff in capturing the spatial variability across HUC8 watersheds over three years. Correlation coefficients above 0.80 for all water years show a reasonable performance of VegET runoff for capturing the spatial dynamics. The VegET runoff values are lower than WaterWatch runoff values for the filtered HUC8s ($R/P \leq 0.40$) for all water years as shown in Figure 12. The underestimations of VegET are within 5% ($\leq$7 mm/yr) for water years 2012 and 2016, and within 15% ($\leq$24 mm/yr) for water year 2018 (Figure 12, Table 7). The root mean square error (RMSE) values are relatively high with an average of 62%, indicating uncertainties over individual HUC8s while capturing the overall spatial dynamics. The performance of the VegET model improved substantially for runoff estimation when the $R/P \leq 0.40$ was applied (Table 7). With the R/P ($\leq$0.4) filter, the model bias reduced for all three water years with the largest reduction of 15.6% (from −20.2% to −4.6%) for 2012 (dry year) and the least of 9.2% (from −23.9% to −14.6%) for 2018 (average year).

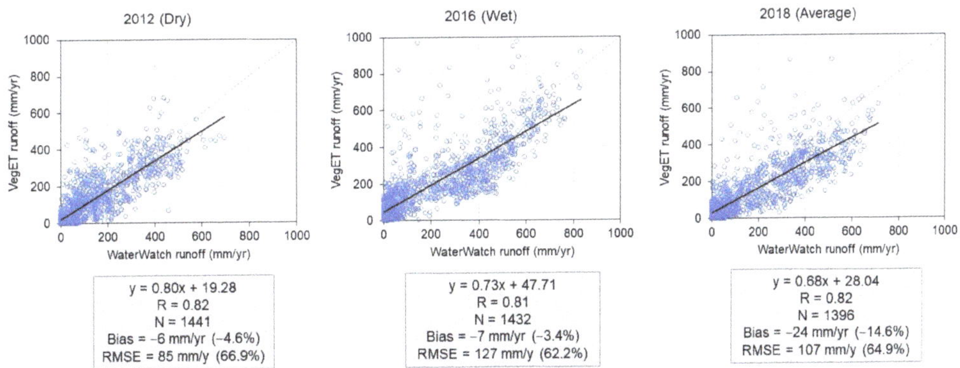

Figure 12. Scatterplot showing the relationship between simulated (VegET) and observed (WaterWatch [49]) annual runoff at eight-digit hydrologic unit code (HUC8) scale for three water years (2012, 2016, 2018) using N number of watersheds across the conterminous United States.

Figure 13 shows major water balance components for the Upper Mississippi River Basin including precipitation, actual evapotranspiration, and runoff, along with their monthly variations for the water year 2012. Areas of higher precipitation correspond with higher VegET-simulated ETa and R, as expected, capturing the general spatial distribution of major fluxes. Additionally, seasonal runoff dynamics compare well between VegET (simulated) and WaterWatch (observed) in relative terms, with higher runoff during spring (March-May) and peak runoff in May (both for VegET and WaterWatch). However, there is a large difference in the monthly R values between the observed and simulated, indicating that calibrating the model and improving the parameterization of the runoff are warranted. The combined ETa and R account for about 93% of water year total precipitation, with the remaining fraction attributed to interception (~8%) and change in storage (~−1%).

Table 7. Summary statistics of eight-digit hydrologic unit code (HUC8) annual runoff comparison between simulated (VegET) and observed (WaterWatch [49]) for three water years (2012, 2016, 2018), without filter and with filter (excluding HUC8s when runoff (R) and precipitation (P) ratio is greater than 0.40.

Statistics	Without Filter			With Filter (R/P ≤ 0.40)		
	2012	2016	2018	2012	2016	2018
N (HUC8)	1762	1762	1762	1441	1432	1396
r (correlation coefficient)	0.90	0.88	0.90	0.82	0.81	0.82
WaterWatch runoff (mm/yr)	216	297	267	128	205	165
VegET runoff (mm/yr)	173	253	203	122	198	141
Bias (mm/yr)	−44	−44	−64	−6	−7	−24
Relative bias (%)	−20.2	−14.8	−23.9	−4.6	−3.4	−14.6
RMSE (mm/yr)	144	163	157	85	127	107
Relative RMSE (%)	66.5	54.9	58.9	66.9	62.2	64.9

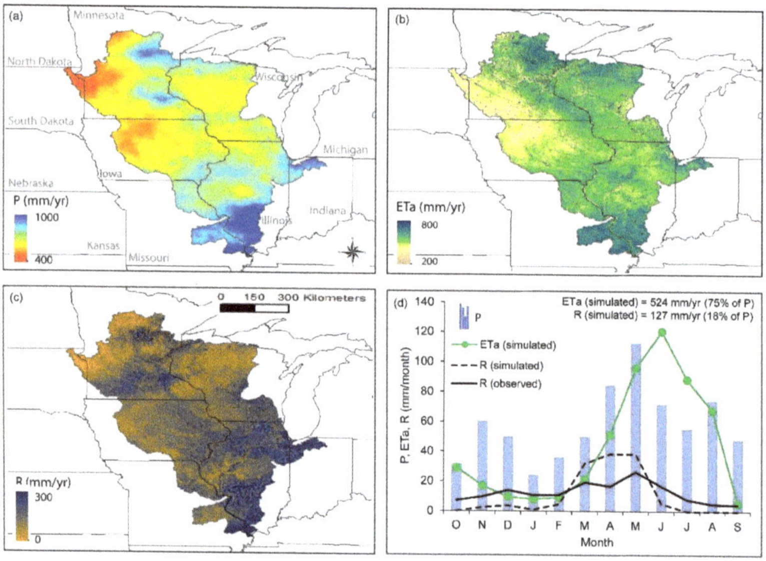

Figure 13. Major water balance components for the Upper Mississippi River Basin for water year 2012 (**a**) precipitation (P) from gridMET [30], (**b**) actual evapotranspiration (ETa) from VegET, (**c**) surface runoff (R) from VegET, and (**d**) monthly P, ETa, and runoff (simulated R (VegET) and observed R (WaterWatch).

Despite the bias and uncertainty, the overall performance of the VegET runoff is satisfactory for an uncalibrated model. Potential sources of errors could be attributed to the gridded precipitation input and model parameters. For example, if gridded precipitation is lower than the actual amount fallen over a basin, the VegET runoff will certainly be lower than the observed. The relative accuracy of VegET runoff is not critical for the estimation of ETa as ETa is assigned a priority in the calculation using the saturation excess principle, i.e., runoff is generated once the root-zone is filled with enough soil moisture.

The annual runoff comparison provided good results with percent bias less than 15% (with $R/P \leq 0.40$ filter) for all water years. The percent bias is acceptable considering these are an uncalibrated results from a simple bucket model to handle complex physical processes that are often unique to each watershed. It is possible that the performance of VegET model varies across HUC8 watersheds, and the single threshold (R/P $\leq$ 0.40) filter applied to the CONUS-scale study may not represent the water balance characteristics of each watershed. However, the VegET model can be calibrated and optimized when finer scale spatial information is needed.

4. Case Study Applications

The spatially explicit Landscape Water Requirement Satisfaction Index (L-WRSI) is an indicator of landscape performance akin to the well-established WRSI for monitoring crop production based on the availability of precipitation and soil moisture to meet crop or landscape water requirements (ETc) during the growing season [22]. L-WRSI can be estimated as the ratio (%) of seasonal ETa to the seasonal ETc. Similar calculations are used for L-WRSI where Kcp is used instead of Kc to define the landscape water requirement phenology as follows:

$$\text{L-WRSI} = \frac{\sum ETa}{\sum ETc} \times 100 \quad (21)$$

$$ET_c = Kcp \times ET_o \quad (22)$$

where $\sum ETa$ is the sum of ETa (mm) for the selected time period (month, season, year); $\sum ETc$ is the sum of the landscape water requirement (mm) for the selected time period and denotes landscape-specific ETo after an adjustment is made to the reference crop ETo by the use of the LSP coefficient (Kcp). Kcp values define the seasonal water requirement patterns of the landscape.

Figure 14 illustrates the concept of the L-WRSI. The gray (ETc) and green (ETa) lines are the two components creating the L-WRSI. The difference between the two lines indicates the water deficit during insufficient precipitation, which leads to the reduction in the L-WRSI from 100%. The annual (January–December) and seasonal (May–September) cumulative deficit are represented by L-WRSI values of 85 and 89, respectively, i.e., 85% and 89% of the median landscape water requirement, met by precipitation, for the year and the season in 2018. The main deficit in the growing season was observed in July with a relatively low amount of precipitation. However, the 11% deficit for the season may not necessarily reflect an actual water deficit that would lead to a proportional yield reduction due to uncertainties in model inputs and assumptions; however, the relative magnitude in space and time could be used for drought monitoring and early warning by comparing the index across years and regions.

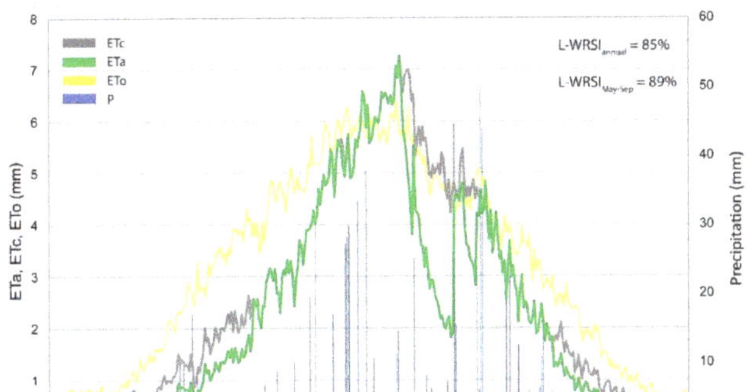

Figure 14. Illustration of the Landscape Water Requirement Satisfaction Index (L-WRSI) concept using daily precipitation (P), reference ET (ETo), actual evapotranspiration (ETa), and landscape water requirement (ETc) for a pixel near the AmeriFlux Station (US-Ne3) for 2018. Seasonal (89%) and annual L-WRSI (85%) indicate some level of dryness during the growing season and through the year.

The L-WRSI values for the CONUS and GHA were calculated and used to illustrate their agro-hydrologic applications for drought monitoring. L-WRSI is an integrated index that includes precipitation, atmospheric demand, phenology, and soil properties.

4.1. CONUS

Figure 15 shows seasonal L-WRSI for three years, namely 2012, 2016, and 2018. L-WRSI less than 100 indicates some form of water stress. Generally, L-WRSI > 95 is considered optimal and less than 80 indicates a serious precipitation shortfall that may lead to a substantial biomass and yield reduction for crops. A crop WRSI < 50 indicates crop failure and need for irrigation to grow crops. It is important to note that L-WRSI is calculated based on availability of moisture in the 1 m root-zone and does not take into account potential access to groundwater by deep-rooted trees and shrubs. This is one explanation why L-WRSI shows lower values (Figure 15) during the growing season in the southeast (e.g., Georgia), where the vegetation demand could be partially met by groundwater resources for the tree-dominated landscapes. It also explains the supplemental irrigation requirement for growing crops during the growing season in the region.

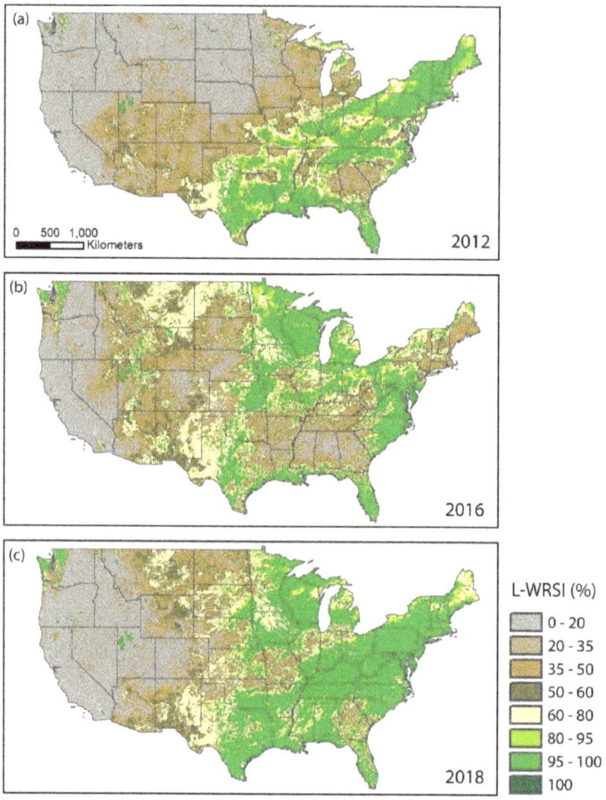

Figure 15. Growing season (May–September) Landscape Water Requirement Satisfaction Index (L-WRSI) for the conterminous United States for (**a**) 2012, (**b**) 2016, and (**c**) 2018. Values close to 100 (green) show availability of enough precipitation to meet crop requirements during the growing season. L-WRSI < 50 (brown tones) indicate severe moisture deficit in the top 1 m root zone to meet the expected water requirement of the landscape. The index does not account for access to groundwater or irrigation water applications.

For the country-wide assessment, L-WRSI was grouped into four qualitative categories of Good (L-WRSI > 95%), Fair (80–95%), Poor (50–80%), and Severe Damage (L-WRSI < 50%). A summary of the L-WRSI by croplands [56] of the CONUS (Figure 16) shows the drought year of 2012 had 66% of the CONUS under severe damage whereas 2016 and 2018 experienced severe damage to a lesser extent (26–27%). The extent observed in 2016 and 2018 may represent the areas that normally require irrigation for crop production. Such kind of metric would allow the expression of the impact of a drought year relative to a normal year. In this case, one could say the 2012 damage was twice as severe as that of 2018 (an average precipitation year).

Figure 16. Summary of seasonal Landscape Water Requirement Satisfaction Index (L-WRSI) for crop areas by four broad categories for the conterminous United States (CONUS). The rectangular charts illustrate the percentage of the CONUS area that falls within the classes of Good (L-WRSI > 95%), Fair (80–95%), Poor (50–80%), and Severe Damage (L-WRSI < 50%) for each year.

4.2. GHA

L-WRSI was generated for the Greater Horn of Africa where frequent droughts create serious food insecurity challenges (Figure 17). In the GHA region, the L-WRSI is combined with other drought monitoring products such as NDVI and hydrologic indicators to develop the convergence of evidence framework needed for food insecurity assessment by FEWS NET. Figure 17 shows 3-month L-WRSI ending on the named month. For example, January 2018 L-WRSI comprises the ratio of ETa to ETc for the months of November 2017, December 2017, and January 2018. The spatial distribution of L-WRSI in the different seasons shows the complex nature of precipitation and vegetation pattern in the region. L-WRSI values can be summarized by district or watershed over a historical period to understand the relative performance of the landscape across regions and time periods.

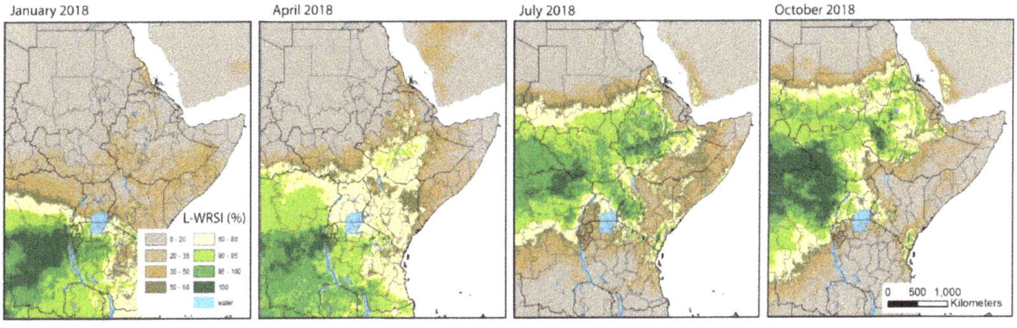

Figure 17. Landscape Water Requirement Satisfaction Index (L-WRSI) distribution in the Greater Horn of Africa using 3-month moving total for ETa and ETc during 2018. L-WRSI spatial patterns reflect the growing season dynamics across the region.

As opposed to the existing WRSI product of FEWS NET [22] for crop monitoring, the current continuous 3-month L-WRSI brings enhanced features of (1) the L-WRSI is continuous in space because the Kcp is generated from the NDVI-based LSP and does not depend on crop types or growing regions where the Kc is applied, (2) L-WRSI does not require estimation of start-of-season and end-of-season layers, which could introduce additional sources of uncertainty, making year-to-year comparison more reliable, and (3) because of the daily, year-round modeling, any desired time period can be simulated in the world instead of pre-specified seasons for a given region.

5. Conclusions

The main objective of this study is to present the updated agro-hydrologic VegET v2.0 model [29] along with performance evaluation results and drought monitoring applications over the conterminous United States and Greater Horn of Africa. A successful integration of a simple temperature-index based snowpack and melt process algorithm has been adapted to work with the VegET model.

Limited evaluation results indicate an encouraging performance in terms of capturing the timing and duration of snow accumulation and melt. Evaluation of soil moisture, ETa, and runoff estimations were reasonable in terms of capturing relative differences in space and time, indicating the usefulness of the model for drought monitoring purposes across diverse ecosystems using the highly integrated L-WRSI product. The operational implementation of the L-WRSI in the Greater Horn of Africa by the Famine Early Warning System Network can be expanded to a global coverage due to the readily available nature of gridded weather datasets and remotely sensed model parameters.

The spatiotemporal patterns of VegET ETa indicate that VegET could be used for the determination of net irrigation water use (blue water) when combined with energy balance models that estimate total ETa by quantifying the green water contribution from precipitation and soil moisture.

With continued evaluation and improvement, the VegET model can also be used to help improve flood forecasting because of the unique inclusion of the readily available land surface phenology (LSP) that accounts for vegetation dynamics in hydrologic modeling, without requiring specification of land cover types.

Author Contributions: Conceptualization, G.B.S.; methodology, G.B.S.; software, S.K. and G.E.L.P.; validation, G.B.S., S.K., G.E.L.P. and K.K.; formal analysis, G.B.S., S.K., G.E.L.P., K.K., O.B. and N.M.V.; resources, G.B.S.; data curation, S.K., G.E.L.P. and K.K.; writing— original draft preparation, G.B.S., S.K., G.E.L.P. and K.K.; writing—review and editing, G.B.S., S.K., G.E.L.P. K.K., O.B. and N.M.V.; visualization, S.K., G.E.L.P. and K.K.; supervision, G.B.S.; project administration, G.B.S.; funding acquisition, G.B.S. All authors have read and agreed to the published version of the manuscript.

Funding: This work was performed under the U.S. Geological Survey (USGS) contract 140G0119C0001 in support of the USGS Land Change Science projects WaterSMART and Landsat Water Balance, and the USGS OpenET activities supported by the USGS Water Mission Area.

Institutional Review Board Statement: Not applicable.

Informed Consent Statement: Not applicable.

Data Availability Statement: The data that support the findings of this study are openly available at https://doi.org/10.5066/P9ILC6RP [54]. The VegET v2.0 model is available at https://doi.org/10.5066/P9ZYT75R [29].

Acknowledgments: We gratefully acknowledge the institutions and individuals who made various geospatial data freely available: GridMET reference evapotranspiration, CHELSA (Climatologies at high resolution for the earth's land surface areas), and SRIC—World Soil Information, runoff by USGS, soil moisture and snow by U.S. Department of Agriculture. We thank individual principal investigators for the use of AmeriFlux tower data in our research. We greatly appreciate the internal reviewer Lei Ji, anonymous journal reviewers, and the USGS Approving Officer for their edits and constructive feedback. Any use of trade, firm, or product names is for descriptive purposes only and does not imply endorsement by the U.S. Government.

Conflicts of Interest: The authors declare no conflict of interest.

References

1. Asante, K.O.; Macuacua, R.D.; Artan, G.A.; Lietzow, R.W.; Verdin, J.P. Developing a flood monitoring system from remotely sensed data for the Limpopo basin. *IEEE Trans. Geosci. Remote Sens.* **2007**, *45*, 1709–1714. [CrossRef]
2. Artan, G.; Gadain, H.; Smith, J.L.; Asante, K.; Bandaragoda, C.J.; Verdin, J.P. Adequacy of satellite derived rainfall data for stream flow modeling. *Nat. Hazards* **2007**, *43*, 167–185. [CrossRef]
3. Senay, G.B. Modeling landscape evapotranspiration by integrating land surface phenology and a water balance algorithm. *Algorithms* **2008**, *1*, 52–68. [CrossRef]
4. Frere, M.; Popov, G. *Early Agrometeorological Crop Yield Assessment*; FAO: Rome, Italy, 1986; p. 144.
5. Molden, D.; Sakthivadivel, R. Water accounting to assess use and productivity of water. *Int. J. Water Resour. Dev.* **1999**, *15*, 55–71. [CrossRef]
6. Leavesley, G.H. *Precipitation-Runoff Modeling System: User's Manual*; US Department of the Interior: Washington, DC, USA, 1984; Volume 83.
7. Arnold, J.G.; Srinivasan, R.; Muttiah, R.S.; Williams, J.R. Large area hydrologic modeling and assessment part I: Model development 1. *JAWRA J. Am. Water Resour. Assoc.* **1998**, *34*, 73–89. [CrossRef]
8. Liang, X.; Lettenmaier, D.P.; Wood, E.F.; Burges, S.J. A simple hydrologically based model of land surface water and energy fluxes for general circulation models. *J. Geophys. Res. Atmos.* **1994**, *99*, 14415–14428. [CrossRef]
9. Niu, G.Y.; Yang, Z.L.; Mitchell, K.E.; Chen, F.; Ek, M.B.; Barlage, M.; Kumar, A.; Manning, K.; Niyogi, D.; Rosero, E. The community Noah land surface model with multiparameterization options (Noah-MP): 1. Model description and evaluation with local-scale measurements. *J. Geophys. Res. Atmos.* **2011**, *116*, D12109. [CrossRef]
10. Miralles, D.G.; Holmes, T.; De Jeu, R.; Gash, J.; Meesters, A.; Dolman, A. Global land-surface evaporation estimated from satellite-based observations. *Hydrol. Earth Syst. Sci.* **2011**, *15*, 453–469. [CrossRef]
11. Herman, A.; Kumar, V.B.; Arkin, P.A.; Kousky, J.V. Objectively determined 10-day African rainfall estimates created for famine early warning systems. *Int. J. Remote Sens.* **1997**, *18*, 2147–2159. [CrossRef]
12. Funk, C.; Peterson, P.; Landsfeld, M.; Pedreros, D.; Verdin, J.; Shukla, S.; Husak, G.; Rowland, J.; Harrison, L.; Hoell, A. The climate hazards infrared precipitation with stations—A new environmental record for monitoring extremes. *Sci. Data* **2015**, *2*, 150066. [CrossRef]
13. Huffman, G.J.; Bolvin, D.T.; Nelkin, E.J.; Wolff, D.B.; Adler, R.F.; Gu, G.; Hong, Y.; Bowman, K.P.; Stocker, E.F. The TRMM multisatellite precipitation analysis (TMPA): Quasi-global, multiyear, combined-sensor precipitation estimates at fine scales. *J. Hydrometeorol.* **2007**, *8*, 38–55. [CrossRef]
14. Abatzoglou, J.T.; Dobrowski, S.Z.; Parks, S.A.; Hegewisch, K.C. TerraClimate, a high-resolution global dataset of monthly climate and climatic water balance from 1958–2015. *Sci. Data* **2018**, *5*, 170191. [CrossRef]
15. Hobbins, M.; Harrison, L.; Blakeley, S.; Dewes, C.; Husak, G.; Shukla, S.; Jayanthi, H.; McNally, A.; Sarmiento, D.; Verdin, J. Drought in Africa: Understanding and exploiting the demand perspective using a new evaporative demand reanalysis. *AGUFM* **2018**, *2018*, GC21D-1121.
16. Senay, G.; Verdin, J.; Lietzow, R.; Melesse, A.M. Global daily reference evapotranspiration modeling and evaluation 1. *JAWRA J. Am. Water Resour. Assoc.* **2008**, *44*, 969–979. [CrossRef]
17. Zomer, R.J.; Xu, J.; Trabucco, A. Version 3 of the global aridity index and potential evapotranspiration database. *Sci. Data* **2022**, *9*, 409. [CrossRef]
18. Milly, P.; Dunne, K. Sensitivity of the global water cycle to the water-holding capacity of land. *J. Clim.* **1994**, *7*, 506–526. [CrossRef]
19. Wang-Erlandsson, L.; Bastiaanssen, W.G.; Gao, H.; Jägermeyr, J.; Senay, G.B.; Van Dijk, A.I.; Guerschman, J.P.; Keys, P.W.; Gordon, L.J.; Savenije, H.H. Global root zone storage capacity from satellite-based evaporation. *Hydrol. Earth Syst. Sci.* **2016**, *20*, 1459–1481. [CrossRef]
20. Adams, T., III. Flood forecasting in the United States NOAA/national weather service. In *Flood Forecasting*; Elsevier: Amsterdam, The Netherlands, 2016; pp. 249–310.
21. McCabe, G.J.; Markstrom, S.L. *A Monthly Water-Balance Model Driven by a Graphical User Interface*; US Geological Survey: Reston, VA, USA, 2007; Volume 1088.
22. Senay, G.B.; Verdin, J. Characterization of yield reduction in Ethiopia using a GIS-based crop water balance model. *Can. J. Remote Sens.* **2003**, *29*, 687–692. [CrossRef]
23. Tercek, M.T.; Thoma, D.; Gross, J.E.; Sherrill, K.; Kagone, S.; Senay, G. Historical changes in plant water use and need in the continental United States. *PLoS ONE* **2021**, *16*, e0256586. [CrossRef]
24. Allen, R.G.; Pereira, L.S.; Raes, D.; Smith, M. *FAO Irrigation and Drainage Paper No. 56*; FAO: Rome, Italy, 1998; p. e.156.
25. Velpuri, N.M.; Senay, G.B. Partitioning evapotranspiration into green and blue water sources in the conterminous United States. *Sci. Rep.* **2017**, *7*, 6191. [CrossRef]
26. Velpuri, N.M.; Senay, G.B. Assessing the potential hydrological impact of the Gibe III Dam on Lake Turkana water level using multi-source satellite data. *Hydrol. Earth Syst. Sci.* **2012**, *16*, 3561–3578. [CrossRef]
27. Beven, K.; Lamb, R.; Quinn, P.; Romanowicz, R.; Freer, J. *Topmodel, Computer Models of Watershed Hydrology*; Water Resources Publications: Littleton, CO, USA, 1995; pp. 627–668.
28. Beven, K.J.; Kirkby, M.J. A physically based, variable contributing area model of basin hydrology/Un modèle à base physique de zone d'appel variable de l'hydrologie du bassin versant. *Hydrol. Sci. J.* **1979**, *24*, 43–69. [CrossRef]

29. Senay, G.; Kagone, S.; Parrish, G.; Butzer, T.; Boiko, O. *Veget_Model*; U.S. Geological Survey software release; U.S. Geological Survey: Reston, VA, USA, 2023. [CrossRef]
30. Abatzoglou, J.T. Development of gridded surface meteorological data for ecological applications and modelling. *Int. J. Climatol.* **2013**, *33*, 121–131. [CrossRef]
31. Didan, K. MOD13Q1 MODIS/Terra vegetation indices 16-day L3 global 250m SIN grid V006. *NASA EOSDIS Land Process. DAAC* **2015**, *10*, 415.
32. NRCS. NRCS Soils: Gridded National Soil Survey Geographic Database (gNATSGO). Available online: https://www.nrcs.usda.gov/resources/data-and-reports/gridded-national-soil-survey-geographic-database-gnatsgo (accessed on 17 May 2023).
33. Boiko, O.; Kagone, S.; Senay, G.B. *Soil Properties Dataset in the United States: U.S. Geological Survey Data Release*; U.S. Geological Survey: Reston, VA, USA, 2021. [CrossRef]
34. DiMiceli, C.; Sohlberg, R.; Townshend, J. MODIS/Terra Vegetation Continuous Fields Yearly L3 Global 250m SIN Grid V061. *NASA EOSDIS Land Process. DAAC* 2022. [CrossRef]
35. Hansen, M.; DeFries, R.; Townshend, J.; Carroll, M.; Dimiceli, C.; Sohlberg, R. Global percent tree cover at a spatial resolution of 500 meters: First results of the MODIS vegetation continuous fields algorithm. *Earth Interact.* **2003**, *7*, 1–15. [CrossRef]
36. Funk, C.C.; Peterson, P.J.; Landsfeld, M.F.; Pedreros, D.H.; Verdin, J.P.; Rowland, J.D.; Romero, B.E.; Husak, G.J.; Michaelsen, J.C.; Verdin, A.P. A quasi-global precipitation time series for drought monitoring. *US Geol. Surv. Data Ser.* **2014**, *832*, 1–12.
37. Hobbins, M.; Dewes, C.; Jansma, T. *Global Reference Evapotranspiration for Food-Security Monitoring: U.S. Geological Survey Data Release*; U.S. Geological Survey: Reston, VA, USA, 2022. [CrossRef]
38. ISRIC. ISRIC—World Soil Information. Available online: https://data.isric.org/geonetwork/srv/eng/catalog.search#/home (accessed on 13 June 2023).
39. Karger, D.N.; Conrad, O.; Böhner, J.; Kawohl, T.; Kreft, H.; Soria-Auza, R.W.; Zimmermann, N.E.; Linder, H.P.; Kessler, M. Climatologies at high resolution for the earth's land surface areas. *Sci. Data* **2017**, *4*, 170122. [CrossRef]
40. Michael, M.G.; Bastiaanssen, W.G. A new simple method to determine crop coefficients for water allocation planning from satellites: Results from Kenya. *Irrig. Drain. Syst.* **2000**, *14*, 237. [CrossRef]
41. Moussav, M.; Wyseure, G.; Feyen, J. Estimation of melt rate in seasonally snow-covered mountainous areas. *Hydrol. Sci. J.* **1989**, *34*, 249–263. [CrossRef]
42. Auch, R.F.; Wellington, D.F.; Taylor, J.L.; Stehman, S.V.; Tollerud, H.J.; Brown, J.F.; Loveland, T.R.; Pengra, B.W.; Horton, J.A.; Zhu, Z. Conterminous United States land-cover change (1985–2016): New insights from annual time series. *Land* **2022**, *11*, 298. [CrossRef]
43. Nemani, R.; Running, S. Land cover characterization using multitemporal red, near-IR, and thermal-IR data from NOAA/AVHRR. *Ecol. Appl.* **1997**, *7*, 79–90. [CrossRef]
44. Allen, R.G.; Pereira, L.S.; Howell, T.A.; Jensen, M.E. Evapotranspiration information reporting: I. Factors governing measurement accuracy. *Agric. Water Manag.* **2011**, *98*, 899–920. [CrossRef]
45. NRCS. UADA NRCS Soil Climate Analysis Network. Available online: https://www.nrcs.usda.gov/resources/data-and-reports/soil-climate-analysis-network (accessed on 24 May 2023).
46. NRCS. USDA NRCS National Water and Climate Center. Available online: https://www.nrcs.usda.gov/wps/portal/wcc/home/ (accessed on 24 May 2023).
47. AmeriFlux. AmeriFlux. Available online: https://ameriflux.lbl.gov/ (accessed on 13 June 2023).
48. Pastorello, G.; Trotta, C.; Canfora, E.; Chu, H.; Christianson, D.; Cheah, Y.-W.; Poindexter, C.; Chen, J.; Elbashandy, A.; Humphrey, M. The FLUXNET2015 dataset and the ONEFlux processing pipeline for eddy covariance data. *Sci. Data* **2020**, *7*, 225. [CrossRef] [PubMed]
49. USGS. USGS WaterWatch. Available online: https://waterwatch.usgs.gov/index.php (accessed on 5 June 2023).
50. Seaber, P.R.; Kapinos, F.P.; Knapp, G.L. *Hydrologic Unit Maps*; US Government Printing Office: Washington, DC, USA, 1987; Volume 2294, p. 1987.
51. Brakebill, J.; Wolock, D.; Terziotti, S. Digital hydrologic networks supporting applications related to spatially referenced regression modeling 1. *JAWRA J. Am. Water Resour. Assoc.* **2011**, *47*, 916–932. [CrossRef] [PubMed]
52. Senay, G.B.; Parrish, G.E.; Schauer, M.; Friedrichs, M.; Khand, K.; Boiko, O.; Kagone, S.; Dittmeier, R.; Arab, S.; Ji, L. Improving the Operational Simplified Surface Energy Balance Evapotranspiration Model Using the Forcing and Normalizing Operation. *Remote Sens.* **2023**, *15*, 260. [CrossRef]
53. Senay, G.B.; Friedrichs, M.; Morton, C.; Parrish, G.E.; Schauer, M.; Khand, K.; Kagone, S.; Boiko, O.; Huntington, J. Mapping actual evapotranspiration using Landsat for the conterminous United States: Google Earth Engine implementation and assessment of the SSEBop model. *Remote Sens. Environ.* **2022**, *275*, 113011. [CrossRef]
54. Senay, G.B.; Kagone, S.; Parrish, G.E.L.; Khand, K. *VegET v2.0 Illustrative Products and Evaluation: U.S. Geological Survey Data Release*; U.S. Geological Survey: Reston, VA, USA, 2023. [CrossRef]

55. Svoboda, M.; LeComte, D.; Hayes, M.; Heim, R.; Gleason, K.; Angel, J.; Rippey, B.; Tinker, R.; Palecki, M.; Stooksbury, D. The drought monitor. *Bull. Am. Meteorol. Soc.* **2002**, *83*, 1181–1190. [CrossRef]
56. Shen, Y.; Zhang, X.; Yang, Z.; Ye, Y.; Wang, J.; Gao, S.; Liu, Y.; Wang, W.; Tran, K.H.; Ju, J. Developing an operational algorithm for near-real-time monitoring of crop progress at field scales by fusing harmonized Landsat and Sentinel-2 time series with geostationary satellite observations. *Remote Sens. Environ.* **2023**, *296*, 113729. [CrossRef]

hydrology

MDPI

Review

Evolution of Tunneling Hydro-Technology: From Ancient Times to Present and Future

Andreas N. Angelakis [1,2,*], Cees W. Passchier [3], Mohammad Valipour [4], Jens A. Krasilnikoff [5], Vasileios A. Tzanakakis [6,*], Abdelkader T. Ahmed [7,8], Alper Baba [9], Rohitashw Kumar [10], Esra Bilgic [11], Andrea G. Capodaglio [12] and Nicholas Dercas [13]

1 School of History and Culture, Hubei University, Wuhan 430061, China
2 National Foundation for Agricultural Research, Institute of Iraklion, 71307 Iraklion, Greece
3 Institute Geosciences, University of Mainz, 55128 Mainz, Germany; cpasschi@uni-mainz.de
4 Department of Engineering and Engineering Technology, Metropolitan State University of Denver, Denver, CO 80217, USA; mvalipou@msudenver.edu
5 Department of History and Classical Studies, School of Culture and Society, Aarhus University, 8000 Aarhus, Denmark; hisjk@cas.au.dk
6 Department of Agriculture, School of Agricultural Science, Hellenic Mediterranean University, 71410 Iraklion, Greece
7 Civil Engineering Department, Faculty of Engineering, Islamic University of Madinah, Medina 42351, Saudi Arabia; dratahmed@aswu.edu.eg
8 KSA & Civil Engineering Department, Faculty of Engineering, Aswan University, Aswan 81528, Egypt
9 Department of International Water Resources, Izmir Institute of Technology, Izmir 35430, Türkiye; alperbaba@iyte.edu.tr
10 College of Agricultural Engineering and Technology, SKUAST—Kashmir, Srinagar 190025, India; rohituhf@rediffmail.com
11 Department of Civil Engineering, Izmir Institute of Technology, Izmir 35430, Türkiye; esrabilgic@iyte.edu.tr
12 Department of Civil Engineering & Architecture, University of Pavia, 27100 Pavia, Italy; andrea.capodaglio@unipv.it
13 Department of Natural Resources Management and Agricultural Engineering, Agricultural University of Athens, 11855 Athens, Greece; ndercas1@aua.gr
* Correspondence: info@a-angelakis.gr (A.N.A.); vtzanakakis@hmu.gr (V.A.T.)

Citation: Angelakis, A.N.; Passchier, C.W.; Valipour, M.; Krasilnikoff, J.A.; Tzanakakis, V.A.; Ahmed, A.T.; Baba, A.; Kumar, R.; Bilgic, E.; Capodaglio, A.G.; et al. Evolution of Tunneling Hydro-Technology: From Ancient Times to Present and Future. *Hydrology* **2023**, *10*, 190. https://doi.org/10.3390/hydrology10090190

Academic Editors: Songhao Shang, Qianqian Zhang, Dongqin Yin, Hamza Gabriel and Magdy Mohssen

Received: 9 August 2023
Revised: 7 September 2023
Accepted: 8 September 2023
Published: 20 September 2023

Abstract: Water tunnels are one of the oldest hydro-technologies for extracting water resources and/or transmitting them through water distribution systems. In the past, human societies have used tunneling for various purposes, including development, as a measure to enable underground resource extraction and the construction of transportation networks in challenging landscapes and topographies. The development of hydro-technology potentially involves the construction of tunnels to feed aqueducts, irrigation and waste water systems. Thus, the ability to make and maintain tunnels became an important component in creating lasting and sustainable water systems, which increased water supply and security, minimized construction costs, and reduced environmental impact. Thus, this review asks how, when and why human societies of the past included tunneling for the development of lasting water supply systems. This review presents a comprehensive overview across time and space, covering the history of tunneling in hydro technology from antiquity to the present, and it ponders how past experiences could impact on future hydro-technological projects involving tunneling. A historical review of tunnel systems enhances our understanding of the potential, performance, challenges, and prospects associated with the use of hydro-techniques. In the past, as the different examples in time and space demonstrate, tunneling was often dedicated to solving local problems of supply and disposal. However, across the world, some features were repeated, including the need for carving through the living rock or digging to create tunnels covered with stone slabs. Also, the world-wide use of extensive and costly tunnel systems indicates the high level of investment which human societies are willing to make for securing control over and with its water resources. This study helps us to gather inspiration from proven technologies of the past and more recent knowledge of water tunnel design and construction. As we face global warming and its derivate problems, including problems of water scarcity and flooding, the ability to create and maintain tunnels remains an important technology for the future.

Keywords: tunnel systems; qanats; Persian Empire; sustainable systems engineering; aqueducts; socio-economical aspects

1. Prolegomena

By studying the past we learn about the present and are planning anything for the future.

Andreas N. Angelakis

Traditional water tunnels were constructed mainly for the exploitation of groundwater in arid and semi-arid regions. These technologies presented major achievements in this scientific field throughout the millennia [1]. It is not easy to study past water tunnels and demonstrate their sustainability. However, Barghouth and Al-Sa'ed [2] presented an overview of the sustainability of ancient water supply systems in Jerusalem from the Chalcolithic period (ca 4500–3200 BC) to the present. Ancient evidence and landscape settings indicated that water resources management in Jerusalem was based on underground hydro-structures. Sustainable water supply facilities were erected, consisting mainly of well-developed water tunnels or other, similar underground hydro-technologies, to supply the town and its agricultural developments, showing that irrigation was practiced for many centuries in that area.

Another example from India demonstrates how traditional water tunnels have been used for centuries to tap into groundwater resources, particularly in arid and semi-arid regions, described in this manuscript in detail. These hydro-technologies have been a significant achievement in the field of water management, and their sustainability can be observed through the ages. The Indus Valley Civilization in ancient India had an extensive network of underground channels, which are called karez there. Details are provided on those underground aqueducts which, in some parts of the world, are named qanats [3]. They are also known as foggaras and khettaras, and were used mainly for irrigation and other purposes [4]. In modern times, India has made significant progress in tunneling engineering. For example, the Mumbai Metro Rail Project has included the construction of a 33.5 km long underground section, which was dug using tunnel boring machines (TBMs) to reduce the cost and duration of tunneling while minimizing environmental impact [5]. Furthermore, the Chenani–Nashri Tunnel, India's longest tunnel, was built using the New Austrian Tunneling Method (NATM), a sustainable and adaptable approach to tunneling that minimizes resource consumption while enhancing worker safety [6,7]. As India continues to invest in infrastructure, it is anticipated that it will make further strides in tunneling engineering, contributing to sustainable development in the country. In addition, tunnels for drainage purposes were developed in central Greece from the end of the Bronze Age. It should also be noted that sometimes tunnels were surface-cut and covered for crossing a watershed, and "valley-side" tunnels were built to pass steep rock walls or to protect an aqueduct in unstable geology. Also, the shafts-and-galley technique was developed, which is known as qanat [8,9]. These are, moreover, discussed in the main text.

The sustainability of ancient water supply systems in India can be seen in the karez system, which is prevalent in the western regions of the country. The karez system dates back to the 2nd century AD at least; it is an underground water management system that collects water from mountain springs and channels it through a series of tunnels to irrigate agricultural land [10]. Similarly, in the southern state of Tamil Nadu, a network of underground tunnels known as "Eri-pattu" has been used since ancient times to provide irrigation to paddy fields. These tunnels collect rainwater during the monsoon season and store it underground, providing a year-round supply of water for irrigation [11]. In recent times, modern technologies such as bore wells and tube wells have become more prevalent in India, but traditional water tunnels are still used in many parts of the country, especially in rural areas. These hydro-technologies have played a crucial role in sustaining agriculture and ensuring the availability of water for domestic use [12].

This paper deals with the construction of water tunnels throughout history [13–15]. It focuses on major water tunnels built as excavation structures in solid rock or sediment, meant to transport flowing water, and it excludes tunnel-like structures built by an excavation of a trench from the surface and the insertion of pipes or masonry-covered channels, such as the main structure of many aqueducts and drains.

This review study is divided into six sections, which include geographical and chronological developments as well as observations on various types of tunneling hydro-technologies and practices. Section 1, the prolegomena, is an introduction to the subject. Section 2 elucidates the distinct histories of tunneling hydro-technologies from the prehistoric to the Medieval Era. Section 3 deals with tunneling hydro-technologies in the Early and Mid-Modern periods, and Section 4 discusses tunneling hydro-technologies in contemporary times. Section 5 deals with emerging trends and possible future challenges of tunneling hydro-technologies and practices. Finally, Section 6, the epilogue, comprises conclusive remarks and highlights.

2. Tunneling: From the Prehistoric to Early Medieval Era (ca 7600 BC–1453 AD)

2.1. Persian and Other Prehistoric Civilizations (ca 7600–110 BC)

Located in an arid and semi-arid region of Asia, ancient Persia (today, Iran) was a dry country that had always faced water shortage problems. Ancient dams, irrigation canals, and qanats show the long-lasting struggle of people to deal with drought. To satisfy the increasing demand for water due to the increasing population, Persians invented a new system to bring groundwater to the surface using gravitational force. This tunneling system, which is called qanat, is still in use and some of them date back 3000 years. Today, there are about 32,000 qanats in Iran, which provide about 10 billion m^3/yr. Qanat was introduced to other regions of the world (e.g., Japan, Egypt, Oman, Spain, and Chile), and it is thus considered the main contribution of Persians to hydraulic practices. Qanat has a main sloping tunnel and many shaft wells, which together bring water from a high mountain region to low-elevation lands. Compared to deep wells, qanats are cost-efficient and long-lasting in transferring water without requiring energy. They also balance natural inflow and outflow [16,17].

Although thousands of years have passed since then, this method is still used in an important proportion of rural, urban, and agricultural water supply. Using this method, Persians/Iranians have been successful in the development of the sustainable exploitation of groundwater and have withstood the drought conditions in Iran [16,18].

The construction of the qanat was undertaken by skilled laborers and exclusively with hand labor. The process was initiated by the search for an appropriate mother well (probably near mountainous areas). For this purpose, some test wells were dug and checked for the groundwater level. After decisions were made about the mother well location, paths towards irrigated lands were defined on the ground. Then, the work team began digging the main tunnel. To be able to work underground (having enough oxygen, sending out the unnecessary soil, and going out and coming back to the tunnel) vertical shafts were established over the path. A schematic process of a qanat construction is shown in Figure 1.

The elements of a qanat can be defined as follows [16,17]:

Appearance: The place where water comes into view on the surface (tunnel reaches the Earth's surface).

Tunnel: The canal, with a section resembling a horseshoe inside the ground, featuring a gentle slope for water conveyance from the aquifer to the appearance.

Wet zone: The infiltrating walls inside the gallery of a qanat. The discharge rate is directly dependent upon the wet zone. Indeed, this is part of the tunnel which goes below the groundwater table.

Dry zone: A portion of the gallery between the wet zone and the appearance. The canal was gradually cut deeper due to the decline of the water table.

Shaft: The dry vertical wells situated across the gallery facilitated soil extraction as well as ventilation and dredging. The distance between the two shafts was based on the depth of the qanat and the air passage. The nearer the shafts were to the mother well, the deeper they were.

Mother well: The furthest, water-infiltrating well is called the mother well.

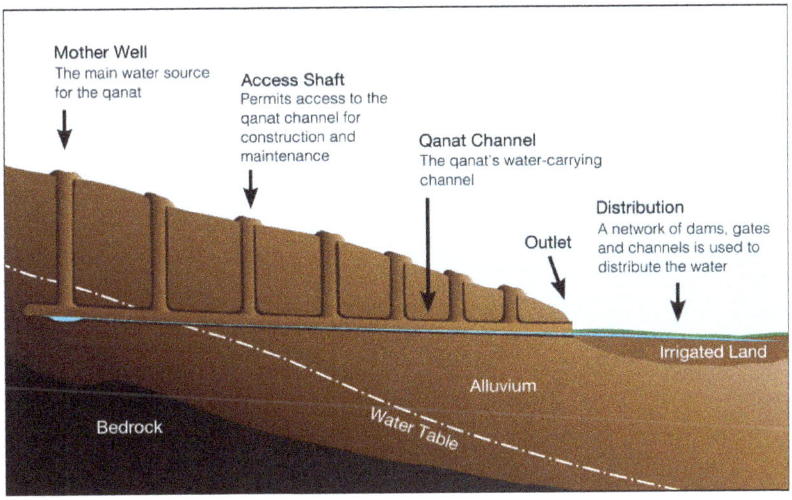

Figure 1. Schematic vertical cross-section of qanat construction (adapted from [3]).

Qanat has many advantages, namely, securing water for irrigation and household consumption in arid regions, balancing the use of groundwater, and low maintenance and operation costs. At the same time, qanats are vulnerable to floods and earthquakes, and they cannot be used for exploiting water from deep layers. Also, in comparison to wells, qanats are more lasting and sustainable and have no energy cost for exploiting water.

Qanat routes need to be regularly cleaned and maintained because they are subjected to damage and destruction by flash floods. To prevent shafts from being filled with sand, they are covered with stone slabs or other objects. One of the famous qanats in Iran is shown in Figure 2 (i.e., Kish qanat).

Figure 2. The Kish qanat: (**a**) the appearance and (**b**) view of the main tunnel.

2.2. Early Ancient Egyptians and Other Civilizations (ca 4000–30 BC)

In Egypt, one of the oldest civilizations, the River Nile has been the main source of freshwater, supplying about 97% of its water resources. Even places far away from the Nile conveyed its water through open and closed aqueducts. The type of aqueduct used in early Egypt was a very basic structure. It consisted of an open canal excavated between the Nile River and the location which required the water, made from stones. Open and closed aqueducts were applied commonly in pyramids that were constructed by pharaohs close to the Nile shoreline. The aqueducts transmitted water to the bottom of these pyramids. They linked the base of the pyramid and the Nile bed with a huge open canal controlled by massive doors of stone that allowed water to pass from the Nile. For example, under the Giza pyramids, the openings and passageways for water transferring are equal to the size of a football playing area. In addition, many vertical openings and aqueducts were used to control the Nile flood, as these openings were lower than the Nile level and sunk the water into the aqueducts underneath the pyramids [19].

Egyptians also used underground aqueducts to deliver the Nile water to the temples. For example, in 57 BC, Ptolemy III built the Edfu temple, in which there was a room called the chamber of the Nile where the priests of the temple obtained the holy water of the Nile. This chamber received Nile water through a stone-built tunnel with a length of one kilometer up to the Nile shoreline. The Dendera temple also featured a similar chamber and stone tunnel [20].

The Persians invaded Egypt in 525 BC and introduced the technology of the long underground aqueducts. For example, they constructed what they called a quant, or aqueduct, to deliver water to the Kharga Oasis 200 km west of the Nile. The aqueduct was constructed from a slightly sloping pathway underground, which connected with many vertical shafts [21]. Another good example of the digging of an underground aqueduct can be found at the Bahariya Oasis, where many sites display remains of this aqueduct. Moreover, in the northeastern part of the Sinai peninsula, there is a spring called Ain EI Gudeirat, which supplied spring water from an aqueduct that was built hundreds of years ago and recently watered olive trees at a daily rate of 1500 m^3 [22,23].

In Alexandria, a city in the northwest part of Egypt, the Greek engineer Archimedes supervised the construction of an overturned (or inverted) siphon to transfer water for kilometers and hundreds of meters of hydraulic heads. These pipes were mostly made of stone and helped transfer Nile water in aqueducts across valleys to the city [24]. A map of ancient Alexandria with a channel of the Nile Delta is shown in Figure 3.

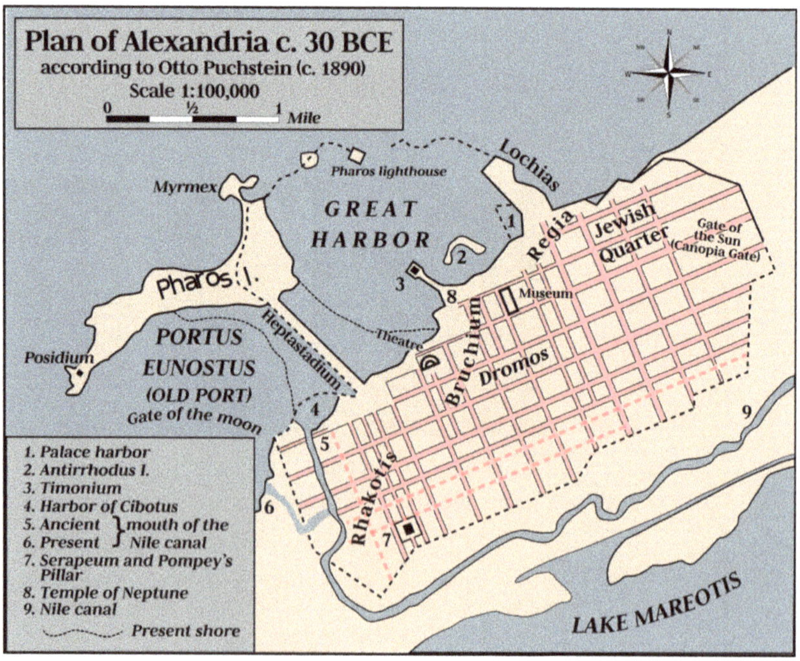

Figure 3. Map of Ancient Alexandria.

It should be noticed that an Inverted siphon Is not a siphon but a term applied to pipes that must dip below an obstruction to form a "U" shaped flow path. Large inverted siphons are used to convey water being carried in canals or flumes across valleys, for irrigation or gold mining. These siphons were developed in Classical times; however, the Romans used inverted siphons of lead pipes to cross valleys that were too big for the construction of an aqueduct (e.g., Aspendos aqueduct) [3].

2.3. Ancient India (ca 3300–185 BC)

India has a rich history of tunneling and hydro-technology dating back to the prehistoric era. Some of the earliest examples of tunneling in India can be found in the Indus Valley Civilization, which existed from approximately 3300 BC to 1300 BC. One of the oldest known tunnels is the Khandagiri–Udayagiri cave complex in Odisha, which was hidden by sandstone cliffs during the Maurya period (321–185 BC), and was used for residential spaces and places of worship [25]. During the Mauryan Empire (321–185 BC), tunnels were used to irrigate farmland and supply water to the growing population. These types of caves are a series of rock-cut Jain and Buddhist temples that were built by carving into the hillside, creating a network of tunnels and chambers.

During the Indus Valley Civilization, underground drainage systems were constructed to manage water supply and mitigate floods. The Great Bath in Mohenjo-Daro is a remarkable example of their expertise in hydro-engineering. This rectangular pool, built around 2600 BC, was constructed using waterproof bricks and a complex system of water channels and drains [26]. The Harappan city of Dholavira also has a sophisticated water management system that included a series of underground tunnels and reservoirs [27] (Figure 4). The Indus Valley people were innovative in their approach to tunneling and used it as a means of managing water supply and creating efficient irrigation systems.

(a)

(b)

Figure 4. Underground tunnels and reservoirs in the Harappan city of Dholavira: (**a**) the southern and (**b**) the eastern views [28].

Finally, there is limited evidence of tunneling in ancient India. However, the Mauryan Empire (321–185 BC) made significant advancements in tunneling technology. Tunnels were used for irrigation, with some examples being the Pataliputra irrigation tunnels in present-day Bihar, India. Additionally, the construction of Emperor Ashoka's rock-cut edicts, dating back to the 3rd century BC, required extensive tunneling and carving into solid rock [29].

2.4. Minoan and Mycenaean Civilizations (ca 3200–1050 BC)

Most Minoan aqueducts transported water through open channels, but a few examples of covered surface channels have survived. In Knossos, water was transported by closed terracotta pipes and/or open or covered channels of various dimensions through a gravity aqueduct about 0.7 km long [30] (Figure 5).

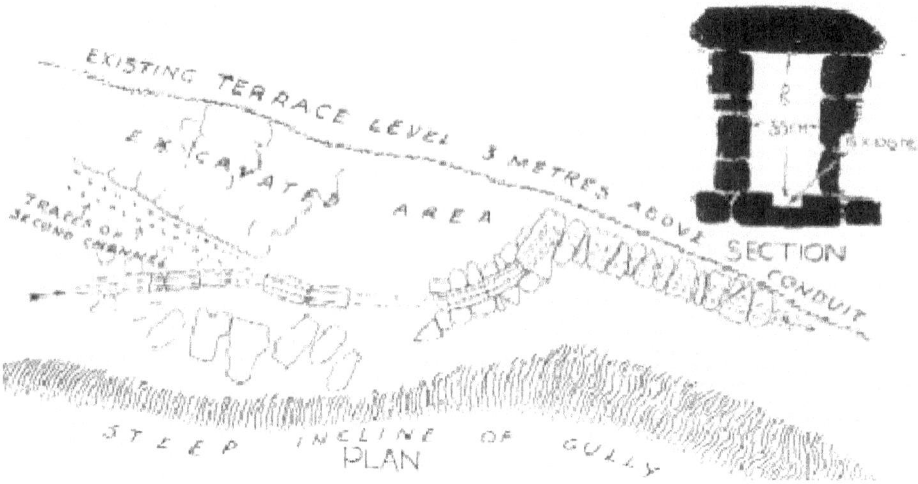

Figure 5. Part of the Knossos palace aqueduct [22].

An advanced hydraulic tunneling technique was introduced by the Minyans of mainland Greece in about 1300 BC, using tunnels for drainage purposes. One prominent example of this is the Akraifnio drainage tunnel, which drained Lake Kopais and used the land for agriculture (Figure 6). The tunnel has a height of 1.8 m and a width of 1.5 m. Sixteen vertical shafts were excavated along the axis of the tunnel, and through those the tunnel was excavated [8].

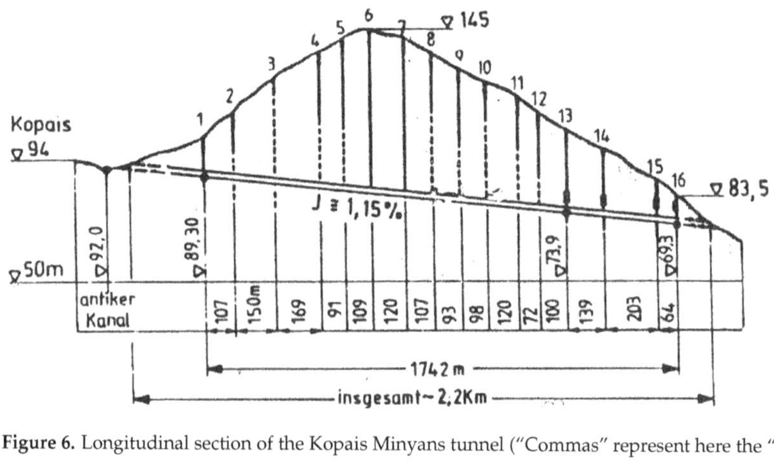

Figure 6. Longitudinal section of the Kopais Minyans tunnel ("Commas" represent here the "dots").

Several studies on the ancient Kopais drainage system have been carried out by several researchers [31–34]. According to Knauss [32], Minyans attempted to gain land from Kopais Lake in two main phases. The first phase used earth dams to protect irrigated polders against floods. After a dam's failure, a second system was developed. The second attempt was based on a 25 km long canal that guided water from the Kopais basin to the natural sinkholes located in the north-eastern part of the area.

The construction of the drainage of Lake Kopais was stopped at the end of the era of Alexander the Great due to the end of the funding of the project. Alexander's engineer had begun the construction of a tunnel that would lead the water of the lake to the sea. The construction method used the technique of that period: the excavation of vertical shafts, followed by horizontal excavation and the connection of the vertical shafts at the bottom level [35].

2.5. Babylonian, Assyrian, and Other Asian Civilizations

Ancient Babylonian and Assyrian civilizations had advanced knowledge of tunneling and hydro-technology, building elaborate underground aqueducts, tunnels, and canals to manage water supply and irrigation.

The Sultanate of Oman is an arid region, and ever since its early history the country has depended on groundwater as a freshwater resource. They used surface and underground tunnels to convey water horizontally via gravity from groundwater into valleys for irrigation and drinking purposes. They called these tunnels and aqueducts 'Aflaj', and they had a long history dating back several thousands of years in Oman [36]. Aflaj is defined as the plural of the term 'Falaj', which refers to a channel supplied by a groundwater source. The term Falaj is Arabic and means 'to divide or split' [37]. Establishing Aflaj helped ancient Omanis to provide freshwater for communities for different purposes. Omanis classify Aflaj as Ghaily, Daudi, and Ayni. Ghaily Falaj is seasonal, as it relies on a shallow underground source that stops in dry periods. The Daudi provides permanent water flow via the top surface of the valley being used as a transferring channel. The Ayni Falaj derives its water from natural springs and the water is usually hot because it comes from very deep layers [38].

2.6. Iron Age (ca 1050–750 BC)

During the Iron Age in India (ca 1200–750 BC), tunneling technology was used primarily for mining and transportation purposes. The Khetri Copper Mines in Rajasthan, India, dating back to the 8th century BC, are an excellent example of ancient Indian mining operations that utilized tunnels. These tunnels were excavated to extract copper ore from the mines and transport it to smelting facilities. The technology used during this period was primitive, with hand tools being the primary means of excavation. However, the expertise of ancient Indian miners and tunnelers cannot be underestimated, as evidenced by the vast network of interconnected tunnels that were constructed during this period [39].

2.7. Archaic, Classical, and Hellenistic Periods (ca 750 BC–31 BC)

One of the oldest tunnels in the world was built below Jerusalem in the 8th century BC from the Gihon karst spring to the Siloam pool [40,41]. Known as Hezekiah's tunnel (Figure 7), this 500 m long structure was built by drilling from the spring and the destination pool in two directions (counter-excavated tunnel), meeting in the middle. The tunnel still carries water.

The technologies of hydraulic tunneling developed by prehistoric civilizations were further developed and improved during historical times. Allegedly, in late Archaic Samos, Greece, the engineer Eupalinus constructed the prestigious and renowned tunnel bearing his name, the Eupalinos or Eupalinian aqueduct (Greek: Ευπαλίνιον όρυγμα, i.e., Efpalinion orygma). The evidence of the historian Herodotus for the construction of the tunnel (*Histories*, 3. 60) potentially connects the construction of the tunnel with the tyrant Polycrates (ruled 540–522 BC). The aqueduct is 1036 m in length and runs through Mount Kastro, and was built to provide fresh water for the island's main city. The tunnel is the second known tunnel in history to have been excavated from both ends (Ancient Greek: ἀμφίστομον, i.e., amphistomon, having two openings), and the first with a geometry-based approach in doing so [19]. The tunnel is inscribed on the UNESCO World Heritage List along with the nearby Pythagoreion and Heraion of Samos, and it was designated as an International Historic Civil Engineering Landmark in 2017 [42]. Today, the tunnel is a popular tourist attraction and can be visited through its southern entrance. A view of a section of the tunnel and a frequently used entrance is depicted below in Figure 8.

Figure 7. Hezekiah's tunnel. This tunnel still carries water (Photo Cees Passschier).

(a) (b)

Figure 8. Eupalinion orygma: (**a**) a view of the orygma and water channel, and (**b**) a frequently used entrance.

Engineer Eupalinos made an effort to have the two construction teams meet either horizontally or vertically by the employment of the following techniques:

(a) In the vertical plane, at the start of work, Eupalinos leveled around the mountain, probably following a contour line to ensure that both tunnels were started at the same altitude. He increased the possibility of the two tunnels meeting each other, by increasing the height of both tunnels at the point near the join. In the north tunnel, he kept the floor horizontal and increased the height of the roof by 2.5 m, while in the south tunnel he kept the roof horizontal and lowered the level of the floor by 0.6 m (Figure 8a). His precautions as to vertical deviation proved unnecessary, since measurements show that there was very little error. At the meeting point, the closing error in altitude for the two tunnels was a few millimeters [43].

(b) In the horizontal plane, Eupalinos calculated the expected position of the meeting point in the mountain. Since two parallel lines never meet, an error of more than 2 m horizontally meant that the north and south tunnels would never meet. Therefore, Eupalinos changed the direction of both tunnels, as shown in the picture (the north tunnel to the left and the south tunnel to the right) (Figure 9b). This gave a catching width that was wider by 17 m so that a crossing point would be guaranteed, even if the tunnels were previously parallel and far away. They thus meet at nearly a right angle [43].

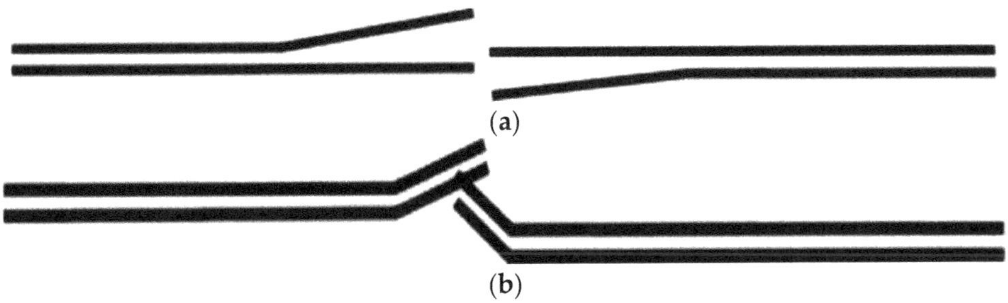

Figure 9. Eupalinos increased the possibility of the two tunnels (right and left sites) meeting each other: (**a**) in the horizontal plane and (**b**) in the vertical plane.

It should be noticed that the water channel used for water transfer was constructed at the bottom of the rock-cut tunnel shown in Figure 8a. The rock-cut tunnel was a working gallery from which the workers could lay the lower-lying terracotta conduit, which still functioned in the time of Herodotus. Later, presumably due to blockage of the pipes, the pipeline was broken open and the water allowed to overflow into the rock-cut channel.

Previously, scholars have used John Camp's study [44] of the wells of the Athenian Agora and their alleged origin in droughts in the ca 8th and 4th centuries as the impetus for the development of underground aqueducts in mainland Greece. This is a likely supposition; however, the builders of the first, major aqueducts, the so-called 'tyrants' of the later Archaic age in the south and eastern parts of mainland Greece and in the Aegean islands, may have had other ambitions as well. The autocratic rulers of the late archaic period probably acknowledged the importance of well-functioning water supplies, both as a means to support the growing populations of cities and as a way to rally support behind their rule 'outside the law'.

Chiotis and Marinos [9] pointed out that the aqueducts from the ca 6th through the 4th centuries fell into versions, which were either surface-cut and covered channels as in the Peisistratean and Acharnian aqueducts or shafts-and-galley techniques, as in the aqueducts of Aegina and Megara. Furthermore, Chiotis and Marinos [9] pointed out the important discussion of whether there might be a link between these aqueducts and the Persian qanats developed during the Achaemenid Empire (538–323 BC). Basically, and unlike the qanats, which collect water from a mother well, Greek aqueducts of the shafts-and-galley type collect water, 'mostly all along their course in temperate areas.' Different types of climate, geology, and topography inspired different strategies of technological development.

The mid-sixth century tyrants, the Peisistratids, who governed Athens after the reform period of Solon in 594 BC, have frequently been associated with improvements in the Athenian water supply. The historian Thucydides (2.15.5) attributed alterations to the fountain 'Enneacrounos', or 'Nine Pipes' to 'the tyrants'. Later this famous fountain appeared under the name of the nymph Callirrhoe—'Fairwater' ([45], 294, et passim). Otherwise, the literature evidence is silent about the construction works of the Peisistratids, and we have to rely on the archaeological evidence for more information about the water supply and construction of tunnels in this period (see further [46–48]).

The city-state of Megara, a western neighbor to Athens, saw an erratic political development in the late archaic age, beginning with the tyrant Theagenes, and an oligarchy followed by democracy in the 5th century. The engineer Eupalinos originated from Megara, and it has been suggested that his water-technological interest may have originated in the city's solutions to water management. A fountain was fed with water from an aqueduct covered with long intersected roof gallery sections ([49]).

The construction of the ancient aqueduct of the Aegean island Naxos, Greece, late in the 6th century BC, may have happened either during the tyranny of Lygdamis or the succeeding brief interval of democracy on the island. The aqueduct ran over 11 km on hillsides at the upper limit of fertile land and consisted of socket-jointed clay pipes of a diameter of ca 0.30 m buried in a ditch ca 1 m underground [50] (Figure 10). Its inclination varied from 0.01 to 0.04%.

Figure 10. Naxos tunnel [50].

Most of the examples of hydraulic tunneling described above were constructed to facilitate the water supply of urban centers, and there seems to have been a keen interest among tyrants, but also later, in democratic Athens, to engage in these projects. As mentioned above, such interest was probably due to the support expected from the population. Furthermore, tunneling associated with water management is probably also found in association with intensive agricultural strategies applied during the Classical Period (ca. 480–323 BC). Some years ago, Moreno [51] argued that an example existed in the Attic deme (local parish) of Euonymon on the southwestern coast of the peninsula. Here, Moreno argued that intensive farming, combining terracing of farmland with extensive tunneling, providing irrigation for the crops and ensured the basis of a lucrative trade in cash crops for nearby Athens. Other types of agriculture of a more extensive nature undoubtedly existed in other locations in Attica, but the southernmost deme of Atene (contra [52], but see [53]), may have shared features with the up-coast example of Euonymon; however, irrigated water of Athens may have been supplied by open conduits.

2.8. Roman Period (31 BC–476 AD)

In the Mediterranean part of the Roman Empire, tunnels were built for (a) the drainage of basins and lakes; (b) the extraction of water from a nappe (spring tunnels); (c) the transport of water in aqueducts; (d) the servicing of harbors, related to silting problems; and (e) for mining gold.

2.8.1. Drainage Tunnels

Some tunnels built in Roman times constitute attempts to drain lakes and use the land for agriculture [14]. The longest Roman tunnel built for drainage is the Lake Fucino tunnel, with a length of 5650 m [54–58]. It was ordered by the emperor Claudius and was built by 30,000 slaves in 11 years in the qanat mode through 40 vertical access shafts. The older drainage tunnels of Lakes Nemi and Albano [13,59–61] were meant to stabilize the level of the lakes rather than drain them. They contained screens to block debris from entering the tunnel. Drainage tunnels from the Roman period are also known in Greece (Lake Kopais), Turkey, and France. In Turkey, the 250 m-long Bezirgan tunnel near Kalkan drained a polje [62]. In France, the Étang de Clausonne tunnel drained a shallow lake [63] (Figure 11). This lake blocked the passage of the Nîmes Roman aqueduct that had to be built below the level of the lake, which therefore had to be drained. The drainage tunnel is adjacent to an aqueduct tunnel (Figure 11).

(a)

(b)

Figure 11. Drainage tunnel of the Étang de Clausonne, France: (**a**) upstream entrance of the drainage tunnel, originally protected by a metal grille; (**b**) downstream aspect of the drainage tunnel (left) built to make the Nîmes Roman aqueduct channel (right) pass below the Étang de Clausonne. Only the aqueduct tunnel has carbonate deposits. Water was flowing in opposite directions in the two tunnels. Sernhac, France (Photos Cees Passchier).

2.8.2. Spring Tunnels

Tunnels dug into solid rock to access groundwater were built to provide water for many ancient aqueducts. They have been thoroughly studied by speleologists in Italy [64], where more than 140 such spring tunnels from Etruscan, Greek, and Roman construction have been described in the "ancient aqueducts of Italy" project [64]. These are complex structures meant to capture enough water to fill an aqueduct downstream. Some are similar in purpose to qanats, but were built by driving a horizontal shaft into the rock without the help of vertical access shafts. Longer spring tunnels, however, were built as proper qanats, with vertical access shafts from which the tunnel was dug in two directions. Examples are tunnels for the Roman aqueducts of Xanthos (Figure 12), Turkey [65]; Sexi, Spain [66–68]; Zadar [69,70] and Novalja, Croatia [71]. There are also several examples in the middle east, especially in Syria [72,73]; in Northern Africa [74]; and, curiously, in western Germany [14,15,75]. A unique case is the tunnel that was excavated to tap the underground water source of Uxellodunum during a siege in the Gallic wars to force the inhabitants to surrender [76].

(a) (b) (c) (d)

Figure 12. Spring tunnel of the Xanthos aqueduct, Turkey. The tunnel was partially dug into the rock and extended with ashlars and cover stones. The channel is deepened, leaving a footpath along the side for access: (**a**) inside, looking to the exterior; (**b**) inside, looking towards the spring; (**c**) exterior. The structure is still in use to provide water for irrigation. (**d**) Branching tunnel of Novalja, Croatia. The tunnel was bifurcated to access two springs (Photos Cees Passchier).

2.8.3. Aqueduct Tunnels

Aqueduct tunnels can be divided into "transfer tunnels" needed to cross below hills and mountains (Figures 13–17), and "valley-side" tunnels built to pass steep rock walls or to protect an aqueduct in unstable geology (Figure 18). Transfer tunnels exist both in the counter-excavated mode, digging from two entrances to a meeting point, or, more commonly, in qanat mode (Figures 13 and 14), starting with vertical shafts dug from the surface downwards to a common level, after which the shafts are connected by horizontal tunnel segments. Counter-excavated tunnels have only one meeting point, while qanat-type tunnels have as many meeting points as there are shafts (Figure 14). Transfer tunnels of both types are among the longest tunnels built in the ancient world. They include the aqueduct tunnel of Bologna (18 km long) [15]; the Vernelles tunnel in the Traconnade tunnel of Aix-en-Provence, which passes below a watershed (>8 km long—[14,77]); the Forino tunnel of the Aqua Augusta near Naples (>6 km long [78,79]); the 4 km long Annio Novus tunnel of Valle Barberini [80–82]; and the 5 km long tunnel of Cella in Spain [83]. Some other tunnel examples are from Jerusalem and Side (both over 2 km long: [14]); Paterno (1903 m: [78]); Syracuse (1385 m: [15]); Lyon (Mornant tunnel in the Gier aqueduct of Lyon, France, 825 m long: [84]); and several shorter tunnels near Naples [78,85]. A famous aqueduct tunnel of 428 m long exists in Saldae, Algeria [14,86]. This tunnel was described on the gravestone of Nonius Datus, a Roman engineer who specialized in the building of water tunnels. He was asked to solve a problem with this counter-excavated tunnel since the workers passed each other without meeting [14,15]. This is one of the few reports we have of Roman tunnel building written by one of the engineers responsible.

Another interesting tunnel is the 230 m long Bullica tunnel of the Marcia aqueduct, Rome, which consisted of a service tunnel wide enough for carts, from which a lower-lying aqueduct tunnel could be accessed, connected to the access tunnel by shafts (Figure 17c) in the sidewall [81].

Transfer tunnels as mentioned above either had water running on the bare rock if the rock was impermeable, or, more commonly, were plastered (Figure 15b) or had a regular excavated or masonry gutter or even a vaulted channel built inside them (Figure 16). Tunnel workers used oil lamps set in niches to light the workforce, and for maintenance. Commonly, a pilot tunnel was dug first, which was then extended and widened downwards (Figure 15a). Tunnels usually have a rounded vault, but they may also have a flat roof (Figures 14d and 17c).

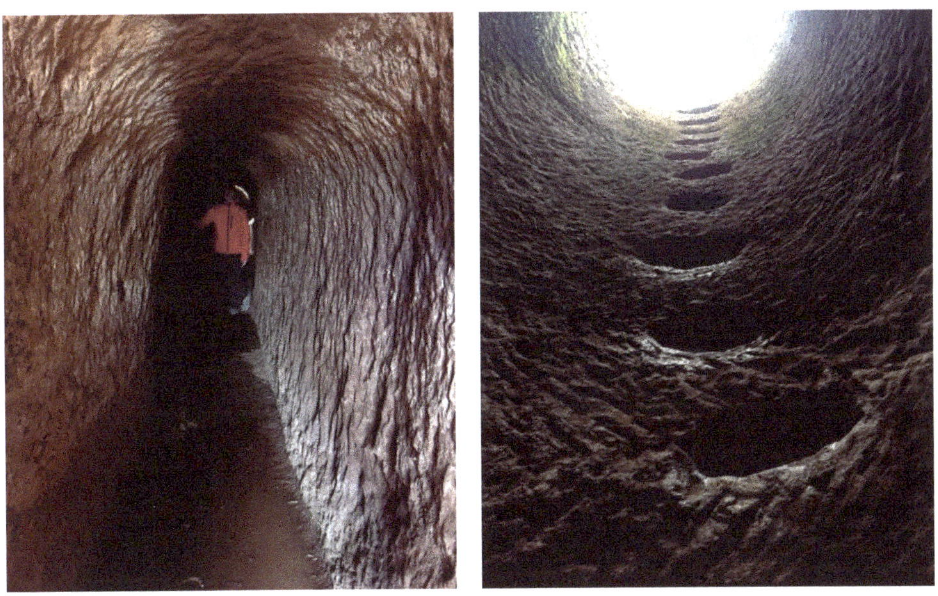

(a) (b)

Figure 13. Qanat-mode tunnels and shafts: (**a**) tunnel and (**b**) vertical access and building shaft with niches for working crews, Tiermes, Spain (Photos Cees Passchier).

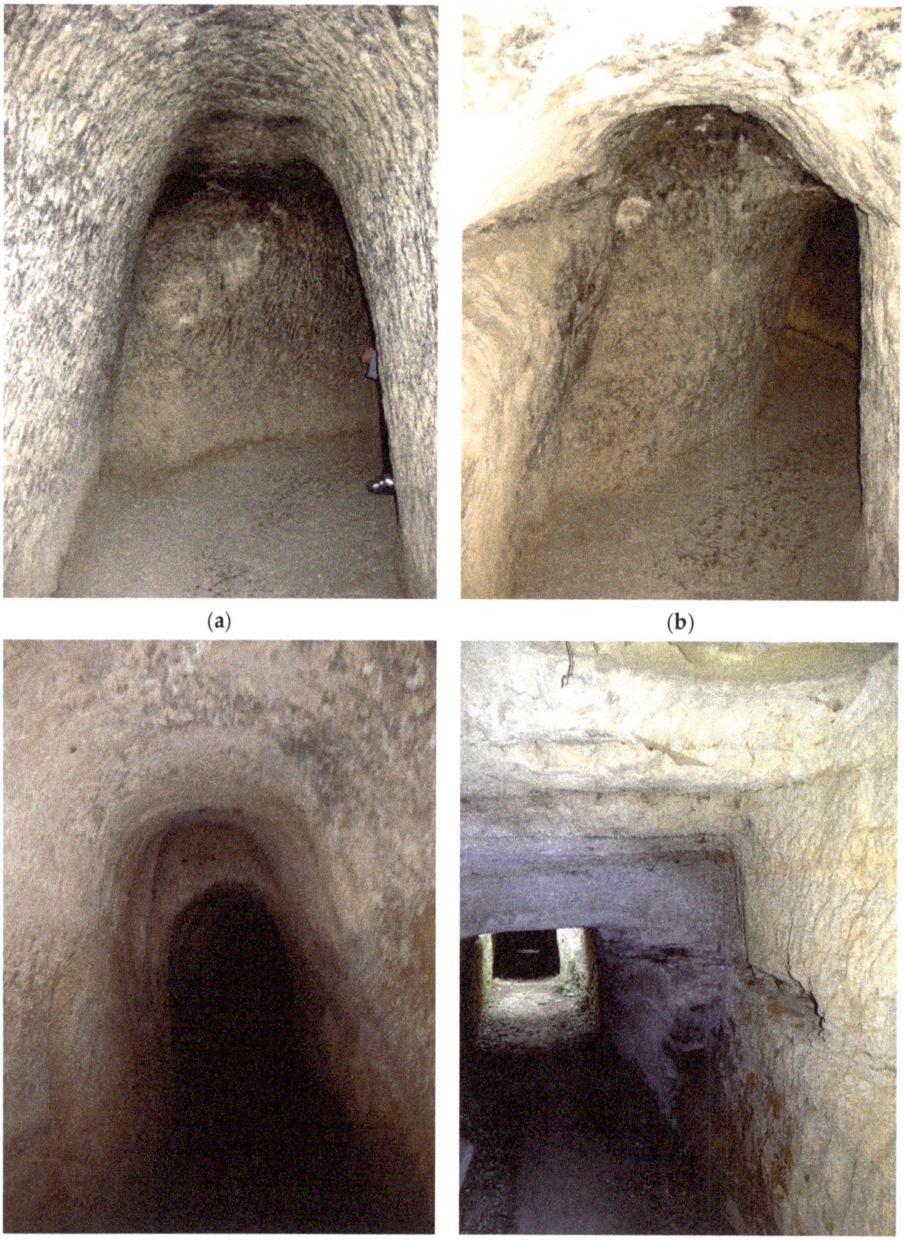

(**a**) (**b**)

(**c**) (**d**)

Figure 14. The meeting point between two excavation sections in several tunnels: (**a**) the Chelva tunnel, Spain, with horizontal offset. The view is towards the end of a section of the gallery that meets another one at the right-hand side (person visible). (**b**) A similar meeting point, with a major horizontal offset and small vertical offset, Chelva aqueduct. (**c**) The meeting point of the aqueduct tunnel of Tiermes, Spain, with a vertical offset. (**d**) Meeting point with vertical offset in the Sernhac tunnel of the Nîmes aqueduct. The tunnel has a flat roof because of the strong horizontal stratification (Photos Cees Passchier).

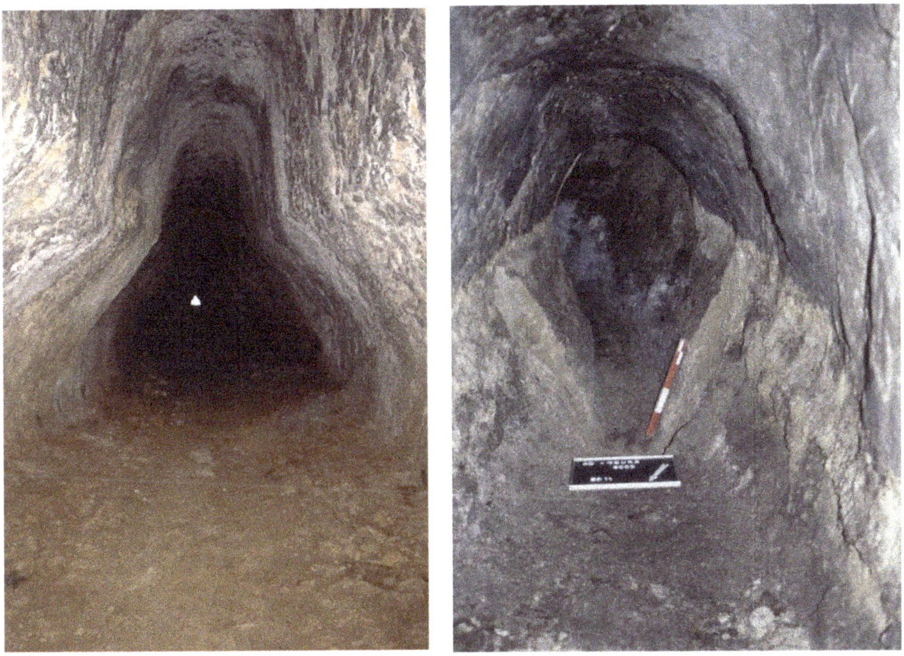

(**a**) (**b**)

Figure 15. Water tunnels: (**a**) tunnel with a pear-shaped profile at Uxama, Spain. This tunnel was probably first dug as a narrow structure, represented by the top, but later widened in its lower part to lower the water level and make access for cleaning crews easier [87]. (**b**) Tunnel of the Cahors aqueduct, Spain, where a tunnel was made narrower and trapezoidal by inserting wedge-shaped masses of mortar (Photos Cees Passchier).

(**a**) (**b**)

Figure 16. Conduits built into an aqueduct tunnel: (**a**) Sernhac tunnel, France; a masonry channel was built into the tunnel for passage of the Nîmes aqueduct (Photo (**a**)—Cees Passchier). (**b**) Cave de Curée, with a vaulted aqueduct channel in the tunnel, part of the Gier aqueduct of Lyon, France—(http://www.romanaqueducts.info/, accessed on 15 August 2023).

(**a**) (**b**) (**c**)

Figure 17. Constructions improving the shape of aqueduct tunnels: (**a**) masonry wall section in a tunnel where the sidewall was broken out. Traconnade aqueduct of Aix-en-Provence (France). (**b**) A vault structure over a tunnel that was dug from the top and then closed with a vault. Traconnade aqueduct, quarry of Santa Anna, Peyrolles. This tunnel is visible in profile, since it was later cut by a quarry. (**c**) Bullica tunnel, a maintenance tunnel of the Marcia aqueduct, with access shafts to the narrow aqueduct tunnel that runs at a deeper level. The metal bars are modern, but in ancient times a wooden beam would have been placed above the shaft to allow workers to descend and clean (Photos Cees Passchier).

(**a**) (**b**)

Figure 18. Valley-side tunnels with "windows" from which the tunnel was excavated: (**a**) Chaves aqueduct, Spain, and (**b**) Galermi aqueduct, Sicily (Photo Cees Passchier).

Since tunnels were not meant to be seen except by maintenance crews, they were purely functional structures and their architectural design was not significant. Therefore, traces of their construction and maintenance are usually well preserved, making it possible to see how they were built. In many tunnels, there are still traces of meeting points where two galleries, dug from opposite sides, met at an angle, or different altitudes (Figure 14).

Although most tunnels have remained unchanged and even lack a constructed water channel, modifications were sometimes made, either widening or narrowing a tunnel (Figure 15). There are also supporting structures, such as a masonry vault or sections of wall-filling in cavities or broken-out sections of tunnel wall (Figure 17a,b).

Many aqueducts have "valley-side tunnels", most built along steep vertical cliffs, which had to be passed (Figure 18). The technique to build them usually involved "horizontal shafts" or windows cut into the wall of the cliff, probably by workers suspended from above, and then connecting the shafts as in qanat construction. Tunnels of this type are known from Chelva, Spain; Galermi, Sicily, and Cella, Spain (Figure 18). A variation of this type of tunnel, the Gadara tunnel in Jordan, was built by excavating sloping shafts with staircases into the side of a valley, which were then connected [88,89]. This produced the longest tunnel of the ancient world, with a minimum length of 107 km and 2900 access shafts, supplying the city of Gadara with water from springs in Syria. This tunnel was probably built instead of a normal aqueduct channel at the surface to avoid problems of land sliding in the local soft, crumbling limestone [85,88].

2.8.4. Harbor-Related Tunnels

Harbor-related tunnels, built to either divert rivers away from a harbor or to regulate the flow of water into a harbor, are known from Seleucea, Turkey, and Cosa, Italy. The Çevlik tunnels of Seleucia Pieria, with a total length of 875 m [90–92], are part of a flood diversion system including dams and channels to keep flood water away from the harbors of Antiochia, the third largest city in the Roman Empire. In Cosa, a smaller structure, the "Tagliata Etrusca", was dug to avoid the silting of the harbor (Figure 19).

Figure 19. Tagliata Etrusca, Ansedonia. A tunnel and gallery cut into the rock to modify currents in the harbor of Cosa, meant to avoid silting (Photo Cees Passchier).

2.8.5. Tunnels Associated with Mining

A special application of water tunnels is those built to support the mining of metals, especially gold. In Spain, the 120 m long Montefurado tunnel (Figure 20a) was built in the time of Trajan to breach a meander of the river Sil and divert it, so that the riverbed could be explored for gold [93]. At the Las Medulas gold mines (Figure 20b), the largest in the Roman Empire [94], tunnels were dug into gold-bearing gravels not to extract the gold, but to assist in the mining process. These tunnels were dug close to the rock wall of the mine, but had no exit; a dammed supply of water upstream was channeled into the tunnels at high velocity, "fracking" the gold-bearing rocks (Figure 20b–d), while continued flow eroded the rock. This is a unique way of using water tunnels in the Roman world.

Figure 20. Tunnels related to gold exploration: (**a**) Tunnel of Montefurado, dug to change the course of the river Sil for gold exploration. (**b**) Water tunnels in Las Medulas, Spain. At the top, one of the original tunnels dug for "fracking" the conglomerate; at the bottom, one of the larger washed-out tunnels. (**c**) Typical wash-out tunnel of the Las Medulas system. (**d**) Tunnel fragments are left in a pillar of conglomerate, while the surrounding area has been mined; Montefurado, Spain (Photos Cees Passchier).

During the Roman period in India (31 BC–476 AD), tunneling technology was used mainly for water management and irrigation purposes. A notable example is the Kaveri Delta system, which dates back to the 1st century AD and features a network of tunnels and canals that were used to divert water from the Kaveri River for agricultural irrigation. The technology used during this period was advanced, with sophisticated engineering techniques being employed for tunnel excavation and maintenance. Additionally, the ancient Indian system of step-wells, such as the Rani ki vav in Gujarat, was also constructed during this period and utilized tunneling techniques for water storage and distribution [95].

2.9. *Byzantine Period (ca 330–1453 AD)*

The Byzantine Empire, which replaced the Roman Empire in Anatolia and the eastern Mediterranean, continued the tradition of building advanced water infrastructure, including water tunnels, to supply fresh water to cities and settlements. Istanbul (formerly known as Constantinople) struggled with water problems throughout its history and made enormous efforts to obtain water from nearby locations. After the city was declared the capital of the Roman Empire, Emperor Constantine built the longest line of tunnels in the Roman Empire, which began in Isırancalar (Figure 21) [96]. It is believed that the construction of this line was started by Constantine between 324 and 337 and completed by later emperors [97]. In a study on this subject, the length of this water supply line, determined through field and map work carried out by [98] between 1993 and 1996, was given as 242 km [99], which is 2.5 times longer than the longest Carthaginian water supply tunnel built by the Romans, with a length of 91 km [100]. Later, Valens (364–378) had a water pipeline built to bring water from the Halkalı area [101]. It is also known that Theodosius (379–395) had an aqueduct built to bring water from the Belgrade Forest [102].

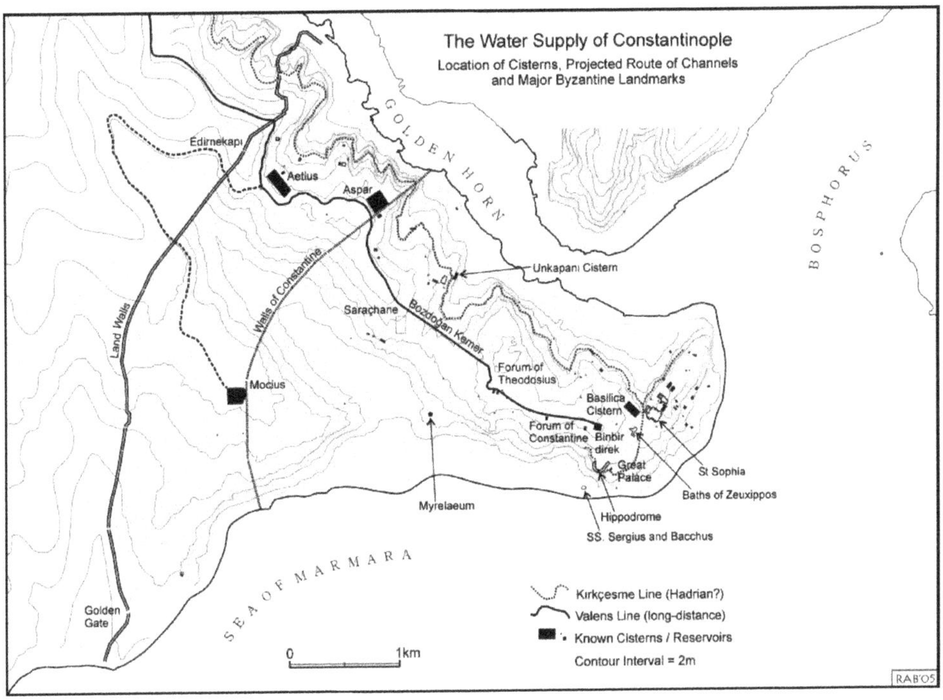

Figure 21. Map showing the locations of cisterns and water channels in Constantinople [97].

Among other water tunnels in this region is the first canal and bridge system, which was completed in 373 AD, and which brought water from the important springs in Danamandra and Pınarca. This system had a total length of 268 km and included an estimated 130 new bridges, varying in size from single-arch to larger-than-average double-arch bridges. All of the enclosed water channels were constructed of mortar-covered stone blocks that were 1 m wide and 1.6 m high, with an arched top. Within the city, the Bozdoğan Bridge (or Valens Aqueduct), with its eighty arches and a length of 971 m, is considered one of the longest water bridges in the Roman world. But that is not all; they built this water channel up to a length of 494 km. With a height of 2 m and a width of 1.6 m, these new tunnels are larger than those built in the first phase [97]. The final water tunnel in Constantine is the Ballıgerme, a wide channel spanning a deep gorge above the Karaman Dere (Figure 22). Both the upper and lower canals then run along the south side of the valley, winding around the elevated ridge crossed by the Anastasian Wall to the southeast [101].

Figure 22. Tunnel near Ballıgerme Hill Farm [101].

Another tunnel in Anatolia is the Kemer dam, including a water tunnel. It is located near the ancient city of Aspendos in the province of Antalya. The Kemer dam and tunnel is a remarkable Byzantine water project. The tunnel was built to carry water from the Koprüçay River to the city of Aspendos. It was built according to the Roman technique, which typically used stone or rock tunnels. The interiors were often lined with a layer of waterproof mortar or concrete to prevent water leakage and erosion. Another water tunnel is the Perge water tunnel [103]. The ancient city of Perge, also in Anatolia, had a Roman water tunnel system. This tunnel, like others in the region, was used to transport water to the city from distant sources. This tunnel was part of a larger aqueduct system that transported water. The tunnel was often dug by hand, with workers using tools such as picks and chisels. The Perge was constructed with a gradual slope or incline. This slope allowed gravity to move the water through the tunnel without the need for pumps or mechanical devices. There are also tunnels from the Byzantine period and underground aqueducts in Cappadocia, which is known for its unique underground cities. These tunnels were used for protection and as a source of water in times of conflict or siege [104–107].

There are many long-distance waterways in the Turkish Aegean and Mediterranean regions. These systems include spring water collection chambers, lead, stone, and clay pipes of various sizes, rock-hewn and masonry canals, tunnels over 2 m high, inverted siphons with pressures up to 190 m for lead pipes and 155 m for stone pipes, and aqueducts up to 40 m high. One of the longest Roman water transport routes, at 100 km, leads to Phoceia (Foça). In addition, lengths of 65 km at Pergamon (Bergama), 30 km at Smyrna (İzmir), and 42 km at Ephesus are among the most fascinating examples of various water supply systems in the ancient world. The 3.3 km long stone pipe siphon of the Karapınar water conduit to Smyrna can withstand water pressure of 155 m, while the lead pipe siphon of the Madradağ water conduit to Pergamon can withstand water pressure of up to 190 m. These siphons, dating from the late decades of the first millennium BC Hellenistic period, functioned at some of the highest pressures ever recorded in antiquity. The stone siphon at Aspendos is the longest in Turkey, on arches at 1.7 km. The Soma transport to the demolished aqueduct of Pergamon across the Karkassos (Ilyas) stream would have been 40 m high, making it the second-highest Roman aqueduct after the Pont-du-Gard of Nîmes. Another tunnel system and river detour from the Roman period is the Çevlik Tunnel. Its construction took place between the first and second centuries AD. The 875 m long system had a capacity of 70 m^3/s. It included two tunnel segments, 90 and 30 m long. The dimensions were in the range of 6–7 m, and the cross sections were either semicircular or trapezoidal. It was the largest structure at that time. The 250 m long Bezirgan tunnel near Kalkan, which is 1.1 m wide and 2.2 m high, serves as a floodwater conduit for the Karst polje [107].

3. Water Tunnelling in Early and Mid-Modern Times (ca 1453–1850 AD)

During medieval times in India (ca 476–1400 AD), tunneling technology was used for water supply, irrigation, and transportation. Notable examples include the Anicut Dam in Tamil Nadu, the Patal Bhuvaneshwar Cave in Uttarakhand, and the Rani-ki-Vav stepwell in Gujarat. These structures and tunnels demonstrate India's engineering capabilities and played an important role in the development of hydro-technology infrastructure during this period.

In Egypt's capital, Cairo, the Citadel Aqueduct was constructed at the beginning of the 13th century by Ayyubid sultans and then completed by the Mamluk State to convert Nile water into a new castle. The aqueduct started from the Nile shoreline, where water was raised by successive waterwheels, and ran into the aqueduct to the Citadel. The aqueduct was raised by a tower of arches constructed from masonry. It was still used during the Ottoman period. Many aqueducts that remain near the Citadel can be seen today, such as the impressive hexagonal tower used for water intake [108,109].

The damage and deterioration of the Constantinople water system that occurred during the Roman period were aggravated by the Latin occupation in 1204, after which the water system became virtually unusable. When Constantinople fell into the hands of the Ottomans in 1453, extensive repairs and additions were made to the system. The Kırkçeşme waterways, originating from the Belgrade forests, are among the most important water sources of Constantinople. In connection with the structures built between 1554 and 1564, 33 arches of different sizes were constructed [110]. The Uzun, the Kovuka, the Moğlova, and the Güzelce aqueducts are the longest ones, with 711 m each. The sixteen different waterways that make up the Halkalı waterways were built between 1453 and 1755. In Constantinople's past, water and the relative architectural structures were considered very valuable, since the city is shaped by water and filled with life through its dams, arches, fountains, water fountains, spa, and cisterns [101,111]. The Mazul aqueduct, the Kara aqueduct, the Turuncluk aqueduct, and the Bozdogan (Valens) aqueduct, which is now known as one of the arches of the Halkalı Canal and was later restored by the Ottoman Empire, are important Roman aqueducts built in Constantinople in the fourth century [96,101]. The Mâzul aqueduct, commonly referred to as the Mâzul aqueduct, was built around the same time in Constantinople and spans the Uzuncaova stream in the Military District [96]. It was the first aqueduct that transported the water of the Halkalı River. It was built of two stories of limestone blocks and has a height of 19 m and a length of 110 m, containing 13 arches in the upper row and 7 arches in the bottom [96]. After its restoration by Fatih Sultan Mehmed, it was used during the Turkish era and again during the time of Constantine V (741–775) [112]. During the Ottoman era, many dams, aqueducts, rivers, fountains, and basins were rebuilt and new fountains and baths were constructed.

The Halkalı waterways, constructed separately in the 16th century, the Kırkçeşme waterways, built during the Kanuni period, the Taksim Waters, built between 1731 and 1839, and the still-operating Hamidiye and Kayışdağı are the four main categories of water facilities built during the Ottoman period in Constantinople [96,98,99]. With these main water supply lines, there is no place in Constantinople where water does not reach [96]. The cargo of the new waterways built or repaired by Fatih Sultan Mehmet consisted of water from the Istranca Mountains, Belgrade Forest, and real sources. Later, new additions were made to the Marmara Region water facilities, which were named Halkalı waterways due to various sources near Halkalı Village, by many rulers and statesmen [113]. These waterways are Fatih (1453–81), Turunçlu (1453–81), Mahmut Paşa (1453–73), Mustafa III (1757–74), Bayezid (1481–1512), Kocamustafa Paşa (1511–12), Süleymaniye (1557), Mihrimah (1565), Ebussuud (1545–74), Cerrahpaşa (1598–99), Sultanahmet (1603–17), Murat IV (Palace fountains) (1623–40), Köprülü (1656–61), Mahmut I (1730–54), Hekimoğlu Ali Paşa (1732–50), Kasım Ağa, and Nuruosmaniye (1748–55). These waterways were used to supply water to mosques, imamates, fountains, and barracks outside the city. The daily output of these facilities is 4335 m^3 [96]. There are four large aqueducts in the Halkalı waterways facilities: Mazul, Kara, Ali Paşa, and Bozdoğan aqueducts [106]. Mazul aqueduct and Bozdoğan aqueduct were built during the Roman era, and later on water aqueducts were constructed over the Halkalı waterways such as Fatih, Turunçluk, and Mahmutpaşa [96,101]. During the reign of Beyazit II, there were 33 water aqueducts, including monumental arches, such as Beyazit waterways, Kırkçeşme waterways, Uzun aqueduct, Kırık aqueduct (Eğri aqueduct, Kovuk aqueduct), Güzelce aqueduct, Moğlova aqueduct, and Paşa aqueduct (Balıkzade aqueduct) [96].

The Kırkçeşme waterways facility collected water from the Alibey and Kağıthane streams, which was then stored in reservoirs and transported to the city through Eğrikapı. Because durable pipes capable of withstanding high pressure were not available at the time, aqueducts were built in valleys and water was transported through them [96]. Uzun aqueduct is the longest arch of all the lines, with a height of 26 m and a length of 711 m. The arches are 4.5–4.6 m wide on the upper row and 3.7–5.2 m wide on the lower row [96]. Kırık aqueduct, also known as Eğri or Kovuk aqueduct, is a three-story arch 35 m high and 342 m long [114]. The Moğlova aqueduct, which is considered an architectural masterpiece of the Kırkçeşme waterway facilities, is a two-story arch that is 35 m high and 258 m long [110]. The geometric structure of the arches is a great engineering achievement. To prevent the arches from tipping over, the base of the legs was widened in the shape of a pyramid so that the arches could be kept unusually thin. To allow the upper part of the lower arch to be used like a bridge, a passage was created through the legs, cleverly connected to the slopes. The legs were given a special shape towards the source to prevent the water from forming a depression in front of them due to the flow [98,99,106]. The Mağlova aqueduct has the largest arch span after the Pont-du-Gard aqueduct in France [114].

The Güzelce aqueduct, across the Cebeciköy Stream, is another aqueduct of the Kırkçeşme waterway infrastructure [96]. Again, Mimar Sinan used a trapezoidal wall system and two-sided buttresses to strengthen the legs. It has 11 openings on the upper floor and 8 openings in the basement, with an opening width of 5.6–6.1 m [114]. It is a two-story building. To resist lateral forces (earthquakes, wind), the legs of this arch were trapezoidal in shape and reinforced with buttresses [98,99,106]. Other important single-story arches include the Kara aqueduct, the Develioğlu aqueduct, the Vâlide aqueduct, and the Alacahamam aqueduct, which was built on a branch of the Cebeciköy stream. The Ali Paşa aqueduct is another arch built under Mimar Sinan. This trapezoidal, two-story aqueduct has a length of 102 m and a height of 16.4 m. The Ali Paşa aqueduct has 13 openings, with a width of 5 m each [96]. During the Ottoman period, aqueducts were built in different parts of the empire. In Constantinople, about forty significant examples are known [110]. The Kırkçeş me water conduit, which is the most important water network in the city, has thirty-five arches, six of which are two- or three-story monumental examples. This water conduit, dating back to the time of Theodosius I (379–395), was destroyed during attacks from the West at the beginning of the 7th century. This facility was almost completely rebuilt by Mimar Sinan between 1554 and 1563 [106,110]. During the reign of Sultan Süleyman the Magnificent, water was brought to the city from sources such as Taşmüsellim, Hıdırağa villages, and the Kurtalçağı stream in the northeast of Edirne in the name of Haseki Hürrem Sultan [115,116].

Although not mentioned in the records, it is accepted that these facilities, including the Hançerli, Ortakçı, Arap, Çifte, Kurt, Yedigöz, Hıdırağa, Üçgöz, Oğlanlı, and Hasanağa aqueducts, were built by Mimar Sinan around 978 (1570–71) during the construction of the Selimiye Mosque and Complex. The Governor of Edirne, İzzet Paşa, repaired these water structures, which were made of cut stone and consisted of pointed arches in a single row, in 1890 after they had been damaged over time [110]. The double-decker aqueduct built in Kavala, Greece, during the reign of Sultan Süleyman the Magnificent is notable among the monumental aqueducts built during the Turkish period in the Balkans. The lower arches were made wider than the upper ones, and lightning holes were drilled between the upper arches. The Mustafa Paşa aqueduct, a 3800 m structure with fifty-five arches in the northwest of Skopje, transports water from Banya Mountain. Bricks were used in the arches, which were constructed with cut stone and sandstone [110]. One of the most important bridges on the Taksim water system, which was built during the reign of Sultan Mahmud I, is the Mahmud I Bridge, which has 21 arches and is 400 m long [96,106,117]. Two rows of arches are only present in the portion that is built along the river.

Another structure with double rows of arches, known as the Ali Paşa Bridge or the Şirin aqueduct, is located on a tributary of the Ayvalı River near the military field. The Avasköy Bridge (also known as Yılanlı aqueduct or Tekaqueduct), was constructed nearby by renowned Ottoman architect Mimar Sinan. Eleven arches make up this bridge, which is constructed of limestone [118]. The Kumrulu Bridge (also known as Akyar Bridge), which has a single arch and is located on the Süleymaniye road, the Kara aqueduct Bridge, with three arches, and the Paşa Bridge, which carries the Turunçlu water to the city at the intersection of the Beyazıt aqueduct, are ordinary water bridges [110]. Also, another aqueduct was constructed by the Grand Vizier Safranbolulu İzzet Mehmed Paşa, and it supplies water to Safranbolu. It is made of mortar and rubble stone, measures 116 m long and 60 m high, and has one major and five tiny arches [110]. The single-pointed arch of the Akdere aqueduct, with a width of 4.10 m and an opening of 1.10 m, traverses the valley as the Kırkgöz water is transported from the Pınarbaşı water source to Kahramanmaraş. The arch, which has a cut stone roof and a base made of rubble stone, is in ruins [110].

India's history of tunneling and hydro-technology continued to evolve and advance throughout the medieval period, with notable examples including the complex water supply system of the Qutb Minar complex in Delhi, built by the Mamluk dynasty in the mid-13th century AD. During the early and mid-modern times in India (ca 1400–1850 AD), tunneling technology continued to play a vital role in the country's hydro-technology infrastructure. A notable example is the Rajon Ki Baoli stepwell in Delhi, built during the 16th century AD. This impressive structure includes a series of underground tunnels and chambers that were used for water storage and purification (Delhi Tourism and Transportation Development Corporation). Another significant example is the Brihadeeswarar Temple in Tamil Nadu, built during the 11th century AD, which features a series of underground channels that collect and distribute water for the temple's use (Archaeological Survey of India).

Furthermore, the Mughal Empire, which ruled over India during the 16th and 17th centuries, contributed considerably to the development of tunneling technology in the country. The Mughals constructed several underground water channels, known as "qanats," to provide water for their gardens, palaces, and cities. One notable example is the Shalimar Bagh garden in Srinagar, which features a network of underground channels that collect and distribute water from a nearby spring (India Water Portal). Overall, during the early and mid-modern periods in India, continued innovation and developments in tunneling technology were achieved, thus supporting the country's growing hydro-technology needs.

4. Tunneling in Contemporary Times (1853 AD–Present)

In contemporary times (1853 AD-present), tunneling technology in India has continued to evolve and expand, playing a significant role in the country's infrastructure development. A major project is the Kaleshwaram Lift Irrigation Project in Telangana, which includes the construction of a network of tunnels to transfer water from the Godavari River to drought-prone regions of the state (India Today). Another notable example is the Chenani–Nashri Tunnel in Jammu and Kashmir, which is the longest road tunnel in India, measuring 9.2 km, and was constructed to provide all-weather connectivity between the two regions (National Highways Authority of India). Additionally, tunneling technology has been used in the construction of metro rail systems in cities such as Delhi, Mumbai, Kolkata, and Bengaluru, to alleviate traffic congestion and provide fast and efficient transportation (The Indian Express Journalism of Encourage, 2017). These projects show that tunneling technology continues to be a critical tool for meeting India's growing infrastructure needs in contemporary times.

A big project was constructed in the area in the Southeastern Anatolia region of Turkey, which includes the provinces of Adıyaman, Batman, Diyarbakır, Gaziantep, Kilis, Mardin, Siirt, Şanlıurfa, and Şırnak, defined as the "GAP Region" (Southeastern Anatolia Project) [119]. The GAP "Southeastern Anatolia" project is one of the most significant water-based development projects in the world in terms of size and impact, and the largest integrated water resources project in Turkey. Irrigation systems and drainage requirements in the Tigris and Euphrates basins have been studied on a project-by-project basis as part of GAP. Numerous studies have been conducted to examine the water resources, irrigation systems, and water distribution methods based on the data and field observations collected during these studies in terms of current demands, as well as drainage requirements and systems, water control structures, covers, and efficient water use [119]. This region, bordering Syria to the south and Iraq to the southeast, comprises 20% of Turkey's irrigable 8.5 million hectares of land and consists of large plains in the river basins of the lower Tigris and Euphrates rivers in the GAP region. Within GAP, 22 dams, 19 hydropower plants, and an area of 1,762,000 hectares have been planned for economically viable irrigated agriculture, with a total installed capacity of over 7476 megawatts and an annual electricity production of 27 billion kilowatt hours [119]. One of these structures is the Şanlıurfa irrigation tunnel. The Şanlıurfa tunnels are located in the Southeast Anatolia region of Turkey. They consist of two parallel tunnels with a total length of 26.4 km, running from the Atatürk Dam reservoir to 5 km northeast of Şanlıurfa. The tunnels are among the longest irrigation tunnels in Turkey and worldwide. Construction began in 1981 and the tunnels, which are among the largest structures in the GAP, were planned to irrigate about 476,000 hectares of land, including about 358,000 hectares of land by gravity and 118,000 hectares of land by pumping. The water tunnels consist of two circular concrete-lined tunnels, each 7.62 m in diameter and 26.4 km long. The total length of the tunnels, including the transport and connecting tunnels, is 57.8 km. The water taken from the Atatürk reservoir through the tunnels, amounting to 328 m^3/s, is to be transferred to the Harran and Mardin plains. In the system consisting of two parallel tunnels, the distance between the tunnels from axis to axis is 40 m. A connecting tunnel has been opened between the tunnels every 500 m so that the excavated material can be easily transported outside, and the excavation and concreting work can be carried out simultaneously. There are 52 connecting tunnels in total. Good ventilation is very important in such a long tunnel. For this purpose, the chimneys in the middle of the connecting tunnels were opened to supply both tunnels. There is a chimney approximately every 1500 m, the depth of which varies between 65.24 m and 207.95 m. The total number of stacks is 23. The tunnels are laid out according to the direction of water flow [120].

In Athens, the capital city of Greece, a contract was signed between the Greek Government, the Bank of Athens, and the American firm ULEN in 1952, for the financing and construction of the new water supply project. The first major work was the construction of the Marathon Dam (1926–1929). The dam is 54 m high and 285 m long and it is considered unique because it is entirely paneled externally with Pentelikon white marble. The Boyati Tunnel, 13.4 km long, 2.6 m wide, and 2.1 m high, was constructed to transport water from the Marathon impounding reservoir to a new water treatment plant in Athens [8]. In 1956, the water from the Yliki Lake was added to the system, and in 1981, the Mornos dam and aqueduct were inaugurated. The Mornos dam is one of the highest earth dams in Europe, with a height of 126 m. The Mornos aqueduct, which transports water from the Mornos reservoir to Athens, is the second longest aqueduct in Europe. It has a total length of 188 km, made up of 15 tunnels of 71 km in length and 3.2 m in diameter, 12 siphons (7 km), and 15 canals (110 km). The first time that a TBM was used in Greece for the excavation of the Gkiona Tunnel, 14.75 km in length [121]. Finally, the last major work, which provided Athens with additional water in 2001, was the Evinos River diversion to the Mornos impounding reservoir, consisting of the Evinos Dam and a diversion tunnel. Works began on the Evinos in 1992 and were completed in 2001. The major structures of the project are a 120 m high earth-fill dam, with a dam volume of 12 million m3, a total barrage capacity of 120 million m^3, and the 29.4 km long Evinos–Mornos tunnel, with a 4.2 m excavation diameter and a 3.50 m internal diameter [8]. The tunnel is one of the longest hydraulic tunnels in the world realized using the TBM method. The adverse geological conditions, the high cover, and the short construction schedule were a great challenge for the successful construction of this tunnel [122]. The tunnel was completed in just two years, which is considered to be a significant achievement given the project scale. The area covered by this major project is shown in Figure 23.

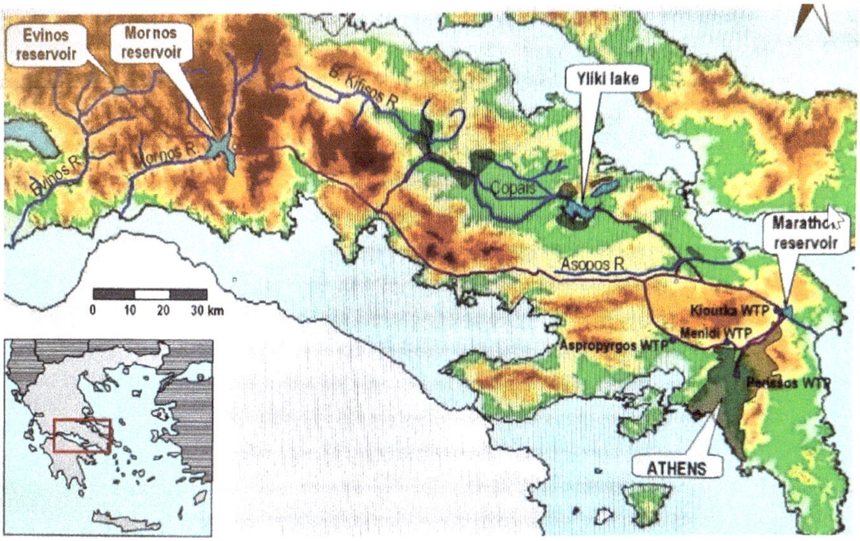

Figure 23. Schematic representation of the Athens water supply system in modern times [123].

In the contemporary time the hydrological tunneling has increased significantly in both size and number. A few more examples, indicating their size, are the following:

(a) The Delaware aqueduct in the New York City water supply system. It was constructed between 1939 and 1945 and carries approximately half of New York City's water supply of 4,900,000 m^3/d. At 4.10 m wide and 137 km long, the Delaware Aqueduct is the world's longest tunnel. It takes water from the Rondout, Cannonsville, Neversink, and Pepacton reservoirs on the west bank of the Hudson River through the Chelsea Pump Station, then into the West Branch, Kensico, and Hillview reservoirs on the east bank, ending at Hillview in Yonkers, New York [124].

(b) The Metropolitan Area Outer Underground Discharge Channel is an underground water infrastructure project in Kasukabe, Saitama, Japan. It is the world's largest underground flood water diversion facility, built to mitigate the overflowing of the city's major waterways and rivers during rain and typhoon seasons [125]. It is located between Showa and Kasukabe in Saitama prefecture, on the outskirts of the city of Tokyo in the Greater Tokyo Area. Construction started in 1992 and was completed by early 2006.

(c) In China, a secretive 500 km long irrigation project being built to divert snowmelt from the Altay Mountains to desert areas in its restive Xinjiang region has developed a too-much-of-a-good-thing problem. Workers keep tapping into gushing flows of groundwater, which has slowed construction to a crawl. It was based in part on the 2000-year-old karez system designed by Uyghurs in Turpan, and China began constructing the 514 km long project years ago, in what is reportedly the longest underground irrigation canal system in the world [126]. The project comprises three deeply dug tunnels, the longest of which is the 280 km long Kashuang Tunnel—twice as long as the Delaware Aqueduct, the main channel supplying water to New York City.

5. Emerging Trends of Tunneling Aqueducts

Governments and municipal authorities, faced with the problems of providing infrastructure within and between densely populated megacities, have acknowledged the importance of tunnels for the installation of underground transport corridors, sewerage systems, and utilities. Nowadays, many tunnels are constructed with advanced mechanical TBMs that have been progressively replacing the older drill and blast methods. TBMs can excavate a full circular face to the diameter of the machine, typically from 2 to 12 m, at astonishingly rapid rates when rock mass conditions are excellent. Even so, and despite dramatic improvements in TBM technology [127,128] TBMs are still not good at coping with rapidly changing or poor geologic conditions that can delay or stop the machines, thus increasing risks and costs to the tunnel project [129].

Studies of tunnel projects in the United States [130] have demonstrated that predesign investigations along the tunnel route using geological mapping and core drilling from the surface can mitigate these risks and reduce costs. These direct exploration methods may be enhanced with appropriate geophysical techniques (e.g., electrical and seismic imaging) to investigate the interval between boreholes or in difficult or complex areas (e.g., [131,132]). These studies may benefit global underground engineering researchers for hazard prediction and in establishing early warning systems [133]. In general, we are seeking to advance the application of geophysical methods to solve problems facing remediation professionals concerning fractured-rock aquifers. To this end, we (a) provide an overview of geophysical methods applied to the characterization and monitoring of fractured-rock aquifers; (b) review case studies showcasing different geophysical methods; and (c) discuss best practices for method selection and rejection based on synthetic modeling and decision support tools [134].

Emerging trends in tunneling aqueducts in India focus on sustainability, cost-effectiveness, and innovation. An interesting example of such trends is the use of micro-tunneling, a trenchless technology used to construct small-diameter tunnels for water supply and drainage systems (Indian Geotechnical Society). This technique has been utilized in the construction of the Ganga–Krishna–Pennar–Link Project, a massive water supply project aimed at transferring water from the Ganges and Godavari rivers to water-deficit regions in the southern part of the country. Another emerging trend is the use of pre-cast tunnel segments, which can be quickly and efficiently assembled on-site, reducing construction time and costs (NBM&CW). These segments have been used in the construction of the Mumbai Metro Line 3 tunnel, a massive underground rail project that will significantly improve transportation in the city. Overall, these emerging trends in tunneling aqueducts demonstrate how innovative technology and sustainable practices are being used to meet India's growing infrastructure needs.

6. Epilogue

It is obvious that the irrigation canals used in modern agriculture still follow the basic technical concepts used in ancient times for the construction of aqueducts. In the past, water sources were usually located outside the cities. Therefore, water was transported through open channels, tunnels, pipes made of various materials, and channels carved into rocks and covered with a lid. Water was transported either under pressure or using gravity in tunnels, galleries, and canals. In gravity conveyance, maintaining the head of the water is critical. In an open channel, gallery, or pipe, water flows freely according to gravity. Since there is no pressure or very little pressure as the elevation rises ahead of the waterway, the pipes are thin. For clay pipes, the thickness is 1–2 cm. In different regions, these pipes are called by different names, including "künk", "pöhrek", and "terracotta" (in Italian).

The "inverted siphon" method, also referred to in the literature as "reverse siphon", was historically used to cross valleys as an alternative to aqueducts. In certain cases, the inverted siphon and the aqueduct were built together, reducing the height of the arch to save costs. These pipelines were built using elements such as earth, stone, and lead. Aspendos in Attalya illustrates these practices: water was transported to the city's reservoir through two towers 65 m high, which were located 924 m apart from each other. Between these towers, there is a water channel, in ruins, 45.00 m high. In Aspendos, both a stone pipe network and an aqueduct were used. The siphon has a depth of about 20 m, and the diameter of the stone pipe is 30 cm.

The United Nations and other organizations encourage the revitalization of traditional water harvesting and supply technologies in arid areas because they consider it important for sustainable water utilization. A qanat as a tunneling system is a gently sloping subterranean conduit, which taps a water-bearing zone at a higher elevation than cultivated lands. It is used to provide a reliable supply of water to human settlements or for irrigation in hot, arid, and semiarid climates and allows the population to live in a desert area. A qanat system has a significant impact on the lives of water users, as it allows those living in a desert environment adjacent to a mountain watershed to create a large oasis in an otherwise stark environment. The advantages of transporting water underground in the qanat system are obvious, given that qanats are subterranean tunnels that tap the groundwater and lead the water entirely by gravity. As they are often dug into the hard subsoil and, when necessary, lined with relatively impermeable clay hoops, there is little seepage, no change in the water table, no water logging, and no evaporation during transit. The rate of water flow in a qanat is controlled by the level of the underground water table and it therefore exploits groundwater as a renewable resource. Thus, qanats are environmentally sustainable water harvesting and conveyance tunneling techniques through which groundwater can be obtained without causing damage to the tapped aquifer in arid regions ([17,135–137]).

The importance of the Kopais project, which was perceived by ancient Greeks, was also recognized in modern times; thus, the drainage of Kopais was among the first land reclamation projects carried out, during the second half of the 19th century, by the newly established Greek state. In this case, a tunnel and a network of ditches were created that sent the waters of the lake to an adjacent lake (Lake Yliki). Modern engineers went further, as they identified the peculiarity of the soil in the Kopais plain, which has an impermeable layer at a shallow depth (~2.00 m) and above it a layer of sand that is under the surface organic layer and a layer of marl. The impervious layer and the sand layer make it possible to apply subirrigation with the use of drainage ditches during the summer. Thus, a network of earthen ditches has been created that drain the area in the winter to Yliki Lake, and in the summer water from Yliki is pumped and led to the ditches. Initially, the system was designed wisely: a suitable level was kept in each ditch, and the plants were irrigated with the sub-irrigation system. Later, pumping with individual pumping stations and sprinkler irrigation was preferred, which is energy-consuming [35].

According to Sir Winston Leonard Spencer-Churchill (1874–1965), *"the more you look back in the past, the more you see into the future"*. Furthermore, an analysis of ancient tunneling techniques and applications can provide many practical solutions invented in the past that can be applied in the modern world. Our ancestors had no access to engines and modern techniques, but they used simple, energy-saving means. Their inventions and practical applications can therefore find a place in a new, environmentally conscious, and energy-conservative world.

In conclusion, tunneling dates back to prehistoric times, with the use of hydro-technology playing a critical role in creating sustainable systems in tunneling. Many ancient civilizations, such as the Ancient Egyptians, Greeks, Romans Persian, and others used tunnelling, from the simplest form of aqueducts, such as ditches cut into the earth, up to complex structures including horizontal and vertical tunnels. These tunnels were at a lower level than the reservoir and relied on a gravity hydraulic system for transferring and distributing water without employing any extra energy, which may cause negative impacts on the environment. From ancient water management systems to modern tunneling engineering, India has made significant strides in enhancing safety, minimizing costs, and reducing environmental impact. As India continues to invest in infrastructure, it is anticipated that it will make further strides in tunneling engineering, contributing to sustainable development in the country.

Author Contributions: A.N.A. had the original idea and wrote the original draft. J.A.K. contributed to the ancient Greek section and reviewed the manuscript. C.W.P. contributed mainly to the Roman section. V.A.T. reviewed and edited the manuscript A.T.A. reviewed and revised the manuscript. A.G.C. reviewed the manuscript. M.V. contributed to the Iranian section and other prehistoric civilizations and Emerging Trends of Tunneling Aqueducts. A.B. and E.B. contributed to the Roman section and Water Tunneling in Early and Mid-Modern Times. R.K. reviewed and revised the manuscript. N.D. reviewed and revised the manuscript. All authors have read and agreed to the published version of the manuscript.

Funding: This research received no external funding.

Data Availability Statement: Not applicable.

Conflicts of Interest: The authors declare no conflict of interest.

References

1. Voudouris, K.S.; Christodoulakos, Y.; Steiakakis, E.; Angelakis, A.N. Hydrogeological characteristics of Hellenic aqueducts-like Qanats. *Water* **2013**, *5*, 1326–1345. [CrossRef]
2. Barghouth, J.M.; Al-Saed, R.M. Sustainability of ancient water supply facilities in Jerusalem. *Sustainability* **2009**, *1*, 1106–1119. [CrossRef]
3. De Feo, G.; Angelakis, A.N.; Antoniou, G.P.; El-Gohary, F.; Haut, B.; Passchier, C.W.; Zheng, X.Y. Historical and technical notes on aqueducts from prehistoric to medieval times. *Water* **2013**, *5*, 1996–2025. [CrossRef]
4. Dutt, S.; Gupta, A.K.; Singh, M.; Jaglan, S.; Saravanan, P.; Balachandiran, P.; Singh, A. Climate variability and evolution of the Indus civilization. *Quat. Int.* **2019**, *507*, 15–23. [CrossRef]
5. Mumbai Metro Rail Corporation Limited. (n.d.). Metro Line 3—Colaba-Bandra-SEEPZ, Mumbai, India. Available online: https://www.railwaygazette.com/infrastructure/mumbai-metro-line-3-breaks-ground/43391.article (accessed on 7 September 2023).
6. Tris, T. TBM Launches on Vishnugad-Pipalkoti Hydropower Project. Available online: https://tunnellingjournal.com/tbm-launches-vishnugad-pipalkoti-hydropower-project/ (accessed on 7 September 2023).

7. Express Web Desk. 10 interesting facts about India's longest Chenani-Nashri tunnel 2 April 2017. In *The Indian Express Jornalism of Encourage (2017)*; The Indian Express: New Delhi, India, 2017. Available online: https://indianexpress.com/article/india/chenani-nashri-tunnel-inauguration-10-facts-about-indias-longest-tunnel-narendra-modi-jammu-srinagar-4596209/ (accessed on 7 September 2023).
8. Tsatsanifos, C.; Michalis, I. Tunnelling in Greece: Past, Present, Future. In Proceedings of the 2nd Eastern European Tunnelling Conference, Athens, Greece, 28 September–1 October 2014.
9. Chiotis, E.; Marinos, P. Geological aspects on the sustainability of ancient aqueducts of Athens. *Bull. Geol. Soc. Greece* **2012**, *46*, 16–38. [CrossRef]
10. Kulkarni, H.; Rao, G. Karez: An Ancient Underground Water Supply System. In *Geospatial Technologies in Urban System Development: Emerging Research and Opportunities*; Mukherjee, S., Ed.; IGI Global: Hershey, PA, USA, 2017; pp. 143–157.
11. Sudhakar, G.J.; Amirthalingam, M.; Sumabala, P. Tank irrigation in South India: A case study of Kolavoy tank. *J. Indian Hist. Cult.* **2007**. Available online: https://www.researchgate.net/publication/312838455_TANK_IRRIGATION_IN_SOUTH_INDIA_A_CASE_STUDY_OF_KOLAVOY_TANK (accessed on 17 September 2023).
12. World Bank. *India—Enhancing Water Resources and Water Efficiency Management Project*; World Bank: Washington, DC, USA, 2018.
13. Castellani, V.; Dragoni, W. Ancient tunnels: From roman outlets back to the early Greek civilization. In Proceedings of the 12th International Congress of Speleology, La-Chaux-de-Fonds, Switzerland, 10–17 August 1997; pp. 265–268.
14. Grewe, K. *Licht am Ende des Tunnels, Planung und Trassierung in Antiker Tunnelbau*; Ant. Welt, Sonderh: Mainz, Germany, 1998.
15. Grewe, K.; Tutlies, P. Der Drover-Berg-Tunnel-Wanderweg bei Düren. Fundgeschichten—Archäologie Nordrh. Westfal. *Mainz* **2010**, *4*, 403–405.
16. Alemohammad, S.H.; Gharari, S. Qanat: An ancient invention for water management in Iran. In Proceedings of the Water History Conference, Delft, The Netherlands, 17–21 July 2010. Available online: https://www.researchgate.net/publication/321914743_Qanat_An_Ancient_Invention_for_Water_Management_in_Iran (accessed on 18 September 2023).
17. Boustani, F. Sustainable Water Utilization in Arid Region of Iran by Qanats. 2008. Available online: chrome-extension://efaidnbmnnnibpcajpcglclefindmkaj/https://www.researchgate.net/profile/Fardin-Boustani-2/publication/242784793_Sustainable_Water_Utilization_in_Arid_Region_of_Iran_by_Qanats/links/5467085a0cf2397f7829f966/Sustainable-Water-Utilization-in-Arid-Region-of-Iran-by-Qanats.pdf (accessed on 21 February 2023).
18. Laureano, P. Water catchment tunnels: Qanat, foggara, falaj. An ecosystem vision. In Proceedings of the EN IWA Specialized Conference on Water & Wastewater Technologies in Ancient Civilizations, Istanbul, Turkey, 22–24 March 2012; pp. 22–24.
19. Angelakis, A.N.; Kavoulaki, E.; Dialynas, M.G. Sanitation and stormwater and wastewater technologies in Minoan era. In *Evolution of Sanitation and Wastewater Management through the Centuries*; Angelakis, A., Rose, J., Eds.; IWA Publishing: London, UK, 2014; Chapter 1; pp. 1–24.
20. TEW. Tour Egypt Website: "Hourse Temple". 2023. Available online: http://www.touregypt.net/edfut.htm#ixzz3pVpWqfyx (accessed on 7 September 2023).
21. Driaux, D. Water supply of ancient Egyptian settlements: The role of the state. Overview of a relatively equitable scheme from the Old to New Kingdom (ca. 2543–1077 BC). *Water Hist.* **2016**, *8*, 43–58. [CrossRef]
22. Evans, S. *The Palace of Minos at Knossos: A Comparative Account of the Successive Stages of the Early Cretan Civilization as Illustrated by the Discoveries*; Reprinted by Biblo and Tannen: New York, NY, USA, 1964; Macmillan and Co.: London, UK, 1921; Volume I–IV.
23. El-Gohary, F.; Angelakis, A.N.; Rose, J. Evolution of sanitation and wastewater technologies in Egypt. In *Evolution of Sanitation and Wastewater Technologies Through Centuries*; IWA Publishing: London, UK, 2014; Chapter 4; pp. 55–68.
24. Jansen, R.B. Dams from the Beginning. *Dams Public Saf.* **1980**, *1*, 1e57.
25. Mishra, U. Shrines as 'monuments': Issues of classification, custody and conflict in Orissa. In *Negotiating Cultural Identity*; Routledge: Delhi, India, 2019; pp. 200–229.
26. Singh, P.K.; Dey, P.; Jain, S.K.; Mujumdar, P.P. Hydrology and water resources management in ancient India. *Hydrol. Earth Syst. Sci.* **2020**, *24*, 4691–4707. [CrossRef]
27. Kenoyer, J.M. *Ancient Cities of the Indus Valley Civilization*; Oxford University Press: Oxford, UK, 1998.
28. Iyer, M. The Best Laid Plans, Deccan Herald. 2019. Available online: https://www.deccanherald.com/sunday-herald/best-laid-plans-713650 (accessed on 7 September 2023).
29. Thapar, R. *Aśoka and the Decline of the Mauryas*; Oxford University Press: Oxford, UK, 2012; p. 90. ISBN 9780199088683.
30. Voudouris, K.; Tsatsanifos, C.; Yannopoulos, S.; Marinos, V.; Angelakis, A. Evolution of underground aqueducts in the Hellenic world. *Water Sci. Technol. Water Supply* **2016**, *16*, 1159–1177. [CrossRef]
31. Mamassis, N.; Moustakas, S.; Zarkadoulas, N. The operation of ancient reclamation works at Lake Copais in Greece. *Water Hist.* **2015**, *7*, 271–287. [CrossRef]
32. Knauss, J. NE Kopais: Technical–Historical Aspects of the Unfinished Ancient Drainage Tunnel. *Proc. Soc. Boeotian Stud.* **1995**, *2*, 83–95.
33. Knauss, J. The prehistoric water management and land reclamation system in the Kopais-Basin, Boiotia, middle Greece. *ICID J.* **2000**, *49*, 39–48.
34. Frazer, J.G. *Pausanias's Description of Greece*; Macmillan and Co. Limited: New York, NY, USA, 1898.
35. Ghembaza, T.; Windell, D. Mysteries of Lake Copais: The Drainage–Massive Bronze Age and Hellenistic Hydraulic Engineering Works. *Open J. Stud. Hist.* **2021**, *4*, 67–84. [CrossRef]

36. Ahmed, A.T. Water quality for irrigation and drinking water use of Aflaj in Oman. *Water Sci. Technol. Water Supply* **2015**, *15*, 421–428. [CrossRef]
37. Al-Marshudi, A.S. Traditional Irrigated Agriculture in Oman. *Water Int.* **2001**, *26*, 259–264. [CrossRef]
38. Al-Marshudi, A.S. The falaj irrigation system and water allocation markets in Northern Oman. *Agric. Water Manag.* **2007**, *91*, 71–77. [CrossRef]
39. Pal, S.; Mukherjee, A.; Senapati, T.; Samanta, P.; Mondal, S.; Ghosh, A. Surface water quality assessment of abandoned opencast coal pit-lakes in Raniganj coalfields area, India. *Ecoscan* **2013**, *4*, 175–188.
40. Frumkin, A.; Shimron, A. Tunnel engineering in the Iron Age: Geoarchaeology of the Siloam Tunnel, Jerusalem. *J. Archaeol. Sci.* **2006**, *33*, 227–237. [CrossRef]
41. Reich, R.; Shukron, E. The date of the Siloam Tunnel reconsidered. *Tel Aviv* **2011**, *38*, 147–157. [CrossRef]
42. UNESCO. Pythagoreion and Heraion of Samos. In *UNESCO World Heritage Convention*; Retrieved 25 November 2022; United Nations Educational Scientific and Cultural Organization: Paris, France, 2022.
43. Kienast, H.J. *The Aqueduct of Eupalinos on Samos*; Archaeological Receipts Fund, Directorate of Publications: Athens, Greece, 2005.
44. Camp, J.M., II. *Drought and Famine in the 4th Century BC*; The American School of Classical Studies at Athens: Athens, Greece, 1982. Available online: http://www.jstor.org/stable/1353941 (accessed on 7 September 2023).
45. Crouch, D.P. *Water Management in Ancient Greek Cities*; Oxford University Press: Cary, NC, USA, 1993.
46. Tölle-Kastenbein, R. *Antike Wasserkultur*; Renate Beck: München, Germany, 1990.
47. Tölle-Kastenbein, R. *Das Archaische Wasserleitungsnetz für Athen und Seine Späteren Bauphasen*; von Zabern: Reverside, CA, USA, 1994.
48. Wikander, Ö. *Handbook of Ancient Water Technology*; Brill: Reverside, CA, USA, 2000; Volume 200.
49. Avgerinou, P. The Archaic Underground Aqueduct in Megara. In *Underground Aqueducts Handbook*; Angelakis, A.N., Chiotis, E., Eslamian, S., Weingartner, H., Eds.; CRC Press: Baton Rouge, LA, USA, 2016; pp. 44–48.
50. Lambrinoudakis, V. Ancient aqueduct in Naxos. In *IWA Regional Symposium on Water, Wastewater and Environment-Traditions and Culture*; Kalavrouziotis, I.K., Angelakis, A.N., Eds.; Hellenic Open University: Patras, Greece, 2014; pp. 453–459.
51. Moreno, A. *Feeding the Democracy: The Athenian Grain Supply in the Fifth and Fourth Centuries BC*; Oxford University Press: Oxford, UK, 2007.
52. Lohmann, H. *Atene: Forschungen zu Siedlungs-und Wirtschaftsstruktur des Klassischen Attika*; Cologne: Böhlau Verlag, Germany, 1993; Volume 1.
53. Krasilnikoff, J. Innovation in ancient Greek agriculture: Some remarks on climate and irrigation in Classical Attica. *Class. Mediaev.* **2014**, *64*, 95–116.
54. Ardito, F. L'emissario del Fucino. In *Città Sotterranee*; Mursia: Milano, Italy, 1990; pp. 44–49.
55. Burri, E. L'emissario Claudio Torlonia. *Speleologia* **1987**, *16*, 33–34.
56. Burri, E. Analisi topografica dell'emissario Claudio-Torlonia'. *Burri* **1994**, 234–261.
57. Burri, E. Problemi di conservazione, tutela e fruizione degli antichi emissari artificiali e sotterranei dei laghi endoreici dell'Italia centrale. In Proceedings of the Acts 1st Intentational Conference "Science and Technology for the Safeguard of Cultural Heritage in the Mediterranean Basin", Catania-Siracusa, Italy, 27 November–2 December 1995; pp. 1595–1601.
58. Döring, M. Der Emissar des Sees von Fucino. *Schr. Front. Ges.* **1995**, *19*, 81–110.
59. Castellani, V.; Caloi, V. L'emissario di Nemi (Roma): Aggiornamenti topografici. *Opera Ipogea* **2000**, *1*, 11–18.
60. Drusiani, R.; Bersani, P.; Penta, P. The ancient Lake Albano tunnel: Origins and considerations regarding the hydraulic regulation achieved. *Water Sci. Technol. Water Supply* **2007**, *7*, 269–276. [CrossRef]
61. Placidi, M. L'emissario del Lago di Nemi. *Archeol. Sotteranea* **2010**, *2*, 3–13.
62. Baykan, O.; Alkan, A.; Bacanli, Ü.G.; Baykan, N.; Öziş, Ü. Testing Flood Estimation Methods on Acient Closed Conduits. In Proceedings of the International Balkans Conference on Challenges of Civil Engineering, BCCCE, EPOKA University, Tirana, Albania, 19–21 May 2011.
63. Fabre, G.; Fiches, J.-L.; Marchand, G.; Mathieu, V.; Pey, J. Entre Gardon et Vistre, clausonne, l'étang, ses drainages et l'aqueduc antique de Nîmes. *Bull. L'école Atl. Nîmes* **2011**, *29*, 147–204.
64. Parise, M.; Galeazzi, C.; Germani, C.; Bixio, R.; Del Prete, S.; Sammarco, M. The map of ancient underground aqueducts in Italy: Updating of the project, and future perspectives. In Proceedings of the International Congress in Artificial Cavities "Hypogea 2015", Italy, Rome, 11–17 March 2015; pp. 235–243.
65. Burdy, J.; Lebouteiller, P. L'aqueduc romain de Xanthos. *Anatol. Atl. Eski Anadolu* **1998**, *6*, 227–248. [CrossRef]
66. Grewe, K. Die römische Wasserleitung nach Almunecar (Spanien). *Der Vermess.* **1983**, *34*, 217–221.
67. Grewe, K. Die römische Wasserleitung von Almuñécar. *Antike Welt* **1991**, *22*, 49–53.
68. López, E.S. El acueducto de sexi firmum iulium (almuñécar, granada). *Cuad. Prehist. Arqueol. Univ. Granada* **2011**, *21*, 127–158.
69. Ilacovac, B. Wasserbauliche Anlagen des Altertums am Kopaissee. *Mitteilungen Leichtweiss-Inst. Wasserbau Tech. Univ. Braunschw.* **1981**, *71*, 275–298.
70. Manenica, H. *Urbanizacija Između Raše i Krke u Vrijeme Ranog Principata*; Faculty of Philosophy in Zagreb Postgraduate Study of Archeology, University of Zagreb: Zagreb, Croatia, 2015.
71. Marasović, K.; Perojević, S.; Margeta, J. Roman Underground Hydraulic Structures in Dalmatia, Croatia. In *Underground Aqueducts Handbook*; CRC Press: Boca Raton, FL, USA, 2016; pp. 19–35.

72. Lightfoot, D.R. Syrian qanat Romani: History, ecology, abandonment. *J. Arid. Environ.* **1996**, *33*, 321–336. [CrossRef]
73. Lightfoot, D.R. Qanats in the Levant: Hydraulic technology at the periphery of early empires. *Technol. Cult.* **1997**, *38*, 432–451. [CrossRef]
74. Wilson, A. Water Management and usage Aksoy T (2015). Aqueducts of Nikomedeia (Izmit). *J. Black Sea Res. Inst.* **1997**, *1*, 187–239. Available online: https://dergipark.org.tr/tr/pub/karen/issue/17526/183261 (accessed on 7 September 2023).
75. Kremer, B. Wasserversorgung aus dem Tunnel: Der römische Qanat von Mehring. *Funde Ausgrab. Bez. Trier.* **1999**, *31*, 37–50.
76. Caesar, J. *The Gallic War*; Edwards, H.J., Ed.; Loeb Classical Library 72; Harvard University Press: Cambridge, MA, USA, 1917; Chapters 41–43.
77. Leveau, P. Les aqueducs d'Aquae Sextiae et la gestion de l'eau sur le territoire de la cité. *Cart. Archéologique Gaule* **2006**, *13*, 93–109.
78. Keenan-Jones, D. The Aqua Augusta and control of water resources in the Bay of Naples. In Proceedings of the Australasian Society for Classical Studies Conference, Perth, Australia, 21–22 May 2009; pp. 1–18.
79. Feo, G.D.; Lorenz, W.F. Route and tunnels of the Aqua Augusta for the water supply of Pompeii. *Int. J. Glob. Environ. Issues* **2015**, *14*, 177–186. [CrossRef]
80. Van Deman, E.B. *The Building of the Roman Aqueducts*; Martino Publications: Bologna, Italy, 1934.
81. Ashby, T.; Richmond, I.A. *The Aqueducts of Ancient Rome*; Oxford University Press: London, UK, 1935.
82. Aicher, P.J. *Guide to the Aqueducts of Ancient Rome*; Bolchazy-Carducci Publishers: Wauconda, IL, USA, 1995; 183p.
83. Moreno, G.I. Análisis técnico y constructivo del acueducto romano de Albarracín a Cella. In *Las Técnicas en la Ingeniería Romana—V Congreso de las Obras Publicas Romanas*; Traianvs Publisher: Madrid, Spain, 2010; pp. 225–248.
84. Burdy, J. L'aqueduc romain du Gier. In *Préinventaire des Monuments et Richesses Artistiques IV: Gier*; Conseil Général du Rhône: Lyon, France, 1996; 407p. [CrossRef]
85. Döring, M. Wasser für Gadara: 94 km langer antiker Tunnel im Norden Jordaniens entdeckt. *Querschnitt* **2007**, *21*, 26–37.
86. Cioli, D. Il tunnel dell'acquedotto romano di Saldae in Algeria. *Archeol. Sotteranea* **2016**, *13*, 3–23.
87. Garcia Merino, C. Avance al estudio del acueducto de Uxama. In *Nuevos Elementos de Ingeniería Romana*; Gallo, I., Ed.; III Congreso de las Obras Públicas Romanas: Madrid, Spain, 2006; pp. 167–194.
88. Kempe, S.; Al-Malabeh, A. A 100-km-Long subterraneous Roman Aqueduct in northern Jordan? In Proceedings of the 17th International Congess of Speleology, Sydney, NSW, Australia, 22–28 July 2017; Volume 2.
89. Döring, M. Qanat Fir'aun: An Underground Roman Water System in Syria and Jordan. In *Underground Aqueducts Handbook*; Angelakis, A.N., Chiotis, E., Eslamian, S., Weingartner, H., Eds.; 6000 BrokenSound Parkway NW, Suoite 300; CRC, Taylor & Francis Group: Boca Raton, FL, USA, 2017; pp. 173–196.
90. Alkan, A.; Özis, Ü. Çevlik historical water tunnel conversion system. In Proceedings of the Historical Water Structures Conference, Proceedings Book, Güneş Ofset, Instabul, Turkey, 26–27 June 2008.
91. Alkan, A.; Özis, Ü. Çevlik historical water tunnel conversion system. In Proceedings of the Historical Water Structures Conference (Regional Meeting for the 5th World Water Forum Preparation); Proceedings Book; Union of Chambers of Turkish Engineers and Architects, Chamber of Civil Engineers İzmir Branch, Istanbul, Turkey, 1–3 October 2015; pp. 93–96.
92. Döring, M. Die antiken Wasserbauten von Antiochia, Türkei. *Wasserwirtsch. J.* **2012**, *102*, 10–16. [CrossRef]
93. Lewis, P.R.; Jones, G.D. Roman gold-mining in north-west Spain. *J. Rom. Stud.* **1970**, *60*, 169–185. [CrossRef]
94. Matías, R. Las Médulas (León-España): El agua en la ingeniería de la mayor explotación minera del mundo antiguo. *Lancia* **2008**, *7*, 17–112.
95. Gupta, A. Stepwells: An architectural legacy of India. In *Prakash Books India*; 113 A Ansari Road, Daryaganj: New Delhi, India, 2007.
96. Arıkan, B. Istanbul's historical water channels and the transmission of water through aqueducts. In Proceedings of the 5th Symposium on Strengthening Historical Artifacts and Handing Them down to the Future, Union of Chambers of Turkish Engineers and Architects, Chamber of Civil Engineers İzmir Branch, Erzurum, Turkey, 1–3 October 2015.
97. Crow, J. The Water Supply System of Byzantine Constantinople. A History of Istanbul. 2023. Available online: https://istanbultarihi.ist/554-the-water-supply-of-byzantine-constantinople (accessed on 21 August 2023).
98. Çeçen, K. *The Longest Roman Water Supply Line*; The Industrial Development Bank: Istanbul, Turkey, 1996; 230p.
99. Çeçen, K.; Kolay, C. *Ottoman Era Waterways of Istanbul*; The Industrial Development Bank: Istanbul, Turkey, 2000; 336p.
100. Avcı, I. *A Prominent Element in the Historical Development Process of Istanbul: Water*; Turkish Engineering News: Instabul, Turkey, 2001.
101. Crow, J.; Bardill, J.; Bayliss, R. *The Water Supply of Byzantine Constantinople*; Society for the Promotion of Roman Studies Publisher: Instabul, Turkey, 2008.
102. ISKI. *Management of Water in Istanbul from Past to Present*; ISKI (Istanbul Water and Sewerage Administration) Publications: Istanbul, Turkey, 2009.
103. Yalgin, E. Perge Ancient City. 2023. Available online: https://www.rotasenin.com/perge-ancient-city/ (accessed on 13 August 2023).
104. Korkanç, M. Tyana aqueducts and stones. *Mavi Gezegen Pop. Earth Sci. J.* **2019**, *26*, 49–53.
105. Kahya, E. Cultural tourism in and around the city of Niğde. In Proceedings of the 13th International Congress on Contemporary Research in Social Sciences, Istanbul, Turkey, 6–8 November 2020; pp. 1568–1592.

106. Çeçen, K. Kırkçeşme Waters. TDV Islamic Encyclopedia. 2022. Available online: https://islamansiklopedisi.org.tr/kirkcesme-sulari (accessed on 7 September 2023).
107. Öziş, Ü. Water Works Through Four Millenia in Turkey. *Environ. Process.* **2015**, *2*, 559–573. [CrossRef]
108. Creswell, K.A.C. *Muslim Architecture of Egypt*; Hacker Art Books: New York, NY, USA, 1978.
109. Williams, C. *Islamic Monuments in Cairo: The Practical Guide*; American University in Cairo Press: Cairo, Egypt, 2008.
110. Karakaya, E. Aqueducts. TDV Islamic Encyclopedia. 2009. Available online: https://islamansiklopedisi.org.tr/su-kemeri (accessed on 1 May 2009).
111. Arıkan, B.; Fuat, A.B. Historical aqueducts of Istanbul and dynamic analysis of the Pasha Kemeri aqueduct. In Proceedings of the International Conference on Earthquake Engineering, Skopje, Republic of Macedonia, 29 May 2013.
112. Zehir, C. The historical development of the ottoman water civilization. In Proceedings of the Ottoman Civilization International Symposium, Istanbul, Turkey, 5–8 May 2000; pp. 173–181.
113. Eroğlu, V. Springs that have supplied water to Istanbul throughout history. In Proceedings of the International Symposium on Ottoman Civilization, Istanbul, Turkey, 5–8 May 2000.
114. Öziş, Ü.; Arısoy, Y.; Alkan, A.; Özdemir, Y. General situation of historical aqueducts in Turkey, Civil Engineers Ankara Branch. In Proceedings of the Consolidation of Historical Artifacts and Their Safe Transfer to the Future Symposium-1, Kardelen Ofset Ltd. Şti, Ankara, Turkey, 27–29 September 2007; pp. 555–565.
115. Özer, M.; Dündar, M. Edirne Sarayı Su Yapıları. *Belleten* **2021**, *85*, 615–644. [CrossRef]
116. Zeybekoğlu, D.; Çakır, H.K.; Özenç, A. Analysis of water scales in Edirne. *Trak. Univ. J. Sci.* **2007**, *8*, 29–33.
117. Şahin, S. Istanbul water structures. In Proceedings of the 4th International Sinan Symposium: Water and Architecturel, Edirne, Turkey, 1–11 April 2008.
118. Cangül, C. Avasköy Aqueduct. Culture Inventory. 2021. Available online: https://kulturenvanteri.com/yer/?P=6952 (accessed on 1 May 2021).
119. Yenigün, K.; Aydoğdu, M.H. General Evaluation of Irrigation and Drainage Systems Within the Scope of GAP. *Turkey's Larg. Integr. Water Resour. Proj. Water Resour.* **2008**, *1*, 12–32.
120. Yeşilnacar, M.İ. *Injection Works in Şanlıurfa Tunnels*; General Directorate of State Hydraulic Works (DSİ) Technical Bulletin: Ankara, Turkey, 2005; Volume 100, pp. 1–8. Available online: https://cdniys.tarimorman.gov.tr/api/File/GetGaleriFile/425/DosyaGaleri/600/tb100.pdf (accessed on 8 May 2005).
121. Wessels, J.I.; Vardakos, S.; Weingartner, H.; Eslamian, S.; Angelakis, A.N. Underground aqueducts: Past, present, and future trends. In *Underground Aqueducts Handbook*; CRC Press: Boca Raton, FL, USA, 2016; pp. 491–509.
122. Grandori, R.; Jaeger, M.; Antonini, F.; Vigl, L. Evinos-Mornos Tunnel-Greece. Construction of a 30 km long hydraulic tunnel in less than three years under the most adverse geological conditions. In Proceedings of the Rapid Excavation and Tunneling ConferenceRECT, San Francisco, Society for Mining, Metallurgy, and Exploration, San Francisco, CA, USA, 18–21 June 1995; pp. 747–767.
123. Xenos, D.; Passios, I.; Georgiades, S.; Parlis, E.; Koutsoyiannis, D. Water demand management and the Athens water supply. In *Water Quality Technologies and Management in Bulgaria, Proceedings of the 7th BNAWQ Scientific and Practical Conference, Sofia, Bulgaria, 20–22 February 2002*; Bulgarian National Association on Water Quality: Sofia, Bulgaria, 2002; pp. 44–50. [CrossRef]
124. Department of Environmental Protection—NYC. Announces Major Milestone for Delaware Aqueduct Repair as Tunneling Machine Completes Excavation. New York City Department of Environmental Protection. 2019. Available online: https://www.nyc.gov/site/dep/news/19-062/dep-major-milestone-delaware-aqueduct-repair-tunneling-machine-completes#/0 (accessed on 7 September 2023).
125. Sheer, J. Metropolitan Area Outer Underground Discharge Channel. AFAR. 2022. Available online: https://web.archive.org/web/20210422165524/https://www.afar.com/places/metropolitan-area-outer-underground-discharge-channel-kasukabe?contex (accessed on 7 September 2023).
126. Kashgary, J. Chinese Irrigation Tunnel Project in Xinjiang Hits Snag: Too Much Water. Radio Free Asia. 2022. Available online: https://www.rfa.org/english/news/uyghur/tunnel-project-02042022142725.html (accessed on 7 September 2023).
127. Robert, J. Gravity Mapping and Seismic Imaging of Paleochannels on Large Tunnel Routes in Sydney, Australia. In *Near-Surface Geophysics Book*; Society of Exploration Geophysicists: Houston, TX, USA, 2005; pp. 503–512. [CrossRef]
128. Biggart, A. A Quiet Revolution. In *Tunnels and Tunnelling International*; Patrick Reynolds; British Tunneling Society: London, UK, 1999; Volume 31, pp. 18–22.
129. Mitani, S. The state of art of TBM excavation and probing ahead technique. In *Engineering Geology: A Global View from the Pacific Rim, Proceedings of the 8th International Congress of the IAEG, Vancouver, BC, Canada, 21–25 September 1998*; CRC Press: Boca Raton, FL, USA, 1998; pp. 3501–3512.
130. USNCTT—US National Committee on Tunnel Technology. *Geotechnical Site Investigations for Underground Projects*; National Academy Press: Washington, DC, USA, 1984.
131. Ishikawa, K.; Feng, S.; Sugiyama, T. Identifying fractured zones using high density electrical prospecting. In Proceedings of the 1st International Workshop on the Application of Geophysics to Rock Engineering, New York, NY, USA, 29 June 1997; Columbia University: New York, NY, USA, 1997; pp. 50–57.

132. Parker, C.; Whiteley, B.; Gee, B. Innovative investigation techniques used for geotechnical modelling of the Northside storage tunnel project, Sydney. In Proceedings of the 10th Australian Tunnelling Conference, Melbourne, VIC, Australia, 21–24 March 1999; pp. 193–205.
133. Khan, M.; He, X.; Farid, A.; Song, D.; Li, Z.; Tian, X.; Ni, M. A novel geophysical method for fractures mapping and risk zones identification in a coalmine, Northeast, China. *Energy Rep.* **2021**, *7*, 3785–3804. [CrossRef]
134. Day-Lewis, F.D.; Slater, L.D.; Robinson, J.; Johnson, C.D.; Terry, N.; Werkema, D. An overview of geophysical technologies appropriate for characterization and monitoring at fractured-rock sites. *J. Environ. Manag.* **2017**, *204*, 709–720. [CrossRef]
135. Golpasand, M.-R.B.; Do, N.A.; Dias, D. Impact of pre-existent Qanats on ground settlements due to mechanized tunneling. *Transp. Geotech.* **2019**, *21*, 100262. [CrossRef]
136. Manuel, M.; Lightfoot, D.; Fattahi, M. The sustainability of ancient water control techniques in Iran: An overview. *Water Hist.* **2018**, *10*, 13–30. [CrossRef]
137. Abouei, R. Conservation of Badgirs and Qanats in Yazd, Central Iran. In Proceedings of the 23rd Conference en Passive and Low Energy Architecture, Geneva, Switzerland, 6–8 September 2006; pp. 6–8.

sustainability

MDPI

Article

Dynamic Simulation Model of Channel Leakage Based on Multiple Regression

Jianqin Ma, Jiangshan Yang *, Xiuping Hao, Bifeng Cui and Shuoguo Yang

Department of Agricultural Hydraulic Engineering, Faculty of Water Resources, North China University of Water Resources and Electric Power, Zhengzhou 450000, China; majianqin@ncwu.edu.cn (J.M.); haoxiuping@ncwu.edu.cn (X.H.)
* Correspondence: z20211010093@stu.ncwu.edu.cn

Abstract: Aiming at the problem that the existing channel leakage calculation methods generally ignore the dynamic changes of influencing factors, which leads to a large calculation error, this study attempts to utilize the machine learning method to accurately calculate the channel leakage loss under the dynamic changes in the influencing factors. By using the machine learning method to analyze the impact of dynamic changes in the flow rate and soil moisture content over time on the channel leakage loss in the water transmission process and quantify the impact of the selected factors on the leakage loss, a dynamic simulation model of the multi-parameter channel leakage loss was constructed, and a test was carried out in the irrigation area to verify the accuracy of the model. The test results are as follows: the actual leakage loss of the U1 channel is 1094.03 m^3, the simulated value of the model is the 1005.24 m^3, and the error between the simulated value and the measured value is 8.12%; the total leakage of the U2 channel is 1111.24 m^3, the simulated value of the model is 1021.1 m^3, and the error between the simulated value and the measured value is 6.31%. The experimental results show that the use of machine learning to construct a dynamic simulation model of channel leakage loss under the comprehensive consideration of the dynamic change in influencing factors over time has a better effect, and the calculation accuracy is high.

Keywords: channel; leakage loss; machine learning; multifactor; leakage test

Citation: Ma, J.; Yang, J.; Hao, X.; Cui, B.; Yang, S. Dynamic Simulation Model of Channel Leakage Based on Multiple Regression. *Sustainability* **2023**, *15*, 14904. https://doi.org/10.3390/su152014904

Academic Editor: Andrea G. Capodaglio

Received: 19 August 2023
Revised: 28 September 2023
Accepted: 12 October 2023
Published: 16 October 2023

1. Introduction

At present, the contradiction of water use in China's irrigation areas is becoming more and more acute, and problems such as serious waste of irrigation water resources and low utilization efficiency still exist. According to statistics in 2020, the water utilization coefficient of the backbone canal system in China's large irrigation districts was only 0.643 [1]. It has been shown that leakage from canals at the dry and branch levels is an important cause of irrigation water losses during the irrigation process [2]. Because the channel leakage rate is affected by flow rate, water depth, wetted perimeter, channel length, lining conditions, soil factor, water transfer time and other factors, and the current practice of using various types of empirical formulas to estimate the leakage loss of different types of channels, but the empirical formulas are generally only selected flow rate, depth of water, wet week, and flow rate of one or more influencing factors as a variable to calculate the leakage amount of the channel. Considering fewer factors and ignoring the dynamic changes of each factor over time, this leads to a large error between the results of empirical formulas and the actual leakage loss, which affects the accurate management of water resources in irrigation districts. Therefore, it is of great significance to construct a multi-parameter dynamic formula for calculating the channel leakage rate to accurately calculate the channel leakage loss and to strengthen the strict management of water resources in irrigation districts in China.

The research methods of channel leakage loss mainly focus on empirical formulas, numerical simulation [3], and two aspects. In the research of empirical formulas of channel leakage, foreign scholars put forward a variety of empirical formulas, including the Davis-Wilson formula, Kosgakov formula, etc., where the Davis-Wilson formula, which uses the wetted perimeter of the channel, water level, and flow rate as variables, is mostly used to calculate channels with liners. However, Kausgakov's formula, which uses flow and soil as variables, is mostly used to make calculations for leakage in earthen channels [4–7]; due to the simplicity and ease of use of empirical formulas, various types of empirical formulas are generally used in practice nowadays to calculate leakage loss losses [8–11]. However, the empirical formulas generally use only one or several channel physical properties and different permeability coefficients as variables for calculation, with fewer factors to consider and poorer calculation accuracy. Shah et al. [9] and Zhang et al. [10] showed that the calculated values of Kausgakov's formula were 1.5- and 2.5-times higher than the actual values, and Akkuzu found that the leakage losses estimated by Moritz's and Davidson-Wilson formulas are much lower than the measured values [11], The empirical formula has been improved by some scholars in China. Men Baohui [12] and others improved the empirical formulas by using the method of integration; Xie Chongbao [13] and others combined different empirical formulas to propose improved empirical formulas that are related to both channel cross-section size and soil type; Wang Bingchuan [14] et al. derived the improved formula by integrating Korsgakov's formula. It is found that the numerical simulation method to calculate the leakage loss of the channel has a better effect, and the object of the study can be either the whole channel or part of the channel section, which has a wider range of applications. The use of computer programming calculations can be quickly obtained as results, the calculation accuracy is higher than the empirical formula, intuition, better expandability than the field test, easy to understand, and later analysis. In the numerical simulation study of channel leakage loss, Zhang Fan [15] et al. proposed a method for estimating channel leakage loss by using a statistical method. The established numerical simulation model is about 10% more accurate than traditional leakage calculation methods. Liao Xiangcheng [16] et al. introduced the concept of pre-influence water content of channel soil, and they proposed a method for calculating the permeability coefficient and index of the soil of the canal bed. By improving the calculation formula, the leakage loss calculation error was reduced to less than 20%. At present, most of the research on channel leakage loss stays in static calculation, and the way and degree of influence of the dynamic changes of each influencing factor on the channel leakage loss have not been clarified, which affects the precise control of the channel water transfer process in irrigation districts.

In this study, we attempted to use a machine learning method to calculate the channel leakage loss under the dynamic change of influencing factors by constructing a dynamic simulation model of multi-parameter channel leakage loss. In this study, we will quantify the influence of each influential factor on leakage loss by comprehensively considering the dynamic changes of some factors in the channel water delivery process, compare with the empirical formulas by constructing a machine learning model, and further explore the changes of channel leakage loss in different time periods by combining numerical simulation and field tests in a long time series, and the results of this study will help the irrigation district to clarify the leakage status of each channel and provide a basis for realizing precise control of channel water transfer.

2. Materials and Methods

2.1. Research Area

Wharf irrigation area district is located in the southeast of Linyi City, Shandong Province (Figure 1). The designed irrigation area is 19,667 hectares, belonging to the temperate monsoon zone of semi-humid transitional climate. The average annual temperature of the irrigation area is 13.3 °C, the annual sunshine hours are 2460 h, and the average rainfall is 840.3 mm per year. The main crops are wheat, rice, corn, and peanuts. There are 3 trunk canals and 24 branch canals in the irrigation area, and the types of canals are unlined earth canals, among which there are 12 branch canals under the first trunk canal, controlling the irrigated area of 13,367 hectares. There are 8 branch canals under the second trunk canal, controlling an irrigated area of 4633 hectares, and there are 4 branch canals under the third trunk canal, controlling an irrigated area of 1667 hectares. The irrigation district canal system project was built a long time ago. The aging and degradation problem is serious, a lot of the canal section channel has collapsed, siltation is present, channel water transfer capacity is seriously insufficient, irrigation water utilization coefficient is only 0.44 or so, and irrigation water resources are seriously wasted.

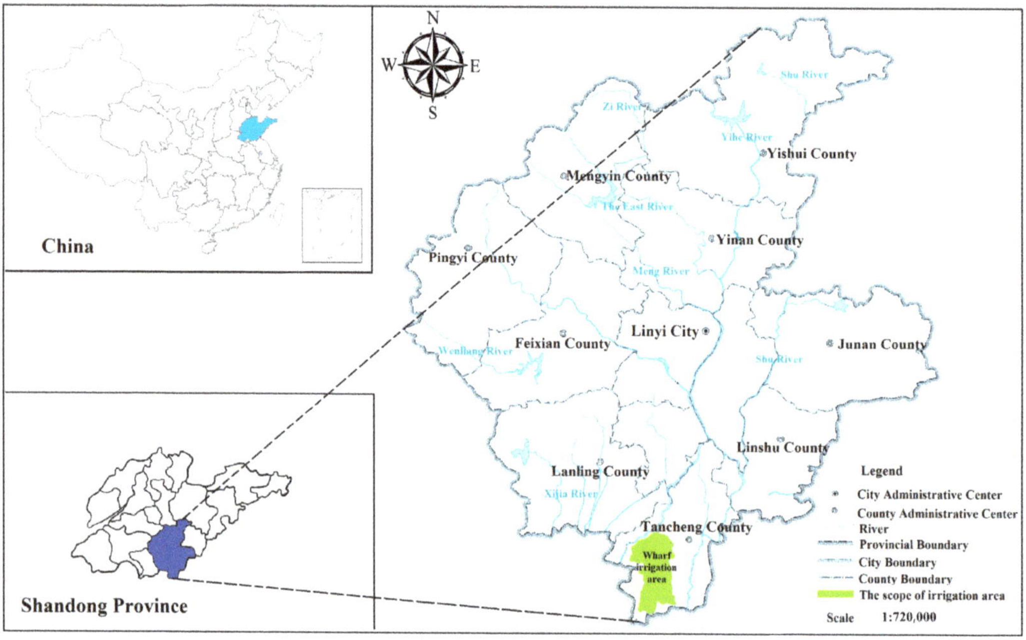

Figure 1. Geographical location of irrigation district.

2.2. Pilot Program

In this study, two sections of channels with pile numbers 7 + 320 to 8 + 520 and 14 + 070 to 15 + 270 were selected as test channels for dynamic simulation of seepage loss in 1 main canal. Among them, the design flow rate of the channel in the section from pile No. 7 + 320 to 8 + 520 is 14.7 m^3/s, the soil type is medium loam with average permeability, and the design flow rate of the channel in the section from pile No. 14 + 070 to 15 + 270 is 13.9 m^3/s, and the soil type is light loam with high permeability. Both sections of the channel are 1 km long, trapezoidal cross-section unlined earth channels; the bottom of the channel is relatively flat, and the length of the channel meets the requirements of the test. Detailed information of the test channel is shown in Table 1. The object of this study is the channel as a whole, without distinguishing the differences between the various segments of the channel, and we only need to measure the leakage loss of the test channel as a whole, so the test selected the dynamic water method to measure the real leakage loss of the channel, through the channel to the test channel, and measure the upstream and downstream flow rate loss within a specified period of time to calculate the actual leakage loss of the channel, and the actual leakage loss of the channel is calculated. This test was set up in the upstream and downstream speed measurement section 8 speed lines, with speed measurement method using the five-point method. Five points were set up near the water surface, 0.2-times the lateral water depth, 0.6-times the lateral water depth, and 0.8-times the lateral water depth. Near the bed of the canal, the length of the flow velocity meter speed measurement was set to 80 s, the test lasted for 15 h of water conveyance, and the measurement interval was 1 h.

Table 1. Detailed information of experimental channels.

Channel Number	Length (m)	Bottom Width (m)	Superelevation (m)	Flow Velocity (m^3/s)	Designed Discharge (m^3/s)	Soil Texture	Water Permeability	Roughness Factor	Gradient	Cross-Section Form
U1	1000	6	0.5	0.852	14.7	medium frequency transformer loam	general	0.02	1/5000	trapezium
U2	1000	6	0.5	0.827	13.9	light flux loam	strong	0.02	1/5000	trapezium

2.3. Data Sources

2.3.1. Soil Data

The soil medium is defined as the soil present in the uppermost layer of the ground. The top weathered portion of the unsaturated zone with substantial biological activity is represented by the soil medium [17]. It has been shown that different soils have different permeability [18,19], and for more clay-heavy soils, the smaller the pore space between the soil particles, the coarser the soil texture; the larger the pores between the soil particles, the stronger the gravitational force on the soil, and the greater the rate of soil leakage [20]. The soil in the irrigation area is mostly sandy loam, with soil particle sizes ranging from 0.02 mm to 0.2 mm, with soil pore ratios ranging from 0.79 to 0.87, vertical permeability coefficients ranging from 0.00019 to 0.00026, and sand content, percolation losses, water retention, and aeration properties being relatively average. The data were obtained from the local irrigation district administration.

2.3.2. Channel Data

Channel data were obtained from local irrigation district authorities (Table 2).

Table 2. Irrigation district channel information.

Name of Branch Canal	Length (m)	Cross-Section Form	Bottom Width (m)	Depth (m)
No.1 branch canal	7320	trapezium	6	0.5
No.2 branch canal	3270	trapezium	6	0.5
No.3 branch canal	2250	trapezium	6	0.5
No.4 branch canal	1100	trapezium	6	0.5
No.5 branch canal	60	trapezium	6	0.5
No.6 branch canal	4550	trapezium	6	0.5
No.7 branch canal	3930	trapezium	6	0.5
No.8 branch canal	1250	trapezium	6	0.5
No.9 branch canal	100	trapezium	6	0.5
No.10 branch canal	2640	trapezium	5	0.5
No.11 branch canal	10	trapezium	5	0.5
No.12 branch canal	470	trapezium	5	0.5

2.4. Model Building

Meta-regression is a statistical method based on mathematical statistics to find an approximate mathematical expression to describe the correlation between several variables. The establishment process includes three parts: regression factor correlation analysis, model establishment, and goodness-of-fit test. The regression model of random variable Y and general variable X can be expressed as:

$$Y = \beta_0 + \beta_1 X_1 + \beta_2 X_2 + \ldots + \beta_k X_k \quad (1)$$

In the formula, β_0 is a constant term; β_1, β_2, ... β_k are regression coefficients; Y is the dependent variable; X_1, X_2, ... X_k are k precisely measurable independent variables.

Correlation analysis is used to determine whether there is a relationship between two or more variable elements, the form of expression of the correlation relationship, the closeness, and the direction of the correlation relationship. It is a method for analyzing causal variables and expressing them with indicators.

$$r_{xy} = \frac{S_{xy}}{S_x S_y} \tag{2}$$

In the formula, r_{xy} is the sample correlation coefficient; S_{xy} is the sample covariance; S_x is the sample standard deviation of X; S_y is the sample standard deviation of y.

The goodness of fit of the model can be tested using the adjusted coefficient of determination $\overline{R^2}$:

$$R^2 = \frac{SSR}{SST} = \frac{SST - SSE}{SST} \tag{3}$$

$$\overline{R^2} = 1 - \frac{\frac{SSR}{n-k-1}}{\frac{SST}{n-1}} = 1 - R^2 \times \left(\frac{n-1}{n-k-1}\right) \tag{4}$$

In the formula, R^2 is the sample determinable coefficient; $\overline{R^2}$ is the adjusted coefficient of determination; SST is the sum of square of total deviation; SSR is the sum of regression squares; SSE is the sum of squared residuals; the value of the coefficient $\overline{R^2}$ is $0 \leq \overline{R^2} \leq 1$. The closer the R^2 value is to 1, the higher the goodness of fit of the equation, and the better the model effect.

2.4.1. Correlation Analysis of Influencing Factors

In the actual water conveyance process, factors, including soil factors, channel characteristics, lining conditions, nature of lining materials [21], groundwater, time factors, specific physical methods [22], and other factors, will have an impact on the channel leakage loss, and these factors are related to each other, interact with each other, and it is difficult to make a clear distinction in practice [23]. Figure 2 shows the results of the correlation analysis of leakage influencing factors selected in this study.

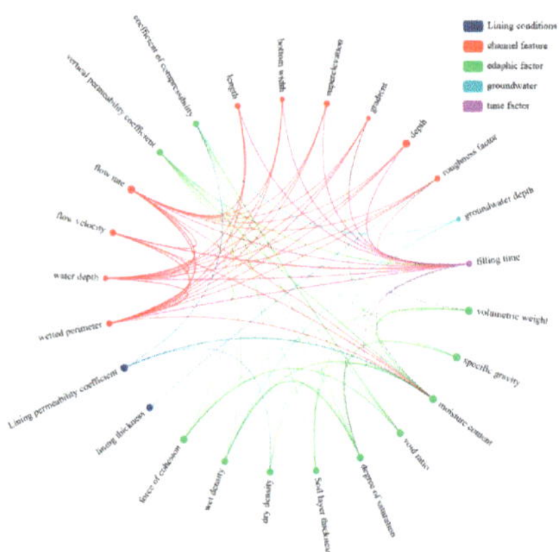

Figure 2. Influencing factors of channel leakage.

In this study, based on the existing channel water transmission data in the irrigation area, the physical properties of the channel were analyzed and selected to include five factors, namely, channel length, bottom width, superelevation, gradient, and roughness. The hydraulic properties included three factors, namely, water level, flow rate, and groundwater level. The soil factors were selected to include three factors, namely, void ratio of the soil, vertical permeability coefficient, and soil water content, with a total of eleven factors to be analyzed in the correlation analysis. The results of the bivariate Pearson test for correlation between factors are shown in Figure 3. The results of correlation analysis show that, except for the two factors of channel length and bottom width, all other factors have strong correlation with seepage loss, so multiple regression analysis can be carried out.

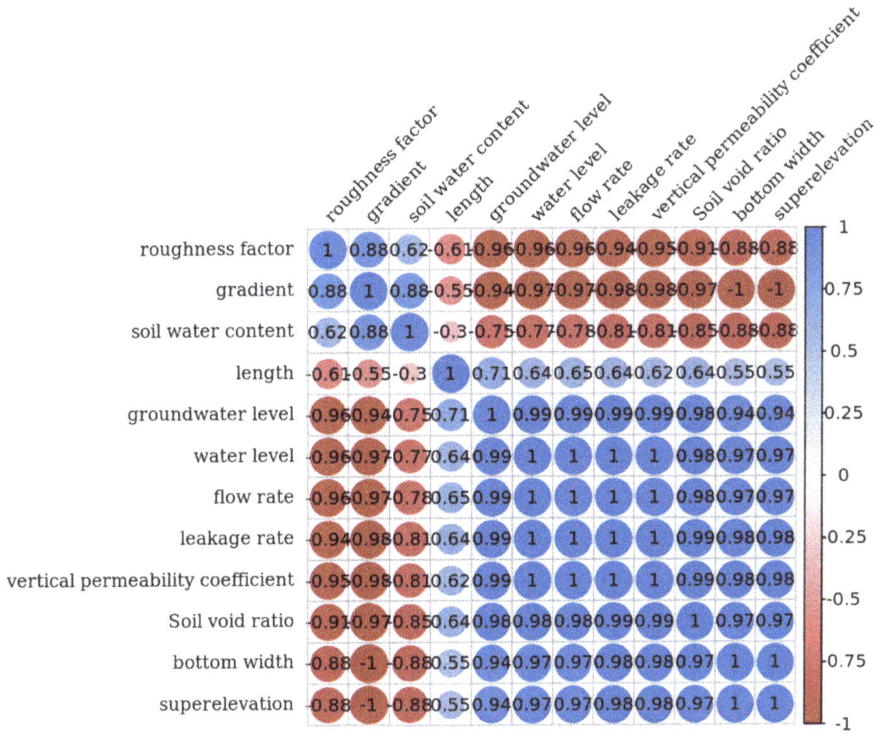

Figure 3. Correlation analysis.

2.4.2. Dynamic Simulation Model of Channel Leakage Rate Based on Multiple Regression

Table 3 shows an attempt to model multiple regressions based on available channel water delivery data. In a multiple regression model, the magnitude of the regression coefficient indicates the magnitude of the effect of the factor on the regressed factor, i.e., the degree of sensitivity of the factor. After analyzing model 3 through the test factor containing channel physical conditions, water factor, and soil moisture, three parts of the factors, in line with the requirements of the selection of factors in this study, selecting the factors with significance less than 0.05, can be seen. The vertical seepage loss coefficient has the greatest impact on seepage loss, the regression coefficient of 89.54, followed by the gradient, roughness factor, the soil void ratio, soil water content, water level, groundwater level, the flow rate, and regression coefficients (46.1030, 0.7470, −0.0370, −0.0320, 0.0070, −0.0010, and 0.0004, respectively). Therefore, in this study, the optimal linear regression equation was established with the channel leakage loss (Y) as the dependent variable, and gradient, roughness factor, water level, flow rate, groundwater level, soil void ratio, vertical permeability coefficient, and soil water content as the regressor as follows: $Y = -0.004 + 89.54X_0 + 46.103X_1 + 0.747X_2 - 0.037X_3 - 0.032X_4 + 0.007X_5 - 0.001X_6 + 0.0004X_7$.

Table 3. Multiple standard regression analysis of channel leakage rate and its influencing factors.

Model		Non-Standardized Coefficient		Standardized Coefficient	t	Significance
		B	STDERR	Beat		
1	(constant)	0.011	0.368	-	28.682	0.689
	length	3.14×10^{-5}	0.598	0.037	0.366	0.723
	flow rate	0	0.211	0.944	9.25	0.587
	soil water content	0.219	0.109	0.293	2.017	0.083
	water level	0.015	0.852	0.188	0.124	0.006
2	(constant)	0.009	0.027	-	0.343	0.764
	length	5.02×10^{-4}	0.397	−0.06	−0.399	0.728
	superelevation	−0.034	0.029	−0.808	−1.188	0.357
	roughness factor	0.47	0.259	0.627	1.816	0.211
	drawdown	−0.023	0.246	−0.032	−0.095	0.933
	flow rate	0.001	0.514	1.603	2.027	0.18
3	(constant)	−0.004	0.01	-	−0.355	0.029
	length	2.54×10^{-4}	0.652	−0.03	−0.273	0.83
	gradient	46.103	22.298	1.081	2.068	0.047
	roughness factor	0.747	0.252	0.997	2.963	0.039
	water level	0.007	0.004	1.138	1.66	0.031
	flow rate	0.0004	0.323	0.847	1.152	0.014
	groundwater level	−0.001	0.981	−0.178	−1.027	0.048
	soil void ratio	−0.037	0.028	−0.557	−1.303	0.028
	vertical permeability coefficient	89.54	41.917	1.218	2.136	0.020
	soil water content	−0.032	0.019	−0.292	−1.652	0.033

3. Results

3.1. Parametric Simulation Results

The reliability of the developed multiple regression model was verified and tested for significance using SPSS 27 software, and Table 4 shows the model test table. From Table 4, the adjusted R^2 is 0.982, and the model Durbin–Watson coefficient is 2.732, which indicates that the model has a good regression effect.

Table 4. Model verification.

Model Summary					
Model	R	R^2	Adjusted R^2	Errors in Standard Estimates	Debin-Watson Coefficient
1	0.999	0.998	0.982	0.000262	2.732

3.2. Experimental Validation

In the actual water transfer process, the two factors of channel flow and soil water content have the characteristics of changing with the change in the water transfer time. This leads to the fact that ignoring the dynamic characteristics of channel leakage losses in the calculation can lead to a large error between the calculation results and the actual leakage rate. Figure 4 shows the flow rate in the channel during the test period measured using the flow meter and the soil moisture content from 10 cm to 20 cm in the channel measured via the real-time monitoring system in the field. The analysis shows that soil water content and flow rate change significantly with water delivery time in the two test channels. The flow varied between 13.14–13.158 $m^3 \cdot s^{-1}$ and 13.139–13.156 $m^3 \cdot s^{-1}$, and the soil water content varied between 0.135–0.241 $m^3 \cdot m^{-3}$ and 0.119–0.24 $m^3 \cdot m^{-3}$, respectively. The results of this measurement are consistent with the results of a study by Ruixuan Li [24] et al.

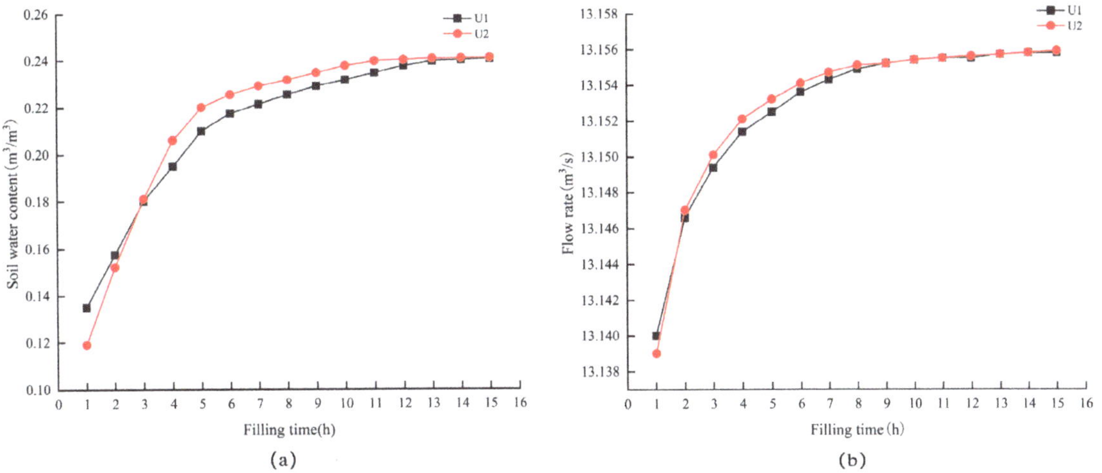

Figure 4. Dynamic monitoring of selected factors: (**a**) dynamic monitoring of soil water content; (**b**) dynamic monitoring of flow.

3.2.1. Dynamic Simulation of Channel Leakage Rate

The data on gradient, roughness, flow rate, and soil moisture content are brought into the regression model to calculate the simulated value of the seepage rate. The actual leakage rate of the test channel was measured using the dynamic water method. In addition, in order to verify the accuracy of the model, the Kausgakov empirical formula to calculate the channel leakage loss was selected as a comparison, and the comparison of the actual leakage rate, the model simulation value, and the calculated value of the empirical formula is shown in Figure 5.

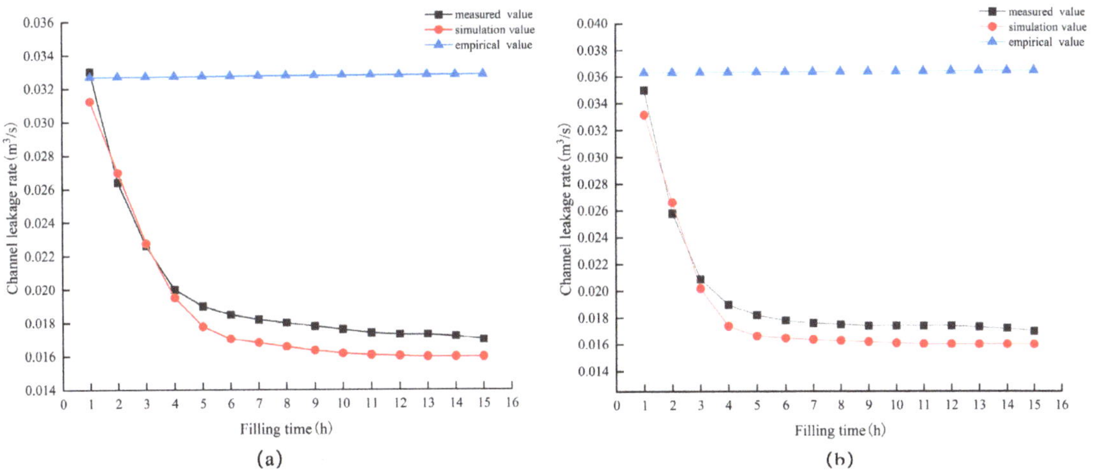

Figure 5. Dynamic simulation of leakage rate: (**a**) dynamic simulation of leakage rate in channel U1; (**b**) dynamic simulation of leakage rate in channel U2.

From the results of the calculations, it is clear that there is a large difference between the channel leakage rates calculated using model simulations and empirical formulas, Kausgakov's empirical formula for calculating seepage rate using soil permeability parameters, soil permeability index, and flow rate. Due to the empirical formulas consideration of fewer factors and ignoring the dynamic changes in the influencing factors, the result was the calculation of its results and the actual leakage rate, being closer only at the beginning of the water transfer. Afterwards, with the increase in the water transmission time, some of the factors affecting leakage change dynamically, and the error between the calculated value of the empirical formula and the actual leakage rate becomes bigger gradually. At the beginning of water transfer U1, the U2 channel empirical formula seepage rate calculation value and the actual seepage rate error are only 1.01% and 3.66%. After that, with the increase in the water transmission time, some of the factors affecting leakage change dynamically, and the error between the calculated value of the empirical formula and the actual leakage rate gradually becomes larger; the error reaches 92.98% and 114.31% at 15 h.

The model simulation value and the actual seepage rate change trend are more consistent. They are presented in the early stage of water transmission seepage rate with a rapid decrease, slowly declining in the middle of the seepage rate and gradually tending to stabilize the change trend. The leakage rate of the U1 channel decreases rapidly from 1 h to 5 h, and after 5 h, the leakage rate decreases gradually and finally reaches a steady seepage state around 10 h. The average relative error between the simulated and measured values was 6.45%. The U2 channel shows a faster decreasing trend in terms of the change in seepage rate than the U1 channel in 1–4 h due to better soil permeability. The seepage loss gradually decreases after 4 h and finally reaches a steady seepage state around 7 h, and the average relative error between the simulated and measured values is 7.04%.

3.2.2. Leakage Loss Simulation

Based on the dynamic simulation model of channel leakage rate, the regression equation was transformed by fitting the flow rate and soil water content elements into a function about time t using the data fitting method. Using SPSS software to fit the data to the flow rate and soil moisture content gives the fitted equation for soil moisture content in channel U1 as follows: $I1 = 0.162 + 0.0315t - 0.0027t^2 + 0.000078079t^3$. The equation fitted to the soil water content of channel U2 is $I2 = 0.0794 + 0.0452t - 0.0042t^2 + 0.00013092t^3$, where I is the soil water content and t is the water delivery time. The U1 channel flow fitting equation is as follows: $Z1 = 13.1362 + 0.0056t - 0.00053505t^2 + 0.000016834t^3$. The U2 channel flow fitting equation is $Z2 = 13.1349 + 0.0065t - 0.00066221t^2 + 0.000021699t^3$, where Z is the channel flow and t is the water delivery time.

Bringing the fitted equation into the U1 channel leakage loss multiple regression equation transforms the regression equation into $Y1 = 0.02014048 + 0.00100576 + 0.000086186t^2 - 0.00000249179t^3$. The multiple regression equation for U2 channel leakage loss can be transformed into $Y2 = 0.02278316 + 0.0014438t + 0.000134135t^2 - 0.00000418076t^3$, where Y is the channel leakage rate and t is the water delivery time.

Establishing the integral model, the data will be brought into the calculation and can be obtained after the model simulation value, the model simulation value, and empirical formula calculated value. The actual leakage loss comparison is shown in Figures 6 and 7.

From the simulation results, it is clear that the simulated value of U1 channel leakage in 1–5 h is 328.8 m^3, and the actual leakage loss is 471.2 m^3; the error of both is 30.22%. The Kausgakov empirical formula results in 587.97 m^3, which is 24.78% error from the actual leakage loss. In 5–10 h, the leakage simulation value is 296.22 m^3, the actual leakage loss is 312.86 m^3, and the error of both is 5.32%. The Kausgakov empirical formula calculates the result as 587.97 m^3, and the actual leakage loss error is 87.93%. At the end of the 10–15 h, the leakage simulation value is 290.76 m^3, the actual leakage loss is 309.97 m^3, and the error of both is 6.2%. The Kausgakov empirical formula calculated results for 587.97 m^3, and the actual leakage loss error is 89.69%. The U2 channel in the 1–4 h leakage simulation value is 295.83 m^3, the actual leakage loss ois425.72 m^3, and the error of the two is 27.8%. The Kausgakov empirical formula calculated results for 491.56 m^3, and the actual leakage loss error is 21.16%. In 4–7 h, the leakage simulation value is 197.14 m^3, the actual leakage loss is 192.03 m^3, and the error of the two is 2.66%. The Kausgakov empirical formula calculates the results as 320.58 m^3, and the actual leakage loss is 66.94% error. At the end of 7–15 h, the leakage simulation value is 505.34 m^3, the actual leakage loss is 513.49 m^3, and the error of both is 1.59%. The Kausgakov empirical formula calculates 854.89 m^3, and the actual leakage loss error is 66.49%. From the simulation results, it can be seen that the established model has a good effect on the simulation of seepage loss in all seepage stages of the channel.

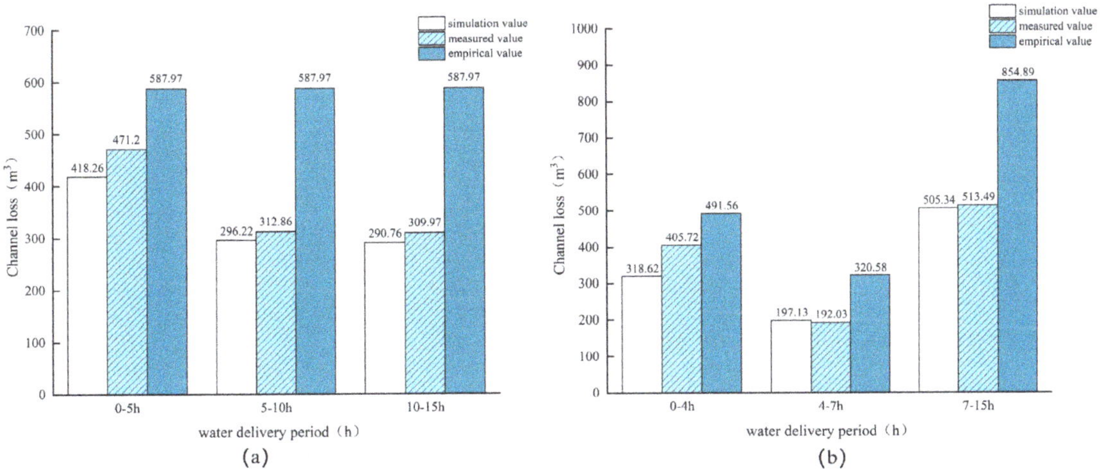

Figure 6. Dynamic simulation of leakage at various stages: (**a**) dynamic simulation of leakage in channel U1; (**b**) dynamic simulation of leakage in channel U2.

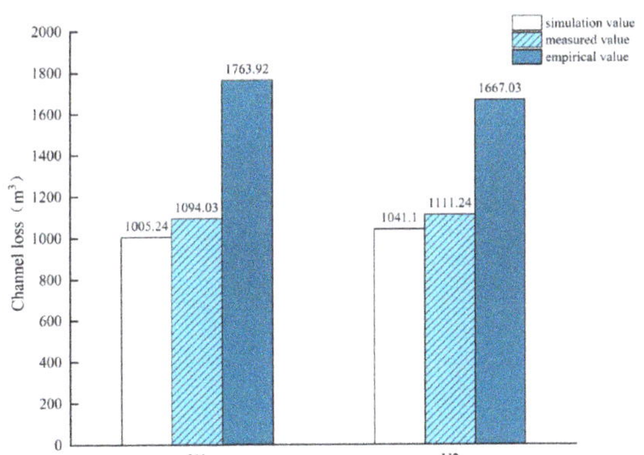

Figure 7. Comparison of total leakage.

During the test period, the total actual leakage of the U1 channel was 1094.03 m^3, the model simulation value was 1005.24 m^3, and the value calculated using the Kausgakov empirical formula was 1763.92 m^3. The error between the simulated and measured values was 8.12%, while the error between the empirical and measured values was 61.23%. The total amount of actual leakage of the U2 channel was 1111.24 m^3, the model simulation value was 1021.1 m^3, and the value calculated using the Korsgakov empirical formula was 1667.03 m^3. The error between the simulated value and measured value was 6.31%, while the error between the empirical value and measured value was 50.02%, and compared with the measured value, the model simulation value was more accurate than the result of the empirical formula calculation.

4. Discussion

The following results were obtained according to the research objectives set at the beginning of this study. On the one hand, the regression results showed that the vertical seepage velocity coefficient and soil pore ratio had a greater influence on the calculation of channel seepage loss, and both factors belonged to soil properties, which indicated that the soil properties had a higher degree of influence on the channel seepage loss, which was consistent with the conclusion in a study by Li Hongxing [18] et al. On the other hand, the test results show that in the actual water conveyance process, the channel bed soil water content and channel flow rate are dynamically changing with time. The dynamic change in the soil water content and channel flow rate will directly affect the infiltration capacity of the channel bed soil, but the channel bed soil water content and channel seepage rate show the opposite change trend, which is similar to the conclusion of the study by Li Mingyang [25] and others. The flow rate, although the change amplitude is small, still has a large impact on the channel seepage loss, which is similar to the conclusion of the study by Kratz et al. [26]. However, in the course of the experiment, there were cases of low precision in individual data, which may have been influenced by the precision of the real-time soil moisture monitoring instrument. Therefore, in future tests, the influence of various monitoring systems in the field for the test should first be considered, and the measurement interval can be set shorter to obtain more continuous real-time data and improve the accuracy of the data. In addition, the precipitation during the test period will affect the channel flow, soil moisture content, and other factors. A subsequent study should fully consider the interference of precipitation on the test. The precipitation factor can be used as a regression factor for regression analysis through the regression coefficient to quantify its impact on the channel seepage.

5. Conclusions

In this study, a dynamic simulation model of multi-factor channel leakage loss was constructed through a machine learning method, and a dynamic simulation test of leakage loss was carried out in the irrigation area to verify the model accuracy, aiming at accurately calculating the channel leakage loss under the dynamic change in the influencing factors, providing effective support for the irrigation area to accurately control the process of channel water conveyance. The research results show that under the conditions of considering the dynamic changes of some influencing factors in the water transfer process and quantifying the degree of influence of multi-factors, the dynamic simulation and calculation model of multi-factor channel leakage loss constructed by using the machine learning method has a better calculation effect than the traditional empirical formula. The model constructed using the machine learning method has higher computational accuracy than the traditional empirical formula. In addition, the model can simulate the dynamic trend of seepage loss more intuitively, which can provide an effective method for irrigation districts to determine the seepage characteristics of the channel, so as to provide strong support for the realization of the precise control of the channel water transfer process.

Author Contributions: Conceptualization, J.M.; methodology, B.C. and J.Y.; software, J.Y., X.H. and B.C.; analysis and validation, J.Y.; data curation, J.Y.; writing—original draft, J.Y.; review, B.C. and S.Y.; visualization, S.Y. All authors have read and agreed to the published version of the manuscript.

Funding: This research was supported by the Henan province university science and technology innovation talent support plan project (No. 15HASTIT046), Henan province science and technology research project (No. 152102110095), and key scientific research projects of colleges and universities in Henan Province (No. 15A570008). Therefore, we thank the Department of Education and the Department of Science and Technology of Henan Province for their strong support.

Institutional Review Board Statement: Not applicable.

Informed Consent Statement: Not applicable.

Data Availability Statement: Derived data supporting the findings of this study are available from the corresponding authors on request.

Conflicts of Interest: The authors declare no conflict of interest.

References

1. Mao, X.; Yao, L.-Q.; Feng, S.-Y.; Wang, Y.-Y. Numerical simulation on canal seepage and soil water distribution for concrete lining canal with layered soil structure. *J. Hydraul. Eng.* **2011**, *42*, 949–955.
2. National Development and Reform Commission, Ministry of Water Resources. Implementation Plan for Supporting Water Saving Renovation of Large and Medium Sized Irrigation Districts in China (2016–2020) [EB/OL] 2017-05-10. Available online: http://www.jsgg.com.cn/Index/Display.asp?NewsID=21849 (accessed on 1 June 2021).
3. Luo, Y.F.; Cui, Y.L.; Zheng, Z.J. Research progress on methods of quantifying seepage from rivers and canals. *Adv. Water Sci.* **2005**, *16*, 444–449.
4. Ye, L.; Ting, Y.; Rong, H.Z.; Yi, B.L.; Wen, J.Z.; Xiao, Y.M. Irrigation Canal System Delivery Scheduling Based on a Particle Swarm Optimization Algorithm. *Water* **2018**, *10*, 1281.
5. Dilini, D.; Hector, M.; Syed, K.S.; Halgamuge, M.N. A novel generic optimization method for irrigation scheduling under multiple objectives and multiple hierarchical layers in a canal network. *Adv. Water Resour.* **2017**, *105*, 188–204.
6. Santhi, C.; Pundarikanthan, N.V. A new planning model for canal scheduling of rotational irrigation. *Agric. Water Manag.* **2000**, *43*, 327–343. [CrossRef]
7. Mishra, A.; Verma, H.C.; Singh, R. Alternative rotational delivery scheduling for better water regime in canal command. *J. Irrig. Drain. Eng.* **2008**, *134*, 175–184. [CrossRef]
8. Li, M.; Fu, Q.; Singh, V.P.; Liu, D.; Li, T.; Zhou, Y. Managing agricultural water and land resources with tradeoff between conomic, environmental, and social considerations: A multi-objective non-linear optimization model under uncertainty. *Agric. Syst.* **2020**, *178*, 102685. [CrossRef]
9. Shah, Z.; Gabriel, H.; Haider, S.; Jafri, T. Analysis of seepage loss from concrete lined irrigation canals in Pun-jab, Pakistan. *Irrig. Drain.* **2020**, *69*, 668–681. [CrossRef]
10. Zhang, Q.; Chai, J.; Xu, Z.; Qin, Y. Investigation of irrigation canal seepage losses through use of four different methods in Hetao Irrigation District, China. *J. Hydrol. Eng.* **2017**, *3*, 05016035. [CrossRef]
11. Akkuzu, E. Usefulness of empirical equations in assessing canal losses through seepage in concrete-lined canal. *J. Irrig. Drain. Eng.* **2012**, *138*, 455–460. [CrossRef]
12. Men, B.H. Discussion on the formula for channel flow loss and water utilization coefficient. *China Rural Water Resour. Hydropower* **2000**, *2*, 33–34.
13. Xie, C.B.; Lance, J.M.; Cui, Y.L.; Bai, M.J.; Huang, B.; Cai, L.G. Exploring the empirical formula for water leakage loss in large and medium-sized irrigation areas' main channels. *China Rural Water Resour. Hydropower* **2003**, *2*, 20–22.
14. Wang, B.C. Improved formula for calculating channel leakage loss and water utilization coefficient. *Agric. Water Conserv. Small Hydropower* **1986**, *08*, 20–21.
15. Zhang, F.; Cai, Y.P.; Guo, P.; Qian, T.; Wang, X. Estimation method of leakage in trunk and branch canals of irrigation areas and its application in optimal allocation of water resources. *J. Agric. Eng.* **2021**, *37*, 140–147.
16. Liao, X.C.; Hu, T.S. Research on the Optimal Water Distribution Model of Canal System Based on the Dynamic Change Characteristics of Water Transmission Loss. *J. Water Resour.* **2021**, *52*, 850–861.
17. Aller, L.; Lehr, J.H.; Petty, R.; Bennett, T. DRASTIC: A Standardized System to Evaluate Ground Water Pollution Potential Using Hydrogeologic Settings. In *Petroleum Hydrocarbons and Organic Chemicals in Ground Water: Prevention, Detection and Restoration*; National Water Well Association: Worthington, OH, USA, 1987; pp. 38–57.
18. Li, H.X.; Fan, G.S. Experimental study on the dominant factors affecting the infiltration capacity of unsaturated soil bed. *J. Water Resour.* **2009**, *40*, 630–634.
19. Hou, H.M.; Dai, X.L.; Wang, Z.J. Study on the effect of different soil structures on water loss in channels. *Gansu Sci. Technol.* **2009**, *25*, 44–47.
20. Fan, Y.; Ma, L.; Wei, H.; Zhu, P. Numerical investigation of wetting front migration and soil water distribution under vertical line source irrigation with different influencing factors. *Water Supply* **2021**, *21*, 2233–2248. [CrossRef]
21. Wang, X.L.; Qu, Z.G.; Lai, T.; Ren, G.F.; Wang, W.K. Enhancing water transport performance of gas diffusion layers through coupling manipulation of pore structure and hydrophobicity. *J. Power Sources* **2022**, *525*, 231121. [CrossRef]
22. Wang, X.L.; Qu, Z.G.; Ren, G.F. Collective enhancement in hydrophobicity and electrical conductivity of gas diffusion layer and the electrochemical performance of PEMFCs. *J. Power Sources* **2023**, *575*, 233077. [CrossRef]
23. Xiao, X.; Wang, X.G.; Tan, D. Comparison of several empirical formulas for calculating channel leakage loss. *J. Wuhan Univ. (Eng. Ed.)* **2016**, *49*, 365–371. [CrossRef]
24. Li, R.X. Comparison of leakage process method and fixed water level method in hydrostatic seepage measurement of channels. *J. Water Resour. Archit. Eng.* **1991**, *2*, 50–53.

25. Li, M.; Liu, T.; Lu, Y.; Duan, L.; Zhang, J.; Zhou, Y. Research on soil infiltration process and conversion function in a semi-arid grassland-type watershed. *J. Water Resour.* **2019**, *50*, 936–946. [CrossRef]
26. Kratz, D.B. *Lining of Irrigation Canals*; He, P., Translator; Water Resources Press: Beijing, China, 1980.

water

MDPI

Article

Water Footprint of Animal Breeding Industry and Driving Forces at Provincial Level in China

Haohan Xing [1], Yun Xie [2], Baoming Li [3,4,5], Hongbin Cong [1], Weichao Zheng [3,4,5,*] and Huan Liu [1]

[1] Key Laboratory of Energy Resource Utilization from Agriculture Residue, Academy of Agricultural Planning and Engineering, Ministry of Agriculture and Rural Affairs, Beijing 100125, China; xhh523059864@126.com (H.X.); conghongbin@163.com (H.C.); liuhuan19851208@163.com (H.L.)

[2] Institute of Food Safety, Chinese Academy of Inspection and Quarantine, Beijing 100176, China; yunxie_whu@163.com

[3] Department of Agricultural Structure and Bioenvironmental Engineering, College of Water Resources and Civil Engineering, China Agricultural University, Beijing 100083, China; libm@cau.edu.cn

[4] Key Lab of Agricultural Engineering in Structure and Environment, Ministry of Agriculture and Rural Affairs, Beijing 100083, China

[5] Beijing Engineering Research Center for Livestock and Poultry Healthy Environment, Beijing 100083, China

* Correspondence: weichaozheng@cau.edu.cn

Abstract: Agriculture significantly contributes to the global water footprint (WF) with the animal breeding industry accounting for over 33% of agricultural water consumption. Since 2000, rapid development in animal breeding has intensified the pressure on water resources. Forecasts indicate a projected 70% increase in freshwater usage in the meat industry by 2025 compared to 2000, particularly in developing countries, such as China, yet comprehensive studies regarding China's animal breeding industry WF remain limited. This study aimed to assess the variations in the green, blue, and gray WF of pork, beef, milk, eggs, and chicken meat across 31 provinces in China from 2000 to 2017. Additionally, a driving force analysis using the Kaya equation and LMDI method was conducted. Findings revealed that the total WF of animal products increased from 1049.67 Gm^3 (in 2000) to 1385.05 Gm^3 (in 2017) in China, and pork exhibited a significantly higher WF compared to other animal products, contributing 64.49% to China's total animal product WF. The sharp rise in the green WF demonstrated regional disparities in water consumption efficiency within the animal breeding industry. The increase in the blue WF was associated with rising livestock numbers and China's efforts to conserve water. The increase in the gray WF indicated that increased consumption of animal products heightened wastewater treatment pressures, particularly in economically developed provinces. The augmentation in China's animal product WF was primarily influenced by policy and economic effects, with increased agricultural equipment funding and enhanced production efficiency identified as effective strategies for WF reduction. This study suggests that the promotion of technology, combined with scientific policies, can alleviate the pressure on water resources in the animal breeding industry in developing countries.

Keywords: water footprint assessment; animal products; agricultural water consumption; logarithmic mean divisia index (LMDI); driving force analysis; Kaya equation

Citation: Xing, H.; Xie, Y.; Li, B.; Cong, H.; Zheng, W.; Liu, H. Water Footprint of Animal Breeding Industry and Driving Forces at Provincial Level in China. *Water* **2023**, *15*, 4264. https://doi.org/10.3390/w15244264

Academic Editors: Songhao Shang, Qianqian Zhang, Dongqin Yin, Hamza Gabriel and Magdy Mohssen

Received: 27 October 2023
Revised: 21 November 2023
Accepted: 27 November 2023
Published: 13 December 2023

1. Introduction

Agriculture represents a major contribution to the water footprint (WF) of humanity, and most of this contribution is linked to the animal breeding industry, which represents more than 33% of agricultural water consumption [1]. There is more pressure on water resources caused by the animal breeding industry. The growing need for livestock products, their substantial water usage, and their significant direct influence on aquatic environments have placed immense pressure on finite water resources. [2]. Predictions show that freshwater consumption in the meat industry will have a 70% increase by 2025 compared to

2000 [3]. Since the beginning of the new century, significant advancements in the animal breeding industry have been witnessed in developing nations. There is an imperative to enhance water productivity within this sector to mitigate its impact on water resources and the environment, thereby fostering the sustainability of animal production.

The WF [4] stands as a comprehensive indicator assessing both water consumption and its environmental ramifications. It quantifies direct and indirect water use by source and identifies polluted water according to the type of pollution, offering distinct spatiotemporal trends that reveal the impact of diverse human activities on water resources [3]. The water usage within the animal breeding industry is elucidated through the evaluation of the WF of animal products. This term encompasses the complete amount of direct and indirect freshwater utilized and contaminated due to livestock feeding practices. Direct water usage incorporates water utilized for drinking and for services like cleaning sheds [5]. Additionally, it encompasses water utilized to mitigate pollutant concentrations arising from animal excreta, commonly referred to as the gray WF. Indirect water consumption is the WF generated in the process of feed production over the lifetime of livestock. The green WF denotes the usage of green water, specifically rainwater that does not transform into runoff. Meanwhile, the blue WF signifies the consumption of blue water resources, including surface and groundwater, throughout the livestock product supply chain. Lastly, the gray WF quantifies water pollution, signifying the volume of freshwater needed to assimilate a pollutant load based on natural background concentrations and established ambient water quality standards [5].

The WF of animal products has been widely studied worldwide and provides the basis for the projections of global future demand for animal products [1,6–9]. Technology and economic development levels and populations are not considered for different countries at the global level. Therefore, the national WF of animal products has recently begun to be intensively analyzed. Most of these studies analyzed the environmental impact of a single WF of animal products in developed countries [2,8–16]. However, the WF of a single animal product has little impact on water resources for a country. The WF of the entire animal breeding industry enables an analysis of the future development of the industry and the impact on the development of a country. There are limited studies that are representative for developing countries, such as China, which is the one of the biggest producers of animal products in the world. China accounted for 46.25% of global pork and 39.13% of egg production in 2017 [17]. The main animal products consumed in China are pork, eggs, milk, poultry meat and beef. The production of these animal products in 2017 was 36.10%, 45.05%, 323.44%, 77.66%, and 25.56% higher than that in 2000, respectively. Water scarcity limits the expansion of animal production. Therefore, large quantities of animal products must be imported into China. China imported 3048.61 thousand tons of meat products in 2018 [17]. However, few studies have examined the driving forces of growth in the WF of the animal breeding industry in China. China's demand for animal products is rapidly increasing with economic development, which corresponds to a highly uneven agricultural development. For its future sustainable development, it is necessary to investigate the impact of the WF of the animal breeding industry on China's water resources.

The driving force of agricultural water consumption has become a major concern. The impact of population, economic, and intensity effects on agricultural water consumption has been widely analyzed [18–25]. Most of these studies ignored the impact of agricultural machinery inputs, the degree of farm automation, and the efficiency of farm production on the WF of animal products. China has continued to increase its investment in agricultural machinery and improve farming automation in recent years. The impact of these factors on the WF of animal products should be examined.

The present paper quantified and evaluated the interannual variations in the blue, green, and gray WFs of pork, eggs, milk, poultry meat, and beef in 31 provinces in China from 2000 to 2017. A driving force analysis was performed by combining a Kaya equation with the logarithmic mean division index (LMDI) method. Clearly, analyzing the variations and driving forces of the WF under changing economic conditions is important for the efficient utilization of regional water resources and for water management and allocation strategies.

2. Materials and Methods

2.1. Animals and Animal Products

In this study, our focus includes pigs, dairy cattle, beef cattle, broiler chickens, and laying hens, along with their respective products such as pork, milk, beef, chicken meat, and eggs. These animal categories represent significant segments within China's livestock industry, and we have extensively analyzed their WF variations and impact on water resources.

2.2. WF of Animal Products

The WFs of pork, beef, milk, eggs and chicken meat are estimated in this paper. The WF of animal products for category *a* in province *p* (WF; m^3) reflects the feed, drinking water, and service water consumed [1],

$$WF[a,p] = WF_{feed}[a,p] + WF_{drink}[a,p] + WF_{serv}[a,p] \quad (1)$$

where WF_{feed} [*a*,*p*], WF_{drink} [*a*,*p*], and WF_{serv} [*a*,*p*] represent the WF of animal products for category *a* in province *p* related to feed, drinking, and service water, respectively. Service water refers to the water used for maintaining cleanliness and cooling in animal buildings. For beef cattle, pigs, and broiler chickens—animals that yield products after slaughter—we ascertain the WF of the animal at the conclusion of its lifetime, subsequently distributing the total WF among the different products (such as meat and leather). In the case of dairy cattle and layer chickens, the process of determining the animal's WF per year (averaged over its lifetime) and relating this annual animal WF to its average annual production (milk, eggs) is more direct [1].

The WF of animal products related to their feed was calculated as follows (the water used to mix the feed was ignored):

$$WF_{feed}[a,p] = \frac{\sum_{i=1}^{n}\left(Feed[a,c,p] \times WF_{prod}[c]\right) + WF_{mixing}[a,p]}{Pop[a,p]} \quad (2)$$

Feed [*a*, *c*, *p*] denotes the yearly consumption of feed ingredient *c* by animal category *a* in province *c* (m^3/y), WF_{prod} [*c*] stands for the WF of feed ingredient *c* (m^3/ton), WF_{mixing} [*a*, *p*] signifies the water volume consumed during the feed mixing process for animal category *a* in province *p* (m^3/y/animal), and Pop [*a*, *p*] represents the annual count of slaughtered animals or the number of milk- or egg-producing animals in a year for animal category *a* in province *p* [1].

The calculation of WF_{drink} and WF_{serv} is as follows:

$$WF_{drink} = \frac{Drink[a,p]}{Pop[a,p]} \quad (3)$$

$$WF_{serv} = \frac{\sum_{i=1}^{n} Serv[a,c,p]}{Pop[a,p]} \quad (4)$$

where Drink [*a*, *p*] represents the annual amount of drinking water consumed by animal category *a* in province *c* (m^3/y) and Serv [*a*, *c*, *p*] is the water consumed to keep the animal building clean and cool *c* for animal category *a* in province *p* (m^3/y).

2.3. Driving Force Analysis

The driving force analysis was extended to identify the factors influencing the increase in the WFs. While the Kaya equation is commonly employed in analyzing the driving forces of carbon emissions and energy consumption [19,25,26], this paper expands its application to decompose the WF as follows.

$$WF = \sum_i \frac{WF_i}{Y_i} \cdot \frac{Y_i}{FARM} \cdot \frac{FARM}{MAC} \cdot \frac{MAC}{EXP_a} \cdot \frac{EXP_a}{EXP} \cdot \frac{EXP}{GDP} \cdot \frac{GDP}{P} \cdot \frac{P}{W} \cdot \frac{W}{W_a} \cdot W_a \tag{5}$$

where WF_i, Y_i, $FARM$, MAC, EXP_a, EXP, GDP, P, W, and W_a represent the WF of animal product i, the yield of animal product i, the number of farms, the number of agricultural machines and China's financial expenditure on agriculture, total financial expenditure, GDP, populations, water resources, and agricultural water, respectively. The ten driving forces are defined in Table 1.

Table 1. Ten driving forces of the WF of animal products in China.

Effect	Decomposition	Symbol	Explanation
Technology Effects	WF_i/Y_i	A	Unit WF content
	$Y_i/FARM$	B	Farm production efficiency
	FARM/MAC	C	Degree of automation on farms
Policy Effects	MAC/EXP_a	D	Funding for agricultural equipment
	EXP_a/EXP	E	National agricultural inputs
	EXP/GDP	F	Scale of national financial expenditure
Economic Effects	GDP/P	G	Agricultural earnings per capita
Endowment Effects	P/W	H	Per capita water resources
	W/W_a	I	Utilization of water resources in agriculture
	W_a	J	Water consumption in agriculture

The study then utilizes the LMDI decomposition method to quantitatively assess the contribution of each influencing factor [27,28]. The LMDI method boasts several advantages, including complete decomposition without residual, a strong theoretical foundation, adaptability, ease of use, and a straightforward interpretation of results [19]. This model aids in analyzing the driving factors within the Kaya equation and their subsequent effects. In accordance with the LMDI method, there exists a variation in the WF between the base year and year t, as illustrated in Equation (6).

$$\Delta WF = WF^t - WF^0 = \Delta WF^A + \Delta WF^B + \Delta WF^C + \Delta WF^D + \Delta WF^E + \Delta WF^F + \Delta WF^G + \Delta WF^H + \Delta WF^I + \Delta WF^J \tag{6}$$

where

$$\Delta WF^A = q \times ln\left(\frac{A^t}{A^0}\right) \tag{7}$$

$$\Delta WF^B = q \times ln\left(\frac{B^t}{B^0}\right) \tag{8}$$

$$\Delta WF^C = q \times ln\left(\frac{C^t}{C^0}\right) \tag{9}$$

$$\Delta WF^D = q \times ln\left(\frac{D^t}{D^0}\right) \tag{10}$$

$$\Delta WF^E = q \times ln\left(\frac{E^t}{E^0}\right) \tag{11}$$

$$\Delta WF^F = q \times ln\left(\frac{F^t}{F^0}\right) \tag{12}$$

$$\Delta WF^G = q \times ln\left(\frac{G^t}{G^0}\right) \tag{13}$$

$$\Delta WF^H = q \times ln\left(\frac{H^t}{H^0}\right) \quad (14)$$

$$\Delta WF^I = q \times ln\left(\frac{I^t}{I^0}\right) \quad (15)$$

$$\Delta WF^J = q \times ln\left(\frac{J^t}{J^0}\right) \quad (16)$$

A, B, C, and D are labeled technology effects, E and F are policy effects, G is the economic effect, and H, I, and J are the endowment effects. After conversion through the application of the LMDI model, the proportion of each of these ten effects is obtained in the explanation for changes in the WF of animal products. If an effect's value is negative, it implies a positive driving factor contributing to the reduction in the WF. The greater the absolute value of this effect, the more pronounced its positive impact as a driving factor. Conversely, a positive effect value indicates a negative driving factor. The magnitude of this effect's value signifies the extent of the impact caused by the driving factor.

q is the logarithmic weight and can be calculated using Equation (17).

$$q = \frac{WF^t - WF^0}{ln\left(WF^T\right) - ln\left(WF^0\right)} \quad (17)$$

2.4. Data

The daily drinking water consumed, daily service water consumed, and feed composition of animals were obtained from Chapagain and Hoekstra [6]. The WF of feed ingredients was obtained from published articles [29,30]. For each province, the annual number of slaughtered animals and production of animal products were obtained, and the number of livestock farms, agricultural machines, financial expenditure on agriculture, total financial expenditure, GDP, populations, water resources, and agricultural water for China were obtained from the National Bureau of Statistics of China.

3. Results and Discussion

3.1. Provincial WF of Animal Products in China

3.1.1. Green, Blue and Gray WF of Animal Products

The green WF of animal products is shown in Figure 1. The results show that a dramatic rise occurred from 2000 to 2017, with a fluctuation of approximately 31.59%. The green WF grew in most provinces, with the exception of Beijing (−45.70%), Shanghai (−63.17%), and Zhejiang Province (−22.57%). The growth rate during the period between 2008 and 2017 (11.53%) was lower than that during 2000–2008 (17.99%), indicating that the demand for animal products is gradually being met as the number of livestock and poultry animals raised in most provinces increases in China.

The regional disparities in the green WF of animal products highlight the diverse efficiency levels in water consumption across provinces within the animal breeding industry. These WFs exhibited notable variations among provinces and underwent significant changes from 2000 to 2017. In 2000, Shandong (67.19 Gm^3), Hunan (64.79 Gm^3) and Hebei (64.45 Gm^3) had higher green WFs, whereas Sichuan (73.95 Gm^3) and Henan (70.67 Gm^3) had the largest green WFs. In 2008, Henan had the largest green WF (86.19 Gm^3). In 2017, Shandong (93.86 Gm^3) surpassed Henan (91.03 Gm^3) and had the largest green WF. Beijing, Shanghai, and Zhejiang had a lower green WF in 2017 than in 2000.

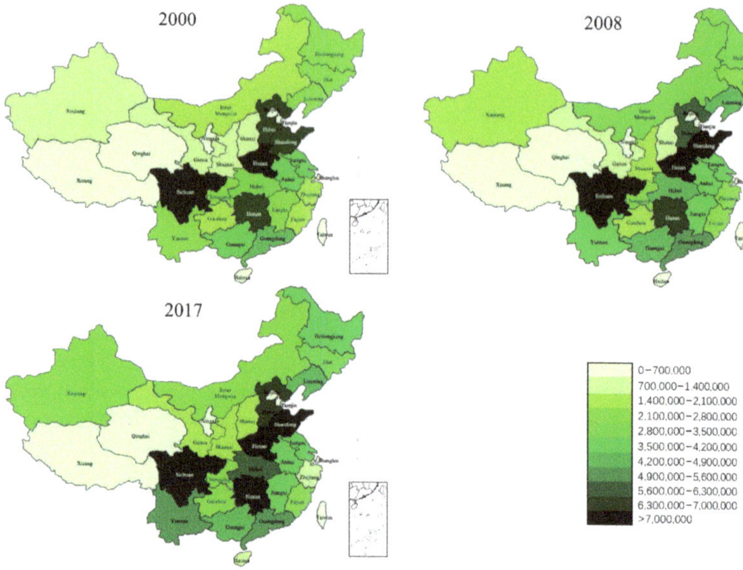

Figure 1. Green WF of animal products in China.

Figure 2 shows the variations in the blue WF of animal products in China over 2000–2017. The blue WF increased by 33.12% during 2000–2017. The blue WF increased in all provinces with the exception of Beijing (−48.69%), Hebei (−0.37%), Shanghai (−67.43%), and Zhejiang (−22.18). The growth rate of the blue WF during 2000-2008 (17.04%) was higher than that during 2008–2017 (13.74%). On the one hand, the increase in the number of livestock and poultry animals led to a larger blue WF, while on the other hand, it can be seen that China made many efforts to save water.

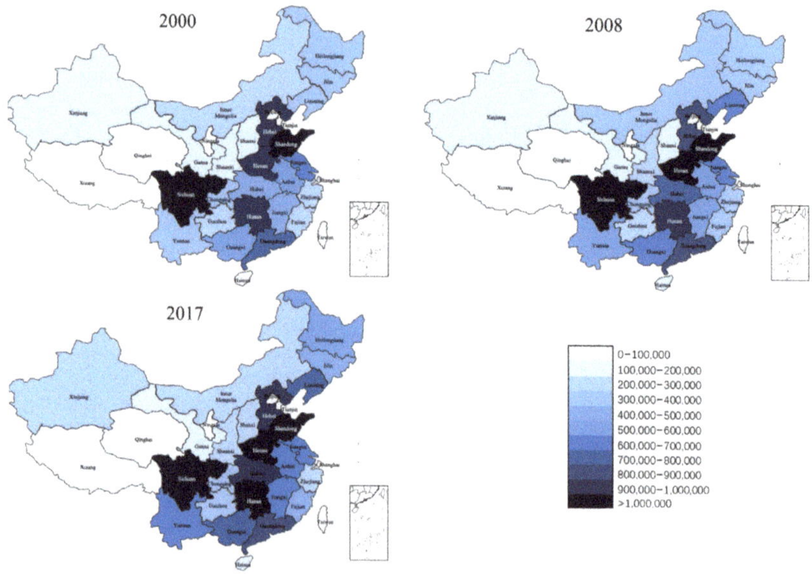

Figure 2. Blue WF of animal products in China.

Differences in the water conservation capacity across provinces are reflected in the depletion of the blue WF in 2000, 2008, and 2017. In 2000, Henan (9.74 Gm^3), Hebei (9.53 Gm^3), and Hunan (9.37 Gm^3) had larger blue WFs, whereas Sichuan (10.82 Gm^3) and Shandong (10.33 Gm^3) had the largest blue WFs. In 2008, Sichuan had the largest blue WF (12.56 Gm^3). In 2017, Shandong (14.95 Gm^3) surpassed Henan (14.26 Gm^3) to have the largest blue WF.

This analysis revealed a notably higher WF for pork compared to other animal products. Pork production contributed significantly to the overall WF of animal products in China, accounting for 64.49%. This dominance of the WF of pork highlights its substantial impact on water resources, emphasizing the importance of reducing its water consumption for ensuring the sustainability of water resources in China. The green and blue pork WF was averaged at 705.9 Gm^3 per year, which is much larger than the 225.8 Gm^3 per year in Xie's study [3]. These numbers may be interpreted as the difference between the number of stocked and slaughtered pigs in China. The pork WF may be overestimated, but the annual water consumption of the stocked pigs was not ignored.

The WFs of all five animal products in this study were higher than those in published studies [1,3,8,12–14]. This study concluded that the total WF of animal products needs to encompass all the water consumed by animals annually. The substantial water consumption by living animals should not be disregarded and needs consideration in assessing water resources.

The green and blue WFs are interconnected. Typically, the green WF from feed crop production does not have a notably adverse environmental impact. Nevertheless, globally, minimizing the green WF might be crucial in decreasing the blue WF in crop production. Enhancing rainwater efficiency, such as augmenting yields per unit of rainwater, holds the potential to decrease the blue WF associated with feed crops.

The large animal population requires a greater capacity for manure treatment, especially in the northern provinces of China (Figure 3). In 2017, Shandong (12.89 Gm^3), Henan (12.60 Gm^3), Sichuan (11.80 Gm^3), and Hunan (10.38 Gm^3) had the highest gray WFs, accounting for 37.39% of the total. In contrast, Tibet had the highest growth rate (227.62%) of the gray WF during 2000–2017.

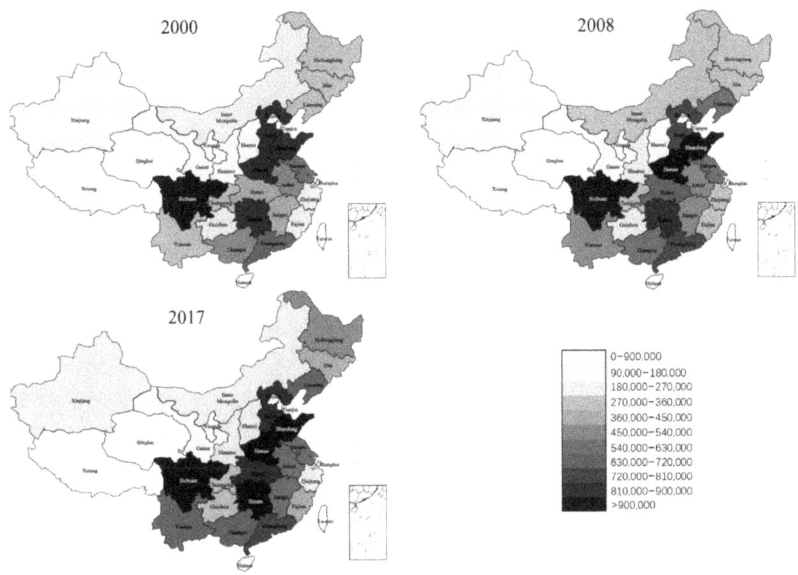

Figure 3. Gray WF of animal products in China.

The total gray WF increased by 33.12% in 2017 compared to 2000. These results are different from the gray WF estimated by Zhang [19], who indicated that the volume of agricultural gray WF remained relatively stable and only slightly increased during 2003–2015. The differences in the gray WF may be explained by the contradiction between technical progress and consumption growth. The increased gray WF of agriculture was curbed with technical progress in pollution treatment, in recent years. The increased gray WF of animal products indicates that the consumption growth of animal products increased the pressure on wastewater treatment while offsetting the benefits of technical progress. Hence, the gray WF of animal products emerges as a pivotal factor in wastewater management across China. It becomes imperative to enhance the discharge norms for pollutants in the animal breeding sector, especially in economically advanced provinces. Implementing effective strategies to diminish wastewater production on farms and to curtail fertilizer usage in feed crop cultivation remains essential.

3.1.2. Changes in the Total WF in Different Regions of China

For descriptive purposes, China was divided into seven regions based on its livestock characteristics (Figure 4): Qinghai–Tibet Plateau (QTP, Tibet, and Qinghai), Inner Mongolia-Xinjiang Region (IXR, Inner Mongolia, and Xinjiang), Loess Plateau (LP, Shaanxi, Ningxia, and Gansu), Southwest Mountain Region (SMR, Yunnan, Guizhou, Sichuan, and Chongqing), Northeast Region (NER, Liaoning, Jilin, and Heilongjiang), Northern China Region (NCR, Beijing, Tianjin, Hebei, Shandong, Henan, Shanxi, Jiangsu, Shanghai, and Anhui), and Southeast region (SER, Zhejiang, Fujian, Guangdong, Guangxi, Hunan, Hubei, Jiangxi, and Hainan).

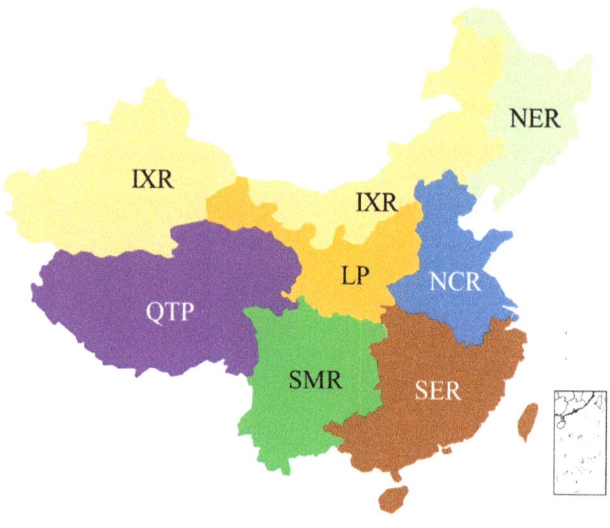

Figure 4. Seven regions in China.

Figure 5 shows the total WF for the different livestock animal compositions for the seven regions in 2000, 2008 and 2017. The NCR had the highest WF. The average total WFs of animal products in 2000, 2008, and 2017 were 15.00 Gm^3 year^{-1}, 17.67 Gm^3 year^{-1}, and 19.50 Gm^3 year^{-1}, respectively. The WF of pork accounted for 54.43%, 51.18%, and 54.49% of the total WF in 2000, 2008 and 2017, respectively. The WF of animal products in most regions showed an increasing trend over the period 2000–2017, except for the IXR region. In 2000–2017, the consumption structure of animal products in the LP, SMR, NER, NCR, and SER regions was dominated by pork, and the total WF of pork increased by 33.28% from 678.5 Gm^3 year^{-1} to 904.3 Gm^3 year^{-1}, whereas beef had the largest WF in the QTP region, accounting for 62.43% on average.

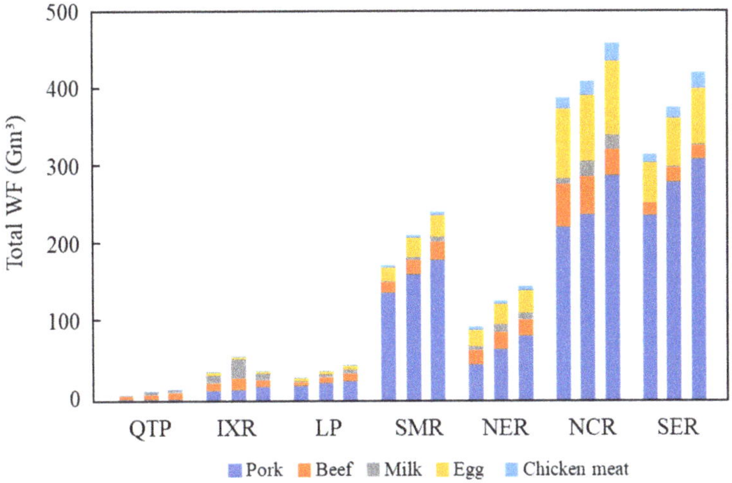

Figure 5. Total WF in different regions of China in 2000, 2008, and 2017.

The contribution of animal products to the total WF varied considerably between different regions. Some provinces produced enormous WFs of animal products and ranked within the top three in 2000, 2008, and 2017. Their enormous WFs were induced by rapid socioeconomic development. These regions are densely populated and have a higher demand for animal products [24,31]. Specifically, a large amount of animal products were produced and consumed in these regions compared to other regions. Provinces in the NCR and NER heavily rely on locally sourced water-intensive animal products and predominantly export feed crops. These areas face severe water stress and grapple with a substantial water scarcity challenge, considering the massive demand for water resources. In these provinces, the animal breeding industry plays a vital role in mitigating the strain on water resources and ensuring sustainable development. Tailoring their development strategies based on local water availability becomes crucial. Therefore, adjusting trade dynamics and relocating a portion of animal product and crop production to regions endowed with abundant water resources could alleviate water scarcity and optimize water utilization in these regions.

The high WF observed in developed areas can be attributed to several factors. Firstly, developed regions often have higher livestock production, leading to increased water consumption due to larger herd sizes and more intensive farming practices. Secondly, these areas tend to rely on more resource-intensive production systems, such as confined animal feeding operations (CAFOs), which require significant water inputs for maintaining livestock health and hygiene. Additionally, the increased use of feed with higher WFs and the reliance on processed feeds might elevate the overall WF in these developed regions. Moreover, the higher demand for animal products in developed areas might drive the need for larger-scale production, contributing to increased water usage throughout the livestock farming process. Lastly, the presence of more advanced but water-consuming technologies and infrastructure in these regions could also contribute to the higher WF associated with livestock farming.

3.2. Driving Force Analysis

In this study, an analysis of the driving forces behind the total WF was conducted. Utilizing the LMDI model, the factors influencing the total WF for each province were categorized into four effects: the technology effect (A, B, C, D), policy effect (E, F), economic effect (G), and endowment effect (H, I, J) (Figure 6).

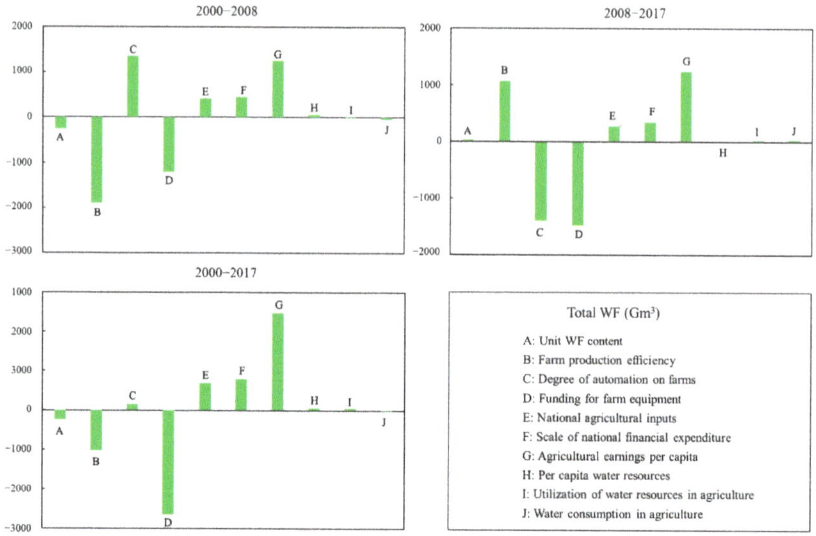

Figure 6. Driving forces of the WFs of animal products in different periods in China.

The results show that G, F, and E were the three most significant driving forces of the total WF increase in the Chinese animal breeding industry. The other driving forces of the increase in the total WF included C, H, and I. The six driving forces contributed 2482.41, 801.97, 696.62, 154.02, 66.29 and 50.93 Gm^3 to the Chinese breeding industry. D was the leading factor in the decline in the total WF, followed by B, A, and J. The four driving forces contributed −2650.51, −1023.00, −237.86, and −5.49 Gm^3 to the total WF of the Chinese breeding industry, respectively.

We decomposed the total WF of the Chinese breeding industry into two different periods to explore the driving forces of the total WF. During 2000–2008, the overall WF increased by 187.11 Gm^3, and C and G had a decisive effect on the increase in WF. The other important driving forces of the increase in total WF were the change in E and F, whereas H and I only marginally affected the total WF. B and D were the leading factors in the decline during this period.

In contrast to the 2000–2008 period, the increase in the total WF of the breeding industry at the national level and the effects of its driving forces differed significantly between 2008 and 2017 in China. B contributed to the increase in total WF, but the positive effects of B notably changed in the two periods. This change may be due to a mismatch between the production of animal products and the increased number of farms. This study finds that D promoted a decline in the total WF during the two periods. The extent of the decrease in 2008–2017 was larger during the two periods since a greater increase in D occurred in 2008–2017. This demonstrates that increased funding for agricultural equipment is effective in reducing WFs. Moreover, G, F, and E were promoted to increase the total WF over the two periods. This promotion could be due to rapid economic growth and a continued rise in the demand for animal products in China. E and F indicate that policy effects may be beneficial to the development of the Chinese breeding industry, while water consumption also increased. The effects of B and C notably changed in the two periods. The change in B occurred because farming productivity was lower during 2008–2017 than in 2000–2008. For example, between 2008 and 2017, the number of farms increased significantly, but the increase in the production of animal products was modest, making 2008–2017 less productive than 2000–2008. The change in C was due to the development of automation on farms. The increase in agro-mechanical equipment has increased the level of automation in farming, and the high level of automation helps to reduce water consumption.

Enhancing farm production efficiency stands as a viable option to alleviate the prevailing pressure on water resources. Namely, the yield of animal products increased under the same conditions of breeding and water consumption, which slowed the expansion of animal breeding scale and reduced water consumption. Despite the advancements witnessed in China's farm production efficiency within the animal breeding industry over the study period, it remained relatively lower in comparison to other industrialized nations, signifying the need for further enhancement. During the past two decades, the substantial expansion of China's animal feeding industry has largely met the demand for animal products among its populace. Efforts are needed create a resource-saving farming model to reduce water consumption while improving production efficiency in the future.

The main drivers behind the increase in the Chinese WF of animal products from 2000 to 2017 were the policy and economic effects, and the technology effects were the main driver of the reduction. Funding for agricultural equipment is increasing via national agricultural inputs and encouraging input from practitioners. The development of the animal breeding industry may be promoted, and the WF of animal products will be reduced. It is more scientific and reasonable to reduce the WF in this manner rather than changing people's consumption habits and continuously expanding the importation of animal products.

3.3. Comparative Analysis

Published studies [2,3,10] broadly addressed the WF of livestock; they mainly concentrated on the overall water resource consumption in specific countries (Korea, the United States) or regions. In contrast, our study focused specifically on the WF of livestock farming in distinct regions of China, undertaking regional and temporal analyses to deeply investigate the variations in WFs across different provinces from 2000 to 2017. This detailed approach provided more specific and concrete data support for managing water resources in the Chinese livestock industry. By conducting an in-depth analysis of the changes in the WF of China's specific regions, including comparisons between different provinces and the impact of various farming scales, our paper provided more precise, practical recommendations and guidance for managing the WF of the livestock industry in specific regions of China.

3.4. Limitations

The study faces certain limitations that could impact the accuracy and comprehensiveness of our findings. Firstly, the diverse regional characteristics of China, including varying climates, economies, and populations, pose a challenge in obtaining representative data for each region, potentially leading to an overestimation of WFs for animal products. Furthermore, our research focused on only five animal species, neglecting other commonly consumed varieties like sheep and duck meat. Expanding the scope of animal products examined could provide a more comprehensive understanding. Additionally, our emphasis on the green WF, while crucial, highlights the necessity to investigate factors influencing water use in the cultivation of feed crops. Future studies should consider delving into the WF of feed crops to offer a more holistic perspective on the overall water impact associated with animal product production.

4. Conclusions

In this study, the WFs of the animal breeding industry were estimated in 31 provinces of China. A driving force analysis of the WF was also conducted by combining a Kaya equation and the LMDI method. The main conclusions were drawn as follows.

The total WF of the animal breeding industry has risen due to the economic development of developing countries. The national total WF of animal products increased 31.95%, from 1049.67 Gm^3 in 2000 to 1385.05 Gm^3 in 2017. The WF of pork constituted the major portion of the total WF in most provinces, consequently being the primary factor contributing to the overall WF increase.

Economically developed regions had higher WFs than less developed regions. The eastern regions of China are more economically developed than the western regions, and the total WF of the animal breeding industry in the eastern regions is higher.

Increasing the funding for agricultural equipment, integrating more small farms, improving production efficiency, and increasing the automation of farms may effectively reduce the WF of animal products.

Author Contributions: Conceptualization, H.X. and W.Z.; methodology, H.X.; software, Y.X.; validation, H.X., B.L.; formal analysis, H.X. and H.C.; investigation, W.Z.; resources, H.X. and Y.X.; data curation, H.X.; writing—original draft preparation, H.X.; writing—review and editing, H.X., Y.X. and H.L.; visualization, H.X.; supervision, B.L.; project administration, W.Z.; funding acquisition, W.Z. and H.C. All authors have read and agreed to the published version of the manuscript.

Funding: This study was supported by the National Key Research and Development Project [Grant number: 2022YFD2300302, 2019YFD1100501], and the Project for R&D of Academy of Agricultural Planning and Engineering, MARA (Project Grant number: QD202109, QNYC-2021-03).

Data Availability Statement: The authors affirm that all data necessary for confirming the conclusions of the article are present within the article, figures, and tables.

Conflicts of Interest: The authors declare no conflict of interest.

References

1. Mekonnen, M.M.; Hoekstra, A.Y. A Global Assessment of the Water Footprint of Farm Animal Products. *Ecosystems* **2012**, *15*, 401–415. [CrossRef]
2. Mekonnen, M.M.; Neale, C.M.U.; Ray, C.; Erickson, G.E.; Hoekstra, A.Y. Water productivity in meat and milk production in the US from 1960 to 2016. *Environ. Int.* **2019**, *132*, 105084. [CrossRef] [PubMed]
3. Xie, D.; Zhuo, L.; Xie, P.; Liu, Y.; Feng, B.; Wu, P. Spatiotemporal variations and developments of water footprints of pig feeding and pork production in China (2004–2013). *Agric. Ecosyst. Environ.* **2020**, *297*, 106932. [CrossRef]
4. Hoekstra, A.Y. Virtual Water Trade. In Proceedings of the Internacional Expert Meeting on Virtual Water Trade, Delft, The Netherlands, 12–13 December 2003.
5. Hoekstra, A.Y.; Chapagain, A.K.; Aldaya, M.M.; Hoekstra, M.M.M. *The Water Footprint Assessment Manual: Setting the Global Standard*; Routledge: London, UK, 2011.
6. Chapagain, A.K.; Hoekstra, A.Y. *Virtual Water Flows between Nations in Relation to Trade in Livestock and Livestock Products*; UNESCO-IHE: Delft, The Netherlands, 2003.
7. Mekonnen, M.; Hoekstra, A.Y. *The Green, Blue and Grey Water Footprint of Animals and Animal Products*; Unesco-IHE Institute for Water Education: Delft, The Netherlands, 2010.
8. Gerbens-Leenes, P.W.; Mekonnen, M.M.; Hoekstra, A.Y. The water footprint of poultry, pork and beef: A comparative study in different countries and production systems. *Water Resour. Ind.* **2013**, *1*, 25–36. [CrossRef]
9. Bosire, C.K.; Ogutu, J.O.; Said, M.Y.; Krol, M.S.; Leeuw, J.D.; Hoekstra, A.Y. Trends and spatial variation in water and land footprints of meat and milk production systems in Kenya. *Agric. Ecosyst. Environ.* **2015**, *205*, 36–47. [CrossRef]
10. Kim, I.; Kim, K.-S. Estimation of water footprint for major agricultural and livestock products in Korea. *Sustainability* **2019**, *11*, 2980. [CrossRef]
11. Kannan, N.; Osei, E.; Gallego, O.; Saleh, A. Estimation of green water footprint of animal feed for beef cattle production in Southern Great Plains. *Water Resour. Ind.* **2017**, *17*, 11–18. [CrossRef]
12. Bosire, C.K.; Lannerstad, M.; Leeuw, J.D.; Krol, M.S.; Ogutu, J.O.; Ochungo, P.A.; Hoekstra, A.Y. Urban consumption of meat and milk and its green and blue water footprints—Patterns in the 1980s and 2000s for Nairobi, Kenya. *Sci Total Environ.* **2017**, *579*, 786–796. [CrossRef]
13. Bosire, C. *The Water and Land Footprints of Meat and Milk Production and Consumption in Kenya: Implications for Sustainability and Food Security*; University of Twente: Twente, The Netherlands, 2016. [CrossRef]
14. de Miguel, Á.; Hoekstra, A.Y.; García-Calvo, E. Sustainability of the water footprint of the Spanish pork industry. *Ecol. Indic.* **2015**, *57*, 465–474. [CrossRef]
15. Bai, X.; Ren, X.J.; Khanna, N.Z.; Zhou, N.; Hu, M.T. Comprehensive water footprint assessment of the dairy industry chain based on ISO 14046: A case study in China. *Resour. Conserv. Recy.* **2018**, *132*, 369–375. [CrossRef]
16. Bai, X.; Ren, X.J.; Khanna, N.Z.; Zhang, G.P.; Zhou, N.; Bai, Y.; Hu, M.T. A comparative study of a full value-chain water footprint assessment using two international standards at a large-scale hog farm in China. *J. Clean. Prod.* **2018**, *176*, 557–565. [CrossRef]
17. FAOSTAT. FAOSTATFAOSTAT Online Database. Available online: https://www.fao.org/faostat/en/#data (accessed on 25 October 2023).
18. Zhang, C.; Zhao, Y.; Shi, C.; Chiu, Y.-H. Can China achieve its water use peaking in 2030? A scenario analysis based on LMDI and Monte Carlo method. *J. Clean. Prod.* **2021**, *278*, 123214. [CrossRef]

19. Zhang, L.; Dong, H.; Geng, Y.; Francisco, M.-J. China's provincial grey water footprint characteristic and driving forces. *Sci. Total Environ.* **2019**, *677*, 427–435. [CrossRef] [PubMed]
20. Long, H.; Lin, B.; Ou, Y.; Chen, Q. Spatio-temporal analysis of driving factors of water resources consumption in China. *Sci. Total Environ.* **2019**, *690*, 1321–1330. [CrossRef] [PubMed]
21. Li, Y.; Wang, S.; Chen, B. Driving force analysis of the consumption of water and energy in China based on LMDI method. *Energy Procedia* **2019**, *158*, 4318–4322. [CrossRef]
22. Zhang, S.; Su, X.; Singh, V.P.; Ayantobo, O.O.; Xie, J. Logarithmic Mean Divisia Index (LMDI) decomposition analysis of changes in agricultural water use: A case study of the middle reaches of the Heihe River basin, China. *Agric. Water Manag.* **2018**, *208*, 422–430. [CrossRef]
23. Zhi, Y.; Hamilton, P.B.; Zhi, C. Analysis of virtual water consumption in China: Using factor decomposition analysis based on a weighted average decomposition model. *Water Environ. J.* **2015**, *29*, 61–70. [CrossRef]
24. Zhao, C.; Chen, B. Driving Force Analysis of the Agricultural Water Footprint in China Based on the LMDI Method. *Environ. Sci. Technol.* **2014**, *48*, 12723–12731. [CrossRef]
25. Jiang, C.H.; Zang, W.R.; Hu, L.L. Influencing factors analysis of carbon dioxide emissions of Liaoning power consumption based on Kaya model. *Adv. Mater. Res.* **2013**, *779*, 1476–1481. [CrossRef]
26. Han, Q.; Sun, C.Z.; Zou, W. Grey water footprint efficiency measure and its driving pattern analysis on provincial scale in China from 1998 to 2012. *Resour. Sci.* **2016**, *38*, 1179–1191.
27. Ang, B.W.; Liu, F.L. A new energy decomposition method: Perfect in decomposition and consistent in aggregation. *Energy* **2001**, *26*, 537–548. [CrossRef]
28. Ang, B.W.; Zhang, F.Q.; Choi, K.-H. Factorizing changes in energy and environmental indicators through decomposition. *Energy* **1998**, *23*, 489–495. [CrossRef]
29. Duan, P.; Qin, L.; Wang, Y.; He, H. Spatial pattern characteristics of water footprint for maize production in Northeast China. *J. Sci. Food Agric.* **2016**, *96*, 561–568. [CrossRef] [PubMed]
30. Mekonnen, M.M.; Hoekstra, A.Y. *The Green, Blue and Grey Water Footprint of Crops and Derived Crops Products*; UNESCO-IHE Institute for Water Education: Delft, The Netherlands, 2010.
31. Hoekstra, A.Y.; Chapagain, A.K. Water footprints of nations: Water use by people as a function of their consumption pattern. *Integr. Assess. Water Resour. Glob. Change North-South Anal.* **2007**, *21*, 35–48.

water

MDPI

Article

Reference Evapotranspiration Estimation Using Genetic Algorithm-Optimized Machine Learning Models and Standardized Penman–Monteith Equation in a Highly Advective Environment

Shafik Kiraga [1], R. Troy Peters [1,*], Behnaz Molaei [2], Steven R. Evett [3] and Gary Marek [3]

[1] Center for Precision and Automated Agricultural Systems, Irrigated Agriculture Research and Extension Center, Washington State University, Prosser, WA 99350, USA; shafik.kiraga@wsu.edu

[2] Department of Agricultural and Environmental Science, College of Agriculture, Tennessee State University, Nashville, TN 37207, USA; bmolaei@tnstate.edu

[3] United States Department of Agriculture-Agricultural Research Service Conservation and Production Research Laboratory, Bushland, TX 79012, USA; steve.evett@usda.gov (S.R.E.); gary.marek@usda.gov (G.M.)

* Correspondence: troy_peters@wsu.edu

Abstract: Accurate estimation of reference evapotranspiration (ET_r) is important for irrigation planning, water resource management, and preserving agricultural and forest habitats. The widely used Penman–Monteith equation (ASCE-PM) estimates ET_r across various timescales using ground weather station data. However, discrepancies persist between estimated ET_r and measured ET_r obtained from weighing lysimeters (ET_r-lys), particularly in advective environments. This study assessed different machine learning (ML) models in comparison to ASCE-PM for ET_r estimation in highly advective conditions. Various variable combinations, representing both radiation and aerodynamic components, were organized for evaluation. Eleven datasets (DT) were created for the daily timescale, while seven were established for hourly and quarter-hourly timescales. ML models were optimized by a genetic algorithm (GA) and included support vector regression (GA-SVR), random forest (GA-RF), artificial neural networks (GA-ANN), and extreme learning machines (GA-ELM). Meteorological data and direct measurements of well-watered alfalfa grown under reference ET conditions obtained from weighing lysimeters and a nearby weather station in Bushland, Texas (1996–1998), were used for training and testing. Model performance was assessed using metrics such as root mean square error (RMSE), mean absolute error (MAE), mean bias error (MBE), and coefficient of determination (R^2). ASCE-PM consistently underestimated alfalfa ET across all timescales (above 7.5 mm/day, 0.6 mm/h, and 0.2 mm/h daily, hourly, and quarter-hourly, respectively). On hourly and quarter-hourly timescales, datasets predominantly composed of radiation components or a blend of radiation and aerodynamic components demonstrated superior performance. Conversely, datasets primarily composed of aerodynamic components exhibited enhanced performance on a daily timescale. Overall, GA-ELM outperformed the other models and was thus recommended for ET_r estimation at all timescales. The findings emphasize the significance of ML models in accurately estimating ET_r across varying temporal resolutions, crucial for effective water management, water resources, and agricultural planning.

Keywords: machine learning; genetic algorithm; advective environments; radiation components; aerodynamic components; reference evapotranspiration

Citation: Kiraga, S.; Peters, R.T.; Molaei, B.; Evett, S.R.; Marek, G. Reference Evapotranspiration Estimation Using Genetic Algorithm-Optimized Machine Learning Models and Standardized Penman–Monteith Equation in a Highly Advective Environment. *Water* **2024**, *16*, 12. https://doi.org/10.3390/w16010012

Academic Editors: Songhao Shang, Magdy Mohssen, Qianqian Zhang, Dongqin Yin and Hamza Gabriel

Received: 29 October 2023
Revised: 8 December 2023
Accepted: 11 December 2023
Published: 20 December 2023

1. Introduction

Evapotranspiration (ET) is a significant factor in the hydrological cycle and is frequently used to calculate hydrological losses through several important processes that take place between the ground and the atmosphere. It is essential for the optimal design of irrigation schedules [1], management of regional water resources [2], and estimation of different hydrological processes [3]. ET has a significant impact on several terrestrial ecosystem processes as well as pertinent characteristics, such as soil water content and energy balances [4]. Considerable progress in estimating ET and understanding the mechanisms of its ongoing variations in daily, annual, and inter-annual timescales has been made through numerous studies, motivated by the early awareness of the importance of water as an essential resource for life sustainability on earth. Despite the findings in these studies, the complex and nonlinear processes that dominate evapotranspiration have made its estimation a great challenge, partly due to several influencing factors, such as landform, geomorphological, soil moisture, and vegetation traits [5–7]. In this context, the precise estimation of ET is still of particular importance for careful water resource management.

An overview of the evolution of ET-estimating methods during the previous century was discussed in Ref. [8], and these methods are often classified as direct and indirect. Direct methods, such as lysimeters and micrometeorological techniques, demand special construction and high maintenance, which is expensive. Indirect methods, on the other hand, are less expensive and time-saving and are in some contexts regarded as suitable alternatives to the direct ones. The indirect methods are commonly classified into water-balance-based, radiation-based, mass-transfer-based, and temperature-based models [9]. Because it considers both aerodynamic and thermodynamic factors, the FAO Penman–Monteith (FAO-PM) model for calculating the ET of a reference short (grass) crop continues to be the most extensively used indirect method for ET_r estimation in a variety of regions and climates [10]. The Food and Agriculture Organization (FAO) approved the equation after determining that it accounts for all the variables influencing evapotranspiration and fixes most of the flaws in the other empirical techniques.

The Penman–Monteith Equation was slightly modified and standardized by the American Society of Civil Engineers (ASCE) for both tall crop (alfalfa) (ET_{rs}) and short crop (clipped grass) surfaces with similar parameterizations as the FAO-PM for computation of the equation components after national and international discussions on the adoption of a taller reference crop [11]. The result was the ASCE-PM equation, and it was formulated to allow calculations for both daily and hourly or shorter time steps [12]. The complexity involved in the calculations of all the required inputs in the ASCE-PM is a disadvantage for its application, which could lead to significant errors [13]. Another drawback in using the ASCE-PM is the difficulty in obtaining the extensive weather data needed and the absence of adequate historical records for each study location, which are crucial aspects for calculating reliable ET_r estimates [4]. This can be particularly challenging in developing countries, where there are limited meteorological stations and a scarcity of weather data records.

Reference evapotranspiration depends on latitude, altitude, and several climatic variables, such as relative humidity, air temperature, soil temperature, wind speed, net radiation, and dew point temperature [14]; several researchers have utilized a combination of these different parameters to model ET_o using machine learning models at daily and monthly timescales to overcome the identified limitations and difficulties of traditional methods [15]. CatBoost, generalized regression neural network (GRNN), and the random forest (RF) models were evaluated for estimating daily ET_o in the arid and semi-arid regions of northern China, using limited meteorological data with eight different combinations of inputs [16]. It was found that CatBoost demonstrated superior performance and was identified as the most effective method for estimating ET_o. The efficiency of extreme learning machines (ELM) was compared to the empirical Penman–Monteith equation and the feedforward backpropagation (FFBP) in predicting ET_o for three meteorological stations in Iraq, utilizing meteorological data from thirteen years (2000–2013) as inputs [5]. They concluded that the ELM model demonstrated efficiency, simplicity, high speed, and good generalization performance for ET_o estimation. Four different variants of an extreme learning machine (ELM) model optimized using bio-inspired search algorithms were evaluated to estimate daily reference evapotranspiration (ET_r) across various regions in China using data from eight meteorological stations [1]. The results highlighted the effectiveness of bio-inspired optimization algorithms, particularly the FPA and CSA algorithms, in enhancing the performance of the conventional ELM model for daily ET_o prediction. Kernel-based (Gaussian process regression (GPR), support vector regression (SVR)) and deep learning methods (Broyden–Fletcher–Goldfarb–Shanno artificial neural network (BFGS-ANN)) were compared to long short-term memory (LSTM) for estimation of monthly reference evapotranspiration using minimal meteorological parameters in ten different combinations [17]. The results showed that all four methods predicted ET_o amounts with acceptable accuracy and error levels.

Due to consistent climatic changes and the complexity of the evapotranspiration process that leads to its high variability in time and space, in this study, modeling the ET_r was extended to quarter-hourly and hourly timescales. Meteorological data at high temporal resolution has become easier to collect due to the recent advancements in the development of automatic weather stations [18]. This, in turn, boosts the estimation of ETr at such resolutions, which is commonly used for calibrating surface energy balance models in the determination of geospatial evapotranspiration from drone imagery or satellite images [19,20]. Therefore, assessing the performance of the ASCE-PM at timescales lower than the daily interval is increasingly becoming necessary.

To the best of our knowledge, there are no studies that have evaluated the use of ML models for the estimation of ET_r at lower timescales, such as hourly or quarter-hourly with varied meteorological input data. In addition, studies that evaluated ET_r at a daily timescale mostly considered parameters directly measured from weather stations. This is desirable for the estimation of reference evapotranspiration in areas with incomplete meteorological data [9]. However, the sensitivity of ET_r to different meteorological variables (directly measured or those calculated from directly measured) has been studied [21,22], and they were all found to influence the energy budget of the surface [23]. Changes in wind speed produced the largest decrease in ET_r, followed by vapor pressure, net radiation, and mean temperature [21]. Also, computed ET_r was found to be most sensitive to net radiation, followed by vapor pressure deficit and wind travel transfers; the contribution of the aerodynamic and net radiation components to the ET_r value varied throughout the year [24]. The difference in the contribution of these components could be attributed to the difference in climates [8]. Therefore, it is evident that for proper evapotranspiration estimation, proper assessment of all meteorological parameters (directly measured or not) is necessary. In this study, we combined directly measured weather parameters with subsequently calculated parameters for ET_r estimation using machine learning models.

Advantages of ML include the ability to use reduced data inputs, capturing non-linearity in the data inputs, and utilizing the computing power of modern-day computer systems to analyze big data. Collectively, these factors have made ML algorithms attractive options for estimating ET_r. The choice of the best possible algorithm and the choice of adequately representative variables are among the challenging aspects of any ML task [25]. Moreover, the performance of ML algorithms strongly depends on the size and structure of available data [25]. To improve performance, several bio-inspired algorithms, such as GA, are coupled with ML models to find the optimal set of parameters during model training [1,5]. This is because default optimization algorithms, such as backpropagation, are often trapped in the local minima, several parameters influence its speed and robustness, and its best parameters appear to vary from problem to problem [26]. Unlike the backpropagation technique that always adjusts weights towards the descending direction of the error function, GA is a parallel stochastic optimization algorithm good at global searching. The drawback of GA is its slowness during model training due to its exploration mechanism through reproduction, crossover, and mutation, as well as searching for optimal solutions from random genes [27,28]. Nonetheless, GA provides multiple optimal solutions from the search space, and has thus gained prominence in recent years. For evapotranspiration studies, ML models coupled with GA were found to perform better than the corresponding single models [29–32]. Therefore, in this paper, GA was utilized to optimize ML models and compared to standardized ASCE-PM for ET_r estimation on daily, hourly, and quarter-hourly timescales considering different input meteorological variables.

2. Materials and Methods

2.1. Lysimetric and Weather Data Collection

A research study was set up for the actual estimation of alfalfa reference evapotranspiration in Bushland, Texas, facility of the USDA Agricultural Research Service Conservation and Production Research Laboratory, in 1996. Bushland has a semi-arid climate that is impacted by local and regional advection events [33]. The site was subdivided into two square fields, designated northeast (NE) and southeast (SE), each measuring ~5.0 ha. Monolithic weighing lysimeters (NE and SE) with 3 m × 3 m surface dimensions and 2.4 m deep were located at the center of each of the two fields. These were used for the direct measurement of ET_r-lys. The details about the study area and the lysimeter's site in Bushland (35°11′ N, 102°6′ W, and 1170 m above MSL) can be found in Refs. [34,35]. Alfalfa was seeded in the two fields, which were irrigated simultaneously by a Lindsay lateral move sprinkler system to maintain a well-watered condition. Experiments were conducted for 4 years from 1996 through 1999 [36], but the data for the year 1999 were omitted in this study because reference conditions for a tall reference crop (alfalfa) were not always met [10]. The leaf area index (LAI), growth stage, and plant height were measured periodically between and at each harvest in each field (Figure 1). The whole planting and agronomic management of alfalfa crops during the four-year growing period were described in Ref. [36], and the data and metadata are available in Ref. [37].

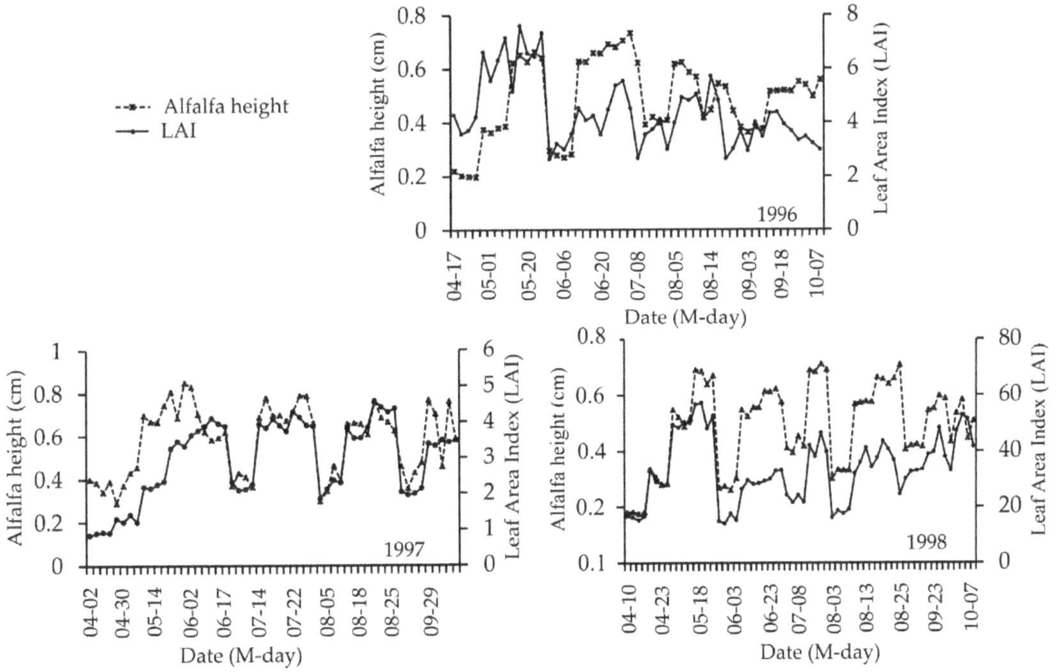

Figure 1. Alfalfa harvesting patterns and days when alfalfa height and LAI measurements were taken for growing seasons from 1996 to 1998.

The methodology for meteorological data acquisition was extensively described in Refs. [35,36] and are available in Ref. [38]; therefore, only the measured variables and other noteworthy points will be presented in this work. Meteorological measurements were made at 5 s intervals and reported as quarter-hourly averages. This meteorological data, including the mean air temperature at 2.0 m height, mean relative humidity at 1.8 m height, mean wind speed at 2.0 m height, and mean solar irradiance were necessary to obtain the ET_r calculated in this study.

2.2. Data Processing

Lysimeter and meteorology data were processed and analyzed under three different timescales: quarter-hourly, hourly, and daily. Only the days when alfalfa height was at least 0.5 m were used in this study in accordance with reference conditions for a tall reference crop (alfalfa) [10]. Also, days were not considered if irrigation or rainfall affected the accuracy of the water balance calculations for measured ET_r or if the crop was lodged or badly watered. Figure 2 shows the K_c values for the days selected in this study. Note that the K_c values were often >1, which indicates that alfalfa ET under reference conditions exceeded the ASCE 2005 PM reference ET in the advective environment of Bushland.

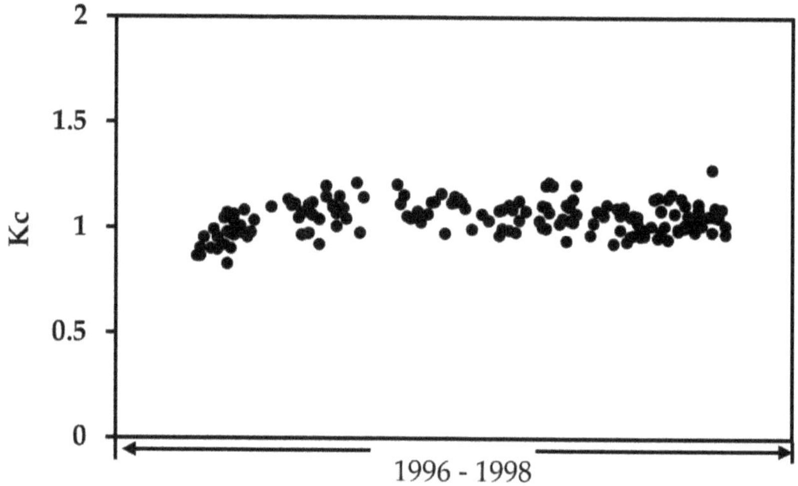

Figure 2. Kc values for the days when reference conditions were met for alfalfa growing period (1996–1998).

ETr-lys values from the NE and SE lysimeters were averaged for each of the data points to obtain the full ET_r-lys dataset under the quarter-hourly timescale. The hourly ET_r-lys was computed as the sum of four (4) consecutive quarter-hourly readings, while the same number of readings were averaged to obtain the mean hourly values of the meteorological variables (air temperature, wind speed, solar radiation, relative humidity). On the daily timescale, the daily ET_r-lys was computed as the sum of all ninety-six (96) consecutive quarter-hourly readings, while the meteorological variables were extracted as follows; the maximum and minimum values of temperature and relative humidity were taken as the maximum and minimum of the 15 min average values for a given day, respectively. The ninety-six (96) consecutive quarter-hourly readings were averaged to obtain the mean daily values of the meteorological variables (wind speed, solar radiation, temperature). This resulted in daily meteorological data, including maximum temperature (T_{max}), minimum temperature (T_{min}), mean wind speed (u_2), maximum relative humidity (RH_{max}), minimum relative humidity (RH_{min}), and mean solar radiation (R_s).

2.3. Calculation of Parameters for Reference Evapotranspiration Estimation

Equation (1) presents the form of the ASCE-PM by Ref. [12] for different time steps. The constants C_d and C_n for each of the timescales are shown in Table 1.

$$ET_{rs} = \frac{0.408\Delta(R_n - G) + \gamma\frac{C_n}{T+273}u_2(e_s - e_a)}{\Delta + \gamma(1 + C_d u_2)} \quad (1)$$

where ET_{rs} is the standardized reference crop evapotranspiration for a tall surface (mm d^{-1} for daily time steps or mm h^{-1} for hourly time steps); R_n is the net radiation at the crop surface (MJ m^{-2} d^{-1} for daily time steps or MJ m^{-2} h^{-1} for hourly time steps); G is the soil heat flux density at the soil surface at the daily (MJ m^{-2} d^{-1}) and hourly (MJ m^{-2} h^{-1}) basis; T is the mean daily or hourly air temperature at 1.5 to 2.5 m height (°C); u_2 is the mean daily wind speed at 2 m height (m s^{-1}); e_s is the saturation vapor pressure at 1.5 to 2.5 m height (kPa); e_a is the mean actual vapor pressure at 1.5 to 2.5 m height (kPa); Δ is the slope of the saturation vapor pressure–temperature curve (kPa $°C^{-1}$); γ is the psychrometric constant (kPa $°C^{-1}$); C_n is the numerator constant that changes with reference type and calculation time step; and C_d is the denominator constant that changes with reference type and calculation time step (s m^{-1}).

Table 1. Parameters of the ASCE-PM at different timescales.

Version	Time Step	C_n	C_d	r_s (m s^{-1})
			Tall reference (0.5 m high)	
ASCE-PM	Daily	1600	0.38	45
	Hourly (daytime)	66	0.25	30
	Hourly (nighttime)	66	1.7	200

The ASCE-PM reference evapotranspiration was calculated using REF-ET software (Version 4.1.4.22) [39], which calculates ET_o and ET_r for grass and alfalfa as short and tall reference crops, respectively, on different timescales. It can be used to calculate reference evapotranspiration on monthly, daily, and hourly or less timescales, and it has been adopted in several studies [40,41]. The calculated ET_{rs} values from the software were compared to the outputs of ML models at daily, hourly, and quarter-hourly timescales.

2.4. Machine Learning Algorithms and Optimization

Four ML models with different operation principles were used in this study to estimate ET_r: random forest (RF), extreme learning machine (ELM), support vector regression (SVR), and artificial neural network (ANN). These models have gained prominence in evapotranspiration studies in recent years [42,43]. ELM is different from ANN and SVR, as it does not require iterative training, and its hidden layer parameters are randomly selected [1]. It was first proposed by Ref. [44] and has received wide applicability due to its fast convergence speed, strong generalization ability, and no local extrema [45]. RF is a tree-based approach that manages high-dimension regression problems, where the final decision results via a bagging procedure. The structure of each of these models is shown in Figure 3, but a further detailed description of the definitions and principle of operation is out of the scope of this paper; it has been extensively discussed elsewhere [46–49].

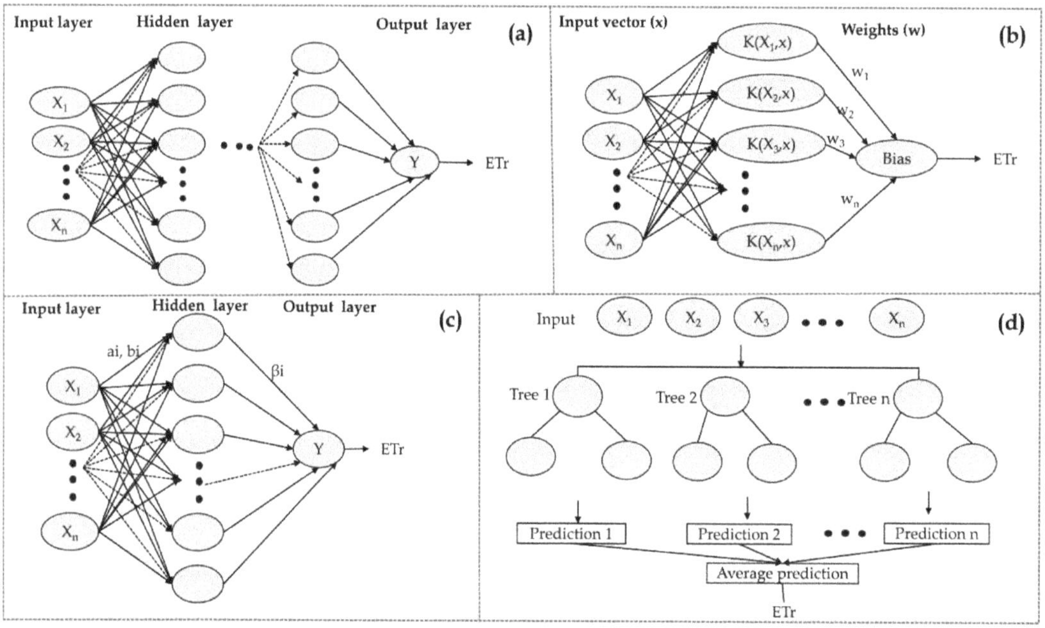

Figure 3. Structure of the machine learning models: ANN (**a**), SVR (**b**), ELM (**c**), and RF (**d**).

2.4.1. Genetic Algorithm

The GA is a widely used optimization technique that has shown promise in agricultural studies for fine-tuning the parameters of ML models [50]. It is an evolutionary algorithm used to search for optimized solutions to the natural evolutionary process through simulation [1]. Several evapotranspiration studies indicated that coupling GA with machine learning models, such as ANN, SVR, and ELM, yields better results than the single models [29–32]. An extensive overview of the implementation of a genetic algorithm including potential integration with ML models was provided in Ref. [51]. In this study, its integration with ML models is demonstrated in Figure 4. Across all models, the population size, mutation probability, crossover probability, and the number of generations of the GA were set to 50, 0.2, 0.3, and 5, respectively. Also, input features were normalized within the range of [−1, 1] using the Mix and Max method [4].

For the SVR model, the radial basis function was used as the kernel function, and the regularization parameter and kernel parameter (gamma) were optimized with them ranging from 5 to 1000, and from 0.05 to 0.99, respectively.

For the RF models, the maximum depth of the tree (max_depth), the number of features (max_features), the minimum number of samples required to be at a leaf node (min_samples_leaf), the minimum number of samples required to split an internal node (min_samples_split), and the number of trees in the forest (n_estimators) were optimized ranging from 5 to 500, from 2 to 6, from 5 to 20, from 5 to 20, and from 1 to 500, respectively.

The weight matrix and bias vector of the ELM were randomly generated and the activation function was set to sigmoid. The hidden units and the regularization parameters were optimized from 5 to 1000 and from 0.02 to 0.9, respectively. Detailed descriptions of the ELM parameters used in this study are extensively discussed in [44].

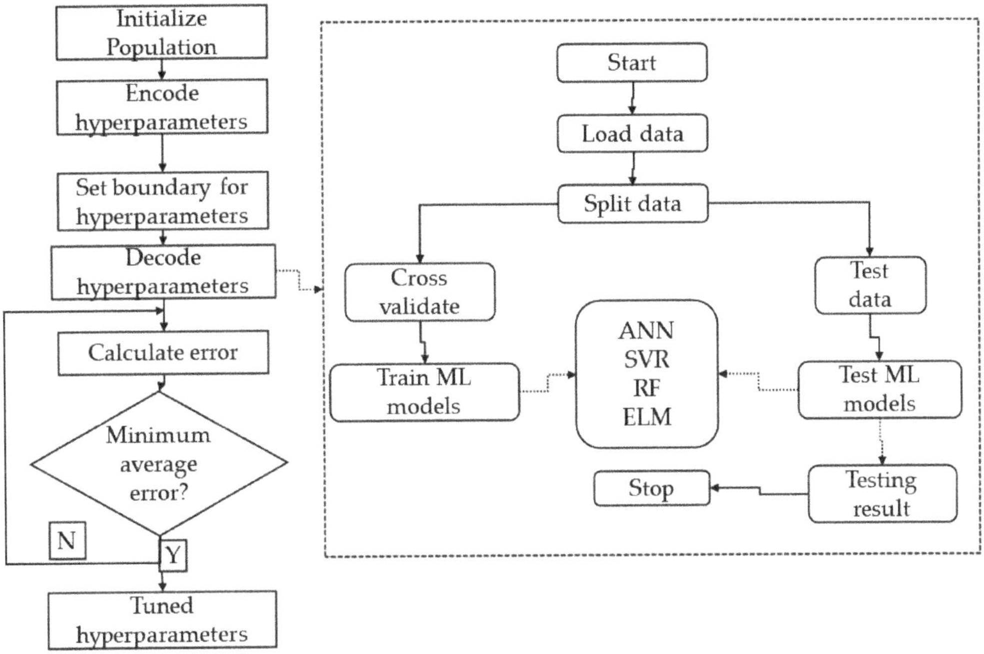

Figure 4. Flow diagram for tuning the hyperparameters of the machine learning models.

The ANN model consisted of an input layer, a hidden layer, and the output layer. The Rectified Linear Unit (ReLu) was taken as the activation function. The neg_mean_squared_error and neg_mean_absolute_error of the scikit-learn package in Pyhtonv3.8 were used as the loss functions. The number of neurons and hidden layers were both optimized ranging from 3 to 100, while the learning rate was optimized from 0.01 to 0.3. Additional parameters were taken as the default values in the scikit-learn package.

2.4.2. k-Folds Cross Validation

The k-fold cross-testing approach was used during this phase to train and test the models. The meteorological dataset was randomly divided into training and testing datasets, each with 80% and 20% of the total data, respectively. The training dataset was then equally divided into five folds, of which four were utilized to train the models and one for model testing. To make sure that each fold was used at least once for model testing, the procedure was carried out five times. The performance of each of the model hyperparameters was assessed by the resulting error. The errors of the five trials were averaged as the expected generalization error. The parameters that provided the minimum average error were returned as the tuned hyperparameters.

2.4.3. Arrangement of Datasets for Machine Learning Models

The datasets for ML models were arranged to reflect the impacts of aerodynamic and radiation components on ET_r (Figure 5). Aerodynamic components are represented by wind speed, relative humidity, and vapor pressure deficit (VPD), while radiation components are represented by net radiation (R_n), solar irradiance (R_s), air temperature, vapor pressure–temperature curve (Δ), and the relative cloudiness (R_s/R_{so}).

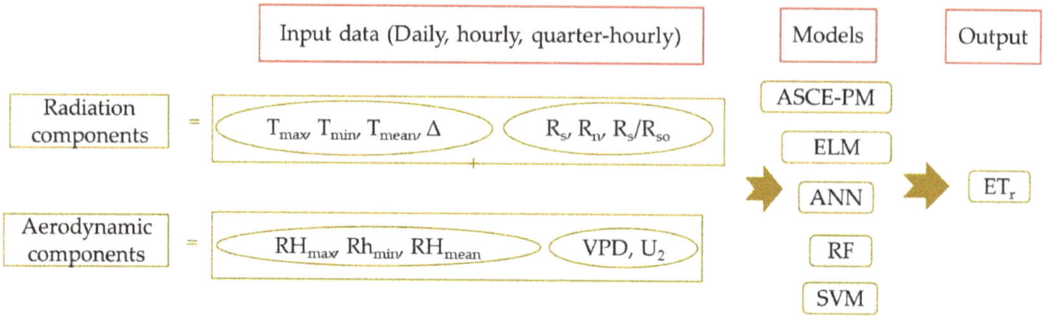

Figure 5. Representation of radiation and aerodynamic components for ET_r estimation.

Further considerations included the simplicity of measuring the weather parameters, the ease of calculation of the intermediate parameters, and the completeness of the climatic variables. Parameters u_2, R_s, and the means, minimums, and maximums of temperature and relative humidity can be directly measured by weather stations. Parameters that are calculated from the directly measurable variables include the VPD, the R_s/R_{so}, the Δ, and the R_n. The requirement for the calculation of all these parameters for ET_r estimation has been cited as one of the challenges of the ASCE-PM that could lead to significant errors [13]. Such errors could be reduced by determining the most crucial parameters that influence evapotranspiration for a given area through machine learning algorithms. For this, eleven DTs were considered for the daily timescale, while seven input datasets (DTs) were considered for the hourly and quarter-hourly timescales, as shown in Table 2. On a daily timescale, DT_3, DT_5, and DT_{10} contain all or the majority of the aerodynamic components, while DT_1, DT_4, DT_7, and DT_8 provide a good mix of both aerodynamic and radiation components. The presence of aerodynamic components is reduced in DT_6, DT_9, and DT_{11}. Among these datasets, all the parameters in DT_8 can be directly measured from a weather station, followed by DT_3 and DT_4, which have additional parameters that require computation. The need for computation is increased in DT_1 and DT_5. A similar arrangement was followed at hourly and quarterly timescales, however, in this case, the minimums and maximums of relative humidity and temperature were replaced by their mean values. The calculation of all the parameters we included in our datasets is defined and extensively explained in Ref. [12]. R_s/R_{so} represents relative cloudiness and can be derived from pyranometer data and calculated R_{so} values [52]. The calculation of VPD requires air temperature and relative humidity, while Δ requires air temperature. The net radiation indicates the amount of solar irradiance absorbed by vegetation, and it is commonly calculated from the short and long-wave radiation components (Equation (2)) at either a daily, hourly, or lower timescale.

$$R_n = R_{ns} - R_{nl} \tag{2}$$

where R_{ns} and R_{nl} are net short-wave and net outgoing long wave radiation in MJ m^{-2} d^{-1}/h^{-1}, respectively.

The equation for R_{ns} (Equation (3)) does not differ based on timescale, while that of R_{nl} (Equations (4) and (5)) differs based on the timescale considered [12].

$$R_{ns}\text{: daily: and hourly } (1-\alpha)Rs \tag{3}$$

$$R_{nl}\text{: daily}: \sigma f_{cd}(0.34 - 0.14\sqrt{e_a})\left[\frac{T_{K\,max}^4 + T_{K\,min}^4}{2}\right] \tag{4}$$

$$R_{nl}\text{: hourly}: \sigma f_{cd}(0.34 - 0.14\sqrt{e_a})T_{K\,hr}^4 \tag{5}$$

where α (0.23) is albedo [dimensionless], R_s is incoming solar radiation [MJ m^{-2} h^{-1}/d^{-1}], σ (2.042×10^{-10}) is Stefan–Boltzmann constant [MJ $K^{-4}m^{-2}$ h^{-1}/d^{-1}], f_{cd} is a cloudiness function [dimensionless] calculated from the relative solar radiation (R_s/R_{so}), e_a is actual vapor pressure [kPa], T_{Khr} is mean absolute temperature during the hourly period [K], and $T_{K\,max}$ and $T_{K\,min}$ are the maximum and minimum absolute temperatures during the 24-h period [K], respectively.

Table 2. Parameter combinations at different timescales used for training machine learning models.

					Daily						
Factor	**DT$_1$**	**DT$_2$**	**DT$_3$**	**DT$_4$**	**DT$_5$**	**DT$_6$**	**DT$_7$**	**DT$_8$**	**DT$_9$**	**DT$_{10}$**	**DT$_{11}$**
VPD	×	×	×		×					×	
Δ	×	×			×						
R_n	×	×	×	×							
u_2	×	×	×	×	×	×	×	×		×	
R_s/R_{so}		×	×		×	×	×		×		×
T_{mean}			×	×			×	×			
RH_{mean}			×	×	×		×	×			
RH_{max}						×					
R_s						×	×	×		×	×
RH_{min}						×					
T_{max}						×			×	×	×
T_{min}						×			×	×	×
					Quarter-hourly and Hourly						
VPD	×	×	×		×						
Δ	×	×			×						
R_n	×	×	×	×							
u_2	×	×	×	×	×	×	×				
R_s/R_{so}		×	×		×	×					
R_s						×	×				
T_{mean}			×	×		×	×				
RH_{mean}			×	×	×	×	×				

2.5. Evaluation Metrics

The competency of the ASCE-PM, GA-SVR, GA-ANN, GA-ELM, and GA-RF models for estimating ET_r-lys at daily, hourly, and quarter-hourly timescales was assessed by comparing the estimated ET from each model with the ET measured using the lysimeters and quantified using four commonly used statistical indices, i.e., the coefficient of determination (R^2) [1], the mean bias error (MBE) [53], the mean absolute error (MAE), and the root mean squared error (RMSE) [54], which can be expressed as follows:

$$R^2: 1 - \frac{\left[\sum_{i=1}^{n} (y_i - y_i)(x_i - x_i)\right]^2}{\sum_{i=1}^{n} (y_i - y_i)^2 \sum_{i=1}^{n} (x_i - x_i)]^2} \quad (6)$$

$$\text{RMSE}: \sqrt{\frac{\sum_{i=1}^{n} (y_i - x_i)^2}{n}} \quad (7)$$

$$\text{MAE}: \frac{\sum_{i=1}^{n} |y_i - x_i|}{n} \quad (8)$$

$$\text{MBE: } \frac{\sum_{i=1}^{n}(y_i - x_i)}{n} \tag{9}$$

Note: y_i denotes estimated ET_r, and x_i denotes the observed ET_r.

3. Results and Discussion

3.1. Comparison of the Estimation Accuracy of the ASCE-PM and Machine Learning Models at a Daily Timescale

The estimated ET_r using ASCE-PM (ET_{rs}) and that from ML models was compared to that from weighing lysimeters (ET_r-lys) on a daily timescale. Table 3 indicates that DT_3, DT_4, and DT_6 provided the best-performing estimators of ET_r-lys across the different ML models. However, Figure 6 illustrates that the ASCE-PM equation performed better than these models (considering R^2) but tended to underestimate ET_r-lys (slope of 0.91, MBE = −0.43 mm/day) for values from 7.5 mm day^{-1} and greater. Also, the residual plot in Figure 7b shows that the largest errors occurred at approximately > 7.5 mm day $^{-1}$ but, generally, the equation seems to persistently underestimate ET_r-lys throughout all estimates. Underestimation or overestimation of ET_r-lys by the different variants of the Penman–Monteith equation due to low and high evaporative demands has been reported in earlier studies, and it was attributed to the difference in local climatic conditions and lysimetric measurement errors. Due to the semi-arid climate and relatively large wind speeds at Bushland [33], the slight deviations between ET_{rs} and ET_r-lys could be attributed to the advective transport that adds energy, thus increasing the ET_r-lys in the reference alfalfa fields, in agreement with Ref. [55]. Such underestimations were reported from other highly advective environments [56].

Table 3. Statistical indicators for machine learning ET_r estimations at a daily timescale (fit linear equations representing the slope and intercept for each respective model are presented in italics).

	Daily																	
	DT_1			DT_2			DT_3			DT_4			DT_5			DT_6		
	R^2	RMSE	MAE	R^2	RMSE	MAE	R^2	RMSE	MAE	R^2	RMSE	MAE	R^2	RMSE	MAE	R^2	RMSE	MAE
ANN	0.83	1.12	0.79	0.76	1.41	0.84	0.83	1.17	0.86	0.91	0.84	0.58	0.82	1.19	0.90	0.87	1.01	0.74
	0.93x + 0.53			*0.94x + 0.52*			*0.97x + 0.31*			*0.90x + 0.70*			*0.83x + 1.10*			*0.95x + 0.26*		
ELM	0.85	1.02	1.03	0.68	0.99	0.68	0.89	0.91	0.61	0.90	0.85	0.56	0.80	1.20	0.90	0.88	0.93	0.61
	0.87x + 1.07			*0.90x + 0.79*			*0.97x − 0.03*			*0.93x + 0.41*			*0.93x + 0.31*			*0.95x + 0.26*		
SVR	0.79	1.24	0.81	0.77	1.28	0.79	0.87	0.95	0.69	0.88	1.92	0.63	0.74	1.39	1.11	0.84	1.07	0.80
	0.92x + 0.50			*0.89x + 0.84*			*0.93x + 0.35*			*0.89x + 0.82*			*0.78x + 1.57*			*0.88x + 0.76*		
RF	0.83	1.06	0.92	0.86	0.99	0.98	0.85	1.12	1.04	0.89	0.83	0.73	0.81	1.21	1.14	0.85	0.90	0.83
	0.82x + 1.22			*0.97x + 0.24*			*1.01x + 0.34*			*0.98x + 0.37*			*0.74x + 2.15*			*0.97x + 0.45*		
	DT_7			DT_8			DT_9			DT_{10}			DT_{11}			ASCE-PM		
ANN	0.83	1.21	0.86	0.89	0.93	0.70	0.62	1.67	1.30	0.77	1.31	0.85	0.83	1.27	0.99	0.94	0.75	0.57
	0.99x + 0.17			*0.94x + 0.34*			*0.63x + 2.76*			*0.86x + 0.98*			*0.83x + 1.43*			*0.91x + 0.30*		
ELM	0.83	1.09	0.79	0.88	0.92	0.69	0.63	1.63	1.31	0.81	1.17	0.73	0.81	1.17	0.91			
	0.96x + 0.31			*0.85x + 1.09*			*0.68x + 2.46*			*0.92x + 0.56*			*0.83x + 1.50*					
SVR	0.81	1.18	0.82	0.87	0.98	0.68	0.43	2.03	1.51	0.73	1.39	0.90	0.66	1.57	1.08			
	0.92x + 0.71			*0.88x + 0.81*			*0.435x + 4.42*			*0.85x + 1.15*			*0.65x + 3.02*					
RF	0.91	0.95	0.86	0.89	0.93	0.89	0.73	1.37	1.40	0.84	1.07	1.12	0.83	1.17	1.06			
	0.94x + 0.38			*0.97x + 0.46*			*0.92x + 0.79*			*0.83x + 0.84*			*0.82x + 1.22*					

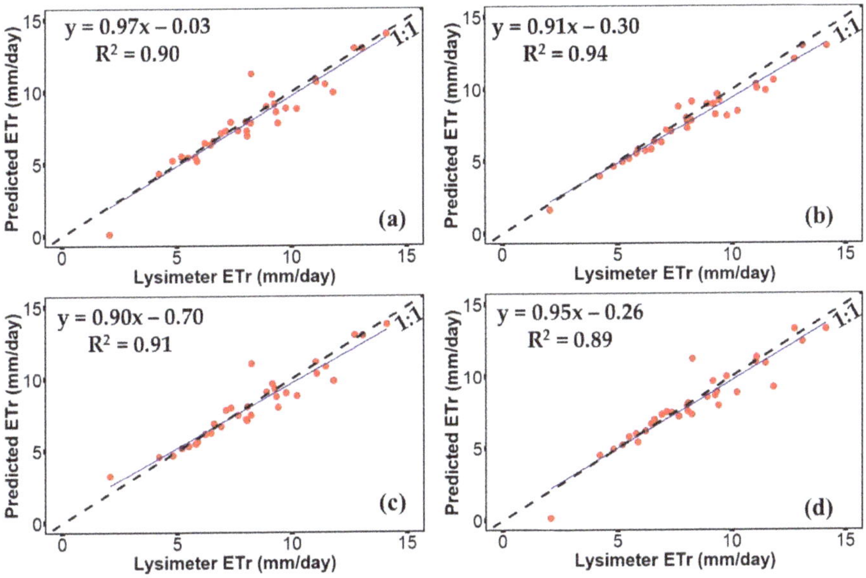

Figure 6. Relationship between ASCE, machine learning algorithms, and ET_r-lys at daily timescale for days when reference conditions were met. (**a**) GA-ELM-DT_3, (**b**) ASCE-PM, (**c**) GA-ANN-DT_4, (**d**) GA-ELM-DT_6.

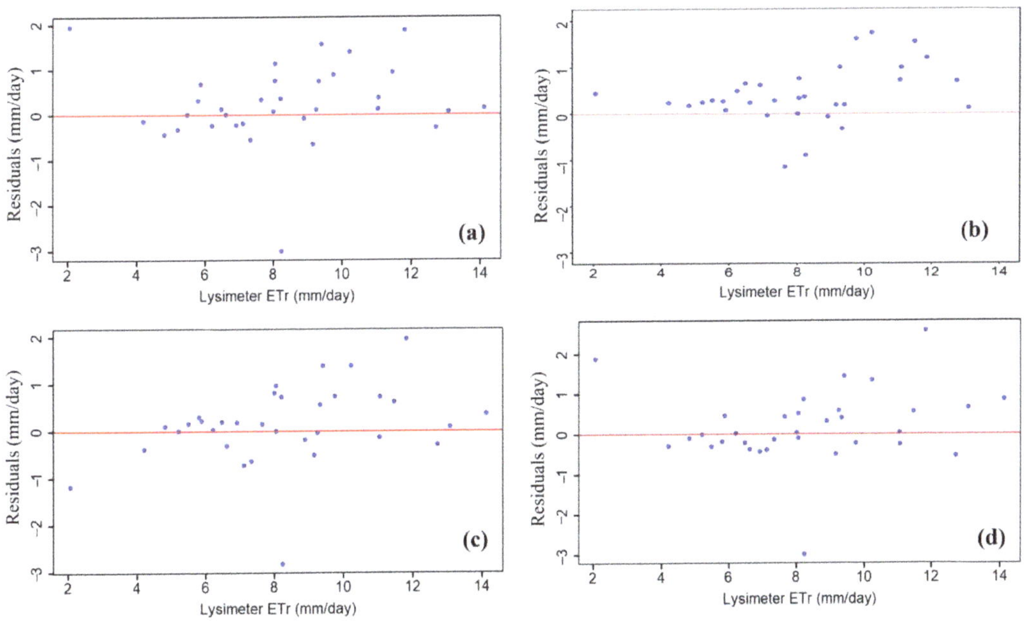

Figure 7. Residual plots for ASCE *and* machine learning algorithms at *a* daily timescale. (**a**) GA-ELM-DT3, (**b**) ASCE-PM, (**c**) GA-ANN-DT4, (**d**) GA-ELM-DT6.

Comparison of the ET_r from ML models with the ETr-lys shows that the underestimation was reduced, most significantly for GA-ELM, when applied to DT_3 (slope = 0.97, MBE = −0.25) and DT_6 (slope = 0.95, MBE = −0.19) (Figure 6a,d). It also demonstrates that GA-ELM can reduce the positive offset and slope, similar to Refs. [22,57], who found ELM to be suitable for reference evapotranspiration estimation. The best-performing model (GA-ANN-DT_4, considering R^2) had a smaller slope (0.90), although it reduced underestimation (MBE = −0.15). According to RMSE values shown in Table 3, ASCE-PM performed better than both the SVR and RF models for the majority of the datasets. These models performed close to the ASCE-PM with RMSE values of 0.98 and 0.93 for the GA-SVR and GA-RF models, respectively, for the DT_8. However, the high positive offsets, 0.81 and 0.46 for the GA-SVR and GA-RF, respectively, indicate the poor performance of the models as compared to that of ASCE-PM (0.30). The statistical parameters shown in Table 3 give a clear idea about the interaction of the meteorological variables used in estimating ET_r-lys across the different models. Based on these results, u_2 was involved in all the best-performing datasets (DT_3, DT_4, and DT_6); therefore, it can be concluded that it is one of the most relevant estimators for daily ET_r-lys in this region and that accurate estimates of ET_r-lys might not be achieved without the inclusion of this parameter. Also, the inclusion of VPD and R_s/R_{so} in DT_3 led to improved slope and offset values compared to those of DT_4 for GA-ELM, GA-ANN, and GA-SVR. In DT_6, when R_n was excluded from the inputs, it did not affect the performance of the ML models. This result presents an opportunity for estimating ET_r-lys without the long R_n calculations and could suggest that in dry and advective environments, radiation components might primarily contribute to pressure deficit through temperature changes, but aerodynamic components play a major role in driving evapotranspiration. This might affirm the relevancy of the inclusion of aerodynamic components in ML models in advective environments for daily ET_r-lys calculations.

Values of RMSE and MAE for GA-ELM-DT_4 and GA-ELM-DT_5 confirm that the inclusion of Δ, VPD, and R_s/R_{so} in DT_5 decreased the model accuracy from a MAE of 0.56 mm/day and RMSE of 0.85 mm/day to a MAE of 0.90 mm/day and RMSE of 1.20 mm/day (Table 3). Removing Δ in GA-ELM-DT_3 and replacing it with T_{mean} and R_n reduced the MAE from 0.90 mm/day to 0.61 mm/day. A similar trend is observed in all the datasets where it was included (DT_1, DT_2). The effect of the inclusion of R_s/R_{so} (an indication of cloudiness) can also be assessed by looking at the values of MAE and RMSE for DT_7 and DT_8 in Table 3. It is evident that across all models, the MAE and RMSE for DT_7 ranged from 0.79 mm/day to 0.86 mm/day and from 0.95 mm/day to 1.21 mm/day, respectively, which are larger than the range for DT_8 (0.68 mm/day–0.89 mm/day and 0.92 mm/day–0.98 mm/day, respectively). Therefore, this suggests that Δ and R_s/R_{so} might not be suitable for daily ETr-lys estimation in the advective environment tested.

It is observed from Table 3 that replacing the maximums and minimums of relative humidity and temperature with their mean values for DT_6 and DT_7 slightly improved the RMSE values. However, when the R_s/R_{so} was excluded as shown for DT_8, the RMSE and MAE were improved. This means R_s/R_{so} might not be suitable as a direct input into ML models for daily ET_r estimation for the conditions of this study. MAE and RMSE values for DT_4 and DT_8 indicate that R_n and R_s might have a similar effect on the daily ET_r-lys estimation and can be used interchangeably. Hence, precautions should be taken into consideration for deploying ML models in locations such as the Bushland, TX station when records of both R_n and R_s are missing. Among all the ML models tested, GA-ANN subjected to DT_4 produced the largest R^2 value (0.91), while the GA-ELM applied to DT_8 produced the slope closest to unity and the smallest offset. Figure 7a,c,d also reveals that residuals for the GA-ELM and GA-ANN were persistently close to the zero line, which indicates that these models can reduce underestimation or overestimation of ET_r-lys.

Overall, the results indicated that on a daily timescale, ML models can give better estimates of ETr-lys than the ASCE-PM under the tested conditions, and we can learn some things about the importance of the various input parameters on the accuracy of the equations. This is important since weather data are normally collected with missing data, which is sometimes estimated or imputed [58]. Therefore, the relevance of the parameters indicates how much effort should be expended to replace or estimate missing data based on the prevailing local conditions. For instance, if accurate estimates of ET_r-lys can be obtained without a particular parameter, no effort would be required to impute its missing values.

3.2. Comparison of the Estimation Accuracy of the ASCE-PM and Machine Learning Models at Hourly and Quarter-Hourly Timescale

Table 4 gives the performance of the ML models across the different datasets for hourly and quarter-hourly timescales. Residual plots for the best-performing ML models and ASCE-PM are shown in Figure 8. Residuals for all ML models in Figure 8a,c,d indicate more points close to the zero line as compared to the ASCE-PM plot (Figure 8b). The errors appear to increase at about 0.6 mm/h. A similar trend is observed in Figure 9b, where the slope of the straight line coincides sensibly with the bisector up to about 0.6 mm/h, indicating the method's success for that set of values. However, the method slightly underestimates ET_r-lys above this value, although the overall MBE was calculated as −0.008 mm/h. Our results agree with the findings of other previous studies [56]. In contrast, some studies have had results slightly different from ours, where grass reference (ET_r) was evaluated and compared with measurements from a lysimeter [59]. In that paper, the ASCE-PM overestimated lysimeter measurements by 4% for values above 0.45 mm/h. The behavior of the ASCE-PM might vary depending on the reference crop considered and the study site [36].

Table 4. Statistical indicators for machine learning ET_r-lys estimations at the hourly and quarter-hourly timescales (fit linear equations representing the slope and intercept for each respective model are presented in italics).

	Hourly																				
	DT_1			**DT_2**			**DT_3**			**DT_4**			**DT_5**			**DT_6**			**DT_7**		
	R^2	**RMSE**	**MAE**	**R^2**	**RMSE**	**MAE**	**R^2**	**RMSE**	**MAE**	**R^2**	**RMSE**	**MAE**	**R^2**	**RMSE**	**MAE**	**R^2**	**RMSE**	**MAE**	**R^2**	**RMSE**	**MAE**
ANN	0.94	0.09	0.06	0.94	0.09	0.06	0.96	0.08	0.05	0.96	0.08	0.05	0.89	0.12	0.08	0.94	0.09	0.06	0.96	0.07	0.04
	0.90x − 0.010			*1.03x + 0.020*			*0.91x + 0.030*			*0.99x − 0.010*			*0.82x + 0.030*			*0.99x + 0.050*			*1.02x + 0.010*		
ELM	0.97	0.06	0.04	0.97	0.06	0.04	0.97	0.06	0.04	0.97	0.06	0.04	0.93	0.10	0.06	0.97	0.06	0.04	0.97	0.06	0.04
	0.98x + 0.011			*0.98x + 0.010*			*0.98x + 0.009*			*0.98x + 0.011*			*0.98x + 0.010*			*0.95x + 0.255*			*0.98x + 0.011*		
SVR	0.97	0.07	0.04	0.97	0.06	0.04	0.97	0.06	0.04	0.97	0.06	0.04	0.93	0.10	0.06	0.97	0.06	0.04	0.97	0.07	0.04
	0.98x + 0.012			*0.98x + 0.009*			*0.98x + 0.009*			*0.984x + 0.012*			*0.94x + 0.023*			*0.98x + 0.011*			*0.98x + 0.011*		
RF	0.97	0.07	0.04	0.97	0.06	0.04	0.97	0.06	0.04	0.97	0.07	0.04	0.93	0.10	0.06	0.97	0.07	0.04	0.97	0.07	0.04
	0.97x + 0.012			*0.97x + 0.011*			*0.97x + 0.012*			*0.97x + 0.012*			*0.91x + 0.030*			*0.97x + 0.013*			*0.97x + 0.013*		
	Quarter-Hourly																				
ANN	0.91	0.03	0.02	0.94	0.02	0.02	0.94	0.02	0.01	0.94	0.02	0.02	0.87	0.03	0.02	0.94	0.02	0.02	0.92	0.03	0.02
	1.030x + 0.0066			*1.004x + 0.0004*			*1.015x − 0.0004*			*0.944x + 0.0084*			*0.892x + 0.0115*			*0.881x + 0.0114*			*0.806x + 0.0089*		
ELM	0.95	0.02	0.01	0.96	0.02	0.01	0.96	0.02	0.01	0.95	0.02	0.01	0.91	0.03	0.02	0.96	0.02	0.01	0.95	0.02	0.01
	0.956x + 0.0038			*0.959x + 0.0034*			*0.959x + 0.0035*			*0.956x + 0.0038*			*0.906x + 0.008*			*0.960x + 0.0035*			*0.955x + 0.004*		
SVR	0.95	0.02	0.01	0.96	0.02	0.01	0.96	0.02	0.01	0.95	0.02	0.01	0.91	0.03	0.02	0.96	0.02	0.01	0.95	0.02	0.01
	0.960x + 0.0034			*0.969x + 0.0031*			*0.969x + 0.0032*			*0.959x + 0.0035*			*0.913x + 0.0067*			*0.969x + 0.0032*			*0.957x + 0.004*		
RF	0.96	0.02	0.01	0.96	0.02	0.01	0.96	0.02	0.01	0.96	0.02	0.01	0.92	0.03	0.02	0.96	0.02	0.01	0.95	0.02	0.01
	0.957x + 0.0036			*0.956x + 0.0037*			*0.953x + 0.0039*			*0.958x + 0.0036*			*0.915x + 0.0072*			*0.957x + 0.0037*			*0.955x + 0.004*		

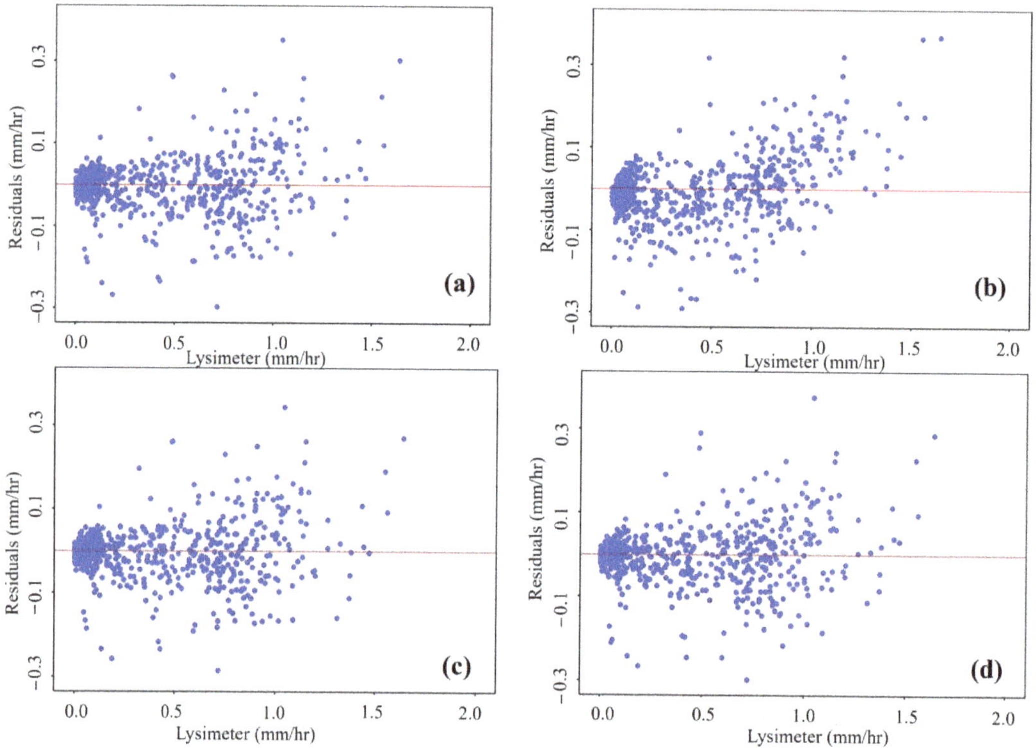

Figure 8. Residual plots for ASCE and machine learning algorithms at hourly timescale. (**a**) GA-ELM-DT_2, (**b**) ASCE-PM, (**c**) GA-ANN-DT_3, (**d**) GA-ELM-DT_7.

For the ML models GA-ELM-DT_2, GA-ELM-DT_3, and GA-ELM-DT_7, the residuals were close to zero and slopes (Figure 9a,c,d) close to unity, indicating that the models performed quite well. It can be perceived from Table 4 that these models performed better than the ASCE-PM, considering all statistical parameters. The MAE and RMSE for ASCE were 0.36 and 0.24, respectively, lower than values for DT_7, where all models performed better than the ASCE, with ranges for MAE and RMSE being 0.04 mm/h–0.05 mm/h and 0.06–0.07 mm/h, respectively. Considering the effect of the different variables on the model's performance, results for DT_4 show that excluding Δ, VPD, and R_s/R_{so} led to a reduction in the slope. However, when the data for the three variables were present in DT_5 and data for R_n and T_{mean} were absent, the slopes were still reduced across all models. This indicates that T_{mean} and R_n could be major contributing factors as direct inputs for hourly ET_r estimation using ML models. The contribution of R_n to hourly ET_r-lys estimation can be further assessed by looking at the results for DT_4 and DT_7. The consistent MAE, RMSE, and slope values for all the evaluated models indicate that R_s and R_n can be used interchangeably as direct inputs to the ML models. In fact, when data for both variables were absent in DT_5, the worst estimates were observed from the ML models. The performance of ML models on DT_4 is encouraging and indicates a potential for estimating hourly ET_r with directly sensed meteorological data.

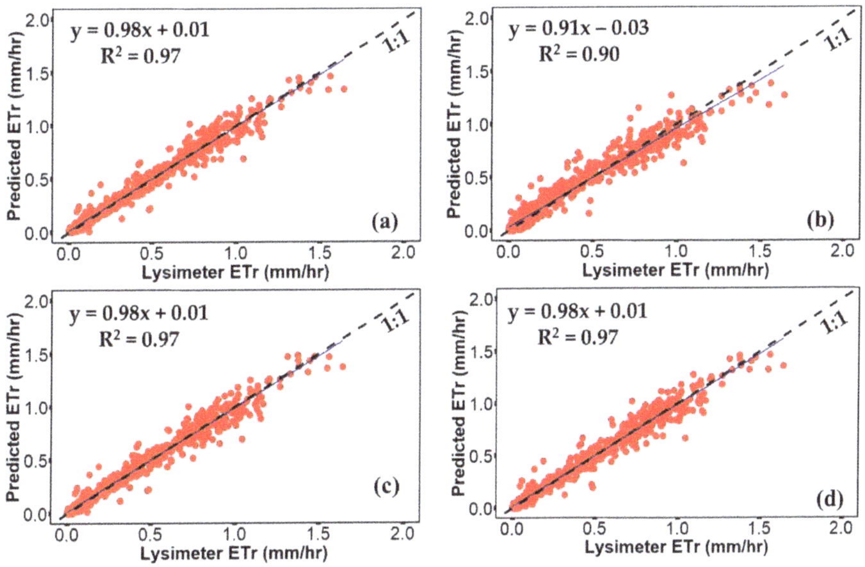

Figure 9. Relationship between ASCE, machine learning algorithms, and ET_r-lys at hourly timescale. (**a**) GA-ELM-DT_2, (**b**) ASCE-PM, (**c**) GA-ELM-DT_3, (**d**) GA-ELM-DT_7.

It can be perceived from both Figures 10a and 11a that the ASCE-PM produced biased estimates at the quarter-hourly timescale (slope of 0.93, MBE = −0.0012 mm/h), tending to underestimate ET_r-lys. On the other hand, the residuals for the best-performing models (Figure 10b,c) appear to be close to the zero line, and the R^2 values in both cases (Figure 11b,c) are larger than that for the ASCE-PM (Figure 11b).

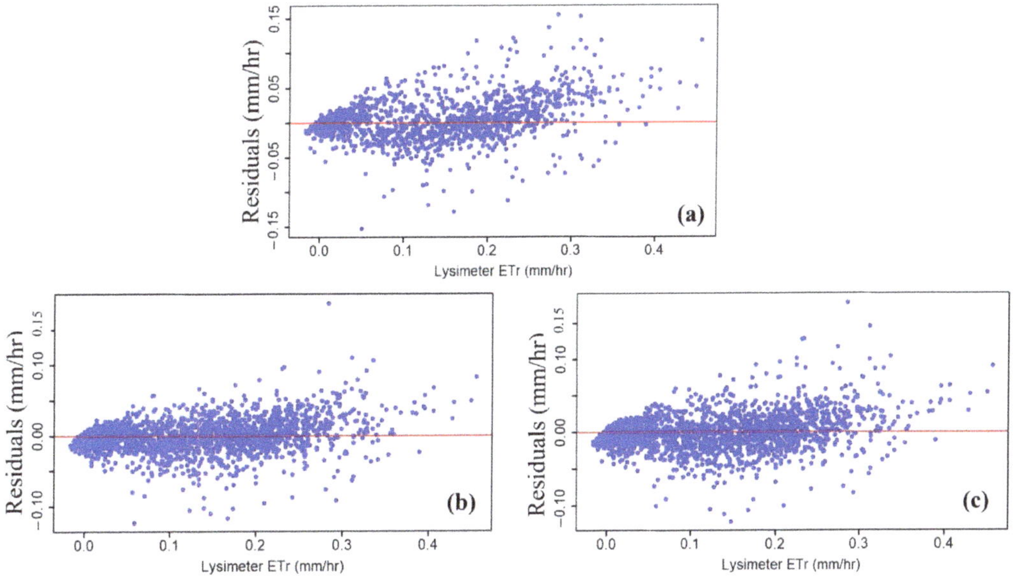

Figure 10. Residual plots for ASCE and machine learning algorithms at quarter-hourly timescale. (**a**) ASCE-PM, (**b**) GA-ANN-DT_2, (**c**) GA-ELM-DT_7.

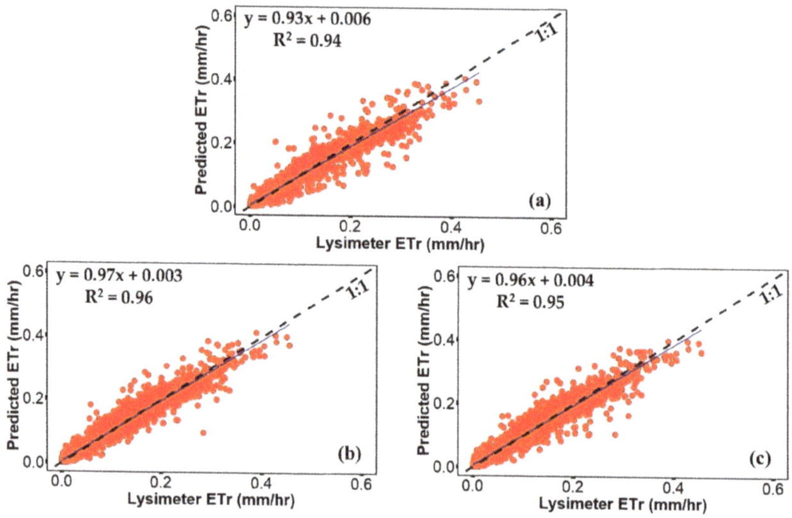

Figure 11. Relationship between ASCE, machine learning algorithms and ET_r-lys at quarter-hourly timescale. (**a**) ASCE-PM, (**b**) GA-ELM-DT_2, (**c**) GA-ELM-DT_7.

An improved performance can be observed from the best-performing ML models (GA-ELM-DT_2, GA-ELM-DT_7) with residuals (Figure 10b,c) close to the zero line and slopes (Figure 11b,c) close to unity. Additionally, all the other models tested (GA-SVR, GA-RF) performed better than ASCE-PM, as shown by the statistical parameters in Table 4. The MAE and RMSE for ASCE were 0.36 and 0.24, respectively, higher than values for all models and across all datasets that ranged from 0.01 to 0.02 mm/h and from 0.02 to 0.03 mm/h, respectively. The superiority of ML models for quarter-hourly ET_r-lys estimation provides an opportunity to use more high temporal resolution data for ET_r-lys estimation.

Some relevant findings have emerged from the results shown above. The adopted machine learning algorithms are a powerful tool for the prediction of reference evapotranspiration at lower timescales. Starting from the parameters that are directly sensed, as in DT_8, to a combination with parameters calculated from them, as in DT_2, and other datasets, it is possible to obtain prediction models characterized by very high accuracy. Mean temperature, as well as net solar radiation, play a significant role in influencing the various processes of the hydrological cycle. These parameters appear to be relevant for modeling at hourly or quarter-hourly timescales; therefore, a data-driven model that considers the above two factors most likely leads to satisfactory results. However, in the absence of net radiation measurements or in case the resources for its long calculations are limiting, it is still possible to build a reliable prediction model of ET_r-lys with the aid of the machine learning algorithms, based only on solar radiation, mean temperature, wind speed, and relative humidity data. This would especially be desirable for remote sensing applications in data-scarce environments. The use of climatic data that are not very recent may appear to be a limitation of this study. However, the ability to interpret patterns in past climatic data will be crucial to climatic forecasting, and ML models have been deemed necessary.

3.3. *Transferability of the Developed Machine Learning Models*

Developing models for reference evapotranspiration estimation requires an understanding of how the different features contribute to the model estimations. This is necessary, especially for data-scarce environments, where the most dominant variables can be utilized to provide real-time ET_r-lys estimates through ML algorithms. Our investigation of the different parameter combinations and contributions to ET_r-lys estimation at different timescales indicated that ML models can perform better than ASCE-PM while utilizing fewer data inputs to achieve reasonable results. If the estimated ET_r-lys values are acceptable representations of ET_{rs}, then the improved accuracy on shorter timescales is especially relevant for remote sensing of ET since daily or weekly ET is extrapolated from ET_{rs} estimates from the small time window that the images were taken, and thus, improved estimates on these shorter timescales could improve the overall accuracy of these methods.

However, the use of the models from this study could be limited to the study region and to places that have similar climatic conditions. Areas with similar climatic conditions might experience similar patterns of the meteorological parameters, producing similar variability as our results. Still, it is necessary to create similar or novel ML models using different parameter combinations to investigate their performance in other regions. Moreover, based on the target application, the timescale of the required data might be different from region to region. The models developed here showed the potential of capturing variability even at small timescales (quarter-hourly), which is a common temporal resolution for weather data collection in many regions. The necessity to test different models at different timescales is paramount for a given region. In fact, different regions have been found to suit different empirical models for ET_r computation, primarily based on local climatic conditions.

In the case that the models of the current study are transferred to other regions, proper training of the models is required as well as paying extra attention to the underfitting and overfitting phenomena commonly experienced during ML model training. Future developments of this study will concern coupling the ML models, surface energy balance models, and satellite imagery to improve the estimation of ET_r. Increasing the spatial scale of ET_r estimation would be necessary to enhance water management and planning.

4. Conclusions

Machine learning models (GA-ELM, GA-SVR, GA-RF, and GA-ANN) were investigated in modeling lysimeter-measured reference evapotranspiration (ET_r-lys) at daily, hourly, and quarter-hourly timescales using various input combinations of radiation and aerodynamic components for a highly advective environment. The results were compared with those of the standardized Penman–Monteith Equation (ASCE-PM). Based on the comparison results, it was observed that machine learning models yielded more accurate ET_r-lys estimates compared to ASCE-PM across all timescales. ASCE-PM consistently underestimated ET_r-lys at all timescales. Radiation components, as well as a combination of radiation and aerodynamic model inputs, demonstrated superior performance at the hourly and quarter-hourly timescales. Conversely, datasets primarily characterized by aerodynamic components performed better at the daily timescale. The results indicate that machine learning models can effectively replace ASCE-PM for ET_r-lys estimation in highly advective environments. Moreover, different climatic variables exert varying influence on model performance at a given timescale based on local weather conditions. This approach can be integrated to enhance the accuracy of ET_r-lys estimation at hourly or quarter-hourly timescales, which is crucial for precise geospatial ET estimation using land surface energy balance models. Additionally, the same approach could be applied in similar advective environments to evaluate the impact of meteorological parameters on ET_r estimation. This is imperative for ensuring precise ET_r estimation for agricultural water management.

Author Contributions: Conceptualization, S.K., R.T.P. and B.M.; methodology, S.K., R.T.P. and S.R.E.; software, S.K.; formal analysis, S.K., R.T.P., B.M. and G.M.; validation, R.T.P., S.R.E. and G.M.; writing: original draft preparation, S.K.; writing: review and editing, S.K., R.T.P., S.R.E. and B.M.; data curation S.K., R.T.P., S.R.E. and B.M.; visualization, S.K., R.T.P. and S.R.E.; investigation, R.T.P.; supervision, R.T.P.; funding acquisition, R.T.P. All authors have read and agreed to the published version of the manuscript.

Funding: This research was supported from USDA-ARS Project 3090-13000-15-00-D, and from the Ogallala Aquifer Program, a Consortium between USDA-Agricultural Research Service, Kansas State University; Texas AgriLife Extension Service & Research, Texas Tech University; and West Texas A&M University, Department of Agriculture. This material is also based upon work supported by the AI

Research Institutes program supported by NSF and USDA-NIFA under the AI Institute: Agricultural AI for Transforming Workforce and Decision Support (AgAID) award No. 2021-67021-35344.

Data Availability Statement: Data are available on the USDA ARS NAL Ag Data Commons at URLs given in Refs. [37,38].

Conflicts of Interest: The authors declare no conflict of interest.

References

1. Wu, L.; Zhou, H.; Ma, X.; Fan, J.; Zhang, F. Daily reference evapotranspiration prediction based on hybridized extreme learning machine model with bio-inspired optimization algorithms: Application in contrasting climates of China. *J. Hydrol.* **2019**, *577*, 123960. [CrossRef]
2. Jia, Y.; Su, Y.; Zhang, R.; Zhang, Z.; Lu, Y.; Shi, D.; Xu, C.; Huang, D. Optimization of an extreme learning machine model with the sparrow search algorithm to estimate spring maize evapotranspiration with film mulching in the semiarid regions of China. *Comput. Electron. Agric.* **2022**, *201*, 107298. [CrossRef]
3. Shiri, J.; Sadraddini, A.A.; Nazemi, A.H.; Kisi, O.; Marti, P.; Fard, A.F.; Landeras, G. Evaluation of different data management scenarios for estimating daily reference evapotranspiration. *Hydrol. Res.* **2013**, *44*, 1058–1070. [CrossRef]
4. Abdullah, S.S.; Malek, M.A.; Abdullah, N.S.; Kisi, O.; Yap, K.S. Extreme Learning Machines: A new approach for prediction of reference evapotranspiration. *J. Hydrol.* **2015**, *527*, 184–195. [CrossRef]
5. Wu, L.; Huang, G.; Fan, J.; Ma, X.; Zhou, H.; Zeng, W. Hybrid extreme learning machine with meta-heuristic algorithms for monthly pan evaporation prediction. *Comput. Electron. Agric.* **2020**, *168*, 105115. [CrossRef]
6. Moazenzadeh, R.; Mohammadi, B.; Shamshirband, S.; Chau, K. Coupling a firefly algorithm with support vector regression to predict evaporation in northern Iran. *Eng. Appl. Comput. Fluid Mech.* **2018**, *12*, 584–597. [CrossRef]
7. Li, X.R.; Jia, R.L.; Zhang, Z.S.; Zhang, P.; Hui, R. Hydrological response of biological soil crusts to global warming: A ten-year simulative study. *Glob. Change Biol.* **2018**, *24*, 4960–4971. [CrossRef] [PubMed]
8. Allen, R.G.; Pereira, L.S.; Howell, T.A.; Jensen, M.E. Evapotranspiration information reporting: I. Factors governing measurement accuracy. *Agric. Water Manag.* **2011**, *98*, 899–920. [CrossRef]
9. Fan, J.; Yue, W.; Wu, L.; Zhang, F.; Cai, H.; Wang, X.; Lu, X.; Xiang, Y. Evaluation of SVM, ELM and four tree-based ensemble models for predicting daily reference evapotranspiration using limited meteorological data in different climates of China. *Agric. For. Meteorol.* **2018**, *263*, 225–241. [CrossRef]
10. Allen, R.G.; Pereira, L.S.; Raes, D.; Smith, M. Crop evapotranspiration-Guidelines for computing crop water requirements-FAO Irrigation and drainage paper 56. *Fao Rome* **1998**, *300*, D05109.
11. Perera, K.C.; Western, A.W.; Nawarathna, B.; George, B. Comparison of hourly and daily reference crop evapotranspiration equations across seasons and climate zones in Australia. *Agric. Water Manag.* **2015**, *148*, 84–96. [CrossRef]
12. Allen, R.G.; Walter, I.A.; Elliott, R.L.; Howell, T.A.; Itenfisu, D.; Jensen, M.E.; Snyder, R.L. *The ASCE Standardized Reference Evapotranspiration Equation*; Water Resources Institute: Reston, VA, USA, 2005.
13. Valiantzas, J.D. Simplified forms for the standardized FAO-56 Penman–Monteith reference evapotranspiration using limited weather data. *J. Hydrol.* **2013**, *505*, 13–23. [CrossRef]
14. Izadifar, Z.; Elshorbagy, A. Prediction of hourly actual evapotranspiration using neural networks, genetic programming, and statistical models. *Hydrol. Process.* **2010**, *24*, 3413–3425. [CrossRef]
15. Kumar, D.; Adamowski, J.; Suresh, R.; Ozga-Zielinski, B. Estimating Evapotranspiration Using an Extreme Learning Machine Model: Case Study in North Bihar, India. *J. Irrig. Drain. Eng.* **2016**, *142*, 04016032. [CrossRef]
16. Zhang, Y.; Zhao, Z.; Zheng, J. CatBoost: A new approach for estimating daily reference crop evapotranspiration in arid and semi-arid regions of Northern China. *J. Hydrol.* **2020**, *588*, 125087. [CrossRef]
17. Sattari, M.T.; Apaydin, H.; Band, S.S.; Mosavi, A.; Prasad, R. Comparative analysis of kernel-based versus ANN and deep learning methods in monthly reference evapotranspiration estimation. *Hydrol. Earth Syst. Sci.* **2021**, *25*, 603–618. [CrossRef]
18. Estévez, J.; Gavilán, P.; Giráldez, J.V. Guidelines on validation procedures for meteorological data from automatic weather stations. *J. Hydrol.* **2011**, *402*, 144–154. [CrossRef]
19. Chandel, A.K.; Molaei, B.; Khot, L.R.; Peters, R.T.; Stöckle, C.O. High Resolution Geospatial Evapotranspiration Mapping of Irrigated Field Crops Using Multispectral and Thermal Infrared Imagery with METRIC Energy Balance Model. *Drones* **2020**, *4*, 52. [CrossRef]
20. Molaei, B.; Peters, R.T.; Khot, L.R.; Stöckle, C.O. Assessing Suitability of Auto-Selection of Hot and Cold Anchor Pixels of the UAS-METRIC Model for Developing Crop Water Use Maps. *Remote Sens.* **2022**, *14*, 4454. [CrossRef]
21. Xie, H.; Zhu, X. Reference evapotranspiration trends and their sensitivity to climatic change on the Tibetan Plateau (1970–2009): Reference evapotranspiration on the tibetan plateau. *Hydrol. Process.* **2013**, *27*, 3685–3693. [CrossRef]
22. Irmak, S.; Kabenge, I.; Skaggs, K.E.; Mutiibwa, D. Trend and magnitude of changes in climate variables and reference evapotranspiration over 116-yr period in the Platte River Basin, central Nebraska–USA. *J. Hydrol.* **2012**, *420–421*, 228–244. [CrossRef]
23. Alexandris, S.; Proutsos, N. How significant is the effect of the surface characteristics on the Reference Evapotranspiration estimates? *Agric. Water Manag.* **2020**, *237*, 106181. [CrossRef]
24. Saxton, K.E. Sensitivity analyses of the combination evapotranspiration equation. *Agric. Meteorol.* **1975**, *15*, 343–353. [CrossRef]

25. Granata, F. Evapotranspiration evaluation models based on machine learning algorithms—A comparative study. *Agric. Water Manag.* **2019**, *217*, 303–315. [CrossRef]
26. Siddique, M.N.H.; Tokhi, M.O. Training neural networks: Backpropagation vs. genetic algorithms. In Proceedings of the IJCNN'01. International Joint Conference on Neural Networks. Proceedings (Cat. No. 01CH37222), Washington, DC, USA, 15–19 July 2001; Volume 4, pp. 2673–2678.
27. Gill, E.J.; Singh, E.B.; Singh, E.S. Training back propagation neural networks with genetic algorithm for weather forecasting. In Proceedings of the IEEE 8th International Symposium on Intelligent Systems and Informatics, Subotica, Serbia, 10–11 September 2010; pp. 465–469.
28. Rozos, E.; Dimitriadis, P.; Mazi, K.; Koussis, A.D. A multilayer perceptron model for stochastic synthesis. *Hydrology* **2021**, *8*, 67. [CrossRef]
29. Huang, G.B.; Ding, X.; Zhou, H. Optimization method based extreme learning machine for classification. *Neurocomputing* **2010**, *74*, 155–163. [CrossRef]
30. Chen, Y.; Chang, F.J. Evolutionary artificial neural networks for hydrological systems forecasting. *J. Hydrol.* **2009**, *367*, 125–137. [CrossRef]
31. Oyebode, O.; Stretch, D. Neural network modeling of hydrological systems: A review of implementation techniques. *Nat. Resour. Model.* **2019**, *32*, 12189. [CrossRef]
32. Liu, Q.; Wu, Z.; Cui, N.; Zhang, W.; Wang, Y.; Hu, X.; Gong, D.; Zheng, S. Genetic Algorithm-Optimized Extreme Learning Machine Model for Estimating Daily Reference Evapotranspiration in Southwest China. *Atmosphere* **2022**, *13*, 971. [CrossRef]
33. Evett, S.R.; Schwartz, R.C.; Howell, T.A.; Louis Baumhardt, R.; Copeland, K.S. Can weighing lysimeter ET represent surrounding field ET well enough to test flux station measurements of daily and sub-daily ET? *Adv. Water Resour.* **2012**, *50*, 79–90. [CrossRef]
34. Evett, S.R.; Howell, T.A.; Schneider, A.D.; Copeland, K.S.; Dusek, D.A.; Brauer, D.K.; Tolk, J.A.; Marek, G.W.; Marek, T.M.; Gowda, P.H. The bushland weighing lysimeters: A quarter century of crop et investigations to advance sustainable irrigation. *Trans. ASABE* **2016**, *59*, 163–179.
35. Evett, S.R.; Marek, G.W.; Copeland, K.S.; Colaizzi, P.D. Quality Management for Research Weather Data: USDA-ARS, Bushland, TX. *Agrosystems Geosci. Environ.* **2018**, *1*, 1–18. [CrossRef]
36. Evett, S.R.; Howell, T.A.; Todd, R.W.; Schneider, A.D.; Tolk, J.A. Alfalfa reference ET measurement and prediction. In Proceedings of the National Irrigation Symposium: Proceedings of the 4th Decennial Symposium, Phoenix, Arizona, 14–16 November 2000; pp. 266–272.
37. Evett, S.R.; Copeland, K.S.; Ruthardt, B.B.; Marek, G.W.; Colaizzi, P.D.; Howell, T.A.; Brauer, D.K.; The Bushland, Texas, Alfalfa Datasets. USDA ARS NAL Ag Data Commons. Available online: https://data.nal.usda.gov/dataset/bushland-texas-alfalfa-datasets (accessed on 6 December 2023).
38. Evett, S.R.; Copeland, K.S.; Ruthardt, B.B.; Marek, G.W.; Colaizzi, P.D.; Howell, T.A.; Brauer, D.K. Standard Quality Controlled Research Weather Data-USDA-ARS, Bushland, Texas. USDA ARS NAL Ag Data Commons. Available online: https://data.nal.usda.gov/dataset/standard-quality-controlled-research-weather-data-%E2%80%93-usda-ars-bushland-texas (accessed on 6 December 2023).
39. Allen, R.G. *REF-ET: Reference Evapotranspiration Calculation Software for FAO and ASCE Standardized Equations*; University of Idaho: Moscow, ID, USA, 2009.
40. DeJonge, K.C.; Thorp, K.R. Implementing Standardized Reference Evapotranspiration and Dual Crop Coefficient Approach in the DSSAT Cropping System Model. *Trans. ASABE* **2017**, *60*, 1965–1981. [CrossRef]
41. Park, J.; Choi, M. Estimation of evapotranspiration from ground-based meteorological data and global land data assimilation system (GLDAS). *Stoch. Environ. Res. Risk Assess.* **2015**, *29*, 1963–1992. [CrossRef]
42. Feng, Y.; Cui, N.; Gong, D.; Zhang, Q.; Zhao, L. Evaluation of random forests and generalized regression neural networks for daily reference evapotranspiration modelling. *Agric. Water Manag.* **2017**, *193*, 163–173. [CrossRef]
43. Yao, Y.; Liang, S.; Li, X.; Chen, J.; Liu, S.; Jia, K.; Zhang, X.; Xiao, Z.; Fisher, J.B.; Mu, Q.; et al. Improving global terrestrial evapotranspiration estimation using support vector machine by integrating three process-based algorithms. *Agric. For. Meteorol.* **2017**, *242*, 55–74. [CrossRef]
44. Huang, G.B.; Zhu, Q.Y.; Siew, C.K. Extreme learning machine: Theory and applications. *Neurocomputing* **2006**, *70*, 489–501. [CrossRef]
45. Feng, Y.; Cui, N.; Hao, W.; Gao, L.; Gong, D. Estimation of soil temperature from meteorological data using different machine learning models. *Geoderma* **2019**, *338*, 67–77. [CrossRef]
46. Raghavendra, N.S.; Deka, P.C. Support vector machine applications in the field of hydrology: A review. *Appl. Soft Comput.* **2014**, *19*, 372–386. [CrossRef]
47. Schonlau, M.; Zou, R.Y. The random forest algorithm for statistical learning. *Stata J. Promot. Commun. Stat. Stata* **2020**, *20*, 3–29. [CrossRef]
48. Eslamian, S.S.; Gohari, S.A.; Zareian, M.J.; Firoozfar, A. Estimating Penman–Monteith reference evapotranspiration using artificial neural networks and genetic algorithm: A case study. *Arab. J. Sci. Eng.* **2012**, *37*, 935–944. [CrossRef]
49. Abdullah, S.S.; Malek, M.A.; Mustapha, A.; Aryanfar, A. Hybrid of artificial neural network-genetic algorithm for prediction of reference evapotranspiration in arid and semiarid regions. *J. Agric. Sci.* **2014**, *6*, 191. [CrossRef]

50. Ikram, R.M.A.; Mostafa, R.R.; Chen, Z.; Islam, A.R.M.T.; Kisi, O.; Kuriqi, A.; Zounemat-Kermani, M. Advanced hybrid meta-heuristic machine learning models application for reference crop evapotranspiration prediction. *Agronomy* **2022**, *13*, 98. [CrossRef]
51. Katoch, S.; Chauhan, S.S.; Kumar, V. A review on genetic algorithm: Past, present, and future. *Multimed. Tools Appl.* **2021**, *80*, 8091–8126. [CrossRef]
52. Lobit, P.; López Pérez, L.; Lhomme, J.P. Retrieving air humidity, global solar radiation, and reference evapotranspiration from daily temperatures: Development and validation of new methods for Mexico. Part II: Radiation. *Theor. Appl. Climatol.* **2018**, *133*, 799–810. [CrossRef]
53. Dombrowski, O.; Hendricks Franssen, H.J.; Brogi, C.; Bogena, H.R. Performance of the ATMOS41 All-in-One Weather Station for Weather Monitoring. *Sensors* **2021**, *21*, 741. [CrossRef] [PubMed]
54. Fu, T.; Li, X.; Jia, R.; Feng, L. A novel integrated method based on a machine learning model for estimating evapotranspiration in dryland. *J. Hydrol.* **2021**, *603*, 126881. [CrossRef]
55. Tolk, J.A.; Evett, S.R.; Howell, T.A. Advection influences on evapotranspiration of alfalfa in a semiarid climate. *Agron. J* **2006**, *98*, 1646–1654. [CrossRef]
56. Berengena, J.; Gavilán, P. Reference Evapotranspiration Estimation in a Highly Advective Semiarid Environment. *J. Irrig. Drain. Eng.* **2005**, *131*, 147–163. [CrossRef]
57. Feng, Y.; Peng, Y.; Cui, N.; Gong, D.; Zhang, K. Modeling reference evapotranspiration using extreme learning machine and generalized regression neural network only with temperature data. *Comput. Electron. Agric.* **2017**, *136*, 71–78. [CrossRef]
58. Doreswamy, I.G.; Manjunatha, B.R. Performance evaluation of predictive models for missing data imputation in weather data. In Proceedings of the 2017 International Conference on Advances in Computing, Communications and Informatics (ICACCI), Kochi, India, 22–24 August 2017; pp. 1327–1334.
59. López-Urrea, R.; de Olalla, F.M.S.; Fabeiro, C.; Moratalla, A. An evaluation of two hourly reference evapotranspiration equations for semiarid conditions. *Agric. Water Manag.* **2006**, *86*, 277–282. [CrossRef]

MDPI
St. Alban-Anlage 66
4052 Basel
Switzerland
www.mdpi.com

MDPI Books Editorial Office
E-mail: books@mdpi.com
www.mdpi.com/books

www.ingramcontent.com/pod-product-compliance
Lightning Source LLC
Chambersburg PA
CBHW061500220926
51850CB00004B/134
* 9 7 8 3 7 2 5 8 0 2 4 7 0 *